Current Topics in Behavioral Neurosciences

Series Editors:
Mark Geyer, La Jolla, CA, USA
Bart Ellenbroek, Hamburg, Germany
Charles Marsden, Nottingham, UK

About this series

Current Topics in Behavioral Neurosciences provides critical and comprehensive discussions of the most significant areas of behavioral neuroscience research, written by leading international authorities. Each volume offers an informative and contemporary account of its subject, making it an unrivalled reference source. Titles in this series are available in both print and electronic formats.

With the development of new methodologies for brain imaging, genetic and genomic analyses, molecular engineering of mutant animals, novel routes for drug delivery, and sophisticated cross-species behavioral assessments, it is now possible to study behavior relevant to psychiatric and neurological diseases and disorders on the physiological level. The *Behavioral Neurosciences* series focuses on "translational medicine" and cutting-edge technologies. Preclinical and clinical trials for the development of new diagnostics and therapeutics as well as prevention efforts are covered whenever possible.

Neal R. Swerdlow
Editor

Behavioral Neurobiology of Schizophrenia and Its Treatment

Springer

Editor
Prof. Neal R. Swerdlow
University of California San Diego
Dept. of Psychiatry – MC 0804
9500 Gilman Drive
La Jolla, California 92093
USA
nswerdlow@ucsd.edu

ISSN 1866-3370 e-ISSN 1866-3389
ISBN 978-3-642-26462-7 ISBN 978-3-642-13717-4 (eBook)
DOI 10.1007/978-3-642-13717-4
Springer Heidelberg Dordrecht London New York

© Springer-Verlag Berlin Heidelberg 2010
Softcover reprint of the hardcover 1st edition 2010
This work is subject to copyright. All rights are reserved, whether the whole or part of the material is concerned, specifically the rights of translation, reprinting, reuse of illustrations, recitation, broadcasting, reproduction on microfilm or in any other way, and storage in data banks. Duplication of this publication or parts thereof is permitted only under the provisions of the German Copyright Law of September 9, 1965, in its current version, and permission for use must always be obtained from Springer. Violations are liable to prosecution under the German Copyright Law.

The use of general descriptive names, registered names, trademarks, etc. in this publication does not imply, even in the absence of a specific statement, that such names are exempt from the relevant protective laws and regulations and therefore free for general use.

Product liability: The publishers cannot guarantee the accuracy of any information about dosage and application contained in this book. In every individual case the user must check such information by consulting the relevant literature.

Cover illustration: Artistic representation of oscillatory synchrony and timing of neurons in networks by Gyorgy Buzsaki

Cover design: WMXDesign GmbH, Heidelberg, Germany

Printed on acid-free paper

Springer is part of Springer Science+Business Media (www.springer.com)

Preface

The Developing Field of Schizophrenia Research

There are developmental milestones in the life of a disorder – the moments that it is defined (or redefined) at a diagnostic level, the moments that it is understood (or better understood) at a scientific level, and the moments that it is effectively treated (or more effectively treated) at a clinical level. Deciding when to pause and take stock of these milestones is a matter of choice, particularly in the absence of a transformative event like the identification of a definitive gene (e.g., BRCA or the Huntington's gene), causative agent (e.g., HIV), enzyme (e.g., Lesch–Nyhan syndrome), or intervention (e.g., the polio vaccine). We do not have such clear transformative milestones to mark our understanding or treatment of schizophrenia; smaller milestones are either part of the distant past (e.g., Bleuler's diagnostic reformulation, or the advent of modern antipsychotics and resulting "deinstitutionalization" of schizophrenia) or perhaps our evolving present (e.g., the growing list of candidate genes).

But just as our field aspires to reject biological determinism in the etiology of schizophrenia, we should hold that the path toward understanding this disorder is not predetermined. For this reason, pausing to assess the field's milestones, even (especially) in the absence of transformative events, affords us the opportunity to better nurture it: to willfully make (or not make) midcourse corrections and thereby alter (or sustain) its developmental trajectory. To do so is not an admission of failure, but to not do so would be a serious omission, and in my opinion, an act of scientific arrogance.

Where on this developmental path do we find ourselves? As the chapters in this volume suggest, we are still in a "learning stage." Diagnostically, the boundaries of the schizophrenias are less clearly marked than we once believed, expanding in some directions toward the bipolar disorders, in others toward the "Cluster A" personality spectrum, and in still others toward "pure" genetic disorders such as Velo-Cardio-Facial Syndrome. In its pathogenesis, we are recognizing a multiplying number of candidate "risk" genes, as well as epigenetic "risk" factors. In its pathophysiology, we have a growing array of increasingly sophisticated experimental tools to characterize its aberrant neural substrates at nano-, micro-, and macro-systems levels, neural information at millisecond–microvolt resolution, and gene networks and neural signaling pathways that seem to interact within two-, three-, and four-dimensions.

Where is this "learning stage" taking our field? As boundaries expand and lists of genes, neural elements, and signal molecules grow, and as the temporal and spatial resolution of our measures increase to reveal more and more about less and less, it can appear that this "learning stage" of schizophreniology is teetering toward a state of fixation, more than of growth. This conclusion would be bolstered by the fact that developments in antipsychotic efficacy, so highly touted by our commercial counterparts, have not withstood the light of data, bringing us full circle, more or less, to where we started 50 years ago, though many pounds heavier. So, are we "fixated" in this learning stage? A closer inspection of our developmental path, described in the chapters in this text, may suggest otherwise.

As Bromley and Brekke describe, our field now has tools to assess and target not merely psychosis but also real-life function and functional outcome in the schizophrenias. This seemingly simple recognition of the importance of "real-life function" in the study of any disorder, but particularly schizophrenia, charts a path away from "learning for learning's sake," and toward a next developmental stage. These real-life metrics will become new benchmarks for assessing the efficacy of current "next generation" interventions, delivered toward different clinical (Barch; Kaur and Cadenhead) or receptor targets (Kim and Stahl), or via different technologies (Rabin and Siegel), even as we better understand and address the failings of the "former generation" interventions (Meyer).

These chapters on schizophrenia neuroimaging (Brown and Thompson; Urban and Abi-Dargham; Levitt et al.), neurophysiology (Rissling and Light; Levy et al.; Braff), neurocognition (Kalkstein et al.), and preclinical models (Young et al; Powell) report that our field has developed a highly advanced ability to submit the neurobiology of the schizophrenias to rigorous experimental analysis. While some of these developments might appear fixated within nitty-gritty experimental issues – finding the most informative ligand, evoked waveform, stimulus condition, scanning state, or neurocognitive domain – they are actually the grist for healthy scientific development: for testing hypotheses under controlled conditions to generate interpretable data and conclusions. And such inquiry across multiple levels of analysis, and across species, creates the opportunity for converging lines of evidence – scientific triangulation – so essential for establishing new knowledge about disorders of brain, mind, and behavior.

At this developmental stage, convergent information has focused our attention on abnormalities in specific brain regions and circuits, including systems within and interconnecting the prefrontal cortex (Volk and Lewis), specific thalamic nuclei (Cronenwett and Csernansky), and mesial temporal lobe (Heckers and Conradi), as causative events in the pathophysiology of the schizophrenias. These abnormalities have been identified and characterized using strategies of volumetric, neurochemical, and functional neuroimaging described in earlier chapters in this volume, and extend to detailed neuropathological studies, and studies of altered developmental and molecular processes. In fact, where the schizophrenias were once characterized pejoratively as "functional" based on the paucity of clear neuropathological findings, it is now characterized pejoratively as "heterogeneous," based on the long list and multiple combinations of such findings across studies. Nobody said development was easy.

Perhaps the most studied (at least in terms of sheer "N") and rapidly evolving facet of our developing field is schizophrenia genetics. Some candidates have emerged as prime "risk gene" suspects (e.g., COMT, NRG1, and DISC1, among others), yet perhaps the bigger developmental advancement is the growing awareness that traditional strategies for identifying disorder genes – even with heroically (and some might say excessively) powered samples – may not be most informative for schizophrenia. Rather, the key to genetic risk in schizophrenia may lie in aberrant patterns of copy number variants (Mantripagada et al.), rare mutations, or DNA methylation (Akbarian) that characterize this disorder. The wide range of different possible genetic disturbances in this disorder might gain coherence via their action within a smaller number of critical molecular signaling pathways (Kvajo et al.) that might ultimately be responsible for downstream disturbances in the development and function of neurons and the limbic-cortical circuits that they populate.

With this pause to assess the developmental trajectory of our field comes the opportunity – and I would say, the obligation – to consider and discuss what lies ahead in our understanding of the neurobiology and treatment of the schizophrenias. Does our "growth chart" suggest that we will ultimately be able to use a molecular toolbox to "fix" this disorder? Or, projecting out some years, will our trajectory make us amenable to other therapeutic approaches? Which paradigms – scientific or therapeutic – once viewed with promise have we now outgrown? And with what will they be replaced? As I note in my chapter, should "midcourse changes" be necessary, this is a sign of growth and not of failure. There can be no question that, with countless dedicated lifetimes of work, our field has learned great amounts about an exquisitely complex biology of schizophrenia; but, I suggest, an equally important question is whether, given all that we know, we can hope to predictably and effectively manipulate this biology within our lifetimes, in a way that will fundamentally change the course of this disorder. If the consensus is "no," or even "who knows?" then our field might consider other approaches (and I raise the speculative example of pharmacologically augmented cognitive therapies), for which the biological and clinical complexity of the intervention is developmentally suited to our ability to deliver it.

The key milestones in schizophrenia research and treatment will be reached only if we maintain a healthy developmental course, and this means that we cannot tolerate fixation. Knowing what we know, and having developed such a rich scientific and clinical knowledge base, to pause and consider whether we are approaching this disorder correctly, is perhaps the best way that we can help our field and the families that we serve.

Summer 2010   Neal R. Swerdlow

Contents

Part I Function, Outcome and Treatment in Schizophrenia

Assessing Function and Functional Outcome in Schizophrenia 3
Elizabeth Bromley and John S. Brekke

Antipsychotics and Metabolics in the Post-CATIE Era 23
Jonathan M. Meyer

**Pharmacological Strategies for Enhancing Cognition
in Schizophrenia** 43
Deanna M. Barch

Treatment Implications of the Schizophrenia Prodrome 97
Tejal Kaur and Kristin S. Cadenhead

Antipsychotic Drug Development 123
Dennis H. Kim and Stephen M. Stahl

Antipsychotic Dosing and Drug Delivery 141
Cara R. Rabin and Steven J. Siegel

**Part II Experimental measures of brain function and dysfunction
in schizophenia**

**Functional Brain Imaging in Schizophrenia: Selected Results
and Methods** 181
Gregory G. Brown and Wesley K. Thompson

Neurochemical Imaging in Schizophrenia 215
Nina Urban and Anissa Abi-Dargham

A Selective Review of Volumetric and Morphometric Imaging in Schizophrenia 243
James J. Levitt, Laurel Bobrow, Diandra Lucia, and Padmapriya Srinivasan

Neurophysiological Measures of Sensory Registration, Stimulus Discrimination, and Selection in Schizophrenia Patients 283
Anthony J. Rissling and Gregory A. Light

Eye Tracking Dysfunction in Schizophrenia: Characterization and Pathophysiology 311
Deborah L. Levy, Anne B. Sereno, Diane C. Gooding, and Gilllian A. O'Driscoll

Prepulse Inhibition of the Startle Reflex: A Window on the Brain in Schizophrenia 349
David L. Braff

Neurocognition in Schizophrenia 373
Solomon Kalkstein, Irene Hurford, and Ruben C. Gur

Animal Models of Schizophrenia 391
Jared W. Young, Xianjin Zhou, and Mark A. Geyer

Models of Neurodevelopmental Abnormalities in Schizophrenia 435
Susan B. Powell

Part III Neural substrates of schizophrenia

Prefrontal Cortical Circuits in Schizophrenia 485
David W. Volk and David A. Lewis

Thalamic Pathology in Schizophrenia 509
Will J. Cronenwett and John Csernansky

Hippocampal Pathology in Schizophrenia 529
Stephan Heckers and Christine Konradi

Integrative Circuit Models and Their Implications for the Pathophysiologies and Treatments of the Schizophrenias 555
Neal R. Swerdlow

Part IV Genetic and molecular substrates of schizophrenia

Experimental Approaches for Identifying Schizophrenia Risk Genes 587
Kiran K. Mantripragada, Liam S. Carroll, and Nigel M. Williams

Epigenetics of Schizophrenia 611
Schahram Akbarian

Molecules, Signaling, and Schizophrenia 629
Mirna Kvajo, Heather McKellar, and Joseph A. Gogos

Index 657

Contributors

Anissa Abi-Dargham
Department of Psychiatry and Radiology, New York State Psychiatric Institute, Columbia University, New York, NY 10032, USA, aa324@columbia.edu

Schahram Akbarian
Department of Psychiatry, Brudnick Neuropsychiatric Research Institute, University of Massachusetts Medical School, Worcester, MA 01604, USA, schahram.akbarian@umassmed.edu

Deanna M. Barch
Departments of Psychology, Psychiatry and Radiology, Washington University, One Brookings Drive, Box 1125, St. Louis, MO 63130, USA, dbarch@artsci.wustl.edu

Laurel Bobrow
Psychiatry Neuroimaging Laboratory, Department of Psychiatry at Brigham and Women's Hospital, Harvard Medical School, Boston, MA, USA, laurel@bwh.harvard.edu

David L. Braff
Department of Psychiatry, University of California, San Diego, 9500 Gilman Drive, La Jolla, CA 92093-0804, USA, dbraff@ucsd.edu

John S. Brekke
School of Social Work, University of Southern California, Los Angeles, CA 90089-0411, USA, brekke@usc.edu

Elizabeth Bromley
Department of Psychiatry and Biobehavioral Sciences, Semel Institute for Health Services and Society, University of California, Los Angeles, 10920 Wilshire Blvd, Suite 300, Los Angeles, CA 90024, USA, ebromley@ucla.edu

Gregory G. Brown
Psychology Service, VA San Diego Healthcare System, 3350 La Jolla Village Drive, San Diego, CA 92161, USA and Department of Psychiatry, University of California, San Diego, 9500 Gilman Dr., La Jolla, CA 92093, USA, gbrown@ucsd.edu

Kristin S. Cadenhead
Department of Psychiatry, University of California, San Diego, 9500 Gilman Drive, La Jolla, CA 92093-0810, USA, kcadenhead@ucsd.edu

Liam S. Carroll
Department of Psychological Medicine and Neurology, MRC Centre in Neuropsychiatric Genetics and Genomics, Cardiff University School of Medicine, Cardiff, UK carrollls@cardiff.ac.uk

Will J. Cronenwett
Psychiatry and Behavioral Sciences and the Stone Institute of Psychiatry, Northwestern University Feinberg School of Medicine and Northwestern Memorial Hospital, 446 E. Ontario, Suite 7-200, Chicago, IL 60611, USA, w-cronenwett@northwestern.edu

John Csernansky
Psychiatry and Behavioral Sciences and the Stone Institute of Psychiatry, Northwestern University Feinberg School of Medicine and Northwestern Memorial Hospital, 446 E. Ontario, Suite 7-200, Chicago, IL 60611, USA, jgc@northwestern.edu

Mark A. Geyer
Department of Psychiatry, University of California, San Diego, 9500 Gilman Drive MC 0804, La Jolla, CA 92093-0804, USA, mgeyer@ucsd.edu

Joseph A. Gogos
Department of Physiology and Cellular Biophysics, and Department of Neuroscience, 630 West 168th Street, P&S 11-159, Columbia University, New York, NY 10032-3702, USA, jag90@columbia.edu

Diane C. Gooding
Departments of Psychology and Psychiatry, University of Wisconsin School of Medicine and Public Health, 1202 W. Johnson Sreet, Madison, WI 53706, USA, dgooding@wisc.edu

Ruben C. Gur
Schizophrenia Research Center, Neuropsychiatry Section, Department of Psychiatry, University of Pennsylvania School of Medicine and Philadelphia Veterans Affairs Medical Center, 10th floor, Gates Building, 3400 Spruce Street, Philadelphia, PA 19104, gur@bblmail.psycha.upenn.edu

Contributors

Stephan Heckers
Department of Psychiatry, Vanderbilt University, 1601 23rd Avenue South, Room 3060, Nashville, TN 37212, USA, stephan.heckers@Vanderbilt.Edu

Irene Hurford
Schizophrenia Research Center, Neuropsychiatry Section, Department of Psychiatry, University of Pennsylvania School of Medicine, 10th Floor, Gates Building, 3400 Spruce Street, Philadelphia, PA 19104, USA and Philadelphia Veterans Affairs Medical Center, University of Pennsylvania School of Medicine, Philadelphia, PA 19104-6021, USA, ihurford@upenn.edu

Solomon Kalkstein
Schizophrenia Research Center, Neuropsychiatry Section, Department of Psychiatry, University of Pennsylvania School of Medicine, 10th Floor, Gates Building, 3400 Spruce Street, Philadelphia, PA 19104, USA and Philadelphia Veterans Affairs Medical Center, University of Pennsylvania School of Medicine, Philadelphia, PA 19104-6021, USA, solomon.kalkstein@va.gov

Tejal Kaur
Department of Psychiatry, University of California, San Diego, 9500 Gilman Drive, La Jolla, CA 92093 and Division of Child and Adolescent Psychiatry, New York Presbyterian Hospital of Columbia, New York, NY, USA and Cornell University, Ithaca, New York, USA, tejalkaur@gmail.com

Dennis H. Kim
Arbor Scientia, 1930 Palomar Point Way, Suite 103, Carlsbad, CA 92008, USA dkim@arborscientia.com

Christine Konradi
Vanderbilt University, Department of Pharmacology, MRB 3, Room 8160, 465 21st Avenue South, Nashville TN 37232, USA, christine.konradi@vanderbilt.edu

Mirna Kvajo
Departments of Physiology and Cellular Biophysics, and Department of Psychiatry, 630 West 168th Street, P&S 11-519, Columbia University, New York, NY 10032-3702, mk2776@columbia.edu

James J. Levitt
Department of Psychiatry at the VA Boston Healthcare System, Harvard Medical School, Brockton Campus, 116A4, 940 Belmont Street, Brockton, MA 02301, USA and Psychiatry Neuroimaging Laboratory, Department of Psychiatry at Brigham and Women's Hospital, Harvard Medical School, Boston, MA, USA, james_levitt@hms.harvard.edu

Deborah L. Levy
Psychology Research Laboratory, McLean Hospital, 115 Mill Street, Belmont, MA 02478, USA, dlevy@mclean.harvard.edu

David A. Lewis
Departments of Psychiatry and Neuroscience, University of Pittsburgh, 3811 O'Hara, BST W1653, Pittsburgh, PA 15213, USA, lewisda@upmc.edu

Gregory A. Light
Department of Psychiatry, University of California, San Diego, 9500 Gilman Drive, La Jolla, CA 92093-0804, glight@ucsd.edu

Diandra Lucia
Psychiatry Neuroimaging Laboratory, Department of Psychiatry at Brigham and Women's Hospital, Harvard Medical School, Boston, MA, USA, diandra@bwh.harvard.edu

Kiran K. Mantripragada
Department of Psychological Medicine and Neurology, MRC Centre in Neuropsychiatric Genetics and Genomics, Cardiff University School of Medicine, Cardiff, UK, MantripragadaKK@cardiff.ac.uk

Heather McKellar
Integrated Program in Cellular, Molecular and Biophysical Studies, 630 West 168th Street, P&S 11-519, Columbia University, New York, NY 10032-3702, hm2126@columbia.edu

Jonathan M. Meyer
Department of Psychiatry, University of California, San Diego, 9500 Gilman Drive, La Jolla, CA 92093; and VA San Diego Healthcare System, 3350 La Jolla Village Drive (116A), San Diego, CA 92161, USA, jmmeyer@ucsd.edu

Gilllian A. O'Driscoll
Department of Psychology, McGill University, Stewart Biological Sciences Bldg, 1205 Dr Penfield Avenue, Montreal QC, Canada H3A 1B1, gillian@psych.mcgill.ca

Susan B. Powell
Department of Psychiatry, University of California, San Diego, 9500 Gilman Dr., La Jolla, CA 92093-0804, USA, sbpowell@ucsd.edu

Cara R. Rabin
Child Psychiatry Branch, National Institute of Mental Health, Bethesda, MD 20892, USA, cara.rabin@gmail.com

Contributors

Anthony J. Rissling
Department of Psychiatry, University of California, San Diego 9500 Gilman Drive, La Jolla, CA 92093-0804, USA, ajrissling@ucsd.edu

Anne B. Sereno
Departments of Neurobiology and Anatomy, University of Texas Medical School at Houston, Houston, TX 77030, USA, anne.b.sereno@uth.tmc.edu

Steven J. Siegel
Translational Neuroscience Program, Department of Psychiatry, 125 So. 31st Street, University of Pennsylvania, Philadelphia, PA 19104, USA, siegels@mail.med.upenn.edu

Padmapriya Srinivasan
Psychiatry Neuroimaging Laboratory, Department of Psychiatry at Brigham and Women's Hospital, Harvard Medical School, Boston, MA, USA, priya@bwh.harvard.edu

Stephen M. Stahl
Department of Psychiatry, School of Medicine, University of California, San Diego, La Jolla, CA 92093, USA and Department of Psychiatry, University of Cambridge, Addenbrooke's Hospital, Hills Road, Cambridge, CB2 2QQ, UK, smstahl@neiglobal.com

Neal R. Swerdlow
Department of Psychiatry, University of California, San Diego, 9500 Gilman Drive, La Jolla, CA 92093-0804, USA, nswerdlow@ucsd.edu

Wesley K. Thompson
Department of Psychiatry, University of California, San Diego, 9500 Gilman Dr., La Jolla, CA 92093, USA and Stein Institute for Research on Aging, University of California, San Diego, La Jolla, CA, USA, wes.stat@gmail.com

Nina B.L. Urban
Department of Psychiatry, New York State Psychiatric Institute, Columbia University, New York, NY 10032, USA, nu2118@columbia.edu

David W. Volk
Department of Psychiatry, University of Pittsburgh, 3811 O'Hara Street, BST W1653, Pittsburgh, PA 15213, USA, volkdw@upmc.edu

Nigel M. Williams
Department of Psychological Medicine and Neurology, MRC Centre in Neuropsychiatric Genetics and Genomics, Cardiff University School of Medicine, Cardiff, UK, WilliamsNM@cardiff.ac.uk

Jared W. Young
Department of Psychiatry, University of California, San Diego, 9500 Gilman Drive, La Jolla, CA 92093-0804, USA, jaredyoung@ucsd.edu

Xianjin Zhou
Department of Psychiatry, University of California, San Diego, 9500 Gilman Drive, La Jolla, CA 92093-0804, USA, xzhou@ucsd.edu

Part I
Function, Outcome and Treatment in Schizophrenia

Assessing Function and Functional Outcome in Schizophrenia

Elizabeth Bromley and John S. Brekke

Contents

1 Introduction 4
   1.1 Functional Dimensions 4
   1.2 Recent Reviews and Overviews of Measures of Functioning
       in Schizophrenia 6
2 Construct Validity 7
   2.1 Functional Outcome as an Experiential Process 8
   2.2 Environmental Moderators of the Functional Dimensions 10
3 Ecological Validity 11
   3.1 Verisimilitude and Veridicality 12
   3.2 Observation in Naturalistic Environments 14
4 Conclusion 16
References 16

Abstract The diagnosis of schizophrenia can only be made in the presence of a loss of functioning in domains such as employment, independent living, and social functioning. Accurately measuring functioning is central to research on the course of the disorder, treatment and rehabilitation outcomes, and biosocial factors in schizophrenia. Assessments of functional disability have described three dimensions of functioning: functional capacity, functional performance, and functional outcome. The "competence/performance" distinction refers to the observation that an individual may demonstrate an ability to perform a functional task (capacity) but may not do so in her own community environment (performance). Functional

E. Bromley
Department of Psychiatry and Biobehavioral Sciences, Semel Institute Health Services Research Center, University of California, Los Angeles, 10920 Wilshire Blvd, Suite 300, Los Angeles, CA 90024, USA
e-mail: ebromley@ucla.edu

J.S. Brekke (✉)
School of Social Work, University of Southern California, Los Angeles, CA 90089-0411, USA
e-mail: brekke@usc.edu

outcomes are the result of both capacity and performance. Several recent reviews have compared the characteristics, reliability, and validity of various functional assessment instruments. Two major initiatives are underway to gather additional comparative data about functional assessment strategies. Recently, both the recovery movement and the recognition of the role of environmental factors in functioning have raised questions about the conceptual content of the functioning construct (construct validity). For instance, several studies have demonstrated that features of functioning need not track together over the course of the illness. In addition, the notion of recovery emphasizes processes like community integration and subjective well-being that are not static outcomes but are continually evolving features of the life course in chronic illness. Findings on the dynamic role of environmental moderators such as support and opportunity also present challenges to scientific constructs. For these reasons and others, the ecological validity of functional assessments has become a central concern. Both the verisimilitude and veridicality of functional assessments can be empirically assessed, but to date very few studies have measured the extent to which functional measures accurately predict individuals' behavior in their usual environments. Observational studies in naturalistic environments are one important area for future research.

1 Introduction

While the diagnosis of schizophrenia can reflect a considerable range in symptomatic presentation, it must include a loss of functioning in crucial psychosocial domains such as employment, independent living, and social functioning. Aside from its diagnostic significance, assessing functional domains has become crucial to a range of studies in schizophrenia. Accurately measuring functional outcomes is central to research on the course of the disorder, treatment responsiveness, rehabilitative outcomes, and the growing literature on biosocial factors in schizophrenia. This chapter provides a brief overview of several recent and important discussions on the measurement of function and functional outcomes in schizophrenia. We will then elaborate three issues that have increasing relevance in this area (1) the degree to which our present functional constructs reflect the theoretical and conceptual breadth that is required, (2) environmental moderators in models that aim to predict functional outcomes, and (3) the degree to which we have confidence that our best measures reflect the everyday living of individuals diagnosed with schizophrenia.

1.1 Functional Dimensions

Functional deficits in schizophrenia are a major cause of disability, accounting for a substantial portion of the indirect costs of the illness (Murray and Lopez 1996).

In the past few years, assessments of functional disability have attempted to delineate three dimensions of functioning in schizophrenia: functional capacity, functional performance, and functional outcomes (Brekke 2003). Functional capacity refers to an individual's ability or competence in performing tasks of daily living (e.g., holding a conversation, preparing a meal, performing job-related tasks, taking public transportation) which are most often assessed in controlled settings such as testing labs or clinics (Deegan 1996; McKibbin et al. 2004b; Harvey and Bellack 2009). Functional capacity tends to reflect the microskills of daily living tasks. While functional capacity is highly variable, it appears to be stable over time (Lieberman et al. 2008).

Second, functional performance refers to the individual's ability to perform or engage in the aforementioned behaviors in the real world, in her natural living environments. Lieberman et al. (2008) note that functional performance has a changeable course with a considerable return of psychosocial functioning after initial episodes, an escalating deterioration after subsequent episodes, and the possibility for a certain degree of functional improvement afterward. While the overall course of functional performance can be documented, the mechanics of functional performance have not been well characterized because so few studies use direct observation of day-to-day functioning in everyday environments. Finally, functional outcomes are the result of both capacity and performance and are typically measured as a level of achievement in work, independent living, and social domains that are occurring in the individual's natural living environments (Bellack et al. 2007). These are macroscopic outcomes such as the amount and type of work and money earned, the independence of living situations, or the breadth and type of social networks.

Evaluating the difference between what individuals are able to do and what they actually do in the real world is referred to as the "competence/performance distinction" (Harvey et al. 2007). The three dimensions of functioning emerged as it became clear that successfully demonstrating functional capacity does not necessarily mean that the individual will be able to perform the tasks in her own community settings. Individual characteristics such as neurocognitive functioning, motivation, confidence, risk-taking, self-evaluative abilities, and environmental factors (e.g., social support, employment, and housing opportunities) may at times have more influence on real-world performance than functional capacity alone. Although the functional capacity measures present logistical advantages over other less direct methods of examining functional outcome, they do not always provide helpful information about community behavior (Harvey et al. 2007).

While the empirical relationships between functional capacity, functional performance, and other outcome constructs have not been widely studied, functional capacity appears to be influenced more by individual factors (such as neurocognition) whereas functional performance and outcome are comparatively more determined by environmental factors (Green et al. 2004; Bowie et al. 2006). Functional capacity measures tend to show good correlations with cognitive performance measures (Leifker et al. 2009; Mausbach et al. 2009) and modest to low correlations with functional outcome measures (Dickerson et al. 1999;

Kurtz et al. 2005; Srinivasan and Tirupati 2005; Marwaha et al. 2007; Perlick et al. 2008). In addition, the strength of the relationship between capacity and outcome is impacted by the identity of the informant. Self-report assessments of functioning tend to show the lowest correlations with functional capacity and objective outcome measures (McKibbin et al. 2004a; Medalia et al. 2008). Informant reports appear to demonstrate stronger relationships with functional outcome (Keefe et al. 2006) than patient reports alone (Ventura et al. 2008). Correspondence between informant reports and patient self-reports tends to be poor, even about objective functional outcomes such as living situation (Bowie et al. 2007), although this is not consistent across all studies (Lecomte et al. 2004).

1.2 Recent Reviews and Overviews of Measures of Functioning in Schizophrenia

Several recent reviews compare the range of functional assessment strategies. Functional capacity measures (McKibbin et al. 2004b; Mausbach et al. 2009), functional outcome measures (Bellack et al. 2007), and measures of social functioning (Bellack et al. 2006) have been recently reviewed. Two conclusions about the relative utility of different assessment strategies are noteworthy. First, functional capacity measures that involve behavioral observation are considered to be the most psychometrically robust of the functional assessment strategies because they avoid the recall bias and informant bias of interview-based ratings or self- and other-report measures (Patterson et al. 2001; Bellack et al. 2007; Mausbach et al. 2007, 2009). Numerous functional capacity measures are available to assess an individual's ability to grocery shop (Hamera and Brown 2000), pay bills (Barrett et al. 2009), arrange medications (Patterson et al. 2002), solve problems with others (Bellack et al. 2006), cook (Semkovska et al. 2004), and accomplish other tasks. Second, self-report approaches are necessary to assess those aspects of functioning that are inherently subjective, such as satisfaction with life and well-being. Many functional outcome measures include subjective aspects of functioning (Lehman 1983; Bengtsson-Tops and Hansson 1999; Naber and Karow 2001; Prieto et al. 2003; Karow et al. 2005), like the semistructured interview-based Heinrichs–Carpenter Quality of Life Scale (Wehmeier et al. 2007) or the self-report Brief Quality of Life Inventory (Fujii et al. 2004). Despite their limitations, self-report strategies remain important because individuals with schizophrenia generally rate their well-being lower than even other socially disadvantaged groups (Lambert et al. 2009).

In addition to the excellent reviews cited earlier, two NIMH-sponsored initiatives are examining the relative utility of functional assessment strategies. MATRICS-CT (http://www.matrics.ucla.edu/matrics-ct/home.html) involves academic and industry scientists in a consensus-oriented process to evaluate functional capacity and interview-based measures. MATRICS-CT investigators have not reported results. The Validation of Everyday Real-World Outcomes (VALERO) project (Leifker et al. 2009) aims to identify optimal functional outcome scales and informants for use in clinical trials of cognition-enhancing medications. In a multi-study process, VALERO researchers will examine the convergence of functional outcome measures and performance-based measures (both functional capacity and neurocognition). VALERO investigators will also assess the validity of reports from various informants (self-report, relative/caregiver, clinician/case manager, prescriber). Following a RAND panel process (Kern et al. 2004) in which nominated scales were discussed and evaluated by a range of experts (Leifker et al. 2009), VALERO investigators selected two scales of everyday living skills (the Life Skills Profile and the Independent Living Skills Survey), two scales of social functioning (the Birchwood Social Functioning Scale and the Social Behavior Schedule), and two scales of both social functioning and everyday living (the Heinrichs–Carpenter Quality of Life Scale and Specific Levels of Functioning Scale) for further study.

Given the comprehensiveness and quality of the work described earlier, we recommend it as fundamental to any discussion of functional outcome. Our understanding of functional deficits in schizophrenia will increase substantially as we accrue empirical data about which existing scales of functioning are most useful. Yet, while ascertaining the relative value of existing scales is important, there is also a growing recognition that current approaches to functional assessment fail to address some crucial aspects of outcome in schizophrenia. Here, we highlight for discussion of two issues for further study and theorization. First, we discuss the degree to which our present functional constructs reflect the theoretical and conceptual breadth that is required; that is, the question of construct validity. Second, the role of environmental factors in models that predict functional outcome and the notion of functional outcome as an experiential process are considered. Finally, we discuss the degree to which we have confidence that our best measures reflect the everyday living of individuals diagnosed with schizophrenia; that is, the question of ecological validity. The advantages and disadvantages of different approaches to achieving ecological validity are addressed.

2 Construct Validity

A crucial question is the degree to which our models and measures of functioning reflect the breadth of the conceptual issues that are encompassed by the terms "function and functional outcome." Here, we discuss two challenges to current conceptualizations of functioning (1) the experiential aspects of functional recovery and (2) the role of environmental factors that can be determinants of functional capacity, performance, and outcome. These experiential factors, including those prioritized by advocates of the recovery movement, present unique psychometric challenges and suggest that a revision of scientific notions of functional outcome in schizophrenia may be warranted. Environment factors are not commonly included in models of functional outcome and their relationship to functional outcome is poorly understood (Hopper and Wanderling 2000).

2.1 Functional Outcome as an Experiential Process

Scientific definitions usually conceptualize functional outcome as a return to normalcy. These definitions focus on the alleviation of clinical symptoms and normalization of manifest behaviors in self-care, social, and work domains. Functional recovery has been defined as a protracted period (e.g., at least 6 months) of minimal symptoms, normal neuropsychiatric functioning, and the ability for people to function independently in the real world (Mausbach et al. 2009). This concept reflects the fact that scientific definitions view mental illness as a predominantly biological illness and view improvement as the absence of the illness (i.e., a cure) and a return to normal life functioning. Though these scientific definitions aim to be value-free, defining what is "normal" functioning involves value judgments because there is no gold standard for functional outcome (Harvey et al. 2009). As a result, the scientific definition of functioning is often recognized as an incomplete, albeit necessary, convention (Bellack et al. 2007; Harvey et al. 2009).

It is apparent that functioning can be usefully understood as composed of distinct domains that need not track together over the course of the illness. For instance, an early model of outcome came from Strauss and Carpenter (1972, 1977). They posited four outcome domains: duration of nonhospitalization, symptoms, work, and social functioning. In an extension of this work, Brekke and Long (2000) posited clinical, functional, and subjective experience outcome domains. Using longitudinal factor analysis, they found evidence for three distinct outcome factors that corresponded to the conceptual domains of clinical, functional, and subjective experience outcomes. There was strong discrimination among the factors with very modest correlations between them. This showed that the variables within the clusters tended to have the same trajectories and to travel together over time, while the three clusters traveled together but to a lesser extent.

It has also become apparent that recovery from schizophrenia need not imply a cure and that enduring symptoms may fluctuate within a certain range (Harvey and Bellack 2009). The notion of functional heterogeneity with regard to recovery is applicable since the relationship between symptoms and functioning in residual or chronic phases of the disorder is modest at best (Green et al. 2000). For example, one individual may experience relatively severe symptoms but function moderately well in employment, while another may have mild symptoms and not function well in their daily activities. Moreover, the sense of well-being experienced by each individual may be strongly shaped by entirely distinct factors. In first-person accounts, consumers describe the importance to their improvement of education and career attainment, self-determination (including choices against medical advice or nonevidence-based treatments), and confrontation of conflicts about shame and identity (Frese et al. 2009; Saks 2009). The significance of the subjective experience domain has been underscored and further elaborated by recent approaches to recovery from mental illness or what might be called the "recovery movement."

Ever since the President's New Freedom Commission on Mental Health report (2003) strongly urged the adoption of the notion of recovery as possible for all and as the guiding vision for mental health services, much effort has been placed in transforming services to achieve recovery outcomes. Although the notion of recovery has begun guiding policies and practices in many mental health systems in the United States, there is little consensus regarding how to define and measure recovery. Definitions of recovery vary between individuals and among groups including consumers, family members, clinicians, and researchers. These definitions have evolved from distinct perspectives, historical contexts, and goals (Onken et al. 2007). Definitions of recovery differ between objective and subjective referents (Lieberman et al. 2008; Harvey and Bellack 2009; McGurk et al. 2009) as well as between recovery as an outcome and recovery as a process (Liberman and Kopelowicz 2005; Bellack 2006; Davidson and Roe 2007).

Contrary to the more scientific approach, consumer definitions of recovery do not view recovery as an outcome but rather as a holistic nonlinear process of adaptation to illness and disability. During the recovery process, one moves beyond disability toward pursuit of a deeply personal and meaningful life that involves hope, empowerment, motivation, personal responsibility, and independent goals (Horan et al. 2006; Frese et al. 2009). Consumer definitions of recovery also encompass human rights, combating stigma, community integration, discrimination, and promoting recovery-oriented practices, services, and policies (Deegan 1996). As evidence of this, Brekke et al. (2009b) describe a stakeholder process for defining the outcomes from rehabilitation services. Consumers were sensitive to the functional aspects of recovery and outcome, but were far more attached to the subjective aspects such as motivation, self-esteem, internalized stigma, hope, and autonomy. In fact, they named these subjective aspects the "core strengths" and argued that improvement in the core strengths was far more important to them than changes in the common functional outcomes of work and independent living.

The recovery movement has also brought attention to issues such as community integration, stigma, and recovery itself. Subjective experiences such as self-efficacy, social isolation, self-esteem, hope, religious and spiritual orientation, motivation, autonomy, self-concept, and satisfaction with life are receiving more study. Outcomes researchers have taken note of the need to reconsider fundamental questions about defining improvement in schizophrenia. Some new conceptualizations of functioning, such as the capabilities approach (Hopper 2007), emphasize that improvements in functioning result from socially brokered processes made possible by specific social and material opportunities. Functioning is defined not as a threshold measure of satisfaction or as a set of normative behaviors, but as a measure of the valued things that individuals are able to do or be in their own environments. Researchers are increasingly recognizing that both emergent processes and subjective experiences can be reliably measured (Corrigan et al. 2006; Gioia and Brekke 2009) and that influential features of the environment can be identified (Brekke et al. 2009a; Frese et al. 2009).

Nonetheless, these revised concepts of outcome in schizophrenia raise considerable challenges. While the scientific definition would describe an individual as recovering "from" mental illness, which suggests an alleviation of the illness, the consumer definition would consider an individual as "being in recovery," which implies progressing with life despite enduring symptoms (Davidson and Roe 2007). Liberman and Kopelowicz (2005) note that process and outcome are always in dynamic interaction with one another. Yet researchers' attempts to separate outcome from process have led to two distinct research foci. The scientific view has attended mainly to clinical and functional recovery (e.g., Bobes et al. 2009) while the consumer view has emphasized the subjective experiences that play a part toward achieving a meaningful life (see Saks 2009 for an example). Some investigators have examined the relationships between functional recovery and subjective experiences (Brekke and Long 2000) but there is a dearth of information on the empirical relationship between functional outcomes and the "process" elements encompassed in consumer definitions of recovery such as hope, motivation, empowerment, and independent goals. Discerning how the subjective and process variables of recovery relate to functional outcomes would add significantly to the knowledge base and clinical interventions for individuals with schizophrenia. Moreover, developing consensus on what aspects are critical to the notion of functional outcome, or whether a novel and broader concept is needed, would help to settle controversies regarding the construct validity of function and functional outcome.

2.2 *Environmental Moderators of the Functional Dimensions*

One way to test the construct validity of the functional dimensions is to examine their performance within a theoretical context. As mentioned earlier, valid and reliable measures of functional capacity have become widely used as predictors of functional outcome. But the relationship between functional capacity and functional outcome is not likely a direct one. There are a host of factors that could intervene as mediators or moderators in models that predict functional outcome. For example, it has been hypothesized that environmental characteristics could moderate the relationship between individual characteristics (like neurocognition) and the functional dimensions (Brekke 2007). In this regard, Brekke (2007) has suggested that three environmental determinants are potentially critical for functional performance to occur and for functional outcome to improve and endure: opportunity, support, and enhancements. Opportunities such as available options for housing, employment, and social engagement must be in place. It is obvious that someone who has the functional capacity for work but who cannot find employment due to a scarcity of options will have poor occupational outcomes. The same is true with regard to social functional and housing. Second, the support of family, friends, peers, and/or staff who encourage adaptive behaviors and behavioral change must also be available. These supports could be particularly significant for individuals who are struggling with long-standing psychiatric challenges and cycles of relapse

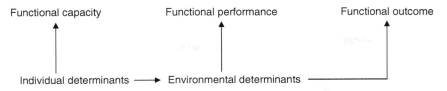

Fig. 1 The relationship between individual and environmental determinants of the functional dimensions

and hospitalization. Finally, enhancements, such as available treatments or services that aim to improve the functional outcome of persons with schizophrenia, are a significant environmental factor as well.

Figure 1 suggests that individual determinants (e.g., cognition, premorbid functioning) will have the most direct impact on indicators of functional capacity because environmental determinants can be controlled and minimized in lab settings. Functional performance and outcome are more environmentally determined; therefore, the influence of individual determinants is modified or moderated by environmental determinants (Green et al. 2004; Brekke 2007; Harvey et al. 2007). While studies are needed to carefully test these notions, there is evidence that neurocognition, for example, is a stronger predictor of functional capacity than functional outcome (Green et al. 2004; Bowie et al. 2006). There is also evidence that social and economic supports affect the strength of the relationship between functional capacity and functional outcome (Harvey et al. 2009) and between neurocognition and functional outcome (Srinivasan and Tirupati 2005).

In general, these moderators can serve to decompose the relationships between predictors (such as cognition, premorbid status, or functional capacity) and functional outcome. For example, the generally modest relationship between neurocognition and functional outcomes in schizophrenia could be well explained by the presence of one or more environmental moderators such that the strength of the relationship between neurocognition and outcomes could vary greatly depending on the presence of moderators (see Nakagami et al. 2008; Brekke et al. 2009a for related examples). Unfortunately, there are not existing scales for many of these moderators. We are currently in a pilot-testing phase of a battery of moderators that could be useful for these purposes. The battery assesses environmental opportunity, social support, social capital, clinical enhancements, and the demand characteristics (i.e., whether functional gains are expected and encouraged) in subjects' usual environments.

3 Ecological Validity

The role of social and cultural factors in the study of functional outcome leads to the final issue that we will address, the ecological validity of our functional measures of capacity and outcome. Sbordone and Long (1996) define ecological validity as the "predictive relationship between the patient's performance on a set of neuropsychological tests and the patient's behavior in a variety of real-world settings" (p. 16). Ecological validity is the degree to which results obtained in controlled experimental conditions are related to those obtained in naturalistic environments (Chaytor and Schmitter-Edgecombe 2003). That is, ecological validity is a judgment regarding the appropriateness of *inferences* that tests allow (Landy 1986; Birchwood et al. 1990; Heinrichs 1990). The ecological validity of functional assessment strategies has become a significant area of research (Silver 2000; Twamley et al. 2003; Chaytor et al. 2006; Koren et al. 2006) because functional assessment strategies are increasingly used to make inferences about the subject's everyday capabilities in usual environments (Heinrichs 1990). Ecological questions concern how functional deficits or strengths identified on structured instruments translate into deficiencies or strengths at home, work, and other social situations. They are often the questions of interest to physicians, families, and patients: How will my loved one do at work? What kinds of supports does she need at home? What kind of social life can I expect him to develop? (Heinrichs 1990)

Validation of assessment tools is a *process* (Bellack et al. 2007) requiring a series of studies using different theoretical approaches, different samples, and different methods over time (Benson 1998). Validation strategies often rely on correlational approaches (Kibby et al. 1998; Brown et al. 2006; Mausbach et al. 2009) in which a capacity or performance measure is empirically associated to a functional outcome measure (Benson 1998; Rempfer et al. 2003) such as housing independence or work performance. However, almost no studies have compared functional measures in schizophrenia with what patients do when they are directly observed in their daily lives in naturalistic community settings (Carpenter 2006; Bellack et al. 2007). To a large extent, we do not know to what degree our best measures of functional outcome correspond to or predict the daily routines of patients in their usual contexts. As a result, current functional assessment strategies have shortcomings in external validity. Moreover, as mentioned above, the field lacks data on dynamic interactions within the environment that improve or impair real-world functioning (Silver 2000; Bellack et al. 2007; Bromley 2007a).

3.1 Verisimilitude and Veridicality

Of the two research strategies available to improve the ecological validity of functional assessment, v*erisimilitude* is most commonly used (Chaytor et al. 2006; Spooner and Pachana 2006). Verisimilitude approaches aim to construct tests that mimic the tasks an individual might perform in life (Silver 2000). For instance, many functional capacity tests assess individuals' capacity to perform cognitive skills in simulated settings (e.g., a hospital kitchen; Aubin et al. 2009a, b) or with prepared props (e.g., a bus map; Patterson et al. 2001); they may assess how well a subject can solve an everyday problem described in a hypothetical scenario (Kee et al. 2009); or they may assess individuals' social functioning skills by engaging them in role-plays (Bellack 2006). The alternative approach, *veridicality*, refers to the degree to which existing tests are empirically related to everyday experiences (Chaytor and Schmitter-Edgecombe 2003). Veridicality approaches use the real world as the ecological standard, assessing the degree to which existing tests are empirically related to actual events in patients' lives. We are aware of only one study involving nine subjects that addresses the veridicality of neuropsychological measures in schizophrenia (Gioia 2006; Gioia and Brekke 2009). Veridicality approaches have been used in the study of childhood behavior and social interaction within the classroom (Lytton 1971; Barkley et al. 1990; Solanto et al. 2001; Ruble and Scott 2002) and in the study of the ecological impact of traumatic brain injury (McDonald et al. 2004; Moseley et al. 2004).

Verisimilitude approaches have limitations in external validity that may come from the tasks assessed, the testing setting, unassessed factors, or subject's ability or willingness to engage in role-play. Testing situations intentionally eliminate obvious deterrents of performance (Chaytor and Schmitter-Edgecombe 2003), and thus may differ markedly from subject's everyday environments. Testing situations are quiet, prompts and instructions are provided for initiation and completion, the subject is supported whether he fails or success (Chaytor and Schmitter-Edgecombe 2003), and tasks are usually short and novel (Sbordone and Long 1996). Tasks may differ from those that subjects are called upon to perform in everyday life, and there is no gold standard that insures that test content matches everyday scenarios (Bellack 1983). Tests may overassess knowledge or cognitive skill and underassess motivation, compensatory strategies, or other environmental factors. For instance, social skills or peer influences may provide crucial support during everyday performance (Silver 2000; Pellegrini 2001). In short, "the external validity of these instruments cannot be automatically assumed" (Bellack et al. 2007, p. 813). Since studies utilizing direct observation in naturalistic environments are lacking, it is unclear to what extent verisimilitude approaches provide a meaningful representation of real-world behaviors (McKibbin et al. 2004b).

For instance, the Test of Grocery Shopping Skills (TOGSS; Hamera and Brown 2000; Rempfer et al. 2003; Brown et al. 2006) observes subjects completing a shopping task in a naturalistic environment (Bellack et al. 2007), a medium-sized grocery store, that is novel to all subjects. Because of the importance of grocery shopping to independent living, the TOGSS appears to be a test with high verisimilitude. Performance on a measure like the TOGSS correlates with cognitive skill (Rempfer et al. 2003; McKibbin et al. 2004b) and with knowledge of grocery shopping (Brown et al. 2006). Yet it is not known how well performance on the TOGSS provides an accurate representation of real-life grocery shopping. Individuals may be adept within a familiar store or may draw upon social skills to ask for help when needed. Similarly, a meal preparation task performed in a standard community grocery store and hospital kitchen (Aubin et al. 2009a) is scored by breaking down tasks into dozens of observable behaviors. Subjects can be scored for their ability to sequence, to perceive problems, and to persist in task completion. The extent to which tests of meal preparation represent how individuals prepare meals in vivo is not known. Individuals may have mastered certain recipes but still have difficulty in following directions efficiently when faced with a new one.

Finally, the extent to which skills on simulated shopping or cooking tasks generalize to other tasks is uncertain.

The most widely used verisimilitude-based measure of functional capacity, the University of California, San Diego Performance-Based Skills Assessment (UPSA; Patterson et al. 2001), shows strong convergent validity with neurocognitive measures and interview-based measures of functioning (McKibbin et al. 2004b; Mausbach et al. 2007), and predicts independent living (Mausbach et al. 2009). But a crosscultural study using the UPSA demonstrates that real-world outcomes are also shaped by social context (Harvey et al. 2009). Swedish and US samples of patients with schizophrenia demonstrated similar ability on the UPSA and received similar functional ratings from case managers. Yet residential outcomes between the two samples were substantially distinct: 80% of the Swedish cohort versus 46% of the New York cohort lived independently. The authors suggest that economic and social support for housing in Sweden resulted in better residential outcomes (Harvey et al. 2009). The Swedish cohort experienced higher disability compensation, generous support for rental costs, and a lower cost of living. In this study, these differences were so consequential that there was no association between performance on the UPSA and residential outcome for people with schizophrenia in Sweden. This study demonstrates that performance on scales with good verisimilitude may predict neither the real-world impact of those skills nor overall outcome.

As this study demonstrates, functional behaviors are embedded within a context (Pellegrini 2004) that can significantly shape functional outcomes (Srinivasan and Tirupati 2005). Clinical and preclinical studies suggest that even basic cognitive processes are context dependent (Delaney and Sahakyan 2007; Bacon and Izaute 2009). The goal properties, motivational content, and cognitive content of a task can shape the nature of neuronal activation (Miller and Cohen 2001; Watanabe and Sakagami 2007). Affective functioning also may be context dependent. While laboratory-based studies show that people with schizophrenia report experiencing similar levels of emotion compared to people without schizophrenia (Kring and Moran 2008), Gard and Kring (2009) use experience sampling methods to show that subjects with schizophrenia experience social interactions as more activating and less pleasant than subjects without schizophrenia. Distinct psychological states such as effort (Gorissen et al. 2005), self-monitoring (Koren et al. 2006; Lysaker et al. 2008), and foresight (Eack and Keshavan 2008) appear to be important determinants of behaviors in context (Lysaker et al. 2007). These studies suggest that social contexts may shape overall functional outcome as well as everyday performance of basic functional tasks.

3.2 *Observation in Naturalistic Environments*

These findings suggest the need for an expanded program of research in ecological validity that assesses the veridicality of functional assessment strategies. Veridicality approaches would empirically assess the degree to which functional measures predict actual behaviors in usual contexts. True veridicality approaches use the real world as the ecological standard. As one group argues:

> Observation in naturalistic settings is considered to be the most robust approach because it allows the rater to evaluate whether the skills are actually implemented in the environment. Such data can also provide a measuring stick by which to evaluate the validity of other measures of everyday functioning. (Bellack et al. 2007, p. 813)

In situ behaviors are so rarely studied that naturalistic observation could generate and test numerous hypotheses about critical unanswered questions. Naturalistic observation can capture intervening variables that hinder or help performance in actual environments. Naturalistic observation can identify factors that account for functional outcomes that are either better or worse than would be predicted by functional capacity. In the study of social processes, naturalistic observation can clarify the phenomenology and everyday impact of social cognitive deficits. Because the real-time manifestations of basic cognitive deficits, such as impaired working memory, are poorly understood, observation can clarify the kinds of behaviors and experiences that should alert patients, clinicians, and caregivers to cognitive impairments (Bromley 2007b). All of the above would suggest potential interventions targeted toward improving functional outcome.

Ethnographic techniques offer well-established strategies for naturalistic observation (Hannerz 2003; Van der Geest and Finkler 2004). While ethnographies traditionally rely on written descriptions (also known as field notes), advances in video and audio technology allow detailed and reliable data to be collected discretely and affordably. Video data are increasingly easy to convert, secure, and store (Walker et al. 1993; Stålberg et al. 2008; Vaskinn et al. 2009). Approaches to quantifying video data, assisted by specialized software programs, can be as reliable as scoring strategies used in performance-based tests. For instance, video data can be rated using coding schemes adapted from performance-based tests (Dickinson et al. 2007), experience sampling methods (Gard and Kring 2009), or diary approaches (Jolley et al. 2006). We would suggest that naturalistic observations are so feasible that they should routinely be used during scale development and validation. For instance, naturalistic observation of shopping or medication management would ascertain the external validity of functional capacity measures already shown to have high internal validity, such as the TOGSS (Hamera and Brown 2000) and the Medication Management Ability Assessment (Patterson et al. 2002).

As an example, the authors have developed a video ethnography method to assess daily living in naturalistic community settings. In pilot testing, subjects with schizophrenia tolerate videotaping and report that daily routines are not altered in the presence of the ethnographer. We have developed coding schemes that quantify video and audio data to capture a range of functional skills. Our approach codes video data for (1) activity level in terms of complexity and initiative; (2) goal pursuit in six functioning areas such as activities of daily living, socialization, and leisure; (3) in vivo problem-solving, which characterizes subjects' ability to respond to contingent circumstances in the moment; and (4) social interaction in terms of frequency, duration, initiation, and relatedness. These features can be reliably coded and they show good range and variability. The four coding schemes capture functional performance, sequencing and multitasking, social competence, social problem-solving, theory of mind skills, emotional processing, and other functional skills. This approach to quantifying naturalistic data can be used to test the ecological validity of neurocognitive and functional measures.

4 Conclusion

Central to the definition of schizophrenia is impairment in functional domains such as employment, self-care, and social functioning. Functional disability can be understood as composed of three distinct dimensions including functional capacity, functional performance, and functional outcome. These dimensions can be in dynamic interaction with one another, but these dimensions may also have quite distinct determinants and a differential impact on overall wellness. Significant strides are being made in developing consensus definitions and in the psychometric validation of functional measures in each domain. However, recent attention to the process of recovery, subjective experience, and social and cultural contextual factors raise challenges for the field. Relationships between subjective process and objective outcomes and the mechanisms by which environmental factors aid functional outcome are poorly understood. We have addressed a set of issues relevant to the construct validity and ecological validity of functional measures that are critical to continuing progress in the field. We have also highlighted the relative scarcity of veridicality approaches and of research utilizing naturalistic observation, despite the fact that functioning research is fundamentally concerned with the everyday realities faced by individuals with schizophrenia. We have argued that an expanded research agenda that includes these issues can identify innovative avenues through which to improve the assessment of function and functional outcomes in schizophrenia.

References

Aubin G, Chapparo C, Gelinas I, Stip E, Rainville C (2009a) Use of the Perceive, Recall, Plan and Perform System of Task Analysis for persons with schizophrenia: a preliminary study. Aust Occup Ther J 56:189–199

Aubin G, Stip E, Gelinas I, Rainville C, Chapparo C (2009b) Daily activities, cognition and community functioning in persons with schizophrenia. Schizophr Res 107:313–318

Bacon E, Izaute M (2009) Metacognition in schizophrenia: processes underlying patients' reflections on their own episodic memory. Biol Psychiatry 66:1031–1037

Barkley RA, DuPaul GJ, McMurray MB (1990) Comprehensive evaluation of attention deficit disorder with and without hyperactivity as defined by research criteria. J Consult Clin Psychol 58:775–789

Barrett JJ, Hart KJ, Schmerler JT, Willmarth K, Carey JA, Mohammed S (2009) Criterion validity of the financial skills subscale of the direct assessment of functional status scale. Psychiatry Res 166:148–157

Bellack AS (1983) Recurrent problems in the behavioral assessment of social skill. Behav Res Ther 21:29–41

Bellack AS (2006) Scientific and consumer models of recovery in schizophrenia: concordance, contrasts, and implications. Schizophr Bull 32:432–442

Bellack AS, Brown CH, Thomas-Lohrman S (2006) Psychometric characteristics of role-play assessments of social skill in schizophrenia. Behav Ther 37:339–352

Bellack AS, Green MF, Cook JA, Fenton W, Harvey PD, Heaton RK, Laughren T, Leon AC, Mayo DJ, Patrick DL, Patterson TL, Rose A, Stover E, Wykes T (2007) Assessment of community functioning in people with schizophrenia and other severe mental illnesses: a white paper based on an NIMH-sponsored workshop. Schizophr Bull 33:805–822

Bengtsson-Tops A, Hansson L (1999) Subjective quality of life in schizophrenic patients living in the community. Relationship to clinical and social characteristics. Eur Psychiatry 14:256–263

Benson J (1998) Developing a strong program of construct validation: a test anxiety example. Educ Meas Issues Pract 17:10–22

Birchwood M, Smith J, Cochrane R, Wetton S, Copestake S (1990) The Social Functioning Scale. The development and validation of a new scale of social adjustment for use in family intervention programmes with schizophrenic patients. Br J Psychiatry 157:853–859

Bobes J, Ciudad A, Alvarez E, San L, Polavieja P, Gilaberte I (2009) Recovery from schizophrenia: results from a 1-year follow-up observational study of patients in symptomatic remission. Schizophr Res 115:58–66

Bowie CR, Reichenberg A, Patterson TL, Heaton RK, Harvey PD (2006) Determinants of real-world functional performance in schizophrenia subjects: correlations with cognition, functional capacity, and symptoms. Am J Psychiatry 163:418–425

Bowie CR, Twamley EW, Anderson H, Halpern B, Patterson TL, Harvey PD (2007) Self-assessment of functional status in schizophrenia. J Psychiatr Res 41:1012–1018

Brekke JS (2003) Functional outcome assessment in schizophrenia. In: NIMH-sponsored conference, measurement and treatment research to improve cognition in schizophrenia (MATRICS), Bethesda, MD, April 14–15, 2003

Brekke J (2007) The relationship between cognitive and environmental determinants of the functional dimensions in schizophrenia. In: NIMH-sponsored conference, measurement and treatment research to improve cognition in schizophrenia (MATRICS-CT), Bethesda, MD, August 21–22, 2007

Brekke J, Long J (2000) Community-based psychosocial rehabilitation and prospective change in functional, clinical, and subjective experience variables in schizophrenia. Schizophr Bull 26:667–680

Brekke JS, Hoe M, Green MF (2009a) Neurocognitive change, functional change and service intensity during community-based psychosocial rehabilitation for schizophrenia. Psychol Med 39:1637–1647

Brekke JS, Phillips E et al (2009b) Implementation practice and implementation research: a report from the field. Res Soc Work Pract 19:592–601

Bromley E (2007a) Barriers to the appropriate clinical use of medications that improve the cognitive deficits of schizophrenia. Psychiatr Serv 58:475–481

Bromley E (2007b) Clinicians' concepts of the cognitive deficits of schizophrenia. Schizophr Bull 33:648–651

Brown CE, Rempfer MV, Hamera E, Bothwell R (2006) Knowledge of grocery shopping skills as a mediator of cognition and performance. Psychiatr Serv 57:573–575

Carpenter WT (2006) Targeting schizophrenia research to patient outcomes. Am J Psychiatry 163:353–355

Chaytor N, Schmitter-Edgecombe M (2003) The ecological validity of neuropsychological tests: a review of the literature on everyday cognitive skills. Neuropsychol Rev 13:181–197

Chaytor N, Schmitter-Edgecombe M, Burr R (2006) Improving the ecological validity of executive functioning assessment. Arch Clin Neuropsychol 21:217–227

Corrigan PW, Watson AC, Miller FE (2006) Blame, shame, and contamination: the impact of mental illness and drug dependence stigma on family members. J Fam Psychol 20: 239–246

Davidson L, Roe D (2007) Recovery from versus recovery in serious mental illness: one strategy for lessening confusion plaguing recovery. J Ment Health 16:459–470

Deegan P (1996) Recovery as a journey of the heart. Psychiatr Rehabil J 19:91–97

Delaney PF, Sahakyan L (2007) Unexpected costs of high working memory capacity following directed forgetting and contextual change manipulations. Mem Cognit 35:1074–1082

Dickerson F, Boronow JJ, Ringel N, Parente F (1999) Social functioning and neurocognitive deficits in outpatients with schizophrenia: a 2-year follow-up. Schizophr Res 37:13–20

Dickinson D, Bellack AS, Gold JM (2007) Social/communication skills, cognition, and vocational functioning in schizophrenia. Schizophr Bull 33:1213–1220

Eack SM, Keshavan MS (2008) Foresight in schizophrenia: a potentially unique and relevant factor to functional disability. Psychiatr Serv 59:256–260

Frese FJ III, Knight EL, Saks E (2009) Recovery from schizophrenia: with views of psychiatrists, psychologists, and others diagnosed with this disorder. Schizophr Bull 35:370–380

Fujii DE, Wylie AM, Nathan JH (2004) Neurocognition and long-term prediction of quality of life in outpatients with severe and persistent mental illness. Schizophr Res 69:67–73

Gard DE, Kring AM (2009) Emotion in the daily lives of schizophrenia patients: context matters. Schizophr Res 115(2):379–380

Gioia D (2006) A contextual study of daily living strategies in neurocognitively impaired adults with schizophrenia. Qual Health Res 16:1217–1235

Gioia D, Brekke J (2009) Neurocognition, ecological validity, and daily living in the community for individuals with schizophrenia: a mixed methods study. Psychiatry 72:94–107

Gorissen M, Sanz J, Schmand B (2005) Effort and cognition in schizophrenia patients. Schizophr Res 78:199–208

Green MF, Kern RS, Braff DL, Mintz J (2000) Neurocognitive deficits and functional outcome in schizophrenia: are we measuring the 'right stuff'? Schizophr Bull 26:119–136

Green MF, Kern RS, Heaton RK (2004) Longitudinal studies of cognition and functional outcome in schizophrenia: implications for MATRICS. Schizophr Res 72:41–51

Hamera E, Brown CE (2000) Developing a context-based performance measure for persons with schizophrenia: the test of grocery shopping skills. Am J Occup Ther 54:20–25

Hannerz U (2003) Being there...and there...and there! Ethnography 4:201–216

Harvey PD, Bellack AS (2009) Toward a terminology for functional recovery in schizophrenia: is functional remission a viable concept? Schizophr Res 35:300–306

Harvey PD, Velligan DI, Bellack AS (2007) Performance-based measures of functional skills: usefulness in clinical treatment studies. Schizophr Bull 33:1138–1148

Harvey PD, Helldin L, Bowie CR, Heaton RK, Olsson AK, Hjärthag F, Norlander T, Patterson TL (2009) Performance-based measurement of functional disability in schizophrenia: a cross-national study in the United States and Sweden. Am J Psychiatry 166:821–827

Heinrichs RW (1990) Current and emergent applications of neuropsychological assessment: problems of validity and utility. Prof Psychol Res Pract 21:171–176

Hopper K (2007) Rethinking social recovery in schizophrenia: what a capabilities approach might offer. Soc Sci Med 65:868–879

Hopper K, Wanderling J (2000) Revisiting the developed versus developing country distinction in course and outcome in schizophrenia: results from ISoS, the WHO collaborative followup project. International Study of Schizophrenia. Schizophr Bull 26:835–846

Horan WP, Kring AM, Blanchard JJ (2006) Anhedonia in schizophrenia: a review of assessment strategies. Schizophr Bull 32:259–273

Jolley S, Garety PA, Ellett L, Kuipers E, Freeman D, Bebbington PE, Fowler DG, Dunn G (2006) A validation of a new measure of activity in psychosis. Schizophr Res 85:288–295

Karow A, Moritz S, Lambert M, Schoder S, Krausz M (2005) PANSS syndromes and quality of life in schizophrenia. Psychopathology 38:320–326

Kee KS, Horan WP, Salovey P, Kern RS, Sergi MJ, Fiske AP, Lee J, Subotnik KL, Nuechterlein K, Sugar CA, Green MF (2009) Emotional intelligence in schizophrenia. Schizophr Res 107:61–68

Keefe RS, Poe M, Walker TM, Kang JW, Harvey PD (2006) The schizophrenia cognition rating scale: an interview-based assessment and its relationship to cognition, real-world functioning, and functional capacity. Am J Psychiatry 163:426–432

Kern RS, Green MF, Nuechterlein KH, Deng BH (2004) NIMH-MATRICS survey on assessment of neurocognition in schizophrenia. Schizophr Res 72:11–19

Kibby MY, Schmitter-Edgecombe M, Long CJ (1998) Ecological validity of neuropsychological tests: focus on the California Verbal Learning Test and the Wisconsin Card Sorting Test. Arch Clin Neuropsychol 13:523–534

Koren D, Seidman LJ, Goldsmith M, Harvey PD (2006) Real-world cognitive – and metacognitive – dysfunction in schizophrenia: a new approach for measuring (and remediating) more "right stuff". Schizophr Bull 32:310–326

Kring AM, Moran EK (2008) Emotional response deficits in schizophrenia: insights from affective science. Schizophr Bull 34:819–834

Kurtz MM, Moberg PJ, Ragland JD, Gur RC, Gur RE (2005) Symptoms versus neurocognitive test performance as predictors of psychosocial status in schizophrenia: a 1- and 4-year prospective study. Schizophr Bull 31:167–174

Lambert M, Schimmelmann BG, Schacht A, Karow A, Wagner T, Wehmeier PM, Huber CG, Hundemer HP, Dittmann RW, Naber D (2009) Long-term patterns of subjective wellbeing in schizophrenia: cluster, predictors of cluster affiliation, and their relation to recovery criteria in 2842 patients followed over 3 years. Schizophr Res 107:165–172

Landy F (1986) Stamp collecting versus science. Am Psychol 41:1183–1192

Lecomte T, Wallace CJ, Caron J, Perreault M, Lecomte J (2004) Further validation of the client assessment of strengths interests and goals. Schizophr Res 66:59–70

Lehman AF (1983) The well-being of chronic mental patients. Arch Gen Psychiatry 40:369–373

Leifker FR, Patterson TL, Heaton RK, Harvey PD (2009) Validating Measures of Real-World Outcome: The Results of the VALERO Expert Survey and RAND Panel. Schizophr Bull [Epub ahead of print, June 12]

Liberman RP, Kopelowicz A (2005) Recovery from schizophrenia: a concept in search of research. Psychiatr Serv 56:735–742

Lieberman JA, Drake RE, Sederer LI, Belger A, Keefe R, Perkins D, Stroup S (2008) Science and recovery in schizophrenia. Psychiatr Serv 59:487–496

Lysaker PH, Dimaggio G, Buck KD, Carcione A, Nicolò G (2007) Metacognition within narratives of schizophrenia: associations with multiple domains of neurocognition. Schizophr Res 93:278–287

Lysaker PH, Warman DM, Dimaggio G, Procacci M, Larocco VA, Clark LK, Dike CA, Nicolò G (2008) Metacognition in schizophrenia: associations with multiple assessments of executive function. J Nerv Ment Dis 196:384–389

Lytton H (1971) Observation studies of parent-child interaction: a methodological review. Child Dev 42:651–684

Marwaha S, Johnson S, Bebbington P, Stafford M, Angermeyer MC, Brugha T, Azorin JM, Kilian R, Hansen K, Toumi M (2007) Rates and correlates of employment in people with schizophrenia in the UK, France and Germany. Br J Psychiatry 191:30–37

Mausbach BT, Harvey PD, Goldman SR, Jeste DV, Patterson TL (2007) Development of a brief scale of everyday functioning in persons with serious mental illness. Schizophr Bull 33:1364–1372

Mausbach BT, Moore R, Bowie C, Cardenas V, Patterson TL (2009) A review of instruments for measuring functional recovery in those diagnosed with psychosis. Schizophr Bull 35:307–318

McDonald S, Flanagan S, Martin I, Saunders C (2004) The ecological validity of TASIT: a test of social perception. Neuropsychol Rehabil 14:285–302

McGurk SR, Mueser KT, DeRosa TJ, Wolfe R (2009) Work, recovery, and comorbidity in schizophrenia: a randomized controlled trial of cognitive remediation. Schizophr Bull 35:319–335

McKibbin C, Patterson TL, Jeste DV (2004a) Assessing disability in older patients with schizophrenia: results from the WHODAS-II. J Nerv Ment Dis 192:405–413

McKibbin CL, Brekke JS, Sires D, Jeste DV, Patterson TL (2004b) Direct assessment of functional abilities: relevance to persons with schizophrenia. Schizophr Res 72:53–67

Medalia A, Thysen J, Freilich B (2008) Do people with schizophrenia who have objective cognitive impairment identify cognitive deficits on a self report measure? Schizophr Res 105:156–164

Miller EK, Cohen JD (2001) An integrative theory of prefrontal cortex function. Annu Rev Neurosci 24:167–202

Moseley AM, Lanzarone S, Bosman JM, van Loo MA, de Bie RA, Hassett L, Caplan B (2004) Ecological validity of walking speed assessment after traumatic brain injury: a pilot study. J Head Trauma Rehabil 19:341–348

Murray CL, Lopez AD (1996) The global burden of disease. Harvard University Press, Cambridge, MA

Naber D, Karow A (2001) Good tolerability equals good results: the patient's perspective. Eur Neuropsychopharmacol 11(Suppl 4):S391–S396

Nakagami E, Xie B, Hoe M, Brekke JS (2008) Intrinsic motivation, neurocognition and psychosocial functioning in schizophrenia: testing mediator and moderator effects. Schizophr Res 105:95–104

Onken SJ, Craig CM, Ridgway P, Ralph RO, Cook JA (2007) An analysis of the definitions and elements of recovery: a review of the literature. Psychiatr Rehabil J 31:9–22

Patterson TL, Goldman S, McKibbin CL, Hughs T, Jeste DV (2001) UCSD Performance-Based Skills Assessment: development of a new measure of everyday functioning for severely mentally ill adults. Schizophr Bull 27:235–245

Patterson TL, Lacro J, McKibbin CL, Moscona S, Hughs T, Jeste DV (2002) Medication management ability assessment: results from a performance-based measure in older outpatients with schizophrenia. J Clin Psychopharmacol 22:11–19

Pellegrini AD (2001) Practitioner review: the role of direct observation in the assessment of young children. J Child Psychol Psychiatry 42:861–869

Pellegrini AD (2004) Observing children in their natural worlds: a methodological primer. Lawrence Erlbaum, Mahwah, NJ

Perlick DA, Rosenheck RA, Kaczynski R, Bingham S, Collins J (2008) Association of symptomatology and cognitive deficits to functional capacity in schizophrenia. Schizophr Res 99:192–199

President's New Freedom Commission on Mental Health Report (2003) Achieving the promise: transforming mental health care in America. DHHS, Rockville, MD

Prieto L, Novick D, Sacristán JA, Edgell ET, Alonso J, SOHO Study Group (2003) A Rasch model analysis to test the cross-cultural validity of the EuroQoL-5D in the Schizophrenia Outpatient Health Outcomes Study. Acta Psychiatr Scand Suppl 416:24–29

Rempfer MV, Hamera EK, Brown CE, Cromwell RL (2003) The relations between cognition and the independent living skill of shopping in people with schizophrenia. Psychiatry Res 117:103–112

Ruble LA, Scott MM (2002) Executive functions and the natural habitat behaviors of children with autism. Autism 6:365–381

Saks ER (2009) Some thoughts on denial of mental illness. Am J Psychiatry 166:972–973

Sbordone RJ, Long CJ (1996) Ecological validity of neuropsychological testing. CRC, Boca Raton, FL

Semkovska M, Bédard MA, Godbout L, Limoge F, Stip E (2004) Assessment of executive dysfunction during activities of daily living in schizophrenia. Schizophr Res 69:289–300

Silver CH (2000) Ecological validity of neuropsychological assessment in childhood traumatic brain injury. J Head Trauma Rehabil 15:973–988

Solanto MV, Abikoff H, Sonuga-Barke E, Schachar R, Logan GD, Wigal T, Hechtman L, Hinshaw S, Turkel E (2001) The ecological validity of delay aversion and response inhibition as measures of impulsivity in AD/HD: a supplement to the NIMH multimodal treatment study of AD/HD. J Abnorm Child Psychol 29:215–228

Spooner DM, Pachana NA (2006) Ecological validity in neuropsychological assessment: a case for greater consideration in research with neurologically intact populations. Arch Clin Neuropsychol 21:327–337

Srinivasan L, Tirupati S (2005) Relationship between cognition and work functioning among patients with schizophrenia in an urban area of India. Psychiatr Serv 56:1423–1428

Stålberg G, Lichtenstein P, Sandin S, Hultman CM (2008) Video-based assessment of interpersonal problem solving skills in patients with schizophrenia, their siblings and non-psychiatric controls. Scand J Psychol 49:77–82

Strauss JS, Carpenter WT Jr (1972) The prediction of outcome in schizophrenia I: characteristics of outcome. Arch Gen Psychiatry 27:739–746

Strauss JS, Carpenter WT Jr (1977) Prediction of outcome in schizophrenia III: five-year outcome and its predictors. Arch Gen Psychiatry 34:159–163

Twamley EW, Jeste DV, Bellack AS (2003) A review of cognitive training in schizophrenia. Schizophr Bull 29:359–382

Van der Geest S, Finkler K (2004) Hospital ethnography: introduction. Soc Sci Med 59:1995–2001

Vaskinn A, Sergi MJ, Green MF (2009) The challenges of ecological validity in the measurement of social perception in schizophrenia. J Nerv Ment Dis 197:700–702

Ventura J, Cienfuegos A, Boxer O, Bilder R (2008) Clinical global impression of cognition in schizophrenia (CGI-CogS): reliability and validity of a co-primary measure of cognition. Schizophr Res 106:59–69

Walker EF, Grimes KE, Davis DM, Smith AJ (1993) Childhood precursors of schizophrenia: facial expressions of emotion. Am J Psychiatry 150:1654–1660

Watanabe M, Sakagami M (2007) Integration of cognitive and motivational context information in the primate prefrontal cortex. Cereb Cortex 17(Suppl 1):i101–i109

Wehmeier PM, Kluge M, Schneider E, Schacht A, Wagner T, Schreiber W (2007) Quality of life and subjective well-being during treatment with antipsychotics in out-patients with schizophrenia. Prog Neuropsychopharmacol Biol Psychiatry 31:703–712

Antipsychotics and Metabolics in the Post-CATIE Era

Jonathan M. Meyer

Contents

1 Introduction 24
2 Sources of Cardiovascular Risk 26
3 The Cardiovascular and Metabolic Risk Profile of Subjects Entering the CATIE Schizophrenia Trial 29
4 The Impact of Antipsychotic Treatment on Cardiovascular and Metabolic Outcomes in the CATIE Schizophrenia Trial 30
    4.1 Metabolic Outcomes 30
    4.2 Framingham Cardiovascular Risk 31
    4.3 Outcomes with Novel Biomarkers 33
5 The Post-CATIE Era 33
    5.1 Clinical Conclusions 33
    5.2 Hypotheses on Schizophrenia and Metabolic Risk, and Adiposity-Independent Drug Effects 35
6 Conclusions 37
References 37

Abstract Schizophrenia patients have high prevalence of cardiovascular (CV) disease risk factors and high CV mortality, with increasing concern over the contribution of antipsychotic medications to cardiometabolic risk. The design of the NIMH-sponsored Clinical Antipsychotic Trials of Intervention Effectiveness (CATIE) Schizophrenia Trial was driven by a need to understand the efficacy and safety differences between atypical antipsychotics, and between atypical and typical antipsychotics. The CATIE data indicated differences between olanzapine and other antipsychotics in phase 1 on the primary outcome measure, time to drug

J.M. Meyer
Department of Psychiatry, University of California, La Jolla, CA 92093, USA

VA San Diego Healthcare System, 3350 La Jolla Village Drive (116A), San Diego, CA 92161, USA
e-mail: jmmeyer@ucsd.edu

discontinuation, yet olanzapine was not superior to risperidone in the phase 2 tolerability arm, and was inferior to clozapine in the phase 2 efficacy arm. However, CATIE provided clear confirmation of the metabolic liability for olanzapine and also quetiapine, particularly on measures associated with insulin resistance: fasting triglycerides and central adiposity. Current research is focused on analyzing the adiposity-independent impact of certain antipsychotics on glucose–insulin homeostasis, and the disease-specific biological factors that predispose schizophrenia patients to metabolic dysfunction. The CATIE data also highlighted the high prevalence of metabolic disorders in chronic schizophrenia patients, and the moderating role of gender and race or ethnicity in antipsychotic-associated metabolic adverse effects. In the post-CATIE era, safety concerns remain the primary driver of antipsychotic prescribing habits. Absent compelling efficacy data that differentiates between antipsychotics for nonrefractory schizophrenia, the CATIE results reinforce the need for additional metabolically neutral antipsychotic treatment options, and the importance of ongoing physical health monitoring for schizophrenia patients.

Keywords Antipsychotic · Cardiovascular · C-reactive protein · Insulin · Metabolic · Metabolic syndrome

1 Introduction

The diagnosis of schizophrenia is associated with early mortality related to suicide and medical illnesses (Allebeck 1989; Brown 1997). Among the medical conditions which are overrepresented in patients with schizophrenia, recent research has convincingly established cardiovascular (CV) disease as the leading natural cause of excess mortality, with standardized mortality ratios for CV causes twofold greater than the general population (Colton and Manderscheid 2006; Osby et al. 2000a, b). As mounting evidence indicates that the mortality gap may be widening between schizophrenia patients and their peers (Saha et al. 2007), the psychiatric community has increasingly focused on means to improve the physical health of patients with schizophrenia (Marder et al. 2004; Meyer and Nasrallah 2009) with the expressed goal of monitoring for cardiometabolic risk, and mitigating medical comorbidity and high CV mortality rates.

Multiple factors contribute to CV risk in schizophrenia patients (Newcomer and Hennekens 2007), including high cigarette smoking prevalence (Brown et al. 1999), undertreatment of medical conditions (Druss et al. 2000; Nasrallah et al. 2006), and inherent metabolic dysfunction associated with schizophrenia (van Nimwegen et al. 2008), yet it is the metabolic effects of antipsychotic treatment that have emerged as one of the most important and contentious elements in the risk equation. By the 1960s, it was apparent that low-potency phenothiazines had deleterious effects on weight, serum lipids, and glucose (Clark et al. 1967; Efron and Balter 1966;

Mefferd et al. 1958; Schwarz and Munoz 1968), but the lessons learned were lost in the ensuing decades as use of metabolically neutral high-potency typical antipsychotics eclipsed that of low-potency agents (Meyer and Koro 2004). In the mid-1990s, the availability of a new generation of antipsychotics, derived from clozapine's pharmacological properties of high $5HT_2$ affinity and low D_2 potency, heralded a new age of schizophrenia treatment with markedly reduced risk for neurological adverse effects. By decade's end, it became apparent that these newer compounds had lower rates of extrapyramidal side effects and tardive dyskinesia, but were at times accompanied by weight gain, and significant derangements in lipid and glucose metabolism (Jin et al. 2002; Meyer 2001a, b). Aside from case reports, the largest data sets with metabolic outcomes were from industry trials, whose restrictive enrollment criteria often precluded generalizability to the broader spectrum of schizophrenia patients. It was in this context that many looked to the NIMH-sponsored Clinical Antipsychotic Trials of Intervention Effectiveness (CATIE) Schizophrenia Trial to provide data on antipsychotic health outcomes from a broad array of atypical antipsychotics, using unbiased entry criteria that promoted enrollment of schizophrenia patients with medical comorbidities (Lieberman et al. 2005) (see Fig. 1). The randomized, double-blind nature of the study, and the large sample size ($n = 1,460$) offered a unique opportunity to resolve many unaddressed issues regarding the relative metabolic impact of atypical antipsychotics; in doing so, the CATIE results confirmed the high prevalence of cardiometabolic risk in

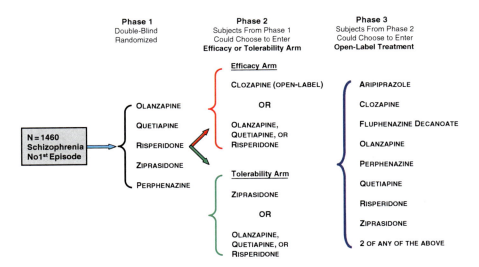

Fig. 1 CATIE Schizophrenia Trial design. *Notes*: (a) Ziprasidone was added to phase 1 after 40% of subjects had been randomized. (b) Subjects with baseline tardive dyskinesia were not randomized to perphenazine in phase 1. (c) Subjects who failed perphenazine in phase 1 were randomized to an atypical (phase 1B) before eligibility for phase 2. (d) Subjects who discontinued phase 1 medication for efficacy reasons were offered treatment in the efficacy arm of phase 2. If they chose not participate due to the possibility of clozapine exposure, they were randomized in the phase 2 tolerability arm. (e) All subjects in phase 2 received a different medication than phase 1

schizophrenia patients, and highlighted those parameters of greatest import to health monitoring during antipsychotic treatment.

The post-CATIE view of antipsychotics is very different from that which engendered the study. Metabolic safety concerns now drive antipsychotic choice, with many providers concluding that there are, at best, limited efficacy differences between atypical antipsychotics (clozapine excepted), and between typical and atypical agents. The purpose of this chapter is to review the current understanding of sources for cardiometabolic risk, and discuss the implications of CATIE data on antipsychotic metabolic outcomes in light of ongoing research into medication and disease mechanisms underlying metabolic dysfunction in schizophrenia patients.

2 Sources of Cardiovascular Risk

The November 20, 2008 issue of the New England Journal of Medicine published findings from a double-blind, placebo-controlled study of rosuvastatin treatment for individuals who did not meet usual criteria for lipid-lowering therapy (Ridker et al. 2008). The findings of this trial, its conceptualization, and its early termination by the safety monitoring board due to markedly lower CV event rates in the rosuvastatin arm represent a paradigm shift in our understanding of CV risk. For decades, clinical CV risk assessment was based on risk factors derived from large longitudinal studies such as the Framingham Heart Study (Wilson et al. 1998). Smoking, hypertension, total cholesterol, and high-density lipoprotein (HDL) cholesterol levels emerged as the most robust predictors of CV events, and were included in empirically derived risk algorithms (Expert Panel on Detection, Evaluation, and Treatment of High Blood Cholesterol in Adults 2001; Wilson et al. 1998), yet there were limitations to these predictive models. Cardiologists were troubled by the fact that 20% of subjects in the Framingham Heart Study experienced major CV events without one major CV risk factor (Wilson et al. 1998). While low-density lipoprotein (LDL) serves as the primary target for lipid-lowering therapy (Expert Panel on Detection, Evaluation, and Treatment of High Blood Cholesterol in Adults 2001), it was perplexing that there was a poor correlation between baseline LDL and future myocardial infarction risk during statin treatment (Heart Protection Study Collaborative Group 2002); moreover, 46% of CV events in large longitudinal studies occurred in those with low serum LDL (<130 mg/dL) (Ridker et al. 2002).

These issues, combined with the knowledge that atherosclerotic disease burden increases serum levels of inflammatory markers such as interleukin-6 (IL-6) and C-reactive protein (CRP), prompted investigators to examine the association between inflammation and CV outcomes. Prospective data from several statin studies confirmed that CRP possessed significant predictive power for CV events, that CRP levels were superior to LDL, HDL, total cholesterol:HDL ratio, and IL-6 for risk prediction, and that CRP was second to systolic blood pressure and ahead of smoking as a predictor of CV risk (Boekholdt et al. 2006; Cook et al. 2006; Ridker et al. 2002). That CRP might be an independent predictor of CV risk was also raised by data, indicating that patients with lower CRP levels after statin treatment sustained fewer CV events than those with higher CRP, regardless of LDL levels (Ridker et al. 2005).

These data served as justification for enrolling 17,802 apparently healthy men and women with normal serum LDL (<130 mg/dL), but with serum CRP above population median levels (>2 mg/L), into a long-term, placebo-controlled trial of the high-potency statin, rosuvastatin. The study was intended as a 4-year trial, but was terminated less than 2 years after commencement by the finding of 37% lower CV events in the rosuvastatin arm, a finding that was consistent across all demographic risk groups (e.g., gender, age, smokers, those with metabolic syndrome, etc.). The appreciation of CRP as a target for CV risk reduction is very recent, but is rapidly assuming relevance for CV monitoring of schizophrenia patients based on the rosuvastatin trial outcome, combined with CATIE data illustrating antipsychotic effects on CRP and other inflammatory markers (Meyer et al. 2009a); moreover, as will be discussed later, there is emerging evidence that elevated levels of systemic inflammation may have deleterious neurocognitive effects, providing another reason to consider CRP reduction as a therapeutic target in schizophrenia patients (Dickerson et al. 2007; Gunstad et al. 2006; Weuve et al. 2006).

Another important source of cardiometabolic risk of relevance to schizophrenia treatment is that related to insulin resistance. Type 2 diabetes mellitus (DM) has long been recognized as a strong predictor of future CV events (Epstein 1967; Kannel and McGee 1979). Current CV risk algorithms deem type 2 DM a coronary heart disease (CHD) risk-equivalent condition, based on findings that the long-term myocardial infarction risk (>20%) is virtually identical to that in patients without DM but with a prior history of myocardial infarction (Expert Panel on Detection, Evaluation, and Treatment of High Blood Cholesterol in Adults 2001). Many are aware of the concerns regarding diabetes risk with atypical antipsychotics, but most are unaware that serum glucose is a relatively insensitive marker of increasing insulin resistance over short periods of time (e.g., <1 year). Given the twofold higher DM prevalence in schizophrenia patients (Bushe and Holt 2004), clinicians are correct in asserting the need to monitor serum glucose, but if the goal is to forestall the onset of diabetes, other laboratory and clinical data provide a more sensitive picture of declining insulin sensitivity.

The metabolic syndrome concept was first elaborated over two decades ago (Reaven 1988), and describes the fact that, in certain susceptible individuals, central (visceral) adiposity results in compensatory hyperinsulinemia, and a cluster of other findings including hypertension, atherogenic dyslipidemia (decreased HDL cholesterol, elevated triglycerides – TGs), and increased serum levels of prothrombotic proteins and inflammatory markers (Table 1). These metabolic parameters define a continuum of risk – those individuals who have more features of this syndrome appear more greatly predisposed to type 2 DM (de Vegt et al. 2001) and CV disease (Ford 2004). There has been an ongoing debate in recent literature over the value of the metabolic syndrome concept (American Heart Association et al. 2005; Kahn et al. 2005), since the metabolic syndrome diagnosis itself confers no greater predictive value for CV events than traditional estimating algorithms

Table 1 National cholesterol education program metabolic syndrome criteria (Expert Panel on Detection, Evaluation, and Treatment of High Blood Cholesterol in Adults 2001)

Criterion	Threshold
Abdominal obesity (waist circumference)	
Men	>40 in.
Women	>35 in.
Fasting triglycerides	≥150 mg/dL
HDL	
Men	<40 mg/dL
Women	<50 mg/dL
Blood pressure	≥130/85 mmHg or use of antihypertensive medication
Fasting glucose[a]	≥110 mg/dL or use of insulin or hypoglycemic medication

[a]Recently lowered to 100 mg/dL (Grundy et al. 2004)

(Wannamethee et al. 2005). Nonetheless, there is a strong association between the number of metabolic syndrome criteria met and increased CHD risk (Girman et al. 2005); moreover, the concept is of value by highlighting clinical findings that, by themselves, may not generate significant attention, and are associated with future risk for DM and CHD, in particular hypertriglyceridemia (Lorenzo et al. 2007). Among laboratory markers of CV risk in schizophrenia patients, hypertriglyceridemia assumes particular importance due to (a) the association with metabolic syndrome and insulin resistance and (b) the fact that elevated TGs may be frequently seen during treatment with certain atypical antipsychotics (Meyer and Koro 2004).

Serum TGs have not been included in CV risk algorithms since, by itself, the predictive power of serum TG levels was inferior to that of total cholesterol and HDL; however, there are data to indicate that, for any level of HDL, elevated TG confers additional CV risk (Jeppesen et al. 1998). Importantly, fasting TG levels highly correlate with insulin resistance among nondiabetics using sophisticated means to measure insulin sensitivity (McLaughlin et al. 2003). This relationship is the direct result of adipocyte insulin resistance to the inhibitory effects on insulin-dependent lipase. As insulin resistance worsens, inappropriately high levels of lipolysis release excess free fatty acids into circulation that are hepatically transformed into TG (Smith 2007). Elevated fasting TG levels thus become a sensitive marker of insulin resistance, but fasting TG to HDL ratios ≥ 3.0 perform better than fasting TGs in identifying prediabetic individuals in the highest tertile of insulin resistance (McLaughlin et al. 2003). Increased TG levels interfere with important regulatory functions governing the production of apolipoprotein B100 (ApoB100), a core lipoprotein in very low, intermediate, and LDL particles (Smith 2007). The overproduction of ApoB100 results in more of these TG-rich particles, and the greater presence of these light TG-rich lipoproteins causes the transfer of TG to HDL at the expense of HDL cholesterol content. After passage through the liver, where TG is cleaved by enzymatic processes, the remaining HDL particle is smaller than normal and more readily cleared in the kidney, resulting in the characteristic low serum HDL levels seen with insulin-resistant states (Smith 2007). The TG:HDL ratio thus reflects the combined effects of low HDL and elevated TG seen in insulin-resistant patients.

While fasting TG values provide important information on insulin resistance, both fasting and nonfasting TGs are associated with CV risk. Nonfasting TG in particular may be more relevant to the development of atherosclerosis and subsequent CV risk than fasting TG levels. The basis of this assertion lies in the concept that arterial injury may occur primarily during the postprandial period, when TG-rich particles reach their highest levels and penetrate arterial intimal cells (Eberly et al. 2003). As serum TG levels require more than 8 h to return to baseline after a meal, individuals are in a nonfasting state most of the day with respect to TG values (Nordestgaard et al. 2007). Supporting clinical data are seen in results from a large ($n = 13,981$) European trial with mean 26 years of follow-up that found a significant correlation between nonfasting TG levels and risk of major CV events (Nordestgaard et al. 2007). Moreover, prospective data from the Women's Health Study ($n = 26,509$) found no relationship between increasing fasting TG values and CV risk over the 11.4 years of follow-up when fully adjusted for CV risk factors; however, using adjusted models, a significant association was found between nonfasting TG and CV risk (Bansal et al. 2007), particularly, TG levels measured 2–4 h postprandially.

There is little debate on the value of established CV risk factors, including older age, low serum HDL, elevated total cholesterol, hypertension, smoking, and the moderating impact of gender (Expert Panel on Detection, Evaluation, and Treatment of High Blood Cholesterol in Adults 2001), but, as the prior discussion illustrates, these factors do not encompass all aspects of CV risk prediction. As will be discussed, CATIE not only provided ample data on traditional CV risk factors and those associated with the metabolic syndrome, but also provided the first introduction for the psychiatric community to the role of nonfasting TG and inflammation in assessing the CV risks of antipsychotic treatment.

3 The Cardiovascular and Metabolic Risk Profile of Subjects Entering the CATIE Schizophrenia Trial

When the CATIE Schizophrenia Trial concluded in fall 2004, it was already evident that schizophrenia patients had 2–4 times higher metabolic syndrome prevalence than general population estimates (Cohn et al. 2004; Heiskanen et al. 2003), but the largest published study at that time was from a Canadian sample of 240 subjects (Cohn et al. 2004). Of 1,460 subjects with baseline data, 689 CATIE subjects had fasting laboratory measures at study baseline to examine metabolic syndrome criteria, and these subjects were matched 1:1 on the basis of age, gender race, and ethnicity with subjects randomly drawn from the Third National Health and Nutrition Examination Survey (NHANES III) to perform comparative analyses. Overall metabolic syndrome prevalence among the CATIE subjects was 40.9% (McEvoy et al. 2005), but was significantly higher in CATIE females (51.6%) than CATIE males (36.0%) ($p = 0.0002$). When compared with the matched NHANES cohort,

Table 2 Comparison of metabolic syndrome data between CATIE subjects and matched NHANES III subjects (McEvoy et al. 2005)

	Males ($n = 509$)			Females ($n = 180$)		
	CATIE (%)	NHANES (%)	p	CATIE (%)	NHANES (%)	p
Metabolic syndrome prevalence	36.0	19.7	0.0001	51.59	25.1	0.0001
Met waist circumference criterion	35.5	24.8	0.0001	76.3	57.0	0.0001
Met triglyceride criterion	50.7	32.1	0.0001	42.3	19.6	0.0001
Met HDL criterion	48.9	31.9	0.0001	63.3	36.3	0.0001
Met BP criterion	47.2	31.1	0.0001	46.9	26.8	0.0001
Met glucose criterion	14.1	14.2	0.9635	21.7	11.2	0.0075

CATIE male subjects were twice as likely to have metabolic syndrome (OR 2.38; 95% CI 1.78–3.18), while CATIE females had three times greater odds for metabolic syndrome (OR 3.51; 95% CI 2.19–5.62). The CATIE subjects also had greater prevalence of every metabolic syndrome criterion when compared to the NHANES sample (Table 2), with fasting glucose among males the sole exception. An additional analysis using the Framingham CV risk algorithm found that 10-year CHD risk was significantly elevated in male (9.4 vs. 7.0%) and female (6.3 vs. 4.2%) schizophrenia patients in the baseline CATIE sample compared to their matched NHANES peers ($p = 0.0001$) (Goff et al. 2005), and significantly higher rates of smoking (68 vs. 35%), diabetes (13 vs. 3%), and hypertension (27 vs. 17%) compared to NHANES subjects ($p < 0.001$). Despite the known cardiometabolic risk factors seen in schizophrenia patients, analysis of the CATIE baseline sample also revealed high rates of nontreatment that ranged from 30.2% for diabetes, to 62.4% for hypertension, and 88.0% for dyslipidemia (Nasrallah et al. 2006).

4 The Impact of Antipsychotic Treatment on Cardiovascular and Metabolic Outcomes in the CATIE Schizophrenia Trial

4.1 Metabolic Outcomes

The initial CATIE publications provided data on changes weight, serum cholesterol, hemoglobin A1C, glucose, and TG (Lieberman et al. 2005; Stroup et al. 2006), and confirmed olanzapine's deleterious effects on weight and TG, but failed to find significant differences between treatments (olanzapine, perphenazine, quetiapine, risperidone, and ziprasidone) for serum glucose. The increased serum TG levels during olanzapine exposure indicate worsening insulin sensitivity, but over the short periods of exposure for subjects in CATIE phase 1, these did not translate into significant between-treatment changes in serum glucose. Ziprasidone exposure was associated with metabolic improvement, as noted in prior switch studies (Weiden et al. 2003, 2008), an effect not related to any weight or lipid-lowering properties of ziprasidone, but rather the removal of prior drug effects combined with ziprasidone's metabolic neutrality. Prior to CATIE, consensus opinion had concluded that risperidone and quetiapine possessed equivalent metabolic profiles, and that both had greater metabolic impact than ziprasidone, but were more benign than olanzapine (American Diabetes Association, American Psychiatric Association, American Association of Clinical Endocrinologists, North American Association for the Study of Obesity 2004); however, the initial CATIE data indicated that quetiapine had an adverse impact on TG not seen with risperidone.

Subsequent detailed analysis of the CATIE phase 1 data focusing on metabolic syndrome components (Meyer et al. 2008b) noted that, after 3 months, there were significant between-drug differences for the change in proportion meeting metabolic syndrome status. The metabolic syndrome prevalence increased for olanzapine-exposed subjects from 34.8 to 43.9%, but decreased for ziprasidone (from 37.7 to 29.9%) ($p = 0.001$). To examine the effect of longer exposure on change in metabolic parameters, an analysis was performed using the last phase 1 data available for each subject. These data were obtained, on average, 9 months from baseline, and the findings were consistent with those at 3 months for mean changes in waist circumference and fasting TG, although between-group differences now emerged for HDL and systolic BP (Table 3). A repeated measures analysis of waist circumference change was also performed which found a significant impact of baseline value on changes in waist circumference: patients who were more centrally obese tended to become thinner. For subjects with baseline values below the median (<39 in.), olanzapine caused greater increases in central adiposity (+1.92 in.) compared to every other medication [range +0.35 in. (ziprasidone) to +0.97 in. (quetiapine)]. Among subjects with baseline waist circumference at or above the median (≥39 in.), only the olanzapine-exposed cohort did not experience an adjusted mean decrease in waist circumference; moreover, perphenazine (−0.97 in.) was found to be significantly superior to both olanzapine (+0.17 in.; $p < 0.0001$) and quetiapine (−0.01 in.; $p = 0.0007$). The exploratory analysis of CATIE phase 1 nonfasting TG samples noted greater increases in median and adjusted mean nonfasting TG levels among those randomized to quetiapine (mean +54.7 mg/dL, median +26 mg/dL) and olanzapine (mean +23.4 mg/dL, median +26.5 mg/dL), with a significant between-group difference for perphenazine versus olanzapine ($p = 0.002$) (Meyer et al. 2008a).

4.2 Framingham Cardiovascular Risk

Since the greatest metabolic impact of atypical antipsychotics is on weight (Newcomer 2005) and serum TG, but much less on cholesterol measures (Meyer and Koro 2004), an important question is whether there would be between-treatment differences in calculated CV risk, given the fact that neither weight, waist circumference, nor

Table 3 Mean changes in individual metabolic criteria at end of CATIE phase 1 visit (Meyer et al. 2008a)

	Waist circumference[a] (in.)		Systolic BP[a] (mmHg)		Diastolic BP (mmHg)	HDL[b,c] (mg/dL) (whites)	HDL[b] (mg/dL) (nonwhites)	Fasting glucose (mg/dL)	Fasting triglycerides[a] (mg/dL)	
	Below median	Above median	Below median	Above median					Below median	Above median
Mean exposure (months)	8.9	8.7	8.7	9.1	8.9	9.2	9.9	9.9	9.9	9.8
OLANZ	1.9 (SE = 0.2) ($n = 146$)	0.4 (SE = 0.3) ($n = 147$)	6.0 (SE = 1.0) ($n = 159$)	−3.6 (SE = 1.2) ($n = 146$)	0.1 (SE = 0.6) ($n = 305$)	−1.7 (SE = 0.6) ($n = 171$)	−0.9 (SE = 0.9) ($n = 115$)	4.5 (SE = 2.3) ($n = 94$)	49.0 (SE = 10.8) ($n = 42$)	5.2 (SE = 17.4) ($n = 51$)
RISP	0.9 (SE = 0.2) ($n = 145$)	−0.7 (SE = 0.3) ($n = 143$)	5.6 (SE = 1.1) ($n = 136$)	−9.0 (SE = 1.1) ($n = 162$)	−1.3 (SE = 0.6) ($n = 298$)	0.1 (SE = 0.6) ($n = 162$)	0.9 (SE = 0.9) ($n = 109$)	−0.4 (SE = 2.6) ($n = 74$)	19.7 (SE = 11.2) ($n = 39$)	−67.1 (SE = 21.2) ($n = 35$)
QUET	0.7 (SE = 0.2) ($n = 142$)	0.0 (SE = 0.2) ($n = 157$)	8.6 (SE = 1.0) ($n = 158$)	−8.0 (SE = 1.2) ($n = 145$)	−0.1 (SE = 0.6) ($n = 303$)	−0.2 (SE = 0.6) ($n = 186$)	0.1 (SE = 1.1) ($n = 85$)	−1.8 (SE = 2.4) ($n = 88$)	29.8 (SE = 10.8) ($n = 42$)	−13.0 (SE = 18.4) ($n = 46$)
ZIP	0.0 (SE = 0.3) ($n = 77$)	−0.4 (SE = 0.3) ($n = 78$)	8.8 (SE = 1.5) ($n = 78$)	−7.6 (SE = 1.6) ($n = 80$)	−0.4 (SE = 0.8) ($n = 158$)	0.6 (SE = 0.9) ($n = 90$)	4.3 (SE = 1.4) ($n = 51$)	0.0 (SE = 3.5) ($n = 39$)	26.0 (SE = 15.6) ($n = 20$)	−96.4 (SE = 28.5) ($n = 19$)
PER	0.6 (SE = 0.2) ($n = 126$)	−1.1 (SE = 0.3) ($n = 110$)	6.0 (SE = 1.2) ($n = 107$)	−6.4 (SE = 1.2) ($n = 137$)	0.0 (SE = 0.6) ($n = 243$)	2.7 (SE = 0.7) ($n = 130$)	−1.3 (SE = 1.0) ($n = 88$)	−1.0 (SE = 2.7) ($n = 68$)	28.7 (SE = 11.6) ($n = 36$)	−27.5 (SE = 22.3) ($n = 31$)
Overall treatment difference	<0.001[d]	0.001[e]	NS	0.017[f]	NS	<0.001[g]	0.012[h]	NS	NS	0.011[i]

Table entries are ANCOVA least-squares adjusted means. All models include time to treatment discontinuation as a covariate, as well as baseline value of outcome. Demographic variables were analyzed, but only age and race entered the models for HDL, SBP, and DBP, and gender for HDL and SBP

Note: NS not significant ($p \geq 0.05$)

[a]Data presented in separate columns due to significant baseline by treatment effect. Median WC = 39 in., median SBP = 122 mmHg, and median TG = 148 mg/dL
[b]Data presented in separate columns due to significant race by treatment effect
[c]There was a significant ziprasidone cohort effect, but the between-group results were not different for cohorts enrolled prior to, or after the introduction of ziprasidone
[d]Between-group comparison significant for olanzapine versus risperidone ($p = 0.001$), quetiapine ($p < 0.001$), ziprasidone ($p < 0.001$), and perphenazine ($p < 0.001$)
[e]Between-group comparison significant for perphenazine versus olanzapine ($p < 0.001$), perphenazine versus quetiapine ($p = 0.003$), and olanzapine versus risperidone ($p = 0.003$)
[f]Between-group comparison significant for olanzapine versus risperidone ($p = 0.001$)
[g]Between-group comparison significant for perphenazine versus olanzapine ($p < 0.001$) and perphenazine versus quetiapine ($p = 0.002$)
[h]Between-group comparisons significant for ziprasidone versus olanzapine ($p = 0.002$) and ziprasidone versus perphenazine ($p = 0.001$)
[i]Between-group comparisons significant for olanzapine versus ziprasidone ($p = 0.003$)

TG are part of Framingham-based CV risk algorithms (Daumit et al. 2008). The other question is whether, over a relatively short time frame, the magnitude of any changes in CHD risk will be significant in those with higher baseline levels of risk (range 8.1–9.1% across antipsychotic treatments). Using the 3-month outcomes, there were significant differences between treatments for covariate-adjusted mean change in 10-year CHD risk (Daumit et al. 2008). Olanzapine was associated with an absolute increase in risk of 0.5% (SE 0.3) and quetiapine, a 0.3% (SE 0.3) increase; whereas risk decreased in patients treated with perphenazine, -0.5% (SE 0.3), risperidone, -0.6% (SE 0.3), and ziprasidone, -0.6% (SE 0.4). The difference in 10-year CHD risk between olanzapine and risperidone was statistically significant ($p = 0.004$).

4.3 Outcomes with Novel Biomarkers

The enormity of the data collected in the CATIE Schizophrenia Trial means that new findings will continue to emerge for years to come. As of this writing, samples are being analyzed to examine various markers associated with cardiometabolic risk, and the first of these to yield important results is the examination of CRP changes during phase 1. As mentioned in Sect. 2, recent clinical data on CRP have greatly influenced our understanding of CV risk, and of possible targets for risk modification. As with waist circumference and HDL, baseline CRP among CATIE subjects was a significant predictor of 3-month CRP change ($p < 0.001$), yet the 3-month analysis still found significant treatment differences in change from baseline after adjustment for baseline CRP ($p = 0.011$) (Meyer et al. 2009a). At 3 months, both olanzapine and quetiapine had the numerically greatest increases. There were no significant treatment differences in those with higher baseline levels of systemic inflammation (CRP ≥ 1 mg/L), but for those with lower baseline CV risk (CRP < 1 mg/L), pairwise comparisons were significantly different for olanzapine versus perphenazine ($p < 0.001$) and versus risperidone ($p = 0.001$). The 12-month repeated measures analysis confirmed the association between baseline CRP and outcomes, and the deleterious impact of olanzapine compared to perphenazine ($p < 0.001$) and ziprasidone ($p = 0.003$) in those with baseline CRP < 1 mg/L.

5 The Post-CATIE Era

5.1 Clinical Conclusions

The CATIE data confirm the fact that metabolic differences exist between various atypical antipsychotics, and pinpoint the metabolic outcomes (adiposity, TGs) most greatly influenced by medications. Among patients with clinical and laboratory findings suggestive of insulin resistance, use of olanzapine is associated with further deterioration in metabolic status, while risperidone and quetiapine have intermediate effects, and ziprasidone appears to be metabolically neutral. Prior consensus panels found equivalent metabolic risk between quetiapine and risperidone (American Diabetes Association, American Psychiatric Association, American Association of Clinical Endocrinologists, North American Association for the Study of Obesity 2004), but the CATIE data suggest that quetiapine, when used at doses >400 mg for schizophrenia treatment, has significant adverse effects on nonfasting TG, HDL (in white subjects), and central adiposity in a manner not seen with risperidone.

Prior literature noted minimal effects of atypical antipsychotic treatment on blood pressure compared to other metabolic parameters, and this is confirmed by CATIE. Weight gain is associated with hypertension, but the time frame of this study may be inadequate to manifest this effect. The absence of a significant signal for HDL changes in prior antipsychotic studies may have been the result of limited duration of exposure and smaller sample sizes (Meyer and Koro 2004), since a deleterious impact of olanzapine and quetiapine was seen for serum HDL in whites, and significant improvement in HDL with ziprasidone in nonwhites. CATIE also provided the first controlled data on the metabolic effects of a medium potency typical antipsychotic published in the past 40 years, with evidence that perphenazine is generally metabolically neutral. The lack of significant between-group differences for glucose should not reassure clinicians that there are no differences in future DM risk between antipsychotics. As previously described, DM develops over 10–20 years, so short-term changes in serum glucose may not be seen despite worsening in other parameters associated with insulin resistance, especially waist circumference and TG. Lastly, the CATIE phase 1 analysis of inflammatory markers and nonfasting TG revealed patterns of changes that paralleled the known metabolic liability for each antipsychotic medication. It should also be noted that aripiprazole was not studied in the first two phases of CATIE, but clinical data indicate a benign metabolic risk profile very similar to ziprasidone (Newcomer 2005).

One of the paramount conclusions from CATIE is that schizophrenia patients possess high levels of CV risk, and that antipsychotic choice may have significant influences on cardiometabolic health over short time frames. One intended outcome for CATIE was to demonstrate whether significant between-drug differences existed for efficacy measures, but the results of CATIE, the British CUtLASS study (Jones et al. 2006), and the recent TEOSS trial in first-episode subjects (Sikich et al. 2008) have caused many to question whether significant efficacy differences exist among atypical antipsychotics (with the exception of clozapine for treatment refractory patients), and between atypical and typical antipsychotics. An unintended outcome, but important for future schizophrenia treatment, is that no novel agent with known metabolic adverse effects is likely to progress very far down the developmental pathway toward approval.

Another useful finding from CATIE relates to demographic factors that moderate metabolic risk. In the general population, gender is associated with differential risk for metabolic syndrome, with the prevalence in US women 27% compared to

22% in men according to recent estimates (Ford et al. 2004). The CATIE baseline analysis not only revealed markedly higher metabolic syndrome prevalence among chronic schizophrenia patients compared to matched peers in the general population, but also a significantly greater gender discrepancy within the CATIE cohort, with CATIE women having 42% higher prevalence than CATIE men (McEvoy et al. 2005). The differential sensitivity to metabolic risk on the basis of race and ethnicity was also seen in the baseline analysis of metabolic syndrome prevalence (McEvoy et al. 2005), and in the analysis of phase 1 metabolic outcomes (Meyer et al. 2008b). The increased vulnerability for black and Hispanic schizophrenia patients to antipsychotic metabolic effects has also been demonstrated in a recent retrospective analysis of a prospective, 26-week, randomized aripiprazole versus olanzapine trial (Meyer et al. 2009b). As treatment of all medical conditions moves into the era of personalized medicine, recognition of sensitivity to antipsychotic adverse effects, and possibly differential psychiatric outcomes on the basis gender and race/ethnicity present significant opportunities for further clinical research.

5.2 Hypotheses on Schizophrenia and Metabolic Risk, and Adiposity-Independent Drug Effects

Patients with schizophrenia exhibit higher metabolic syndrome prevalence than the general population (Cohn et al. 2004; De Hert et al. 2006; Hagg et al. 2006; Heiskanen et al. 2003; Mackin et al. 2007; McEvoy et al. 2005; Meyer et al. 2006; Saari et al. 2005; Srisurapanont et al. 2007; Suvisaari et al. 2007; Tirupati and Chua 2007), and the age of metabolic syndrome onset occurs much earlier than the general population (McEvoy et al. 2005). As these data are primarily accrued in antipsychotic treated individuals, there has been an ongoing debate whether there is any biological contribution of mental illness toward metabolic disease risk that exists independently of other disease-related risks such as medications, inactivity, smoking (Willi et al. 2007), and dietary habits.

Early data from samples of neuroleptic-naïve and drug-free schizophrenia patients showed contradictory findings regarding the presence of fasting hyperinsulinemia, increased central adiposity, and other abnormalities (e.g., hypercortisolemia) (Arranz et al. 2004; Ryan et al. 2003, 2004; Thakore et al. 2002; Zhang et al. 2004), but recent studies performed in neuroleptic-naïve first-episode schizophrenia patients have found evidence of pretreatment metabolic dysfunction. In a Canadian study of nine predominantly drug-naïve schizophrenia patients compared to nine matched controls, mean insulin sensitivity was 42% lower in the schizophrenia patients ($p = 0.026$), although acute insulin response to glucose did not differ between the groups ($p = 0.752$) (Cohn et al. 2006). Oral glucose tolerance testing of 38 drug-naïve, first-episode Irish schizophrenia patients and 38 matched controls found that the prevalence of impaired glucose tolerance was 10.5% in the schizophrenia patients, compared to 0% in the controls (Spelman et al. 2007).

The investigators found no significant differences on any lipid measure, leptin, hemoglobin A1C, or fasting glucose between patients and controls, but there were differences for fasting cortisol, baseline insulin, and 2-h postload insulin. Fasting plasma glucose, insulin, insulin-like growth factor-1 (IGF-1), and cortisol levels were examined in 44 neuroleptic-naïve schizophrenia patients (mean age 33.3 years) in India, and 44 matched control subjects, with schizophrenia patients exhibiting significantly higher fasting insulin and cortisol levels, but lower IGF-1 levels (Venkatasubramanian et al. 2007). Most recently, a study of seven first-episode neuroleptic-naïve Dutch schizophrenia patients and seven matched controls found no between-group differences on insulin-mediated peripheral glucose uptake; the schizophrenia patients had significantly greater endogenous glucose production (van Nimwegen et al. 2008). The failure of insulin to suppress hepatic glucose production (the primary source of endogenous glucose) indicates a level of hepatic insulin resistance existing prior to the effects of treatment.

One major research question not addressed by CATIE relates to adiposity-independent effects of atypical antipsychotics on glucose–insulin homeostasis. An early detailed analysis of new onset diabetes cased culled from the FDA Med-Watch database showed that 78% of those on clozapine or olanzapine exhibited reversal of their metabolic derangements after antipsychotic discontinuation (Koller et al. 2001; Koller and Doraiswamy 2002). This was followed by a review (Jin et al. 2002) of the published literature that noted a significant proportion of atypical antipsychotic-associated new onset DM cases occurred literally within weeks of treatment initiation, again suggesting a direct antipsychotic effect on glycemic control.

These data were indicative of direct antipsychotic impact on glucose–insulin homeostasis, but the case literature is always clouded by confounding variables, including prior drug exposure, baseline obesity, and health habits which increase diabetes risk. To address these issues, Houseknecht et al. (2007) performed a series of euglycemic clamps in laboratory rats exposed to single doses of clozapine, olanzapine, ziprasidone, or risperidone. In the euglycemic clamp, a fixed intravenous insulin infusion is administered, and a simultaneous infusion of 20% glucose is also given, with the glucose infusion rate (GIR) adjusted to achieve a predetermined serum glucose level in the euglycemic range (typically 90 mg/dL in humans). The GIR thus becomes a measure of insulin effectiveness, and therefore the extent of insulin resistance, with low GIR representing a high degree of insulin resistance. In animal models, GIR equilibrium is reached after 90 min, at which point Houseknecht administered a single parenteral antipsychotic dose to each rat. This single dose design examined the acute antipsychotic impact on insulin sensitivity, completely eliminating the confounding effects of prior drug exposure and any possible effect of weight gain (Houseknecht et al. 2007). A significant impact of olanzapine on whole body insulin sensitivity was seen after injections at $t = 100$ min, as evidenced by a mean 43% decrease in GIR. A similar impact on GIR was also seen with clozapine (mean decrease of 65.2%), but not with risperidone or ziprasidone. The investigators followed this discovery by looking for evidence of impaired insulin action on target tissues. A main effect of insulin is to suppress de novo glucose production by the liver, and individuals who are insulin resistant continue to have inappropriately high levels of endogenous hepatic glucose production. In animals and humans, hepatic glucose production can be studied using labeled glucose tracers (e.g., [6,6-^{2}H$_2$]-glucose) that are infused prior to and during the clamp procedure. Changes in hepatic glucose production during the clamp procedure are measured as percentage increase in endogenous glucose production from baseline. Following a single dose of ziprasidone, there was no change in hepatic glucose production compared to baseline, but a single dose of olanzapine or clozapine significantly impaired the ability of insulin to inhibit hepatic glucose production, with an 18.5-fold (olanzapine) and 22.7-fold (clozapine) increase in hepatic glucose production, a result consistent with severe hepatic insulin resistance.

This constellation of findings provided the first compelling biological data to support the concept of adiposity-independent effects of certain antipsychotics on glucose–insulin homeostasis. As this research is extended into human models, it may provide useful information on the molecular basis of antipsychotic induced disturbances in metabolic functioning, and potentially elucidate specific markers to predict differential risk among patients exposed to medication with known metabolic liabilities (e.g., clozapine and olanzapine). The ability to identify individuals at low risk for metabolic dysfunction from clozapine in particular might permit certain treatment-resistant schizophrenia patients to be transitioned more quickly to clozapine, after having failed adequate trials of other antipsychotics.

6 Conclusions

Safety concerns exert considerable influence over antipsychotic prescribing practices, particularly in the absence of compelling effectiveness differences between agents. Results of the CATIE trial confirmed the high prevalence of cardiometabolic risk among chronic schizophrenia patients, and the immediate impact of antipsychotics on certain parameters related to insulin resistance, especially central adiposity and triglyceride levels. Emerging data have augmented the information provided by CATIE on demographic factors that increase risk for adverse metabolic effects from antipsychotic treatment, even with metabolically more benign medications. Regardless of medication choice, the burden remains on clinicians to monitor all metabolic parameters associated with increased CV risk in schizophrenia patients. In the post-CATIE era it is important that psychiatrists are mindful that many patients with severe mental illnesses receive limited primary care services, and psychotropic medications are one of many contributors to increased CV risk.

References

Allebeck P (1989) Schizophrenia: a life-shortening disease. Schizophr Bull 15:81–89
American Diabetes Association, American Psychiatric Association, American Association of Clinical Endocrinologists, North American Association for the Study of Obesity (2004)

Consensus development conference on antipsychotic drugs and obesity and diabetes. J Clin Psychiatry 65:267–272

American Heart Association, National Heart, Lung, and Blood Institute, Grundy SM, Cleeman JI, Daniels SR, Donato KA, Eckel RH, Franklin BA, Gordon DJ, Krauss RM, Savage PJ, Smith SC Jr, Spertus JA, Costa F (2005) Diagnosis and management of the metabolic syndrome. An American Heart Association/National Heart, Lung, and Blood Institute Scientific Statement. Executive summary. Cardiol Rev 13:322–327

Arranz B, Rosel P, Ramirez N, Duenas R, Fernandez P, Sanchez JM, Navarro MA, San L (2004) Insulin resistance and increased leptin concentrations in noncompliant schizophrenia patients but not in antipsychotic-naive first-episode schizophrenia patients. J Clin Psychiatry 65: 1335–1342

Bansal S, Buring JE, Rifai N, Mora S, Sacks FM, Ridker PM (2007) Fasting compared with nonfasting triglycerides and risk of cardiovascular events in women. JAMA 298:309–316

Boekholdt SM, Hack CE, Sandhu MS, Luben R, Bingham SA, Wareham NJ, Peters RJG, Jukema JW, Day NE, Kastelein JJP, Khaw K-T (2006) C-reactive protein levels and coronary artery disease incidence and mortality in apparently healthy men and women: the EPIC-Norfolk prospective population study 1993–2003. Atherosclerosis 187:415–422

Brown S (1997) Excess mortality of schizophrenia. A meta-analysis. Br J Psychiatry 171:502–508

Brown S, Birtwistle J, Roe L, Thompson C (1999) The unhealthy lifestyle of people with schizophrenia. Psychol Med 29:697–701

Bushe C, Holt R (2004) Prevalence of diabetes and glucose intolerance in patients with schizophrenia. Br J Psychiatry 184:67–71

Clark ML, Ray TS, Paredes A, Ragland RE, Costiloe JP, Smith CW, Wolf S (1967) Chlorpromazine in women with chronic schizophrenia: the effect on cholesterol levels and cholesterol–behavior relationships. Psychosom Med 29:634–642

Cohn T, Prud'homme D, Streiner D, Kameh H, Remington G (2004) Characterizing coronary heart disease risk in chronic schizophrenia: high prevalence of the metabolic syndrome. Can J Psychiatry 49:753–760

Cohn TA, Remington G, Zipursky RB, Azad A, Connolly P, Wolever TMS (2006) Insulin resistance and adiponectin levels in drug-free patients with schizophrenia: a preliminary report. Can J Psychiatry 51:382–386

Colton CW, Manderscheid RW (2006) Congruencies in increased mortality rates, years of potential life lost, and causes of death among public mental health clients in eight states. Prev Chronic Dis 3:1–14

Cook NR, Buring JE, Ridker PM (2006) The effect of including C-reactive protein in cardiovascular risk prediction models for women. Ann Intern Med 145:21–29

Daumit GL, Goff DC, Meyer JM, Davis VG, Nasrallah HA, McEvoy JP, Rosenheck RA, Davis SM, Hsiao JK, Stroup TS, Lieberman JA (2008) Antipsychotic effects on estimated 10-year coronary heart disease risk in the CATIE schizophrenia study. Schizophr Res 105:175–187

De Hert MA, van Winkel R, van Eyck D, Hanssens L, Wampers M, Scheen A, Peuskens J (2006) Prevalence of the metabolic syndrome in patients with schizophrenia treated with antipsychotic medication. Schizophr Res 83:87–93

de Vegt F, Dekker JM, Jager A, Hienkens E, Kostense PJ, Stehouwer CD, Nijpels G, Bouter LM, Heine RJ (2001) Relation of impaired fasting and postload glucose with incident type 2 diabetes in a Dutch population: the Hoorn study. JAMA 285:2109–2113

Dickerson F, Stallings C, Origoni A, Boronow J, Yolken R (2007) C-reactive protein is associated with the severity of cognitive impairment but not of psychiatric symptoms in individuals with schizophrenia. Schizophr Res 93:261–265

Druss BG, Bradford DW, Rosenheck RA, Radford MJ, Krumholz HM (2000) Mental disorders and use of cardiovascular procedures after myocardial infarction. JAMA 283:506–511

Eberly LE, Stamler J, Neaton JD, Multiple Risk Factor Intervention Trial Research Group (2003) Relation of triglyceride levels, fasting and nonfasting, to fatal and nonfatal coronary heart disease. Arch Intern Med 163:1077–1083

Efron HY, Balter AM (1966) Relationship of phenothiazine intake and psychiatric diagnosis to glucose level and tolerance. J Nerv Ment Dis 142:555–561

Epstein FH (1967) Hyperglycemia: a risk factor in coronary heart disease. Circulation 36:609–619

Expert Panel on Detection, Evaluation, and Treatment of High Blood Cholesterol in Adults (2001) Executive summary of the third report of the national cholesterol education program (NCEP) expert panel on detection, evaluation, and treatment of high blood cholesterol in adults (Adult Treatment Panel III). JAMA 285:2486–2497

Ford ES (2004) The metabolic syndrome and mortality from cardiovascular disease and all-causes: findings from the National Health and Nutrition Examination Survey II Mortality Study. Atherosclerosis 173:309–314

Ford ES, Giles WH, Mokdad AH (2004) Increasing prevalence of the metabolic syndrome among US adults. Diabetes Care 27:2444–2449

Girman CJ, Dekker JM, Rhodes T, Nijpels G, Stehouwer CDA, Bouter LM, Heine RJ (2005) An exploratory analysis of criteria for the metabolic syndrome and its prediction of long-term cardiovascular outcomes: the Hoorn study. Am J Epidemiol 162:438–447

Goff DC, Sullivan L, McEvoy JP, Meyer JM, Nasrallah HA, Daumit GL, Lamberti S, D'Agnostino RB, Stroup TS, Davis S, Lieberman JA (2005) A comparison of ten-year cardiac risk estimates in schizophrenia patients from the CATIE Study and matched controls. Schizophr Res 80:45–53

Grundy SM, Brewer B, Cleeman JI, Smith SC, Lenfant C (2004) Definition of metabolic syndrome: report of the National Heart, Lung, and Blood Institute/American Heart Association conference on scientific issues related to definition. Circulation 109:433–438

Gunstad J, Bausserman L, Paul RH, Tate DF, Hoth K, Poppas A, Jefferson AL, Cohen RA (2006) C-reactive protein, but not homocysteine, is related to cognitive dysfunction in older adults with cardiovascular disease. J Clin Neurosci 13:540–546

Hagg S, Lindblom Y, Mjorndal T, Adolfsson R (2006) High prevalence of the metabolic syndrome among a Swedish cohort of patients with schizophrenia. Int Clin Psychopharmacol 21:93–98

Heart Protection Study Collaborative Group (2002) MRC/BHF Heart Protection Study of cholesterol lowering with simvastatin in 20, 536 high-risk individuals: a randomised placebo-controlled trial. Lancet 360:7–22

Heiskanen T, Niskanen L, Lyytikainen R, Saarinen PI, Hintikka J (2003) Metabolic syndrome in patients with schizophrenia. J Clin Psychiatry 64:575–579

Houseknecht KL, Robertson AS, Zavadoski W, Gibbs EM, Johnson DE, Rollema H (2007) Acute effects of atypical antipsychotics on whole body insulin resistance in rats: implications for adverse metabolic effects. Neuropsychopharmacology 32:289–297

Jeppesen J, Hein HO, Suadicani P, Gyntelberg F (1998) Triglyceride concentration and ischemic heart disease: an eight-year follow-up in the Copenhagen Male Study. Circulation 97:1029–1036

Jin H, Meyer JM, Jeste DV (2002) Phenomenology of and risk factors for new-onset diabetes mellitus and diabetic ketoacidosis associated with atypical antipsychotics: an analysis of 45 published cases. Ann Clin Psychiatry 14:59–64

Jones PB, Barnes TRE, Davies L, Dunn G, Lloyd H, Hayhurst KP, Murray RM, Markwick A, Lewis SW (2006) Randomized controlled trial of the effect on quality of life of second- vs first-generation antipsychotic drugs in schizophrenia: cost utility of the latest antipsychotic drugs in schizophrenia study (CUtLASS 1). Arch Gen Psychiatry 63:1079–1087

Kahn R, Buse J, Ferrannini E, Stern M (2005) The metabolic syndrome: time for a critical appraisal. Diabetes Care 28:2289–2304

Kannel WB, McGee DL (1979) Diabetes and cardiovascular risk factors: the Framingham study. Circulation 59:8–13

Koller EA, Doraiswamy PM (2002) Olanzapine-associated diabetes mellitus. Pharmacotherapy 22:841–852

Koller E, Schneider B, Bennett K, Dubitsky G (2001) Clozapine-associated diabetes. Am J Med 111:716–723

Lieberman JA, Stroup TS, McEvoy JP, Swartz MS, Rosenheck RA, Perkins DO, Keefe RSE, Davis SM, Davis CE, Lebowitz BD, Severe J, Hsiao JK (2005) Effectiveness of antipsychotic drugs in patients with chronic schizophrenia. N Engl J Med 353:1209–1223

Lorenzo C, Williams K, Hunt KJ, Haffner SM (2007) The National Cholesterol Education Program – Adult Treatment Panel III, International Diabetes Federation, and World Health Organization definitions of the metabolic syndrome as predictors of incident cardiovascular disease and diabetes. Diabetes Care 30:8–13

Mackin P, Bishop D, Watkinson H, Gallagher P, Ferrier IN (2007) Metabolic disease and cardiovascular risk in people treated with antipsychotics in the community. Br J Psychiatry 191:23–29

Marder SR, Essock SM, Miller AL, Buchanan RW, Casey DE, Davis JM, Kane JM, Lieberman JA, Schooler NR, Covell N, Stroup S, Weissman EM, Wirshing DA, Hall CS, Pogach L, Pi-Sunyer X, Bigger JT Jr, Friedman A, Kleinberg D, Yevich SJ, Davis B, Shon S (2004) Physical health monitoring of patients with schizophrenia. Am J Psychiatry 161:1334–1349

McEvoy JP, Meyer JM, Goff DC, Nasrallah HA, Davis SM, Sullivan L, Meltzer HY, Hsiao J, Scott Stroup T, Lieberman JA (2005) Prevalence of the metabolic syndrome in patients with schizophrenia: baseline results from the Clinical Antipsychotic Trials of Intervention Effectiveness (CATIE) Schizophrenia Trial and comparison with national estimates from NHANES III. Schizophr Res 80:19–32

McLaughlin T, Abbasi F, Cheal K, Chu J, Lamendola C, Reaven G (2003) Use of metabolic markers to identify overweight individuals who are insulin resistant. Ann Intern Med 139: 802–809

Mefferd RB, Labrosse EH, Gawienowski AM, Williams RJ (1958) Influence of chlorpromazine on certain biochemical variables of chronic male schizophrenics. J Nerv Ment Dis 127:167–179

Meyer JM (2001a) Effects of atypical antipsychotics on weight and serum lipid levels. J Clin Psychiatry 62(Suppl 27):27–34, discussion 40–41

Meyer JM (2001b) Novel antipsychotics and severe hyperlipidemia. J Clin Psychopharmacol 21:369–374

Meyer JM, Koro CE (2004) The effects of antipsychotic therapy on serum lipids: a comprehensive review. Schizophr Res 70:1–17

Meyer JM, Nasrallah HA (2009) Medical illness and schizophrenia, 2nd edn. American Psychiatric Press, Washington, DC

Meyer J, Loh C, Leckband SG, Boyd JA, Wirshing WC, Pierre JM, Wirshing D (2006) Prevalence of the metabolic syndrome in veterans with schizophrenia. J Psychiatr Pract 12:5–10

Meyer JM, Davis VG, Goff DC, McEvoy JP, Nasrallah HA, Davis SM, Daumit GL, Hsiao J, Swartz MS, Stroup TS, Lieberman JA (2008a) Impact of antipsychotic treatment on nonfasting triglycerides in the CATIE Schizophrenia Trial phase 1. Schizophr Res 103:104–109

Meyer JM, Davis VG, Goff DC, McEvoy JP, Nasrallah HA, Davis SM, Rosenheck RA, Daumit GL, Hsiao J, Swartz MS, Stroup TS, Lieberman JA (2008b) Change in metabolic syndrome parameters with antipsychotic treatment in the CATIE Schizophrenia Trial: prospective data from phase 1. Schizophr Res 101:273–286

Meyer JM, McEvoy JP, Davis VG, Goff DC, Nasrallah HA, Davis SM, Hsiao J, Swartz MS, Stroup TS, Lieberman JA (2009a) Inflammatory markers in schizophrenia: comparing antipsychotic effects in phase 1 of the CATIE effective study. Biol Psychiatry 66:1013–1022

Meyer JM, Rosenblatt LC, Kim E, Baker RA, Whitehead R (2009b) The moderating impact of ethnicity on metabolic outcomes during treatment with olanzapine and aripiprazole in patients with schizophrenia. J Clin Psychiatry 70:318–325

Nasrallah HA, Meyer JM, Goff DC, McEvoy JP, Davis SM, Stroup TS, Lieberman JA (2006) Low rates of treatment for hypertension, dyslipidemia and diabetes in schizophrenia: data from the CATIE Schizophrenia Trial sample at baseline. Schizophr Res 86:15–22

Newcomer JW (2005) Second-generation (atypical) antipsychotics and metabolic effects: a comprehensive literature review. CNS Drugs 19(Suppl 1):1–93

Newcomer JW, Hennekens CH (2007) Severe mental illness and risk of cardiovascular disease. JAMA 298:1794–1796

Nordestgaard BG, Benn M, Schnohr P, Tybjærg-Hansen A (2007) Nonfasting triglycerides and risk of myocardial infarction, ischemic heart disease, and death in men and women. JAMA 298:299–308

Osby U, Correia N, Brandt L, Ekbom A, Sparen P (2000a) Mortality and causes of death in schizophrenia in Stockholm county, Sweden. Schizophr Res 45:21–28

Osby U, Correia N, Brandt L, Ekbom A, Sparen P (2000b) Time trends in schizophrenia mortality in Stockholm county, Sweden: cohort study. BMJ 321:483–484

Reaven GM (1988) Banting lecture 1988: role of insulin resistance in human disease. Diabetes 37:1595–1607

Ridker PM, Rifai N, Rose L, Buring JE, Cook NR (2002) Comparison of C-reactive protein and low-density lipoprotein cholesterol levels in the prediction of first cardiovascular events. N Engl J Med 347:1557–1565

Ridker PM, Cannon CP, Morrow D, Rifai N, Rose LM, McCabe CH, Pfeffer MA, Braunwald E (2005) Pravastatin or atorvastatin evaluation and infection therapy-thrombolysis in myocardial infarction investigators. C-reactive protein levels and outcomes after statin therapy. N Engl J Med 352:20–28

Ridker PM, Danielson E, Fonseca FAH, Genest J, Gotto AM, Kastelein JJP, Koenig W, Libby P, Lorenzatti AJ, MacFayden JG, Nordestgaard BG, Shepherd J, Willerson JT, Glynn RJ, Group JTS (2008) Rosuvastatin to prevent vascular events in men and women with elevated C-reactive protein. N Engl J Med 359:2195–2207

Ryan MCM, Collins P, Thakore JH (2003) Impaired fasting glucose tolerance in first-episode, drug-naive patients with schizophrenia. Am J Psychiatry 160:284–289

Ryan MCM, Flanagan S, Kinsella U, Keeling F, Thakore JH (2004) The effects of atypical antipsychotics on visceral fat distribution in first episode, drug-naive patients with schizophrenia. Life Sci 74:1999–2008

Saari KM, Lindeman SM, Viilo KM, Isohanni MK, Jarvelin M-R, Lauren LH, Savolainen MJ, Koponen HJ (2005) A 4-fold risk of metabolic syndrome in patients with schizophrenia: the Northern Finland 1966 Birth Cohort study. J Clin Psychiatry 66:559–563

Saha S, Chant D, McGrath J (2007) A systematic review of mortality in schizophrenia: is the differential mortality gap worsening over time? Arch Gen Psychiatry 64:1123–1131

Schwarz L, Munoz R (1968) Blood sugar levels in patients treated with chlorpromazine. Am J Psychiatry 125:253–255

Sikich L, Frazier JA, McClellan J, Findling RL, Vitiello B, Ritz L, Ambler D, Puglia M, Maloney AE, Michael E, De Jong S, Slifka K, Noyes N, Hlastala S, Pierson L, McNamara NK, Delporto-Bedoya D, Anderson R, Hamer RM, Lieberman JA (2008) Double-blind comparison of first- and second-generation antipsychotics in early-onset schizophrenia and schizo-affective disorder: findings from the treatment of early-onset schizophrenia spectrum disorders (TEOSS) study. Am J Psychiatry 165:1420–1431

Smith DA (2007) Treatment of the dyslipidemia of insulin resistance. Med Clin North Am 91:1185–1210

Spelman LM, Walsh PI, Sharifi N, Collins P, Thakore JH (2007) Impaired glucose tolerance in first-episode drug-naive patients with schizophrenia. Diabet Med 24:481–485

Srisurapanont M, Likhitsathian S, Boonyanaruthee V, Charnsilp C, Jarusuraisin N (2007) Metabolic syndrome in Thai schizophrenic patients: a naturalistic one-year follow-up study. BMC Psychiatry 7:14

Stroup TS, Lieberman JA, McEvoy JP, Swartz MS, Davis SM, Rosenheck RA, Perkins DO, Keefe RS, Davis CE, Severe J, Hsiao JK (2006) Effectiveness of olanzapine, quetiapine, risperidone, and ziprasidone in patients with chronic schizophrenia following discontinuation of a previous atypical antipsychotic. Am J Psychiatry 163:611–622

Suvisaari JM, Saarni SI, Perala J, Suvisaari JVJ, Harkanen T, Lonnqvist J, Reunanen A (2007) Metabolic syndrome among persons with schizophrenia and other psychotic disorders in a general population survey. J Clin Psychiatry 68:1045–1055

Thakore JH, Mann JN, Vlahos I, Martin A, Reznek R (2002) Increased visceral fat distribution in drug-naive and drug-free patients with schizophrenia. Int J Obes Relat Metab Disord 26: 137–141

Tirupati S, Chua L-E (2007) Obesity and metabolic syndrome in a psychiatric rehabilitation service. Aust NZ J Psychiatry 41:606–610

van Nimwegen LJ, Storosum JG, Blumer RM, Allick G, Venema HW, de Haan L, Becker H, van Amelsvoort T, Ackermans MT, Fliers E, Serlie MJ, Sauerwein HP (2008) Hepatic insulin resistance in antipsychotic naive patients with schizophrenia, a detailed study of glucose metabolism with stable isotopes. J Clin Endocrinol Metab 93:572–577

Venkatasubramanian G, Chittiprol S, Neelakantachar N, Naveen MN, Thirthall J, Gangadhar BN, Shetty KT (2007) Insulin and insulin-like growth factor-1 abnormalities in antipsychotic-naive schizophrenia. Am J Psychiatry 164:1557–1560

Wannamethee SG, Shaper AG, Lennon L, Morris RW (2005) Metabolic syndrome vs Framingham risk score for prediction of coronary heart disease, stroke, and type 2 diabetes mellitus. Arch Intern Med 165:2644–2650

Weiden PJ, Daniel DG, Simpson GM, Romano SJ (2003) Improvement in indices of health status in outpatients with schizophrenia switched to ziprasidone. J Clin Psychopharmacol 23:1–6

Weiden PJ, Newcomer JW, Loebel A, Yang R, Lebovitz H (2008) Long-term changes in weight and plasma lipids during maintenance treatment with ziprasidone. Neuropsychopharmacology 33:985–994

Weuve J, Ridker PM, Cook NR, Buring JE, Grodstein F (2006) High-sensitivity C-reactive protein and cognitive function in older women. Epidemiology 17:183–189

Willi C, Bodenmann P, Ghali WA, Faris PD, Cornuz J (2007) Active smoking and the risk of type 2 diabetes: a systematic review and meta-analysis. JAMA 298:2654–2664

Wilson PW, D'Agostino RB, Levy D, Belanger AM, Silbershatz H, Kannel WB (1998) Prediction of coronary heart disease using risk factor categories. Circulation 97:1837–1847

Zhang Z-J, Yao Z-J, Liu W, Fang Q, Reynolds GP (2004) Effects of antipsychotics on fat deposition and changes in leptin and insulin levels. Magnetic resonance imaging study of previously untreated people with schizophrenia. Br J Psychiatry 184:58–62

Pharmacological Strategies for Enhancing Cognition in Schizophrenia

Deanna M. Barch

Contents

1. Introduction 44
2. Cholinergic Agents 46
   2.1 Cholinesterase Inhibitors 47
   2.2 Nicotine, Nicotinic Receptor Agonists, and Muscarinic Receptor Agonists 51
3. Glutamatergic Agents 60
   3.1 Glycine Allosteric Modulators 60
   3.2 AMPA Receptor Modulators 61
   3.3 Phosphodiesterase 5 Inhibitors 66
   3.4 NMDA Receptor Antagonists 66
4. Gamma-Aminobutyric Acid Modulating Agents 67
5. Dopaminergic Agents 70
   5.1 Indirect Dopamine Agonists 71
   5.2 Atomoxetine and Amantadine 71
   5.3 Selective Dopamine Agonists 76
6. Modafinil 77
7. Other Agents 80
8. Conclusions 83
References 85

Abstract Researchers have long recognized that individuals with schizophrenia experience challenges in a wide range of cognitive domains, and research on cognitive impairment in schizophrenia is not a recent phenomena. However, the past 10–20 years have seen an increasing recognition of the central importance of cognition to understanding function and outcome in this illness (Green et al. in Schizophr Bull 26:119–136, 2000), an awareness that has shifted the emphasis of at least some work on schizophrenia. More specifically, there has been a rapidly growing body of work on methods of enhancing cognition in schizophrenia, as

---

D.M. Barch
Washington University in St. Louis, One Brookings Drive, Box 1125, St. Louis, MO 63130, USA
e-mail: dbarch@artsci.wustl.edu

a means to potentially facilitate improved outcome and quality of life for individuals with this debilitating illness. The current chapter reviews the results of a range of studies examining adjunctive pharmacological treatments to enhance cognition in schizophrenia using a range of designs, including single-dose studies, open-label repeated dosing studies, and double-blind parallel group and crossover designs with repeated dosing. Although many of the single-dose and open-label studies have suggested positive cognitive effects from a range of agents, few of the larger-scale double-blind studies have generated positive results. The current state of results may reflect the need to identify alternative molecular mechanisms for enhancing cognition in schizophrenia or the need to reconceptualize the ways in which pharmacological agents may improve cognition in this illness, with a concomitant change in the traditional clinical trial study design used in prior studies of cognitive enhancement in schizophrenia.

Keywords Cognition · Control · Executive control · Improvement · Pharmacological · Schizophrenia · Working memory

1 Introduction

Researchers and theorists as far back as Bleuler and Kraeplin have recognized that abnormalities in cognitive function are a key component of schizophrenia, one of the most debilitating psychiatric disorders (Bleuler 1950; Kraeplin 1950). There are some data to suggest that the degree of impairment in certain aspects of cognition predicts the subsequent onset of schizophrenia (Cornblatt et al. 1999; Niendam et al. 2003; Sorensen et al. 2006). Further, individuals who share unexpressed genetic components of vulnerability to schizophrenia also experience impairments in cognitive function (Delawalla et al. 2006; Seidman et al. 2006; Snitz et al. 2006; Toulopoulou et al. 2003). In addition, some evidence suggests that the stronger the genetic risk, the greater the impairment in cognitive function in first-degree relatives (Glahn et al. 2003; Tuulio-Henriksson et al. 2003). As such, the attempt to understand the specific nature and sources of cognitive deficits in this disorder has a long and well-established history in the schizophrenia literature. However, in many ways, the last two decades have witnessed a relative explosion of research on cognition in schizophrenia, much of it couched within the framework of understanding the cognitive neuroscience of schizophrenia. There are at least two major forces driving the current wave of research on the psychological and neurobiological mechanisms that give rise to cognitive deficits in schizophrenia. One of these forces is the fact that over the past 20 years, the field of basic cognitive neuroscience has generated a wealth of new information on the neural systems that support specific processes involved of a range of cognitive functions, including the domains of working memory (WM), episodic memory (EM), and other aspects of "executive" control. As such, clinical scientists have been able to use this information to guide the search for the neural mechanisms that give rise to cognitive deficits in schizophrenia and a wealth of information is now available about the neurobiological abnormalities associated with a range of cognitive impairments in schizophrenia (Barch 2005; Minzenberg et al. 2009; Tamminga 2006).

A second force is that a growing body of research suggests that cognitive function in schizophrenia is one of the most critical determinants of social and occupational function in schizophrenia, potentially more so than the severity of other aspects/symptoms of schizophrenia such as hallucinations, delusions, or even negative symptoms (Cervellione et al. 2007; Gold et al. 2002; Green et al. 2000; Heinrichs et al. 2008; McClure et al. 2007; Williams et al. 2007). Such findings have led even the most applied of researchers and clinicians to become more invested in understanding the nature and source of cognitive deficits in schizophrenia, as such information may help to identify the treatment approaches that may be most effective in ameliorating such cognitive deficits in schizophrenia. This renewed emphasis on understanding cognitive impairment in schizophrenia has contributed to a rapidly growing body of work on methods of enhancing cognition in schizophrenia, as a means to potentially facilitate improved outcome and quality of life for individuals with this debilitating illness. This work has included major initiatives within the field designed to improve the measurement of cognition in schizophrenia during the conduct of clinical trials, such as the Measurement and Treatment Research to Improve Cognition in Schizophrenia (MATRICS) program (Marder and Fenton 2004) and the Cognitive Neuroscience Treatment Research to Improve Cognition in Schizophrenia (CNTRICS) program (Carter and Barch 2007), as well as major initiatives to identify and test novel pharmacological and behavioral approaches for improving cognition and functional outcome in schizophrenia (Buchanan et al. 2007a; Floresco et al. 2005; Stover et al. 2007a, b).

The goal of the current chapter is to provide an overview of the current status of results from treatment trials conducted to investigate the ability of pharmacological agents to enhance cognitive function in schizophrenia. This review focuses on studies that were either open-label, single-blind, or double-blind studies of adjunctive therapy (not monotherapy) designed to improve cognition in schizophrenia, with the obvious hope that such improvement would translate into improved life function and outcome in this illness. For good or for bad, the wealth of different neurobiological abnormalities identified in individuals with schizophrenia and which have been associated with cognitive impairment provides a wealth of potential targets for enhancing cognitive function in schizophrenia. This "embarrassment of riches" is not necessarily positive, as it suggests that the complexity of the neurobiology of cognitive impairment in schizophrenia may hamper the potential effectiveness of agents targeting a single mechanism, and that more creative approaches may be necessary for this complex disorder. Nonetheless, the MATRICS Neuropharmacology Committee identified a number of promising targets for the treatment of cognitive impairment in schizophrenia, including cholinergic, dopaminergic, and glutamatergic agents (Buchanan et al. 2007a). The field has been pushing forward with a range of clinical trials using agents targeting these mechanisms as well as many others, and there are now over 70 published studies of trials of examining pharmacological enhancement of cognitive function in schizophrenia.

2 Cholinergic Agents

A large body of human and animal research has highlighted the critical role that acetylcholine plays in a range of cognitive functions. Much of the animal work has focused on its role in memory (Gold 2003, 2004; Hasselmo 2006; Power et al. 2003), as research has demonstrated that impairing the septohippocampal and nucleus basalis of Meynert cortical cholinergic projections can impair memory function (Mandel et al. 1989). Further, supplementing the cholinergic system pharmacologically (e.g., via physostigmine administration) can help reverse the impairments causes by the disruption of these pathways through lesions, other pharmacological agents, or even aging (Arendt et al. 1990; Kamei et al. 1990; Mandel et al. 1989). Further, much research points to a consistent impairment in the cholinergic system in dementias such as Alzheimer's disease that involve impairments in memory as well as other cognitive functions (Schliebs and Arendt 2006). Not surprisingly, drugs that enhance the function of the cholinergic system have been a major focus of treatment for Alzheimer's disease and other neurological disorders involving memory impairment (Birks et al. 2009; Cincotta et al. 2008; Mohan et al. 2009; Razay and Wilcock 2008; Reingold et al. 2007).

The pattern of cognitive impairment in schizophrenia is not identical to that found in Alzheimer's disease and there is not the same type of evidence for cholinergic impairments in schizophrenia as there is in Alzheimer's. Nonetheless, there are studies suggesting a range of impairments in the cholinergic system in schizophrenia, including alterations in choline acetyltransferase (Powchik et al. 1998), and both nicotinic and muscarinic receptors (Breese et al. 2000; Crook et al. 1999, 2000, 2001; Deng and Huang 2005; Griffith et al. 1998; Scarr et al. 2009) in the brains of individuals with schizophrenia (for reviews, see Adams and Stevens 2007; Berman et al. 2007; Martin and Freedman 2007; Terry 2008).

There are a number of pathways by which one could attempt to enhance or regulate cholinergic function in schizophrenia. These include the use of acetylcholinesterase inhibitors (a focus for dementia work) that serve to inhibit acetylcholinesterase, the primary enzyme serving to break down acetylcholine in the synaptic cleft. Blocking the function of acetylcholinesterase should enhance cholinergic function by making more acetylcholine available for both nicotinic and muscarinic receptors in a nonspecific fashion (Buchanan et al. 2007a). An alternative would be to use direct nicotinic and muscarinic agonists that could target a range of receptor types that are either thought to be impaired in schizophrenia or involved in cognitive function. Both of these approaches have been tried in individuals with schizophrenia.

2.1 Cholinesterase Inhibitors

The first studies focusing on cholinergic enhancement in schizophrenia used the cholinesterase inhibitor donepezil (trade name Aricept), which is currently approved for the treatment of Alzheimer's disease and has shown evidence of efficacy for stabilizing cognition and functional outcome in this illness (Tsuno 2009). As shown in Table 1, results of these studies in schizophrenia have overall been negative. Several open-label studies found improvements in either memory (Chung et al. 2009) or motor function (Buchanan et al. 2003), and one double-blind crossover study found evidence for improved verbal memory (Erickson et al. 2005). However, the vast majority of the double-blind, placebo-controlled studies have not found any evidence for cognitive enhancement with donepezil compared to placebo in schizophrenia (Akhondzadeh et al. 2008; Fagerlund et al. 2007; Freudenreich et al. 2005; Friedman et al. 2002; Kohler et al. 2007; Tugal et al. 2004), including a recent large-scale multisite study with over 200 patients (Keefe et al. 2008). Keefe et al. offered a number of speculations as to why donepezil may not be effective in schizophrenia, including the hypothesis that a 10-mg dose of donepezil (the final dose in the Keefe et al. study) may be too high and may actually impair cognitive function.

A number of smaller-scale studies have also examined rivastigmine as a cognitive enhancer in schizophrenia Table 2. Rivastigmine is also a cholinesterase inhibitor, but one that leads to longer-lasting inhibition than donepezil, which is considered a short-acting inhibitor (Polinsky 1998). There have been fewer studies with rivastigmine than donepezil and none had the large sample size of the Keefe et al. study. As shown in Table 2, an early open-label study did provide evidence of improvement in memory function among individuals with schizophrenia taking rivastigmine (Lenzi et al. 2003), and two early 12-week, double-blind, parallel group studies provided some trend-level evidence for improved memory and executive control, as well as altered functional brain activity during cognitive task performance in individuals with schizophrenia (Aasen et al. 2005; Kumari et al. 2006). However, three subsequent 24-week double-blind studies did not provide any evidence for cognitive enhancement on any working memory, long-term memory, executive or speed measures (Chouinard et al. 2007; Guillem et al. 2006; Sharma et al. 2006).

In the last 4 years, a number of additional studies have also examined galantamine as a cognitive enhancer in schizophrenia. Like donepezil and rivastigmine, galantamine is a cholinesterase inhibitor (Table 3). However, it is also an allosteric modulator for both α_7 and $\alpha_4\beta_2$ nicotinic receptors (Dajas-Bailador et al. 2003; Schilstrom et al. 2007; Wang et al. 2007). Further, galantamine may indirectly augment dopamine function, which could have beneficial effects on some aspects of cognitive function, though with a risk for psychosis augmentation (Schilstrom et al. 2007; Wang et al. 2007). The studies with galantamine have had mixed results. Of the five double-blind, placebo-controlled, multiweek trials, four have provided some evidence of improvements in cognitive function with galantamine compared to placebo

Table 1 Donepezil studies in schizophrenia

Agent and dose	Authors	Sample size	Antipsychotics	Additional criteria	Cognitive assessment	Design	Outcome
Donepezil 5 and 10 mg	Friedman et al. (2002)	36 (18/18); ages 48.8/50.3	Risperidone	Clinically stable; 2+ SD below normal on the CLVT	Spatial Working Memory; CPT-IP; TMT; WCST; DSPT; VF; RAVLT	12-week, double-blind, placebo-controlled, parallel group	No significant improvements on any cognitive measure
Donepezil 5 mg → 10 mg	Buchanan et al. (2003)	15; age 43.1	Olanzapine		P50; RAVLT; BVRT; PEG; Digit Symbol; GDS CPT	6-week, open-label	Significant improvement on PEG
Donepezil 5 mg	Tugal et al. (2004)	12; ages 18–45	Conventional	Duration of schizophrenia or at least 2 years; clinically stable; high school graduate; no anticholinergics	WMS-R Figural Memory, Visual Reproduction, Visual Paired Associates, Logical Memory, and Verbal Paired Associates; VF; TMT; WCST	12-week, double-blind, placebo-controlled, crossover design	No significant improvements on any cognitive measure
Donepezil 5 mg → 10 mg	Freudenreich et al. (2005)	36 (19/17); age 48.7; 80% smokers	Conventional and second generation	Clinically stable; <20 on MMSE; no antipsychotic with strong anticholinergic properties	Digit Span; HVLT-R; TMT; Benton Oral Word Association Test; PEG	8-week, double-blind, parallel group	No significant improvements on any cognitive measure
Donepezil	Erickson et al. (2005)	15; age 43	Conventional and second generation	Stable doses of antipsychotics for at least 4 weeks	RAVLT; TMT	18 double-blind, placebo-controlled, crossover	Significantly improved verbal learning
Donepezil 5 mg → 10 mg	Fagerlund et al. (2007)	11 (7/4); ages 23–43; 71/50% smokers	Ziprasidone	No treatment refractory patients; no anticholinergics	CANTAB; Buschke Selective Reminding; Rey Complex Figure; TMT; Symbol Digit; VF	16-week, double-blind, placebo-controlled, parallel group	Impairments in planning efficiency; No improvement in verbal recall compared to placebo
Donepezil 5 mg	Lee et al. (2007a)	24 (12/12); ages 42.2/44.2; 66.7/50% smokers	Haloperidol	Score between 15 and 24 on K-MMSE	K-MMSE; HVLT; RAVLT; Digit Span; Digit Symbol; Stroop; TMT; VF; Boston Naming Test	24-week, double-blind, placebo-controlled, parallel group	Significant improvement in HVLT; Trend-level improvements in K-MMSE, RAVLT; Digit Span
Donepezil 5 mg → 10 mg	Kohler et al. (2007)	26 (13/13); ages 31.7/30.0; 54% smokers	Second generation	Clinically stable; BPRS <35; no anticholinergics	University of Pennsylvania Computerized Neurocognitive Battery	12-week, double-blind, placebo-controlled parallel group	No significant improvements on any cognitive measure

Donepezil 5 mg → 10 mg	Akhondzadeh et al. (2008)	30 (15/15); ages 32.3/33.9	Risperidone	Clinically stable; >19 on MMSE	WCST; WMS-R Figural Memory, Visual Reproduction, Verbal and Visual Paired Associates, Logical Memory: Digit Span; Block Design	12-week, double-blind, placebo-controlled parallel group	No significant improvements on any cognitive measure
Donepezil 5 mg → 10 mg	Keefe et al. (2008)	250 (124/121); ages 40.9/39.7; 73/72% smokers	Second generation	<5 on CGI-S; <81 on PANSS; −0.5 to −2.5 SD on BACS composite; >5th grade reading level on WRAT-3; no anticholinergics or antiparkinsonians	CATIE Neurocognitive Battery: COWAT; Category Instances; HVLT: Digit Symbol; Letter–Number Auditory Working Memory; CPT-IP; PEG; Visual–Spatial Working Memory; WISC-R Mazes	12-week, double-blind, placebo-controlled parallel group	No significant improvements on any cognitive measure
Donepezil 5 mg → 10 mg	Chung et al. (2009)	13; age 36.6; 23% smokers	Second generation	Clinically stable; at least 1 SD below norm on Computerized Neurocognitive Function Test	SCoRS; Computerized Neurocognitive Function Test (Digit span, visual span, auditory and visual CPT; Stroop, Trail Making, verbal and visual learning, hypothesis formation and finger tapping	12-week, open-label	Follow-up testing after 12 weeks of donepezil showed improved backward digit span, CVLT performance; Trail Making Performance (both A and B)

CPT-IP Continuous Performance Test – Identical Pairs, *COWAT* Controlled Oral Word Association Test, *DSPT* Digit Span Distraction Test, *GDS CPT* Gordon Diagnostic System Continuous Performance Test, *HVLT* Hopkin's Verbal Learning Test, *K-MMSE* Korean Mini-Mental Status Exam, *MMSE* Mini-Mental Status Exam, *PEG* Grooved Pegboard, *RAVLT* Rey Auditory Visual Learning Test, *SCoRS* Schizophrenia Cognition Rating Scale, *TMT* Trail Making Test, *VF* Verbal Fluency, *WMS-R* Wechsler Memory Scale-Revised, *WCST* Wisconsin Card Sorting Test

Table 2 Rivastigmine studies in schizophrenia

Agent and dose	Authors	Sample size	Antipsychotics	Additional criteria	Cognitive assessment	Design	Outcome
Rivastigmine 3 mg → 12 mg	Lenzi et al. (2003)	16; age 32	Second generation	Clinically stable; at least 2-year duration of schizophrenia	MMSE; CPT; WMS	12-month, open-label	Significant improvement in MMSE, WMS
Rivastigmine 3 mg → 12 mg	Aasen et al. (2005)	20 (11/9); age 42.6	Second generation	Clinically stable; <41 errors on the NART; −0.5 to −2 SDs below normal on CVLT	CPT with fMRI	12-week, double-blind, parallel group	Trend for more correct responses in rivastigmine group in control condition of CPT; increase in cerebellar activity
Rivastigmine 3 mg → 12 mg	Kumari et al. (2006)	21 (11/10); 73 and 90% smokers	Second generation	Medication Stable; no anticholinergics; 1–2 SDs below normal on CVLT	N-back task (0-, 1-, and 2-back) with fMRI	12-week, double-blind, parallel group	Trend for accuracy improvement across loads in rivastigmine group; increased occipital gyrus activation in rivastigmine group
Rivastigmine 3 mg → 18 mg	Guillem et al. (2006)	18 (9/9); ages 32.7/25.1; Fagerstrom 3.8/3.9	Second generation	<90 on RBANS composite	Continuous face recognition task with ERPs	24-week, crossover, comparing rivastigmine + antipsychotic to antipsychotic alone	No significant improvements in cognitive data, some significant modulation of various ERP components
Rivastigmine 3 mg → 12 mg	Sharma et al. (2006)	21 (11/10); ages 42.6/46.8; some overlap with prior studies	Second generation	Clinically stable; no anticholinergics; >41 errors on NART; −1 to −2 SD on CVLT	CVLT; NART; WCST; TMT; Digit Symbol; Dot Test; CPT-IP; Finger Tapping	24-week, double-blind parallel group	No significant improvements on any of the cognitive measures
Rivastigmine 3 mg → 12 mg	Chouinard et al. (2007)	20 (9/11); ages 32.7/25.7; 55/82% smokers	Conventional or second generation	<75 on the immediate or delayed memory indices of the RBANS	CANTAB Paired Associates Learning, Reaction Time, Rapid Visual Processing; Stockings of Cambridge; Spatial Working Memory	24-week, crossover, comparing rivastigmine + antipsychotic to antipsychotic alone	No significant improvements on any of the cognitive measures

BACS Brief Assessment of Cognition Scale, *CPT-IP* Continuous Performance Test – Identical Pairs, *DSPT* Digit Span Distraction Test, *GDS CPT* Gordon Diagnostic System Continuous Performance Test, *HVLT* Hopkin's Verbal Learning Test, *K-MMSE* Korean Mini-Mental Status Exam, *MMSE* Mini-Mental Status Exam, *RBANS* Repeatable Battery for the Assessment of Neuropsychological Status, *RAVLT* Rey Auditory Visual Learning Test, *SCoRS* Schizophrenia Cognition Rating Scale, *TMT* Trail Making Test, *WMS-R* Wechsler Memory Scale-Revised, *WCST* Wisconsin Card Sorting Test on both total battery scores and measure of memory and processing speed (Buchanan et al. 2008; Lee et al. 2007b; Noren et al. 2006; Schubert et al. 2006). One of these was a relatively large-scale study (86 patients) that found greater improvements on galantamine versus placebo for Digit Symbol and California Verbal Learning Performance, though more improvement on placebo for a Continuous Performance Test measure (Buchanan et al. 2008). However, one was a very small study (only 12 patients) which did not conduct formal statistical tests and provided only qualitative evidence for the benefits of galantamine (Noren et al. 2006). The most recent double-blind study was of a shorter duration (8 weeks) than the majority of the previous studies (12 weeks), and found that instead of improving cognitive function, galantamine impaired Continuous Performance Test – Identical Pairs (CPT-IP) performance, Stroop Interference, and some aspects of working memory (Dyer et al. 2008). This last study also used a higher dose of galantamine than prior studies (32 mg compared to 24 mg). As discussed by the authors of this negative galantamine study (Dyer et al. 2008), the mechanism of action of galantamine also differs at lower versus higher doses. At lower doses, galantamine acts as an allosteric modulator of α_7 and $\alpha_4\beta_2$ nicotinic receptors, increases burst-firing activity of dopamine cells in the ventral tegmental area (VTA), and increases prefrontal dopamine (Schilstrom et al. 2007; Wang et al. 2007). At higher doses, galantamine primary acts primarily as a cholinesterase inhibitor. Thus, the positive results at lower doses may reflect the allosteric modulatory effects and the dopaminergic effects. In contrast, the negative results at higher doses of galantamine are consistent with the generally negative results with donepezil and rivastigmine, both of which primarily operate as cholinesterase inhibitors.

Taken together, the results of the studies with donepezil, rivastigmine, and galantamine suggest that cholinergic agents that primarily act as cholinesterase inhibitors (which would include high-dose galantamine) are not particularly effective at improving cognitive function in schizophrenia. However, the somewhat more positive results with lower dose galantamine, which operates as an allosteric modulator of α_7 and $\alpha_4\beta_2$ nicotinic receptors and which has dopaminergic effects as well, suggest that an alternative mechanism for modulating cholinergic function in schizophrenia may have more positive effects of cognition. This latter suggestion is consistent with the body of evidence implicating abnormalities in the α_7 receptor as a potential pathophysiological mechanism in schizophrenia, and thus as a potential treatment target, as discussed in more detail in the next section.

2.2 Nicotine, Nicotinic Receptor Agonists, and Muscarinic Receptor Agonists

This body of work implicating nicotinic receptor abnormalities includes evidence that individuals with schizophrenia and their relatives show a failure to inhibit or filter responses to sensory stimuli, as evidenced by reduced P50 suppression

Table 3 Galantamine studies in schizophrenia

Agent and dose	Authors	Sample size	Antipsychotics	Additional criteria	Cognitive assessment	Design	Outcome
Galantamine 8 mg → 24 mg	Schubert et al. (2006)	16 (8/8); ages 48.3/46.8; 94% smokers	Risperidone	No anticholinergics	RBANS; Conner's CPT; Object Matching Memory Test; Tower of Toronto Puzzle	8-week, double-blind, placebo-controlled, parallel group	Significant improvement in RBANS total with galantamine
Galantamine 8 mg → 16 mg	Lee et al. (2007b)	24 (12/12); ages 39.5/41.5	Conventional	18–24 on K-MMSE	K-MMSE: HVLT: RCFT; Digit Span; Digit Symbol; Stroop; TMT; VF; Boston Naming Test	12-week, double-blind, placebo-controlled parallel group	Significant improvement in RCFT recognition for galantamine compared to placebo
Galantamine 8 mg → 24 mg	Noren et al. (2006)	12 (9/3)	Second generation		RAVLT; TMT; Letter–Number Sequencing; Vocabulary; WCST	12-week, parallel group comparing galantamine + antipsychotic to antipsychotic alone	No formal statistical tests conducted
Galantamine 0, 4, and 8 mg doses	Sacco et al. (2008)	21; 9 nonsmokers, 6 satiated, and 6 nonsatiated smokers; ages 44.7/47.0/48.6	Conventional and second generation	Deficits in Visual–Spatial Working Memory	CPT; TMT; Stroop; Digit Span; Simple Auditory Attention	Acute dose, double-blind, placebo-controlled	No significant improvements on any cognitive measure
Galantamine 8 mg → 24 mg	Buchanan et al. (2008)	86 (42/44); ages 49.9/49.5	Second generation other than clozapine	Clinically stable; <90 on RBANS; no anticholinergic medications	WAIS-III Letter–Number Sequencing; BACS number sequencing; CVLT; Brief Visual Memory Test; PEG; Digit Symbol; GDS CPT	12-week, double-blind, placebo-controlled, parallel group	Galantamine improved Digit Symbol and CVLT more than placebo, placebo improved GDS CPT more than galantamine

| Galantamine 8 mg → 32 mg | Dyer et al. (2008) | 18 (9/9); ages 44.3/50.5 | Second generation | No anticholinergic medications | CPT-IP; Stroop; WAIS-III Letter–Number Span; PEG | 8-week, double-blind, placebo-controlled, parallel group | Galantamine impaired CPT-IP, Stroop Interference, and LNS without reordering |

BACS Brief Assessment of Cognition Scale, *CPT-IP* Continuous Performance Test – Identical Pairs, *DSPT* Digit Span Distraction Test, *GDS CPT* Gordon Diagnostic System Continuous Performance Test, *HVLT* Hopkin's Verbal Learning Test, *K-MMSE* Korean Mini-Mental Status Exam, *MMSE* Mini-Mental Status Exam, *PEG* Grooved Pegboard, *RBANS* Repeatable Battery for the Assessment of Neuropsychological Status, *RAVLT* Rey Auditory Visual Learning Test, *RCFT* Rey Complex Figure Test, *SCoRS* Schizophrenia Cognition Rating Scale, *TMT* Trail Making Test, *VF* Verbal Fluency, *WMS-R* Wechsler Memory Scale-Revised, *WCST* Wisconsin Card Sorting Test

(Adler et al. 1998; Cullum et al. 1993; Griffith et al. 1998; Waldo et al. 1995; Young et al. 1996). Animal work suggests that activation of α_7 nicotinic receptors is critical for the presence of sensory gating in P50 paradigms (Luntz-Leybman et al. 1992). Further, there is evidence for the reduced expression of the α_7 nicotinic receptors in the hippocampus (Freedman et al. 1995) and cingulate cortex (Marutle et al. 2001) in postmortem brains of individuals with schizophrenia. The smoking behavior of individuals with schizophrenia has also been interpreted as evidence for nicotinic receptor abnormalities in this illness, as a high percentage of individuals with schizophrenia smoke (Smith et al. 2006), and when they smoke, they tend to extract more nicotine than a smoker without schizophrenia (Olincy et al. 1997). Consequently, a number of researchers have interpreted the smoking behavior in schizophrenia from a "self-medication" perspective, suggesting that nicotine may have positive effects on both symptoms and cognitive function in schizophrenia (Kumari and Postma 2005; Leonard et al. 2007).

The hypothesis that smoking may improve cognitive function in schizophrenia has led to the conduct of a number of controlled trials of nicotine in schizophrenia. As shown in Table 4, these studies have used the nicotine patch, nicotine spray, and nicotine gum in both smoking and nonsmoking individuals with schizophrenia to examine a range of cognitive functions. All of the studies have essentially been single-dose studies examining the acute effects of nicotine on cognitive function, typically following withdrawal of some length if the individuals were smokers. The vast majority of these studies found some positive benefit of nicotine on some aspect of cognitive function, including faster reaction times in a range of tasks (AhnAllen et al. 2008; Barr et al. 2008; Hong et al. 2009; Jubelt et al. 2008; Levin et al. 1996; Smith et al. 2002, 2006), as well as improved EM (Jacobsen et al. 2004; Jubelt et al. 2008; Myers et al. 2004; Smith et al. 2002), working memory (Levin et al. 1996; Myers et al. 2004; Smith et al. 2006), and attentional/executive control functions (AhnAllen et al. 2008; Barr et al. 2008; Depatie et al. 2002; Harris et al. 2004; Hong et al. 2009; Smith et al. 2006). This was true for both smokers and nonsmokers, and for individuals taking both conventional and second-generation antipsychotics. Although nicotine has had relatively consistent positive effects across studies, these effects should be considered modest in that nicotine did not improve every measure examined in each study and the magnitude of the effects are not large. Further, nicotine as a therapeutic agent may be limited by tachyphylaxis. For example, a 6-mg nicotine gum dose improved the Repeatable Battery for the Assessment of Neuropsychological Status (RBANS) attentional index in nonsmokers, but impaired it in smokers. However, in this study, smokers were only required to be abstinent for 2 h prior to testing, as compared to overnight in many of the other studies. The authors suggest that this relatively short withdrawal period may have meant that the agonist effects of nicotine were still subject to tachyphylaxis and that longer periods of abstinence may be necessary to reinstate positive agonist effects in smokers (Harris et al. 2004). However, taken together, the results of these studies do suggest some benefit of nicotine on cognitive function in schizophrenia, consistent with a self-medication hypothesis.

Table 4 Nicotinic and muscarinic receptor agonist studies in schizophrenia

Agent and dose	Authors	Sample size	Antipsychotics	Additional criteria	Cognitive assessment	Design	Outcome
Nicotine							
Nicotine patch (0, 7, 14, 21 mg/day) combined with haldol	Levin et al. (1996)	15; age 38.9; 100% smokers	Haldol starting at 2 mg/day and incremented at 2 mg every 2–3 days until bradykinesia detected; high/med/lose doses of haldol	Smoker	Automated Neuropsychological Assessment Metrics (reaction time, verbal, and visual memory)	Before smoking each day, given patch of 0, 7, 14, and 21 mg of nicotine in randomized counterbalanced design	Nicotine significantly improved reaction time, delayed match to sample performance in medium- and high-dose haldol. Quadratic effects of nicotine on CPT reaction time
Nicotine nasal spray (10 mg/ml) and research cigarettes (0.1 vs. 1.9 mg)	Smith et al. (2002)	31; age 40.8; 100% smokers	Conventional and second generation	Smoker	Automated Neuropsychological Assessment Metrics (reaction time, verbal, and visual memory); Verbal Memory from RANDT; VF	Single-dose, double-blind, crossover (following abstinence)	Nasal spray significantly improved verbal memory; higher dose nasal spray (2 puffs) improved two-choice reaction time and spatial rotation accuracy
Nicotine patch (14 mg)	Depatie et al. (2002)	15; age 36.7	Conventional and second generation	Smoker; FSIQ > 80	CPT-IP; Visually guided saccades and antisaccades; Smooth pursuit	Single-dose, double-blind, placebo-controlled, crossover study	Nicotine increased hits in on CPT, decreased antisaccade errors, and increased gain in no-monitoring condition compared to placebo
Nicotine nasal spray (1.0 mg)	Myers et al. (2004)	29 (15 smokers, 14 nonsmokers); ages 39.4/41.6	Conventional and second generation		Delay recognition task; Working memory task	Single-dose, crossover design, no drug versus nasal spray in randomized order at least 1 h apart	Recognition and d' was significantly better for smoking individuals with schizophrenia with nicotine than in the control condition

(*continued*)

Table 4 (continued)

Agent and dose	Authors	Sample size	Antipsychotics	Additional criteria	Cognitive assessment	Design	Outcome
Nicotine gum 6 mg	Harris et al. (2004)	20 (10 smokers, 10 nonsmokers); ages 44.5/43.1	Conventional and second generation		RBANS	Single-dose, double-blind, placebo-controlled, crossover study (after 2 h withdrawal)	Smoker by nicotine interaction for the RBANS attention index, such that nonsmokers got better on nicotine and smokers got worse
Nicotine patch 28 mg or 35 mg depending on BMI	Jacobsen et al. (2004)	13; age 42.9; 100% smoker	Conventional and second generation	Smoker	N-back (1- and 2-back), with either monaural or binaural presentation (high selective attention demand) with fMRI	Single-dose, double-blind, placebo-controlled, crossover study (after 15 h withdrawal)	On dichotic 2-back, higher plasma nicotine concentration was associated with better accuracy; greater activity in left insula and right putamen at high memory load under nicotine; greater activity in right thalamus, right globus pallidus and left lingual gyrus with high attention demand under nicotine
Nicotine nasal spray	Smith et al. (2006)	27; age 37.6	Conventional and second generation	Smoker; male	Conner's CPT; Spatial Rotation Test; Dot Memory Test; Verbal Memory from RANDT	Double-blind, placebo-controlled (active or placebo) nicotine spray on each of 2 days	Nicotine reduced CPT reaction time, reduced difference in error between immediate and delay memory on Dot

Drug/Dose	Study	N; age	Medication	Smoking status	Measures	Design	Results
Nicotine patch 14 mg	Barr et al. (2008)	28; age 47.7	Conventional and second generation	Nonsmoker; >35 on WRAT	CPT-IP; Stroop; Letter–Number Sequencing; PEG	Single-dose, double-blind, placebo-controlled, crossover study	Nicotine reduced reaction time, SD of hit RT and random and commission errors on CPT-IP, reduced interference score on Stroop
Nicotine patch	AhnAllen et al. (2008)	22; age 47.59; 100% smokers	Conventional and second generation		Attention Network Test	Baseline, withdrawal and patch rescue conditions	Executive network function and overall reaction time improved on patch compared to baseline
Nicotine patch 14 mg	Jubelt et al. (2008)	10; age 46; 0% smoker	Second generation	Nonsmoker; <35 on WRAT-3	Source Monitoring Task	Single-dose, placebo-controlled, crossover study	Nicotine reduced false alarm rates and reaction times for new items
Nicotine patch 21 or 35 mg depending on smoking level	Hong et al. (2009)	20; age 35; 100% smoker	Second generation	Smoker	Rapid Visual Information Processing Task with fMRI	Single-dose, double-blind, placebo-controlled, crossover study	Nicotine improved hit rate and reduced reaction time; increased BOLD in a range of regions on nicotine
DMXB-A							
DMXB-A 75 mg (+37.5 mg) and 150 mg (+75 mg)	Olincy et al. (2006)	12; ages 20–58	Conventional and second generation	Nonsmoking	RBANS; P50	Single-dose, placebo-controlled, double-blind, crossover	Low-dose DMXB-A improved total RBANS compared to placebo and increased P50 suppression
DMXB-A 75 and 150 mg	Freedman et al. (2008)	31; 22–60	Conventional and second generation	Clinically stable; nonsmoker	MCCB (no social cognition); P50	4-week (for each arm), double-blind, placebo-controlled, crossover	No differential improvement with either DMXB-A dose compared to placebo on full analysis. In first arm only, attention

(*continued*)

Table 4 (continued)

Agent and dose	Authors	Sample size	Antipsychotics	Additional criteria	Cognitive assessment	Design	Outcome
							and working memory significantly improved with DMXB-A, but not with placebo
Xanomeline							
Xanomeline 75 mg → 225 mg	Shekhar et al. (2008)	20 (10/10); ages 34.4/42.1	Unmedicated	Total PANSS score > 60	CPT-IP; Stroop; WMS; WAIS; TMT; Word list memory and recall; COWAT; Shipley Vocabulary Test; Finger Tapping	4-week, double-blind, placebo-controlled, parallel group	Significant improvements on xanomeline for List Learning, Story Recall and Delayed Memory

BACS Brief Assessment of Cognition Scale, *CPT-IP* Continuous Performance Test – Identical Pairs, *COWAT* Controlled Oral Word Association Test, *DSPT* Digit Span Distraction Test, *GDS CPT* Gordon Diagnostic System Continuous Performance Test, *HVLT* Hopkin's Verbal Learning Test, *K-MMSE* Korean Mini-Mental Status Exam, *MCCB* MATRICS Consensus Cognitive Battery, *MMSE* Mini-Mental Status Exam, *PEG* Grooved Pegboard, *RBANS* Repeatable Battery for the Assessment of Neuropsychological Status, *RAVLT* Rey Auditory Visual Learning Test, *RCFT* Rey Complex Figure Test, *SCoRS* Schizophrenia Cognition Rating Scale, *TMT* Trail Making Test, *VF* Verbal Fluency, *WMS-R* Wechsler Memory Scale-Revised, *WCST* Wisconsin Card Sorting Test

Given the potential therapeutic benefit of nicotinic receptor modulation, researchers have also been exploring the development of α_7 nicotinic receptor agonists as a treatment for cognitive impairment in schizophrenia. One such drug is 3-[(2,4-dimethoxy)benzylidene]anabaseine (DMXB-A) developed by the Kem laboratory (Kem et al. 2004; Walker et al. 2006). At lower concentrations, DMXB-A is an α_7-selective partial agonist, though at higher doses it is also a weak antagonist at $\alpha_4\beta_2$ receptors (Kem et al. 2004). As shown in Table 4, the first study with DMXB-A was a single-dose proof-of-concept study with two doses of DMXB-A in individuals with schizophrenia who were nonsmokers. The low dose improved total RBANS scores and increased P50 suppression (Olincy et al. 2006). The fact that the high dose did not improve the RBANS may be an indication of tachyphylaxis or the involvement of other receptors besides α_7 (Olincy et al. 2006). However, despite this initial positive proof-of-concept study, the subsequent 4-week, double-blind placebo-controlled, crossover study did not provide clear evidence for cognitive benefits of DMXB-A over and above placebo, though there was an significant though modest improvement in negative symptoms at the high dose (Freedman et al. 2008). This second study with DMXB-A used the MATRICS Consensus Cognitive Battery (MCCB), and one of the issues raised by the authors was the presence of unexpected practice effects in some of the MCCB domains. For example, speed of processing showed a significant effect of testing session, and a trend-level repetition effect was found for the attention/vigilance domain. Thus, the authors compared only the first arm of the study for each individual, providing essentially a small-sample size parallel-group design. Both the attention/vigilance and working memory domain scores showed either a significant or trend-level improvement with both doses of DXMB-A and not placebo, and numerically the visual learning and reasoning domains showed the same pattern. However, the opposite trend was apparent for speed of processing and verbal learning, with numerically more improvement under placebo than under either DXMB-A dosage.

In sum, the work on nicotine effects on cognitive function in schizophrenia suggests that this is a potentially promising avenue for cognitive enhancement, though clearly nicotine itself will not be an effective approach due to tachyphylaxis. The initial trials with DMXB-A suggest some promise for selective α_7 nicotinic receptor agonists, though the support is not strong at this point. There is also additional positive evidence from a study with Tropisetron (Koike et al. 2005), which is a strong 5-HT3 antagonist that is also a high-affinity partial agonist for α_7 nicotinic receptors (Koike et al. 2005). Tropisetron improved P50 suppression among individuals with schizophrenia, though only in nonsmokers. Thus, while modulation of nicotinic receptors remains a viable approach for cognitive enhancement in schizophrenia, the extent studies suggest that issues related to dosing, tachyphylaxis, and sustainability of response will need careful attention.

Although much of the attention in terms of cholinergic function in schizophrenia has focused on nicotinic receptors for the reasons outlined earlier, acetylcholine also acts at muscarinic receptors that would also be a potential cognition-enhancing target. Further, modulation of muscarinic receptors can also modulate dopamine function (Gomeza et al. 1999, 2001; Zhang et al. 2002). Xanomeline is a selective M1 and M4 muscarinic agonist that exerts functional dopamine antagonism without having affinity for any dopamine receptor (Shekhar et al. 2008). In a recent 4-week double-blind, placebo-controlled, parallel-group study in unmedicated individuals, xanomeline improved total Brief Psychiatric Rating Scale (BPRS) and Positive and Negative Symptom Scale (PANNS) scores more than placebo, and improved a number of measures of EM (see Table 4). However, this was a small-scale study, and further work will be needed to determine whether agents such as xanomeline are effect cognition either as primary treatments or as adjunctive treatments for cognitive impairment in schizophrenia.

3 Glutamatergic Agents

Glutamate is the primary neurotransmitter mediating excitatory neurotransmission in the human brain and as such plays a role in numerous cognitive functions. Glutamate can exert its excitatory effects through either ionotropic or metabotropic receptors. The ionotropic glutamate receptors include NMDA, AMPA, and kainite receptors, and there are at least three classes of metabotropic glutamate receptors, with one group (mGluR1 and mGluR4) serving to strengthen presynaptic glutamate release and postsynaptic NMDA neurotransmission. In contrast, the other two groups limit glutamate release (Buchanan et al. 2007a; Meldrum 2000). Much of the attention in terms of schizophrenia has focused on the NMDA receptor, given the fact that NMDA receptor antagonists such as PCP or ketamine can induce psychosis and some aspects of cognitive impairment (Javitt and Zukin 1991; Newcomer et al. 1999; Newcomer and Krystal 2001). This has led to an NMDA receptor-hypofunction hypothesis for schizophrenia (Olney et al. 1999).

3.1 Glycine Allosteric Modulators

NMDA receptors have a number of different binding sites. Given that direct agonism of the glutamate-binding site can cause excitotoxicity (Camon et al. 2001), much of the research on cognitive enhancing effects has focused on the glycine allosteric modulatory site (Buchanan et al. 2007a). In terms of cognitive function, studies have mainly focused on the amino acids glycine and D-serine, as well as D-cycloserine, an antituberculosis drug that easily crosses the blood–brain barrier and has partial agonist properties in a specific dose range (Goff et al. 1995). A number of early small-sample studies have provided evidence for a positive effect of glycine on negative symptoms (Table 5). (Heresco-Levy et al. 1996, 1998, 1999, 2004). In addition, several studies have found evidence for improvements in negative symptoms when D-cycloserine was added to conventional antipsychotics (e.g., Goff et al. 1999), though not when added to clozapine (e.g., Goff et al. 1996).

Despite these initial positive results in regards to negative symptoms, neither glycine nor D-cycloserine has proven to have any consistent positive benefits on any aspects of cognitive function as assessed by a formal test. Table 5 outlines all of the glycine and D-cycloserine studies in schizophrenia that have used cognitive measures, and only one out of seven studies found any positive benefit. The 1995 dose-finding study by Goff et al. (1995) found improved reaction time on the Sternberg Item Recognition test, but only at the 50-mg dose. However, this finding was not replicated in subsequent larger-scale studies with patients taking either conventional neuroleptics (Goff et al. 1999) or clozapine (Goff et al. 1996). These null results cannot readily be explained by factors such as type of antipsychotic medication, trial length, dosage, or baseline negative symptom levels, as the majority of the studies have required individuals to have significant negative symptoms, have studied patients on both conventional and second-generation medications, and have used a range of trial lengths. The largest trial – The Cognitive and Negative Symptoms in Schizophrenia Trial (CONSIST) study – directly compared glycine and D-cycloserine to placebo in individuals taking any antipsychotic but clozapine and did not find any overall significant improvements in either negative symptoms or cognition as a function of either adjunctive treatment (Buchanan et al. 2007b). One concern with the CONSIST study and the prior ones is that daily dosing of agents such as D-cycloserine can be subject to tachyphylaxis, which would limit effectiveness. Thus, Goff and colleagues also explored the efficacy of once-weekly D-cycloserine. However, although there was some improvement in Logical Memory thematic recall after 1 week, there were no significant improvements on any cognitive measure compared to placebo after 8 weeks (Goff et al. 2008a). Taken together, the consistent negative findings for cognitive function with either glycine or D-cycloserine suggest that these are not promising pathways for cognitive enhancement in schizophrenia.

3.2 AMPA Receptor Modulators

Although activation of the glycine allosteric modulatory site has not yet proven effective at enhancing cognitive function in schizophrenia, there are other approaches being examined for enhancing glutamate function. One such alternative is to modulate AMPA receptors rather than NMDA receptors. Although much of the focus on glutamate dysfunction in schizophrenia has been on NMDA receptor abnormalities, there is also evidence for decreased AMPA receptor density in the hippocampus of individuals with schizophrenia (Meador-Woodruff and Healy 2000). AMPA plays an important role in learning and memory via its impact on long-term potentiation. Activation of AMPA receptors facilitates a level of depolarization that eliminates the magnesium block in NMDA channels, allowing activation of NMDA receptors and setting in motion intracellular processes that induce LTP and changes in genetic expression that are thought to facilitate memory processes. Goff et al. (2001, 2008b) (see Table 5) have used CX516, which they

Table 5 Glutamatergic modulation studies in schizophrenia

Agent and dose	Authors	Sample size	Antipsychotics	Additional criteria	Cognitive assessment	Design	Outcome
Glycine							
Glycine 60 g	Evins et al. (2000)	27 (14/13); ages 52/58	Olanzapine or risperidone	Clinically stable	Stroop; WAIS Vocabulary, Information, digit span, and block design; CVLT, JOLO	8-week, double-blind, placebo-controlled, parallel group	No significant effects on any cognitive measure
D-cycloserine							
D-cycloserine 5, 15, 50, and 250 mg	Goff et al. (1995)	9; age 43	Conventional	Stable medication dose for at least 4 months	SIRP	Consecutive 2-week trials of placebo, 5, 15, 50, and 250 mg D-cycloserine	Significantly improved reaction times on the SIRP at the 50 mg dose
D-cycloserine 5, 15, 50, and 250 mg	Goff et al. (1996)	10; age 40.4	Clozapine	Meet primary deficit syndrome criteria	SIRP	Consecutive 2-week trials of placebo, 5, 15, 50, and 250 mg D-cycloserine	No significant effects on Sternberg
D-cycloserine 50 mg	Goff et al. (1999)	47 (23/24)	Conventional	Meet primary deficit syndrome criteria; stable dose of conventional narcoleptic for 4 weeks; 30+ on SANS	SIRP; Stroop; Miller-Selfridge; VF; Digit Span; Finger Tapping	8-week, double-blind, placebo-controlled, parallel group	No significant effects on any cognitive measure
D-cycloserine 5, 15, 50, and 250 mg	Evins et al. (2002)	10; age 42	Risperidone	Meet primary deficit syndrome criteria; 30+ on SANS	Word List Generation; Finger Tapping; Digit Span and Stroop	Consecutive 2-week trials of placebo, 5, 15, 50, and 250 mg D-cycloserine	No significant effects on any cognitive measure
D-cycloserine 50 mg	Goff et al. (2005)	55 (27/28); only 26 completed; ages 45.9/47	Conventional monotherapy	40+ on SANS	CVLT; Vocabulary; Information; Digit Span; Block Design; ANART; Stroop; Category Fluency; Finger Tapping and WCST	6-month, double-blind, placebo-controlled, parallel group	No significant effects on any cognitive measure

Drug/Dose	Study	N (sample); ages	Antipsychotic	Inclusion criteria	Cognitive measures	Design	Results
D-cycloserine 50 mg	Buchanan et al. (2007b)	157 (52/53/52); ages 42.6/ 44.4/43.4	Any antipsychotic but clozapine	Persistent moderate to severe negative symptoms; BRPS positive symptoms >19; Simpson–Angus EPS <9	WAIS Digit Symbol; WAIS Digit Search; Phonemic and Category Fluency; CPT; RAVLT; Brief Visual–Spatial Memory Test; Letter–Number Span; WAIS Letter–Number Sequencing; Visual–Spatial Working Memory; WCST	16-week, double-blind, double-dummy, placebo-controlled, parallel group	No significant effects on any cognitive measure
D-cycloserine 50 mg (1× per week)	Goff et al. (2008a)	38 (19/19); ages 50.1/48.0	Any antipsychotic but clozapine	Stable dose of antipsychotic for 4 weeks	NAART; HVLT; WCST: TMT Part B: Phonemic and Category Fluency; WMS Face Memory; Letter–Number Span; WAIS Letter–Number Sequencing; Logical Memory	8-week, double-blind, placebo-controlled, parallel group	D-cycloserine improved thematic recall on Logical Memory at 1 week; no significant effects on any cognitive measure at week 8
Ampakine CX516 (Ampakine) 2,700 mg	Goff et al. (2001)	13 (8/5); age 39.8	Clozapine	Stable medication dose for at least 6 months	RANDT 5-item acquisition; WCST; Rey/Taylor Figure; GDS CPT; TMT (B); VF; Finger Tapping	4-week, double-blind, placebo-controlled, parallel group	No formal statistical tests; large positive effect size for CX516 on WCST, RANDT 5-item acquisition and moderate positive effects on GDS CPT, VF, and Rey/Taylor Figure

(*continued*)

Table 5 (continued)

Agent and dose	Authors	Sample size	Antipsychotics	Additional criteria	Cognitive assessment	Design	Outcome
CX516 (Ampakine) 2,700 mg	Goff et al. (2008b)	95 (54/51); ages 43.7/42	Clozapine, olanzapine, or risperidone	At least 6 months on stable dose of antipsychotic	Degraded Stimulus CPT; CVLT; WCST; TMT; Letter–Number Span; Grooved Peg Board; Phonemic and Category Fluency	4-week, double-blind, placebo-controlled, parallel group	No significant effects on any cognitive measure
Lamotrigine							
Lamotrigine 50 mg → 400 mg	Goff et al. (2007)	Study 1: 209; age 41 Study 2: 212; age 41.6	One or more of clozapine, aripiprazole, olanzapine, quetiapine, risperidone, ziprasidone	Persistent positive symptoms for at least 3 months; >4 CGI; stable optimal dose of antipsychotic for at least 1 month	BACS + delayed verbal recall; Stroop	14-week, double-blind, placebo-controlled, parallel group	In study 2, but not study 1, the BACS composite score improved more in the lamotrigine group than in the placebo group
Sildenafil							
Sildenafil 50 and 100 mg	Goff et al. (2009)	15; age 49.7	Conventional or second generation	Stable dose of all medications for 4 weeks	HVLT; Logical Memory; Letter–Number Sequencing; Digit Symbol; Category Fluency; CPT-IP; Spatial Span	Double-blind, placebo-controlled, random-order, single-dose crossover design (placebo 50 and 100 mg)	No significant effects on any cognitive measure
Memantine							
Memantine 5 mg → 20 mg	Krivoy et al. (2008)	7; ages 20–56	Conventional or second generation	Stable antipsychotic dose for at least 6 weeks	Neurobehavioral Cognitive Scale Examination; Clock Drawing Test	6-week, open-label	No significant effects on any cognitive measure

| Memantine 5 mg → 20 mg | Lieberman et al. (2009) | 138 (70/68); ages 40.9/40.1 | Second-generation monotherapy | >25 on BPRS; >3 on at least one BRPS psychosis symptom; stable antipsychotic dose for 4 weeks | BACS | 8-week, double-blind, placebo-controlled, parallel group | No significant effects on any cognitive measure |

CPT-IP Continuous Performance Test – Identical Pairs, *COWAT* Controlled Oral Word Association Test, *DSPT* Digit Span Distraction Test, *GDS CPT* Gordon Diagnostic System Continuous Performance Test, *HVLT* Hopkin's Verbal Learning Test, *JOLO* Judgment of Line Orientation, *K-MMSE* Korean Mini-Mental Status Exam, *MMSE* Mini-Mental Status Exam, *RAVLT* Rey Auditory Visual Learning Test, *SCoRS* Schizophrenia Cognition Rating Scale, *SIRP* Sternberg Item Recognition Performance Task, *TMT* Trail Making Test, *VF* Verbal Fluency, *WMS-R* Wechsler Memory Scale-Revised, *WCST* Wisconsin Card Sorting Test refer to as an "Ampakine." CX516 binds to an allosteric site of the AMPA receptor. This enhances depolarization by prolonging channel opening in the presence of glutamate (Goff et al. 2008b). The first small-sample size study did not conduct formal statistical tests, but did see positive effects of CX516 on a number of cognitive measures, as well as on negative symptoms (Goff et al. 2001). Unfortunately, however, a relatively large sample size follow-up study did not find any positive benefits for CX516 on either cognitive function or symptoms in the overall analysis. This was true even when subjects taking clozapine were examined separately from those taking either olanzapine or risperidone (Goff et al. 2008b). The authors suggest that at least three factors could have influenced this null result, including too small of a dose (though this dose did produce insomnia and fatigue), too short of a trial (although fast effects were seen in animal studies), practice effects under placebo (though these were small), and potentially negative interactions with antipsychotics. Further, the authors note that the negative results with CX516 should not be taken as evidence again the potential effectiveness of any pathway toward modulating AMPA receptors, as there are a number of alternative mechanisms that might be more efficacious (Goff et al. 2008b).

3.3 Phosphodiesterase 5 Inhibitors

As another potential option for enhancing glutamate function, Goff et al. (2009) have also explored the use of phosphodiesterase 5 (PDE5) inhibitors. These agents increase cyclic guanosine monophosphate (cGMP) that would augment an NMDA–nitric oxide synthase (NOS)–cGMP pathway thought to be involved in long-term potentiation. Goff et al. (2009) used a PDE5 inhibitor called *sildenafil* that has been approved for the treatment of erectile dysfunction. There were no positive effects of single doses of either 50 or 100 mg of sildenafil on any cognitive measure in individuals with schizophrenia taking either conventional or second-generation antipsychotics (Goff et al. 2009). However, this was a small-sample size study and it is possible that repeated dosing is necessary for individuals with schizophrenia to experience a cognitive benefit from a PDE5.

3.4 NMDA Receptor Antagonists

In a different approach to regulating glutamate function in schizophrenia, Goff and colleagues have also examined the effects of lamotrigine, an anticonvulsant thought to inhibit excessive glutamate release related to impaired inhibitory feedback from GABAergic interneurons (Anand et al. 2000). The logic behind the use of lamotrigine is as follows. There is evidence that ketamine produces symptoms analogous to schizophrenia by blocking NMDA receptors on GABAergic inhibitory interneurons. This results in excessive glutamate release at non-NMDA glutamate receptors, as the blockage of the NMDA receptors on the gamma-aminobutyric acid (GABA) interneurons reduces inhibitory feedback that would normally serve to limit glutamate release. If this mechanism of action is at play for NMDA-mediated glutamatergic function in schizophrenia, as suggested by Olney et al. (1999), then blocking excessive glutamate release resulting from chronic NMDA hypofunction could enhance cognition and reduce negative symptoms in schizophrenia. Goff and colleagues reported on two separate large-scale clinical trials with this agent. Lamotrigine did not have any greater benefit than placebo on any clinical measure in either study. In one study, but not the other, the composite score from the Brief Assessment of Cognition Scale (BACS) improved more with lamotrigine than with placebo. However, the magnitude of this effect was relatively small, and in study 1 there were more cognitive responders on placebo than on lamotrigine (Goff et al. 2007). To test a similar hypothesis, several groups have also examined the effectiveness of memantine as an adjunctive treatment for cognitive impairment in schizophrenia (see Table 5). Memantine is currently licensed for use in Alzheimer's and is a noncompetitive low-affinity NMDA receptor antagonist. In an initial small-scale open-label study, memantine improved negative symptoms, but did not alter cognitive function (Krivoy et al. 2008). However, in a subsequent much larger placebo-controlled double-blind study, memantine did not improve either clinical symptoms or cognitive function (Lieberman et al. 2009).

In sum, the extent studies examining a variety of mechanisms designed to enhance or regulate glutamate function in schizophrenia as a means of improving negative symptoms and cognitive function have not shown much positive evidence for efficacy in either regard. This is particularly disappointing given the range of different mechanisms studied to date and the fact that a number of the trials had relatively large samples and good power to detect clinically relevant effects. This is not to say that we should give up hope for mechanisms that might enhance cognitive function via modulation of the glutamate systems. As recently reviewed by Javitt (2008), glycine transport inhibitors might represent a promising pathway to augment NMDA receptor activation. In addition, as noted by Goff, other means of modulating AMPA receptor function that differ from the mechanism of action of CX516 might very well end up being more effective at regulating glutamate function and enhancing cognition in schizophrenia.

4 Gamma-Aminobutyric Acid Modulating Agents

In recent years, much interest has also centered on the role of GABA dysfunction in the pathophysiology of schizophrenia, due in large part to the work of David Lewis and colleagues (Hashimoto et al. 2008; Konopaske et al. 2006; Lewis and Hashimoto 2007; Lewis et al. 2005; Lewis and Moghaddam 2006) (see chapter, this text). Lewis and others have shown that mRNA for GAD67 interneurons is reduced in postmortem brains of individuals with schizophrenia (Volk et al. 2000, 2001).

These GABA neurons express the calcium-binding protein parvalbumin (Lewis and Moghaddam 2006). These parvalbumin-expressing neurons include chandelier neurons whose axons terminate on the axon initial segments of pyramidal neurons (Melchitzky and Lewis 2003). Importantly, animal research has demonstrated that GABA neurons in prefrontal cortex are critical for intact working memory function, potentially because inhibitory GABA interneurons help to regulate gamma-activity in cortical pyramidal neurons, a mechanism that may serve to support working memory representations (Barr et al. 2009; Farzan et al. 2009). Individuals with schizophrenia showed reduced gamma-band activity in prefrontal cortex during cognitive control tasks, a finding which Lewis and colleagues have hypothesized to be related to altered GABAergic function in prefrontal cortex in this illness (Cho et al. 2006).

To test this hypothesis, Menzies and colleagues (see Table 6) examined the influence of lorazepam (a GABA agonist for the benzodiazepine receptor site that allosterically enhances postsynaptic inhibitory effects) and flumazenil (antagonist or partial inverse agonist at the GABA receptor benzodiazepine site) on working memory and functional brain activation as measured by fMRI. The task used was an N-back working memory task with multiple memory loads (0-, 1-, 2-, and 3-back). As hypothesized, lorazepam impaired working memory performance in schizophrenia, though not in controls. This effect was significant in individuals with schizophrenia at the 1-back (with a trend at the 2-back load). In contrast, flumazenil improved working memory performance at both the 1- and 2-back condition in patients, though not in controls (Menzies et al. 2007).

In a subsequent study (see Table 6), Lewis et al. (2008) conducted a 4-week small-sample double-blind placebo-controlled study with MK-0777. MK-0777 is a benzodiazepine-like drug with selective activity at GABA receptors containing α_2- or α_3-subunits. The logic of targeting these subunits is that it would avoid the negative cognitive effects and sedation that occur with GABA receptors containing α_1- or α_5-subunits are activated (Lewis et al. 2008). The authors used a number of different cognitive measures in this study, including the RBANS and both the N-back and the AX-CPT, as well as using EEG to measure γ-activity during a cognitive control task. On the RBANS, only the delayed memory subtest showed a positive effect of MK-0777 over placebo. However, on the AX-CPT, there was a trend for improvement in the discriminability index, and some trends for improved N-back performance and γ-activity during cognitive control performance. The effect sizes were relatively large, but the small-sample size precluded clear statistical interpretations of the results. However, a subsequent larger-scale study with a very similar design (4-week, double-blind, placebo-controlled) using either a 6- or 16-mg dose failed to replicate these results (Buchanan et al. 2010). Specifically, the Buchanan et al. study did find any significant positive effects of MK-0777 on the MCCB battery (total score or subscale scores) or on either the AX-CPT or the N-back. Thus, although the approach of targeting GABA receptors has received some positive support in small-scale and single-dose study, the more recent larger study was not supportive and additional work is needed to establish the efficacy of this or related approaches.

Table 6 GABAergic modulation studies in schizophrenia

Agent and dose	Authors	Sample size	Antipsychotics	Additional criteria	Cognitive assessment	Design	Outcome
Lorazepam 2 mg and flumazenil 0.9 mg	Menzies et al. (2007)	12; age 44.4	Second generation	Male; no anticholinergics or benzodiazepines	N-back (0-, 1-, 2-, and 3-back) working memory with fMRI	Single-dose, double-blind, placebo-controlled	Lorazepam impaired d' (at 1-back) and flumazenil improved d' (at 1- and 2-back)
MK-0777 6 mg → 16 mg	Lewis et al. (2008)	15 (6/9); ages 24–50	Second generation	Clinically stable; unemployed; IQ > 79; RBANS composite < 90	RBANS; N-back (0-, 1-, and 2-back); AX-CPT; Preparing to Overcome Prepotency (POP) Task	4-week, double-blind, placebo-controlled, parallel group	On RBANS, only delayed memory index showed MK-0777-related improvement; MK-0777 improved POP task performance, trend for d'-context on AX-CPT
MK-0777 6 and 16 mg	Buchanan et al. (2010)	60 (19/21/21); ages 43.3/44.9/40	Second generation	Clinically stable; BPRS Hallucinatory Behavior or Unusual Thought Content > 6; Conceptual Disorganization > 5; SAS > 7; CDS > 11; Maximum MCCB performance below 1.0 SD from perfect; HVLT < 32; >4th grade on WTAR	MCCB; AX-CPT; N-back (0-, 1-, and 2-back)	4-week, double-blind, placebo-controlled, parallel group	No significant effects on any cognitive measure

AX-CPT AX Continuous Performance Test, *BPRS* Brief Psychiatric Rating Scale, *MCCB* MATRICS Consensus Cognitive Battery, *RBANS* Repeatable Battery for the Assessment of Neuropsychological Status

5 Dopaminergic Agents

Early and simple forms of the dopamine hypothesis of schizophrenia – focused primarily on enhanced subcortical dopamine function – are no longer considered tenable. However, there continues to be evidence that dopamine is critically involved in the pathophysiology of schizophrenia, but that the nature of dopaminergic abnormalities is much more complex than originally thought. Current theories focus on a dysregulation of the dopamine system that involves both disrupted phasic dopamine function in subcortical regions, as well as hypodopaminergic function in prefrontal cortex (Lisman et al. 2008). It has been hypothesized that the heightened dopamine neurotransmission in subcortical regions contributes to the positive psychotic symptoms of schizophrenia, while hypoactive dopamine neurotransmission in cortical regions contributes to negative symptoms and cognitive impairment (Toda and Abi-Dargham 2007).

Further, there is a wealth of work implicating the dopamine system in many of the cognitive domains that are impaired in schizophrenia with the most evidence in regards to working memory and cognitive control (Goldman-Rakic et al. 2000). For example, working memory function is impaired in nonhuman primates following 6-hydroxydopamine lesions in PFC (Brozoski et al. 1979), or administration of dopamine antagonists (Sawaguchi and Goldman-Rakic 1994). In addition, administration of low-dose DA agonists can improve working memory in monkeys (Williams and Goldman-Rakic 1995), especially those with impaired performance (Arnsten et al. 1994; Cai and Arnsten 1997; Castner et al. 2000). Current models of the role of dopamine in working memory emphasize the important interactions between multiple dopamine receptors (e.g., D1 vs. D2) as well as interactions with other neurotransmitter systems (Gonzalez-Burgos et al. 2005; Seamans and Yang 2004).

There is also growing evidence that the administration of dopamine agonists can improve cognition in humans, including working memory. Methylphenidate (Clark et al. 1986; Elliott et al. 1997; Mehta et al. 2000), amphetamine (Mattay et al. 1996, 2000), bromocriptine (Kimberg et al. 1997; Luciana and Collins 1997; Luciana et al. 1992, 1998), and pergolide (Kimberg and D'Esposito 2003; Muller et al. 1998) have all been shown to improve working memory in healthy human participants. Interestingly, there is also research to suggest that dopamine agonists may be particularly effective for those individuals with the worst performance in the absence of drug (Kimberg and D'Esposito 2003; Kimberg et al. 1997; Mattay et al. 2000, 2003; Mehta et al. 2001). For example, individuals with the high-activity form of the COMT gene (leading to more catabolism of dopamine) have worse working memory performance than individuals with the low-activity form of the COMT gene (Egan et al. 2001; Malhotra et al. 2002), and also show the greatest positive benefit of amphetamine (Mattay et al. 2003). Although several of these agents are not selective for dopamine, and it is likely that all of these drugs influence neurotransmitter systems other than the dopamine system, such results are generally consistent with the hypothesis that administration of dopamine agonists can improve working memory. Further, there is evidence that levodopa can improve working memory and related cognitive functions in individuals with impaired dopamine function, such as those with Parkinson's disease (Cools et al. 2002; Cooper et al. 1992; Costa et al. 2003; Kulisevsky et al. 1996, 2000; Lange et al. 1995).

5.1 Indirect Dopamine Agonists

Given these lines of evidence suggesting a role for dopamine in cognitive function in schizophrenia, a number of studies have examined the influence of various dopaminergic agents in schizophrenia, though most have not been in the form of a traditional clinical trial (see Table 7). These studies have provided evidence that individuals with schizophrenia taking haloperidol show improved performance on the Wisconsin Card Sorting Task with the administration of amphetamine, despite minimal or no exacerbation of positive symptoms (Daniel et al. 1991; Goldberg et al. 1991). In more recent work, Barch and Carter found that individuals with schizophrenia on stable doses of haloperidol or fluphenazine showed improvement in a number of cognitive domains following a single dose of amphetamine (0.25 mg/kg), including increased accuracy and faster reaction times on spatial working memory, increased language production, and decreased reaction time on the Stroop (with no loss of accuracy). These cognitive improvements occurred without an exacerbation of the positive symptoms of psychosis. The interpretation of these results has been that cognition is improved in schizophrenia with the coadministration of haloperidol and amphetamine because treatment with a typical antipsychotic blocks D2 receptors in subcortical regions. This blockage is thought to prevent a negative impact of dopamine agonists of positive symptoms, leaving D1 receptors in regions such as prefrontal cortex free to benefit for enhanced cholinergic transmission (Goldberg et al. 1991).

Further, individuals with schizotypal personality disorder also show improved performance on the Wisconsin Card Sorting Task, on a spatial working memory task, and on reaction time in an antisaccade task with the administration of amphetamine (Kirrane et al. 2000; Siegel et al. 1996; Wonodi et al. 2006), even in the absence of stable treatment with an antipsychotic. In contrast, one additional study used methylphenidate in young individuals with schizophrenia, once off medication and once after the patients were stabilized on medication (Szeszko et al. 1999). These researchers found that this indirect dopamine agonist reduced word production, increased redundant errors on a verbal fluency tasks, and increased disorganization symptoms, both when patients were on medication and when they were off medication (Szeszko et al. 1999).

5.2 Atomoxetine and Amantadine

Two additional agents that have been used in schizophrenia are also thought to have at least an indirect effect on increasing dopamine function (see Table 7). One such

Table 7 Dopaminergic modulation studies in schizophrenia

Agent and dose	Authors	Sample size	Antipsychotics	Additional criteria	Cognitive assessment	Design	Outcome
Dextroamphetamine							
Dextroamphetamine 0.25 mg/kg	Goldberg et al. (1991)	21; age 32	Haldol		WCST; VF; Selective Reminding Test; CTP; Stroop; Understanding Communication Test; Finger Tapping; TMT (B)	Double-blind, single-dose, placebo-controlled, crossover	Amphetamine significantly improved motor speed on Trails and Finger Tapping, trend-level improvement in correct responses on WCST and on Understanding Communication Test
Dextroamphetamine 0.25 mg/kg	Daniel et al. (1991)	19; ages 20–40	Haldol		WCST and Sensorimotor Control with SPECT	Double-blind, single-dose, placebo-controlled, crossover	Amphetamine improved number of correct responses and percentage of conceptual-level responses on WCST; Amphetamine increased DLPFC activity during WCST
Dextroamphetamine 0.25 mg/kg	Siegel et al. (1996)	9; age 43	Unmedicated	Schizotypal personality disorder	WCST	Double-blind, single-dose, placebo-controlled, crossover	Fewer errors on amphetamine compared to placebo, after controlling for placebo performance
Dextroamphetamine 30 mg	Kirrane et al. (2000)	12; age 39	Unmedicated	Schizotypal personality disorder;	DOT Test of Spatial Working Memory	Double-blind, single-dose, placebo-	Improved delay condition

Drug/dose	Study	N; age	Medication status	Inclusion criteria	Cognitive measures	Design	Results
Dextroamphetamine 0.25 mg/kg	Barch and Carter (2005)	10; age 36.6	Stable dose haldol or prolixin	Clinically stable	Spatial Working Memory; Language Production; Category Monitoring CPT; Stroop	Double-blind, single-dose, placebo-controlled, crossover	Amphetamine increased accuracy and speeded RT on spatial working memory, increased language production, and decreased RT on the Stroop (with no loss of accuracy)
Dextroamphetamine 30 mg	Wonodi et al. (2006)	11; age 34.8	Unmedicated	Schizophrenia spectrum personality disorder no lifetime exposure to antipsychotic medications	Antisaccade Task	Double-blind, single-dose, placebo-controlled, crossover, with two amphetamine sessions and one placebo session	Amphetamine reduced antisaccade latency, but not errors
Methylphenidate							
Methylphenidate 0.5 mg/kg	Szeszko et al. (1999)	11; age 24.5	Unmedicated at active phase	<12 weeks of lifetime antipsychotic exposure; 4 or more on at least one psychotic symptom of SAPS	Word Production	Pre- and postmethyl-phenidate perfusion at unmedicated baseline and after treatment stabilization criteria met for 8 weeks	Methylphenidate reduced word production and increased redundant word production

(*continued*)

Table 7 (continued)

Agent and dose	Authors	Sample size	Antipsychotics	Additional criteria	Cognitive assessment	Design	Outcome
Atomoxetine							
Atomoxetine 40 mg → 80 mg	Friedman et al. (2008)	20 (10/10)	Second generation	Stable antipsychotic doses for at least 4 weeks	BACS; N-back (0-, 2-, and 3-back) with fMRI	4-week, double-blind, placebo-controlled, parallel group	No significant improvements on any cognitive measure
Atomoxetine 0, 40, and 80 mg	Sacco et al. (2009)	12 (3/4/5)	Conventional and second generation	Smoker; must have deficit in visuospatial working memory	Attention; Cognitive Switching; Working memory; COWAT; Fine Motor; Learning and Memory	1-week, double-blind, placebo-controlled, crossover, with two doses of atomoxetine and placebo	Individuals in 80 mg group showed improvement in Visual–Spatial Working Memory, and COWAT compared to placebo, but no changes were statistically significant
Amantadine							
Amantadine 100 mg	Fayen et al. (1988)	9; age 30	Conventional	Required antiparkinsonian daily for 4 weeks	RAVLT	6-week, double-blind, crossover design, amantadine versus trihexy-phenidyl	First four trials of RAVLT significantly better on amantadine compared to trihexyphenidyl; trends for recognition to be better as well

Drug	Study	N; age	Antipsychotic	Inclusion criteria	Cognitive measures	Design	Results
Amantadine 200 mg	Silver and Geraisy (1995)	26; age 36.7	Conventional or second generation	Stable medication doses for at least 2 months	BVRT; WMS; MMSE	2-week, double-blind, crossover design, amantadine versus biperiden	Logical Memory and visual reproduction from the WMS significantly better on amantadine versus biperiden
Amantadine 200 mg	Silver et al. (2005)	29; age 36.86	Conventional or second generation	Stable medication dose for at least 1 month	WAIS Digit Span; DOT Test; Finger Tapping; BVRT; MMSE; Abstraction Inhibition and Working Memory Task; Penn CPT; Penn Face Memory Test; Visual Object Learning Test; JOLO	3-week, double-blind, placebo-controlled, parallel group	No significant improvements on any cognitive measure
Dihydrexidine							
Dihydrexidine 20 mg	George et al. (2007)	13; age 39.45	Conventional and second generation	PANSS score >50 but less than 90, at least one PANSS negative item >4; stable antipsychotic dose for at least 2 weeks	TMT; HVLT; COWAT	Double-blind, single-dose, placebo-controlled, crossover	No significant improvements on any cognitive measure

CPT-IP Continuous Performance Test – Identical Pairs, *COWAT* Controlled Oral Word Association Test, *DSPT* Digit Span Distraction Test, *GDS CPT* Gordon Diagnostic System Continuous Performance Test, *HVLT* Hopkin's Verbal Learning Test, *JOLO* Judgment of Line Orientation, *K-MMSE* Korean Mini-Mental Status Exam, *MMSE* Mini-Mental Status Exam, *RAVLT* Rey Auditory Visual Learning Test, *SCoRS* Schizophrenia Cognition Rating Scale, *SIRP* Sternberg Item Recognition Performance Task, *TMT* Trail Making Test, *VF* Verbal Fluency, *WMS-R* Wechsler Memory Scale-Revised, *WCST* Wisconsin Card Sorting Test agent is atomoxetine, which is a selective norepinephrine reuptake inhibitor that also serves to increase extracellular dopamine in prefrontal cortex (Friedman et al. 2008), but not in subcortical regions (Bymaster et al. 2002). Neither study found evidence for statistically significant improvements in cognitive performance on atomoxetine among individuals with schizophrenia (Friedman et al. 2008; Sacco et al. 2009), though Sacco et al. (2009) reported relatively large effect size improvements that did not reach statistical significance given the small-sample size. Several additional studies have also examined the effects of amantadine, which has both NMDA antagonist properties and indirect dopamine agonist properties. Two of these studies compared amantadine to an anticholinergic medication, and found better memory performance on amantadine versus either biperiden (Silver and Geraisy 1995) or trihexyphenidyl (Fayen et al. 1988). However, given the absence of a placebo condition, it is difficult to tell whether this reflected actual benefits of amantadine, or impairments due to the anticholinergic medications. A third study did compare amantadine to placebo, but did not find any significant improvements in cognitive performance associated with amantadine, though it did improve visual motor coordination (Silver et al. 2005).

5.3 Selective Dopamine Agonists

As described earlier, single-dose studies of an indirect dopamine agonist have provided evidence of cognitive improvements among individuals with schizophrenia taking stable doses of high-potency antipsychotics, and among individuals with schizotypal personality disorder not taking any medication. However, concerns about amphetamine sensitization and the potential negative effects of more global enhancement of dopamine availability even among medicated patients with schizophrenia have really prevented agents such as amphetamine or methylphenidate from being seen as viable long-term adjunctive treatments for the enhancement of cognitive function in schizophrenia. To be specific, this is because indirect (and nonselective) dopamine agonists will enhance dopamine function in both subcortical and cortical regions, and will modulate neurotransmitters other than dopamine, making it more likely to generate negative effects alongside any potential positive effects. As such, much interest has centered on the possibility of developing selective dopamine agonists that target receptors thought to mediate the positive effects of dopamine on cognitive function in prefrontal cortex, such as D1 receptors. One such agent is dihydrexidine (DAR-0100), which is the first full D1 agonist (Zhang et al. 2009). This agent has been used in a single-dose crossover design, which found that DAR-0100 increased perfusion in prefrontal cortex, as well as in temporal and parietal regions (Mu et al. 2007). However, DAR-0100 did not produce any significant positive effects on cognition, though the sample size was clearly too small to have any power to detect significant effects, and no means or standard deviations were presented that would allow computation of an effect size (George et al. 2007). Nonetheless, the field remains optimistic that agents similar to DAR-0100 may end up providing evidence of cognitive enhancement in the absence of psychosis exacerbation, should it become clinically feasible to conduct a larger-scale, longer-term studies with a direct D1 agonist.

6 Modafinil

Although the choice of many agents as potential cognitive enhancers in schizophrenia has been driven by theoretical considerations of potential pathophysiological mechanisms and ways to target these mechanisms, some choices have been based on more practical considerations of proven efficacy in other disorders. For example, a number of studies have examined the use of modafinil as a potential cognitive enhancing agent in schizophrenia. Modafinil has the trade name Provigil and has been approved for use in various sleep disorders (Didato and Nobili 2009; Kumar 2008). Modafinil can improve cognition as well as mood and fatigue in sleep-deprived individuals, and there is evidence for beneficial effects on cognition even in nonsleep-deprived individuals (Kumar 2008; Minzenberg and Carter 2008). It is not yet clear exactly how modafinil works. Many hypotheses about the effects of modafinil on cognitive function have centered on dopamine. However, modafinil is structurally different than amphetamine, and it is clear that modafinil has effects on many neurotransmitter systems (Minzenberg and Carter 2008), including the ability to inhibit the function of both the dopamine transporter and the norepinephrine transporter, leading to functionally higher levels of both dopamine and norepinephrine. An additional hypothesis is that modafinil can act as a hypocretin/orexin agonist with excitatory influences on locus coeruleus adrenergic system (Minzenberg et al. 2008; Morein-Zamir et al. 2007).

Given that modafinil had good efficacy as a cognitive enhancer in sleep-deprived and even healthy individuals, as well as positive evidence for cognitive enhancement with modafinil in a number of psychiatric disorders (Turner 2006; Turner et al. 2003, 2004a), a number studies have examined its effects on cognitive and brain function in schizophrenia (see Table 8). Early studies using a single-dose design provided evidence for improvement in either cognition (Turner et al. 2004b) or brain function (Hunter et al. 2006; Spence et al. 2005) and a 4-week open-label study also showed evidence for some cognitive improvement among individuals with schizophrenia (Rosenthal and Bryant 2004). However, despite this early promise, subsequent longer-term, double-blind, placebo-controlled studies have not provided any consistent evidence for cognitive enhancement in schizophrenia as a function of modafinil (Freudenreich et al. 2009; Pierre et al. 2007; Sevy et al. 2005). Though these more recent studies are not encouraging, Freudenreich et al. (2009) have suggested that more definitive larger-scale studies are still needed, particularly if they examine the effects of modafinil on a range of cognitive, motor, and fatigue parameters in patients treated with different types of antipsychotics.

Table 8 Modafinil studies in schizophrenia

Agent and dose	Authors	Sample size	Antipsychotics	Inclusion/exclusion criteria	Cognitive assessment	Design	Outcome
Modafinil 200 mg	Turner et al. (2004b)	20; age 43	Conventional and second generation	Clinically stable; stable doses of neuroleptics; >25 on MMSE	CANTAB: Digit Span; PRM; DMTS; SWM; SSP; NTOL; IDED; STOP	Double-blind, placebo-controlled, single-dose, crossover	Modafinil improved Digit Span forward and backward, extradimensional shift performance on the IDED, but slowed response latency on NTOLS
Modafinil 100 mg → 200 mg	Rosenthal and Bryant (2004)	11; age 38.8	Conventional and second generation	Illness duration of at least 2 years; stable doses of antipsychotics for at least 1 month; a maximum score of 4 on no more than 1 PANSS positive item	WAIS-III Letter–Number Sequencing	4-week, open-label	Significant improvement on raw Letter–Number Sequencing scores
Modafinil 200 mg	Sevy et al. (2005)	20 (10/10); 35.9/38.9	Conventional and second generation	Stable doses of antipsychotics for at least 1 month; 4+ on CGI fatigue	CPT-IP; Oculomotor Delayed Response Task; DMTS; COWAT; RAVLT	8-week, double-blind, Placebo-controlled, parallel group	No significant improvements on any cognitive measures
Modafinil 100 mg	Spence et al. (2005)	21; age 37.7	Conventional and second generation	>70 NART; >2 on at least one SANS item; no prominent positive symptoms	2-Back working memory task with fMRI	Double-blind, placebo-controlled, single-dose, crossover	No significant improvement on cognitive measures; significantly greater working memory-related activity in anterior cingulated on modafinil

Modafinil 100 mg	Hunter et al. (2006)	12; age 37	Conventional and second generation	>70 NART; >2 on at least one SANS item; no prominent positive symptoms	VF with fMRI; Sheffield Activity in Time Task	Double-blind, placebo-controlled, single-dose, crossover	Modafinil associated with increased left dorsolateral PFC activity; Worse letter fluency performance at baseline associated with greater increase in activity
Modafinil 100 mg	Pierre et al. (2007)	20 (10/10); ages 49.8/48.7	Conventional and second generation	Stable doses of antipsychotics for at least 1 month; <15 on BRPS Psychosis; >19 on SANS total and >1 on either affective flattening or alogia	TMT; Degraded Stimulus CPT; CVLT	8-week, double-blind, Placebo-controlled, parallel group	No significant improvements on any cognitive measures
Modafinil 100 mg → 300 mg	Freudenreich et al. (2009)	35 (19/16); ages 44.2/46.4; 50/75% smoker	Clozapine	Clinically stable; clozapine for at least 6 months	NAART; Degraded Stimulus CPT; HVLT; WMS-III Faces and Family Pictures; WCST; TMT; WAIS-III Letter–Number Sequencing; VF; PEG	8-week, double-blind, Placebo-controlled, parallel group	No significant improvements on any cognitive measures

CPT-IP Continuous Performance Test – Identical Pairs, *COWAT* Controlled Oral Word Association Test, *DSPT* Digit Span Distraction Test, *GDS CPT* Gordon Diagnostic System Continuous Performance Test, *HVLT* Hopkin's Verbal Learning Test, *K-MMSE* Korean Mini-Mental Status Exam, *MMSE* Mini-Mental Status Exam, *PEG* Grooved Pegboard, *RAVLT* Rey Auditory Visual Learning Test, *SCoRS* Schizophrenia Cognition Rating Scale, *SIRP* Sternberg Item Recognition Performance Task, *WMS-R* Wechsler Memory Scale-Revised, *WCST* Wisconsin Card Sorting Test, *DMTS* Delayed Match to Sample Size, *SWM* Spatial Working Memory, *SSP* Spatial Span Performance, *NTOL* One-Touch Tower of London, *PRM* Pattern Recognition Memory, *IDED* Intradimensional/Extradimensional Shift Task, *STOP* Stop Signal Task, *VF* Verbal Fluency

7 Other Agents

In addition to the different classes of agents described in the sections earlier, there are a number of additional novel approaches that have been tried in only one or two studies each (Table 9). For example, two studies have examined the influence of mirtazapine on cognition in schizophrenia. Mirtazapine is a dual-acting antidepressant that has antagonist effects at the α_2-adrenergic, 5-HT$_2$ and 5-HT$_3$ receptors, as well as indirect agonist effects at the 5-HT$_{1a}$ receptor. The logic for the use of this drug as a potential enhancer of cognition in schizophrenia stems from the argument that clozapine (which may also have adrenergic and serotonin effects) has shown evidence of enhanced efficacy for negative symptoms in schizophrenia (Berk et al. 2009; Delle Chiaie et al. 2007). The first open-label study with mirtazapine showed beneficial effects on RBANS total scale scores as well as immediate and delayed memory scores (Delle Chiaie et al. 2007). However, a subsequent 6-week double-blind placebo-controlled study did not replicate the initial positive findings of the open-label study, failing to find any significant improvement as a function of mirtazapine in any cognitive measure among individuals with schizophrenia.

One study has also examined mifepristone (RU-486), which is a progesterone receptor antagonist and an antagonist of glucocorticoid receptors at high doses. The logic behind the use of mifepristone is that there is evidence that chronic elevations of endogenous cortisol levels (as is found in Cushing's syndrome) are associated with cognitive impairment (Gallagher et al. 2005) and the fact that HPA axis dysfunction may play a role in schizophrenia (Walker and Diforio 1997). However, a 1-week double-blind, placebo-controlled crossover study did not find any significant positive effects on any cognitive measures in schizophrenia. There has also been one study examining the potential benefits of pregnenolone, a neurosteriod shown to have beneficial effects on learning and memory in animal models (Marx et al. 2009). However, a small-sample 8-week, double-blind, placebo-controlled, parallel-group study did not find any significant improvements on either the BACS or the MCCB, though it did show significant improvements on SANS negative symptom scores compared to placebo (Marx et al. 2009). In addition, increases in serum pregnenolone levels predicted improvements in BACS scores at the end of the trial. Thus, although this small study did not find positive overall effects on the BACS or MCCB, it may be that with appropriate dosing, more positive effects might be found. Lastly, one additional study examined minocycline, a tetracycline. Minocycline has reduced cognitive impairments induced by PCP in animal models (Levkovitz et al. 2010). It has a number of effects and its mechanism of action in regards to cognitive enhancement is currently unclear. These effects include an ability to block nitric oxide-induced neurotoxicity, its influence on dopamine neurotransmission, and its influence on microglia that may impact apoptosis (Levkovitz et al. 2010). Interestingly, this 24-week, single-blind, placebo-controlled study showed positive benefits on SANS negative symptom ratings as well as positive effects on a number of cognitive measures, including an executive function

Table 9 Mixed or unique agent studies in schizophrenia

Agent and dose	Authors	Sample size	Antipsychotics	Additional criteria	Cognitive assessment	Design	Outcome
Mirtazapine 30 mg	Delle Chiaie et al. (2007)	15; age 32.3	Clozapine	Clinically Stable	RBANS	8-week, open-label	Significant improvement on RBANS immediate memory, delayed memory and total scale
Mirtazapine 30 mg	Berk et al. (2009)	38 (18/20); ages 37.8/35.9	Second generation	Clinically Stable	Digit Span; Word Learning; TMT and VF	6-week, double-blind, placebo-controlled, parallel group	No significant effects on any cognitive measure
Mifepristone 600 mg	Gallagher et al. (2005)	19; age 43.1	Conventional and second generation	Clinically stable	CANTAB Spatial Working Memory; RAVLT; CANTAB Spatial Span; WAIS Digit Span; CANTABN Pattern and Spatial Recognition Tasks; VF; CPT	1-week, double-blind, placebo-controlled, crossover	No significant effects on any cognitive measure
Pregnenolone 500 mg	Marx et al. (2009)	18 (9/9); ages 52.7/49/4	Second generation	Stable antipsychotic doses for at least 4 weeks	BACS and MCCB	8-week, double-blind, placebo-controlled, parallel group	No significant effects on any cognitive measure

(continued)

Table 9 (continued)

Agent and dose	Authors	Sample size	Antipsychotics	Additional criteria	Cognitive assessment	Design	Outcome
Minocycline 200 mg	Levkovitz et al. (2010)	54 (36/18); ages 24.7/25.1	Risperidone, olanzapine, quetiapine, or clozapine	Within 5 years of diagnosis; had not received antipsychotics for 6 months prior to current symptom exacerbation; had been initiated on treatment 14 days before	CANTAB Psychomotor Speed; Rapid Visual Processing; Pattern Recognition Memory; Spatial Recognition Memory; Spatial Working Memory; ID/ED; Stockings of Cambridge	24-week, single-blind, placebo-controlled, parallel group	Significant positive effects of minocycline on executive functioning composite score, Spatial Recognition Memory, ED errors, Spatial Working Memory

CPT-IP Continuous Performance Test – Identical Pairs, *COWAT* Controlled Oral Word Association Test, *DSPT* Digit Span Distraction Test, *GDS CPT* Gordon Diagnostic System Continuous Performance Test, *HVLT* Hopkin's Verbal Learning Test, *K-MMSE* Korean Mini-Mental Status Exam, *MMSE* Mini-Mental Status Exam, *RAVLT* Rey Auditory Visual Learning Test, *SCoRS* Schizophrenia Cognition Rating Scale, *SIRP* Sternberg Item Recognition Performance Task, *WMS-R* Wechsler Memory Scale-Revised, *WCST* Wisconsin Card Sorting Test, *VF* Verbal Fluency composite score, spatial recognition memory, extradimensional shift errors, and spatial working memory (Levkovitz et al. 2010). These initial positive results suggest a need for a more definitive double-blind placebo-controlled study to more fully delineate the potential cognitive enhancing effects of this novel agent.

8 Conclusions

The studies reviewed earlier represent a concerted effort on the part of the field to identify adjunctive treatments that could potentially improve cognitive function in schizophrenia, with the hope that such cognitive improvement would subsequently lead to enhancements in social, occupational, and educational achievement. The various studies represent diverse approaches that span many different neurotransmitter systems thought to be impaired in schizophrenia, as well as different mechanisms for modulating the function of those systems. Many single-dose and open-label studies have shown promising positive effects of a number of different agents on cognitive function in schizophrenia. In particular, many of the single-dose studies with nicotine and the single-dose studies with amphetamine have shown relatively consistent positive effects on measures of both accuracy and reaction time, though it is not always clear that these effects are of a clinically significant magnitude. However, the holistic view of the results of the more definitive or larger-scale studies is not nearly so encouraging. Specifically, relatively few large-scale, well-controlled, double-blind studies have shown any robust evidence for improvement in any domain of cognitive function in schizophrenia. This is true regardless of whether the individuals in the study were limited to those taking conventional antipsychotics, second-generation antipsychotics, or even limited to clozapine alone. The studies with galantamine are perhaps overall the most encouraging, but even galantamine has at least one null result in a double-blind study (Dyer et al. 2008), and a mixed result in another (Buchanan et al. 2008).

There are two responses one could have to the observation that few large-scale studies have generated positive evidence for cognitive enhancement effects in schizophrenia. One response is that we have just not yet found the right agent or mechanism, and that with continued drug development, we will hone in on an effective approach. It is certainly true that there are many theoretically motivated attractive targets that are currently in development. As described earlier, although glycine agonists or partial agonists have not been effective, glycine transporter inhibitors are another alternative approach to enhancing glutamate function (Javitt 2008). Although DMXB-A, an α_7-selective partial agonist, did not reveal positive results in its first double-blind study, xanomeline (a muscarinic agonist) did show positive effects and there are other approaches to muscarinic and nicotinic receptor activation that may be more effective, including targeting M_1 or $\alpha_4\beta_2$ receptors (Lieberman et al. 2008). In addition, another potential target is the histamine H_3 receptor (Esbenshade et al. 2008). In particular, H_3 receptor antagonists are attractive because of the fact that they regulate the release of many other neurotransmitters relevant for cognition, including dopamine, acetylcholine, and norepinephrine, and have shown positive effects on a range of cognitive functions in animal models (Esbenshade et al. 2008). Thus, H_3 antagonists may be able to modulate multiple neurotransmitter systems simultaneously in a way that may end up having more efficacy than mechanisms that focus on a single neurotransmitter system, although many such putatively selective agents also result in modulation of additional neurotransmitter systems.

A corollary to the above argument is that we may also need more predictive animal models in order to identify promising agents with a higher hit rate. Many of the drugs that come to early clinical trials have shown some evidence of improving cognitive function in animal studies, typically with rodents. These studies have used paradigms thought to capture core aspects of the cognitive functions impaired in individuals with schizophrenia or individuals with other cognitive disorders. However, many of these animal models lack construct validity, in the sense that they may not be tapping into the same cognitive or neurobiological processes operating in humans (Geyer 2008), because of species differences both in behavioral repertoires and in neural systems. Such rodent models are attractive for their ease of use and practicality, but these qualities become less helpful if the results of such studies lack predictive utility for knowing how drugs will influence cognition in humans. Thus, either the development of rodent models of cognition with more construct validity for human cognition, and/or the greater use of primate models that may have better predictive utility may be necessary advancements in order to move the field forward.

An alternative response, however, is to suggest that the approaches we have been using to test the ability of various agents to enhance cognition in schizophrenia are part of the problem, and that novel approaches are needed. The primary approach has been to simply add some type of potentially procognitive molecule to ongoing antipsychotic treatment, and then to test individuals prior to the start of the trial and at the end (as well as at some intermediate points in various studies). This approach assumes that the procognitive effects of the molecule will occur via relatively fast acting changes in neurotransmission that may or may not remain stable over the course of the trial. However, the logic for targeting many of the neurotransmitter systems upon which we have focused is their role in plasticity, learning, and memory. Individuals with schizophrenia may have experienced impairments in the wiring or developmental connectivity of various neural systems in the brain because of long-standing impairments in one or more of these neurotransmitter systems. As such, it may not be reasonable to expect that administration of a drug over a matter of weeks or months will be sufficient to reverse this damage or to help "rewire" the systems that support many cognitive functions that are impaired in schizophrenia. Instead, it may be that a combination of pharmacological enhancements with systematic and potentially intensive cognitive rehabilitation or training may be necessary to engender more robust and potentially longer-lasting change. Obviously, this may not be true for all molecules or for all cognitive functions. However, it is not unreasonable to expect that truly clinically significant cognitive change may require more than just either pharmacological enhancement or cognitive rehabilitation alone, but rather a combination of approaches that use the scaffolding provided by regulation of impaired neurotransmitter systems to engender plasticity and learning in a way that leads to more profound cognitive change among individuals with schizophrenia. This is of course an empirical question, and such studies are clearly time consuming and costly. However, such efforts will be well worth the cost should they reveal new avenues for robust, clinically significant and long-lasting cognitive change in schizophrenia, particularly if this translates into enhanced life function for individuals with this illness.

Acknowledgments Dr. Barch has received grants from the NIMH, NIA, NARSAD, Novartis, and the McDonnel Center for Systems Neuroscience.

References

Aasen I, Kumari V, Sharma T (2005) Effects of rivastigmine on sustained attention in schizophrenia: an FMRI study. J Clin Psychopharmacol 25:311–317
Adams CE, Stevens KE (2007) Evidence for a role of nicotinic acetylcholine receptors in schizophrenia. Front Biosci 12:4755–4772
Adler LE, Olincy A, Waldo M, Harris JG, Griffith J, Stevens K, Flach K, Nagamoto H, Bickford P, Leonard S, Freedman R (1998) Schizophrenia, sensory gating, and nicotinic receptors. Schizophr Bull 24:189–202
AhnAllen CG, Nestor PG, Shenton ME, McCarley RW, Niznikiewicz MA (2008) Early nicotine withdrawal and transdermal nicotine effects on neurocognitive performance in schizophrenia. Schizophr Res 100:261–269
Akhondzadeh S, Gerami M, Noroozian M, Karamghadiri N, Ghoreishi A, Abbasi SH, Rezazadeh SA (2008) A 12-week, double-blind, placebo-controlled trial of donepezil adjunctive treatment to risperidone in chronic and stable schizophrenia. Prog Neuropsychopharmacol Biol Psychiatry 32:1810–1815
Anand A, Charney DS, Oren DA, Berman RM, Hu XS, Cappiello A, Krystal JH (2000) Attenuation of the neuropsychiatric effects of ketamine with lamotrigine: support for hyperglutamatergic effects of N-methyl-D-aspartate receptor antagonists. Arch Gen Psychiatry 57:270–276
Arendt T, Schugens MM, Bigl V (1990) The cholinergic system and memory: amelioration of ethanol-induced memory deficiency by physostigmine in rat. Acta Neurobiol Exp (Wars) 50:251–261
Arnsten AF, Cai JX, Murphy BL, Goldman-Rakic PS (1994) Dopamine D1 receptor mechanisms in the cognitive performance of young adult and aged monkeys. Psychopharmacology 116:143–151
Barch DM (2005) The cognitive neuroscience of schizophrenia. In: Cannon T, Mineka S (eds) Annual review of clinical psychology, vol 1. American Psychological Association, Washington, DC, pp 321–353
Barch DM, Carter CS (2005) Amphetamine improves cognitive function in medicated individuals with schizophrenia and in healthy volunteers. Schizophr Res 77:43–58
Barr RS, Culhane MA, Jubelt LE, Mufti RS, Dyer MA, Weiss AP, Deckersbach T, Kelly JF, Freudenreich O, Goff DC, Evins AE (2008) The effects of transdermal nicotine on cognition in nonsmokers with schizophrenia and nonpsychiatric controls. Neuropsychopharmacology 33:480–490
Barr MS, Farzan F, Rusjan PM, Chen R, Fitzgerald PB, Daskalakis ZJ (2009) Potentiation of gamma oscillatory activity through repetitive transcranial magnetic stimulation of the dorsolateral prefrontal cortex. Neuropsychopharmacology 34:2359–2367

Berk M, Gama CS, Sundram S, Hustig H, Koopowitz L, D'Souza R, Malloy H, Rowland C, Monkhouse A, Monkhouse A, Bole F, Sathiyamoorthy S, Piskulic D, Dodd S (2009) Mirtazapine add-on therapy in the treatment of schizophrenia with atypical antipsychotics: a double-blind, randomised, placebo-controlled clinical trial. Hum Psychopharmacol 24:233–238

Berman JA, Talmage DA, Role LW (2007) Cholinergic circuits and signaling in the pathophysiology of schizophrenia. Int Rev Neurobiol 78:193–223

Birks J, Grimley Evans J, Iakovidou V, Tsolaki M, Holt FE (2009) Rivastigmine for Alzheimer's disease. Cochrane Database Syst Rev CD001191

Bleuler E (1950) Dementia praecox, or the group of schizophrenias. International Universities Press, New York, NY

Breese CR, Lee MJ, Adams CE, Sullivan B, Logel J, Gillen KM, Marks MJ, Collins AC, Leonard S (2000) Abnormal regulation of high affinity nicotinic receptors in subjects with schizophrenia. Neuropsychopharmacology 23:351–364

Brozoski TJ, Brown RM, Rosvold HE, Goldman PS (1979) Cognitive deficit caused by regional depletion of dopamine in prefrontal cortex of rhesus monkey. Science 205:929–931

Buchanan RW, Summerfelt A, Tek C, Gold J (2003) An open-labeled trial of adjunctive donepezil for cognitive impairments in patients with schizophrenia. Schizophr Res 59:29–33

Buchanan RW, Freedman R, Javitt DC, Abi-Dargham A, Lieberman JA (2007a) Recent advances in the development of novel pharmacological agents for the treatment of cognitive impairments in schizophrenia. Schizophr Bull 33:1120–1130

Buchanan RW, Javitt DC, Marder SR, Schooler NR, Gold JM, McMahon RP, Heresco-Levy U, Carpenter WT (2007b) The cognitive and negative symptoms in schizophrenia trial (CONSIST): the efficacy of glutamatergic agents for negative symptoms and cognitive impairments. Am J Psychiatry 164:1593–1602

Buchanan RW, Conley RR, Dickinson D, Ball MP, Feldman S, Gold JM, McMahon RP (2008) Galantamine for the treatment of cognitive impairments in people with schizophrenia. Am J Psychiatry 165:82–89

Buchanan RW, Marder S, Group TT (2010) MK-0777 for the treatment of cognitive impairments in people with schizophrenia. Schizophr Res

Bymaster FP, Katner JS, Nelson DL, Hemrick-Luecke SK, Threlkeld PG, Heiligenstein JH, Morin SM, Gehlert DR, Perry KW (2002) Atomoxetine increases extracellular levels of norepinephrine and dopamine in prefrontal cortex of rat: a potential mechanism for efficacy in attention deficit/hyperactivity disorder. Neuropsychopharmacology 27:699–711

Cai JX, Arnsten AFT (1997) Dose-dependent effects of the dopamine D1 receptor agonists A77636 or SKF81297 on spatial working memory in aged monkeys. J Pharmacol Exp Ther 283:183–189

Camon L, de Vera N, Martinez E (2001) Polyamine metabolism and glutamate receptor agonists-mediated excitotoxicity in the rat brain. J Neurosci Res 66:1101–1111

Carter CS, Barch DM (2007) Cognitive neuroscience-based approaches to measuring and improving treatment effects on cognition in schizophrenia: the CNTRICS initiative. Schizophr Bull 33:1131–1137

Castner SA, Williams GV, Goldman-Rakic PS (2000) Reversal of antipsychotic induced working memory deficits by short term dopamine D1 receptor stimulation. Science 287:2020–2022

Cervellione KL, Burdick KE, Cottone JG, Rhinewine JP, Kumra S (2007) Neurocognitive deficits in adolescents with schizophrenia: longitudinal stability and predictive utility for short-term functional outcome. J Am Acad Child Adolesc Psychiatry 46:867–878

Cho RY, Konecky RO, Carter CS (2006) Impairments in frontal cortical gamma synchrony and cognitive control in schizophrenia. Proc Natl Acad Sci USA 103:19878–19883

Chouinard S, Stip E, Poulin J, Melun JP, Godbout R, Guillem F, Cohen H (2007) Rivastigmine treatment as an add-on to antipsychotics in patients with schizophrenia and cognitive deficits. Curr Med Res Opin 23:575–583

Chung YC, Lee CR, Park TW, Yang KH, Kim KW (2009) Effect of donepezil added to atypical antipsychotics on cognition in patients with schizophrenia: an open-label trial. World J Biol Psychiatry 10:156–162

Cincotta SL, Yorek MS, Moschak TM, Lewis SR, Rodefer JS (2008) Selective nicotinic acetylcholine receptor agonists: potential therapies for neuropsychiatric disorders with cognitive dysfunction. Curr Opin Investig Drugs 9:47–56

Clark CR, Geffen GM, Geffen LB (1986) Role of monoamine pathways in the control of attention: effects of droperidol and methylphenidate on normal adult humans. Psychopharmacology 90:28–34

Cools R, Stefanova E, Barker RA, Robbins TW, Owen AM (2002) Dopaminergic modulation of high-level cognition in Parkinson's disease: the role of the prefrontal cortex revealed by PET. Brain 125:584–594

Cooper JA, Sagar JH, Doherty SM, Jordan N, Tidswell P, Sullivan EV (1992) Different effects of dopaminergic and anticholinergic therapies on cognitive and motor function in Parkinson's disease: a follow-up study of untreated patients. Brain 115:1701–1725

Cornblatt B, Obuchowski M, Roberts S, Pollack S, Erlenmeyer-Kimling L (1999) Cognitive and behavioral precursors of schizophrenia. Dev Psychopathol 11:487–508

Costa A, Peppe A, Dell'Agnello G, Carlesimo GA, Murri L, Bonuccelli U, Caltagirone C (2003) Dopaminergic modulation of visual–spatial working memory in Parkinson's disease. Dement Geriatr Cogn Disord 15:55–66

Crook JM, Dean B, Pavey G, Copolov D (1999) The binding of [3H]AF-DX 384 is reduced in the caudate-putamen of subjects with schizophrenia. Life Sci 64:1761–1771

Crook JM, Tomaskovic-Crook E, Copolov DL, Dean B (2000) Decreased muscarinic receptor binding in subjects with schizophrenia: a study of the human hippocampal formation. Biol Psychiatry 48:381–388

Crook JM, Tomaskovic-Crook E, Copolov DL, Dean B (2001) Low muscarinic receptor binding in prefrontal cortex from subjects with schizophrenia: a study of Brodmann's areas 8, 9, 10, and 46 and the effects of neuroleptic drug treatment. Am J Psychiatry 158:918–925

Cullum CM, Harris JG, Waldo MC, Smernoff E, Madison A, Nagamoto HT, Griffith J, Adler LE, Freedman R (1993) Neurophysiological and neuropsychological evidence for attentional dysfunction in schizophrenia. Schizophr Res 10:131–141

Dajas-Bailador FA, Heimala K, Wonnacott S (2003) The allosteric potentiation of nicotinic acetylcholine receptors by galantamine is transduced into cellular responses in neurons: Ca^{2+} signals and neurotransmitter release. Mol Pharmacol 64:1217–1226

Daniel DG, Weinberger DR, Jones DW, Smernoff E, Madison A, Nagamoto HT, Griffith J, Adler LE, Freedman R (1991) The effect of amphetamine on regional cerebral blood flow during cognitive activation in schizophrenia. J Neurosci 11:1907–1917

Delawalla Z, Barch DM, Fisher Eastep JL, Thomason ES, Hanewinkel MJ, Thompson PA, Csernansky JG (2006) Factors mediating cognitive deficits and psychopathology among siblings of individuals with schizophrenia. Schizophr Bull 32:525–537

Delle Chiaie R, Salviati M, Fiorentini S, Biondi M (2007) Add-on mirtazapine enhances effects on cognition in schizophrenic patients under stabilized treatment with clozapine. Exp Clin Psychopharmacol 15:563–568

Deng C, Huang XF (2005) Decreased density of muscarinic receptors in the superior temporal gyrusin schizophrenia. J Neurosci Res 81:883–890

Depatie L, O'Driscoll GA, Holahan AL, Atkinson V, Thavundayil JX, Kin NN, Lal S (2002) Nicotine and behavioral markers of risk for schizophrenia: a double-blind, placebo-controlled, cross-over study. Neuropsychopharmacology 27:1056–1070

Didato G, Nobili L (2009) Treatment of narcolepsy. Expert Rev Neurother 9:897–910

Dyer MA, Freudenreich O, Culhane MA, Pachas GN, Deckersbach T, Murphy E, Goff DC, Evins AE (2008) High-dose galantamine augmentation inferior to placebo on attention, inhibitory control and working memory performance in nonsmokers with schizophrenia. Schizophr Res 102:88–95

Egan MF, Goldberg TE, Kolachana BS, Callicott JH, Mazzanti CM, Straub RE, Goldman D, Weinberger DR (2001) Effect of COMT Val108/158 Met genotype on frontal lobe function and risk for schizophrenia. Proc Natl Acad Sci USA 98:6917–6922

Elliott R, Sahakian BJ, Matthews K, Bannerjea A, Rimmer J, Robbins TW (1997) Effects of methylphenidate on spatial working memory and planning in healthy young adults. Psychopharmacology 131:196–206

Erickson SK, Schwarzkopf SB, Palumbo D, Badgley-Fleeman J, Smirnow AM, Light GA (2005) Efficacy and tolerability of low-dose donepezil in schizophrenia. Clin Neuropharmacol 28:179–184

Esbenshade TA, Browman KE, Bitner RS, Strakhova M, Cowart MD, Brioni JD (2008) The histamine H3 receptor: an attractive target for the treatment of cognitive disorders. Br J Pharmacol 154:1166–1181

Evins AE, Fitzgerald SM, Wine L, Rosselli R, Goff DC (2000) Placebo-controlled trial of glycine added to clozapine in schizophrenia. Am J Psychiatry 157:826–828

Evins AE, Amico E, Posever TA, Toker R, Goff DC (2002) D-Cycloserine added to risperidone in patients with primary negative symptoms of schizophrenia. Schizophr Res 56:19–23

Fagerlund B, Soholm B, Fink-Jensen A, Lublin H, Glenthoj BY (2007) Effects of donepezil adjunctive treatment to ziprasidone on cognitive deficits in schizophrenia: a double-blind, placebo-controlled study. Clin Neuropharmacol 30:3–12

Farzan F, Barr MS, Wong W, Chen R, Fitzgerald PB, Daskalakis ZJ (2009) Suppression of gamma-oscillations in the dorsolateral prefrontal cortex following long interval cortical inhibition: a TMS-EEG study. Neuropsychopharmacology 34:1543–1551

Fayen M, Goldman MB, Moulthrop MA, Luchins DJ (1988) Differential memory function with dopaminergic versus anticholinergic treatment of drug-induced extrapyramidal symptoms. Am J Psychiatry 145:483–486

Floresco SB, Geyer MA, Gold LH, Grace AA (2005) Developing predictive animal models and establishing a preclinical trials network for assessing treatment effects on cognition in schizophrenia. Schizophr Bull 31:888–894

Freedman R, Hall M, Adler LE, Leonard S (1995) Evidence in postmortem brain tissue for decreased numbers of hippocampal nicotinic receptors in schizophrenia. Biol Psychiatry 38:22–33

Freedman R, Olincy A, Buchanan RW, Harris JG, Gold JM, Johnson L, Allensworth D, Guzman-Bonilla A, Clement B, Ball MP, Kutnick J, Pender V, Martin LF, Stevens KE, Wagner BD, Zerbe GO, Soti F, Kem WR (2008) Initial phase 2 trial of a nicotinic agonist in schizophrenia. Am J Psychiatry 165:1040–1047

Freudenreich O, Herz L, Deckersbach T, Evins AE, Henderson DC, Cather C, Goff DC (2005) Added donepezil for stable schizophrenia: a double-blind, placebo-controlled trial. Psychopharmacology (Berl) 181:358–363

Freudenreich O, Henderson DC, Macklin EA, Evins AE, Fan X, Cather C, Walsh JP, Goff DC (2009) Modafinil for clozapine-treated schizophrenia patients: a double-blind, placebo-controlled pilot trial. J Clin Psychiatry 70(12):1674–1680

Friedman JI, Adler DN, Howanitz E, Harvey PD, Brenner G, Temporini H, White L, Parrella M, Davis KL (2002) A double blind placebo controlled trial of donepezil adjunctive treatment to risperidone for the cognitive impairment of schizophrenia. Biol Psychiatry 51:349–357

Friedman JI, Carpenter D, Lu J, Fan J, Tang CY, White L, Parrella M, Bowler S, Elbaz Z, Flanagan L, Harvey PD (2008) A pilot study of adjunctive atomoxetine treatment to second-generation antipsychotics for cognitive impairment in schizophrenia. J Clin Psychopharmacol 28:59–63

Gallagher P, Watson S, Smith MS, Ferrier IN, Young AH (2005) Effects of adjunctive mifepristone (RU-486) administration on neurocognitive function and symptoms in schizophrenia. Biol Psychiatry 57:155–161

George MS, Molnar CE, Grenesko EL, Anderson B, Mu Q, Johnson K, Nahas Z, Knable M, Fernandes P, Juncos J, Huang X, Nichols DE, Mailman RB (2007) A single 20 mg dose of dihydrexidine (DAR-0100), a full dopamine D1 agonist, is safe and tolerated in patients with schizophrenia. Schizophr Res 93:42–50

Geyer MA (2008) Developing translational animal models for symptoms of schizophrenia or bipolar mania. Neurotox Res 14:71–78

Glahn DC, Therman S, Manninen M, Huttunen M, Kaprio J, Lönnqvist J, Cannon TD (2003) Spatial working memory as an endophenotype for schizophrenia. Biol Psychiatry 53:624–626

Goff DC, Tsai G, Manoach DS, Coyle JT (1995) Dose-finding trial of D-cycloserine added to neuroleptics for negative symptoms in schizophrenia. Am J Psychiatry 152:1213–1215

Goff DC, Tsai G, Manoach DS, Flood J, Darby DG, Coyle JT (1996) D-Cycloserine added to clozapine for patients with schizophrenia. Am J Psychiatry 153:1628–1630

Goff DC, Tsai G, Levitt J, Amico E, Manoach D, Schoenfeld DA, Hayden DL, McCarley R, Coyle JT (1999) A placebo-controlled trial of D-cycloserine added to conventional neuroleptics in patients with schizophrenia. Arch Gen Psychiatry 56:21–27

Goff DC, Leahy L, Berman I, Posever T, Herz L, Leon AC, Johnson SA, Lynch G (2001) A placebo-controlled pilot study of the ampakine CX516 added to clozapine in schizophrenia. J Clin Psychopharmacol 21:484–487

Goff DC, Herz L, Posever T, Shih V, Tsai G, Henderson DC, Freudenreich O, Evins AE, Yovel I, Zhang H, Schoenfeld D (2005) A six-month, placebo-controlled trial of D-cycloserine co-administered with conventional antipsychotics in schizophrenia patients. Psychopharmacology (Berl) 179:144–150

Goff DC, Keefe R, Citrome L, Davy K, Krystal JH, Large C, Thompson TR, Volavka J, Webster EL (2007) Lamotrigine as add-on therapy in schizophrenia: results of 2 placebo-controlled trials. J Clin Psychopharmacol 27:582–589

Goff DC, Cather C, Gottlieb JD, Evins AE, Walsh J, Raeke L, Otto MW, Schoenfeld D, Green MF (2008a) Once-weekly D-cycloserine effects on negative symptoms and cognition in schizophrenia: an exploratory study. Schizophr Res 106:320–327

Goff DC, Lamberti JS, Leon AC, Green MF, Miller AL, Patel J, Manschreck T, Freudenreich O, Johnson SA (2008b) A placebo-controlled add-on trial of the Ampakine, CX516, for cognitive deficits in schizophrenia. Neuropsychopharmacology 33:465–472

Goff DC, Cather C, Freudenreich O, Henderson DC, Evins AE, Culhane MA, Walsh JP (2009) A placebo-controlled study of sildenafil effects on cognition in schizophrenia. Psychopharmacology (Berl) 202:411–417

Gold PE (2003) Acetylcholine modulation of neural systems involved in learning and memory. Neurobiol Learn Mem 80:194–210

Gold PE (2004) Coordination of multiple memory systems. Neurobiol Learn Mem 82:230–242

Gold JM, Goldberg RW, McNary SW, Dixon LB, Lehman AF (2002) Cognitive correlates of job tenure among patients with severe mental illness. Am J Psychiatry 159:1395–1402

Goldberg TE, Bigelow LB, Weinberger DR, Daniel DG, Kleinman JE (1991) Cognitive and behavioral effects of the coadministration of dextroamphetamine and haloperidol in schizophrenia. Am J Psychiatry 148:78–84

Goldman-Rakic PS, Muly EC, Williams GV (2000) D1 receptors in prefrontal cells and circuits. Brain Res Rev 31:295–301

Gomeza J, Zhang L, Kostenis E, Felder C, Bymaster F, Brodkin J, Shannon H, Xia B, Deng C, Wess J (1999) Enhancement of D1 dopamine receptor-mediated locomotor stimulation in M (4) muscarinic acetylcholine receptor knockout mice. Proc Natl Acad Sci USA 96:10483–10488

Gomeza J, Zhang L, Kostenis E, Felder CC, Bymaster FP, Brodkin J, Shannon H, Xia B, Duttaroy A, Deng CX, Wess J (2001) Generation and pharmacological analysis of M2 and M4 muscarinic receptor knockout mice. Life Sci 68:2457–2466

Gonzalez-Burgos G, Kroener S, Seamans JK, Lewis DA, Barrionuevo G (2005) Dopaminergic modulation of short-term synaptic plasticity in fast-spiking interneurons of primate dorsolateral prefrontal cortex. J Neurophysiol 94:4168–4177

Green MF, Kern RS, Braff DL, Mintz J (2000) Neurocognitive deficits and functional outcome in schizophrenia: are we measuring the "right stuff"? Schizophr Bull 26:119–136

Griffith JM, O'Neill JE, Petty F, Garver D, Young D, Freedman R (1998) Nicotinic receptor desensitization and sensory gating deficits in schizophrenia. Biol Psychiatry 44:98–106

Guillem F, Chouinard S, Poulin J, Godbout R, Lalonde P, Melun P, Bentaleb LA, Stip E (2006) Are cholinergic enhancers beneficial for memory in schizophrenia? An event-related potentials (ERPs) study of rivastigmine add-on therapy in a crossover trial. Prog Neuropsychopharmacol Biol Psychiatry 30:934–945

Harris JG, Kongs S, Allensworth D, Martin L, Tregellas J, Sullivan B, Zerbe G, Freedman R (2004) Effects of nicotine on cognitive deficits in schizophrenia. Neuropsychopharmacology 29:1378–1385

Hashimoto T, Arion D, Unger T, Maldonado-Avilés JG, Morris HM, Volk DW, Mirnics K, Lewis DA (2008) Alterations in GABA-related transcriptome in the dorsolateral prefrontal cortex of subjects with schizophrenia. Mol Psychiatry 13:147–161

Hasselmo ME (2006) The role of acetylcholine in learning and memory. Curr Opin Neurobiol 16:710–715

Heinrichs RW, Goldberg JO, Miles AA, McDermid Vaz S (2008) Predictors of medication competence in schizophrenia patients. Psychiatry Res 157:47–52

Heresco-Levy U, Javitt DC, Ermilov M, Mordel C, Horowitz A, Kelly D (1996) Double-blind, placebo-controlled, crossover trial of glycine adjuvant therapy for treatment-resistant schizophrenia. Br J Psychiatry 169:610–617

Heresco-Levy U, Javitt DC, Ermilov M, Silipo G, Shimoni J (1998) Double-blind, placebo-controlled, crossover trial of D-cycloserine adjuvant therapy for treatment-resistant schizophrenia. Int J Neuropsychopharmacol 1:131–135

Heresco-Levy U, Javitt DC, Ermilov M, Mordel C, Silipo G, Lichtenstein M (1999) Efficacy of high-dose glycine in the treatment of enduring negative symptoms of schizophrenia. Arch Gen Psychiatry 56:29–36

Heresco-Levy U, Ermilov M, Lichtenberg P, Bar G, Javitt DC (2004) High-dose glycine added to olanzapine and risperidone for the treatment of schizophrenia. Biol Psychiatry 55:165–171

Hong LE, Schroeder M, Ross TJ, Buchholz B, Salmeron BJ, Wonodi I, Thaker GK, Stein EA (2009) Nicotine enhances but does not normalize visual sustained attention and the associated brain network in schizophrenia. Schizophr Bull [Epub ahead of print, Aug 27]

Hunter MD, Ganesan V, Wilkinson ID, Spence SA (2006) Impact of modafinil on prefrontal executive function in schizophrenia. Am J Psychiatry 163:2184–2186

Jacobsen LK, D'Souza DC, Mencl WE, Pugh KR, Skudlarski P, Krystal JH (2004) Nicotine effects on brain function and functional connectivity in schizophrenia. Biol Psychiatry 55:850–858

Javitt DC (2008) Glycine transport inhibitors and the treatment of schizophrenia. Biol Psychiatry 63:6–8

Javitt DC, Zukin SR (1991) Recent advances in the phencyclidine model of schizophrenia. Am J Psychiatry 148:1301–1308

Jubelt LE, Barr RS, Goff DC, Logvinenko T, Weiss AP, Evins AE (2008) Effects of transdermal nicotine on episodic memory in non-smokers with and without schizophrenia. Psychopharmacology (Berl) 199:89–98

Kamei C, Tsujimoto S, Tasaka K (1990) Effects of cholinergic drugs and cerebral metabolic activators on memory impairment in old rats. J Pharmacobiodyn 13:772–777

Keefe RS, Malhotra AK, Meltzer HY, Kane JM, Buchanan RW, Murthy A, Sovel M, Li C, Goldman R (2008) Efficacy and safety of donepezil in patients with schizophrenia or schizoaffective disorder: significant placebo/practice effects in a 12-week, randomized, double-blind, placebo-controlled trial. Neuropsychopharmacology 33:1217–1228

Kem WR, Mahnir VM, Prokai L, Papke RL, Cao X, LeFrancois S, Wildeboer K, Prokai-Tatrai K, Porter-Papke J, Soti F (2004) Hydroxy metabolites of the Alzheimer's drug candidate 3-[(2,4-dimethoxy)benzylidene]-anabaseine dihydrochloride (GTS-21): their molecular properties, interactions with brain nicotinic receptors, and brain penetration. Mol Pharmacol 65:56–67

Kimberg DY, D'Esposito M (2003) Cognitive effects of the dopamine receptor agonist pergolide. Neuropsychologia 41:1020–1027

Kimberg DY, D'Esposito M, Farah MJ (1997) Effects of bromocriptine on human subjects depend on working memory capacity. NeuroReport 8:381–385

Kirrane RM, Mitropoulou V, Nunn M, New AS, Harvey PD, Schopick F, Silverman J, Siever LJ (2000) Effects of amphetamine on visuospatial working memory performance in schizophrenia spectrum personality disorder. Neuropsychopharmacology 22:14–18

Kohler CG, Martin EA, Kujawski E, Bilker W, Gur RE, Gur RC (2007) No effect of donepezil on neurocognition and social cognition in young persons with stable schizophrenia. Cogn Neuropsychiatry 12:412–421

Koike K, Hashimoto K, Takai N, Shimizu E, Komatsu N, Watanabe H, Nakazato M, Okamura N, Stevens KE, Freedman R, Iyo M (2005) Tropisetron improves deficits in auditory P50 suppression in schizophrenia. Schizophr Res 76:67–72

Konopaske GT, Sweet RA, Wu Q, Sampson A, Lewis DA (2006) Regional specificity of chandelier neuron axon terminal alterations in schizophrenia. Neuroscience 138:189–196

Kraeplin E (1950) Dementia praecox and paraphrenia. International Universities Press, New York, NY

Krivoy A, Weizman A, Laor L, Hellinger N, Zemishlany Z, Fischel T (2008) Addition of memantine to antipsychotic treatment in schizophrenia inpatients with residual symptoms: a preliminary study. Eur Neuropsychopharmacol 18:117–121

Kulisevsky J, Avila A, Barbanoj M, Antonijoan R, Berthier ML, Gironell A (1996) Acute effects of levodopa on neuropsychological performance in stable and fluctuating Parkinson's disease patients at different levodopa plasma levels. Brain 199:2121–2132

Kulisevsky J, Garcia-Sanchex C, Berthier ML, Barbanoj M, Pascual-Sedano B, Gironell A, Estévez-González A (2000) Chronic effects of dopaminergic replacement on cognitive function in Parkinson's disease: a two year follow-up study of previously untreated patients. Mov Disord 15:613–626

Kumar R (2008) Approved and investigational uses of modafinil: an evidence-based review. Drugs 68:1803–1839

Kumari V, Postma P (2005) Nicotine use in schizophrenia: the self medication hypotheses. Neurosci Biobehav Rev 29:1021–1034

Kumari V, Aasen I, ffytche D, Williams SC, Sharma T (2006) Neural correlates of adjunctive rivastigmine treatment to antipsychotics in schizophrenia: a randomized, placebo-controlled, double-blind fMRI study. NeuroImage 29:545–556

Lange KW, Paul GM, Naumann M, Gsell W (1995) Dopaminergic effects on cognitive performance in patients with Parkinson's disease. J Neural Transm 46:423–432

Lee BJ, Lee JG, Kim YH (2007a) A 12-week, double-blind, placebo-controlled trial of donepezil as an adjunct to haloperidol for treating cognitive impairments in patients with chronic schizophrenia. J Psychopharmacol 21:421–427

Lee SW, Lee JG, Lee BJ, Kim YH (2007b) A 12-week, double-blind, placebo-controlled trial of galantamine adjunctive treatment to conventional antipsychotics for the cognitive impairments in chronic schizophrenia. Int Clin Psychopharmacol 22:63–68

Lenzi A, Maltinti E, Poggi E, Fabrizio L, Coli E (2003) Effects of rivastigmine on cognitive function and quality of life in patients with schizophrenia. Clin Neuropharmacol 26:317–321

Leonard S, Mexal S, Freedman R (2007) Smoking, genetics and schizophrenia: evidence for self medication. J Dual Diagn 3:43–59

Levin ED, Wilson W, Rose JE, McEvoy J (1996) Nicotine–haloperidol interactions and cognitive performance in schizophrenics. Neuropsychopharmacology 15:429–436

Levkovitz Y, Mendlovich S, Riwkes S, Braw Y, Levkovitch-Verbin H, Gal G, Fennig S, Treves I, Kron S (2010) A double-blind, randomized study of minocycline for the treatment of negative and cognitive symptoms in early-phase schizophrenia. J Clin Psychiatry 71(2):138–149

Lewis DA, Hashimoto T (2007) Deciphering the disease process of schizophrenia: the contribution of cortical GABA neurons. Int Rev Neurobiol 78:109–131

Lewis DA, Moghaddam B (2006) Cognitive dysfunction in schizophrenia: convergence of gamma-aminobutyric acid and glutamate alterations. Arch Neurol 63:1372–1376

Lewis DA, Hashimoto T, Volk DW (2005) Cortical inhibitory neurons and schizophrenia. Nat Rev Neurosci 6:312–324

Lewis DA, Cho RY, Carter CS, Eklund K, Forster S, Kelly MA, Montrose D (2008) Subunit-selective modulation of GABA type A receptor neurotransmission and cognition in schizophrenia. Am J Psychiatry 165:1585–1593

Lieberman JA, Javitch JA, Moore H (2008) Cholinergic agonists as novel treatments for schizophrenia: the promise of rational drug development for psychiatry. Am J Psychiatry 165:931–936

Lieberman JA, Papadakis K, Csernansky J, Litman R, Volavka J, Jia XD, Gage A, MEM-MD-29 Study Group (2009) A randomized, placebo-controlled study of memantine as adjunctive treatment in patients with schizophrenia. Neuropsychopharmacology 34:1322–1329

Lisman JE, Coyle JT, Green RW, Javitt DC, Benes FM, Heckers S, Grace AA (2008) Circuit-based framework for understanding neurotransmitter and risk gene interactions in schizophrenia. Trends Neurosci 31:234–242

Luciana M, Collins PF (1997) Dopamine modulates working memory for spatial but not object cues in normal humans. J Cogn Neurosci 4:58–68

Luciana M, Depue RA, Arbisi P, Leon A (1992) Facilitation of working memory in humans by a D_2 dopamine receptor agonist. J Cogn Neurosci 4:58–68

Luciana M, Collins PF, Depue RA (1998) Opposing roles for dopamine and serotonin in the modulation of human spatial working memory functions. Cereb Cortex 8:218–226

Luntz-Leybman V, Bickford PC, Freedman R (1992) Cholinergic gating of response to auditory stimuli in rat hippocampus. Brain Res 587:130–136

Malhotra AK, Kestler LJ, Mazzanti CM, Bates JA, Goldberg TE, Goldman D (2002) A functional polymorphism in the COMT gene and performance on a test of prefrontal cognition. Am J Psychiatry 159:652–654

Mandel RJ, Chen AD, Connor DJ, Thal LJ (1989) Continuous physostigmine infusion in rats with excitotoxic lesions of the nucleus basalis magnocellularis: effects on performance in the water maze task and cortical cholinergic markers. J Pharmacol Exp Ther 251:612–619

Marder SR, Fenton W (2004) Measurement and treatment research to improve cognition in schizophrenia: NIMH MATRICS initiative to support the development of agents for improving cognition in schizophrenia. Schizophr Res 72:5–9

Martin LF, Freedman R (2007) Schizophrenia and the alpha7 nicotinic acetylcholine receptor. Int Rev Neurobiol 78:225–246

Marutle A, Zhang X, Court J, Piggott M, Johnson M, Perry R, Perry E, Nordberg A (2001) Laminar distribution of nicotinic receptor subtypes in cortical regions in schizophrenia. J Chem Neuroanat 22:115–126

Marx CE, Keefe RS, Buchanan RW, Hamer RM, Kilts JD, Bradford DW, Strauss JL, Naylor JC, Payne VM, Lieberman JA, Savitz AJ, Leimone LA, Dunn L, Porcu P, Morrow AL, Shampine LJ (2009) Proof-of-concept trial with the neurosteroid pregnenolone targeting cognitive and negative symptoms in schizophrenia. Neuropsychopharmacology 34:1885–1903

Mattay VS, Berman KF, Ostrem JL, Esposito G, Van Horn JD, Bigelow LB, Weinberger DR (1996) Dextroamphetamine enhances "neural network-specific" physiological signals: a positron-emission tomography rCBF study. J Neurosci 15:4816–4822

Mattay VS, Callicott JH, Bertolino A, Heaton I, Frank JA, Coppola R, Berman KF, Goldberg TE, Weinberger DR (2000) Effects of dextroamphetamine on cognitive performance and cortical activation. NeuroImage 12:268–275

Mattay VS, Goldberg TE, Fera F, Hariri AR, Tessitore A, Egan MF, Kolachana B, Callicott JH, Weinberger DR (2003) Catechol O-methyltransferase val158-met genotype and individual variation in the brain response to amphetamine. Proc Natl Acad Sci USA 100:6186–6191

McClure MM, Bowie CR, Patterson TL, Heaton RK, Weaver C, Anderson H, Harvey PD (2007) Correlations of functional capacity and neuropsychological performance in older patients with schizophrenia: evidence for specificity of relationships? Schizophr Res 89:330–338

Meador-Woodruff JH, Healy DJ (2000) Glutamate receptor expression in schizophrenic brain. Brain Res Brain Res Rev 31:288–294

Mehta MA, Owen AM, Sahakian BJ, Mavaddat N, Pickard JD, Robbins TW (2000) Methylphenidate enhances working memory by modulating discrete frontal and parietal lobe regions in the human brain. J Neurosci 20:1–6

Mehta MA, Swainson R, Gogilvie AD, Sahakian BJ, Robbins TW (2001) Improved short-term spatial memory but impaired reversal learning following the dopamine D2 agonist bromocriptine in human volunteers. Psychopharmacology 159:10–20

Melchitzky DS, Lewis DA (2003) Pyramidal neuron local axon terminals in monkey prefrontal cortex: differential targeting of subclasses of GABA neurons. Cereb Cortex 13:452–460

Meldrum BS (2000) Glutamate as a neurotransmitter in the brain: review of physiology and pathology. J Nutr 130:1007S–1115S

Menzies L, Ooi C, Kamath S, Suckling J, McKenna P, Fletcher P, Bullmore E, Stephenson C (2007) Effects of gamma-aminobutyric acid-modulating drugs on working memory and brain function in patients with schizophrenia. Arch Gen Psychiatry 64:156–167

Minzenberg MJ, Carter CS (2008) Modafinil: a review of neurochemical actions and effects on cognition. Neuropsychopharmacology 33:1477–1502

Minzenberg MJ, Watrous AJ, Yoon JH, Ursu S, Carter CS (2008) Modafinil shifts human locus coeruleus to low-tonic, high-phasic activity during functional MRI. Science 322: 1700–1702

Minzenberg MJ, Laird AR, Thelen S, Carter CS, Glahn DC (2009) Meta-analysis of 41 functional neuroimaging studies of executive function in schizophrenia. Arch Gen Psychiatry 66:811–822

Mohan M, Carpenter PK, Bennett C (2009) Donepezil for dementia in people with Down syndrome. Cochrane Database Syst Rev CD007178

Morein-Zamir S, Turner DC, Sahakian BJ (2007) A review of the effects of modafinil on cognition in schizophrenia. Schizophr Bull 33:1298–1306

Mu Q, Johnson K, Morgan PS, Grenesko EL, Molnar CE, Anderson B, Nahas Z, Kozel FA, Kose S, Knable M, Fernandes P, Nichols DE, Mailman RB, George MS (2007) A single 20 mg dose of the full D1 dopamine agonist dihydrexidine (DAR-0100) increases prefrontal perfusion in schizophrenia. Schizophr Res 94:332–341

Muller U, von Cramon Y, Pollmann S (1998) D1- versus D2-receptor modulation of visuospatial working memory in humans. J Neurosci 18:2720–2728

Myers CS, Robles O, Kakoyannis AN, Sherr JD, Avila MT, Blaxton TA, Thaker GK (2004) Nicotine improves delayed recognition in schizophrenic patients. Psychopharmacology (Berl) 174:334–340

Newcomer JW, Krystal JH (2001) NMDA receptor regulation of memory and behavior in humans. Hippocampus 11:529–542

Newcomer JW, Farber NB, Jevtovic-Todorovic V, Selke G, Melson AK, Hershey T, Craft S, Olney JW (1999) Ketamine-induced NMDA receptor hypofunction as a model of memory impairment and psychosis. Neuropsychopharmacology 20:106–118

Niendam TA, Bearden CE, Rosso IM, Sanchez LE, Hadley T, Nuechterlein KH, Cannon TD (2003) A prospective study of childhood neurocognitive functioning in schizophrenic patients and their siblings. Am J Psychiatry 160:2060–2062

Noren U, Bjorner A, Sonesson O, Eriksson L (2006) Galantamine added to antipsychotic treatment in chronic schizophrenia: cognitive improvement? Schizophr Res 85:302–304

Olincy A, Young DA, Freedman R (1997) Increased levels of the nicotine metabolite cotinine in schizophrenic smokers compared to other smokers. Biol Psychiatry 42:1–5

Olincy A, Harris JG, Johnson LL, Pender V, Kongs S, Allensworth D, Ellis J, Zerbe GO, Leonard S, Stevens KE, Stevens JO, Martin L, Adler LE, Soti F, Kem WR, Freedman R (2006) Proof-of-concept trial of an alpha7 nicotinic agonist in schizophrenia. Arch Gen Psychiatry 63:630–638

Olney JW, Newcomer JW, Farber NB (1999) NMDA receptor hypofunction model of schizophrenia. J Psychiatr Res 33:523–533

Pierre JM, Peloian JH, Wirshing DA, Wirshing WC, Marder SR (2007) A randomized, double-blind, placebo-controlled trial of modafinil for negative symptoms in schizophrenia. J Clin Psychiatry 68:705–710

Polinsky RJ (1998) Clinical pharmacology of rivastigmine: a new-generation acetylcholinesterase inhibitor for the treatment of Alzheimer's disease. Clin Ther 20:634–647

Powchik P, Davidson M, Haroutunian V, Gabriel SM, Purohit DP, Perl DP, Harvey PD, Davis KL (1998) Postmortem studies in schizophrenia. Schizophr Bull 24:325–341

Power AE, Vazdarjanova A, McGaugh JL (2003) Muscarinic cholinergic influences in memory consolidation. Neurobiol Learn Mem 80:178–193

Razay G, Wilcock GK (2008) Galantamine in Alzheimer's disease. Expert Rev Neurother 8:9–17

Reingold JL, Morgan JC, Sethi KD (2007) Rivastigmine for the treatment of dementia associated with Parkinson's disease. Neuropsychiatr Dis Treat 3:775–783

Rosenthal MH, Bryant SL (2004) Benefits of adjunct modafinil in an open-label, pilot study in patients with schizophrenia. Clin Neuropharmacol 27:38–43

Sacco KA, Creeden C, Reutenauer EL, George TP (2008) Effects of galantamine on cognitive deficits in smokers and non-smokers with schizophrenia. Schizophr Res 103:326–327

Sacco KA, Creeden C, Reutenauer EL, Vessicchio JC, Weinberger AH, George TP (2009) Effects of atomoxetine on cognitive function and cigarette smoking in schizophrenia. Schizophr Res 107:332–333

Sawaguchi T, Goldman-Rakic PS (1994) The role of D1-dopamine receptor in working memory: local injections of dopamine antagonists into the prefrontal cortex of rhesus monkeys performing an oculomotor delayed-response task. J Neurophysiol 71:515–528

Scarr E, Cowie TF, Kanellakis S, Sundram S, Pantelis C, Dean B (2009) Decreased cortical muscarinic receptors define a subgroup of subjects with schizophrenia. Mol Psychiatry 14:1017–1023

Schilstrom B, Ivanov VB, Wiker C, Svensson TH (2007) Galantamine enhances dopaminergic neurotransmission in vivo via allosteric potentiation of nicotinic acetylcholine receptors. Neuropsychopharmacology 32:43–53

Schliebs R, Arendt T (2006) The significance of the cholinergic system in the brain during aging and in Alzheimer's disease. J Neural Transm 113:1625–1644

Schubert MH, Young KA, Hicks PB (2006) Galantamine improves cognition in schizophrenic patients stabilized on risperidone. Biol Psychiatry 60:530–553

Seamans JK, Yang CR (2004) The principal features and mechanisms of dopamine modulation in the prefrontal cortex. Prog Neurobiol 74:1–58

Seidman LJ, Giuliano AJ, Smith CW, Stone WS, Glatt SJ, Meyer E, Faraone SV, Tsuang MT, Cornblatt B (2006) Neuropsychological functioning in adolescents and young adults at genetic risk for schizophrenia and affective psychoses: results from the Harvard and Hillside adolescent high risk studies. Schizophr Bull 32:507–524

Sevy S, Rosenthal MH, Alvir J, Meyer S, Visweswaraiah H, Gunduz-Bruce H, Schooler NR (2005) Double-blind, placebo-controlled study of modafinil for fatigue and cognition in schizophrenia patients treated with psychotropic medications. J Clin Psychiatry 66:839–843

Sharma T, Reed C, Aasen I, Kumari V (2006) Cognitive effects of adjunctive 24-weeks Rivastigmine treatment to antipsychotics in schizophrenia: a randomized, placebo-controlled, double-blind investigation. Schizophr Res 85:73–83

Shekhar A, Potter WZ, Lightfoot J, Lienemann J, Dubé S, Mallinckrodt C, Bymaster FP, McKinzie DL, Felder CC (2008) Selective muscarinic receptor agonist xanomeline as a novel treatment approach for schizophrenia. Am J Psychiatry 165:1033–1039

Siegel BV, Trestman RL, O'Faithbheartaigh S, Mitropoulou V, Amin F, Kirrane R, Silverman J, Schmeidler J, Keefe RS, Siever LJ (1996) D-Amphetamine challenge effects on Wisconsin Card Sort Test. Performance in schizotypal personality disorder. Schizophr Res 20:29–32

Silver H, Geraisy N (1995) Effects of biperiden and amantadine on memory in medicated chronic schizophrenic patients. A double-blind cross-over study. Br J Psychiatry 166:241–243

Silver H, Goodman C, Isakov V, Knoll G, Modai I (2005) A double-blind, cross-over comparison of the effects of amantadine or placebo on visuomotor and cognitive function in medicated schizophrenia patients. Int Clin Psychopharmacol 20:319–326

Smith RC, Singh A, Infante M, Khandat A, Kloos A (2002) Effects of cigarette smoking and nicotine nasal spray on psychiatric symptoms and cognition in schizophrenia. Neuropsychopharmacology 27:479–497

Smith RC, Warner-Cohen J, Matute M, Butler E, Kelly E, Vaidhyanathaswamy S, Khan A (2006) Effects of nicotine nasal spray on cognitive function in schizophrenia. Neuropsychopharmacology 31:637–643

Snitz BE, Macdonald AW III, Carter CS (2006) Cognitive deficits in unaffected first-degree relatives of schizophrenia patients: a meta-analytic review of putative endophenotypes. Schizophr Bull 32:179–194

Sorensen HJ, Mortensen EL, Parnas J, Mednick SA (2006) Premorbid neurocognitive functioning in schizophrenia spectrum disorder. Schizophr Bull 32:578–583

Spence SA, Green RD, Wilkinson ID, Hunter MD (2005) Modafinil modulates anterior cingulate function in chronic schizophrenia. Br J Psychiatry 187:55–61

Stover E, Marder S, Carpenter W (2007a) Progress on NIMH initiatives (memorial theme for Wayne Fenton, MD). Schizophr Bull 33:1084–1085

Stover EL, Brady L, Marder SR (2007b) New paradigms for treatment development. Schizophr Bull 33:1093–1099

Szeszko PR, Bilder RM, Dunlop JA, Walder DJ, Lieberman JA (1999) Longitudinal assessment of methylphenidate effects on oral word production and symptoms in first-episode schizophrenia at acute and stabilized phases. Biol Psychiatry 45:680–686

Tamminga CA (2006) The neurobiology of cognition in schizophrenia. J Clin Psychiatry 67:e11

Terry AV Jr (2008) Role of the central cholinergic system in the therapeutics of schizophrenia. Curr Neuropharmacol 6:286–292

Toda M, Abi-Dargham A (2007) Dopamine hypothesis of schizophrenia: making sense of it all. Curr Psychiatry Rep 9:329–336

Toulopoulou T, Rabe-Hesketh S, King H, Murray RM, Morris RG (2003) Episodic memory in schizophrenic patients and their relatives. Schizophr Res 63:261–271

Tsuno N (2009) Donepezil in the treatment of patients with Alzheimer's disease. Expert Rev Neurother 9:591–598

Tugal O, Yazici KM, Anil Yagcioglu AE, Gogus A (2004) A double-blind, placebo controlled, cross-over trial of adjunctive donepezil for cognitive impairment in schizophrenia. Int J Neuropsychopharmacol 7:117–123

Turner D (2006) A review of the use of modafinil for attention-deficit hyperactivity disorder. Expert Rev Neurother 6:455–468

Turner DC, Robbins TW, Clark L, Aron AR, Dowson J, Sahakian BJ (2003) Cognitive enhancing effects of modafinil in healthy volunteers. Psychopharmacology (Berl) 165:260–269

Turner DC, Clark L, Dowson J, Robbins TW, Sahakian BJ (2004a) Modafinil improves cognition and response inhibition in adult attention-deficit/hyperactivity disorder. Biol Psychiatry 55:1031–1040

Turner DC, Clark L, Pomarol-Clotet E, McKenna P, Robbins TW, Sahakian BJ (2004b) Modafinil improves cognition and attentional set shifting in patients with chronic schizophrenia. Neuropsychopharmacology 29:1363–1373

Tuulio-Henriksson A, Arajarvi R, Partonen T, Haukka J, Varilo T, Schreck M, Cannon T, Lönnqvist J (2003) Familial loading associates with impairment in visual span among healthy siblings of schizophrenia patients. Biol Psychiatry 54:623–628

Volk DW, Austin MC, Pierri JN, Sampson AR, Lewis DA (2000) Decreased glutamic acid decarboxylase67 messenger RNA expression in a subset of prefrontal cortical gamma-aminobutyric acid neurons in subjects with schizophrenia. Arch Gen Psychiatry 57:237–245

Volk D, Austin M, Pierri J, Sampson A, Lewis D (2001) GABA transporter-1 mRNA in the prefrontal cortex in schizophrenia: decreased expression in a subset of neurons. Am J Psychiatry 158:256–265

Waldo M, Myles-Worsley M, Madison A, Byerley W, Freedman R (1995) Sensory gating deficits in parents of schizophrenics. Am J Med Genet 60:506–511

Walker E, Diforio D (1997) Schizophrenia: a neural diathesis-stress model. Psychol Rev 104:667–685

Walker DP, Wishka DG, Piotrowski DW, Jia S, Reitz SC, Yates KM, Myers JK, Vetman TN, Margolis BJ, Jacobsen EJ, Acker BA, Groppi VE, Wolfe ML, Thornburgh BA, Tinholt PM, Cortes-Burgos LA, Walters RR, Hester MR, Seest EP, Dolak LA, Han F, Olson BA, Fitzgerald L, Staton BA, Raub TJ, Hajos M, Hoffmann WE, Li KS, Higdon NR, Wall TM, Hurst RS, Wong EH, Rogers BN (2006) Design, synthesis, structure-activity relationship, and *in vivo* activity of azabicyclic aryl amides as alpha7 nicotinic acetylcholine receptor agonists. Bioorg Med Chem 14:8219–8248

Wang D, Noda Y, Zhou Y, Mouri A, Mizoguchi H, Nitta A, Chen W, Nabeshima T (2007) The allosteric potentiation of nicotinic acetylcholine receptors by galantamine ameliorates the cognitive dysfunction in beta amyloid25–35 i.c.v.-injected mice: involvement of dopaminergic systems. Neuropsychopharmacology 32:1261–1271

Williams GV, Goldman-Rakic PS (1995) Modulation of memory fields by dopamine D1 receptors in prefrontal cortex. Nature 376(17):572–575

Williams LM, Whitford TJ, Flynn G, Wong W, Liddell BJ, Silverstein S, Galletly C, Harris AW, Gordon E (2007) General and social cognition in first episode schizophrenia: identification of separable factors and prediction of functional outcome using the IntegNeuro test battery. Schizophr Res 99:182–191

Wonodi I, Cassady SL, Adami H, Avila M, Thaker GK (2006) Effects of repeated amphetamine administration on antisaccade in schizophrenia spectrum personality. Psychiatry Res 141:237–245

Young DA, Waldo M, Rutledge JH III, Freedman R (1996) Heritability of inhibitory gating of the P50 auditory-evoked potential in monozygotic and dizygotic twins. Neuropsychobiology 33:113–117

Zhang W, Yamada M, Gomeza J, Basile AS, Wess J (2002) Multiple muscarinic acetylcholine receptor subtypes modulate striatal dopamine release, as studied with M1–M5 muscarinic receptor knock-out mice. J Neurosci 22:6347–6352

Zhang J, Xiong B, Zhen X, Zhang A (2009) Dopamine D1 receptor ligands: where are we now and where are we going. Med Res Rev 29:272–294

Treatment Implications of the Schizophrenia Prodrome

Tejal Kaur and Kristin S. Cadenhead

Contents

1 Introduction 98
   1.1   Early Identification of Psychotic Illness 100
   1.2   Duration of Untreated Psychosis: Individual and Public Health Concern 100
   1.3   Identifying and Predicting Risk for Psychotic Illnesses 101
   1.4   Review of Treatment Studies in the Psychotic Prodrome 104
   1.5   Nonpharmacologic Interventions 105
   1.6   Psychopharmacologic Interventions 106
   1.7   Pharmacologic Potential for Neuroprotection 107
   1.8   Preliminary Treatment Recommendations 109
   1.9   Ethical Implications 109
   1.10 Development of Clinical Staging Criteria 111
   1.11 Recommended Treatment Guidelines 112
   1.12 General Summary 115
References 115

Abstract Schizophrenia is a debilitating neurodevelopmental disorder that strikes at a critical period of a young person's life. Early identification of individuals in the prodromal phase of a psychotic illness can lead to earlier treatment and perhaps prevention of many of the devastating effects of a first psychotic episode. International research efforts have demonstrated the success of community outreach and education regarding the schizophrenia prodrome and it is now possible to use empirically defined clinical and demographic criteria to identify individuals at a substantially increased risk for a psychotic illness. The development of clinical staging criteria for

T. Kaur
Department of Psychiatry, University of California, San Diego
Division of Child and Adolescent Psychiatry, New York Presbyterian Hospital of Columbia and Cornell Universities
e-mail: kcadenhe@ucsd.edu

K.S. Cadenhead (✉)
Department of Psychiatry, University of California, San Diego

psychosis that incorporates type and severity of clinical symptoms, level of global and social functioning, family history, substance use, neurocognitive functioning, and perhaps neurobiological information, could help to specify appropriate treatment for vulnerable individuals at different phases of the prodrome. Preliminary psychosocial and pharmacologic treatment studies report initial success in reducing severity of prodromal symptoms in "at-risk" samples, but further work is needed to refine the prodromal criteria and perform well controlled treatment studies in adequately powered samples. Treatment algorithms can then be tailored to presenting symptoms, number of risk factors present, and evidence of progression of the illness, to assure appropriate, safe and effective interventions in the early stages of psychosis.

1 Introduction

Schizophrenia has been conceptualized as a chronic and debilitating disease with ongoing cognitive, social, and functional losses since dementia praecox was first described at the turn of the twentieth century. Up until the 1950s, those with psychotic illnesses were locked away, imprisoned by the notion that life could not exist beyond the desolate spaces of state hospitals. Then with deinstitutionalization, while many patients with psychotic illness relished new found freedoms, others were relegated to lives of poverty and homelessness. In the past two decades, schizophrenia researchers and clinicians have challenged this notion of inevitable decline, demonstrating that early intervention and treatment of psychotic illness can mitigate losses in psychotic illness, improve functional outcomes, and provide hope to patients and families (Addington et al. 2004; Hafner and an der Heiden 1999; Hegarty et al. 1994; McGlashan et al. 2005).

Despite the more favorable prognosis for individuals with psychotic illness compared to those of Kraeplin's time, up to two-thirds of patients with schizophrenia never return to their previous level of functioning (Addington et al. 2003). College students may never again return to school after a first psychotic episode, young adults early in their careers may find themselves unable to maintain the cognitive demands required in their jobs, and most young people encounter persistent difficulties in interpersonal relationships (Ballon et al. 2007; Grant et al. 2001). Current research supports these clinical observations as studies show significant cognitive (Bilder et al. 2006; Eastvold and Cadenhead 2003) and social functioning (Ballon et al. 2007) deficits in the first episode of psychosis that may have been present before the illness began (Caspi et al. 2003).

Clearly, schizophrenia cannot be best characterized as beginning with the onset of frank psychosis, as biological changes occur long before this period (Niendam et al. 2006). Rather, it is better conceptualized as an illness characterized by premorbid, prodromal, acute, and chronic phases that correspond to neurodevelopmental changes (Lieberman 1999). The premorbid phase is characterized by a period of stable social and cognitive deficits, alongside frequent subtle neurological abnormalities which long precede the first episode of psychosis (Davidson et al. 1999). In contrast, the "prodromal" period is defined by its lack of stability, worsening positive and negative symptoms, and a deteriorating course of psychosocial impairment culminating in the onset of frank psychosis (Keith and Matthews 1991; Yung and McGorry 1996). With the onset of frank psychosis comes a period of recurrent exacerbations and remissions from psychotic symptoms as well as ongoing functional decline until an individual settles into the chronic phase of illness where deficits and symptoms reach a level of symptomatic stability.

While in the latter two phases (acute and chronic), symptoms are more externally identifiable, the subtle, nonspecific symptoms that first emerge during the prodrome are often overlooked. The failure to recognize these early changes is particularly concerning as the duration of untreated psychosis (DUP) corresponds to further functional decline (Melle et al. 2005). Yet, the idea of early identification itself challenges the notion that downward decline is inevitable. Although researchers have studied changes in the first episode of psychosis in the hopes that intervention at the onset of psychosis may prevent further decline and morbidity, recent research has delved further back to the prodrome. However, as the prodrome can only be accurately defined retrospectively, research efforts strive to develop measures which may predict risk of future psychosis with increasing sensitivity and specificity (Cannon et al. 2008; Ruhrmann et al. 2003).

In 1996, Yung and her colleagues heralded the field of early identification by establishing "prodromal" criteria which appeared to predict conversion to psychosis in 40% of individuals at 1 year follow-up. Individuals between the ages of 16–30 were recruited if they had developed subsyndromal psychotic symptoms within the last year or had a familial risk for schizophrenia plus a recent functional decline (Yung et al. 2003). As many of these individuals may not convert to psychosis, and to avoid early stigmatization, the prodromal literature uses terms such as "at-risk," "ultra high risk," or "clinically high risk" to better describe a population who meets such "at-risk" criteria, but whose individual prognosis is yet unknown.

In order to measure dimensions of psychopathology as well as to operationally define at-risk criteria, the Melbourne Australia group developed the Comprehensive Assessment of At-Risk Mental State (CAARMS; Yung et al. 2002). The CAARMS operationally defined the at-risk state by creating three distinct at-risk criteria. The vulnerability group, defined by genetic risk and recent functional deterioration, includes those who experienced a significant functional loss in the past year and who either meet DSM-IV criteria for schizotypal personality disorder and/or have a first-degree relative with a diagnosed psychotic disorder. The attenuated positive symptom group includes individuals with recent onset (<1 year) subsyndromal psychotic symptoms which do not reach psychotic intensity. The third group includes those with brief limited intermittent psychotic symptoms (BLIPS), which do not meet full criteria for psychotic disorder due to limited frequency or duration. Subsequently, McGlashan developed the Scale of Prodromal Symptoms (SOPS) which also utilizes a structured assessment, the Structured Interview of Prodromal Symptoms (SIPS) to elicit whether individuals meet criteria for three distinct at-risk criteria, similar to those created by Yung's group (Miller et al. 2004). In original reports, both the CAARMS and the SOPS predicted conversion to psychosis at rates as high as 40–54% over the course of 6 months to 1 year, implying these instruments have increased sensitivity in late phases of the psychosis prodrome.

However, as numerous "basic symptoms," such as subtle disturbances of thought, speech, and perception, may be altered years prior to onset of frank psychosis, the Bonn Scale for the Assessment of Basic Symptoms (BSABS; Huber and Gross 1989) has shown that the presence of such basic symptoms predicted schizophrenia with a probability of 70% over 10 years, while the absence of such symptoms excluded schizophrenia with a probability of 96% (Klosterkotter et al. 2001). Particular disturbances, such as thought interference, disturbances of receptive language, or visual distortions, predicted schizophrenia with a probability up to 91% (Klosterkotter et al. 2001). In the current literature, the BSABS is thought to define individuals in an "early" prodromal phase that is characterized by the more negative or deficit-like symptoms as well as neurocognitive deficits.

In the last decade, early identification centers worldwide have adapted similar prodromal criteria and instruments, establishing themselves both as resources for help-seeking youth as well as critical centers of academic research (Cornblatt et al. 2002; Haroun et al. 2006; Klosterkotter et al. 2001; Larsen et al. 2000; Miller et al. 2002). The centers can be classified generally as either descriptive, treatment, or neurobiological translational studies, with many centers employing a combination of strategies. Descriptive studies monitor at-risk youth for conversion to psychosis in order to validate and further identify risk. In addition to monitoring at-risk youth, treatment studies apply either pharmacologic or nonpharmacologic treatment strategies during the putative prodrome in the hopes of delaying or preventing psychosis. Lastly, neurobiological translational studies follow neurobiological markers in order to better understand the biological processes contributing to the development of psychosis, as well as to validate biological measures that may indicate vulnerability towards psychosis (McGlashan et al. 2007).

1.1 Early Identification of Psychotic Illness

As international research efforts have demonstrated the success of community outreach and education regarding the schizophrenia prodrome, it is now possible to use empirically defined criteria to identify individuals at a substantially increased risk for a psychotic illness (Addington et al. 2007).Yet, combining "at-risk" criteria with objective biological markers, may further improve predictive potential, and thereby open up a window of opportunity for primary prevention (Cannon et al. 2008; Haroun et al. 2006; McGorry et al. 2006).

1.2 Duration of Untreated Psychosis: Individual and Public Health Concern

Imagine the patient who comes into the hospital, handsome, conversant, and even charming at times, but starkly psychotic. Clinical experience paints a portrait of steady decline with each subsequent hospitalization for psychotic decompensation. It is possible that untreated psychosis is "toxic," with longer durations of untreated illness corresponding to depression, anxiety, negative symptoms, positive symptoms, and overall poor functioning (Marshall et al. 2005; Perkins et al. 2005). Preliminary studies suggest that early intervention can decrease DUP and improve short-term clinical outcomes (Marshall et al. 2005; Melle et al. 2004).

In the real world, however, most individuals do not receive early intervention, and the onset of positive symptoms often occurs 2 years prior to initial treatment, while negative symptoms date back about 5 years prior (Häfner et al. 1992; Salokangas and McGlashan 2008). Furthermore, biological correlates of illness appear to coincide with this period of untreated symptoms. Studies show a greater temporal and frontal gray matter reduction in patients with a long duration of illness suggesting either a progressive pathological process prior to treatment or a more insidious onset of illness and a later presentation to services (Lappin et al. 2006; Takahashi et al. 2007).

As schizophrenia occurs during late adolescence and early adulthood, the illness sets in during a period of critical development, thwarting normal brain processes that sustain cognition, function, and independence. Even beyond an individual loss, functional loss occurs at a community level, impairing an individual's potential contribution to society, making schizophrenia among the leading causes of disability in the United States (Murray and Lopez 1996). An analysis in 2002 showed the overall cost of schizophrenia in the United States to be $62.7 billion, with total indirect excess costs due to unemployment of $32.4 billion (Wu et al. 2005). Although the cost of early identification may initially appear higher, it is suggested that even at 24-month follow-up, early identification may be cost-saving (Valmaggia et al. 2009). But now, beyond individual centers in particular communities, efforts at community education show potential in decreasing DUP, even to a matter of weeks. The Norwegian TIPS study examined whether specialized community education and a mobile detection program could affect DUP (Johannessen et al. 2005). When comparing two health service regions in which an early psychosis detection program was introduced with two areas without such a program, DUP was reduced to a matter of weeks, with a greater reduction in the experimental regions where community education and mobile detection teams were provided. Positive clinical differences were maintained at all 3-month, 1-year, and 2-year follow-ups (Larsen et al. 2006; McGorry et al. 2007).

1.3 *Identifying and Predicting Risk for Psychotic Illnesses*

As we stand now over a decade following the development of Yung's initial criteria, it appears that as the number of early identification centers has grown, the conversion rates at these centers has diminished (Haroun et al. 2006; Yung et al. 2007). From 1 to 2 years follow-up rates of 40–50% seen in earlier studies (Yung et al. 2006); subsequent studies have yielded rates closer to 25% per year using the same prodromal criteria (Cannon et al. 2008; Haroun et al. 2006; Yung et al. 2007). It is possible that variability in conversion rates may be attributable to the development of such centers, such that knowledge regarding at-risk states has actually increased the process of inherent treatment, subsequently delaying, or even preventing the onset of psychosis (Yung et al. 2007).

As current centers strive to intervene early, it is very possible that simple measures such as regular assessments, frequent clinical contact, and referrals addressing emerging affective and psychotic symptoms may be ameliorating current symptoms. It becomes clear in clinical practice that those who continue to present for such research studies are likely seeking more than financial compensation. Individuals followed in at-risk studies often present as a unique population characterized not only by being putatively prodromal, but by being distressed, help seeking, and open to clinical contact (Cornblatt et al. 2001; Haroun et al. 2006). As comorbid affective, anxiety and attentional symptoms are often prominent, these individuals come to expect a level of concern and advisement from early identification centers (Haroun et al. 2006). However, to identify each individual who meets at-risk criteria as "prodromal" would be misleading as simply meeting at-risk criteria does not indicate the certainty of future psychosis, especially considering the phenomenon of decreasing conversion rates in areas of early identification (Yung et al. 2007). To improve the predictive validity of the established criteria and better identify which individuals are at highest risk of conversion to psychosis, schizophrenia researchers strive to strengthen the sensitivity of current at-risk criteria to best predict imminent risk.

For instance, particular clinical symptoms have been identified across studies that show an increased association with later conversion to psychosis. Klosterkotter et al. (2001) reported that the basic symptoms of thought blocking, disturbances of receptive speech, and sensory perceptual disturbances were found more often in those who converted to psychosis. In the Cognitive Assessment and Risk Evaluation (CARE) sample in San Diego, individuals meeting at-risk criteria who converted to psychosis at 1 year follow-up were more likely to have a history of cannabis abuse or dependence than the nonconverted group and similarly to other samples, higher ratings on measures of delusional-like experiences, suspiciousness, and thought disorder (Haroun et al. 2006).

A number of prospective population studies have suggested that cannabis use, likely in those displaying vulnerability for psychosis, confers a clear increase in the relative risk of subsequently developing psychosis (Arseneault et al. 2004; Caspi et al. 2005). Furthermore, by identifying specific clinical factors as conferring particularly high risk, such studies allow research to inform clinical practice by encouraging psychoeducation regarding high risk behaviors.

In an attempt to increase the power of prodromal studies to predict psychotic conversion, the NAPLS (the North American Prodromal Longitudinal Studies) Consortium was created as a collaboration between seven early identification centers with NIMH funding (Addington et al. 2007). A recent analysis of the NAPLS dataset revealed that individuals who were most likely to convert to psychosis had a family history of psychosis, symptoms of suspiciousness or delusional-like experiences, a decline in social functioning, and/or a history of drug abuse. A combination of any three of these criteria increased the positive predictive power to 80% suggesting that it may be possible to develop an algorithm for treatment that will target those at highest risk (Cannon et al. 2008). Yung et al. (2003) also report that combining certain predictive variables, such as long duration of prodromal symptoms, poor functioning at intake, low-grade psychotic symptoms, depression, and disorganization, can yield a strategy for psychosis prediction with good sensitivity (86%), specificity (91%), positive predictive value (80%), and negative predictive value (94%) at 6-month follow-up. Other groups suggest that further specificity could be reached by applying the concept of basic symptoms to current at-risk criteria (Meng et al. 2009). Klosterkotter et al. (2001) found that basic symptoms, defined as subtle, subclinical self-experienced disturbances in thought, speech, and perception processes, were highly sensitive indicators of subsequent psychosis. While the study is limited by its retrospective design, it suggests that basic symptoms may be used as complementary approach to current at-risk criteria, perhaps even by aiding in timing the onset of a psychotic disorder by assessing onset of basic symptoms as well as attenuated psychotic symptoms (Schultze-Lutter et al. 2010).

While clinical criteria aid in identifying risk, subsequent gains in schizophrenia research will likely come from understanding the neurobiological mechanisms of psychotic processes and incorporating such biological markers into current at-risk criteria. Numerous deficits in information processing which serve as endophenotypes have been clearly defined in the schizophrenia spectrum including prepulse inhibition of the startle response (Cadenhead et al. 2000; Maier et al. 2008; Tenn et al. 2005), P50 event related potentials (Cadenhead 2005), and mismatch negativity (Umbricht and Krljes 2005). Longitudinal follow-up will determine whether these deficits are predictive of later psychosis in addition to providing insight into the mechanism of neuropathological change in the early stages of the illness.

Neurocognitive deficits, appear to be quite predictive of later conversion to psychosis in the at-risk sample, especially measures of verbal learning and general intelligence (Brewer et al. 2005; Cosway et al. 2000; Eastvold et al. 2007; Jahshan et al. 2010). In addition, one early study (Brewer et al. 2003) demonstrated that at-risk subjects with olfactory identification deficits prior to the onset of illness were more likely to later convert to psychosis. It is conceivable that measures of neurocognition or olfactory identification performed during initial assessment could add to the clinical assessment in determining which individuals are at greatest risk and might benefit from early cognitive remediation or pharmacologic intervention that enhances cognitive functioning.

A number of small scale neuroimaging studies have suggested that those at-risk individuals who progress to psychosis appear to display both structural and functional deficits prior to the onset of illness (Cannon et al. 2007, 2008; Pantelis et al. 2003; Wood et al. 2003). Specifically, progressive cortical volume loss may be associated with the onset of psychosis, indicating ongoing pathological processes during the progression to illness (Sun et al. 2009). Moreover, white matter development may be altered in individuals at-risk for psychosis, possibly due to disrupted developmental mechanisms (Karlsgodt et al. 2009).

With the advent of the Human Genome Project and subsequent advances in the field of genetics, it is foreseeable that the addition of genetic markers could further increase the positive predictive power of current at-risk criteria (Cannon et al. 2003). Ultimately, by understanding the neuropathological processes in the early phase of psychosis, treatment may be targeted earlier based on identified neurobiological deficits, ideally preventing some of the devastating aspects of a first psychotic episode.

1.4 Review of Treatment Studies in the Psychotic Prodrome

Early identification and treatment in the early phase of psychosis has been an area of increased interest and ethical debate over the last decade. Various studies have been conducted worldwide that highlight the potential of intervening in the early phase of illness with both pharmacologic and psychosocial treatment (see Table 1). According to a recent Cochrane review, while at this time there are likely still insufficient trials to draw definitive conclusions (Marshall and Rathbone 2006),

Table 1 Treatment studies in the psychotic prodrome

Study	Modality	Result
PRIME (USA)	Olanzapine vs. placebo ($N = 60$, double blind)	No difference in rate of psychotic conversion. Significant weight gain
PACE (Australia)	Risperidone + CBT vs. specialized care ($N = 59$)	Reduced psychotic conversion rate at 6 months but not at 12 months
EDIE (UK)	CBT vs. monitoring ($N = 58$)	Reduced rate of antipsychotic use in CBT group; no difference in psychotic conversion or symptoms
German Research Network	Amisulpride vs. needs-focused intervention ($N = 124$, open-label)	Reductions in attenuated psychotic, negative, and basic symptoms, but no reduction in conversion. Greater improvement in GAF scores. No weight gain or EPS but increased prolactin levels
RAPP (USA)	Naturalistic study of antidepressants vs. antipsychotics ($N = 48$)	No conversions among antidepressant-treated adolescents. The antipsychotic group was more disorganized at entry, was less compliant with medication, and more likely to become psychotic
Prime (USA)	Aripiprizole ($N = 15$, open-label)	Improvement in the total number of prodromal symptoms and none of the participants converted to psychosis. Neuropsychological measures showed no consistent improvement. There was a mean weight gain of 1.2 kg and akathesia emerged in over half of the participants

recent and ongoing studies continue to build a foundation of evidence based medicine to guide preliminary treatment guidelines in at-risk youth.

1.5 Nonpharmacologic Interventions

The validation of nonpharmacologic interventions in the treatment of chronic schizophrenia has lead to similar studies in earlier stages of illness. As rigorous trials conducted in at-risk youth are limited, psychosocial, randomized controlled trials conducted in first episode patients will also be briefly discussed because of their applicability in this population. Future studies are necessary to validate whether such interventions can be effective in individuals meeting at-risk criteria.

A UK-based study (EDIE) randomized 58 people with prodromal symptoms to 6 months of cognitive behavioral therapy (CBT) or a monitoring group. Morrison et al. (2004) report that the CBT group had a lower risk for conversion to psychosis with 1-year longitudinal follow-up (Morrison et al. 2004). The CBT group displayed fewer indicators towards conversion on all measures. However, an analysis of the EDIE data in the Cochrane report stated that of the two outcomes reported, neither the rate of psychotic conversion nor the rate of subjects leaving the study was significantly different between groups (Marshall and Rathbone 2006).

Other trials focused on first episode psychosis have shown promise, but are also similarly limited in applicability. LifeSPAN-Australia, a randomized controlled trial of a phase-specific brief individual cognitively orientated therapy for people in the first episode of psychosis with suicidal ideation (Power et al. 2003), noted benefits in the treatment group on indirect measures of suicidality such as hopelessness; however, these findings did not reach statistical significance. A Chinese trial by Zhang et al. (1994) also utilized a phase-specific intervention (family therapy) by randomizing 78 males diagnosed with schizophrenia to either family therapy or treatment as usual upon discharge from their first psychiatric hospitalization. Over the 18-month follow-up, there was a significantly lower rate of hospital readmission in the family intervention group than in the control group, and the mean hospital-free period for those who were readmitted was significantly longer in the experimental group than in the control (Marshall and Rathbone 2006). A Dutch randomized controlled trial ($N = 76$; Linszen et al. 2001) comparing phase-specific intervention including family therapy plus a specialized team for individuals in their first episode of schizophrenia found no difference between intervention and control groups at 12 months for the outcome of relapse (Marshall and Rathbone 2006). Finally, the OPUS-Scandinavia study included 547 people with a diagnosis of first episode schizophrenia in which 275 were randomly assigned to integrated treatment (consisting of an assertive community treatment, family therapy, social skills training, and modifications of medication regime) and 272 to standard treatment (Petersen et al. 2005). At 1 year follow-up, global GAF significantly favored integrated treatment by 1 year, but neither group differed significantly at 2 year follow-up (Marshall and Rathbone 2006).

While psychosocial interventions often complement treatment in chronic schizophrenia, further studies are necessary to validate such strategies in both the prodrome and first episode of psychosis.

1.6 Psychopharmacologic Interventions

Relatively few psychopharmacologic studies have been performed in individuals meeting prodromal criteria (see Table 1). The absence of information is in part related to the short history of prodromal research but also the difficulty in recruiting sufficient numbers of subjects who are willing to enter into a clinical trial for nonspecific symptoms and diagnosis. It is also difficult to obtain approval for large scale studies when the prodromal phase of a psychotic illness is not a defined diagnostic entity, the majority of at-risk youth are adolescents and there are many ethical concerns on the part of the pharmaceutical industry and federal granting agencies in performing such studies.

The PRIME (Prevention through Risk Identification, Management and Education) study is the only placebo-controlled double blind study of an antipsychotic in the prodromal phase of illness. In the PRIME study, olanzapine vs. placebo was compared in 60 at-risk individuals over the course of 1 year with subsequent 1-year follow-up (McGlashan et al. 2006). No significant differences were found between olanzapine and placebo in preventing conversion to psychosis by 12 months or improving symptoms. In the placebo group, 11 out of 29 converted to psychosis compared to 5 out of 31 in the olanzapine group, yielding a nonsignificant difference. It is interesting, however, that all of the psychoses in the olanzapine group occurred in the first 4 weeks of the clinic trial when doses of olanzapine were relatively low, implying that perhaps those who converted in this group may not have had sufficient time on olanzapine for it to affect the active processes leading to psychosis in these particular youth. Furthermore, by week 8, it appeared that prodromal symptoms had significantly decreased further in the group treated with olanzapine compared to that given placebo. It is noteworthy to mention that the weight gain in the olanzapine group as 8.79 kg compared to the 0.30 kg weight gain in the placebo group.

The PACE (Personal Assessment and Crisis Evaluation) study ($N = 59$) randomized individuals meeting at-risk criteria with low dose risperidone and CBT vs. "needs-based intervention" and found that individuals who received the treatment were significantly less likely to develop psychosis at a 6-month follow-up than people who only received need-based treatment (McGorry et al. 2002). After the 6-month treatment period, 10 of 28 people who received needs-based intervention converted to psychosis in comparison to 3 of 31 from the specific preventive intervention group. However, using intention-to-treat analysis, the difference was no longer significant at 12-month follow-up ($N = 59$, 1 RCT, RR 0.54 CI 0.2–1.3; Marshall and Rathbone 2006). It is interesting to note, however, that in the subsequent 6-month follow-up period, it appeared that those who were not adherent to the risperidone were those who were most likely to convert to psychosis.

A report from the PRIME clinic (Woods et al. 2007) included 15 participants in an open-label, single site trial with fixed-flexible dosing of aripiprazole (5–30 mg/day) for 8 weeks. There was improvement from baseline in the total number of prodromal symptoms and none of the participants converted to psychosis. Neuropsychological measures showed no consistent improvement. There was a mean weight gain of 1.2 kg and akathesia emerged in over half of the participants. While results are promising, the significance of these findings are complicated by small sample size as well as lack of control group and blinding.

A study by the German Research Network compared treatment with amisulpride in combination with needs-focused intervention to needs-focused intervention alone in a sample of 124 people in an open-label 3-month design (Ruhrmann et al. 2007). Individuals in the antipsychotic group showed reductions in attenuated psychotic symptoms, negative symptoms and basic symptoms as well as a greater improvement in GAF scores. Amisulpride did not cause weight gain or extrapyramidal symptoms but it did increase prolactin levels. Again, applicability is limited due to open-label design, but can inform further rigorous studies.

One study conducted by RAPP (Recognition and Prevention of Psychological Problems) examined the effects of antidepressants using a prospective naturalistic treatment study of clinical high risk adolescents comparing individuals who received antidepressants ($N = 20$) vs. second-generation antipsychotics ($N = 28$; Cornblatt et al. 2007a, b). The group who received antipsychotics showed evidence of more disorganized thinking at entry, was less compliant with medication, and more likely to become psychotic at follow-up. Interestingly, there were no conversions among antidepressant-treated adolescents. It is possible that the adolescents treated with antidepressants were less ill at entry and the authors acknowledge that a substantial number of false positives may have been present among the antidepressant-treated subgroup. However, as animal models have demonstrated increased neurogenesis, dendritic arborization, and synaptogenesis with SSRI treatment (Richtand and McNamara 2008), it is plausible that treating these young people with psychotropic agents may provide an element of neuroprotection even during the initial stages of a psychotic illness. While the naturalistic nature of this study prevented generalization of these results, such studies invite subsequent randomized controlled trials to further evaluate treatment efficacy. Future studies ideally would also integrate larger sample sizes such that more definitive conclusions may be reached regarding the efficacy of pharmacologic intervention in early identification.

1.7 Pharmacologic Potential for Neuroprotection

Ideally, understanding brain changes that accompany early psychosis can help to inform not only treatment strategies but ultimately neuroprotection. In a comprehensive review, Berger et al. (2003) outline what is known about the early and late neurodevelopmental abnormalities in early psychosis and how altered regulatory mechanisms of progenitor cell generation and death in some brain areas could explain these changes.

Although researchers previously believed that stem and progenitor cell generation in mammals was only possible in early life, recent research suggests that the hippocampi (Kornack and Rakic 1999), periventricular zone (Steindler and Pincus 2002), and olfactory bulbs (Byrd and Brunjes 2001) retain the capacity to generate progenitor cells which differentiate into neurons. Therefore, disequilibrium of pruning or trophic processes during the prodrome may account for manifestation of psychosis above and beyond pre-existing neurodevelopment injury (Feinberg 1982). The dysregulation of mechanisms for apoptosis in individuals with schizophrenia likely further contributes to the longitudinal biological changes seen in early psychosis on magnetic resonance imaging (Catts and Catts 2000; Jarskog et al. 2000; Pantelis et al. 2003).

A number of compounds that are under consideration for their neuroprotective qualities include agents that modulate apoptosis pathways (lithium, sodium valproate, BDNF, clozapine, quetiapine, lamotrogine, omega-3 fatty acid), block necrosis pathways (vitamin E), increase synaptogenesis (selective serotonin reuptake inhibitors), or block the inflammatory response (Cox-2 inhibitors; Jacobs et al. 2000; Malberg et al. 2000; Vaidya et al. 1997).

Atypical antipsychotics appear to display neuroprotective effects by inducing BDNF mRNA expression which serves to increase the threshold for apoptosis (Bai et al. 2003). Patients treated with antipsychotics near the time of death display increased levels of bcl-2 compared to those not exposed to antipsychotics (Angelucci et al. 2000), while postmortem analysis of patients with schizophrenia shows a 25% reduction in bcl-2 transcription factor in the temporal cortex (Jarskog et al. 2000). Furthermore, inducers of apoptosis, such as TNF-α, also correlate with illness and appear elevated in the serum of antipsychotic-naïve patients (Erbagci et al. 2001). Interestingly, clozaril appears to normalize TNF-α levels (Monteleone et al. 1997).

Other novel therapies such eicosapentaenoic acid, an omega-3 fatty acid, or estrogen are also suggested as adjunctive agents due to neuroprotective properties. One randomized, double-blinded, placebo-controlled trial by Berger et al. (2007) suggests that augmenting antipsychotic medication with omega-3 fatty acids may improve tolerability of, and accelerate response to, antipsychotic medications. Protective effects of omega-3 fatty acids may be modulated by phospoinositide-protein kinase C (PI-PKC) signal transduction as omega-3 fatty acids antagonize PI-PKC and are found to be deficient in the peripheral tissue of individuals with schizophrenia (McNamara et al. 2006). Estrogen is postulated to indirectly interfere with mitochondrial properties, and therefore protect against apoptosis (Arnold and Beyer 2009). Applying such treatments at early stages of illness when neuronal pathways may not be irrevocable damaged may maximize the neuroprotective potential of such agents.

Neuroprotective strategies should thus be instituted at early stages of illness when such biological developmental abnormalities could be averted. Further research will elucidate the mechanisms by which particular pharmacologic treatments induce neuroprotection. Understanding the pathophysiology of such changes will inform the literature as to when and what to implement during specific stages of early intervention.

1.8 Preliminary Treatment Recommendations

Although the literature is mixed, the majority of studies support the notion that shortening the DUP improves outcome and course of illness (Addington et al. 2004; Harrigan et al. 2003; Larsen et al. 2000; Malla et al. 2002; Melle et al. 2005). This can lead to the assumption that starting treatment even before the onset of psychosis may further improve outcomes. However, current research does not adequately inform us *which* treatment modalities are most indicated in the prodromal period. Although clinicians and researchers differ on thoughts regarding treatment protocols, it is universally agreed upon that the prodromal population is ill and necessitates some type of treatment. Ballon et al. (2007) revealed that quality of life and everyday functioning in at-risk youth are nearly as low as patients who have already developed psychosis, with numerous other studies confirming the presence of significant social deficits (Cornblatt et al. 2007a, b; Niendam et al. 2007). Even in those who fit at-risk criteria, but may never develop psychosis, research is needed to determine whether the initiation of proposed treatments can ameliorate the course of other psychiatric disorders which may arise in this high risk population. Yet, even when such treatments are proposed, there are no guidelines that can inform us *when* in the course of illness to initiate particular interventions.

1.9 Ethical Implications

Although understanding risk and protective factors can help to inform potential preventative and treatment strategies, the ethical implications of such treatments also remain complex. Considering the large number of at-risk youth who may never convert to psychosis, the ethical considerations including potential stigmatization, loss of confidentiality, and insurability become particularly significant. Furthermore, as Corcoran (2005) eloquently points out, it is not safe to assume that individuals or institutions will appreciate the difference between a risk assessment and a diagnosis. Subsequently, even the act of providing medical information in the form of risk assessments can affect wellbeing in unintended and potentially unexpected ways (Corcoran et al. 2005). When keeping in mind this potential group of false positives, the need for an exit strategy is critically important (Cornblatt et al. 2001). If prophylactic treatments are begun, how long does one continue to administer them and what factors do we consider in deciding to stop? How do we weigh risks and benefits of providing unnecessary treatment, and yet, how to do we weigh the risk that stopping may elicit psychosis in true positives (Cornblatt et al. 2001; Haroun et al. 2006)?

Specifically, if clinicians begin to treat at-risk youth with atypical antipsychotic medications, they run not only the risk of potentially treating false positives with unnecessary medication (McGlashan 2001; Schaffner and McGorry 2001), but they will also unnecessarily expose individuals to the known metabolic risks of such medications. Atypical antipsychotics are known to cause weight gain of more than 10 pounds in 3 months in young people, a risk significant not only for long-term medical health but also for psychosocial development (Correll and Malhotra 2004). Furthermore, as the long-term effects of these medications on the adolescent brain is yet speculative, implications of dopamine blockade on psychosexual development are also unknown. Risks and benefits will need to be weighed if incorporating into clinical practices.

And yet, we cannot assume that treating false positives is unwarranted. It is possible that "false positives," are individuals who may yet convert to psychosis but had not converted during the study design. Conversely, it may be that indirect treatment administered as a result of early identification may have delayed or prevented the onset of psychosis. As the putatively prodrome group is well documented to display significant clinical symptoms and functional decline (Haroun et al. 2006), it is possible that individuals could have been offered a variety of psychosocial intervention or even treatment indicated psychotropic medications which may have had neuroprotective effects. Or this "false positive" group could truly be comprised of individuals whose clinical symptomatology only masqueraded as subsyndromal symptoms of psychosis and they would never have developed psychosis (Haroun et al. 2006).

As the prodrome can only truly be identified retrospectively, those considered *putatively* prodromal are a heterogenous group, most of whom present with elevations on a variety of psychiatric symptom scales. Even those who do not convert to psychosis within the window of such research studies appear to remain quite ill, such that by follow-up, many meet the criteria for a number of affective Axis I conditions (Haroun et al. 2006).

When we advise at-risk patients of their risks, however, certain individuals who display multiple risk factors may be at particularly high risk for psychosis (Cannon et al. 2008). Given that degree of risk likely varies even within a high risk cohort, clinicians must also consider the ethical consideration of *not* informing of risk. McGlashan discusses that although the risks of stigma and treating false positives are widely discussed, we often fail to discuss the greater stigma involving the social consequences of psychotic behavior which leads to social losses, hospitalizations, and even incarceration (McGlashan et al. 2007).

At-risk subjects are symptomatic and present with a range of symptoms and comorbid anxiety, mood, attentional, and substance abuse diagnoses that do not necessarily require a "one size fits all" treatment approach but are better suited to needs-based treatment that includes a comprehensive multidimensional approach (Haroun et al. 2006).

1.10 Development of Clinical Staging Criteria

While the concept of clinical staging is not new in the field of medicine, its application to psychiatric illness remains largely untouched. Clinical staging is a more refined form of diagnosis in that it attempts to define where an individual may lie in the course of an illness. It can be a useful algorithm in disorders which either tend to *or* may progress.

In the field of psychiatry, naturalistic studies inform us of the course of psychiatric illness, but the next step would be for research to then translate this into clinical stages, thereby facilitating early recognition and therefore early intervention. Ideally, staging would be defined by a system of clinicopathological indicators in which endophenotypic markers can be integrated into clinical profiles, lending to the further refinement of the validity of clinical staging.

McGorry et al. (2006) have promoted the concept of clinical staging for psychotic disorders much like what is done in the treatment of illnesses such as cancer or diabetes. The fundamental assumptions to clinical staging are two-fold. First, patients in earlier stages of an illness are more likely to respond to treatment than those in the later stages of illness. Second, treatments offered in the earlier stages should be more benign as well as more effective in order to be in accordance with ethical risk-benefit considerations. These assumptions are based on lessons learned from other areas of medicine where aggressive treatment in early stages of illness carries significant complications which may negate the potential benefits of that particular early intervention (McGorry et al. 2006).

If we were to apply such an algorithm to at-risk individuals, those individuals with milder symptoms and/or fewer risk factors would be treated with psychosocial treatments such as CBT, crisis intervention, or supportive psychotherapy, while those who have more severe symptoms and risk factors would be treated with pharmacotherapy in addition to psychosocial treatment.

Clearly, biological research is needed to continue to elucidate appropriate treatment strategies. While critics of the clinical staging model for psychotic illness would argue that staging should begin with biological correlates, the complexity of psychotic disorders precludes a systematic approach as we do not yet know which biological changes are inherent to the disorder and which are merely epiphenomenon. Moreover, it is entirely unclear which endophenotypes are static and which reflect disease progression (Jahshan et al. 2010; McGorry et al. 2006). Yet, the syndromal and help-seeking nature of at-risk youth preclude the luxury of waiting until definite algorithms are created, especially as clinical practice currently already employs a variety of treatments, not limited to antipsychotic medication. As such, it is imperative that the guidelines we do develop as a community are conservative in nature to reflect the preliminary nature of the current evidence.

To date, a variety of treatment algorithms have been proposed based on our current knowledge of the psychotic prodrome. McGorry et al. have developed an algorithm that begins with individuals with increased risk (e.g., first-degree teenage relatives) who do not have symptoms currently (Stage 0), and progresses through the prodrome (Stage 1a and 1b) to the acute (Stage 2–3) and chronic stages of psychosis (Stage 4). Potential intervention at each stage would include family education, substance abuse reduction, and psychosocial treatments such as CBT. Neuroprotective strategies such as omega-3 fatty acids as well as atypical antipsychotics and/or antidepressants and mood stabilizers would be initiated at Stage 1b when the individual meets full prodromal criteria. The proposed algorithm also proposes that neurocognitive deficits be incorporated as part of the definition of the Ultra High Risk or prodromal state (McGorry et al. 2006).

One clear benefit of using clinical staging models is that it encourages clinicians to realize that functional disability and early symptoms occur long before the onset of frank psychosis. As such, by intervening early, patients and family can be offered consultation for symptoms comorbid with the prodromal state, such as affective symptoms or substance abuse. Furthermore, patients may begin to develop a therapeutic alliance with clinicians at a stage in illness where trust, cognition, and decisional capacity are not yet compromised by frank psychosis (McGlashan et al. 2007). In clinical practice, patients often enter care at the time of a first psychotic break, often coming to the attention of clinicians due to hospitalizations or incarceration, clearly situations less conducive to building rapport and establishing long-term treatment alliances. However, when conversion occurs in the context of early identification centers providing education, support, and treatment, clinical outcomes appear better. In one published report by McGlashan, among the New Haven clinical trial sample of prodromal patients who converted to schizophrenia, no patient required hospitalization. All but one continued their daily schedule at work or school, maintained social relationships, and were characterized by an average medical compliance pill count of 93% (McGlashan et al. 2007).

Beyond improving function and symptom severity if conversion occurs, it appears that such early intervention and effective treatment may also be contributing to a decline in psychotic conversion rates (Yung et al. 2007). Therefore, with even earlier identification of the prodrome, the treatment algorithm might involve a period free of specific psychotropic medication and instead a period of observation, monitoring, and treatment of psychiatric disorders, such as depression, anxiety disorders, and substance use problems. By identifying young people with mental health problems earlier, it is likely that a range of disorders would be identified close to their onset and early intervention could reduce the burden of these other illnesses as well.

1.11 Recommended Treatment Guidelines

Individuals who meet the prodromal syndrome criteria or individuals at clinical high risk group for psychosis are clinically heterogeneous but help-seeking. Less than 40% of those who meet the SIPS criteria are likely to become psychotic, but those with additional risk factors such as a family history of psychosis, more severe ratings on delusional-like symptoms, social functioning deficits, or substance abuse

Table 2 Treatment guidelines for at-risk individuals

Stage	Treatment
• All stages	• Psychoeducation regarding the presenting symptoms and risk of ongoing substance abuse
• Mild symptoms ("3" on SIPS positive or disorganized items) • Recent functional decline	• Psychosocial treatment, including crisis intervention, reduction of stress, and ongoing support. CBT to target specific symptoms. Social skills training in a group setting • Stop or reduce agents that might worsen symptoms (e.g., stimulants or antidepressants causing hypomania)

• Moderate symptoms ("4–5" on SIPS positive or disorganized items)
• Functional impairment
• Significant risk factors for psychosis
  Family history
  Drug abuse
  Decline in social functioning
  Neurocognitive deficits
• Axis I anxiety or mood disorder
	• Psychosocial treatment including multifamily groups
	• Case management
	• School support/vocational rehabilitation
	• Pharmacologic intervention to target specific symptoms
	  Depression/anxiety – antidepressant, anxiolytic
	  Hypomania – mood stabilizer
	  Worsening auditory hallucinations, ideas of reference, paranoia – antipsychotic
	• Neuroprotection (e.g., omega-3 fatty acid)

• Severe symptoms ("6" on SIPS positive or disorganized items) • Intermittent psychosis • Severe depression, mania, suicidality, homicidality • Poor functioning	• Crisis intervention • Inpatient hospitalization • Pharmacologic intervention

are even more likely to develop schizophrenia or an affective disorder (Cannon et al. 2008). Additional risk factors such as neurocognitive deficits will likely improve the positive predictive power of current prediction algorithms.

As delineated in Table 2, treatment needs to be tailored to the presenting symptoms and risk factors. An important step in determining the most appropriate treatment requires a full diagnostic assessment, including differential diagnosis by a well trained clinician. It is important to recognize that presenting symptoms could be emerging anxiety, affective, substance use or the iatrogenic effects of medications. Pervasive developmental disorder, learning disabilities, and attention deficit disorders should also be carefully considered as comorbid conditions. A full medical work up should rule out metabolic, endocrine, or neurological etiologies.

Psychosocial treatment including crisis intervention, stress reduction, CBT, supportive therapy, social skills training, cognitive remediation, psychoeducation, dual diagnosis, multifamily groups, case management, school intervention, or vocational rehabilitation should be considered at all stages of illness but especially in the early stage of illness that are more mild or characterized by nonspecific symptoms.

Individuals who are considered "high risk" based on family history of a psychotic disorder would benefit not only from the above psychosocial treatments, but also from regular follow-up and assessment. In those individuals who are considered at clinical high risk, with or without family history of psychosis, our staging criteria recommends that SIPS rating scales, in conjunction with clinical risk factors, can be used to categorize risk severity and specific treatment interventions.

For instance, for those without a family history but who now present with functional decline and mild symptoms corresponding to a "3" on SIPS positive or disorganized items, specific symptom clusters may be targeted with nonpharmacologic treatment such as CBT while current pharmacology may be reassessed to ensure medications may not be exacerbating or causing symptoms. On the other hand, for those with functional decline and moderate symptoms corresponding to a "4–5" on SIPS, who may also have other risk factors, such as family history of psychosis, drug abuse, or neurocognitive and social deficits, more aggressive treatment plans are recommended, employing strategies such as neuroprotective and pharmacologic treatments as delineated in Table 2. Finally, when individuals reach the stage of severe symptoms, corresponding with a "6" on the SIPS, when they no longer have insight, intervention may entail crisis intervention, pharmacologic intervention as well as possible hospitalization.

When pharmacologic intervention is considered, it should be tailored to the differential diagnosis and target the presenting symptoms. Often, even on initial presentation to prodromal clinics, patients are more often than not already suffering from a variety of complaints for which pharmacotherapy may have already been initiated. In these cases, it is essential to recognize that medication could be exacerbating symptoms as antidepressants might induce a hypomanic/manic or mixed episode in a depressed individual prone to bipolar illness, just as high doses or abuse of stimulants might induce subsyndromal psychosis.

Usually, however, if the individual is already being treated with medication for mood, anxiety, or attentional problems, it is most prudent to continue such treatment as long as it is clinically indicated and risk–benefit ratios are thoroughly assessed. Although such treatment may be termed "symptomatic" in nature, it is far from clear that such treatment does not also contribute to declining conversion rates (McGlashan et al. 2007). Thus, the use of antidepressants and mood stabilizers should be considered early if indicated by presenting symptoms. Also, given the potential neuroprotective properties of the SSRIs and mood stabilizers as well as omega-3 fatty acids, they are good first line pharmacologic interventions. Brief trials of antipsychotics should be used to target worsening subsyndromal psychotic or disorganized symptoms if the above interventions are not effective. More severe symptoms such as suicidal or homicidal thoughts as well as agitation will likely require an inpatient stay along with intensive crisis and pharmacologic intervention.

The development of clinical staging criteria for psychosis that incorporates type and severity of clinical symptoms, level of global and social functioning, family history, substance abuse, neurocognitive functioning, and perhaps neurobiological information could help to specify appropriate treatment for vulnerable individuals at different phases of the prodrome and first episode of psychosis. If we could use knowledge gained from neurobiological markers in these populations to not only improve our ability to predict who will develop psychosis but also target specific deficits identified with these markers; it might be possible to better personalize treatment in the early stages of psychosis. Treatment algorithms can then be tailored to presenting symptoms, number of risk factors present, and evidence of progression of the illness to assure appropriate, safe, and effective interventions in the early stages of psychosis.

Yet, as we discuss such treatment strategies it is imperative to keep in mind that since the at-risk state is a new clinical syndrome; all treatments to date are still considered preliminary and "off-label." Ideally, randomized controlled trials of large samples will be necessary before clear and convincing treatment guidelines may be codified. However, in the absence of the strictest evidence based medicine, clinicians faced with complexity of the prodrome must be armed with guidelines based on clinical practice across prodromal centers internationally.

1.12 General Summary

Given the range of possible presentations of the psychotic prodrome, ethical issues, and potential treatments, it is not surprising that clinicians do not know what to do with individuals who present in an at-risk state. Clearly, translational studies are needed to better inform treatment, but it is also apparent that more work is needed to characterize the psychotic prodrome and develop clinical, demographic, and vulnerability marker assessment tools to better identify those who are at greatest risk. The predictive data can then inform preventive treatment and the development of treatment algorithms based on clinical staging. Finally, treatment studies are needed that target specific stages of early psychosis and select individuals at greatest risk for pharmacologic trials and those with more mild symptoms or few risk factors for psychosocial interventions. For now the best recommendation is to treat at-risk youth using the best clinical wisdom recognizing that it may be transient or evolve into a more serious condition.

References

Addington J, Leriger E, Addington D (2003) Symptom outcome 1 year after admission to an early psychosis program. Can J Psychiatry 48:204–207

Addington J, Van Mastrigt S, Addington D (2004) Duration of untreated psychosis: impact on 2-year outcome. Psychol Med 34:277–284

Addington J, Cadenhead KS, Cannon TD, Cornblatt B, McGlashan TH, Perkins DO, Seidman LJ, Tsuang M, Walker EF, Woods SW, Heinssen R (2007) North American Prodrome Longitudinal Study: a collaborative multisite approach to prodromal schizophrenia research. Schizophr Bull 33:665–672

Angelucci F, Mathe AA, Aloe L (2000) Brain-derived neurotrophic factor and tyrosine kinase receptor TrkB in rat brain are significantly altered after haloperidol and risperidone administration. J Neurosci Res 60:783–794

Arnold S, Beyer C (2009) Neuroprotection by estrogen in the brain: the mitochondrial compartment as presumed therapeutic target. J Neurochem 110:1–11

Arseneault L, Cannon M, Witton J, Murray RM (2004) Causal association between cannabis and psychosis: examination of the evidence. Br J Psychiatry 184:110–117

Bai O, Chlan-Fourney J, Bowen R, Keegan D, Li XM (2003) Expression of brain-derived neurotrophic factor mRNA in rat hippocampus after treatment with antipsychotic drugs. J Neurosci Res 71:127–131

Ballon JS, Kaur T, Marks II, Cadenhead KS (2007) Social functioning in young people at risk for schizophrenia. Psychiatry Res 151:29–35

Berger GE, Wood S, McGorry PD (2003) Incipient neurovulnerability and neuroprotection in early psychosis. Psychopharmacol Bull 37:79–101

Berger GE, Proffitt TM, McConchie M, Yuen H, Wood SJ, Amminger GP, Brewer W, McGorry PD (2007) Ethyl-eicosapentaenoic acid in first-episode psychosis: a randomized, placebo-controlled trial. J Clin Psychiatry 68:1867–1875

Bilder RM, Reiter G, Bates J, Lencz T, Szeszko P, Goldman RS, Robinson D, Lieberman JA, Kane JM (2006) Cognitive development in schizophrenia: follow-back from the first episode. J Clin Exp Neuropsychol 28:270–282

Brewer WJ, Wood SJ, McGorry PD, Francey SM, Phillips LJ, Yung AR, Anderson V, Copolov DL, Singh B, Velakoulis D, Pantelis C (2003) Impairment of olfactory identification ability in individuals at ultra-high risk for psychosis who later develop schizophrenia. Am J Psychiatry 160:1790–1794

Brewer WJ, Francey SM, Wood SJ, Jackson HJ, Pantelis C, Phillips LJ, Yung AR, Anderson VA, McGorry PD (2005) Memory impairments identified in people at ultra-high risk for psychosis who later develop first-episode psychosis. Am J Psychiatry 162:71–78

Byrd CA, Brunjes PC (2001) Neurogenesis in the olfactory bulb of adult zebrafish. Neuroscience 105:793–801

Cadenhead K (2005) The stability of prepulse inhibition of the startle response in individuals at-risk for schizophrenia and in the early phases of illness. Schizophr Bull 31:450

Cadenhead K, Light GA, Geyer MA, Braff DL (2000) Sensory gating deficits assessed by the P50 event-related-potential in subjects with schizotypal personality disorder. Am J Psychiatry 157:55–59

Cannon TD, van Erp TG, Bearden CE, Loewy R, Thompson P, Toga AW, Huttunen MO, Keshavan MS, Seidman LJ, Tsuang MT (2003) Early and late neurodevelopmental influences in the prodrome to schizophrenia: contributions of genes, environment, and their interactions. Schizophr Bull 29:653–669

Cannon R, Lubar J, Congedo M, Thornton K, Towler K, Hutchens T (2007) The effects of neurofeedback training in the cognitive division of the anterior cingulate gyrus. Int J Neurosci 117:337–357

Cannon TD, Cadenhead K, Cornblatt B, Woods SW, Addington J, Walker E, Seidman LJ, Perkins D, Tsuang M, McGlashan T, Heinssen R (2008) Prediction of psychosis in youth at high clinical risk: a multisite longitudinal study in North America. Arch Gen Psychiatry 65:28–37

Caspi A, Reichenberg A, Weiser M, Rabinowitz J, Kaplan Z, Knobler H, Davidson-Sagi N, Davidson M (2003) Cognitive performance in schizophrenia patients assessed before and following the first psychotic episode. Schizophr Res 65:87–94

Caspi A, Moffitt TE, Cannon M, McClay J, Murray R, Harrington H, Taylor A, Arseneault L, Williams B, Braithwaite A, Poulton R, Craig IW (2005) Moderation of the effect of adolescent-onset cannabis use on adult psychosis by a functional polymorphism in the catechol-O-methyltransferase gene: longitudinal evidence of a gene X environment interaction. Biol Psychiatry 57:1117–1127

Catts VS, Catts SV (2000) Apoptosis and schizophrenia: is the tumour suppressor gene, p53, a candidate susceptibility gene? Schizophr Res 41:405–415

Corcoran C, Malaspina D, Hercher L (2005) Prodromal interventions for schizophrenia vulnerability: the risks of being at "risk". Schizophr Res 73:173–184

Cornblatt BA, Lencz T, Kane JM (2001) Treatment of the schizophrenia prodrome: is it presently ethical? Schizophr Res 51:31–38

Cornblatt B, Lencz T, Obuchowski M (2002) The schizophrenia prodrome: treatment and highrisk perspectives. Schizophr Res 54:177–186

Cornblatt BA, Auther AM, Niendam T, Smith CW, Zinberg J, Bearden CE, Cannon TD (2007a) Preliminary findings for two new measures of social and role functioning in the prodromal phase of schizophrenia. Schizophr Bull 33(3):688–702

Cornblatt BA, Lencz T, Smith CW, Olsen R, Auther AM, Nakayama E, Lesser ML, Tai JY, Shah MR, Foley CA, Kane JM, Correll CU (2007b) Can antidepressants be used to treat the schizophrenia prodrome? Results of a prospective, naturalistic treatment study of adolescents. J Clin Psychiatry 68:546–557

Correll CU, Malhotra AK (2004) Pharmacogenetics of antipsychotic-induced weight gain. Psychopharmacology (Berl) 174:477–489

Cosway R, Byrne M, Clafferty R, Hodges A, Grant E, Abukmeil SS, Lawrie SM, Miller P, Johnstone EC (2000) Neuropsychological change in young people at high risk for schizophrenia: results from the first two neuropsychological assessments of the Edinburgh High Risk Study. Psychol Med 30:1111–1121

Davidson M, Reichenberg A, Rabinowitz J, Weiser M, Kaplan Z, Mark M (1999) Behavioral and intellectual markers for schizophrenia in apparently healthy male adolescents. Am J Psychiatry 156:1328–1335

Eastvold AD, Cadenhead K (2003) Neurocognitive markers in prodromal schizophrenia. Schizophr Res 60:133

Eastvold AD, Heaton RK, Cadenhead KS (2007) Neurocognitive deficits in the (putative) prodrome and first episode of psychosis. Schizophr Res 93:266–277

Erbagci AB, Herken H, Koyluoglu O, Yilmaz N, Tarakcioglu M (2001) Serum IL-1beta, sIL-2R, IL-6, IL-8 and TNF-alpha in schizophrenic patients, relation with symptomatology and responsiveness to risperidone treatment. Mediators Inflamm 10:109–115

Feinberg I (1982) Schizophrenia: caused by a fault in programmed synaptic elimination during adolescence? J Psychiatr Res 17:319–334

Grant C, Addington J, Addington D, Konnert C (2001) Social functioning in first- and multiepisode schizophrenia. Can J Psychiatry 46:746–749

Hafner H, an der Heiden W (1999) The course of schizophrenia in the light of modern follow-up studies: the ABC and WHO studies. Eur Arch Psychiatry Clin Neurosci 249(Suppl 4):14–26

Häfner H, Riecher-Rössler A, Maurer K, Fätkenheuer B, Löffler W (1992) First onset and early symptomatology of schizophrenia. A chapter of epidemiological and neurobiological research into age and sex differences. Eur Arch Psychiatry Clin Neurosci 242:109–118

Haroun N, Dunn L, Haroun A, Cadenhead KS (2006) Risk and protection in prodromal schizophrenia: ethical implications for clinical practice and future research. Schizophr Bull 32:166–178

Harrigan SM, McGorry PD, Krstev H (2003) Does treatment delay in first-episode psychosis really matter? Psychol Med 33:97–110

Hegarty JD, Baldessarini RJ, Tohen M, Waternaux C, Oepen G (1994) One hundred years of schizophrenia: a meta-analysis of the outcome literature. Am J Psychiatry 151:1409–1416

Huber G, Gross G (1989) The concept of basic symptoms in schizophrenic and schizoaffective psychoses. Recenti Prog Med 80:646–652

Jacobs BL, Praag H, Gage FH (2000) Adult brain neurogenesis and psychiatry: a novel theory of depression. Mol Psychiatry 5:262–269

Jahshan C, Heaton RK, Golshan S, Cadenhead KS (2010) Course of neurocognitive deficits in the prodrome and first episode of schizophrenia. Neuropsychology 24:109–120

Jarskog LF, Gilmore JH, Selinger ES, Lieberman JA (2000) Cortical bcl-2 protein expression and apoptotic regulation in schizophrenia. Biol Psychiatry 48:641–650

Johannessen JO, Larsen TK, Joa I, Melle I, Friis S, Opjordsmoen S, Rund BR, Simonsen E, Vaglum P, McGlashan TH (2005) Pathways to care for first-episode psychosis in an early detection healthcare sector: part of the Scandinavian TIPS study. Br J Psychiatry 48: s24–s28

Karlsgodt KH, Niendam TA, Bearden CE, Cannon TD (2009) White matter integrity and prediction of social and role functioning in subjects at ultra-high risk for psychosis. Biol Psychiatry 66:562–569

Keith SJ, Matthews SM (1991) The diagnosis of schizophrenia: a review of onset and duration issues. Schizophr Bull 17:51–67

Klosterkotter J, Hellmich M, Steinmeyer EM, Schultze-Lutter F (2001) Diagnosing schizophrenia in the initial prodromal phase. Arch Gen Psychiatry 58:158–164

Kornack DR, Rakic P (1999) Continuation of neurogenesis in the hippocampus of the adult macaque monkey. Proc Natl Acad Sci USA 96:5768–5773

Lappin JM, Morgan K, Morgan C, Hutchison G, Chitnis X, Suckling J, Fearon P, McGuire PK, Jones PB, Leff J, Murray RM, Dazzan P (2006) Gray matter abnormalities associated with duration of untreated psychosis. Schizophr Res 83:145–153

Larsen TK, Moe LC, Vibe-Hansen L, Johannessen JO (2000) Premorbid functioning versus duration of untreated psychosis in 1 year outcome in first-episode psychosis. Schizophr Res 45:1–9

Larsen TK, Melle I, Auestad B, Friis S, Haahr U, Johannessen JO, Opjordsmoen S, Rund BR, Simonsen E, Vaglum P, McGlashan T (2006) Early detection of first-episode psychosis: the effect on 1-year outcome. Schizophr Bull 32:758–764

Lieberman JA (1999) Is schizophrenia a neurodegenerative disorder? A clinical and neurobiological perspective. Biol Psychiatry 46:729–739

Linszen D, Dingemans P, Lenior M (2001) Early intervention and a five year follow up in young adults with a short duration of untreated psychosis: ethical implications. Schizophr Res 51:55–61

Maier W, Mossner R, Quednow BB, Wagner M, Hurlemann R (2008) From genes to psychoses and back: the role of the 5HT2alpha-receptor and prepulse inhibition in schizophrenia. Eur Arch Psychiatry Clin Neurosci 258(Suppl 5):40–43

Malberg JE, Eisch AJ, Nestler EJ, Duman RS (2000) Chronic antidepressant treatment increases neurogenesis in adult rat hippocampus. J Neurosci 20:9104–9110

Malla AK, Norman RM, Manchanda R, Ahmed MR, Scholten D, Harricharan R, Cortese L, Takhar J (2002) One year outcome in first episode psychosis: influence of DUP and other predictors. Schizophr Res 54:231–242

Marshall M, Rathbone J (2006) Early intervention for psychosis. Cochrane Database Syst Rev CD004718

Marshall M, Lewis S, Lockwood A, Drake R, Jones P, Croudace T (2005) Association between duration of untreated psychosis and outcome in cohorts of first-episode patients: a systematic review. Arch Gen Psychiatry 62:975–983

McGlashan TH (2001) Psychosis treatment prior to psychosis onset: ethical issues. Schizophr Res 51:47–54

McGlashan TH, Vaglum P, Friis S, J. Johannessen O, Simonsen E, Larsen TK, Melle I, Haahr U, Opjordsmoen S, Zipursky R, Perkins D, Addington J, Miller T, Woods S, Hoffman R, Preda A, Epstein I, Addington D, Lindborg S, Trzaskoma Q, M. Tohen M, Breier A (2005) Early detection and intervention in first episode psychosis: empirical update of the TIPS and PRIME projects. Schizophr Bull 31:496

McGlashan TH, Zipursky RB, Perkins D, Addington J, Miller T, Woods SW, Hawkins KA, Hoffman RE, Preda A, Epstein I, Addington D, Lindborg S, Trzaskoma Q, Tohen M, Breier A (2006) Randomized, double-blind trial of olanzapine versus placebo in patients prodromally symptomatic for psychosis. Am J Psychiatry 163:790–799

McGlashan TH, Addington J, Cannon T, Heinimaa M, McGorry P, O'Brien M, Penn D, Perkins D, Salokangas RK, Walsh B, Woods SW, Yung A (2007) Recruitment and treatment practices for help-seeking "prodromal" patients. Schizophr Bull 33:715–726

McGorry PD, Yung AR, Phillips LJ, Yuen HP, Francey S, Cosgrave EM, Germano D, Bravin J, McDonald T, Blair A, Adlard S, Jackson H (2002) Randomized controlled trial of interventions designed to reduce the risk of progression to first-episode psychosis in a clinical sample with subthreshold symptoms. Arch Gen Psychiatry 59:921–928

McGorry PD, Hickie IB, Yung AR, Pantelis C, Jackson HJ (2006) Clinical staging of psychiatric disorders: a heuristic framework for choosing earlier, safer and more effective interventions. Aust NZ J Psychiatry 40:616–622

McGorry PD, Killackey E, Yung AR (2007) Early intervention in psychotic disorders: detection and treatment of the first episode and the critical early stages. Med J Aust 187:S8–S10

McNamara RK, Ostrander M, Abplanalp W, Richtand NM, Benoit SC, Clegg DJ (2006) Modulation of phosphoinositide-protein kinase C signal transduction by omega-3 fatty acids: implications for the pathophysiology and treatment of recurrent neuropsychiatric illness. Prostaglandins Leukot Essent Fatty Acids 75:237–257

Melle I, Larsen TK, Haahr U, Friis S, Johannessen JO, Opjordsmoen S, Simonsen E, Rund BR, Vaglum P, McGlashan T (2004) Reducing the duration of untreated first-episode psychosis: effects on clinical presentation. Arch Gen Psychiatry 61:143–150

Melle I, Haahr U, Friis S, Hustoft K, Johannessen JO, Larsen TK, Opjordsmoen S, Rund BR, Simonsen E, Vaglum P, McGlashan T (2005) Reducing the duration of untreated first-episode psychosis – effects on baseline social functioning and quality of life. Acta Psychiatr Scand 112:469–473

Meng H, Schimmelmann BG, Koch E, Bailey B, Parzer P, Günter M, Mohler B, Kunz N, Schulte-Markwort M, Felder W, Zollinger R, Bürgin D, Resch F (2009) Basic symptoms in the general population and in psychotic and non-psychotic psychiatric adolescents. Schizophr Res 111:32–38

Miller TJ, McGlashan TH, Rosen JL, Somjee L, Markovich PJ, Stein K, Woods SW (2002) Prospective diagnosis of the initial prodrome for schizophrenia based on the Structured Interview for Prodromal Syndromes: preliminary evidence of interrater reliability and predictive validity. Am J Psychiatry 159:863–865

Miller TJ, McGlashan TH, Rosen JL, Cannon TD, Ventura J, Cadenhead K, McFarlane W, Perkins DO, Pearlso GD, Woods SW (2004) Prodromal assessment using the SIPS and SOPS. Schizophr Res 70:74

Monteleone P, Fabrazzo M, Tortorella A, Maj M (1997) Plasma levels of interleukin-6 and tumor necrosis factor alpha in chronic schizophrenia: effects of clozapine treatment. Psychiatry Res 71:11–17

Morrison AP, French P, Walford L, Lewis SW, Kilcommons A, Green J, Parker S, Bentall RP (2004) Cognitive therapy for the prevention of psychosis in people at ultra-high risk: randomised controlled trial. Br J Psychiatry 185:291–297

Murray CJ, Lopez AD (1996) Evidence-based health policy – lessons from the Global Burden of Disease Study. Science 274:740–743

Niendam TA, Bearden CE, Johnson JK, McKinley M, Loewy R, O'Brien M, Nuechterlein KH, Green MF, Cannon TD (2006) Neurocognitive performance and functional disability in the psychosis prodrome. Schizophr Res 84:100–111

Niendam TA, Bearden CE, Zinberg J, Johnson JK, O'Brien M, Cannon TD (2007) The course of neurocognition and social functioning in individuals at ultra high risk for psychosis. Schizophr Bull 33:772–781

Pantelis C, Velakoulis D, McGorry PD, Wood SJ, Suckling J, Phillips LJ, Yung AR, Bullmore ET, Brewer W, Soulsby B, Desmond P, McGuire PK (2003) Neuroanatomical abnormalities before and after onset of psychosis: a cross-sectional and longitudinal MRI comparison. Lancet 361:281–288

Perkins DO, Gu H, Boteva K, Lieberman JA (2005) Relationship between duration of untreated psychosis and outcome in first-episode schizophrenia: a critical review and meta-analysis. Am J Psychiatry 162:1785–1804

Petersen L, Nordentoft M, Jeppesen P, Ohlenschaeger J, Thorup A, Christensen TØ, Krarup G, Dahlstrøm J, Haastrup B, Jørgensen P (2005) Improving 1-year outcome in first-episode psychosis: OPUS trial. Br J Psychiatry Suppl 48:s98–s103

Power PJ, Bell RJ, Mills R, Herrman-Doig T, Davern M, Henry L, Yuen HP, Khademy-Deljo A, McGorry PD (2003) Suicide prevention in first episode psychosis: the development of a randomised controlled trial of cognitive therapy for acutely suicidal patients with early psychosis. Aust NZ J Psychiatry 37:414–420

Richtand NM, McNamara RK (2008) Serotonin and dopamine interactions in psychosis prevention. Prog Brain Res 172:141–153

Ruhrmann S, Schultze-Lutter F, Klosterkotter J (2003) Early detection and intervention in the initial prodromal phase of schizophrenia. Pharmacopsychiatry 36(Suppl 3):S162–S167

Ruhrmann S, Bechdolf A, Kühn KU, Wagner M, Schultze-Lutter F, Janssen B, Maurer K, Häfner H, Gaebel W, Möller HJ, Maier W, Klosterkötter J, LIPS study group (2007) Acute effects of treatment for prodromal symptoms for people putatively in a late initial prodromal state of psychosis. Br J Psychiatry Suppl 51:s88–s95

Salokangas RK, McGlashan TH (2008) Early detection and intervention of psychosis. A review. Nord J Psychiatry 62:92–105

Schaffner KF, McGorry PD (2001) Preventing severe mental illnesses – new prospects and ethical challenges. Schizophr Res 51:3–15

Schultze-Lutter F, Ruhrmann S, Berning J, Maier W, Klosterkötter J (2010) Basic symptoms and ultrahigh risk criteria: symptom development in the initial prodromal state. Schizophr Bull 36:182–191

Steindler DA, Pincus DW (2002) Stem cells and neuropoiesis in the adult human brain. Lancet 359:1047–1054

Sun D, Phillips L, Velakoulis D, Yung A, McGorry PD, Wood SJ, van Erp TG, Thompson PM, Toga AW, Cannon TD, Pantelis C (2009) Progressive brain structural changes mapped as psychosis develops in 'at risk' individuals. Schizophr Res 108:85–92

Takahashi T, Suzuki M, Tanino R, Zhou SY, Hagino H, Niu L, Kawasaki Y, Seto H, Kurachi M (2007) Volume reduction of the left planum temporale gray matter associated with long duration of untreated psychosis in schizophrenia: a preliminary report. Psychiatry Res 154:209–219

Tenn CC, Fletcher PJ, Kapur S (2005) A putative animal model of the "prodromal" state of schizophrenia. Biol Psychiatry 57:586–593

Umbricht D, Krljes S (2005) Mismatch negativity in schizophrenia: a meta-analysis. Schizophr Res 76:1–23

Vaidya VA, Marek GJ, Aghajanian GK, Duman RS (1997) 5-HT2A receptor-mediated regulation of brain-derived neurotrophic factor mRNA in the hippocampus and the neocortex. J Neurosci 17:2785–2795

Valmaggia LR, McCrone P, Knapp M, Woolley JB, Broome MR, Tabraham P, Johns LC, Prescott C, Bramon E, Lappin J, Power P, McGuire PK (2009)Economic impact of early intervention in people at high risk of psychosis. Psychol Med 39:1617-1626

Wood SJ, Berger G, Velakoulis D, Phillips LJ, McGorry PD, Yung AR, Desmond P, Pantelis C (2003) Proton magnetic resonance spectroscopy in first episode psychosis and ultra high-risk individuals. Schizophr Bull 29:831–843

Woods SW, Tully EM, Walsh BC, Hawkins KA, Callahan JL, Cohen SJ, Mathalon DH, Miller TJ, McGlashan TH (2007) Aripiprazole in the treatment of the psychosis prodrome: an open-label pilot study. Br J Psychiatry Suppl 51:s96–s101

Wu EQ, Birnbaum HG, Shi L, Ball DE, Kessler RC, Moulis M, Aggarwal J (2005) The economic burden of schizophrenia in the United States in 2002. J Clin Psychiatry 66:1122–1129

Yung AR, McGorry PD (1996) The prodromal phase of first-episode psychosis: past and current conceptualizations. Schizophr Bull 22:353–370

Yung A, Phillips L, McGorry P, Ward J, Donovan K, Thompson K (2002) Comprehensive Assessment of At Risk Mental States (CAARMS). The PACE Clinic, University of Melbourne, Department of Psychiatry, Melbourne, Australia, In

Yung AR, Phillips LJ, Yuen HP, Francey SM, McFarlane CA, Hallgren M, McGorry PD (2003) Psychosis prediction: 12-month follow up of a high-risk ("prodromal") group. Schizophr Res 60:21–32

Yung AR, Stanford C, Cosgrave E, Killackey E, Phillips L, Nelson B, McGorry PD (2006) Testing the ultra high risk (prodromal) criteria for the prediction of psychosis in a clinical sample of young people. Schizophr Res 84:57–66

Yung AR, Yuen HP, Berger G, Francey S, Hung TC, Nelson B, Phillips L, McGorry P (2007) Declining transition rate in ultra high risk (prodromal) services: dilution or reduction of risk? Schizophr Bull 33:673–681

Zhang M, Wang M, Li J, Phillips MR (1994) Randomised-control trial of family intervention for 78 first-episode male schizophrenic patients: an 18-month study in Suzhou, Jiangsu. Br J Psychiatry Suppl 24:96–102

Antipsychotic Drug Development

Dennis H. Kim and Stephen M. Stahl

Tell me, what sort of thing is an elephant?
– Udana 68–69, Buddhist canon

Contents

1	Introduction	124
2	Dopamine D_2 Receptors: Antagonism, Inverse Agonism, and Partial Agonism	125
3	Dopamine D_3 Receptors: Cognition and "Optimized Antagonism" at D_3 Versus D_2 Receptors	129
4	Serotonin 5-HT_{2A} Receptors: Dopamine Brakes	130
5	Serotonin 5-HT_{1A} Receptors: Dopamine Accelerators	130
6	Serotonin 5-HT_{2C} Receptors	131
7	Serotonin 5-HT_7 Receptors	132
8	Glutamatergic Receptors: NMDA and mGluR	132
9	Glycine Agonists	134
10	Receptors That Mediate Side Effects	134
11	Conclusion	135
References		135

Abstract Schizophrenia typically manifests itself with a wide array of symptoms – positive, negative, cognitive, and affective – and may also involve neurodevelopmental and neurodegenerative aspects. Each of these symptom dimensions may be derived from pathology at one or more receptor types, localized in different regions of the brain. The absence of a single therapeutic target for schizophrenia has therefore prompted the de-emphasis of selective "magic bullets" and a critical re-examination of the intramolecular polypharmacy afforded by antipsychotics.

D.H. Kim
Arbor Scientia, 1930 Palomar Point Way, Suite 103, Carlsbad, CA 92008, USA

S.M. Stahl (✉)
Arbor Scientia, 1930 Palomar Point Way, Suite 103, CA 92008, USA
Department of Psychiatry, School of Medicine, University of California, San Diego, La Jolla, CA 92093, USA
Honorary Visiting Senior Fellow, Department of Psychiatry University of Cambridge, Addenbrooke's Hospital, Hills Road, Cambridge, CB2 2QQ
e-mail: smstahl@neiglobal.com

In this chapter, we present a review of some of the receptor targets that are currently thought to mediate symptoms of schizophrenia, and discuss their possible implications for future antipsychotic drug development. Therapeutic strategies for schizophrenia that successfully exploit the multifunctionality of antipsychotics will take into account the entire receptor activity "portfolio" of the agent and provide a total therapeutic response that, like the elephant of the Buddhist parable, is greater than the sum of its parts.

Keywords Antipsychotics · Intramolecular polypharmacy · Multifunctional drugs · Receptors

1 Introduction

In a well-known parable attributed to Buddhist, Hindu, and other traditions, an elephant is presented to several blind men who each feel a different part of the animal – foot, tusk, body, tail, etc. (Fig. 1). On the basis of their limited analyses, the blind men arrive at very different conclusions regarding the nature of the elephant, variously likening it to a pillar, plowshare, granary, and broom. The moral of the story, of course, is that the whole is more than the sum of its parts.

Fig. 1 "Blind monks examining an elephant" by Hanabusa Itchō (1652–1724) (courtesy of Library of Congress, Prints and Photographs Division, Washington, DC 20540, USA)

This tale serves as a metaphor for the development of pharmacotherapies for schizophrenia. Attempts to develop selective "magic bullets" that possess a single therapeutic action and factor out side effects have been frustrated by the complexity of schizophrenia and the fact that the symptomatology of the disease does not appear to arise from a single neurobiological entity. Schizophrenia has not only psychotic features (positive symptoms) but also negative, cognitive, and affective symptoms, and probably neurodevelopmental and neurodegenerative aspects. Each of these symptom dimensions may be derived from pathology at one or more receptor types. The absence of a single therapeutic target for schizophrenia has prompted a critical re-examination of the multifunctionality of "dirty drugs," so termed for their combination of both therapeutic actions and unwanted side effects. From the perspective of rational therapeutic design, polypharmacy aspires to take advantage of potential synergies among independent therapeutic mechanisms, whether they are present in a single drug or in multiple drugs working in tandem. In particular, the term "intramolecular polypharmacy" has been coined to describe the capacity of a single drug to affect multiple receptor types (Stahl 2008a).

Antipsychotics fit squarely in this category, with each agent in the class having a unique portfolio of activity at various receptors. The most robust action of any antipsychotic is generally its ability to reduce the positive symptoms of psychosis (e.g., delusions and hallucinations), and these symptoms are often the first targets of treatment of schizophrenia. It is important to note, however, that in current therapeutic strategies for schizophrenia, the greatest unmet needs are in treatment of negative and cognitive symptoms. In 2006, the National Institute of Mental Health issued a consensus statement that "persistent and clinically significant negative symptoms are an unmet therapeutic need" that should be a focus of future drug development efforts (Kirkpatrick et al. 2006). Unfortunately, to date the actions of most therapeutics on negative, cognitive, and affective symptom domains have been theoretically interesting but clinically disappointing. In this chapter, we present a review of some of the receptor targets that are currently thought to mediate symptoms of schizophrenia, and discuss their possible implications for future multifunctional antipsychotics. Figure 2 summarizes the receptors, which are depicted iconographically on a hypothetical multifunctional drug.

Not surprisingly, the highly varied receptor profiles of different antipsychotics give rise to a number of other therapeutic applications, both on and off label; for example, some antipsychotics may have uses as mood stabilizers and anxiolytics. A question that is germane to this discussion and worthwhile to consider is: what constitutes an effective antipsychotic? Is it the tusk, the trunk, or the whole elephant?

2 Dopamine D_2 Receptors: Antagonism, Inverse Agonism, and Partial Agonism

Conventional and atypical antipsychotics alike are dopamine D_2 antagonists or inverse agonists – when they bind to D_2 receptors, they either *inhibit* basal activity in the D_2 receptor-linked G protein messenger system (inverse agonism) or *have no effect* on basal activity (silent antagonism) (Stahl 2008a; Strange 2008). In either

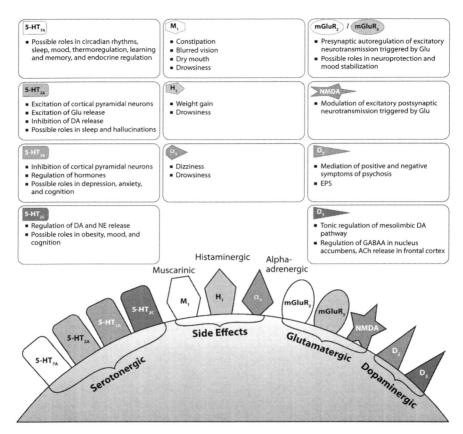

Fig. 2 Potential functions of selected CNS receptors that are currently thought to mediate symptoms of schizophrenia. The various symptoms of schizophrenia are not only mediated by specific receptor sites, but also localized in unique regions of the brain. The functions ascribed to the receptors here are therefore dependent upon the location of the specific receptors in the brain. Furthermore, whether therapeutic effects are the result of full or partial agonism, antagonism, or inverse agonism depends on each receptor type

case, the antipsychotic blocks postsynaptic receptors and normalizes the hyperactivity theoretically arising from excessive dopamine at the synapse. In the mesolimbic dopamine pathway, the result of the hypothetical normalization is a reduction in positive psychotic symptoms. While the rubric of D_2 antagonism/inverse agonism has been widely accepted, subtleties of a D_2-based mechanism of antipsychotic action have been suggested by several researchers. For example, Kapur and Seeman (2001) have proposed that atypicality arises specifically in antipsychotics that bind to D_2 postsynaptic receptors with relatively low affinity, and therefore more rapid dissociation rates. Theoretically, such an agent is able to occupy D_2 receptors long enough to exert an antipsychotic action, but not long enough for extrapyramidal symptoms (EPS), prolactin elevation, or secondary negative symptoms to develop. Currently, it is still unclear how long an atypical must reside at D_2 receptors to achieve the balance of therapeutic efficacy and minimal EPS (Miyamoto et al. 2005). Although rapid dissociation from the D_2 receptor in vitro is a good predictor of low EPS potential in patients, this "hit-and-run" action may not be acceptable as a general mechanism of atypicality because it does not account for the fact that the atypical antipsychotics olanzapine, risperidone, and ziprasidone all have high D_2 affinities and slow dissociation rates from the receptor (Meltzer 2004; Schmidt et al. 2001).

Another D_2-centered mechanism that has gained currency is *dopamine D_2 partial agonism*. In the spectrum of pharmacological actions that ranges from agonism to inverse agonism, the actions of dopamine partial agonists (DPAs) on D_2 receptors lie somewhere between full agonism and silent antagonism (Fig. 3). As their name implies, DPAs activate the D_2 receptor's second messenger system, but in a manner that is less than the activation produced by full agonists (including dopamine itself). Pharmacologically, DPAs activate at low dopaminergic tone and inhibit at high dopaminergic tone. In other words, in the presence of excessive agonist action by dopamine, DPAs *reduce* signal transduction in the D_2-linked G protein messenger system. Conversely, when dopamine is deficient, DPAs *increase* D_2-receptor-mediated signal transduction (Stahl 2008b). By mediating between the extremes of hyperdopaminergia and hypodopaminergia, partial agonists are thus sometimes called "stabilizers" (Burris et al. 2002; Carlsson et al. 1999).

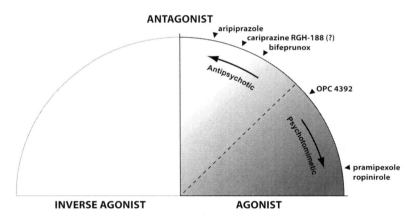

Fig. 3 The agonist spectrum. Naturally occurring neurotransmitters stimulate receptors and thus are agonists. Some drugs also stimulate receptors and therefore are agonists as well. It is possible for drugs to stimulate receptors to a lesser degree than the natural neurotransmitter; these are called *partial agonists* or *stabilizers*. Antagonists have no intrinsic activity of their own in the absence of an agonist; for this reason, antagonists are sometimes called "silent." Inverse agonists, on the other hand, have the opposite effect as agonists. With respect to D_2 partial agonists (DPAs), some DPAs have too much agonism and may actually be psychotomimetic (an example is the agent OPC 4293). In contrast, partial agonists that are closer to the antagonist range of the spectrum are thought to have more favorable antipsychotic profiles (examples include aripiprazole and bifeprunox)

Partial agonism has unique functional and clinical consequences. In theory, DPAs used to treat schizophrenia reduce D_2 hyperactivity in mesolimbic dopamine neurons to a degree that is sufficient to exert an antipsychotic action on positive symptoms, even though they do not completely shut down D_2 receptors as do conventional antipsychotics. At the same time, DPAs reduce dopamine activity in the nigrostriatal system to an extent that is insufficient to induce EPS. However, clinical experience with DPAs suggests that only those that are close to the antagonist end of the spectrum may have antipsychotic efficacy. For example, the partial agonist OPC 4293 developed by Otsuka Pharmaceutical was found to be too activating, improving negative symptoms but worsening positive symptoms (Stahl 2008b). Another example of a DPA that is "too full" is bifeprunox (developed by Solvay Pharmaceuticals), which exhibited a slow onset of action due to activating side effects and was associated with nausea and vomiting (Stahl 2008b). Furthermore, bifeprunox showed no significant efficacy at high doses, while efficacy at lower doses was inferior to that of a D_2 antagonist (risperidone); accordingly, bifeprunox was given a nonapproval by the Food and Drug Administration (Stahl 2008b).

On the other hand, another Otsuka compound, aripiprazole, was found to be "less full" as a D_2 partial agonist and demonstrated effectiveness in treating psychotic symptoms as well as symptoms of acute mania. Actions at serotonin receptors may also contribute to aripiprazole's clinical properties, but are not as potent as aripiprazole's actions at dopamine receptors, and may only be relevant at higher doses (Stahl 2008a; Lawler et al. 1999; Mamo et al. 2007). Aripiprazole can still be activating in some patients, however, and unlike D_2 antagonists, aripiprazole can show an inverted U-shaped dose response for its antipsychotic actions, consistent with partial agonist actions at D_2 receptors (Stahl 2008b). In fact, dosing of aripiprazole for use in augmenting antidepressants in major depression may be even lower than the ideal dose of aripiprazole for use as an antipsychotic, possibly optimizing its agonist rather than its antagonist actions (Stahl 2008b; Berman et al. 2007; Thase et al. 2008). Aripiprazole is apparently most effective for depression at very low doses (<10 mg), and perhaps most effective for mania and psychosis at higher doses (>10 mg), which suggests the possibility that *increasing* receptor activation at low doses may be what is needed in depression, while *blocking* at high doses may be what is needed in psychosis. Another D_2 partial agonist, cariprazine (RGH-188), is currently in development and has also shown some early signs of efficacy as an antipsychotic.

The upshot of these explorations of the role of the D_2 receptor in psychosis is that there are theoretically *several ways* in which an antipsychotic can have efficacy in treating positive symptoms. The antipsychotic can be an antagonist, inverse agonist, or even a partial agonist. Full antagonism or full inverse agonism may come at the price of EPS, hyperprolactinemia, and negative symptoms, unless the mitigating influence of 5-HT_{2A} antagonism is also present (see below). Partial agonism may provide an intermediate effect, as long as the action falls closer to silent antagonism than to full agonism on the agonist spectrum. In general, the closer the partial agonist is to the full agonist end of the spectrum, the less effective it becomes as an antipsychotic, the more difficult it is to find the optimal dose, and the more activating and psychotomimetic the drug becomes (Stahl 2008a, b). Lastly, "hit-and-run" receptor binding involves the drug occupying D_2 receptors long enough to have an antipsychotic effect, but not long enough for EPS or other adverse effects to arise.

3 Dopamine D_3 Receptors: Cognition and "Optimized Antagonism" at D_3 Versus D_2 Receptors

Dopamine D_3 receptors belong to the D_2 receptor family and are often coexpressed with D_2, but are far less abundant than D_2 receptors (Joyce and Millan 2005). The concentration of D_3 receptors is greatest in the nucleus accumbens, suggesting that the role of D_3 may be related to the mesolimbic dopaminergic system (Schwartz et al. 2000). Indeed, two features of D_3 receptors combine to support a role for these receptors in tonic regulation of the limbic system (1) the affinity of D_3 receptors for dopamine is greater than 60-fold higher than that of D_2 receptors and (2) postsynaptic D_3 receptors are localized away from the synapse (Joyce 2001). It has therefore been hypothesized that D_3 serves as an extrasynaptic sensor of low endogenous dopamine concentrations. Recent research also shows that D_3 autoreceptors possibly regulate extracellular dopamine levels via an interaction with the dopamine transporter (Zapata and Shippenberg 2002).

Until recently, it was difficult to study the specific function of D_3 receptors because of the absence of D_3-selective agents. Now, with the development of such agents (e.g., the selective D_3 antagonist SB277011), preclinical results suggest that the D_3 receptor may be a useful target for therapeutics that ameliorate not only negative and cognitive symptoms of schizophrenia, but also substance abuse disorders (Reavill et al. 2000; Laszy et al. 2005; Richtand et al. 2001). Selective D_3 antagonists have been found to robustly enhance acetylcholine release in the frontal cortex, resulting in improvements in attention and working memory in rats (Dekeyne et al. 2004; Lacroix et al. 2003; Laszy et al. 2005). Research has also revealed that D_3 receptor agonists regulate γ-aminobutyric acid ($GABA_A$) receptor function, affecting inhibitory synaptic neurotransmission in the nucleus accumbens (Chen et al. 2006).

Interestingly, the antipsychotic agent S33138 has been characterized as an "optimized" antagonist at D_3 versus D_2 receptors, with 25-fold higher affinity for D_3 over D_2 receptors (Millan et al. 2008). It has been proposed that S33138 behaves as a pure D_3 antagonist at low concentrations of the agent, potentially improving cognitive and negative symptom dimensions of schizophrenia. At higher doses, S33138 blocks D_2 receptors, alleviating positive symptoms (Joyce and Millan 2005).

4 Serotonin 5-HT$_{2A}$ Receptors: Dopamine Brakes

Therapeutic agents for schizophrenia must target dysfunction in neurotransmission simultaneously in multiple brain regions, and in doing so must balance potentially competing therapeutic needs. For example, the treatment of positive psychotic symptoms requires a reduction of dopamine hyperactivity in the mesolimbic dopamine pathway. Unfortunately, a pure D_2 antagonist will act indiscriminately at D_2 receptors throughout the brain, thereby giving rise to unwanted hypoactivity in other dopamine pathways. In the mesocortical dopamine pathways, this hypoactivity may spell negative, cognitive, and affective symptoms, while in the nigrostriatal pathway, EPS may result. As pure D_2 antagonists, conventional antipsychotics therefore face a conundrum of "robbing Peter to pay Paul" – efficacy in treating positive symptoms comes at the price of negative symptom exacerbation and side effects.

Multifunctional agents that possess 5-HT$_{2A}$ antagonism among their roster of receptor activities may be able to sidestep this problem. Serotonergic neurons that innervate dopaminergic neurons modulate dopamine release either directly via somatodendritic and axodendritic 5-HT$_{2A}$ receptors on dopamine neurons or indirectly via 5-HT$_{2A}$ receptors on GABA interneurons. Antagonism of these 5-HT$_{2A}$ receptors can stimulate dopamine release in certain brain areas, including the prefrontal cortex and the striatum. Antipsychotics with both D_2 and 5-HT$_{2A}$ antagonist properties therefore have the ability to not only *decrease* dopamine activity by blocking D_2 receptors but also *increase* dopamine release by blocking 5-HT$_{2A}$. The latter action thus mitigates to some extent the absolute blockade associated with pure D_2 antagonism. Which action predominates is the subject of intense current investigation and appears to be specific to different regions of the brain. It is also still unclear whether the activity at 5-HT$_{2A}$ receptors is antagonism or inverse agonism.

Despite the questions surrounding the 5-HT$_{2A}$ receptor and its precise role in the mechanism of action of antipsychotics, it is already clear that the addition of 5-HT$_{2A}$ antagonist properties to D_2 antagonist properties yields a very different type of drug – an atypical antipsychotic with therapeutic actions not only on positive symptoms but also on negative, cognitive, and affective symptoms, with a significant reduction in the incidence of EPS and hyperprolactinemia. In essence, the 5-HT$_{2A}$ antagonism of atypical antipsychotics can be conceptualized as "fine-tuning" dopamine output in malfunctioning neuronal circuits. Clinicians can exploit these properties by individualizing drug selection and dosage to individual patients, since the exact balance of 5-HT$_{2A}$ antagonism versus D_2 antagonism differs between drugs, regions of the brain, and, indeed, between patients.

5 Serotonin 5-HT$_{1A}$ Receptors: Dopamine Accelerators

Another pharmacological mechanism that may contribute to the therapeutic profile of antipsychotics is the ability of some agents to act at 5-HT$_{1A}$ receptors either as full agonists or as partial agonists (serotonin partial agonists – SPAs).

The regulatory effect of 5-HT_{1A} receptors on dopamine release can be thought of as the converse of the influence exerted by 5-HT_{2A} receptors: whereas 5-HT_{2A} agonism acts as a dopamine brake, 5-HT_{1A} agonism acts as a dopamine accelerator. Activation of 5-HT_{1A} autoreceptors disinhibits dopamine release; in the striatum, this action lessens hypodopaminergia resulting from D_2 blockade and theoretically reduces the risk of EPS, while in the pituitary the same action potentially lowers the risk of hyperprolactinemia. In a similar vein, enhanced dopamine release in the prefrontal cortex could potentially improve negative, cognitive, and affective symptoms (Ichikawa et al. 2001). Thus, 5-HT_{1A} agonism has similar net effects to 5-HT_{2A} antagonism. In principle, if the two effects were copresent in a single pharmacological entity, the resulting therapeutic actions of the agent could be additive or synergistic (Stahl 2008a).

Currently, there is debate as to whether the ideal 5-HT_{1A} actions are agonist, partial agonist, or antagonist. Postsynaptic 5-HT_{1A} receptors in the cortex are most concentrated on the pyramidal neurons of the cortex and hippocampus, which suggests a role of the receptor in cognitive function (Azmitia et al. 1996). However, conflicting data have emerged from clinical experiments on the effects of SPAs on cognitive function. For example, Sumiyoshi et al. (2001) administered the SPA tandospirone as an adjunctive treatment to patients with schizophrenia who were receiving stable doses of typical antipsychotics and found that executive function and verbal memory were significantly improved in patients receiving tandospirone over patients who did not receive the drug. On the other hand, Yasuno et al. (2003) found that the administration of tandospirone dose-dependently impaired explicit verbal memory while leaving other cognitive functions intact, although the subjects in this experiment were healthy volunteers with no neuropsychiatric disorders. The results of preclinical studies in animals are also divided on the question of whether 5-HT_{1A} agonism or antagonism enhances cognition (Gray and Roth 2007a). In any event, the net long-term consequences of antagonism and agonism may be similar since antagonists block and chronic agonists downregulate 5-HT_{1A} receptors (Stahl 2008a).

6 Serotonin 5-HT_{2C} Receptors

Serotonin 5-HT_{2C} receptors are able to tonically regulate dopamine release via terminal receptors in the mesolimbic and nigrostriatal pathways (Alex and Pehek 2007). 5-HT_{2C} receptors are also found at high density on cell bodies in the ventral tegmentum, where they are believed to tonically inhibit dopamine release from dopaminergic neurons projecting to the frontal cortex (Millan et al. 1998). 5-HT_{2C} therefore presents an intriguing but challenging target for therapeutics: agonism at 5-HT_{2C} in the mesolimbic dopamine pathway may reduce dopaminergic transmission and improve positive symptoms of schizophrenia. A concern, however, is that dopaminergic suppression via 5-HT_{2C} agonism in the mesocortical pathway could potentially worsen cognition, while in the nigrostriatal pathway the same action might cause EPS (Gray and Roth 2007b; Alex et al. 2005; Gunes et al. 2008; Reavill et al. 1999). On the other hand, 5-HT_{2C} antagonism at inhibitory GABA interneurons in the brainstem may be expected to reduce depression and improve cognition by disinhibiting dopamine and norepinephrine release in the prefrontal cortex (Stahl 2007c, 2008a).

Receptors associated with weight gain are H_1 histamine receptors and 5-HT_{2C} receptors in the hypothalamus. When these receptors are blocked, particularly at the same time, patients may experience enhanced appetite in hypothalamic eating centers, and this increase in appetite may lead to weight gain. Antipsychotics associated with the greatest degree of weight gain are those that have the most potent antagonist actions simultaneously at H_1 and 5-HT_{2C} receptors.

Features such as the ones we have just described underscore the fact that, as with 5-HT_{1A} receptors, there is no consensus yet on the ideal actions of antipsychotics at 5-HT_{2C} receptors. 5-HT_{2C} agonists and antagonists both present potential rewards as well as pitfalls.

7 Serotonin 5-HT_7 Receptors

Serotonin 5-HT_7 receptors were discovered only 16 years ago, and their physiological role is still being elucidated (Bard et al. 1993; Lovenberg et al. 1993; Ruat et al. 1993). The relatively high density of 5-HT_7 in the limbic and thalamocortical regions hints at a possible role of the receptor in mediating depressive disorders. Downregulation of 5-HT_7 has been observed after chronic antidepressant treatment, and several antipsychotics and antidepressants (such as clozapine, zotepine, risperidone, pimozide, and fluoxetine) have been found to bind with high affinity at 5-HT_7 receptors (Mullins et al. 1999; Roth et al. 1994). It is still not conclusively known, however, whether and to what extent these actions contribute to therapeutic effects. Other potential roles for the 5-HT_7 receptor include regulation of circadian rhythms (Lovenberg et al. 1993) and sleep (Thomas et al. 2003), thermoregulation (Guscott et al. 2003; Hagan et al. 2000; Hedlund et al. 2003), learning and memory (Manuel-Apolinar and Meneses 2004; Roberts et al. 2004), and endocrine regulation (Jorgensen et al. 2003; Laplante et al. 2002).

8 Glutamatergic Receptors: NMDA and mGluR

In recent years, glutamate has attained a key theoretical role in the pathophysiology of schizophrenia. One hypothesis of schizophrenia suggests that early in the illness, excessive glutamate activity could lead to excitotoxicity, which might interfere with normal neurodevelopment and be linked to disease progression. However, it is now also widely hypothesized that once schizophrenia has developed, ionotropic *N*-methyl-D-aspartate (NMDA) glutamate receptors are actually hypofunctional. In any event, a multiplicity of glutamatergic receptors has provided a number of potential therapeutic targets from which to choose. The question of whether to use glutamate agonists or antagonists, however, is not a simple one to answer, and may depend on what stage of the disease is being treated and what symptoms are being targeted (Stahl 2008a). For example, antagonists of the NMDA receptor block the opening of the NMDA-coupled channel; in principle, this action has the potential to produce a neuroprotective effect by blocking excitotoxic downstream neurotransmission (Stahl 2008a). The consequences of this blockade, however, are complex and depend on the location of the NMDA receptor. For example, the blockade of NMDA receptors located postsynaptically on cortical GABA interneurons prevents the release of inhibitory GABA from the interneuron, which may result in the *disinhibition* of downstream glutamatergic neurons. In this instance, the action of an NMDA antagonist (e.g., ketamine) blocks glutamate receptors and yet *increases* glutamate release. This action has the potential short-term benefit of "rebooting" cortical glutamate regulation in patients with treatment-resistant depression; however, the same action also poses the risk of interfering with prefrontal cortical functioning and, over time, causing symptoms similar to the positive and negative symptoms of schizophrenia (Stahl 2008a).

Complicating this scenario is the fact that glutamatergic neurons also modulate dopaminergic neurotransmission. In the mesolimbic pathway, a descending cortico-brainstem glutamate pathway acts as a brake on dopamine release through an inhibitory GABA interneuron in the ventral tegmental area. NMDA hypoactivation in this circuit may therefore result in dopamine hyperactivity and give rise to positive symptoms of schizophrenia (Stahl 2007a). Similarly, blockade of NMDA receptors on glutamatergic neurons that synapse directly on dopaminergic neurons in the mesocortical pathway may produce dopamine hypoactivity in the dorsolateral and ventromedial prefrontal cortices, leading to negative, affective, and cognitive symptoms (Stahl 2007a). These mechanisms have given rise to the theory that NMDA receptors may be pathologically hypofunctional in untreated schizophrenia, similar to the effects produced by psychotomimetic agents such as phencyclidine and ketamine (Newcomer et al. 1999; Olney et al. 1999). The use of direct agonists of NMDA receptors to rectify NMDA hypoactivity, however, may not be clinically feasible because of the aforementioned risk of excess excitation of NMDA-coupled channels and resulting neurotoxicity (Gray and Roth 2007a).

Another class of glutamate receptors known as metabotropic glutamate receptors (mGluRs) regulates neurotransmission at glutamate synapses. Agents that act as agonists at presynaptic mGluR2 and mGluR3 (group II) autoreceptors may reduce glutamate release and be potentially useful as anticonvulsants and mood stabilizers, in addition to being neuroprotective against glutamate neurotoxicity (Stahl 2008a). Group II mGluRs also have been linked to attenuation of cognitive symptoms, such as deficits in working memory (Moghaddam and Adams 1998). A possible limitation to mGluR as a therapeutic target, however, is the suggestion that group II receptors may desensitize during chronic treatment (Javitt 2004).

9 Glycine Agonists

Much of the current targeting of the glutamate system in schizophrenia is aimed at increasing glutamatergic neurotransmission to compensate for NMDA hypoactivation. The challenge is to do so without producing neurotoxic effects; pharmacologic therapies for schizophrenia that involve direct enhancement of glutamate run the risk of excitotoxicity if glutamatergic neurotransmission becomes excessive (Stahl 2008a). A potentially safer way to enhance glutamate is to exploit the fact that NMDA glutamate receptors require allosteric binding by glycine (Johnson and Ascher 1987). In addition to glycine itself, the glycine site of the NMDA receptor also binds the endogenous agonist D-serine (Mothet et al. 2000), the partial agonist D-cycloserine (Goff et al. 2005; Heresco-Levy and Javitt 2004), and D-alanine (Tsai et al. 2006). These coagonists appear to modulate, but not activate, the NMDA receptor, and therefore seem to be free of excitotoxic effects associated with glutamate (Javitt 2006). Small-scale clinical trials of these agents in the treatment of schizophrenia have begun, and some evidence has emerged that glycine agonists may be able to reduce negative and/or cognitive symptoms. Of the four agents, D-cycloserine has been the least promising, possibly due to its partial agonism and the fact that it acts as an antagonist at high (100-mg) doses (van Berckel et al. 1999).

Another proposed approach to rectifying NMDA hypoactivity via glycine allosterism involves the glial glycine transporter GLYT1. GLYT1 is a reuptake pump that terminates the synaptic action of glycine by shuttling glycine back into glial cells; inhibition of this transporter should theoretically increase the synaptic availability of glycine, thus enhancing NMDA neurotransmission (Bergeron et al. 1998). In this manner, GLYT1 inhibitors are analogous to drugs that inhibit reuptake of other neurotransmitters such as serotonin-selective reuptake inhibitors (Stahl 2007b). Several GLYT1 inhibitors are now in clinical or preclinical testing, including the natural agent *N*-methylglycine (also known as *sarcosine*).

10 Receptors That Mediate Side Effects

In addition to dopaminergic, serotonergic, and glutamatergic effects, antipsychotics typically block a number of other receptors, including muscarinic cholinergic, histaminic, and adrenergic receptors. Blockade of these receptors can give rise to certain undesirable side effects: constipation, blurred vision, dry mouth, sedation, and cognitive dysfunction (via M_1 muscarinic cholinergic receptors); weight gain and sedation (via H_1 histaminic receptors); and orthostatic hypotension and sedation (via α_1-adrenergic receptors; Stahl 2008a). If antagonism of these receptors causes side effects, it is germane to the discussion of therapeutic design to ask whether the action of agonists at these same receptors might have therapeutic effects. This tack already has been followed by researchers looking for new therapeutic targets in schizophrenia. For example, the major active metabolite of clozapine, *N*-desmethylclozapine, has been reported to be a potent M_1 agonist and dopamine partial agonist with the ability to potentiate hippocampal NMDA activity as well (Burstein et al. 2005; Sur et al. 2003). As a result of these findings, N-desmethylclozapine (ACP-104) and other M_1 receptor agonists are in clinical trials as potential treatments of the cognitive dysfunction of schizophrenia (Gray and Roth 2007b).

11 Conclusion

Advances in neurobiology, particularly brain imaging techniques, have revealed that the various symptoms of schizophrenia and other psychiatric disorders are not only localized in unique regions of the brain, but also mediated by specific receptor sites. Treating the broad spectrum of symptoms of schizophrenia therefore needs to be viewed as an endeavor to treat the *whole* brain. In order to optimize and individualize treatments to address the unique symptoms of each patient, it is important to consider which symptoms are being expressed, and therefore which areas of the brain are hypothetically malfunctioning (Stahl 2008a). Each brain region has characteristic neurotransmitters, receptors, enzymes, and genes that regulate its activity, with substantial overlap and communication occurring with other brain regions. A solid understanding of these interrelationships can help the clinician in choosing appropriate medications that address the specific needs of the patient. Therapeutic strategies for schizophrenia that exploit the pharmacological multifunctionality of antipsychotics therefore represent the rational outcome of our improved understanding of the disease. Once a therapeutic agent (or agents) have been selected, it will furthermore be essential to remember that, in general, multifunctional drugs do not have the same potencies at each of their functions, and as we have seen with aripiprazole and certain agents that have been reported to be optimized antagonists at D_3 versus D_2 receptors, they may actually be very different drugs at low doses versus high doses. Dosage must therefore be adjusted properly in order to take full advantage of the pharmacological receptor profile of the antipsychotic and to achieve the therapeutic goals. When successful, this approach will take into account the entire receptor activity "portfolio" of the agent and provide a total therapeutic response that, like the elephant of the Buddhist parable, is greater than the sum of its parts.

Acknowledgments The authors wish to thank Daniel Lara Rios and Jahon Jabali for preparation of the figures.

References

Alex KD, Pehek EA (2007) Pharmacologic mechanisms of serotonergic regulation of dopamine neurotransmission. Pharmacol Ther 113:296–320
Alex KD, Yavanian GJ, McFarlane HG, Pluto CP, Pehek EA (2005) Modulation of dopamine release by striatal 5-HT2C receptors. Synapse 55:242–251

Azmitia EC, Gannon PJ, Kheck NM, Whitaker-Azmitia PM (1996) Cellular localization of the 5-HT1A receptor in primate brain neurons and glial cells. Neuropsychopharmacology 14:35–46

Bard JA, Zgombick J, Adham N, Branchek TA, Weinshank RL (1993) Cloning of a novel human serotonin receptor (5-HT7) positively linked to adenylate cyclase. J Biol Chem 268:23422–23426

Bergeron R, Meyer TM, Coyle JT, Greene RW (1998) Modulation of N-methyl-D-aspartate receptor function by glycine transport. Proc Natl Acad Sci USA 95:15730–15734

Berman RM, Marcus RN, Swanink R, McQuade RD, Carson WH, Corey-Lisle PK, Khan A (2007) The efficacy and safety of aripiprazole as adjunctive therapy in major depressive disorder: a multicenter, randomized, double-blind, placebo-controlled study. J Clin Psychiatry 68:843–853

Burris KD, Molski TF, Xu C, Ryan E, Tottori K, Kikuchi T, Yocca FD, Molinoff PB (2002) Aripiprazole, a novel antipsychotic, is a high-affinity partial agonist at human dopamine D2 receptors. J Pharmacol Exp Ther 302:381–389

Burstein ES, Ma J, Wong S, Gao Y, Pham E, Knapp AE, Nash NR, Olsson R, Davis RE, Hacksell U, Weiner DM, Brann MR (2005) Intrinsic efficacy of antipsychotics at human D2, D3, and D4 dopamine receptors: identification of the clozapine metabolite N desmethylclozapine as a D2/D3 partial agonist. J Pharmacol Exp Ther 315:1278–1287

Carlsson A, Waters N, Carlsson ML (1999) Neurotransmitter interactions in schizophrenia – therapeutic implications. Biol Psychiatry 46:1388–1395

Chen G, Kittler JT, Moss SJ, Yan Z (2006) Dopamine D3 receptors regulate GABAA receptor function through a phospho-dependent endocytosis mechanism in nucleus accumbens. J Neurosci 26:2513–2521

Dekeyne A, Di Cara B, Gobert A, Millan MJ (2004) Blockade of dopamine D3 receptors enhances frontocortical cholinergic transmission and cognitive function in rats. Soc Neurosci Abstr 29:776.4

Goff DC, Herz L, Posever T, Shih V, Tsai G, Henderson DC, Freudenreich O, Evins AE, Yovel I, Zhang H, Schoenfeld D (2005) A six-month, placebo-controlled trial of D-cycloserine co-administered with conventional antipsychotics in schizophrenia patients. Psychopharmacology (Berl) 179:144–150

Gray JA, Roth BL (2007a) Molecular targets for treating cognitive dysfunction in schizophrenia. Schizophr Bull 33:1100–1119

Gray JA, Roth BL (2007b) The pipeline and future of drug development in schizophrenia. Mol Psychiatry 12:904–922

Gunes A, Dahl ML, Spina E, Scordo MG (2008) Further evidence for the association between 5-HT2C receptor gene polymorphisms and extrapyramidal side effects in male schizophrenic patients. Eur J Clin Pharmacol 64:477–482

Guscott MR, Egan E, Cook GP, Stanton JA, Beer MS, Rosahl TW, Hartmann S, Kulagowski J, McAllister G, Fone KC, Hutson PH (2003) The hypothermic effect of 5-CT in mice is mediated through the 5-HT7 receptor. Neuropharmacology 44:1031–1037

Hagan JJ, Price GW, Jeffrey P, Deeks NJ, Stean T, Piper D, Smith MI, Upton N, Medhurst AD, Middlemiss DN, Riley GJ, Lovell PJ, Bromidge SM, Thomas DR (2000) Characterization of SB-269970-A, a selective 5 HT7 receptor antagonist. Br J Pharmacol 130:539–548

Hedlund PB, Danielson PE, Thomas EA, Slanina K, Carson MJ, Sutcliffe JG (2003) No hypothermic response to serotonin in 5-HT7 receptor knockout mice. Proc Natl Acad Sci USA 100:1375–1380

Heresco-Levy U, Javitt DC (2004) Comparative effects of glycine and D-cycloserine on persistent negative symptoms in schizophrenia: a retrospective analysis. Schizophr Res 66:89–96

Ichikawa J, Ishii H, Bonaccorso S, Fowler WL, O'Laughlin IA, Meltzer HY (2001) 5-HT2A and D2 receptor blockade increases cortical DA release via 5-HT1A receptor activation: a possible mechanism of atypical antipsychotic-induced cortical dopamine release. J Neurochem 76:1521–1531

Javitt DC (2004) Glutamate as a therapeutic target in psychiatric disorders. Mol Psychiatry 9:984–997

Javitt DC (2006) Is the glycine site half saturated or half unsaturated? Effects of glutamatergic drugs in schizophrenia patients. Curr Opin Psychiatry 19:151–157

Johnson JW, Ascher P (1987) Glycine potentiates the NMDA response in cultured mouse brain neurons. Nature 325:529–531

Jorgensen H, Riis M, Knigge U, Kjaer A, Warberg J (2003) Serotonin receptors involved in vasopressin and oxytocin secretion. J Neuroendocrinol 15:242–249

Joyce JN (2001) Dopamine D3 receptor as a therapeutic target for antipsychotic and antiparkinsonian drugs. Pharmacol Ther 90:231–259

Joyce JN, Millan MJ (2005) Dopamine D3 receptor antagonists as therapeutic agents. Drug Discov Today 10:917–925

Kapur S, Seeman P (2001) Does fast dissociation from the dopamine D2 receptor explain the action of atypical antipsychotics? A new hypothesis. Am J Psychiatry 158:360–369

Kirkpatrick B, Fenton WS, Carpenter WT Jr, Marder SR (2006) The NIMH-MATRICS consensus statement on negative symptoms. Schizophr Bull 32:214–219

Lacroix LP, Hows MEP, Shah AJ, Hagan JJ, Heidbreder CA (2003) Selective antagonism at dopamine D3 receptors enhances monoaminergic and cholinergic neurotransmission in the rat anterior cingulate cortex. Neuropsychopharmacology 28:839–849

Laplante P, Diorio J, Meaney MJ (2002) Serotonin regulates hippocampal glucocorticoid receptor expression via a 5-HT7 receptor. Brain Res Dev Brain Res 139:199–203

Laszy J, Laszlovszky I, Gyertyán I (2005) Dopamine D3 receptor antagonists improve the learning performance in memory-impaired rats. Psychopharmacology 179:567–575

Lawler CP, Prioleau C, Lewis MM, Mak C, Jiang D, Schetz JA, Gonzalez AM, Sibley DR, Mailman RB (1999) Interactions of the novel antipsychotic aripiprazole (OPC-14597) with dopamine and serotonin receptor subtypes. Neuropsychopharmacology 20:612–627

Lovenberg TW, Baron BM, de Lecea L, Miller JD, Prosser RA, Rea MA, Foye PE, Racke M, Slone AL, Siegel BW, Danielson PE, Sutcliff JG, Erlander MG (1993) A novel adenylyl cyclase-activating serotonin receptor (5-HT7) implicated in the regulation of mammalian circadian rhythms. Neuron 11:449–458

Mamo D, Graff A, Mizrahi R, Shammi CM, Romeyer F, Kapur S (2007) Differential effects of aripiprazole on D2, 5-HT2, and 5-HT1A receptor occupancy in patients with schizophrenia: a triple tracer PET study. Am J Psychiatry 164:1411–1417

Manuel-Apolinar L, Meneses A (2004) 8-OH-DPAT facilitated memory consolidation and increased hippocampal and cortical cAMP production. Behav Brain Res 148:179–184

Meltzer HY (2004) What's atypical about atypical antipsychotic drugs? Curr Opin Pharmacol 4:53–57

Millan MJ, Dekeyne A, Gobert A (1998) Serotonin (5-HT)2C receptors tonically inhibit dopamine (DA) and noradrenaline (NA), but not 5-HT, release in the frontal cortex in vivo. Neuropharmacology 37:953–955

Millan MJ, la Cour CM, Novi F, Maggio R, Audinot V, Newman-Tancredi A, Cussac D, Pasteau V, Boutin JA, Dubuffet T, Lavielle G (2008) S33138 [N-[4-[2-[(3aS,9bR)-8-cyano-1,3a,4,9b-tetrahydro[1]-benzopyrano[3,4-c]pyrrol-2(3H)-yl)-ethyl]phenylacetamide], a preferential dopamine D3 versus D2 receptor antagonist and potential antipsychotic agent. I. Receptor-binding profile and functional actions at G-protein-coupled receptors. J Pharmacol Exp Ther 324:587–599

Miyamoto S, Duncan GE, Marx CE, Lieberman JA (2005) Treatments for schizophrenia: a critical review of pharmacology and mechanisms of action of antipsychotic drugs. Mol Psychiatry 10:79–104

Moghaddam B, Adams BW (1998) Reversal of phencyclidine effects by a group II metabotropic glutamate receptor agonist in rats. Science 281:1349–1352

Mothet JP, Parent AT, Wolosker H, Brady RO Jr, Linden DJ, Ferris CD, Rogawski MA, Snyder SH (2000) D-Serine is an endogenous ligand for the glycine site of the N-methyl-D-aspartate receptor. Proc Natl Acad Sci USA 97:4926–4931

Mullins UL, Gianutsos G, Eison AS (1999) Effects of antidepressants on 5-HT7 receptor regulation in the rat hypothalamus. Neuropsychopharmacology 21:352–367

Newcomer JW, Farber NB, Jevtovic-Todorovic V, Selke G, Melson AK, Hershey T, Craft S, Olney JW (1999) Ketamine-induced NMDA receptor hypofunction as a model of memory impairment and psychosis. Neuropsychopharmacology 20:106–118

Olney JW, Newcomer JW, Farber NB (1999) NMDA receptor hypofunction model of schizophrenia. J Psychiatr Res 33:523–533

Reavill C, Kettle A, Holland V, Riley G, Blackburn TP (1999) Attenuation of haloperidol-induced catalepsy by a 5-HT2C receptor antagonist. Br J Pharmacol 126:572–574

Reavill C, Taylor SG, Wood MD, Ashmeade T, Austin NE, Avenell KY, Boyfield I, Branch CL, Cilia J, Coldwell MC, Hadley MS, Hunter AJ, Jeffrey P, Jewitt F, Johnson CN, Jones DN, Medhurst AD, Middlemiss DN, Nash DJ, Riley GJ, Routledge C, Stemp G, Thewlis KM, Trail B, Vong AK, Hagan JJ (2000) Pharmacological actions of a novel, high-affinity, and selective human dopamine D3 receptor antagonist, SB-277011-A. J Pharmacol Exp Ther 294(3):1154–1165

Richtand NM, Woods SC, Berger SP, Strakowski SM (2001) D3 dopamine receptor, behavioral sensitization, and psychosis. Neurosci Biobehav Rev 25:427–443

Roberts AJ, Krucker T, Levy CL, Slanina KA, Sutcliffe JG, Hedlund PB (2004) Mice lacking 5 HT7 receptors show specific impairments in contextual learning. Eur J Neurosci 19:1913–1922

Roth BL, Craigo SC, Choudhary MS, Uluer A, Monsma FJ Jr, Shen Y, Meltzer HY, Sibley DR (1994) Binding of typical and atypical antipsychotic agents to 5-hydroxytryptamine-6 and 5-hydroxytryptamine-7 receptors. J Pharmacol Exp Ther 268:1403–1410

Ruat M, Traiffort E, Leurs R, Tardivel-Lacombe J, Diaz J, Arrang JM, Schwartz JC (1993) Molecular cloning, characterization, and localization of a high-affinity serotonin receptor (5-HT7) activating cAMP formation. Proc Natl Acad Sci USA 90:8547–8551

Schmidt AW, Lebel LA, Howard HR Jr, Zorn SH (2001) Ziprasidone: a novel antipsychotic agent with a unique human receptor binding profile. Eur J Pharmacol 425:197–201

Schwartz J, Diaz J, Pilon C, Sokoloff P (2000) Possible implications of the dopamine D3 receptor in schizophrenia and in antipsychotic drug actions. Brain Res Brain Res Rev 31:277–287

Stahl SM (2007a) Beyond the dopamine hypothesis to the NMDA glutamate receptor hypofunction hypothesis of schizophrenia. CNS Spectr 12:265–268

Stahl SM (2007b) Novel therapeutics for schizophrenia: targeting glycine modulation of NMDA glutamate receptors. CNS Spectr 12:423–427

Stahl SM (2007c) Novel mechanism of antidepressant action: norepinephrine and dopamine disinhibition (NDDI) plus melatonergic agonism. Int J Neuropsychopharmacol 10:575–578

Stahl SM (2008a) Stahl's essential psychopharmacology: neuroscientific basis and practical applications, 3rd edn. Cambridge University Press, New York, NY

Stahl SM (2008b) Do dopamine partial agonists have partial efficacy as antipsychotics? CNS Spectr 13:279–282

Strange PG (2008) Antipsychotic drug action: antagonism, inverse agonism or partial agonism. Trends Pharmacol Sci 29:314–321

Sumiyoshi T, Matsui M, Nohara S, Yamashita I, Kurachi M, Sumiyoshi C, Jayathilake K, Meltzer HY (2001) Enhancement of cognitive performance in schizophrenia by addition of tandospirone to neuroleptic treatment. Am J Psychiatry 158:1722–1725

Sur C, Mallorga PJ, Wittmann M, Jacobson MA, Pascarella D, Williams JB, Brandish PE, Pettibone DJ, Scolnick EM, Conn PJ (2003) N-desmethylclozapine, an allosteric agonist at muscarinic 1 receptor, potentiates N-methyl-D-aspartate receptor activity. Proc Natl Acad Sci USA 100:13674–13679

Thase ME, Jonas A, Khan A, Bowden CL, Wu X, McQuade RD, Carson WH, Marcus RN, Owen R (2008) Aripiprazole monotherapy in nonpsychotic bipolar I depression: results of 2 randomized, placebo-controlled studies. J Clin Psychopharmacol 28:13–20

Thomas DR, Melotto S, Massagrande M, Gribble AD, Jeffrey P, Stevens AJ, Deeks NJ, Eddershaw PJ, Fenwick SH, Riley G, Stean T, Scott CM, Hill MJ, Middlemiss DN, Hagan

JJ, Price GW, Forbes IT (2003) SB-656104-A, a novel selective 5-HT7 receptor antagonist, modulates REM sleep in rats. Br J Pharmacol 139:705–714

Tsai GE, Yang P, Chang YC, Chong MY (2006) D-Alanine added to antipsychotics for the treatment of schizophrenia. Biol Psychiatry 59:230–234

van Berckel BN, Evenblij CN, van Loon BJ, Maas MF, van der Geld MA, Wynne HJ, van Ree JM, Kahn RS (1999) D-Cycloserine increases positive symptoms in chronic schizophrenic patients when administered in addition to antipsychotics: a double-blind, parallel, placebo-controlled study. Neuropsychopharmacology 21:203–210

Yasuno F, Suhara T, Nakayama T, Ichimiya T, Okubo Y, Takano A, Ando T, Inoue M, Maeda J, Suzuki K (2003) Inhibitory effect of hippocampal 5-HT1A receptors on human explicit memory. Am J Psychiatry 160:334–340

Zapata A, Shippenberg TS (2002) D3 receptor ligands modulate extracellular dopamine clearance in the nucleus accumbens. J Neurochem 81:1035–1042

Antipsychotic Dosing and Drug Delivery

Cara R. Rabin and Steven J. Siegel

Contents

1. Background 142
2. History 143
3. Dopamine Hypothesis 143
4. D$_2$ Mechanism of Antipsychotic Medication 144
5. Clozapine 145
6. Genesis and Interpretation of "Atypicality" 145
7. Current State of Affairs 146
8. Dosing 146
    - 8.1 Chlorpromazine Equivalents 146
    - 8.2 Other Contributions (Non-D$_2$) to Efficacy 149
9. Adherence 153
    - 9.1 Background and Clinical Perspective 153
    - 9.2 Development of Depots 154
    - 9.3 Development of Polymer-Based Microsphere Systems to Overcome Limitations of Chemistry 155
    - 9.4 Potential Extension to Implants 156
    - 9.5 Transdermal Delivery Systems 165
    - 9.6 Barriers to Development of Novel Delivery Systems 168
10. Summary and Conclusions 170
References 170

Abstract This chapter addresses the current state of affairs regarding proposed mechanism of action for antipsychotic medications and how this mechanism relates to dosing and delivery strategies. The initial portion describes the history of antipsychotic medication, including key discoveries that contribute to the dopamine

C.R. Rabin
Child Psychiatry Branch, National Institute of Mental Health, Bethesda, MD 20892, USA
S.J. Siegel (✉)
Translational Neuroscience Program, Department of Psychiatry, University of Pennsylvania, Philadelphia, PA 19104, USA
e-mail: siegels@mail.med.upenn.edu

hypothesis of schizophrenia and provide evidence that dopamine D_2 receptor antagonism remains the most copasetic explanation for both determination of dose and degree of efficacy for current antipsychotic medications. Early observations regarding the unique properties of clozapine and how those observations led to the misconception and misnomer of atypicality are also discussed. Subsequent sections relate the dosing of available medications using chlorpromazine equivalents, with a discussion of non-D_2-related mechanisms to antipsychotic effects. The balance of the chapter explores the temporal pattern of receptor occupancy as a key determinant of antipsychotic effectiveness, noting that continuous infusion would present the optimal method of treatment. In addition to the pharmacodynamic benefits of continuous long-term delivery systems, the incidence, causes, and clinical consequences of poor adherence are addressed. These observations are then discussed in the context of clinical studies and meta-analyses, demonstrating superiority of long-term depot preparations over oral administration. However, despite overwhelming evidence in favor of long-term delivery systems, few options are available to provide such ideal medication delivery profiles. Barriers to creating traditional depot preparations for a large number of antipsychotic agents, as well as efforts to address these limitations with polymer-based microspheres are described. The potential extension of current formulations to very long-term delivery implants using biodegradable and nonbiodegradable platforms is then described. Benefits as well as limitations of such systems are discussed with respect to clinical and ethical issues as well as a brief description of potential regulatory and logistic barriers to developing better delivery options. In summary, this chapter describes the basis for relating the dose of all existing antipsychotic medications to dopamine D_2 receptor affinity and the potential contribution of continuous occupancy to enhanced efficacy through superior biological effects and improved adherence.

Keywords Schizophrenia · Medication · Dopamine · D2 · Depot · Delivery

1 Background

Schizophrenia is among the most devastating of all illnesses, affecting approximately 1–2% of the world population. Age of onset is generally between 20 and 30 years with a chronic, unremitting disease course for the duration of the patient's life. Schizophrenia is characterized by two main symptom clusters, termed positive and negative, as well as lasting cognitive and functional disability. Positive symptoms include hallucinations, delusions, paranoia, and disorganized thinking. Negative symptoms include loss of interest and motivation, as well as loss of normal ability to both feel and express emotion. Lasting functional decline, a hallmark of the illness, leads to enormous personal, familial, and societal costs due to disability and lost productivity.

Although there have been effective medications for the positive symptoms of schizophrenia since the mid-1950s, there has been relatively little progress on effective treatments for the negative, cognitive, and functional domains. Recent large-scale multicenter studies and meta-analyses demonstrate that newer, formerly called atypical, medications have had little impact on improved efficacy, tolerability, or compliance with treatment. Additionally, nonadherence to existing medications significantly complicates treatment because the majority of patients do not take medication consistently. Future treatments will focus on both new mechanisms of action and improved delivery systems to target unmet needs in schizophrenia.

2 History

This chapter addresses historical and current approaches to the treatment of schizophrenia. Although schizophrenia remains among the most severe and debilitating illnesses known to medicine, its treatment has remained virtually unchanged for over 50 years due to a paucity of significant advances to our understanding of pharmacological targets. This is in part due to a lack of adequate animal models and preclinical screening tools to identify biological, mechanistic, or molecular therapeutic targets that address the pathophysiology of negative symptoms and cognitive deficits. Major concepts in this chapter include the use of dopamine receptor type 2 (D_2) antagonists for the treatment of psychosis and the limitation of the concept of "typical" versus "atypical" classifications for antipsychotic drugs. The final portion of the chapter will discuss several promising areas for future therapeutic advances, including improved drug delivery strategies that have the potential to effect rapid and dramatic improvements in schizophrenia outcomes.

The use of medicine for the treatment of schizophrenia dates back to 1952 with the serendipitous observation that chlorpromazine yielded symptomatic improvement when given as a preanesthetic agent (Delay et al. 1952a, b; Shen 1999). This was followed by the empiric discovery of other chemical entities with similar therapeutic effects. Approximately 10 years later, a seminal article by Seeman and colleagues documented that the clinically derived dose of all antipsychotic medications was highly correlated with the activity (IC_{50}) at striatal dopamine receptors (Fig. 1) (Seeman et al. 1975; Seeman and Lee 1975). A similarly important discovery resulted from complimentary evidence that agents that improved psychosis in schizophrenia resulted in increased dopamine metabolites in brain and cerebrospinal fluid (Carlsson and Lindqvist 1963; Sedvall 1980).

3 Dopamine Hypothesis

These observations in total were the basis for the dopamine hypothesis of schizophrenia, which remains as valid today as it was then. The addition of *in vivo* imaging with positron emission tomography (PET) allowed further confirmation

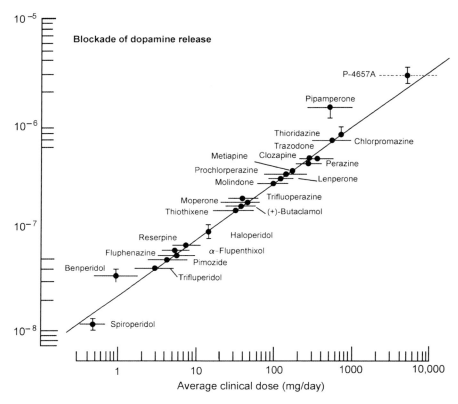

Fig. 1 The neuroleptic IC_{30} values (the neuroleptic concentrations which inhibited the stimulated release of dopamine by 50%) correlate with the average clinical doses (in mg/day) for controlling schizophrenia. The *horizontal bars* indicate the range of clinical doses. The *vertical bars* show the 20% variation in the IC_{30} values. The correlation includes such diverse compounds as phenothiazines, butyrophenones, reserpine, pimozide, trazodone, clozapine, and (+)-butaclamol. *Trans*-thiothixene (P-4657A) is of the order of one-hundredth the potency of its *cis*-isomer (Seeman 1975, p. 5)

that therapeutic doses of antipsychotic agents resulted in approximately 70–80% dopamine D_2 receptor occupancy (Farde et al. 1986, 1988). It must be noted, however, that all of these observations remain correlations and dose finding for these medications is empirical rather than based on receptor affinity.

4 D_2 Mechanism of Antipsychotic Medication

The most obvious characteristic of schizophrenia is the positive symptom cluster, so named for the presence of behaviors or sensations that should not be present. However, these symptoms could also be called "positive" because there is a positive response to medication. That is, most available medications will dramatically reduce these symptoms in most patients, most of the time. This is true for both the majority of older, "typical" agents and to a variable degree for newer, "atypical" ones (Davis et al. 2003; Geddes et al. 2000). The treatment of positive symptoms appears to be related to antagonism of the dopamine D_2-type receptor. The D_2 receptor is a membrane-bound G protein (guanine nucleotide-binding protein) that is linked to inhibition of adenylyl cyclase through its α-subunit. When dopamine binds to this receptor subtype, it leads to a cascade of intracellular events that decrease adenylyl cyclase activity, which in turn decreases cyclic adenosine monophosphate (cAMP) formation and reduces activation of protein kinase A (PKA, cAMP-dependent protein kinase). This in turn reduces phosphorylation of many intracellular substrates including ion channels and DNA-binding proteins. Thus, in the presence of a D_2 receptor antagonist, this process is reduced and there is an increase in intracellular cAMP and its downstream effects.

5 Clozapine

The next major advance in the treatment of psychosis resulted from the observation by Kane et al. (1988) that clozapine appeared to have superior efficacy as compared with other agents among a population of treatment-resistant patients. This observation was validated during the CATIE study, in which the time to treatment discontinuation for any reason was significantly longer for clozapine than for the other newer antipsychotics (Tandon et al. 2006). One notable feature of clozapine was that it also appeared to achieve significant improvement without the typical motor side effects seen with most other agents.

6 Genesis and Interpretation of "Atypicality"

This lack of motor side effects at therapeutic doses led to the term "atypical antipsychotic" for clozapine. Unfortunately, the term "atypical" medication took on broader meanings as it was applied to newer medications over time. For example, many theories have been put forward to suggest why clozapine displays efficacy without the same degree of dopamine blockade as other drugs. However, a close inspection of the data presented in the landmark Seeman and Lee article clearly show that clozapine sits squarely on the same correlation line as all other drugs for dopamine blockade and dose. Thus, the observation is more accurately stated that clozapine appears to have a lower degree of motor side effects despite the same level of dopamine blockade as other medications. Much attention was then placed on decomposing the receptor-binding profile for clozapine to discover which receptor-binding properties might contribute to the relative lack of motor side effects. This led to initial hypotheses that the ratio of serotonin type 2 ($5HT_2$) to dopamine type 2 (D_2) receptor binding was a key feature of atypicality (Ereshefsky et al. 1989; Nordstrom et al. 1993, 1995).

7 Current State of Affairs

However, the aforementioned theory has failed to yield any medications that match clozapine in efficacy. Additionally, chlorpromazine (the first antipsychotic medication) actually meets this and many other definitions of "atypicality" since it has significant $5HT_2$ and D_2 binding at clinical doses (Trichard et al. 1998). A number of other theories have been proposed in an attempt to retain the idea that newer medications are categorically, mechanistically, or clinically distinguished from older ones. For example, one theory poses that "atypical" agents have less prolactin elevation than typical ones (Rauser et al. 2001). However, risperidone, which is considered atypical, has among the highest prolactin elevation among all antipsychotic medications, and chlorpromazine, the original agent, has only moderate increases (Cheer and Wagstaff 2004; Gruen et al. 1978; Kinon et al. 2003; Langer et al. 1977).

Similarly, many authors have hypothesized that "atypical" agents might improve cognition (Meltzer and McGurk 1999). However, several studies have not found cognitive benefits for newer agents over older ones and a recent large-scale comparison of newer agents versus the "typical" agent perphenazine showed that the older, typical agent resulted in a slight but significant advantage over newer drugs for cognitive improvement (Keefe et al. 2005, 2006; Siegel et al. 2006). Other definitions have similarly failed to distinguish any class-like properties among older (prerisperidone) and newer (risperidone and newer) agents. Table 1 lists a summary of antipsychotic agents available in the United States and Table 2 lists the receptor affinity for a subset of these medications (Seeman et al. 1997).

8 Dosing

8.1 Chlorpromazine Equivalents

The term "chlorpromazine equivalents" indicates that doses of all antipsychotic medications can be normalized to a single scale based on potency. The scale is defined by how the clinically efficacious dose of each medication compares to the efficacious dose of all others. By convention, chlorpromazine is used as the standard because it was the first agent. For example, the system correctly notes that 100 mg of chlorpromazine is about as efficacious as 0.5 mg of haloperidol. Thus, haloperidol is 0.02 (2%) chlorpromazine equivalents (e.g., 500 mg of chlorpromazine will work about as well as 5 mg of haloperidol). All antipsychotic

Table 1 List of antipsychotic medications

Medication	Category	Class	Metabolism	Prominent side effects	Prolactin elevation	Daily dose range (mg)	Frequency	Chlorpromazine equivalents
Aripiprazole	Newer	Quinolinone	CYP450 2D6, 3A4	Insomnia, agitation, anxiety	Mild	10–15	Daily	0.02
Chlorpromazine	Older	Phenothiazine	CYP450 2D6	Sedation, hypotension, shuffling gait	Mild	200–800	Three times daily	1.00
Clozapine	Older	Dibenzodiazepine	CYP450 1A2, 2D6, 3A4	Sedation, tachycardia, sialorrhea	Low	300–600	Twice daily	0.75
Fluphenazine	Older	Phenothiazine	CYP450 2D5 2D6	Nausea, EPS	Moderate	0.5–10	Daily	0.01
Fluphenazine decanoate	Long-acting	Phenothiazine with 10 carbon ester linkage	CYP450 2D6	Nausea, EPS*	Moderate	*12.5–100 per month	Monthly	0.00
Haloperidol	Older	Butyrophenone	CYP450 2D6, 3A4, 2D6 inhibitor	EPS, akethesia	Moderate	1–15	Twice daily	0.02
Haloperidol decanoate	Long-acting	Butyrophenone with 10 carbon ester linkage	CYP450 2D6, 3A4	EPS, akethesia*	Moderate	50–100 monthly	Monthly	0.005
Loxapine	Older	Dibenzodiazepine	CYP450 unknown	EPS, sedation	Moderate	60–100	Twice daily	0.13
Molindone	Older	Dihydroindolene	CYP450 unknown	Sedation, EPS	Mild	50–200	Three times daily	0.25
Olanzapine	Newer	Thienobenzodiazepine	CYP450 1A2, 2D6, 2C18	Weight gain, constipation	Mild	10–20	Daily	0.03
Olanzapine – rapidly dissolving	Newer	Thienobenzodiazepine	CYP450 1A2, 2D6, 2C19	Weight gain, constipation	Mild	10–20	Daily	0.03
Perphenazine	Older	Phenothiazine	CYP450 2D6	EPS, sedation	Mild	8–64	Twice daily	0.08

(continued)

Table 1 (continued)

Medication	Category	Class	Metabolism	Prominent side effects	Prolactin elevation	Daily dose range (mg)	Frequency	Chlorpromazine equivalents
Pimozide	Older	Diphenylbutylpiperidine	CYP450 1A2, 3A4	Akethesia, EPS	Moderate	2–10	Twice daily	0.01
Quetiapine	Newer	Dibenzothiazepine	CYP450 3A4	Sedation	Mild	150–750	Twice to three daily	0.94
Risperidone	Newer	Benzisoxazole	CYP450 2D6	EPS, hypotension, hyperprolactinema	Severe	1–4	Daily	0.01
Risperidone microspheres	Long-acting	Risperidone in PLGA microspheres	CYP450 2D6	EPS, hypotension, hyperprolactinema*	Severe	25–50 every 2 weeks	Bimonthly	0.00
Thioridazine	Older	Phenothiazine	CYP450 2D6	Sedation, dry mouth, constipation	Mild	200–800	Twice to three daily	1.00
Thiothixene	Older	Thioxanthene	CYP450 unknown	Akethesia, EPS	Moderate	4–15	Twice daily	0.02
Trifluoperazine	Older	Phenothiazine	CYP450 unknown	Hypotension, EPS	Moderate	4–10	Twice daily	0.01
Ziprasidone	Newer	Benzothiazolylpiperazine	CYP450 3A4	Sedation, headache	Mild	80–160	Twice daily	0.20

Table 2 Radioligand-independent dissociation constants

	D_2	D_4	$5HT_{2A}$	D_2:$5HT_{2A}$ ratio	D_2:D_4 ratio
Chlorpromazine	0.66 ± 0.05(12)	1.15 ± 0.04(9)	3.5 ± 0.06(9)	0.19	0.58
Clozapine	44 ± 8(27)	1.6 ± 0.4(96)	11 ± 3.5(9)	4.00	28.00
Fluphenazine	0.32 ± 0.03(7)	50 ± 10(11)	80 ± 19(6)	0.004	0.0064
Haloperidol	0.35 ± 0.05(18)	0.84 ± 0.05(54)	25 ± 8(5)	0.014	0.42
Molindone	6 ± 3(9)	2,400 ± 800(11)	5,800 ± 1,300(6)	0.001	0.0025
Olanzapine	3.7 ± 0.6(12)	2 ± 0.4(22)	5.8 ± 0.7(14)	0.64	1.85
Risperidone	0.3 ± 0.1(19)	0.25 ± 0.1(17)	0.14 ± 0.1(5)	2.14	1.2
Seroquel	78 ± 28(13)	3,000 ± 300(14)	2,500 ± 600(5)	0.03	0.026
Thioridazine	0.4 ± 0.12(12)	1.5 ± 0.5(16)	60 ± 15(6)	0.007	0.27
Trifluoperazine	0.96 ± 0.2(11)	44 ± 6(11)	135 ± 50(6)	0.007	0.022

In nM ± SE (*n* experiments in duplicate)

medications can be described in this manner using the maximum or average doses. There is a very high correlation between chlorpromazine equivalents and binding affinity at the dopamine D_2 receptor because antipsychotic efficacy is essentially a function of D_2 antagonism. However, it is a misconception to think that equivalents are based on, or determined by binding affinities instead of clinical, empirical data. This system allows clinicians to estimate the dose of a new medication based on the dose a patient took of a previous one. For example, a patient switching from 4 to 6 mg of risperidone would need 15–20 mg of olanzapine or 8–12 mg of haloperidol for equivalent effect.

Although there were some efforts to move away from using the term chlorpromazine equivalents for newer agents, there is no rationale to distinguish older or newer medications on this scale. The movement to create a new system was driven by several factors including the misconception that equivalents reflected something other than the empirical dose used in practice; a misconception that D_2 antagonism was less relevant in newer agents than older ones; and an agenda to support the illusion that new agents acted through a fundamentally different mechanism than older ones. Since there are now overwhelming data that newer and older antipsychotic medications are indistinguishable in efficacy or mechanism, the system of equivalents is just as valid today as it was 30 years ago. Table 1 shows chlorpromazine equivalents for available antipsychotic medications.

8.2 Other Contributions (Non-D_2) to Efficacy

As noted above, early efforts to recreate clozapine focused on D_2 to $5HT_2$ binding ratio. However, this approach has not yielded medications of comparable efficacy (Davis et al. 2003). Therefore, it may be prudent to examine other receptor-binding properties that contribute to the overall effect of antipsychotic medications and clozapine in particular. Clozapine displays significant affinity for non-D_2 dopaminergic receptors, α-adrenergic receptors, serotonergic, and cholinergic receptors. In particular, clozapine possesses a high affinity for α_2-adrenceptors (Ashby and Wang 1996). Activity at α-adrenergic receptors has been suggested to improve negative and cognitive symptoms via α_2-receptor antagonism and positive symptoms through α_1-adrenoceptor antagonism. α_2-adrenoceptor blockage may act by augmenting and improving prefrontal dopaminergic functioning, which some authors have suggested may contribute to proposed efficacy for negative symptoms (Svensson 2003).

Because the prominent α-adrenoceptor antagonistic action of clozapine may contribute to its efficacy in schizophrenia, some agents currently in the premarketing stage of development demonstrate similar α-adrenergic receptor antagonism. One such drug, asenapine, appears to have relatively high affinity for adrenergic receptors and binds $5HT_{2A}$ and $5HT_{2C}$ receptors with higher affinity than D_2, which has been proposed to have potential benefits for negative symptoms (Bishara and Taylor 2008). It must, however, be noted that this remains an untested hypothesis in real-world applications, since neither clozapine nor asenapine have demonstrated clinically meaningful effects on negative symptoms. Another compound in development, iloperidone, a broad-spectrum dopamine, serotonin, and noradrenaline antagonist, binds with a high affinity to α_1-adrengeric receptors and $5HT_{2A}$ receptors, as well as having moderate affinity for D_2 receptors, resulting in doses of 24–48 mg/day in clinical trials (Kalkman et al. 2001). This constellation of binding profiles at multiple receptor systems is similar to existing midpotency agents such as perphenazine and trifluoperazine.

Observations that noncompetitive antagonists of the NMDA receptor can induce a psychosis similar to schizophrenia in healthy subjects imply that dysfunction of the glutamatergic system could play a role in the disease pathophysiology (Javitt et al. 1987). While direct glutamate agonists are unlikely to be suitable as antipsychotic agents because of their excitotoxic effects, agents with agonist activity at the glycine site (an obligatory coagonist) have been proposed to have potential benefits by enhancing glutamatergic function (Sodhi et al. 2008). Several clinical trials of coagonists at the glycine site suggest promise for this approach as adjunctive therapy (Heresco-Levy et al. 2005; Tsai et al. 2006). Similarly, metabotropic glutamate receptors have emerged as another potential target for the treatment of schizophrenia based on the hypothesis that these receptors act presynaptically to influence glutamate release. Preliminary results from a clinical study using the mGluR2/3 agonist LY2140023 support this approach. When evaluated in a randomized, three-armed, double-blind, placebo-controlled study of 196-Russian patients, LY2140023 was found to be safe and well tolerated. Patients showed statistically significant improvements in both positive and negative symptoms compared with placebo, though the traditional D_2 antagonist olanzapine performed better when overall clinical response was measured (Patil et al. 2007). Although these data serve as early indication that mGlu2/3 receptor agonists may have novel antipsychotic properties, it is yet to be confirmed that LY2142003 does not act, at least in part, through an indirect downstream dopaminergic mechanism (Harrison 2008).

Although there is some exploration of non-D_2 contributions to antipsychotic efficacy, the majority of emerging drugs have not been proven to act through mechanisms that are distinct from the familiar $D_2/5HT_2$ binding or to satisfy the current clinical need for negative and cognitive symptom alleviation in schizophrenia. Thus, while the field searches for improved drug targets and mechanisms, developments in medication delivery can provide a complementary avenue toward fulfilling unmet needs in schizophrenia.

8.2.1 Potential Benefits of Continuous Infusion

We propose that continuous infusion from antipsychotic implants may have biological benefits leading to superior clinical outcomes. Multiple studies indicate that intermittent adherence to antipsychotic medication leads to reduced efficacy (Keith and Kane 2003). Dosing an antipsychotic medication less frequently than its plasma half-life yields wide fluctuations in drug levels with reduced efficacy (Borgman 1982; Ereshefsky and Mascarenas 2003). This is compounded by interindividual variation in plasma half-life for each drug, complicating the ability to achieve steady-state levels with a unitary dosing regimen (i.e., once a day or twice a day for every person). The average serum half-life for risperidone is approximately 14 h, suggesting that optimal dosing should occur approximately twice a day to maintain steady state (Borison et al. 1994). However, many patients choose to take their medication once a day, resulting in wider peaks and troughs than are ideal. Higher peak values result in excess side effects, and lower troughs lead to periods of unopposed disease process (i.e., increased dopamine availability at the D_2 receptor). Indeed, fluctuations in drug level may lead to a continuously dynamic state, with alternating periods of hyperdopaminergic tone at the receptor during drug-level troughs and excessive D_2 blockade during drug-level peaks. We propose that such fluctuations at the D_2 receptor impede the ability of the postsynaptic cell to arrive at a stable state of intracellular signal transduction. As noted above, the dopamine D_2 receptor is an inhibitory G protein-coupled receptor (G_i). Binding of dopamine to D_2 receptors reduces adenylyl cyclase activity, resulting in downregulation of cAMP production and a reduction in PKA activity. PKA is responsible for phosphorylation of multiple intracellular substrates with resulting early functional changes (e.g., receptor phosphorylation) and long-term adaptations in gene expression through the cAMP responsive element (CRE) (Fig. 2). Changes in gene expression then allow the postsynaptic cell to modify its activity and connectivity. This cascade of intra- and intercellular events allows for therapeutic remodeling throughout the brain. Thus, variation in antipsychotic drug level from oral administration and resulting fluctuations in D_2 blockade impedes the ability of the postsynaptic cell to establish homeostasis with respect to dopamine tone. Alternatively, steady-state infusion of antipsychotic medication may provide constant modulation of dopaminergic tone, thereby allowing the brain to remodel in a stable therapeutic environment.

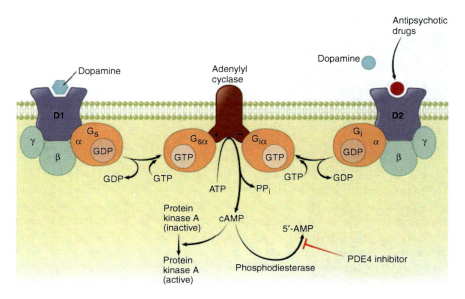

Fig. 2 The DA receptor-coupled adenylyl cyclase (AC)–cyclic adenosine monophosphate (cAMP) pathway. All current antipsychotic drugs are believed to work via antagonism at the G protein-coupled D_2 dopamine receptor. Whereas D1/5 type DA receptors increase the activity of AC, D_2 dopamine receptors decrease AC activity. AC is an enzyme that transforms adenosine triphosphate (ATP) to cAMP, which is a ubiquitous intracellular second messenger. Furthermore, cAMP levels are tightly controlled by phosphodiesterase (PDE) activity because PDEs degrade the cAMP molecule to 5′-AMP

8.2.2 Alternative Hypothesis

Although the authors propose that continuous administration of antipsychotic medication will have biological benefits relative to intermittent oral administration, the counterargument has been advocated by Kapur and colleagues (Kapur and Seeman 2001; Kapur et al. 2000; Samaha et al. 2008). According to this line of reason, chronic antipsychotic treatment leads to long-term upregulation of postsynaptic D_2 receptors, which results in both motor side effects and breakthrough psychosis. Ginovart and colleagues modeled clinically salient chronic administration of antipsychotics in cats through the continuous infusion of haloperidol at 0.25 mg/kg to produce high and stable (exceeding 78%) levels of D_2 receptor occupancy. They observed that this dosing paradigm created a robust upregulation of striatal D_2 receptors and induced the development of behavioral tolerance to the effect of haloperidol on spontaneous locomotor activity (Ginovart et al. 2009). The authors postulated that continuously high levels of D_2 receptor occupancy result in D_2 receptor upregulation, which in turn is responsible for the development of drug tolerance, loss of antipsychotic efficacy, and the development of tardive dyskinesia. Additionally, Kapur and colleagues suggest that whereas tolerance develops following continuous administration, reverse tolerance – sensitization – follows intermittent antipsychotic administration because the drug is allowed to wear off between repeated administrations (Barnes et al. 1990; Csernansky et al. 1990; Ezrin-Waters and Seeman 1977; See and Ellison 1990). Future studies will examine these two opposing perspectives using animal models of antipsychotic efficacy and side effects, as well as molecular measures of drug response.

9 Adherence

9.1 Background and Clinical Perspective

9.1.1 Adherence Rates

Nonadherence with medication remains a primary obstacle to better treatment outcomes in schizophrenia. Although the acceptance of a prescribed pharmacological regimen often dictates treatment outcome, more than 35% of patients demonstrate adherence problems during the first 4–6 weeks of treatment and within 2 years, up to 74% of patients are unable to adhere to their original prescribed treatment (Gilmer et al. 2004; Lieberman et al. 2005; Valenstein et al. 2004). The short- and long-term consequences of nonadherence to pharmacological treatment include increased relapses, more frequent and longer hospitalizations, cumulative deteriorations in functioning, and a diminished capacity to maintain employment and relationships. The more relapses a patient has, the more difficult it is to achieve remission after the next episode, creating a negative feedback loop that severely hinders long-term functioning and quality of life (Turner and Stewart 2006).

9.1.2 Relapse Rates as a Function of Adherence

Nonadherence to antipsychotic medication treatment leads to several clinically relevant outcomes, including psychotic relapse, increased clinic and emergency department visits, rehospitalization, and deterioration in social function (Adams and Howe 1993; Bergen et al. 1998; Casper and Regan 1993; Cramer 1998; Rittmannsberger et al. 2004; Valenstein et al. 2001; Viguera et al. 1997; Weiden and Olfson 1995). There is overwhelming evidence for an increased risk of recurrent psychotic episodes with interruptions in antipsychotic therapy, even in the early course of the disorder. Nonadherence to pharmacological intervention, then, leads to a fivefold chance of increased relapse. In a study performed by Johnson (1981), patients who suffered a relapse did not return to their prelapse level of social adjustment 1 year after recovery. This finding emphasizes that the clinical and societal costs of relapse are much greater than just the cost of rehospitalization (Marder 2003). Furthermore, the functional gap present between the adherent and nonadherent groups at baseline widens over time as a result of sustained improvements among adherent patients and decline among nonadherent patients. These differential outcome trajectories bode well for the adherent group, but signal alarm for nonadherent patients (Ascher-Svanum et al. 2006).

9.1.3 Causes of Poor Adherence

Although there are numerous factors underlying nonadherence in schizophrenia, the primary reasons that determine the degree to which a patient adheres to treatment include comorbid substance abuse, cognitive impairment, and forgetfulness, as well as unclear understanding of one's illness associated with lack of insight (Nasrallah and Lasser 2006). Approximately, 60% of patients with schizophrenia have a comorbid substance abuse disorder (Kane et al. 2003). Substance abuse is a significant factor precipitating psychotic episode relapse and is also positively correlated with poor treatment adherence (Castle and Pantelis 2003). Another considerable factor in the degree of patient adherence arises from inherent cognitive deficits and working memory impairment of patients with schizophrenia. As many as 75% of schizophrenia patients demonstrate some form of deficiency on neuropsychological test batteries (Kasper 2006). Cognitive domains implicated in schizophrenia include working memory, attention, learning of new information, and cognitive flexibility, all of which may be barriers to adhering to a complicated oral antipsychotic regimen (Palmer and Jeste 2006).

Comorbid substance abuse and cognitive impairment may impact a patient's ability to adhere to treatment, but most psychiatrists contend that lack of insight is the most salient factor in patient nonadherence (McEvoy et al. 1989a). Recent studies suggest that poor insight is causally related to medication nonadherence, stressing the direct linear relationship between the two variables. Poor insight in psychosis is described as lacking awareness of the deficits, consequences of the disorder, and need for treatment (David 1990; McEvoy et al. 1989b). Essentially, the patient's acknowledgment of his illness is significantly associated with the level of medication adherence (McEvoy et al. 1989b).

9.2 Development of Depots

9.2.1 Clinical Studies and Meta-Analyses Comparing Depots to Oral Administration

In an effort to improve medication adherence, researchers in the 1960s introduced long-lasting depot antipsychotic drugs, consisting of an ester of antipsychotic agent in a fat-soluble solution (Adams et al. 2001). Because they provide lower steady-state therapeutic drug concentrations and more sustained release characteristics compared with oral medications, depot formulations are associated with a lower propensity to induce side effects (Keith 2006). Thus, depot formulations augment the impact of oral formulations by providing further reductions in morbidity and mortality (Kane et al. 2003). Because they bypass the gastrointestinal tract, depot treatments decrease the amount of medication needed and may minimize certain peripheral side effects including hepatotoxicity and hyperprolactinemia (Knox and Stimmel 2004).

9.2.2 Barriers to Creating Depots for More Agents

However, depot formulations have limitations that restrict more significant improvements in adherence and efficacy. Decanoate treatment is limited by chemistry, as many compounds are unable to make the required ester linkages. Additionally, depot injections are irreversible and can result in prolonged pain at the injection site, which increases the likelihood that patients will discontinue treatment (Bloch et al. 2001; Kane et al. 1998).

9.3 Development of Polymer-Based Microsphere Systems to Overcome Limitations of Chemistry

Therefore, additional drug delivery systems have been developed to overcome the limitations of depots, improve the efficiency of drug delivery, and minimize toxic side effects (Johnson 1984; Lambert et al. 2003). Biodegradable microparticles have received a great deal of attention because of their ability to administer a wide range of drugs over longer periods of time than conventional drug delivery and depots. Their initial promise for oral delivery was dampened by the fact that there was a size limit of less than 10 mm for the particles to cross the intestinal lumen into the lymphatic system (Lemoine and Preat 1998; Torche et al. 2000). However, there are several other routes through which microparticles are a useful drug delivery system. Subcutaneous or intramuscular administration of microparticles is less constrained in terms of the size limits than those required to be suitable for oral delivery or intravenous delivery of particles (Rafati et al. 1997). In these injectable depot formulations, one can take advantage of the sustained release from microparticles to release small molecular weight drugs, peptides, and proteins (Berkland et al. 2002; Oh et al. 2006). Drugs such as gentamicin, and vaccines for mucosal immunity, can be delivered intranasally (Lim et al. 2002; Spiers et al. 2000). Microparticles have also proven effective for pulmonary administration (Sharma et al. 2001).

Pharmacokinetic studies have found that risperidone depot produces total drug exposure (the area under the time–plasma concentration curve) that is similar to that of oral risperidone, but with lower peak plasma concentration and less peak-to-trough variability in plasma drug concentration (Love 2002). The efficacy and safety of extended-release risperidone were recently examined in a large, randomized, double-blind, placebo-controlled clinical trial (Kane et al. 2003). A total of 400 patients with schizophrenia were randomized to biweekly injections of either placebo or one of three doses of long-acting risperidone (25, 50, or 75 mg). Placebo-treated patients exhibited a slight worsening from baseline (an increase of 2.6 points) on the positive and negative symptom scale (PANSS), the primary study endpoint. In contrast, patients in the three risperidone groups improved from baseline by an average of 6.2, 8.5, and 7.4 points for the 25, 50, and 75 mg dose groups, respectively. Treatment with long-acting risperidone was well tolerated by the patients, and local injection-site reactions were rarely rated as painful. Motor symptoms were no more common in the low-dose risperidone group (10%) of patients than with placebo (13%). The incidence of such side effects was greater in the two higher-dose risperidone groups.

More recent research suggests that it may be possible to combine the more favorable efficacy and tolerability profile of risperidone with a more convenient long-lasting dosing formulation. A new formulation of risperidone was recently developed for biweekly administration. This extended-release risperidone formulation uses biodegradable drug-containing microspheres that are manufactured from poly(lactic-*co*-glycolic acid) (PLGA). PLGA is a copolymer used for a wide variety of medical applications, due to its biocompatibility and nontoxicity. The United States Food and Drug Administration approved this product for clinical use in late 2003.

9.4 Potential Extension to Implants

9.4.1 Biodegradable Versus Nonbiodegradable

Nondegradable

One of the goals for the development of a long-term delivery system is to reduce the side effects associated with the peaks that result from conventional oral dosing. For the antipsychotic haloperidol, this is particularly important due to the emergence of motor side effects at higher plasma concentrations. Ideally, a continuous, low concentration of haloperidol would decrease these side effects. To produce a long-term delivery system for psychoactive compounds, several nonbiodegradable polymer systems were previously developed. Although these systems have not yielded therapeutic products, some are worth mentioning in part due to their similarity to biodegradable delivery systems currently in development. Implants were made in a previous study using the nondegradable ethylene vinyl acetate (EVA) copolymer with 37.5% haloperidol (wt/wt) (Kohler et al. 1994). After 3 weeks of *in vivo* studies in rats, similar dopamine supersensitivity was seen in the rats with haloperidol implants as well as the rats receiving daily doses of haloperidol. The major limitation to such nondegradable systems is the need for implant removal at the end of the delivery interval. Thus, biodegradable systems may offer more flexibility due to the ability to provide similar intervals of delivery without the need for eventual removal (Siegel et al. 2002).

Biodegradable Systems

Biodegradable polymers retain their properties for a limited period of time *in vivo* and then gradually degrade into materials that can become soluble or are metabolized and excreted from the body. Biodegradable polymers have some advantages for the purpose of drug delivery since there is no need for surgical removal of the exhausted polymer matrix. In order to be used for *in vivo* applications, the polymers used for such systems must have favorable properties for biocompatibility, processability, sterilization capability, and shelf life.

9.4.2 PLGA/PLA

Many biodegradable systems rely on the polymers of PLGA or poly(lactic acid) (PLA). These classes of polymers are highly biocompatible and have good mechanical properties for drug delivery applications (Kitchell and Wise 1985). The biomedical use of PLGA has been reported since the 1960s (Kulkarni et al. 1966). Numerous systems already utilize PLGA and PLA to successfully achieve long-term delivery, including several microparticle and nanoparticle systems, as well as devices to control thyrotropin-releasing hormone in controlling metabolism (Okada and Toguchi 1995), L-dopa to treat Parkinson's disease (Sabel et al. 1990), and naltrexone in treating narcotic addiction (Sharon and Wise 1981). Several intraocular systems, including Vitrasert (Bausch and Lomb), offer biocompatible delivery systems with controlled-release drug therapy for periods ranging from several days up to 1 year. In addition, PLA and PLGA have been approved by the FDA for clinical applications, such as sutures, bone plates, and abdominal mesh. Indeed, the major rationale for using PLGA in a variety of existing products such as Risperdal Consta, as well as implants in development, is that it is the only biodegradable polymer currently approved for drug delivery. Having this material on the FDA inactive ingredient list dramatically reduces development costs. Because of this, some equally safe polymers with superior physical and chemical properties such as polycaprolactone (PCL) cannot be used without very significant additional financial investment.

PLGA degrades chemically by hydrolytic cleavage of the ester bonds in the polymer backbone (Fig. 3). Its degradation products, lactic acid and glycolic acid, are water-soluble, nontoxic products of normal metabolism that are either excreted or further metabolized to carbon dioxide and water in the Krebs cycle (Jain 2000; Okada and Toguchi 1995). Polymer molecular weight, while being an important determinant of mechanical strength, is also a key factor in determining the

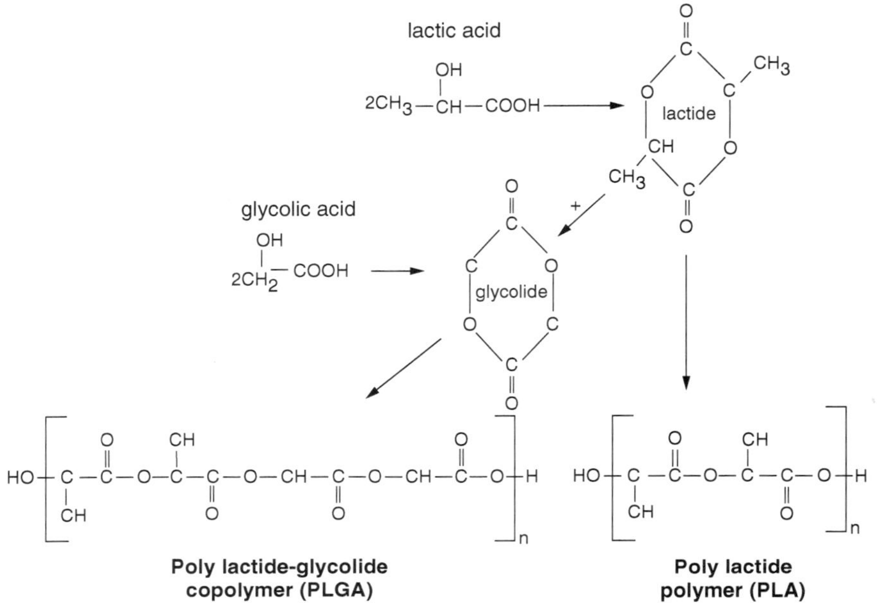

Fig. 3 PLGA and PLA synthesis. Biodegradable polymers of poly(lactide-*co*-glycolide) are formed from units of individual monomers of lactic and glycolic acids. The resulting polymer degrades by simple hydrolysis and the rate of degradation is proportional to the size of the polymer chain

degradation rate of biodegradable polymers. Low molecular weight polymers degrade faster and lose their structural integrity more quickly than high molecular weight polymers. As chain scission occurs over time, the small polymer chains that result become more soluble in the aqueous environment of the body. This introduces "holes" into the polymer matrix. Consequently, lower molecular weight polymers release drug molecules more quickly (Blanco and Alonso 1998). This knowledge can be exploited to engineer a system to control the release rate. A combination of molecular weights might be used to tailor a system to meet the demands of specific release profiles (Metzger et al. 2007; Rabin et al. 2008).

9.4.3 Poly(ε-Caprolactone)

As noted above, PCL is biodegradable, nontoxic polymer that could be used for drug delivery. Advantages of PCL include extremely high durability and strength while remaining flexible. These materials are also very easily processed at low temperatures, facilitating the ability to incorporate a wide variety of drugs without degradation. Delivery interval, as well as physical properties such as flexibility and strength, are dictated by the molecular weight of the PCL chosen. These materials are also very hydrophobic and therefore do not allow a high degree of implant hydration. As such, this material is well suited for release of hydrophilic drugs that are less amenable to delivery by PLGA.

9.4.4 Drug Release Mechanisms

Biodegradable polymers release drug in one of two ways: erosion and diffusion. Drug release from biodegradable polymers *in vivo* is governed by a combination of both mechanisms, and therefore depends on the relative rates of erosion and diffusion. The specific drug used and concentration of that drug will effect the mechanism of release as well as polymer degradation (Siegel et al. 2006). In short, hydrophobic drugs draw water into the polymeric delivery system and accelerate diffusion and degradation. Highly hydrophobic drugs retard hydration of the delivery material (implant or particle) and therefore reduce degradation of the polymer and diffusion of drug through the matrix. Thus, highly hydrophobic drugs such as risperidone and haloperidol are ideal for such delivery systems since higher drug loads cause longer delivery intervals and therefore can produce comparable release rates for longer periods of time as well as smaller systems. A third mechanism of release results from the fact that degradation of PLGA causes a reduction in pH within the polymeric matrix. This can facilitate release of drugs with pH-dependent water solubility. For example, both haloperidol and risperidone are more water soluble at low pH. Thus, degradation of the polymer increases solubility, which increases diffusion of remaining drug. This allows for pH-dependent control of release as a third mechanism to control delivery (Rabin et al. 2008).

9.4.5 Erosion

Erosion is defined as the physical dissolution of a polymer as a result of its degradation (Edlund and Albertsson 2003). Most biodegradable polymers used for drug delivery are degraded by hydrolysis. Hydrolysis is a reaction between water molecules and bonds in the polymer backbone, typically ester bonds, that repeatedly cuts the polymer chain until it is returned to monomers (Fig. 4). There are two possible mechanisms of erosion. When water is confined to the surface of the matrix, as in the case of a hydrophobic polymers like PCL, polymer hydrolysis will occur only on the surface and drug will be released as the surface of the polymer matrix erodes (Fig. 4a), which is called *surface erosion*. If, however, water penetrates the polymer matrix faster than it hydrolyzes the bonds on the surface, then erosion will occur throughout the entire material, which is also called *bulk erosion* (Fig. 4b, c). In many cases, the erosion of a polymer matrix *in vivo* is some combination of these mechanisms.

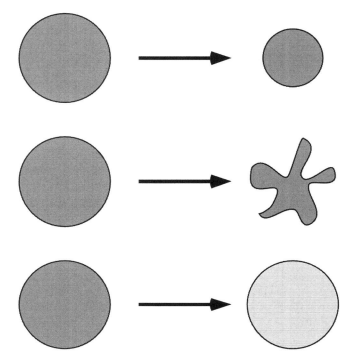

Fig. 4 Types of polymeric delivery system drug release. (**a**) Surface erosion of a homogeneous polymer matrix. Note that the matrix degrades and drug is released only from the surface, while the internal regions remain unchanged. (**b**) Heterogeneous surface erosion. Note that the matrix degrades and the drug is released from the surface, but since the polymer matrix is not homogeneous the surface degradation is not evenly distributed. (**c**) Bulk erosion of a polymer matrix. Note that the matrix is degraded and drug is released from the entire volume of the system. As the polymer matrix is eroded, drug molecules are free to be released via diffusion as well

9.4.6 Diffusion

In the case of diffusion controlled release, the drug's concentration gradient in the polymer matrix is the driving force for the molecules to diffuse into the surrounding medium. The diffusion of a drug molecule through the polymer matrix is dependent upon the solubility of the drug in the polymer matrix and the surrounding environment (subcutaneous space), its concentration throughout the polymer matrix, and the distance necessary for diffusion.

When a drug is dissolved in the matrix and the mechanism for delivery is diffusion, then the driving force for release is the concentration gradient.

Frequently, diffusion-controlled release is important in the early stages of drug release. For many of the polymeric delivery systems, there is some concentration of drug molecules entrapped near the surface of the matrix and adsorbed onto the surface of the matrix. Upon immersion into a medium, the release of these drug molecules is controlled by the rate of diffusion of the drug into the surrounding environment. This can cause a problem referred to as the "burst effect" that can potentially release a toxic amount of drug within the first 24 h (Huang and Brazel 2001). In many drug delivery applications, the burst effect must be overcome before the system is suitable for clinical trials.

The specific chemical and biological characteristics of the drug and the polymer are crucial in designing a polymeric delivery system. For example, drugs with greater hydrophilicity can increase the overall release rate by promoting polymer swelling and degradation, which in turn increases drug diffusion. Additionally, certain drug molecules may potentially react with the polymer matrix (Liu et al. 2003). The drug's molecular weight, solubility in biological fluids, and its miscibility in the polymer matrix also influence the drug's diffusivity from the system and the concentration profile of the drug throughout the matrix.

9.4.7 Advantages

Duration

Successes with decanoate technology in schizophrenia treatment indicate that the introduction of long-acting medications to prevent relapse could significantly improve the pharmacologic treatment of many psychiatric conditions. Furthermore, despite the advent of newer antipsychotic medications with the promise of less severe and disabling side effects, there is little evidence that progress has been made in increasing adherence to these compounds (Zygmunt et al. 2002). For example, Conlon et al. (2002) switched patients from depot antipsychotics to treatment with the newer antipsychotic risperidone. A total of 33 patients who consented to the treatment change discontinued their depot antipsychotics and began treatment with open-label risperidone at a maximum daily dose of 8 mg (mean dose 4 mg). The acute study period was 6 months with additional follow-up for as long as 2 years. Compared with baseline, risperidone treatment was associated with small but statistically significant improvements in psychotic symptoms and quality of life during the 6-month treatment phase. However, the rate of treatment discontinuation by the patients was very high during long-term follow-up; 40% of risperidone-treated patients discontinued treatment over 2 years, whereas none of the patients who continued on depot antipsychotics discontinued treatment. These findings suggest that even though risperidone may be more tolerable for patients than older antipsychotics, the benefits of the superior medication were outweighed in long-term treatment by the disadvantages of a less convenient dosing regimen.

With an understanding of these factors, researchers have investigated new methods of long-term medication delivery to improve adherence using implantable systems (Siegel et al. 2002). Such long-term drug delivery systems may complement the development of new molecules not yet available for clinical application or not yet demonstrating an impact on adherence rates. In this way, implants could provide a complementary approach for improved adherence while bypassing the limiting irreversibility of depot formulations.

The type and molar ratio of individual monomer components (lactide and glycolide) in the copolymer chain partially determine the implant's degradation rate. Polyglycolic (PGA) acid is more hydrophilic than PLA (Dorta et al. 2002). Subsequently, higher glycolic acid molar ratio polymers degrade more quickly through hydrolysis by allowing more biological fluids to penetrate and swell the polymer matrix (Mundargi et al. 2008). Essentially, the proportion of glycolide is a critical parameter to control the hydrophilicity of the polymer and thus the degradation and drug release rate. Geometry also affects implant degradation and drug delivery rates. Several geometries for implants are possible, including films, rods, disks, and gels. As in the case of microparticles, the surface area of the implant will affect release rate. Although previous studies indicate that the shape of an implant (i.e., rod vs. film) does not have a significant effect on release rate independent of surface area, it does affect the induction time for drug release (Witt and Kissel 2001; Witt et al. 2000). Various methods are available for implant fabrication, such as solvent casting, compression molding, and extrusion (Coombes and Heckman 1992; Siegel et al. 2002; Witt and Kissel 2001; Witt et al. 2000).

Reversibility

Implantable systems can be easily removed in case of adverse side effects, offering a degree of reversibility not available with depot injections. Specifically, in the case of an emergency, an overdose, or a change of consent, the remaining drug and matrix can be removed from the body. The reversibility of implants also means that patients maintain the ability to discontinue a specific treatment if they chose to stop or switch medications. However, this would be an active decision by the patient, as part of a process involving their physician, rather than a passive or unintentional act of not taking a daily pill.

Because medication adherence is poor across many persistent disorders, the implant approach has now been extended for the treatment of substance abuse using the opiate antagonist naltrexone. Long-term delivery of such agents is less dependent on patient participation to maintain efficacy, and is therefore more likely to ensure that patients continue to receive their medication even when confronted by cues that increase their risk of relapse. Additionally, evidence suggests that a more gradual delivery of naltrexone could improve substance abuse treatment. A pilot study conducted in the United Kingdom found that naltrexone implants reduced the reinforcing properties of opiate drugs and reduced the incidence of early relapse in a supervised opiate detoxification program (Foster et al. 2003).

9.4.8 Limitations

Physical Drug Characteristics

Stability of the Molecule in a Physiological Environment
Long-term implantable delivery systems require pharmaceutical compounds that are chemically stable under physiological conditions for the entire delivery interval. This is important to ensure that the drug molecules that are released at the end of the delivery interval remain bioactive and effective despite sequestration in a biological environment. Previous studies demonstrate that both haloperidol and risperidone remain stable in physiologic and low pH environments, as may occur inside of biodegradable implants for greater than 12 months (Metzger et al. 2007; Rabin et al. 2008). Less is known about the stability of other agents in biodegradable polymeric systems.

Water Solubility. Hydrophobicity of a pharmaceutical agent has a large effect on how well it can be delivered from implantable long-term delivery systems. Ideal agents have very low water solubility as this retards hydration of the implant, which results in prolongation of both implant degradation and drug release.

Tolerability in the Subcutaneous Environment. A lack of local irritation with a resulting high degree of tolerability in the subcutaneous space is a key requirement for any implantable system. Several reports have shown that PLGA implants are well tolerated in the subcutaneous environment (Metzger et al. 2007; Okada and Toguchi 1995; Rabin et al. 2008; Siegel et al. 2002). However, the tolerability of any specific drug implant system will vary substantially based on the biocompatibility of the pharmaceutical compound in the subcutaneous environment. For example, studies in the authors' laboratory indicate that implants made with the atypical agent quetiapine result in local irritation and small cutaneous lesions, similar to but less severe than those caused by subcutaneous injection of the compound alone. Risperidone and haloperidol implants are well tolerated in the subcutaneous space in mice (Metzger et al. 2007; Rabin et al. 2008; Siegel et al. 2002). Similarly, risperidone is well tolerated in humans when injected intramuscularly, supporting the hypothesis that risperidone implants would be well tolerated in rats and eventually humans. Additionally, ziprasidone is currently available in an acute injectable formulation, suggesting that it too would likely be well tolerated in the subcutaneous space. The authors are not aware of any data available for the parenteral tolerability of aripiprazole, obscuring the ability to make informed predictions of its tolerability in implants. Although implants can be easily removed, they do require at least an initial implantation procedure that may be unfavorable to some patients. There is also a greater risk of local infection and immunogenic reaction at the implant site where there is a higher concentration of both released drug and degradation products.

Dose/Potency. The potency and resulting daily dose of each antipsychotic agent are an important characteristic when considering the feasibility of sequestering a 6-month supply in the subcutaneous space through long-term delivery implants.

Table 3 Estimated implant mass for each atypical medication

Agent	Daily oral dose (mg)	Daily parenteral dose (mg/kg/day)	Monthly parenteral dose (mg/kg/month)	Monthly parenteral dose (mg/month)	Semiannual implant dose (mg)
Risperidone	4–8	0.03–0.06	1–2	50–100 (Consta)	300–600
Haloperidol	6–12	0.03–0.06	1–2	50–100 (decanoate)	300–600
Olanzapine	10–20	0.05–0.10	2–4	100–200	600–1,200
Aripiprazole	10–30	0.05–0.15	2–6	100–300	600–1,800
Ziprasidone	120–160	0.6–1.2	20–40	1,000–2,000	6,000–12,000
Quetiapine	400–600	3–6	100–200	5,000–10,000	30,000–60,000

Specifically, lower-dose requirements result in smaller and therefore more easily tolerated implants for any duration. For example, haloperidol decanoate is traditionally given at approximately 0.03 mg/kg/day (1 mg/kg/month = 50–100 mg/month). Similarly, the biweekly depot preparation of risperidone (Consta, Janssen) requires about the same dose (25 mg/14 days = 50 mg/month). The feasibility for each of the newer antipsychotic agents can be assessed based on the comparable oral doses required to achieve symptom remission. Using these ratios, the required parenteral dose in a 6-month implant for each agent can be estimated based on package inserts (Table 3) from Rabin et al. (2008). As shown, risperidone results in the lowest amount of drug required in an implantable system with resulting reduction in size of the proposed implants. For example, ziprasidone and quetiapine would require implants that are 20 and 100 times larger than those using risperidone. Alternatively, olanzapine and aripiprazole would require only a two- to threefold increase in size relative to risperidone.

Ethical Considerations

Ethical considerations related to the use of implantable delivery systems in psychiatric populations may arise as the concept gains clinical momentum. Ethical factors include careful attention to protect patient autonomy, safeguards to ensure patients' ability to provide informed consent, and assessment of the patient's comprehension of the intervention and desire to continue or discontinue throughout the course of implantation. By allowing patients to make decisions regarding long-term treatment during periods of relative health, this method of delivery may minimize interruptions in antipsychotic medication. A recent study indicates that the majority of psychiatric patients are able to comprehend the issues related to long-term drug delivery. Furthermore, comprehension was highly correlated with the decision to endorse a long-term delivery system for schizophrenia medication (Dankert et al. 2008).

Extension to Other Areas

The widespread applicability of PLGA polymer implants is especially pertinent given the significant medication nonadherence in all chronic illnesses, such as acquired immunodeficiency syndrome (AIDS), asthma, diabetes, and hypertension. In studies of patients with diabetes and hypertension, the rates of medication adherence were only 25 and 53%, respectively (Hudson et al. 2004). Like these medical conditions, schizophrenia requires long-term medication management. Despite rapid advances in the development of antipsychotic drugs, nonadherence remains a pernicious block to long-term improvements in the quality of life of patients with schizophrenia. Furthermore, nonadherence does not appear to be significantly ameliorated by the advent of new agents in clinical trials (Jones et al. 2006). This finding was quantified in the NIH-sponsored Clinical Antipsychotic Trials of Intervention Effectiveness, during which 74% of patients discontinued their initial medication despite being enrolled in a clinical trial (Lieberman 2006). In addition to the clinical consequences of nonadherence, researchers estimate that the economic burden of health-care costs for nonadherence in mental illness is at least \$2.3 billion annually in the United States (Menzin et al. 2003). Implantable antipsychotic medication may complement future development of new molecules in the service of improved adherence and long-term clinical improvement and cost effectiveness.

9.5 Transdermal Delivery Systems

9.5.1 Potential Benefits

Transdermal drug delivery systems – "patches" – represent another method of delivering long-term medication to individuals with chronic conditions that optimizes dosing and convenience while minimizing side effects (Table 4). The first transdermal patch for systemic delivery, a 3-day patch that delivers scopolamine to treat motion sickness, was FDA approved for use in 1979 (Prausnitz and Langer 2008). Today, there are 19 such transdermal delivery systems available in the US (Tanner and Marks 2008). Transdermal patches are designed to deliver therapeutically effective amounts of drug across a patient's skin to exert a systemic effect. The rationale behind patches echoes the reasoning behind both injectable and implantable delivery systems. Reasons include improved patient adherence; avoidance of first-pass metabolism by the liver, which can prematurely metabolize drugs; maintenance of sustained therapeutic plasma concentrations of drugs; and greater flexibility in dosage, such that dosing can be easily terminated by removal of the patch (Chan et al. 2008; Tandon et al. 2006).

Table 4 Patches for treatment of neuropsychiatric conditions

Indication	Medication	Mechanism of action	Adverse effects	Pharmacokinetics
Attention deficit hyperactivity disorder	Methylphenidate (Daytrana®)	Norepinephrine and dopamine reuptake inhibitor	Anorexia and insomnia (comparable to oral medications)	Overcomes enteric and hepatic presystemic de-esterification, which may limit oral bioavailability. A substantial amount of drug remains in the patient after patch removal (Patrick et al. 2009)
Chronic pain	Fentanyl (Duragesic®)	Opiate receptor antagonist	Application site reaction; sweating	Forms a depot within the upper skin layers before entering the microcirculation. Therapeutic blood levels are attained 12–16 h after patch application and decrease slowly with a half-life of 16–22 h following removal (Vadalouca et al. 2006)
Dementia of Alzheimer's or Parkinson's disease	Rivastigmine (Exelon®)	Acetylcholinesterase inhibitor	Application site reaction (local erythema, pruritus, and dermatitis) Nausea and vomiting (less than oral administration)	Provides consistent, controlled drug delivery with a substantially lowered maximum concentration and greater interval to maximum concentration following drug administration (Chan et al. 2008; Tanner and Marks 2008) Reduces plasma fluctuation compared with a similar level of exposure obtained with oral administration (Cummings et al. 2007)
Major depressive disorder	Selegiline (Emsam®)	Monoamine oxidase inhibitor	Application site reaction; insomnia	Steady-state plasma levels are reached after 5 days of treatment. 6 mg/24 h once-daily dose maintains stable plasma levels (Pae et al. 2007)
Parkinson's disease	Rotigotine (Neupro®) (currently not available in US; available in EU)	Dopamine agonist	Application site reaction	Release is constant over 24 h and proportional to patch surface area. 6 mg/24 h dose maintains stable plasma rotigotine concentrations (Baldwin and Keating 2007)
Smoking cessation	Nicotine (Nicoderm®), (Habitrol®), (ProStep®)	Nicotine receptor agonist	Application site reaction	Continuous release of nicotine over 16–24 h (Nides 2008)

This table represents a portion of the transdermal delivery systems approved by the FDA. Other patches are indicated for contraception, menopausal symptoms, and hypertension

9.5.2 Passive Systems

Patches are usually designed such that the drug is stored in a reservoir which is enclosed on one side with an impermeable backing and an adhesive in contact with skin on the other side (Venkatraman and Gale 1998). Some patches utilize drug dissolved in a liquid or gel-based reservoir. This design is composed of four layers: an impermeable membrane, a drug reservoir, a semipermeable membrane, and an adhesive layer. Other transdermal patches incorporate the drug into a solid polymer matrix, which may simplify manufacturing (Prausnitz and Langer 2008). Patches may be designed to deliver medication for periods of a single day (methylphenidate), 1 week (clonidine), or even 4 weeks (ethinyl estradiol with norelgestromin) (Tanner and Marks 2008).

9.5.3 Active Iontophoretic Systems

Iontophoretic transdermal drug delivery (patches) involves the transport of ionized drug molecules into the skin using an applied electric current. Commercially available iontophoretic drug delivery systems typically employ an isolated stimulator producing one or two channels of direct current, with an output of 4–5 mA and a voltage of 50–100 V. A drug delivery reservoir holds the ionized drug solution on the electrode and is positioned on the skin directly over the treatment site. The electrical circuit is completed by a dispersive electrode placed on the nearby skin. The quantity of drug transported into the skin is proportional to the total charge delivered and is dependent upon a number of criteria, including the molecular weight of the drug ion, drug concentration, solution pH, and buffer concentration. The treatment dosage is quantified by the total charge delivered and is often expressed in units of mA min. Adverse events due to the iontophoretic process include local erythema, irritation, and pruritus. The extent to which these events occur is dependent on total charge as well as the specific properties of the pharmaceutical agent being delivered.

Relatively, fewer randomized controlled trials have investigated the use of iontophoresis for systemic delivery of active pharmaceutical agents, and none are currently available for psychiatric applications:

- *Single agents.* A double-blind study of 60 patients demonstrated efficacious iontophoretic administration of the nonsteroidal anti-inflammatory drug ketorolac for patients with pain from rheumatic disease (Saggini et al. 1996). This study utilized five sessions of 30 min at 2 mA, which resulted in no adverse reactions noted. Additionally, iontophoretic delivery of morphine has been tested for postoperative analgesia (Ashburn et al. 1992). This study used an external power supply to deliver 1.3 mg of morphine/mA/h for 6 h but did not disclose the current or voltages applied. All patients who were iontophoresed with morphine demonstrated a wheal and flare under and surrounding the applied electrode. Within 10–20 min of beginning iontophoretic morphine, a

distinct red flare developed, which extended beyond the rim of the electrode and biased in the direction of venous drainage, and a raised wheel appeared under the electrode. After 2–3 h of treatment, the red flare subsided but the wheel remained for up to 12 h after iontophoresis. This reaction was not present among the group that received placebo solution via iontophoresis, suggesting that it was due to the morphine rather than the applied current. However, one patient in the placebo group and two in the morphine group also displayed arc-shaped erythema under the electrode that lasted for 72 h. Similar reactions, described as skin irritation and painless electrical burns, were observed in a controlled trial that evaluated iontophoresis of vincristine and saline (15–30 mA for 60 min) in postherpetic neuralgia (Layman et al. 1986).
- *Combination therapy.* One study evaluated the ability to reduce local irritation following transdermal iontophoresis (0.5 mA for 240 min) by performing codelivery of hydrocortisone with the antiemetic agent metoclopramide (Cormier et al. 1999). This study determined that coadministration of hydrocortisone did not change the pharmacokinetics of metoclopramide but did significantly decreased erythema and edema scores produced by the iontophoresis of metoclopramide.

9.5.4 Limitations

The most significant barrier to the production of transdermal delivery systems is that only a limited number of drugs are amenable to administration via the skin. The criteria are a relatively low molecular mass, hydrophobicity, and dosages of mg/day or less (Prausnitz et al. 2004). Despite these challenges, it is likely that transdermal drug delivery, alongside implantable drug delivery, will have an increasingly widespread impact on medicine in the future.

9.6 Barriers to Development of Novel Delivery Systems

There are several unique challenges to the clinical development of long-term delivery systems. Specifically, there is little or no precedent for an antipsychotic medication product that could last 6 months or more. As such, pharmaceutical companies need to create clinical trial designs that allow the field to traverse additional barriers to bring forward this approach.

9.6.1 Long Clinical Trial Length

The length of a clinical trial for novel long-term delivery systems would need to exceed the intended interval for the system. For example, an implant that lasts 6 months would likely require a Phase I clinical trial lasting at least 6 month to demonstrate the full pharmacokinetic profile, stability, as well as reversibility over that time course. Because these systems would also need to be replaced at the end of each delivery interval in patients, clinical trial would likely need to last twice the delivery interval to establish the optimal crossover timing.

9.6.2 Inclusion of Controls in Studies of 6-Month Exposure to Antipsychotic Medications

Inclusion of normal controls in Phase I trials would present another unique issue for very long-term trials of antipsychotic medication. Specifically, it may not be possible to expose nonpatients to antipsychotic medications for 6 months to examine the pharmacokinetic and safety profile of the delivery system. As such, Phase I trials may need to be performed with patients that normally take antipsychotic medication. This would reduce the risks associated with long-term exposure to antipsychotic medications in nonpsychotic individuals. A similar approach was used in the testing of implants for prostate cancer, and may provide precedence for antipsychotic medications (Blom et al. 1989; Fowler et al. 2000).

9.6.3 Placebo Arm in Patients for Length of Time Needed to Truly Test a 6-Month Delivery System

The use of placebo arms in Phase III clinical trials always presents significant risk to patients. However, the Food and Drug Administration has traditionally insisted that a placebo arm is required to demonstrate antipsychotic efficacy for new chemical entities. However, new long-term delivery systems using existing medications may present unique challenges and solutions to this dilemma. One possible solution would be to have a traditional Phase III clinical trial design with active agent compared to placebo for a short initial period of 6–8 weeks, comparable to designs for oral medications. This could be followed by a longer trial of 6–12 months in which active implants could be compared to oral medication for noninferiority rather than compared to placebo. Regardless of interval, all patients in such trials would likely need to have both an implant and oral medication using a double-dummy design in order to retain the blinded conditions. Specifically, participants on oral medication would need placebo implants (polymer without drug), and participants on implanted medication would need placebo pills. An alternative design could test novel delivery systems to active comparators only. This approach has been used for implants to treat prostate cancer. Phase III trials of LHRH-Hydrogel implants were conducted using open-labeled, randomized, parallel studies that compared the hydrogel implant with the active comparators, leuprorelin acetate 22.5 mg depot (TAP Pharmaceutical's Lupron Depot), and a 3-month implant of goserelin acetate (Astra Zeneca's Zoladex 2005). Similarly, testing of antipsychotic implants for 6–12 months could incorporate both the ethical contraindication to lengthy placebo arms and the exceedingly high replace rates over that interval.

Specifically, there is a substantial amount of data indicating that 60–80% of patients would relapse with severe psychotic exacerbations if not treated for 12 months. Thus, it is possible significantly reduced relapse rates over that interval could be shown without having to include a placebo group.

10 Summary and Conclusions

"Schizophrenia is a terrible illness. It most commonly strikes people at a young age. It severely limits their potential to work and support themselves, to marry, and to have healthy relationships" (McEvoy 2004). If left untreated or undertreated, schizophrenia marginalizes the individual, who continues to follow a deteriorating trajectory for the duration of their life. Despite recent efforts to create improved medications, antipsychotics have not changed in their basic mechanism, dopamine D_2 receptor antagonism, since the advent of chlorpromazine in 1952. These medications effectively treat the positive symptom cluster of schizophrenia but demonstrate minimal efficacy for the negative, cognitive, and functional domains of the illness. Furthermore, discontinuation or irregular dosing of antipsychotic medication is correlated with worse long-term functioning, quality of life, ability to maintain employment and relationships (Gilmer et al. 2004; Nasrallah and Lasser 2006).

To ameliorate the problems associated with nonadherence, novel mechanisms of drug delivery have been developed that offer continuous infusion of antipsychotic medication. Depot, polymer-based microsphere, implantable, and transdermal systems are long-term methods of delivery that may have biological and psychological benefits compared to conventional dosing, leading to superior clinical outcomes. These methods can provide consistent, controlled delivery that avoids wide fluctuations in drug levels. Additionally, these novel delivery systems avoid first-pass metabolism by the liver, and avoid the peaks and troughs of conventional oral delivery, which are potentially detrimental in psychiatric conditions.

Drug development for schizophrenia continues to focus on novel mechanisms of action including positive allosteric modulators at NMDA and metabotropic glutamate receptors, selective nicotine receptor agonists, and modulators of intracellular cAMP. As we wait for these new mechanisms of action to be validated, current and future advancements in long-term delivery provide a complementary approach to reduce morbidity and mortality from the consequences of untreated psychosis.

References

Adams SG Jr, Howe JT (1993) Predicting medication compliance in a psychotic population. J Nerv Ment Dis 181:558–560

Adams CE, Fenton MK, Quraishi S, David AS (2001) Systematic meta-review of depot antipsychotic drugs for people with schizophrenia. Br J Psychiatry 179:290–299

Adis International Limited (2005) Histrelin Hydrogel Implant – Valera: Histrelin implant, LHRH-Hydrogel implant, RL 0903, SPD 424. Drugs RD 6:53–55

Ascher-Svanum H, Faries DE, Zhu B, Ernst FR, Swartz MS, Swanson JW (2006) Medication adherence and long-term functional outcomes in the treatment of schizophrenia in usual care. J Clin Psychiatry 67:453–460

Ashburn MA, Stephen RL, Ackerman E, Petelenz TJ, Hare B, Pace NL, Hofman AA (1992) Iontophoretic delivery of morphine for postoperative analgesia. J Pain Symptom Manage 7: 27–33

Ashby CR Jr, Wang RY (1996) Pharmacological actions of the atypical antipsychotic drug clozapine: a review. Synapse 24:349–394

Baldwin CM, Keating GM (2007) Rotigotine transdermal patch: a review of its use in the management of Parkinson's disease. CNS Drugs 21:1039–1055

Barnes DE, Robinson B, Csernansky JG, Bellows EP (1990) Sensitization versus tolerance to haloperidol-induced catalepsy: multiple determinants. Pharmacol Biochem Behav 36:883–887

Bergen J, Hunt G, Armitage P, Bashir M (1998) Six-month outcome following a relapse of schizophrenia. Aust N Z J Psychiatry 32:815–822

Berkland C, King M, Cox A, Kim K, Pack DW (2002) Precise control of PLG microsphere size provides enhanced control of drug release rate. J Control Release 82:137–147

Bishara D, Taylor D (2008) Upcoming agents for the treatment of schizophrenia: mechanism of action, efficacy and tolerability. Drugs 68:2269–2292

Blanco D, Alonso MJ (1998) Protein encapsulation and release from poly(lactide-*co*-glycolide) microspheres: effect of the protein and polymer properties and of the co-encapsulation of surfactants. Eur J Pharm Biopharm 45:285–294

Blom JH, Hirdes WH, Schroder FH, de Jong FH, Kwekkeboom DJ, van't Veen AJ, Sandow J, Krauss B (1989) Pharmacokinetics and endocrine effects of the LHR analogue buserelin after subcutaneous implantation of a slow release preparation in prostatic cancer patients. Urol Res 17:43–46

Bloch Y, Mendlovic S, Strupinsky S, Altshuler A, Fennig S, Ratzoni G (2001) Injections of depot antipsychotic medications in patients suffering from schizophrenia: do they hurt? J Clin Psychiatry 62:855–859

Borgman RJ (1982) Bioavailability, dosage regimens, and new delivery systems. In: Craig CR, Stitzel RE (eds). Modern pharmacology. Little, Brown and Company, Boston pp 63–74

Borison RL, Diamond B, Pathiraja A, Meibach RC (1994) Pharmacokinetics of risperidone in chronic schizophrenic patients. Psychopharmacol Bull 30:193–197

Castle DJ, Pantelis C (2003) Comprehensive care for people with schizophrenia living in the community. Med J Aust 178(Suppl 9):S45–S46

Carlsson A, Lindqvist M (1963) Effect of chlorpromazine or haloperidol on formation of 3methoxytyramine and normetanephrine in mouse brain. Acta Pharmacol Toxicol (Copenh) 20:140–144

Casper ES, Regan JR (1993) Reasons for admission among six profile subgroups of recidivists of inpatient services. Can J Psychiatry 38:657–661

Chan AL, Chien YW, Jin Lin S (2008) Transdermal delivery of treatment for Alzheimer's disease: development, clinical performance and future prospects. Drugs Aging 25:761–775

Cheer SM, Wagstaff AJ (2004) Quetiapine. A review of its use in the management of schizophrenia. CNS Drugs 18:173–199

Coombes AG, Heckman JD (1992) Gel casting of resorbable polymers. 1. Processing and applications. Biomaterials 13:217–224

Conlon L, Fahy TJ, OT R, Gilligan J, Prescott P (2002) Risperidone in chronic schizophrenia: a detailed audit, open switch study and two-year follow-up of patients on depot medication. Eur Psychiatry 17:459–465

Cormier M, Chao ST, Gupta SK, Haak R (1999) Effect of transdermal iontophoresis codelivery of hydrocortisone on metoclopramide pharmacokinetics and skin-induced reactions in human subjects. J Pharm Sci 88:1030–1035

Cramer JR, Cramer JA, Rosenheck R (1998) Compliance with medication regimens for mental and physical disorders. Psychiatr Serv 49:196–201

Csernansky JG, Bellows EP, Barnes DE, Lombrozo L (1990) Sensitization versus tolerance to the dopamine turnover-elevating effects of haloperidol: the effect of regular/intermittent dosing. Psychopharmacology (Berl) 101:519–524

Cummings J, Lefevre G, Small G, Appel-Dingemanse S (2007) Pharmacokinetic rationale for the rivastigmine patch. Neurology 69:S10–S13

Dankert ME, Brensinger CM, Metzger KL, Li C, Koleva SG, Mesen A et al (2008) Attitudes of patients and family members towards implantable psychiatric medication. Schizophr Res 105:279–286

David AS (1990) Insight and psychosis. Br J Psychiatry 156:798–808

Davis JM, Chen N, Glick ID (2003) A meta-analysis of the efficacy of second-generation antipsychotics. Arch Gen Psychiatry 60:553–564

Delay J, Deniker P, Harl JM, Grasset A (1952a) [N-dimethylamino-prophylchlorophenothiazine (4560 RP) therapy of confusional states]. Ann Med Psychol (Paris) 110:398–403

Delay J, Deniker P, Harl JM (1952b) [Therapeutic use in psychiatry of phenothiazine of central elective action (4560 RP)]. Ann Med Psychol (Paris) 110:112–117

Dorta MJ, Santovena A, Llabres M, Farina JB (2002) Potential applications of PLGA film-implants in modulating *in vitro* drugs release. Int J Pharm 248:149–156

Edlund U, Albertsson AC (2003) Polyesters based on diacid monomers. Adv Drug Deliv Rev 55:585–609

Ereshefsky L, Mascarenas CA (2003) Comparison of the effects of different routes of antipsychotic administration on pharmacokinetics and pharmacodynamics. J Clin Psychiatry 16(Suppl 64): 18–23

Ereshefsky L, Watanabe MD, Tran-Johnson TK (1989) Clozapine: an atypical antipsychotic agent. Clin Pharm 8:691–709

Ezrin-Waters C, Seeman P (1977) Tolerance of haloperidol catalepsy. Eur J Pharmacol 41: 321–327

Foster J, Brewer C, Steele T (2003) Naltrexone implants can completely prevent early (1-month) relapse after opiate detoxification: a pilot study of two cohorts totalling 101 patients with a note on naltrexone blood levels. Addict Biol 8:211–217

Fowler JE Jr, Gottesman JE, Reid CF, Andriole GL Jr, Soloway MS (2000) Safety and efficacy of an implantable leuprolide delivery system in patients with advanced prostate cancer. J Urol 164:730–734

Farde L, Hall H, Ehrin E, Sedvall G (1986) Quantitative analysis of D2 dopamine receptor binding in the living human brain by PET. Science 231:258–261

Farde L, Wiesel FA, Jansson P, Uppfeldt G, Wahlen A, Sedvall G (1988) An open label trial of raclopride in acute schizophrenia. Confirmation of D2-dopamine receptor occupancy by PET. Psychopharmacology (Berl) 94:1–7

Ginovart N, Wilson AA, Hussey D, Houle S, Kapur S (2009) D2-receptor upregulation is dependent upon temporal course of D2-occupancy: a longitudinal [11C]-raclopride PET study in cats. Neuropsychopharmacology 34:662–671

Geddes J, Freemantle N, Harrison P, Bebbington P (2000) Atypical antipsychotics in the treatment of schizophrenia: systematic overview and meta-regression analysis. BMJ 321:1371–1376

Gilmer TP, Dolder CR, Lacro JP, Folsom DP, Lindamer L, Garcia P et al (2004) Adherence to treatment with antipsychotic medication and health care costs among Medicaid beneficiaries with schizophrenia. Am J Psychiatry 161:692–699

Gruen PH, Sachar EJ, Langer G, Altman N, Leifer M, Frantz A et al (1978) Prolactin responses to neuroleptics in normal and schizophrenic subjects. Arch Gen Psychiatry 35:108–116

Harrison PJ (2008) Metabotropic glutamate receptor agonists for schizophrenia. Br J Psychiatry 192:86–87

Heresco-Levy U, Javitt DC, Ebstein R, Vass A, Lichtenberg P, Bar G, Catinari S, Ermilov M (2005) D-Serine efficacy as add-on pharmacotherapy to risperidone and olanzapine for treatment-refractory schizophrenia. Biol Psychiatry 57:577–585

Hudson TJ, Owen RR, Thrush CR, Han X, Pyne JM, Thapa P et al (2004) A pilot study of barriers to medication adherence in schizophrenia. J Clin Psychiatry 65:211–216

Huang X, Brazel CS (2001) On the importance and mechanisms of burst release in matrix-controlled drug delivery systems. J Control Release 73:121–136

Javitt DC, Jotkowitz A, Sircar R, Zukin SR (1987) Non-competitive regulation of phencyclidine/sigma-receptors by the N-methyl-D-aspartate receptor antagonist D-($-$)-2-amino-5-phosphonovaleric acid. Neurosci Lett 78:193–198

Jain RA (2000) The manufacturing techniques of various drug loaded biodegradable poly(lactide-co-glycolide) (PLGA) devices. Biomaterials 21:2475–2490

Johnson DA (1981) Long-term maintenance treatment in chronic schizophrenia. Some observations on outcome and duration. Acta Psychiatr Belg 81:161–172

Johnson DA (1984) Observations on the use of long-acting depot neuroleptic injections in the maintenance therapy of schizophrenia. J Clin Psychiatry 45:13–21

Jones PB, Barnes TR, Davies L, Dunn G, Lloyd H, Hayhurst KP et al (2006) Randomized controlled trial of the effect on quality of life of second- vs first-generation antipsychotic drugs in schizophrenia: Cost Utility of the Latest Antipsychotic Drugs in Schizophrenia Study (CUtLASS 1). Arch Gen Psychiatry 63:1079–1087

Kalkman HO, Subramanian N, Hoyer D (2001) Extended radioligand binding profile of iloperidone: a broad spectrum dopamine/serotonin/norepinephrine receptor antagonist for the management of psychotic disorders. Neuropsychopharmacology 25:904–914

Kane JM, Aguglia E, Altamura AC, Ayuso Gutierrez JL, Brunello N, Fleischhacker WW et al (1998) Guidelines for depot antipsychotic treatment in schizophrenia. European neuropsychopharmacology consensus conference in Siena, Italy. Eur Neuropsychopharmacol 8:55–66

Kane J, Honigfeld G, Singer J, Meltzer H (1988) Clozapine for the treatment-resistant schizophrenic. A double-blind comparison with chlorpromazine. Arch Gen Psychiatry 45:789–796

Kane JM, Leucht S, Carpenter D, Docherty JP (2003) The expert consensus guideline series. Optimizing pharmacologic treatment of psychotic disorders. Introduction: methods, commentary, and summary. J Clin Psychiatry 64(Suppl 12):5–19

Kapur S, Seeman P (2001) Does fast dissociation from the dopamine d(2) receptor explain the action of atypical antipsychotics?: a new hypothesis. Am J Psychiatry 158:360–369

Kapur S, Zipursky R, Jones C, Shammi CS, Remington G, Seeman P (2000) A positron emission tomography study of quetiapine in schizophrenia: a preliminary finding of an antipsychotic effect with only transiently high dopamine D2 receptor occupancy. Arch Gen Psychiatry 57:553–559

Kasper S (2006) Optimisation of long-term treatment in schizophrenia: treating the true spectrum of symptoms. Eur Neuropsychopharmacol 16(Suppl 3):S135–S141

Keith S (2006) Advances in psychotropic formulations. Prog Neuropsychopharmacol Biol Psychiatry 30:996–1008

Keith SJ, Kane JM (2003) Partial compliance and patient consequences in schizophrenia: our patients can do better. J Clin Psychiatry 64:1308–1315

Keefe RS (2006) Neurocognitive effects of antipsychotic medications in patients with chronic schizophrenia in the CATIE trial. Biol Psychiatry 59:965

Keefe R, Gu H, Perkins D, McEvoy J, Hamer R, Lieberman JA (2005) A comparison of the effects of olanzapine, quetiapine, and risperidone on neurocognitive function in first-episode psychosis. ACNP annual meeting, 127

Kitchell JP, Wise DL (1985) Poly(lactic/glycolic acid) biodegradable drug–polymer matrix systems. Methods Enzymol 112:436–448

Kinon BJ, Stauffer VL, McGuire HC, Kaiser CJ, Dickson RA, Kennedy JS (2003) The effects of antipsychotic drug treatment on prolactin concentrations in elderly patients. J Am Med Dir Assoc 4:189–194

Knox ED, Stimmel GL (2004) Clinical review of a long-acting, injectable formulation of risperidone. Clin Ther 26:1994–2002

Kohler U, Schroder H, Augustin W, Sabel BA (1994) A new animal model of dopamine supersensitivity using s.c. implantation of haloperidol releasing polymers. Neurosci Lett 170:99–102

Kulkarni RK, Pani KC, Neuman C, Leonard F (1966) Polylactic acid for surgical implants. Arch Surg 93:839–843

Layman PR, Argyras E, Glynn CJ (1986) Iontophoresis of vincristine versus saline in post-herpetic neuralgia. A controlled trial. Pain 25:165–170

Lambert T, Brennan A, Castle D, Kelly DL, Conley RR (2003) Perception of depot antipsychotics by mental health professionals. J Psychiatr Pract 9:252–260

Langer G, Sachar EJ, Gruen PH, Halpern FS (1977) Human prolactin responses to neuroleptic drugs correlate with antischizophrenic potency. Nature 266:639–640

Lemoine D, Preat V (1998) Polymeric nanoparticles as delivery system for influenza virus glycoproteins. J Control Release 54:15–27

Liu FI, Kuo JH, Sung KC, Hu OY (2003) Biodegradable polymeric microspheres for nalbuphine prodrug controlled delivery: *in vitro* characterization and *in vivo* pharmacokinetic studies. Int J Pharm 257:23–31

Lieberman JA, Stroup TS, McEvoy JP, Swartz MS, Rosenheck RA, Perkins DO et al (2005) Effectiveness of antipsychotic drugs in patients with chronic schizophrenia. N Engl J Med 353:1209–1223

Lim ST, Forbes B, Berry DJ, Martin GP, Brown MB (2002) In vivo evaluation of novel hyaluronan/chitosan microparticulate delivery systems for the nasal delivery of gentamicin in rabbits. Int J Pharm 231:73–82

Love RC (2002) Strategies for increasing treatment compliance: the role of long-acting antipsychotics. Am J Health Syst Pharm 59:S10–S15

Marder SR (2003) Overview of partial compliance. J Clin Psychiatry 64(Suppl 16):3–9

McEvoy J (2004) The relationship between insight into psychosis and compliance with medications. In: Amador XF, David AS (eds) Insight and psychosis. Oxford University Press, Oxford, pp 311–334

McEvoy JP, Applebaum PS, Apperson LJ, Geller JL, Freter S (1989) Why must some schizophrenic patients be involuntarily committed? The role of insight. Compr Psychiatry 30:13–17

McEvoy JP, Apperson LJ, Appelbaum PS, Ortlip P, Brecosky J, Hammill K, Geller JL, Roth L (1989a) Insight in schizophrenia. Its relationship to acute psychopathology. J Nerv Ment Dis 177:43–47

McEvoy JP, Freter S, Everett G, Geller JL, Appelbaum P, Apperson LJ, Roth L (1989b) Insight and the clinical outcome of schizophrenic patients. J Nerv Ment Dis 177:48–51

Meltzer HY, McGurk SR (1999) The effects of clozapine, risperidone, and olanzapine on cognitive function in schizophrenia. Schizophr Bull 25:233–255

Menzin J, Boulanger L, Friedman M, Mackell J, Lloyd JR (2003) Treatment adherence associated with conventional and atypical antipsychotics in a large state Medicaid program. Psychiatr Serv 54:719–723

Metzger KL, Shoemaker JM, Kahn JB, Maxwell CR, Liang Y, Tokarczyk J et al (2007) Pharmacokinetic and behavioral characterization of a long-term antipsychotic delivery system in rodents and rabbits. Psychopharmacology (Berl) 190:201–211

Mundargi RC, Babu VR, Rangaswamy V, Patel P, Aminabhavi TM (2008) Nano/micro technologies for delivering macromolecular therapeutics using poly(D,L-lactide-*co*-glycolide) and its derivatives. J Control Release 125:193–209

Nasrallah HA, Lasser R (2006) Improving patient outcomes in schizophrenia: achieving remission. J Psychopharmacol 20:57–61

Nasrallah HA (2008) Atypical antipsychotic-induced metabolic side effects: insights from receptor-binding profiles. Mol Psychiatry 13:27–35

Nides M (2008) Update on pharmacologic options for smoking cessation treatment. Am J Med 121:S20–S31

Nordstrom AL, Farde L, Halldin C (1993) High 5-HT2 receptor occupancy in clozapine treated patients demonstrated by PET. Psychopharmacology (Berl) 110:365–367

Nordstrom AL, Farde L, Nyberg S, Karlsson P, Halldin C, Sedvall G (1995) D1, D2, and 5-HT2 receptor occupancy in relation to clozapine serum concentration: a PET study of schizophrenic patients. Am J Psychiatry 152:1444–1449

Oh SH, Lee JY, Ghil SH, Lee SS, Yuk SH, Lee JH (2006) PCL microparticle-dispersed PLGA solution as a potential injectable urethral bulking agent. Biomaterials 27:1936–1944

Okada H, Toguchi H (1995) Biodegradable microspheres in drug delivery. Crit Rev Ther Drug Carrier Syst 12:1–99

Pae CU, Lim HK, Han C, Neena A, Lee C, Patkar AA (2007) Selegiline transdermal system: current awareness and promise. Prog Neuropsychopharmacol Biol Psychiatry 31: 1153–1163

Palmer BW, Jeste DV (2006) Relationship of individual cognitive abilities to specific components of decisional capacity among middle-aged and older patients with schizophrenia. Schizophr Bull 32:98–106

Patil ST, Zhang L, Martenyi F, Lowe SL, Jackson KA, Andreev BV et al (2007) Activation of mGlu2/3 receptors as a new approach to treat schizophrenia: a randomized Phase 2 clinical trial. Nat Med 13:1102–1107

Patrick KS, Straughn AB, Perkins JS, Gonzalez MA (2009) Evolution of stimulants to treat ADHD: transdermal methylphenidate. Hum Psychopharmacol 24:1–17

Prausnitz MR, Langer R (2008) Transdermal drug delivery. Nat Biotechnol 26:1261–1268

Prausnitz MR, Mitragotri S, Langer R (2004) Current status and future potential of transdermal drug delivery. Nat Rev Drug Discov 3:115–124

Rabin C, Liang Y, Ehrlichman RS, Budhian A, Metzger KL, Majewski-Tiedeken C, Winey KI, Siegel SJ (2008) *In vitro* and *in vivo* demonstration of risperidone implants in mice. Schizophr Res 98:66–78

Rauser L, Savage JE, Meltzer HY, Roth BL (2001) Inverse agonist actions of typical and atypical antipsychotic drugs at the human 5-hydroxytryptamine(2C) receptor. J Pharmacol Exp Ther 299:83–89

Rabin C, Liang Y, Ehrlichman RS, Budhian A, Metzger KL, Majewski-Tiedeken C et al (2008) In vitro and in vivo demonstration of risperidone implants in mice. Schizophr Res 98:66–78

Rafati H, Lavelle EC, Coombes AG, Stolnik S, Holland J, Davis SS (1997) The immune response to a model antigen associated with PLG microparticles prepared using different surfactants. Vaccine 15:1888–1897

Rittmannsberger H, Pachinger T, Keppelmuller P, Wancata J (2004) Medication adherence among psychotic patients before admission to inpatient treatment. Psychiatr Serv 55:174–179

Sabel BA, Dominiak P, Hauser W, During MJ, Freese A (1990) Levodopa delivery from controlled-release polymer matrix: delivery of more than 600 days *in vitro* and 225 days of elevated plasma levels after subcutaneous implantation in rats. J Pharmacol Exp Ther 255:914–922

Saggini R, Zoppi M, Vecchiet F, Gatteschi L, Obletter G, Giamberardino MA (1996) Comparison of electromotive drug administration with ketorolac or with placebo in patients with pain from rheumatic disease: a double-masked study. Clin Ther 18:1169–1174

Samaha AN, Reckless GE, Seeman P, Diwan M, Nobrega JN, Kapur S (2008) Less is more: antipsychotic drug effects are greater with transient rather than continuous delivery. Biol Psychiatry 64:145–152

Sedvall G (1980) Relationships among biochemical, clinical, and pharmacokinetic variables in neuroleptic-treated schizophrenic patients. Adv Biochem Psychopharmacol 24:521–528

Seeman P, Lee T (1975) Antipsychotic drugs: direct correlation between clinical potency and presynaptic action on dopamine neurons. Science 188:1217–1219

Seeman P, Chau-Wong M, Tedesco J, Wong K (1975) Brain receptors for antipsychotic drugs and dopamine: direct binding assays. Proc Natl Acad Sci USA 72:4376–4380

See RE, Ellison G (1990) Intermittent and continuous haloperidol regimens produce different types of oral dyskinesias in rats. Psychopharmacology (Berl) 100:404–412

Seeman P, Corbett R, Van Tol HH (1997) Atypical neuroleptics have low affinity for dopamine D2 receptors or are selective for D4 receptors. Neuropsychopharmacology 16:93–110, discussion 111–135

Sharma R, Saxena D, Dwivedi AK, Misra A (2001) Inhalable microparticles containing drug combinations to target alveolar macrophages for treatment of pulmonary tuberculosis. Pharm Res 18:1405–1410

Sharon AC, Wise DL (1981) Development of drug delivery systems for use in treatment of narcotic addiction. NIDA Res Monogr 28:194–213

Shen WW (1999) A history of antipsychotic drug development. Compr Psychiatry 40:407–414

Siegel SJ, Winey KI, Gur RE, Lenox RH, Bilker WB, Ikeda D et al (2002) Surgically implantable long-term antipsychotic delivery systems for the treatment of schizophrenia. Neuropsychopharmacology 26:817–823

Siegel SJ, Irani F, Brensinger CM, Kohler CG, Bilker WB, Ragland JD et al (2006) Prognostic variables at intake and long-term level of function in schizophrenia. Am J Psychiatry 163:433–441.

Siegel SJ, Kahn JB, Metzger K, Winey KI, Werner K, Dan N (2006) Effect of drug type on the degradation rate of PLGA matrices. Eur J Pharm Biopharm 64:287–293

Sodhi M, Wood KH, Meador-Woodruff J (2008) Role of glutamate in schizophrenia: integrating excitatory avenues of research. Expert Rev Neurother 8:1389–1406

Spiers ID, Eyles JE, Baillie LW, Williamson ED, Alpar HO (2000) Biodegradable microparticles with different release profiles: effect on the immune response after a single administration via intranasal and intramuscular routes. J Pharm Pharmacol 52:1195–1201

Svensson TH (2003) Alpha-adrenoceptor modulation hypothesis of antipsychotic atypicality. Prog Neuropsychopharmacol Biol Psychiatry 27:1145–1158

Tandon R, Targum SD, Nasrallah HA, Ross R (2006) Strategies for maximizing clinical effectiveness in the treatment of schizophrenia. J Psychiatr Pract 12:348–363

Tanner T, Marks R (2008) Delivering drugs by the transdermal route: review and comment. Skin Res Technol 14:249–260

Torche AM, Le Corre P, Albina E, Jestin A, Le Verge R (2000) PLGA microspheres phagocytosis by pig alveolar macrophages: influence of poly(vinyl alcohol) concentration, nature of loaded-protein and copolymer nature. J Drug Target 7:343–354

Trichard C, Paillere-Martinot ML, Attar-Levy D, Recassens C, Monnet F, Martinot JL (1998) Binding of antipsychotic drugs to cortical 5-HT2A receptors: a PET study of chlorpromazine, clozapine, and amisulpride in schizophrenic patients. Am J Psychiatry 155:505–508

Tsai GE, Yang P, Chang YC, Chong MY (2006) D-Alanine added to antipsychotics for the treatment of schizophrenia. Biol Psychiatry 59:230–234

Turner MS, Stewart DW (2006) Review of the evidence for the long-term efficacy of atypical antipsychotic agents in the treatment of patients with schizophrenia and related psychoses. J Psychopharmacol 20:20–37

Vadalouca A, Siafaka I, Argyra E, Vrachnou E, Moka E (2006) Therapeutic management of chronic neuropathic pain: an examination of pharmacologic treatment. Ann NY Acad Sci 1088:164–186

Valenstein M, Copeland LA, Owen R, Blow FC, Visnic S (2001) Adherence assessments and the use of depot antipsychotics in patients with schizophrenia. J Clin psychiatry 62:545–551

Valenstein M, Blow FC, Copeland LA, McCarthy JF, Zeber JE, Gillon L et al (2004) Poor antipsychotic adherence among patients with schizophrenia: medication and patient factors. Schizophr Bull 30:255–264

Venkatraman S, Gale R (1998) Skin adhesives and skin adhesion. 1. Transdermal drug delivery systems. Biomaterials 19:1119–1136

Viguera AC, Baldessarini RJ, Hegarty JD, van Kammen DP, Tohen M (1997) Clinical risk following abrupt and gradual withdrawal of maintenance neuroleptic treatment. Arch Gen Psychiatry 54:49–55

Weiden PJ, Olfson M (1995) Cost of relapse in schizophrenia. Schizophr Bull 21:419–429

Witt C, Kissel T (2001) Morphological characterization of microspheres, films and implants prepared from poly(lactide-co-glycolide) and ABA triblock copolymers: is the erosion controlled by degradation, swelling or diffusion? Eur J Pharm Biopharm 51:171–181

Witt C, Mader K, Kissel T (2000) The degradation, swelling and erosion properties of biodegradable implants prepared by extrusion or compression moulding of poly(lactide-co-glycolide) and ABA triblock copolymers. Biomaterials 21:931–938

Zygmunt A, Olfson M, Boyer CA, Mechanic D (2002) Interventions to improve medication adherence in schizophrenia. Am J Psychiatry 159:1653–1664

Part II
Experimental measures of brain function and dysfunction in schizophenia

Functional Brain Imaging in Schizophrenia: Selected Results and Methods

Gregory G. Brown and Wesley K. Thompson

Contents

1   Introduction 182
2   Critical Regions 183
    2.1   Positive Symptoms 184
    2.2   Negative Symptoms 190
3   Brain Systems 199
    3.1   Functional Connectivity 199
    3.2   Summary and Comments 205
4   Final Comments and Emerging Trends 206
References 207

Abstract Functional brain imaging studies of patients with schizophrenia may be grouped into those that assume that the signs and symptoms of schizophrenia are due to disordered circuitry within a critical brain region and studies that assume that the signs and symptoms are due to disordered connections among brain regions. Studies have investigated the disordered functional brain anatomy of both the positive and negative symptoms of schizophrenia. Studies of spontaneous hallucinations find that although hallucinations are associated with abnormal brain activity in primary and secondary sensory areas, disordered brain activation associated with hallucinations is not limited to sensory systems. Disordered activation in non-sensory regions

G.G. Brown (✉)
Psychology Service, VA San Diego Healthcare System, 3350 La Jolla Village Drive, San Diego, CA 92161, USA
Department of Psychiatry, University of California, San Diego, 9500 Gilman Dr., La Jolla, CA 92093, USA
e-mail: gbrown@ucsd.edu

W.K. Thompson
Department of Psychiatry, University of California, San Diego, 9500 Gilman Dr., La Jolla, CA 92093, USA
Stein Institute for Research on Aging, University of California, San Diego, La Jolla, CA, USA

appear to contribute to the emotional strength and valence of hallucinations, to be a factor underlying an inability to distinguish ongoing mental processing from memories, and to reflect the brain's attempt to modulate the intensity of hallucinations and resolve conflicts with other processing demands. Brain activation studies support the view that auditory/verbal hallucinations are associated with an impaired ability of internal speech plans to modulate neural activation in sensory language areas. In early studies, negative symptoms of schizophrenia were hypothesized to be associated with impaired function in frontal brain areas. In support of this hypothesis meta-analytical studies have found that resting blood flow or metabolism in frontal cortex is reduced in schizophrenia, though the magnitude of the effect is only small to moderate. Brain activation studies of working memory (WM) functioning are typically associated with large effect sizes in the frontal cortex, whereas studies of functions other than WM generally reveal smaller effects. Findings from some functional connectivity studies have supported the hypothesis that schizophrenia patients experience impaired functional connections between frontal and temporal cortex, although the nature of the disordered connectivity is complex. More recent studies have used functional brain imaging to study neural compensation in schizophrenia, to serve as endophenotypes in genetic studies and to provide biomarkers in drug development studies. These emerging trends in functional brain imaging research are likely to help stimulate the development of a general neurobiological theory of the complex symptoms of schizophrenia.

1 Introduction

In 1919, Kraepelin published a comprehensive description of a "series of states" that shared the destruction of the "emotional and volitional aspects of mental life." He believed that these states were manifestations of a common disease process that justified their being classified into a single disease, which he called dementia praecox – a form of dementia that often begins in late adolescence and early adulthood.[1] Given how successfully postmortem methods of the time had been in identifying the neuropathological changes of some forms of severe intellectual and personality loss, Kraepelin was optimistic that the brain substrate of dementia praecox would be similarly identified. In his review of the postmortem literature on dementia praecox, some of which was done by Alzheimer, Nissl and Wada, Kraepelin reports "widespread disease of the *nerve-tissue*" (italics in original). These nerve cell changes included swollen nuclei, wrinkling of cell membranes, shrunken cell bodies, and diffuse loss of cortical cells (Kraepelin 1919/1971, p. 213). Despite this promising beginning, what we learned from these early studies

---

[1]Kraepelin and his colleagues used the term dementia broadly to mean a loss of mental functions, a phrase encompassing emotion, personality, and intelligence. Authors of the time used the terms manic-depressive dementia, delusional depressive dementia, paranoid gravis dementia, in addition to senile dementia.

of the neuropathology of schizophrenia did not provide the same etiological and diagnostic specificity as neuropathological studies of Alzheimer's disease or neurosyphilis provided for those disorders (Adams and Victor 1985; David 1957; Harrison and Lewis 2003). More recent neuropathological studies, however, have produced several well replicated findings, which suggest that schizophrenia is characterized by altered neural circuits (Harrison and Lewis 2003).

Evidence that the fundamental disorder in schizophrenia occurs at the level of brain circuits implies that data from functional brain imaging should be critical for tests of neurobiological theories of schizophrenia (Andreasen 1997; Brown and Eyler 2006). As a group of techniques, in vivo functional brain imaging maps the brain substrate of specific cognitive functions, assesses the distribution of metabolic and blood flow changes associated with healthy and disordered brain states, and monitors neurotransmitter changes related to disease and treatment (Andreasen 1989). In this chapter, we present a selective review of recent studies using functional brain imaging to investigate abnormal regional brain activity and disordered brain systems associated with schizophrenia. To help organize these results we divided studies into those that investigated the neural functioning of schizophrenia patients in critical brain regions and studies that investigated disordered brain systems. Other chapters in this volume will discuss metabolic and structural imaging. We also refer readers to Frankle's recent review of neuroreceptor imaging studies in schizophrenia (Frankle 2007; see also Urban and Abi-Dargham, this text). Readers interested in the technical details of the functional imaging methods used in the studies reviewed are referred to Andreasen (1989) especially for its excellent chapters on single-photon emission computed tomography (SPECT) and positron emission tomography (PET), to Wood (1987) for descriptions of cerebral blood flow (CBF) methods, and to Buxton (2009) and Huettel et al. (2009) for descriptions of functional magnetic resonance imaging (fMRI).

2 Critical Regions

For abnormal neural functioning in a brain region to be a critical cause of schizophrenia, its dysfunction would need to be a biologically plausible mechanism that was empirically associated with some of the core symptoms and signs of schizophrenia in a majority of patients. Abnormal neural function in the critical region might be the primary cause of a set of symptoms and signs or the primary cause of the diathesis that places an individual at risk for particular signs and symptoms. Given the symptom heterogeneity of schizophrenia, disrupted neural functioning in the critical region would not need to be casually related to all of the core symptoms of schizophrenia in order to be theoretically useful. In a strong form of the critical region account, the appearance of characteristic symptoms and signs would necessarily imply dysfunction in the critical region. This strong concept of a critical region justifies the concept of anatomical diagnosis that has been used successfully to localized brain disease in neurology (Adams and Victor 1985, p. 3). Historically, most imaging studies of localization of function in schizophrenia seem to have assumed the strong form of the critical region account, although more recent papers have also focused on connectivity models of brain dysfunction in schizophrenia.

One of the challenges for a critical region theory is to explain the scope of the symptoms and signs composing the syndrome to be explained. The classical approach to the problem of symptom scope developed by early behavioral neurologists assumes the existence of a critical neurobehavioral function that is a necessary component of each symptom and sign defining the syndrome (Hecaen and Albert 1978, p. 14; Wernicke 1874/1969). A lesion in the brain substrate where the critical function is computed impairs the function, which in turn causes all of those observable behaviors that depend on the critical function to be disrupted. Wernicke (1874/1969), for example, argued that the neural representation of speech sounds was stored in the posterior portion of the left superior temporal gyrus (STG), and that damage to this area caused a syndrome that included difficulties comprehending speech, impaired repetition of spoken words, phonemic and semantic paraphasia, alexia, and agraphia (Wernicke 1874/1969). In Wernicke's analysis of this form of receptive aphasia, the storage of neural representations of speech sounds was the critical neurobehavioral function whose damage would lead to the observable symptoms of impaired speech comprehension, impaired repetition, etc. The strong form of the critical region account, as represented by Wernicke's aphasia, will be our starting point when discussing the brain substrate of schizophrenia.

2.1 Positive Symptoms

Functional brain imaging studies of the hallucinations, delusions, and disorganized speech observed among individuals with schizophrenia have examined brain activity during the spontaneous generation of symptoms and during controlled neurobehavioral experiments. There are several imaging hypotheses about the functional brain systems underlying hallucinations in schizophrenia. One hypothesis draws an analogy to focal seizures originating in primary and secondary sensory association areas. According to this account, hallucinations would be caused by excessive neural activity originating in primary or secondary sensory areas, for example, the auditory cortex during auditory/verbal hallucinations (Lennox et al. 1999). When considering this hypothesis it is worthwhile to recognize that auditory hallucinations are an uncommon initial manifestation of seizures, even though auditory/verbal hallucinations are common among patients with schizophrenia (Adams and Victor 1985, p. 240; Lewis et al. 2009). A variant of the excessive activation hypothesis asserts that the abnormal neural activity in sensory areas leading to hallucinations is due to disordered inhibition of sensory areas caused by neural dysfunction in a remote inhibitory center. Investigators have also tested the hypothesis that auditory–verbal hallucinations arise because schizophrenia patients experience a disorder in the monitoring of internal speech. Other investigators have argued that hallucinations are caused by an inability to distinguish current experience from previous memories.

2.1.1 Spontaneous Presentation

To investigate the neural basis of hallucinations among patients with schizophrenia, imaging researchers have compared spontaneous brain activity during hallucinations with brain activity during non-hallucinatory states, using within-session or between-session designs. Table 1 presents the main results from functional imaging studies of adult schizophrenia patients experiencing spontaneous hallucinations. All studies except one, reported only increased brain activity during hallucinations, with the Hoffman and colleagues' (2008) study finding a mixture of increased and decreased brain activity. Several studies reported increased brain activation in the primary or secondary auditory cortex during periods when patients signaled they were experiencing auditory/verbal hallucinations (Copolov et al. 2003; Dierks et al. 1999; Silbersweig et al. 1995; van de Ven et al. 2005). One paranoid schizophrenia patient with visual hallucinations displayed increased activity in the "higher visual areas" (Oertel et al. 2007). Areas in the left temporal lobe, such as the left superior and middle temporal gyri, which are involved in speech sound perception, the processing of sound based word representations, and naming, were found to be activated during hallucinations in four additional studies (Hecaen and Angelergues 1964). The finding of increased activity in auditory and speech areas support the view that abnormal neural activity in higher order sensory and speech processing areas contribute to the genesis of hallucinations.

About 70% of the studies listed in Table 1 found that auditory/verbal hallucinations increased neural activity in the right middle temporal gyrus and/or in the right superior temporal cortex. Investigators have hypothesized that this increased activation in the right hemisphere was due to abnormal speech lateralization, to excessive left-handedness in the samples studied, or to the engagement of right hemispheric regions involved in processing the emotional prosody of speech (Hoffman et al. 2008; Lennox et al. 1999).

Studies summarized in Table 1 that show activation during hallucinations in the left inferior frontal cortex along with activation in the left middle and STG are in line with the view that hallucinations reflect disordered internal speech. Simons and colleagues (2010) found that ratings by schizophrenia patients of the reality of auditory–verbal hallucinations correlated with the magnitude of hallucination-related activation in the inferior frontal gyri, which are activated during overt speech (Troiani et al. 2008). Moreover, judgments of the reality of hallucinations was correlated with the extent of coupling within a network composed of the inferior frontal gyrus, ventral striatum, auditory cortex, anterior cingulate and right posterior temporal lobe. These results imply that the perceived reality of hallucinations is related to activity in classical frontotemporal speech circuits that are modulated by activity in reward circuitry and by activity in conflict resolution brain regions. In other studies, increased brain response to hallucinations was observed in the thalamus, caudate, amygdala, and insula, regions involved in the processing and modulation of arousal, emotion, and reward (Balleine et al. 2007; Haber and Calzavara 2009; Van der Werf et al. 2002). Activation in these regions might contribute to the emotional strength and valence of hallucinations. Activation

Table 1 Imaging studies of adult schizophrenia or psychotic patients with spontaneous hallucinations

First author (year published)	Modality	N	Age years (SD)	Gender M/F	Chronicity years (SD)	Imaging modality	Areas of increased activity	Medication mean (SD)
Copolov (2003)	Auditory/verbal	8	30 (8)	8/0	Not reported	[$H_2^{15}O$] PET	Auditory cortex, R medial frontal cortex (extending to anterior cingulate), R prefrontal regions, L parahippocampal gyrus/hippocampus	Chlorpromazine equivalents 640 (740) mg
Dierks (1999)	Auditory	3	36 (8)	2/1	12(7)	BOLD fMRI	Transverse temporal gyrus, frontoparietal operculum, posterior superior temporal gyrus, hippocampus, amygdala	One haloperidol, two clozapine
Hoffman (2008)	Auditory/verbal	6	Not reported	Not reported	Not reported	BOLD fMRI	L anterior insula, R middle temporal site, superior temporal gyrus	Three clozapine, one haloperidol, one olanzapine, one no concurrent medications
Jardri (2009)	Complex: auditory, visual	1	32	1/0	19	BOLD fMRI	GLM: insula, superior temporal gyrus; ICA: auditory, visual and multisensory components	No concurrent medications
Lennox (1999)	Auditory/verbal	1	26	1/0	Not reported	BOLD fMRI	R middle temporal gyrus, R superior and L superior temporal gyri, R middle and inferior frontal gyri, R anterior cingulate, R cuneus	Stable medication
McGuire (1993)	Auditory/verbal	12	34	12/0	10	HMPAO SPECT	L inferior frontal regions (Broca's area)	Chlorpromazine equivalents 667 (446) mg

Functional Brain Imaging in Schizophrenia

Study	Modality	N	Age (SD)	M/F	Duration of illness	Method	Regions	Medications
Oertel (2007)	Visual	1	27	1/0	12	BOLD fMRI	Higher visual areas, hippocampus	Clozapine
Parellada (2008)	Auditory/ verbal	9	25 (4)	4/5	Estimated 1 (1.25)	FDG PET	Supplementary motor area, anterior cingulate, medial superior frontal area, cerebellum	
Shergill (2000)	Auditory/ verbal	6	35 (11)	6/0	11 (8)	BOLD fMRI	L inferior frontal cortex, insular region, anterior cingulate, L middle temporal cortex, R superior temporal gyrus, R thalamus and inferior colliculus, L hippocampal and parahippocampal cortex	Five taking 2nd generation antipsychotics, one taking haloperidol injection
Shergill (2004)	Auditory/ verbal	2	36	2/0	14	BOLD fMRI	L inferior frontal gyrus, middle temporal gyrus, superior temporal gyrus, R middle frontal gyrus	Clozapine and olanzepine
Sommer (2009)	Auditory/ verbal	24	37 (10)	17/7	15 years (est.)	BOLD fMRI	R inferior frontal cortex, insula, supramarginal gyrus, R superior temporal gyrus	13 clozapine, remaining medications a mixture of 1st and 2nd generation antipsychotics
Silbersweig (1995)	Auditory/ visual	5/1	32	Group study: not reported; case study 2 years		H$_2^{15}$O PET	Thalamus, striatum, hippocampus, cingulate gyri, orbitocortex, auditory cortex, visual cortex	Not reported
Woodruff (1994)	Auditory/ verbal	1[a]	48	1/0	Not reported	BOLD fMRI	R middle temporal gyrus	Not reported

(continued)

Table 1 (continued)

First author (year published)	Modality	N	Age years (SD)	Gender M/F	Chronicity years (SD)	Imaging modality	Areas of increased activity	Medication mean (SD)
van de Ven (2005)	Auditory/verbal	6	Not reported	3/3	Not reported	BOLD fMRI independent component analysis of auditory cortex activity	Bilateral auditory cortex in one patient, unilateral auditory cortex activity in two patients	Three on 1st generation antipsychotics; three on 2nd generation antipsychotics

R Right hemisphere, *L* Left Hemisphere; when no hemisphere is listed the effect is bilateral. *fMRI* functional magnetic resonance imaging, *FDG* 18F-fluorodeoxyglucose, *PET* positron emission tomography, *HMPAO* hexamethylpropyleneamineoxime, *SPECT* single-photon emission computed tomography, *BLOOD* oxygen level dependent, *GLM* general linear model

[a]Patient was mixed-handed

in orbitofrontal, medial frontal cortex, prefrontal cortex, and anterior cingulate has also been observed among schizophrenia patients during hallucinations. Activation in these regions might reflect the brain's attempt to modulate hallucinations and resolve conflicts with other processing demands. Finally, findings of hippocampal and parahippocampal activation support a contribution of disordered memory functioning to the genesis of hallucinations (Seal et al. 2004). Studies of spontaneous hallucinations reveal that although hallucinations are associated with abnormal activity in higher order sensory systems in many schizophrenia patients, brain activation associated with hallucinations is not limited to sensory systems.

2.1.2 Brain Activation Studies

If hallucinations activate brain regions involved in speech perception, hallucinations might interfere with brain activation elicited by listening to external speech. In support of this prediction, studies of brain activation to external speech while patients experience auditory/verbal hallucinations have found reduced activation in the temporal lobe to speech perception compared to periods of diminished hallucinations (David et al. 1996; Woodruff et al. 1997). These results indicate that for some schizophrenia patients hallucinations appear to compete with external speech for neural processing resources in the temporal lobe.

Internal speech is one source of hallucinations that could compete for neural resources usually allocated to process external speech. This view would be compatible with the hypothesis that auditory/verbal hallucinations arise when patients characterize internal speech as originating from an external source rather than being self-generated. Schizophrenia patients with a strong history of hallucinations do display abnormal activation in the left temporal lobe while engaged in controlled speech imagery compared with either schizophrenia patients without a history of auditory/verbal hallucinations or with healthy controls (McGuire et al. 1996). Moreover, direct comparison of internal and external speech has found that patients show a greater activation in the left STG than do controls during internal speech, even though both groups show very similar levels of temporal activation to external speech (Simons et al. 2010). These data have been interpreted as supporting the idea of disordered self-monitoring of internal speech in schizophrenia. In this account of silent speech generation, efferent copies of the internal speech plan generated in frontal regions, probably in the left inferior frontal gyri, are compared with the internal afferent copy of the silent speech (Ford et al. 2001; Simons et al. 2010; Stephan et al. 2009). If the copies match, the neural activity in the sensory area is reduced (Simons et al. 2010). In schizophrenia, there is a disruption between the planned and perceived speech, giving the perceived speech an alien quality (Ford and Mathalon 2004; Heinks-Maldonado et al. 2007). Because the left inferior frontal cortex is normally activated in schizophrenia patients when they engage in internal speech, the disordered monitoring appears to occur at comparison or inhibitory stages, perhaps due to impaired functional connectivity (FC) between left inferior frontal and left superior temporal cortex during internal speech (McGuire et al. 1996;

Shergill et al. 2000; Simons et al. 2010; Stephan et al. 2009). This model is supported by the finding of altered activation in the left STG when schizophrenia patients misattribute self-generated speech (Allen et al. 2007). Evidence that schizophrenia patients experience impaired activation in the pulvinar while listening to their own voice, suggests that impaired modulation of auditory attention might also contribute to disordered self-monitoring of the patient's internal speech. This idea has, however, not been directly tested.

2.1.3 Comments

Functional imaging studies of hallucinations have established that perceptions occurring without external stimulation are associated with neural activation in specific brain regions. The brain dysfunction underlying hallucinations is complex, involving abnormal activity in primary and secondary sensory regions, in language cortex, in systems involved with the processing of reward and emotion, in the brain substrate of executive functioning, and in memory systems. An open question is whether more than one of these systems must be involved to generate hallucinations as experienced by schizophrenia patients. If so, the analogy to focal seizures described above would be inappropriate. Alternatively, subgroups of schizophrenia patients might exist each with a distinct neural basis for their hallucinations – a hypothesis that requires more direct testing. Evidence is increasing that one common mechanism underlying hallucinations is the reduced ability of an internal speech plan to modulate language activation in sensory speech areas, such as the STG. The commonly reported activation of the right middle and superior temporal gyri during hallucinations implies that the prosodic and emotional aspects of language are as poorly modulated as are the phonological and lexical aspects of language, typically processed in the left temporal lobe. The disordered modulation of neural processing related to sensory, emotional and lexical aspects of internal speech generation does not appear to be due to impaired activation of fronto-executive systems that are activated by internal speech among healthy individuals. Rather, the current data suggests that the abnormal modulation of temporal lobe activity during the internal speech of schizophrenia patients might be due to impaired functional connectivity between frontal and temporal regions. The definition of frontal connectivity and the contribution that abnormal connectivity might make to genesis of the symptoms of schizophrenia is discussed in more detail in Sect. 3.

2.2 Negative Symptoms

Since the time of Kraepelin, investigators have attributed the deficits that schizophrenia patients experience in "higher intellectual abilities" to neural abnormalities in the frontal lobe (Kraepelin 1919/1971, p. 219). More recently, investigators have ascribed a broad range of negative symptoms of schizophrenia, such as anhedonia, poverty of speech and affective flattening to reduced resting metabolism, or impaired activation in the frontal lobe (Andreasen et al. 1992; Volkow et al. 1987; Wolkin et al. 1992). This frontal lobe-deficit hypothesis is another instance of the critical brain region hypothesis described above. Because of the influence that studies of frontal lobe dysfunction have had on theories of negative symptoms in schizophrenia, we focus on the evidence of impaired frontal lobe function below (Weinberger et al. 1994; Weinberger and Berman 1996).

2.2.1 Studies of Uncontrolled Mental State

Many early brain imaging studies acquired scans from patients in a "resting" condition. Operationally, these were conditions where the mental activity of patients was uncontrolled or where control was limited to reducing ambient stimulation to one or two sensory modalities. In the earliest reported study of regional CBF (rCBF) in schizophrenia, Ingvar and Franzen (1974a, b) observed an altered distribution of CBF among schizophrenia patients with reduced CBF seen in the dorsal frontal cortex and increased CBF in postcentral regions. Even though the patients studied were quite chronic, with mean symptom duration about 30 years, and many patients were severely ill, including some with catatonia, schizophrenia patients showed normal mean hemispheric CBF. Although symptom severity did not influence the magnitude of the hypofrontal effect, the increased flow observed in occipital–parietal and occipital–temporal areas was apparent only in the high symptom group.

Ingvar and Franzen's papers stimulated a large number of studies aimed at replicating and extending the original findings. Subsequent studies, however, have not consistently found reduced blood flow or reduced metabolic activity in the frontal cortex of schizophrenia patients (Goldstein et al. 1990; Gur and Gur 1995; Liddle 1996). A review of studies published from 1982 to 1997 using PET to measure glucose metabolism is representative of the empirical basis of the mixed findings (Buchsbaum and Hazlett 1998). Of the 22 papers where data were available, 15 reported lower mean frontal/occipital ratio values in schizophrenia patients than among healthy controls, six reported larger values among patients, and one reported equal values (Buchsbaum and Hazlett 1998; Table 1). If only studies where significant differences in mean frontal/occipital ratios were considered, seven found lower glucose metabolic ratios among patients, whereas three found larger ratios. When the ratio of frontal glucose metabolism to whole slice was calculated, schizophrenia patients had lower values in 15 of 18 studies, although only eight of these studies yielded significant differences (Buchsbaum and Hazlett 1998). Even though the preponderance of studies reported reduced frontal metabolism in schizophrenia, the results were only moderately consistent.

The finding of non-significant group differences in resting frontal flow or metabolism for a sizable minority of functional imaging studies suggests that if the hypofrontal effect exists, the effect size is only small to moderate. Two meta-analytic studies investigated the between group effect size associated with resting hypofrontality (Davidson and Heinrichs 2003; Hill et al. 2004). Their results are

Table 2 Pooled effect sizes reported in functional imaging studies of resting cerebral metabolism and blood flow comparing schizophrenia patients and healthy volunteers

Brain region	Mean Cohen's d^a	Confidence interval
Davidson and Heinrichs (2003) Meta-analysis[b]		
Left frontal	−0.48	−0.80 to −0.15
Right frontal	−0.43	−0.74 to −0.12
Total frontal	−0.65	−0.88 to −0.42
Hill et al. (2004) Meta-analysis[c]		
Whole brain	−0.26	−0.40 to −0.11
Frontal resting state (relative)	−0.32	−0.43 to −0.21
Frontal resting state (absolute)	−0.55	−0.68 to −0.41

[a]Mean difference between schizophrenia patients and healthy volunteers divided by the pooled standard deviation. Negative values imply lower values for the schizophrenia group
[b]Data represented glucose metabolism or blood flow measured with positron emission tomography (PET) or with single-photon emission computed tomography (SPECT). Rates, that is, absolute values, were analyzed. Period of review: 1980–2001
[c]Data represented glucose metabolism or blood flow measure with PET, SPECT or xenon inhalation. Period of review: 1974–2003

presented in Table 2. These meta-analyses confirm that schizophrenia patients experience reduced CBF or metabolism in the frontal cortex compared with healthy volunteers. The standardized mean difference in resting absolute metabolism or flow between schizophrenia patients and controls – the effect size – falls into the medium range, with a mean frontal lobe effect size of −0.55 in the Hill et al. (2004) study and −0.65 in the Davidson and Heinrichs (2003) study. The effect size magnitude for absolute resting metabolism or flow is composed of two components: one reflecting the impact of schizophrenia on global values and the other reflecting regional effects. As Table 2 shows, when frontal values are normalized by whole-brain or hemispheric values, the mean effect size for relative hypofrontal metabolism or flow was reduced to the small to moderate effect size range ($d = -0.32$; Hill et al. 2004). A portion of the absolute resting effect size appears to be related to the small but measurable reduction of whole-brain blood flow among schizophrenia patients.

One implication of the finding of small to medium effect sizes is that most functional imaging studies of resting frontal flow and metabolism in schizophrenia are underpowered. To use a 2-tailed t-test to detect a medium effect size of −0.6 at $p = 0.05$ and with good statistical power = 0.8 requires a sample size of 45 per group (Cohen 1988). If the p-value is increased to 0.01, the required sample size is 67 per group. Less than one-third of the 29 studies of absolute resting frontal metabolism or flow reviewed by Hill and colleagues (2004) had sample sizes large enough to detect differences at $p = 0.05$ with good statistical power (> 0.80), and only about one-sixth involved sample sizes sufficient to detect differences at $p = 0.01$ with good statistical power. Not surprisingly, therefore, sample size has emerged as a significant moderator of functional brain-imaging effect sizes, with larger studies associated with smaller effect sizes (Davidson and Heinrichs 2003). An additional implication of the finding of moderate effect sizes is that resting flow or metabolism is unlikely to be a powerful marker separating schizophrenia patients from healthy controls. If resting flow and metabolism are normally distributed across individuals with equal variance in patient and control groups, even a mean effect size of 0.60 corresponds to a 76% overlap of the two groups.

Both meta-analytic studies found significant moderators of effect size magnitude. Handedness, gender, age, and duration of illness moderated the magnitude of the resting metabolism/flow effect size, in addition to sample size (Davidson and Heinrichs 2003; Hill et al. 2004). Older, more chronic, right-handed, male schizophrenia patients experienced larger reductions in frontal metabolism and blood flow (Davidson and Heinrichs 2003; Hill et al. 2004). Hill and colleagues (2004) found that neuroleptic treatment had a significant impact on the magnitude of the patient/control difference in resting whole-brain blood flow (mean Cohen's d for untreated patients = −0.08 versus mean d for treated patients = −0.63), a finding that had been reported in an early study of CBF in patients with schizophrenia or major affective disorder (Goldstein et al. 1990). The presence of neuroleptic medication was also significantly associated with larger hypofrontality effects measured in absolute flow units (Hill et al. 2004). Neuroleptic medication, however, did not contribute to variation in effect size for relative measures of resting frontal metabolism and flow in the study of Hill and colleagues (2004). The Davidson and Heinrichs (2003) review did not find any impact of medication (on versus off medications) on the size of the hypofrontality effect, although they did not provide separate data for absolute and relative hypofrontality effects.

Differences in imaging methods, interestingly, did not greatly influence the consistency of results. Scanning method (xenon inhalation, PET, SPECT) was not significantly associated with effect size, although the mean SPECT effect size was about one-third the size of the effect sizes obtained from xenon and PET studies (Hill et al. 2004). Whether metabolism or flow was measured, scanner resolution, and use of an individuals' anatomical scan or a standardized template to localize the region of interest did not significantly contribute to variation in effect sizes (Davidson and Heinrich 2003). More recent studies, interestingly, found smaller hypofrontal effect sizes than initial studies (Hill et al. 2004).

2.2.2 Studies of Brain Activation: Working Memory

Studies imaging CBF and metabolism at rest have found only small to moderate differences in neural functioning when comparing schizophrenia patients with healthy individuals. Some investigators have criticized resting studies for their lack of control over psychological and physiological sources of unwanted variation (Weinberger and Berman 1996). Cognitive challenge tasks, it is argued, might produce larger between group effects and provide more specific links between disordered behavior and disordered brain function (Brown and Eyler 2006; Weinberger and Berman 1996). We will examine below the evidence that cognitive challenge tasks have led to more robust group differences and to stronger theories of neurobehavioral dysfunction in schizophrenia by examining functional brain imaging studies of frontal lobe activation, especially when participants perform working memory (WM) tasks (Table 3).

Table 3 Pooled effect sizes comparing schizophrenia patients and healthy volunteers reported in functional imaging studies of cerebral metabolism and blood flow response to an activating task

Brain region	Mean Cohen's d[a]	Confidence interval
Davidson and Heinrichs (2003) Meta-analysis		
Left frontal activation	−0.54	−0.78 to −0.31
Right frontal activation	−0.54	−0.90 to −0.18
Total frontal activation	−0.81	−1.06 to −0.57
Hill et al. (2004) Meta-analysis		
Frontal activation (relative)	−0.37	−0.53 to −0.22
Frontal activation (absolute)	−0.42	−0.65 to −0.20

[a]See notes to Table 1 for details

In 1994, Goldman-Rakic published one of the most systematic statements of a critical region account of the functional neuropathology of schizophrenia. In this theory, the cardinal symptoms of schizophrenia are caused by a disorder in neural systems by which the prefrontal cortex maintains and accesses "on line" knowledge representations of the external world (Goldman-Rakic 1994). A disorder in these neural systems was posited to disrupt cortical feedback important to bringing knowledge representations in line with external reality. Such disordered feedback models were described above in the review of brain imaging studies of hallucinations. Goldman-Rakic's account, however, was unusual in attributing the varied symptoms of schizophrenia to the impairment of a single cognitive function, namely to impaired WM.

WM, though variously described, is often defined as a multi-component memory system that stores information over the short-term to support the cognitive manipulations involved in mental work (Baddeley 1976, pp. 169–187; Cowan 2008). Of the various cognitive components involved in WM, Goldman-Rakic emphasized the importance of maintaining information from moment to moment, updating the contents of short-term memory and accessing these contents as core WM components that when impaired could cause the symptoms of schizophrenia (Goldman-Rakic 1994).

Much of the evidence supporting Goldman-Rakic's 1994 theory was based on animal studies of the delayed response task. To use delayed response paradigms to study human WM, investigators need to account for the use of rehearsal to maintain the to-be-remembered item during the delay (Peterson and Peterson 1959). One method of reducing the impact of rehearsal is to introduce intervening items to shift attention from rehearsing the initial items to processing the intervening items (Brown 1958; Posner and Rossman 1965). If intervening items are presented in a continuous manner, they will serve the dual functions of being new memory targets and preventing rehearsal of previous items (Brown and Eyler 2006). To date, the N-back task has been the most commonly used continuous memory task in functional imaging studies of the WM abnormalities experienced by schizophrenia patients (Callicott et al. 2000; Glahn et al. 2005). The N-back memory task requires individuals to remember items presented separately in a continuous stream and to respond when a repeated item was presented N-trials previously (Callicott et al. 2000; Gevins and Cutillo, 1993). Although little formal construct validity research has been done with this task (Kane et al. 2007), the N-back task provides a face valid assessment of an individual's ability to update WM with newly presented sequential information on a moment-by-moment basis, to maintain the sequential information in the face of distraction, to monitor and manipulate online information, and to access sequential information when responding (Cohen et al. 1997; Glahn et al. 2005). The task might also involve the encoding of item familiarity and appears to require participants to decide on how to allocate attention between updating and retrieving from WM when new items are presented.

Glahn and colleagues used activation likelihood estimation to find concordances in the localization of N-back effects in functional imaging studies of schizophrenia patients. Their meta-analytic review of 12 studies found consistently greater activation during N-back performance among healthy participants than among schizophrenia patients in dorsolateral-prefrontal cortex (DLPFC) [Brodmann areas (BA) 9], in inferior prefrontal cortex (BA 11) and in the right ventrolateral/insular cortex (BA 13). They also found consistently greater activation among patients than controls in the left frontal pole (BA 10) and in anterior cingulate (BA 32).

Although the study by Glahn and colleagues provided evidence about the consistency of the location of the hypofrontality effect associated with schizophrenia, it did not provide data on effect sizes. We examined the studies presented in the Glahn paper, as well as additional studies found in a Medline search of papers using the N-back task to study schizophrenia patients. For our analysis to correspond to the time frame covered in the meta-analyses of Glahn and colleagues, we examined papers published before 2006. Because of challenges due to the wide difference in methods and reporting of results, our results include only 18 patient–control comparisons taken from half-dozen studies. Our estimate of Cohen's d when schizophrenia patients activated less than controls in the DLPFC ranged from 0.77 to 2.30, with a median of 1.26. Corresponding percentages of overlap in patient and control distributions ranged from 25 to 70% with a median of 53%. Although these effect size results support the hypothesis that hypofrontality effects are larger for behavioral challenge studies than for resting studies, this conclusion should be tempered by the tendency for authors who found no effect to provide no data useful for calculating effect sizes. Our findings are, therefore, biased towards larger effect sizes. Perhaps more troubling, several authors reported larger activation effects for schizophrenia patients than for healthy controls on N-back tasks. Effect sizes associated with findings of hyperfrontality ranged from 0.78 to 1.82 with a median value of 1.02.

A second approach to dealing with rehearsal during a delayed response task is to permit rehearsal to occur and then study the impact of experimental manipulations of component WM functions on brain activity. The typical dependent variable in the behavioral analysis of such tasks is response time. The item memory task developed by Sternberg is a delayed response paradigm where memory load is manipulated; it has been used in several studies investigating the brain substrate of abnormal WM functioning in schizophrenia (Sternberg 1969; Manoach et al. 1999). In the two initial studies using this paradigm, schizophrenia patients displayed greater activation to increasing memory load in the DLPFC (Manoach et al. 1999, 2000). Recently, the

Function Biomedical Informatics Research Network reported results from a large sample study of Sternberg's item recognition test involving 106 schizophrenia patients and 111 healthy controls. Schizophrenia patients experienced greater activation than healthy volunteers in bilateral dorsolateral-prefrontal regions of interest for memory probe events, during which item information in WM is accessed (Potkin et al. 2009). The group difference was largest at intermediate memory loads, a finding that was reflected in a significant quadratic trend (Potkin et al. 2009). In a subset of fBIRN patients who had response time and functional MRI data for the Sternberg task, memory scanning speed was estimated by fitting a mathematical model to each individual's response time data (Brown et al. 2009). The correlation between memory scanning speed and the linear brain response to increasing memory load differed between patients and controls in left inferior and left middle frontal gyrus, bilateral caudate, and right precuneus. The pattern of findings in these regions indicated that among healthy volunteers high scanning capacity was associated with high neural capacity, whereas among patients memory scanning speed was uncoupled from brain response to increasing memory load (Brown et al. 2009). Group differences in correlation of memory scanning speed with the quadratic trend in DLPFC response to memory load suggested inefficient or disordered neural inhibition among individuals with schizophrenia, especially in the left perirhinal and entorhinal cortices (Brown et al. 2009).

As with the N-back task, some investigators have failed to find any group difference in the degree of neural activation in the prefrontal cortex during performance of the Sternberg task (Koch et al. 2008). When group differences between patients and controls were found, the differences tended to be smaller for the Sternberg task than the N-back task with Cohen d ranging from 0.36 to 1.27.

Investigators studying WM impairment in schizophrenia patients have found no differences between patients and controls in neural activation in the DLPFC, greater activation among patients, or greater activation among healthy comparison subjects. Several hypotheses have been offered to account for the heterogeneity of results (Callicott et al. 2003; Manoach 2003). Patients might experience less spatial coherence to their activation patterns than controls (Manoach 2003). In the extreme case, patients might show a mixture of hyper- and hypo-activation (Callicott et al. 2003; Johnson et al. 2006). Patients and controls might also differ in the temporal pattern of activation. Differences in the spatial and temporal patterns of activation might reflect impaired cognitive control of processing resources among schizophrenia patients (Johnson et al. 2006; Koch et al. 2008; Manoach 2003). WM tasks involve component processes that might be differentially affected by subgroups of schizophrenia patients (Brown et al. 2009; Manoach 2003). Moreover, the activation of a particular component of a WM task might be related to the task strategy adopted by participants. Differences in processing strategy might, therefore, contribute to the variation in WM findings. Perhaps the most commonly advanced hypothesis about the heterogeneity of WM results is that the relationship between performance and neural activation in the DLPFC is an inverted U (Callicott et al. 2003; Manoach 2003). If this curve is shifted leftwards in the schizophrenia group, those patients achieving intermediate levels of performance success will display greater neural activation than healthy comparison subjects. When this performance–activation pattern is observed, some investigators conclude that the neurocognitive functioning of patients is inefficient (Callicott et al. 2000). Finally, variations in the differences between schizophrenia patients and healthy controls in resting CBF and blood volume could contribute to some of the variability seen in the literature reporting on blood oxygen level dependent (BOLD) functional MRI findings. Indeed, several investigators have reported increased resting cerebral blood volume among schizophrenia patients (Cohen et al. 1995; Loeber et al. 1999). In Hill's meta-analysis schizophrenia patients taking neuroleptic medications experienced larger reductions of global resting CBF than did medication free patients (Hill et al. 2004). According to a standard model of the BOLD fMRI signal, changes in resting blood volume and flow will alter the maximum activation values observable in a particular fMRI study (Buxton 2009; Davis et al. 1998). Differences in resting vascular status between patients and controls, among patients themselves, and within patients across time might be important sources of uncontrolled variation in fMRI studies of individuals with schizophrenia.

Some researchers have attributed reduced brain activation in cognitive challenge studies of schizophrenia to the poorer performance of the patients (Ebmeier et al. 1995; Weinberger and Berman 1996). This criticism needs to be considered in light of the aim of many functional brain imaging studies, namely to identify disordered brain systems that underlie impaired cognitive and behavioral functioning. Two general approaches to studying neurobehavioral impairment seem possible, both potentially flawed. Investigators might use behavioral tasks on which patients perform as well as controls to investigate impaired neurocognitive functioning in schizophrenia. Such tasks, however, would have poor face validity as indicators of the impaired neurocognitive function that investigators are trying to explain. Alternatively, investigators might use cognitive tasks that produce impaired performance of the neurocognitive function of interest. Although ideally the impaired performance of schizophrenia patients would be due to dysfunction in the neurobehavioral function of interest, poor performance might be caused by factors, such as reduced motivation, altered time on task, misinterpreted expectancies, or distractibility, which would reduce performance on any effortful task. The possibility that several factors in addition to the neurobehavioral function of interest might cause impairment would place the task's construct validity under question.

Early papers using functional MRI to study the brain's response to behavioral tasks did find that the magnitude of brain activation can be altered by manipulating performance level, even among healthy individuals (Rao et al. 1996). Rao and colleagues (1996), for example, instructed subjects to move the fingers of the right hand at various rates to match a metronome and found the BOLD response to be monotonically related to the rate of movement. Investigators studying schizophrenia have broadly interpreted performance–activation data from studies like Rao et al. (1996) in two different ways. One interpretation is that activation data from patients cannot be unambiguously interpreted unless the performance of patient and control groups has been rendered comparable. Several methods to achieve comparability have been proposed including retrospectively selecting subgroups of patients and healthy volunteers whose performance is matched, prospectively screening and selecting patients based on performance, developing tasks on which the performance of schizophrenia patients and controls are comparable, relying on event-related designs to separate trials representing intact and impaired performance, and using analysis of covariance to statistically control for performance differences. The strengths and weaknesses of each of these strategies to mitigate the performance–activation dilemma faced by clinical functional imaging investigators have been reviewed elsewhere (Brown and Eyler 2006). The other interpretation of performance–activation data is that when groups are mismatched on performance, general factors that might cause performance decrements, such as poor motivation, lack of cooperativeness or inattentiveness, must be excluded as explanations of impaired performance (Crespo-Facorro et al. 2001). In the study by Rao and colleagues the additional factor was the instruction to keep pace with the metronome. As long as the impact of general performance factors on a schizophrenia patient's ability to engage in the task at hand can be ruled out, studies of abnormally responding brain systems associated with impaired patient performance should provide useful insight into the disordered brain systems underlying the cognitive and behavioral deficits that schizophrenia patients experience.

2.2.3 Summary and Comments

The bulk of the evidence indicates that schizophrenia patients experience both absolute and relative deficits in CBF or metabolism when scanned at rest. The magnitude of the group difference, though, is small – perhaps explaining why a large minority of the studies analyzed in a previous review found no evidence of diminished resting flow (Andreasen et al. 1992). Functional brain imaging studies using cognitive challenges have produced evidence of aberrant neural activity in the prefrontal cortex of schizophrenia patients, though the pattern of the abnormality is complexly determined. When found, abnormal frontal activity elicited by the N-back WM task appears to produce large between group differences. Other frontal lobe tasks, however, do not seem to yield such large effect sizes. In Hill and colleague's (2004) meta-analysis, effect sizes (Cohen's d) for cognitive challenge tests of frontal lobe abnormality were found to be in the small to moderate range (~ 0.45). In their meta-analysis, Davidson and Heinrichs (2003) reported a larger mean Cohen d value (0.81) that was still below the levels we found for N-back tasks. Although there is evidence to support the hypothesis that cognitive challenge tasks produce larger group differences than resting studies, at least for some tasks, a direct comparison of resting and activation frontal values using a priori defined regions of interest was not reported in any of the studies we reviewed.

Activation studies have produced increasing evidence that neural activity can be greater in some brain regions among patients with schizophrenia compared with healthy participants. These findings support the hypothesis that schizophrenia patients might be able to compensate for inefficient brain functioning. Monitoring neural activity while participants perform a cognitive challenge task provides some of the strongest evidence in support of a neural compensation hypothesis The successful use of functional brain imaging data to establish the existence of neural compensation in schizophrenia is one of the most exciting developments in the field, and strongly supports the use of cognitive activation studies to identify specific abnormalities of brain activity that underlie the disordered functioning of schizophrenia patients.

3 Brain Systems

Theorists have long believed that the human brain uses distributed networks of anatomically distinct and functionally-specialized brain regions that communicate with each other to process information (Cohen and Tong 2001; Toga and Mazziotta 2002; Wernicke 1874/1969). It has only been with recent developments in functional neuroimaging, however, that investigators have been able to test hypotheses about how disordered communication or connectivity among brain regions might contribute to the pathophysiology of schizophrenia (Honey and Fletcher 2006). Connectivity analyses involving human brain functional data were first performed using PET (Clark et al. 1984; Horwitz et al. 1984; Friston et al. 1993), and have since expanded into fMRI, magnetoencephalography (MEG), electroencephalography (EEG), and peripheral physiological measures (Friston 1994; Sun et al. 2004; Salvador et al. 2005; David et al. 2004; Chen et al. 2008; Jeong et al. 2001).

3.1 Functional Connectivity

FC denotes the observed temporal associations in functional neuroimaging time series data obtained from anatomically-distinct, spatially-separated brain regions (Friston et al. 1993; Horwitz 2003; Salvador et al. 2005). FC has been operationalized utilizing a broad variety of statistical techniques, each tailored to a particular experimental design and imaging modality (Zhou et al. 2009). These techniques quantify the notion of FC in different ways, and it is unclear to what extent they index the same or different constructs. In this section we will generally limit discussion to applications in schizophrenia research that specifically attempt to characterize time or frequency domain relationships between neuroimaging time series extracted from two or more regions or voxels. FC techniques which fulfill this requirement include zero-order (Biswal et al. 1995) and lagged cross-correlational analyses (Siegle et al. 2007), approaches based on singular value decompositions of time series (Friston and Frith 1995), cross-coherence-type frequency domain measures (Peled et al. 2001; Koenig et al. 2001; Spencer et al. 2003), and independent components analysis (ICA; Calhoun et al. 2003).

Whereas FC analyses are agnostic with respect to potential causal directions in the observed associations, *effective connectivity* (EC) refers to relationships in functional neuroimaging time series which can be ascribed to causal interactions among regions (Friston 1994). EC analyses are generally hypothesis based, whereas FC analyses are often more exploratory. Analytical methods which have been employed for EC analyses in schizophrenia research include structural equation models (SEMs; McIntosh and Gonzalez-Lima 1994), dynamic causal models (Friston et al. 2003; Benetti et al. 2009), and Granger causality (Goebel et al. 2003; Demirci et al. 2009). To date, EC analyses have been applied much less frequently than FC analyses to examine connectivity in schizophrenia functional imaging data, perhaps due to the diffuse nature of observed connectivity abnormalities in schizophrenia subjects.

There is a host of evidence from both structural and functional studies that individuals with schizophrenia have abnormal cortical connectivity. This has given rise to the *dysconnectivity hypothesis* (Friston and Frith 1995; Friston 1998) that the pathophysiology of schizophrenia is characterized by impaired cortical FC. Building on early postmortem data, recent investigators have been especially interested in investigating the hypothesis of disordered frontotemporal connectivity in schizophrenia (Friston and Frith 1995; Kraepelin 1919/1971). Burns (2004), in part evaluating evidence from both structural and functional studies of schizophrenia, concluded that both frontotemporal and frontoparietal networks were compromised in schizophrenia. Weinberger et al. (1992) proposed an impaired prefrontal–limbic pathway as the source of memory problems in patients with schizophrenia. Andreasen et al. (1996) hypothesized that schizophrenia involved a dysfunction of cognitive coordination (*cognitive dysmetria*) in neural circuitry involving perception, retention, retrieval, and response. The underlying source of this dysfunction was further hypothesized to involve compromised networks involving prefrontal cortex, thalamus, and cerebellum.

Over the last 15 years, considerable research has focused on the relationship between cognitive deficits in schizophrenia and disturbed cortical FC. Pioneering applications of FC analysis to the pathophysiology of schizophrenia involved data obtained from PET studies (Friston 1993). Friston and Frith (1995) presented PET data from 18 schizophrenia subjects and six control subjects with word production tasks. Each subject was scanned six times, and the resulting images were subjected to a singular value decomposition analysis, decomposing variation into spatial patterns (eigenimages) representing the major modes of covariation. Left prefrontal–bilateral superior temporal regions were negatively related in the normal control group but not so in the schizophrenia group. Moreover, there were left prefrontal–left temporal positive correlations present in the schizophrenia group but not in the control group. The authors concluded that schizophrenia subjects failed to modulate activity in temporal lobes due to aberrant connectivity with the prefrontal cortex.

A subsequent PET study of verbal fluency (Spence et al. 2000) in ten schizophrenia patients, ten controls, and ten at-risk subjects used left DLPFC as a seed region and performed zero-order temporal correlations of rCBF with other frontotemporal regions. They found no differences in FC between left DLPFC and left

STG but did find reduced FC between left DLPFC and anterior cingulate cortex (ACC). Josin and Liddle (2001) also examined FC utilizing a verbal fluency task PET study of 17 schizophrenia subjects and six normal control subjects. Using left LPFC as a seed region, voxel-wise time series were correlated with this seed region to produce connectivity maps. They found that left ACC, temporal, and parietal regions exhibited greater positive correlation in the schizophrenia group, whereas left cuneus and lingual gyrus exhibited greater negative correlations.

While not technically an FC analysis, Yurgelun-Todd et al. (1995) presented evidence for abnormal frontotemporal networks in an fMRI study of 12 schizophrenia subjects and 11 controls using a verbal fluency task, finding increased PFC activation and reduced activation in STG among schizophrenia subjects. Lawrie et al. (2002) conducted a block design fMRI trial employing a sentence completion task in eight schizophrenia subjects and an equal number of controls. These authors examined FC only between left DLPFC and temporal cortex, finding that correlations were lower in the schizophrenia group and that lower FC was associated with more severe auditory hallucinations. Whalley et al. (2005) also administered a sentence completion task in an fMRI study with 69 subjects at high risk for schizophrenia and 21 normal controls. They specified seed voxels in bilateral DLPFC, superior/middle frontal gyrus, medial frontal gyrus, thalamus, and inferior parietal lobule. Whole-brain voxel time series were then cross-correlated with these seed voxels. They found decreased FC between right medial prefrontal regions and contralateral cerebellum, and increased FC between left parietal and left prefrontal regions.

Schizophrenia-related FC abnormalities in frontotemporal networks have also been found from studies employing auditory oddball tasks. Winterer et al. (2003) conducted an EEG auditory oddball study of 64 schizophrenia patients, 79 unaffected siblings, and 88 unrelated normal controls. FC was determined by coherence of EEG channels in the frequency domain during a P300 peak activation time-window. Normal controls exhibited negative frontotemporal coherence in the delta band during this time-window, whereas schizophrenia subjects had positive coherence; unaffected siblings were in-between. Kiehl and Liddle (2001) employed an auditory oddball task in an event-related fMRI design with 11 schizophrenia subjects and 11 matched controls. While not an FC analysis, patients showed smaller and less extensive activations in STG, left supramarginal gyrus, right superior and inferior parietal lobule, anterior and posterior cingulate, thalamus, and right LPFC.

Kim et al. (2009) analyzed auditory oddball data from a large fMRI study (fBIRN) of 109 schizophrenia subjects and 114 control subjects. They employed ICA to decompose fMRI whole-brain time series into spatially-independent and temporally-coherent time courses. Thus, each independent component (IC) represents a network of temporally-coherent regions. They found eight such ICs involved in schizophrenia/normal control group differences. One IC consisted of a bilateral temporal network containing the superior and middle temporal gyrus; several others involved networks containing the DLPFC. Another network contained regions considered to be members of the so-called "default mode network" (DMN; described in more detail below), including posterior cingulate, precuneus, and middle frontal gyrus.

WM experiments have also been an important source of information regarding the relationship between cognitive deficits and FC abnormalities in persons with schizophrenia. Fletcher et al. (1999) used a learning and recall of words task in a PET study of 12 schizophrenia patients and 7 normal controls. Using a hypothesis driven approach, rCBF of voxels in the temporal cortex across 12 scans were regressed on ACC and PFC and their interaction, an FC analysis approach termed *psychophysiological interactions* (PPI) (Friston et al. 1997a, b). A significant interaction term was interpreted as a modulation of PFC-temporal FC by ACC activity. Results from this analysis showed a significant decrease in superior temporal rCBF as a function of the PFC–ACC interaction for normal controls but not for schizophrenia subjects. They conclude that schizophrenia involves disruption of ACC modulation of frontotemporal networks.

Meyer-Lindenberg et al. (2001) implemented an N-back WM paradigm in a PET study of 13 schizophrenia patients and 13 normal controls. Eigenimage analysis obtained through partial least squares (PLS) methodology. The first eigenimage, accounting for more than half of the overall variance, was task independent and significantly different across diagnostic groups: schizophrenia subjects loaded on inferotemporal, hippocampal, and cerebellar regions, whereas normal controls loaded on DLPFC and cingulate gyrus bilaterally. Kim et al. (2003) also utilized an N-back design in a PET study of 12 schizophrenia subjects and 12 normal control subjects. Lateral PFC was used as a seed region by correlating its time course with other activated voxels. Significantly lower prefrontal–parietal FC was discovered in the schizophrenia subjects.

Schloesser et al. (2003) performed a 2-back WM task in an fMRI study with 12 schizophrenia subjects and 6 controls. These data were entered into a SEM examining EC relationships among bilateral parietal association cortex (PAC), ventrolateral prefrontal cortex (VLPFC), DLPFC, thalamus, and cerebellum. Schizophrenia subjects exhibited reduced EC within prefrontal-cerebellar and cerebellar-thalamic limbs, but increased EC in the thalamo-cortical limb.

Wolf et al. (2007) used several WM paradigms, including a letter N-back task in an fMRI study of 14 schizophrenia subjects and 14 normal controls. Using seed regions in the STG and parahippocampal gyrus (PHG), whole-brain zero-order correlations found reduced DLPFC-temporal lobe FC, including PHG and STG. Patients also showed increased FC between these same temporal lobe regions and VLPFC. They concluded that reduced DLPFC-temporal lobe FC could account for encoding deficits in schizophrenia subjects. Henseler et al. (2009) examined subcomponents of WM through the application of several WM tasks in an fMRI study of 12 schizophrenia subjects and matched controls using right frontal opercular cortex and superior parietal lobule as seed regions in a PPI analysis. Compared to controls, schizophrenia subjects demonstrated reduced FC of the PFC with the intraparietal cortex and the hippocampus and abnormal negative FC between DLPFC and VLPFC. They also found these FC abnormalities were correlated with positive symptoms in the schizophrenia subjects.

Calhoun et al. (2006) present an fMRI study of 15 schizophrenia subjects and matched controls using both an auditory oddball task and a Sternberg WM task.

ICA was applied to both experimental tasks simultaneously to extract eight joint temporally-coherent networks. Patients demonstrated reduced loadings in an IC comprising regions of the temporal lobe, cerebellum, thalamus, basal ganglia, and lateral frontal regions. This can be interpreted as reduced FC between entire networks of regions (*functional network connectivity*; Calhoun et al. 2009). Moreover, correlation between voxels in this network across the two tasks was significantly higher in schizophrenia subjects, possibly indicating less specialization of brain networks when processing distinct cognitive tasks.

Kim et al. (2009) analyzed data from two fMRI trials (fBIRN and MCIC) of 115 schizophrenia subjects and 130 normal controls. The subjects were administered the Sternberg item recognition paradigm (SIRP). Group spatial ICA was applied, detecting six ICs with significant differences between patients and controls. Two of these ICs were positively correlated with task (*task-positive* networks): the first IC loaded on the cerebellum, whereas the second IC loaded strongly on left DLPFC, IPL, and cingulate gyrus. Patients had significantly higher loadings than controls during the probe phase of the SIRP task, possibly indicating processing differences between the two groups. The remaining four ICs to show differences between groups were negatively correlated with task presentation (*task-negative* networks), and loaded on regions including posterior cingulate, precuneus, medial PFC, ACC, IPL, and parahippocampus. For two of these ICs, patients tended to show less deactivation of these networks compared to controls during the probe phase of the SIRP task, but more deactivation during the encoding phase; this pattern was also observed in the task-positive IC loading on cerebellum. This may be indicative of abnormal interaction between task-positive and task-negative networks.

FC abnormalities in schizophrenia patients within task-positive and task-negative networks have received growing attention over the last few years. Research in this area has been driven by the discovery of functionally connected regions which tend to activate more strongly in resting state and deactivate upon presentation of an attention-demanding task (Biswal et al. 1995; Greicus and Menon 2004). This network has been termed *DMN*, though some evidence indicates these regions may in fact be involved in several functionally-related task-negative sub-networks (Kim et al. 2009). Regions implicated in the DMN include anterior and posterior cingulate, medial prefrontal cortex, inferior temporal regions, and cerebellum (Williamson 2007). Possible functions of the DMN include attending to external and internal stimuli, and self-referential and reflective activity including episodic memory retrieval, inner speech, mental images, emotions, and planning future events (Garrity et al. 2007 and references therein). Many of these mental processes are abnormal in persons with schizophrenia, which has engendered interest in determining whether DMN dysfunction could partially account for positive and negative symptoms. One hypothesized pathway for these symptoms is abnormal interaction of the DMN with a network (or networks) which tends to become more active during attention-demanding tasks; regions involved in this so-called *task-activated network* include DLPFC, supplemental motor area, IPL, and the middle temporal region (Williamson 2007).

Evidence for abnormal DMN connectivity in schizophrenia has recently been found in a few studies. Garrity et al. (2007) conducted an fMRI study using an auditory oddball task with 21 schizophrenia subjects and 22 normal controls. Spatial ICA was used to obtain the IC most strongly correlated with pre-determined DMN regions. This identification was unambiguous, in that one IC correlated strongly with DMN regions for each subject, and this IC was strongly negatively correlated with task presentation. However, controls had stronger correlation with the DMN template and showed greater task deactivation on the DMN than did schizophrenia subjects. Schizophrenia subjects also had more temporal variability in DMN components at higher frequencies (0.13 Hz) compared with controls, who had higher variability in lower frequencies (0.067 Hz).

Bluhm et al. (2007) conducted a resting-state fMRI study with 17 schizophrenia subjects and an equal number of controls. They computed correlations of spontaneous fluctuations across voxels with a seed region identified in the posterior cingulate. Schizophrenia patients exhibited less correlation between posterior cingulate and lateral parietal, medial prefrontal, and cerebellar regions of the DMN. Attenuation of DMN connectivity was correlated with both positive and negative symptoms in the schizophrenia group. Jafri et al. (2008) examined data from another resting-state fMRI study with 29 schizophrenia subjects and 25 normal controls, using ICA to extract functional network time courses. Patients showed higher correlation than controls among the most of the dominant extracted networks; patients also had slightly more variability in FC than controls.

Finally, resting-state studies have also been used to investigate the graph-theoretical properties of large scale brain networks (Watts and Strogatz 1998). Graph-theoretical analyses summarize aspects of functional networks in terms of the graphs derived from them. These FC graphs consist of *nodes* (regions) connected by *edges* between pairs of regions indicating significant FC relationships. From the resulting FC graphs it is possible to compute summary measures such as the *clustering coefficient* (a measure of how functionally interconnected the regions are) and *characteristic path length* (the average length of the shortest number of edges to traverse between pairs of regions), an index of global integration (Micheloyannis et al. 2006). It has been found that the most efficient distributed networks tend to have high clustering coefficients and short characteristic path lengths (Sporns and Zwi 2004). Networks which exhibit these optimal graph-theoretical properties are said to be *small-world networks*. Since small-world networks are highly efficient in processing information, departures from small-worldness may indicate lack of functional integration (Liu et al. 2008). For example, *ordered networks* have high clustering but long characteristic path lengths, while *random networks* have short characteristic path lengths but low clustering coefficients.

Liu et al. (2008) performed a resting-state fMRI study with 31 schizophrenia subjects and an equal number of normal controls. They parcellated the brain into 90 pre-determined anatomical regions and computed partial correlations for all pairs of regions; this partial correlation matrix was thresholded at various levels to determine significant FC relationships between pairs of regions controlling for relationships with the remaining 88. Clustering coefficients and characteristic path lengths were computed for each threshold. Small-world properties of the FC graphs over a broad range of thresholds were disrupted in schizophrenia subjects compared to normal control subjects, particularly in prefrontal, parietal, and temporal lobes. Specifically, clustering values were significantly smaller and characteristic path lengths were significantly longer in these regions. Furthermore, it was found that longer duration of illness correlated with smaller clustering coefficients and longer characteristic path lengths. Rubinov et al. (2009) report on a resting-state EEG study of 40 schizophrenia subjects and 40 normal control subjects. Using graphs derived from nonlinear correlation matrices, they also found lower clustering coefficients and higher characteristic path lengths in the schizophrenia sample.

Small-world analysis has also been applied to functional data from task-related designs. For example, Micheloyannis et al. (2006) investigated small-world properties in an EEG study with a 2-back WM test in 20 schizophrenia subjects and an equal number of normal controls. They used a variation of coherence which measures both linear and nonlinear interdependencies in EEG time series to construct FC graphs. During the 2-back tasks normal subjects exhibited small-world properties in alpha, beta, and gamma bands, whereas the schizophrenia sample did not. This may indicate dysfunctional attention-demanding networks in persons with schizophrenia.

3.2 Summary and Comments

To date, studies of FC have typically been exploratory, with investigators employing a wide variety of statistical models and behavioral tasks to identify the distributed brain systems involved in schizophrenia. Although the sample sizes studied have often been small and little work has been done to assess the robustness of the reported findings, this exploratory work has been heuristically valuable. The findings from some FC studies have supported the hypothesis that schizophrenia patients experience impaired functional connections between frontal and temporal cortex (Burns 2004; Friston and Frith 1995; Wolf et al. 2007). Yet the nature of the disordered connectivity does not support the hypothesis of a simple disconnection of frontotemporal pathways. The abnormality, rather, appears to be due to aberrant connectivity that might be expressed as altered patterns of correlation in the neural activity of frontal and temporal regions, disordered modulation of frontotemporal connectivity by other brain regions, or abnormal interactions of frontotemporal systems with other brain systems.

Not all studies reporting abnormal connectivity of the dorsal lateral prefrontal cortex in schizophrenia have found abnormal DLPFC-temporal connectivity. Rather abnormal DLPFC connectivity has been reported to involve VLPFC, anterior cingulate, inferior parietal lobule, and cerebellum, among other regions. Moreover, aberrant connectivity among schizophrenia patients has been reported in brain systems not involving the DLPFC. Whether or not investigators find abnormal frontotemporal connectivity among schizophrenia patients depends in part on what behavioral task was used, with word production or word generation tasks more commonly generating findings of aberrant frontotemporal connectivity.

Studies of the FC of resting networks have recently gained considerable momentum. Initial interest in resting-state connectivity initially stemmed from the observation by Raichle and colleagues (2001) that low or no stimulation behavioral states are typically associated with a uniform oxygen extraction fraction across brain regions. These authors argued that low stimulation states could be used as a physiological baseline for functional brain imaging studies (Raichle et al. 2001). Since the publication of this seminal paper, investigators have found coherent spatio-temporal patterns of functional brain signals during low stimulation or resting conditions (Williamson 2007). As discussed above, some investigators have reported that schizophrenia experience disordered resting connectivity. The finding of several spatially independent, task-negative sub-networks suggests, however, that functional connections studied during low stimulation conditions might be more than a single system that becomes disengaged during goal-directed behavior, as originally proposed. As investigators, such as Garrity and colleagues, begin to parse the functions of critical nodes of the resting-state networks, it might become possible to develop experimental methods to manipulate neural systems involved in self-reflective functions, leading to better validated theories of how internal mental states are related to the observed symptoms of schizophrenia. The potential for an integrated theory of the distributed brain systems involved in the aberrant neural processing of goal-directed and self-reflective states in schizophrenia appears within reach, a prospect that would be hard to envision without FC methods.

4 Final Comments and Emerging Trends

Functional brain imaging studies have produced a wealth of suggestive findings. What settled science can it offer? There is a consensus among investigators that the magnitude and regional pattern of neural activity observed among schizophrenia patients differs from the neural activity observed among healthy individuals. Although altered brain response can be seen in studies of resting blood flow and metabolism, the effect size is at best small to moderate in frontal regions where resting state has been most carefully studied. Altered neural response to cognitive challenge tasks has often produced larger effect sizes than have most resting studies. Yet this increased effect size has most often been observed for a narrow range of tasks – those assessing WM. Even for WM tasks the magnitude and direction of the effects vary from study to study with little consensus about the precise determinants of these variations.

When moving from settled science to clear trends in the literature, the findings become more exciting. The field has developed several intriguing hypotheses about the brain substrate of hallucinations. Whether these competing hypotheses will give way to a single consensus hypothesis or whether they represent different subsets of patients or different states within the same patient requires additional studies that compare these possibilities directly. Studies of neurocognitive deficits in schizophrenia show similar promise. Evidence from several types of experimental paradigms now shows that schizophrenia patients who achieve normal performance often display a larger neural response during task performance. These findings suggest that schizophrenia patients can compensate for inefficient neurocognitive processing at least for intermediate processing loads. A study by Quintana and colleagues has found that the location of compensatory responses can be altered in WM tasks by manipulating cue anticipation or memory processing aspects of the task (Quintana et al. 2003). Experimental studies manipulating neural patterns of compensation will likely lead to treatment methods that focus on potentiating these native compensatory mechanisms. Early cognitive rehabilitation studies focusing on cognitive adaptation strategies in schizophrenia show that it is possible to change neural activity in task relevant brain systems (Kurtz 2003).

There is also considerable interest in the use of functional brain imaging methods to develop new pharmacological treatments, to monitor treatment outcome, and to predict treatment response (Davis et al. 2005). Imaging investigators have argued that functional brain imaging can identify specific neural systems as treatment targets and provide a pathway for direct translational research between studies of animal models of human disease and clinical research (Carter 2005). A consensus study group has been formed to develop neurocognitive constructs relevant to translational and pharmacotherapy studies of cognitive deficit in schizophrenia (Carter et al. 2008).

Functional brain imaging data are being studied as intermediate phenotypes to identify genes that regulate the neural substrate of complex brain functions (Hariri and Weinberger 2003; Windemuth et al. 2008). Schizophrenia investigators have argued that both magnitude activation data and connectivity data might serve as useful intermediate phenotypes (Turner et al. 2006; Whalley et al. 2009). Imaging studies of genes regulating dopamine and serotonin neurotransmission and genes related to schizophrenia susceptibility are active areas of research (Gallinat et al. 2008; Lawrie et al. 2008). Functional brain imaging methods are now investigated as potential links to integrate neurogenetic information into drug development and, ultimately, to guide tailored treatment (Di Giorgio et al. 2009). Such studies should move the field towards a general neurobiological theory of the complex symptoms that trouble individuals with schizophrenia.

Acknowledgments Support for the preparation of this chapter was provided by the VA VISN 22 Mental Illness, Research, Education and Clinical Center and by NIH grant 2K25 MH076981 awarded to the second author. This chapter is dedicated to Gerald Rosenbaum whose enthusiasm for the study of schizophrenia survives in the students and colleagues whom he greatly influenced.

References

Adams RD, Victor M (1985) Principles of neurology, 3rd edn. McGraw Hill, New York
Allen P, Aleman A, McGuire PK (2007) Inner speech models of auditory verbal hallucinations: evidence from behavioural and neuroimaging studies. Int Rev Psychiatry 19:407–415
Andreasen NC (1989) Brain imaging: applications in psychiatry. American Psychiatric Press, Washington DC

Andreasen NC (1997) Linking mind and brain in the study of mental illnesses: a project for a scientific psychopathology. Science 275:1586–1593

Andreasen NC, Rezai K, Alliger R, Swayze VW 2nd, Flaum M, Kirchner P, Cohen G, O'Leary DS (1992) Hypofrontality in neuroleptic-naive patients and in patients with chronic schizophrenia. Assessment with xenon 133 single-photon emission computed tomography and the Tower of London. Arch Gen Psychiatry 49:943–958

Andreasen NC, O'Leary DS, Cizadlo T, Arndt S, Rezai K, Boles Ponto LL, Watkins GL, Hichwa R (1996) Schizophrenia and cognitive dysmetria: a positron-emission tomography study of dysfunctional prefrontal-thalamic-cerebellar circuitry. Proc Natl Acad Sci USA 93:9985–9990

Baddeley AD (1976) The psychology of memory. Basic Books, Inc, New York

Balleine BW, Delgado MR, Hikosaka O (2007) The role of the dorsal striatum in reward and decision-making. J Neurosci 27:8161–8165

Benetti S, Mechelli A, Picchioni M, Broome M, Williams S, McGuire P (2009) Functional integration between the posterior hippocampus and prefrontal cortex is impaired in both first episode schizophrenia and the at risk mental state. Brain 132:2426–2436

Biswal B, Yetkin FZ, Haughton VM, Hyde JS (1995) Functional connectivity in the motor cortex of resting human brain using echo-planar MRI. Magn Reson Med 34:537–541

Bluhm RL, Miller J, Lanius R, Osuch EA, Boksman K, Neufeld RWJ, Theberge J, Schaefer B (2007) Spontaneous low-frequency fluctuations in the BOLD signal in schizophrenic patients: anomalies in the default network. Schizophr Bull 33:1004–1012

Brown J (1958) Some tests of the decay theory of immediate memory. Q J Exp Psychol 10:12–21

Brown GG, Eyler LT (2006) Methodological and conceptual issues in functional magnetic resonance imaging: applications to schizophrenia research. Annu Rev Clin Psychol 2:51–81

Brown GG, McCarthy G, Bischoff-Grethe A, Ozyurt B, Greve D, Potkin SG, Turner JA, Notestine R, Calhoun VD, Ford JM, Mathalon D, Manoach DS, Gadde S, Glover GH, Wible CG, Belger A, Gollub RL, Lauriello J, O'Leary D, Lim KO (2009) Brain-performance correlates of working memory retrieval in schizophrenia: a cognitive modeling approach. Schizophr Bull 35:32–46

Buchsbaum MS, Hazlett EA (1998) Positron emission tomography studies of abnormal glucose metabolism in schizophrenia. Schizophr Bull 24:343–364

Burns JK (2004) An evolutionary theory of schizophrenia: cortical connectivity, metarepresentation, and the social brain. Behav Brain Sci 27:831–885

Buxton RB (2009) Introduction to functional magnetic resonance imaging: principles and techniques, 2nd edn. Cambridge University Press, New York

Calhoun VD, Adali T, Hansen LK, Larsen J PJJ (2003) ICA of functional MRI data: an overview: in Proceedings of the International Workshop on Independent Component Analysis and Blind Signal Separation. Nara, Japan, pp 921–926

Calhoun VD, Adali T, Kiehl KA, Astur RS, Pekar JJ, Pearlson GD (2006) A method for multi-task fMRI data fusion applied to schizophrenia. Hum Brain Mapp 27:598–610

Calhoun VD, Eichele T, Pearlson G (2009) Functional brain networks in schizophrenia: a review. Front Hum Neurosci 3:1–12

Callicott JH, Bertolino A, Mattay VS, Langheim FJ, Duyn J, Coppola R, Goldberg TE, Weinberger DR (2000) Physiological dysfunction of the dorsolateral prefrontal cortex in schizophrenia revisited. Cereb Cortex 10:1078–1092

Callicott JH, Mattay VS, Verchinski BA, Marenco S, Egan MF, Weinberger DR (2003) Complexity of prefrontal cortical dysfunction in schizophrenia: more than up or down. Am J Psychiatry 160:2209–2215

Carter CS (2005) Applying new approaches from cognitive neuroscience to enhance drug development for the treatment of impaired cognition in schizophrenia. Schizophr Bull 31:810–815

Carter CS, Barch DM, Buchanan RW, Bullmore E, Krystal JH, Cohen J, Geyer M, Green M, Nuechterlein KH, Robbins T, Silverstein S, Smith EE, Strauss M, Wykes T, Heinssen R (2008) Identifying cognitive mechanisms targeted for treatment development in schizophrenia: an overview of the first meeting of the Cognitive Neuroscience Treatment Research to Improve Cognition in Schizophrenia Initiative. Biol Psychiatry 64:4–10

Cates W Jr (1998) Syphilis. In: Evans AS, Brachman PS (eds) Bacterial infections of humans, 3rd edn. Plenum, New York, pp 713–740

Chen C-C, Hsieh J-C, Wu Y-Z, Lee P-L, Chen S-S, Niddam DM, Yeh T-C, Wu Y-T (2008) Mutual-information-based approach for neural connectivity during self-paced finger lifting task. Hum Brain Mapp 29:265–280

Clark CM, Kessler R, Buchsbaum MS, Margolin RA, Holcomb HH (1984) Correlational methods for determining regional coupling of cerebral glucose metabolism: a pilot study. Biol Psychiatry 19:663–678

Cohen J (1988) Statistical power analysis for the behavioral sciences. Lawrence Erlbaum, Hillsdale, NJ

Cohen JD, Tong F (2001) The face of controversy. Science 293:2405–2407

Cohen BM, Yurgelun-Todd D, English CD, Renshaw PF (1995) Abnormalities of regional distribution of cerebral vasculature in schizophrenia detected by dynamic susceptibility contrast MRI. Am J Psychiatry 152:1801–1803

Cohen JD, Perlstein WM, Braver TS, Nystrom LE, Noll DC, Jonides J, Smith EE (1997) Temporal dynamics of brain activation during a working memory task. Nature 386:604–608

Copolov DL, Seal ML, Maruff P, Ulusoy R, Wong MTH, Tochon-Danguy HJ, Egan GF (2003) Cortical activation associated with the experience of auditory hallucinations and perception of human speech in schizophrenia: a PET correlation study. Psychiatry Res 122:139–152

Cowan N (2008) What are the differences between long-term, short-term, and working memory? Prog Brain Res 169:323–338

Crespo-Facorro B, Wiser AK, Andreasen NC, O'Leary DS, Watkins GL, Boles Ponto LL, Hichwa RD (2001) Neural basis of novel and well-learned recognition memory in schizophrenia: a positron emission tomography study. Hum Brain Mapp 12:219–231

David GB (1957) The pathological anatomy of the schizophrenias. In: Richter D (ed) Schizophrenia: somatic aspects. Pergamon, Oxford, pp 93–130

David AS, Woodruff PW, Howard R, Mellers JD, Brammer M, Bullmore E, Wright I, Andrew C, Williams SC (1996) Auditory hallucinations inhibit exogenous activation of auditory association cortex. NeuroReport 7:932–936

David O, Cosmelli D, Friston KJ (2004) Evaluation of different measures of functional connectivity using a neural mass model. Neuroimage 21:659–673

Davidson LL, Heinrichs RW (2003) Quantification of frontal and temporal lobe brain-imaging findings in schizophrenia: a meta-analysis. Psychiatry Res 122:69–87

Davis TL, Kwong KK, Weisskoff RM, Rosen BR (1998) Calibrated functional MRI: mapping the dynamics of oxidative metabolism. Proc Natl Acad Sci USA 95:1834–1839

Davis CE, Jeste DV, Eyler LT (2005) Review of longitudinal functional neuroimaging studies of drug treatments in patients with schizophrenia. Schizophr Res 78:45–60

Demirci O, Stevens MC, Andreasen NC, Michael A, Liu J, White T, Pearlson GD, Clark VP, Calhoun VD (2009) Investigation of relationships between fMRI brain networks in the spectral domain using ICA and Granger causality reveals distinct differences between schizophrenia patients and healthy controls. NeuroImage 46:419–431

Di Giorgio A, Sambataro F, Bertolino A (2009) Functional imaging as a tool to investigate the relationship between genetic variation and response to treatment with antipsychotics. Curr Pharm Des 15:2560–2572

Dierks T, Linden DE, Jandl M, Formisano E, Goebel R, Lanfermann H, Singer W (1999) Activation of Heschl's gyrus during auditory hallucinations. Neuron 22:615–621

Ebmeier KP, Lawrie SM, Blackwood DH, Johnstone EC, Goodwin GM (1995) Hypofrontality revisited: a high resolution single photon emission computed tomography study in schizophrenia. J Neurol Neurosurg Psychiatry 58:452–456

Fletcher P, McKenna PJ, Friston KJ, Frith CD, Dolan RJ (1999) Abnormal cingulate modulation of fronto-temporal connectivity in schizophrenia. Neuroimage 9:337–342

Ford JM, Mathalon DH (2004) Electrophysiological evidence of corollary discharge dysfunction in schizophrenia during talking and thinking. J Psychiatr Res 38:37–46

Ford JM, Mathalon DH, Heinks T, Kalba S, Faustman WO, Roth WT (2001) Neurophysiological evidence of corollary discharge dysfunction in schizophrenia. Am J Psychiatry 158:2069–2071

Frankle WG (2007) Neuroreceptor imaging studies in schizophrenia. Harv Rev Psychiatry 15:212–232

Friston KJ (1993) Fronto-limbic integration in schizophrenia: abnormal functional connectivity measured with PET. J Neurol Sci 120:6

Friston KJ (1994) Functional and effective connectivity in neuroimaging: a synthesis. Hum Brain Mapp 2:56–78

Friston KJ (1998) The disconnection hypothesis. Schizophr Res 30:115–125

Friston KJ, Frith CD (1995) Schizophrenia: a disconnection syndrome? Clin Neurosci 3:89–97

Friston KJ, Frith CD, Liddle PF, Frackowiak RSJ (1993) Functional connectivity: the principal-component analysis of large(PET) data sets. J Cereb Blood Flow Metab 13:5–14

Friston KJ, Buechel C, Fink GR, Morris J, Rolls E, Dolan RJ (1997) Psychophysiological and modulatory interactions in neuroimaging. Neuroimage 6:218–229

Friston KJ, Harrison L, Penny W (2003) Dynamic causal modeling. Neuroimage 19:1273–1302

Gallinat J, Bauer M, Heinz A (2008) Genes and neuroimaging: advances in psychiatric research. Neurodegener Dis 5:277–285

Garrity AG, Pearlson GD, McKiernan K, Lloyd D, Kiehl KA, Calhoun VD (2007) Aberrant 'default mode' functional connectivity in schizophrenia. Am J Psychiatry 164:450–457

Gevins A, Cutillo B (1993) Spatiotemporal dynamics of component processes in human working memory. Electroencephalogr Clin Neurophysiol 87:128–143

Glahn DC, Ragland JD, Abramoff A, Barrett J, Laird AR, Bearden CE, Velligan DI (2005) Beyond hypofrontality: a quantitative meta-analysis of functional neuroimaging studies of working memory in schizophrenia. Hum Brain Mapp 25:60–69

Goebel R, Roebroeck A, Kim D-S, Formisano E (2003) Investigating directed cortical interactions in time-resolved fMRI data using vector autoregressive modeling and Granger causality mapping. Magn Reson Imaging 21:1251–1261

Goldstein PC, Brown GG, Marcus A, Ewing JR (1990) Effects of age, neuropsychological impairment, and medication on regional cerebral blood flow in schizophrenia and major affective disorder. Henry Ford Hosp Med J 34:202–206

Greicus MD, Menon V (2004) Default-mode activity during a passive sensory task: uncoupled from deactivation but impacting activation. J Cogn Neurosci 16:1484–1492

Gur RC, Gur RE (1995) Hypofrontality in schizophrenia: RIP. Lancet 345:1383–1384

Haber SN, Calzavara R (2009) The cortico-basal ganglia integrative network: the role of the thalamus. Brain Res Bull 78:69–74

Hariri AR, Weinberger DR (2003) Imaging genomics. Br Med Bull 65:259–270

Harrison PJ, Lewis DA (2003) Neuropathology of schizophrenia. In: Hirsch SR, Weinberger D (eds) Schizophrenia. Blackwell, Malden, MA, pp 310–325

Hecaen H, Albert ML (1978) Human neuropsychology. Wiley, New York

Hecaen H, Angelergues R (1964) Localization of symptoms in aphasia. In: de Reuck AUS, O'Conner M (eds) Disorders of language. Churchill, London, pp 223–246

Heinks-Maldonado TH, Mathalon DH, Houde JF, Gray M, Faustman WO, Ford JM (2007) Relationship of imprecise corollary discharge in schizophrenia to auditory hallucinations. Arch Gen Psychiatry 64:286–296

Henseler I, Falkai P, Gruber O (2009) Disturbed functional connectivity within brain networks subserving domain-specific subcomponents of working memory in schizophrenia: relation to performance and clinical symptoms. J Psychiatr Res [Epub ahead of print, Oct 16]

Hill K, Mann L, Laws KR, Stephenson CM, Nimmo-Smith I, McKenna PJ (2004) Hypofrontality in schizophrenia: a meta-analysis of functional imaging studies. Acta Psychiatr Scand 110:243–256

Hoffman RE, Anderson AW, Varanko M, Gore JC, Hampson M (2008) Time course of regional brain activation associated with onset of auditory/verbal hallucinations. Br J Psychiatry 193:424–425

Honey GS, Fletcher PC (2006) Investigating principles of human brain function underlying working memory: what insights from schizophrenia? Neuroscience 139:59–71

Horwitz B (2003) The elusive concept of brain connectivity. Neuroimage 19:466–470

Horwitz B, Duara R, Rapoport SI (1984) Intercorrelations of glucose metabolic rates between brain regions: application to healthy males in a state of reduced sensory input. J Cereb Blood Flow Metab 4:484–499

Huettel SA, Song AW, McCarthy G (2009) Functional magnetic resonance imaging, 2nd edn. Sinauer Associates, Inc, Sunderland, MA

Ingvar DH, Franzén G (1974a) Abnormalities of cerebral blood flow distribution in patients with chronic schizophrenia. Acta Psychiatr Scand 50:425–62

Ingvar DH, Franzén G (1974b) Distribution of cerebral activity in chronic schizophrenia. Lancet 21(2):1484–1486

Jafri M, Pearlson GD, Stevens M, Calhoun VD (2008) A method for functional network connectivity among spatially independent resting-state components in schizophrenia. Neuroimage 39:1666–1681

Jardri R, Pins D, Bubrovszky M, Lucas B, Lethuc V, Delmaire C, Vantyghem V, Despretz P, Thomas P (2009) Neural functional organization of hallucinations in schizophrenia: multisensory dissolution of pathological emergence in consciousness. Conscious Cogn 18:449–457

Jeong J, Gore JC, Peterson BS (2001) Mutual information analysis of the EEG in patients with alzheimer's disease. Clin Neurophysiol 112:827–835

Johnson MR, Morris NA, Astur RS, Calhoun VD, Mathalon DH, Kiehl KA, Pearlson GD (2006) A functional magnetic resonance imaging study of working memory abnormalities in schizophrenia. Biol Psychiatry 60:11–21

Josin GM, Liddle PF (2001) Neural network analysis of the pattern of functional connectivity between cerebral areas in schizophrenia. Biol Cybern 84:117–122

Kane MJ, Conway AR, Miura TK, Colflesh GJ (2007) Working memory, attention control, and the N-back task: a question of construct validity. J Exp Psychol Learn Mem Cogn 33:615–622

Kiehl KA, Liddle PF (2001) An event-related functional magnetic resonance imaging study of an auditory oddball task in schizophrenia. Schizophr Res 48:159–171

Kim JJ, Kwon JS, Park HJ, Youn T, Kang DH, Kim MS, Lee DS, Lee MC (2003) Functional disconnection between the prefrontal and parietal cortices during working memory processing in schizophrenia: a [^{15}O]H$_2$O PET Study. Am J Psychiatry 160:919–923

Kim DI, Mathalon DH, Ford JM, Mannell M, Turner JA, Brown GG, Belger A, Gollub R, Lauriello J, Wible C, O'Leary DO, Kim K, Toga A, Potkin SG, Birn F, Calhoun VD (2009) Auditory oddball deficits in schizophrenia: an independent components analysis of the fMRI multisite function BIRN study. Schizophr Bull 35:67–81

Koch K, Wagner G, Nenadic I, Schachtzabel C, Schultz C, Roebel M, Reichenbach JR, Sauer H, Schlösser RG (2008) Fronto-striatal hypoactivation during correct information retrieval in patients with schizophrenia: an fMRI study. Neuroscience 153:54–62

Koenig T, Lehmann D, Saito N, Kuginuki T, Kinoshita T, Koukkou M (2001) Decreased functional connectivity of EEG theta-frequency activity in first-episode, neuroleptic-naïve patients with schizophrenia: preliminary results. Schizophr Res 50:55–60

Kraepelin E (1971) Dementia praecox and paraphrenia (trans: Barclay RM, Robertson GM). Original publication date 1919. Robert E. Krieger Publishing Co Inc, Huntington, New York

Kurtz MM (2003) Neurocognitive rehabilitation for schizophrenia. Curr Psychiatry Rep 5: 303–310

Lawrie SM, Buechel C, Whalley HC, Frith CD, Friston KJ, Johnstone EC (2002) Reduced frontotemporal functional connectivity in schizophrenia associated with auditory hallucinations. Biol Psychiatry 51:1008–1011

Lawrie SM, Hall J, McIntosh AM, Cunningham-Owens DG, Johnstone EC (2008) Neuroimaging and molecular genetics of schizophrenia: pathophysiological advances and therapeutic potential. Br J Pharmacol 153(Suppl 1):S120–S124

Lennox BR, Park SB, Jones PB, Morris PG (1999) Spatial and temporal mapping of neural activity associated with auditory hallucinations. Lancet 353:644, Erratum in: Lancet 1999 Aug 7;354 (9177):518. Park G [corrected to Park SB]

Lewis L, Escalona R, Keith SJ (2009) Phenomenology of schizophrenia. In: Sadock BJ, Sadock VA, Ruiz P (eds) Kaplan & Sadock's comprehensive textbook of psychiatry, vol 1, 9th edn. Wolters Kluwer/Lippincott Williams & Wilkins, Philadelphia, pp 1433–1451

Liddle PF (1996) Functional imaging – schizophrenia. Br Med Bull 52:486–494

Liu YL, Liang M, Zhou Y, He Y, Hao Y, Song M, Yu C, Liu Z, Jiang T (2008) Disrupted small-world networks in schizophrenia. Brain 131:945–961

Loeber RT, Sherwood AR, Renshaw PF, Cohen BM, Yurgelun-Todd DA (1999) Differences in cerebellar blood volume in schizophrenia and bipolar disorder. Schizophr Res 37:81–89

Manoach DS (2003) Prefrontal cortex dysfunction during working memory performance in schizophrenia: reconciling discrepant findings. Schizophr Res 60:285–298

Manoach DS, Press DZ, Thangaraj V, Searl MM, Goff DC, Halpern E, Saper CB, Warach S (1999) Schizophrenic subjects activate dorsolateral prefrontal cortex during a working memory task, as measured by fMRI. Biol Psychiatry 45:1128–1137

Manoach DS, Gollub RL, Benson ES, Searl MM, Goff DC, Halpern E, Saper CB, Rauch SL (2000) Schizophrenic subjects show aberrant fMRI activation of dorsolateral prefrontal cortex and basal ganglia during working memory performance. Biol Psychiatry 48:99–109

McGuire PK, Shah GM, Murray RM (1993) Increased blood flow in Broca's area during auditory hallucinations in schizophrenia. Lancet 342:703–706

McGuire PK, Silbersweig DA, Murray RM, David AS, Frackowiak RS, Frith CD (1996) Functional anatomy of inner speech and auditory verbal imagery. Psychol Med 26:29–38

McIntosh AR, Gonzalez-Lima F (1994) Structural equation modeling and its applications to network analysis of functional brain imaging. Hum Brain Mapp 2:2–22

Meyer-Lindenberg AS, Poline J-B, Kohn PD, Holt JL, Egan MF, Weinberger DR, Berman KF (2001) Evidence for abnormal cortical functional connectivity during working memory in schizophrenia. Am J Psychiatry 158:1809–1817

Micheloyannis S, Pachou E, Jan Stam C, Breakspear M, Bitsios P, Vourkas M, Erimaki S, Zervakis M (2006) Small-world networks and disturbed functional connectivity in schizophrenia. Schizophr Res 87:60–66

Oertel V, Rotarska-Jagiela A, van de Ven VG, Haenschel C, Maurer K, Linden DE (2007) Visual hallucinations in schizophrenia investigated with functional magnetic resonance imaging. Psychiatry Res 156:269–273

Parellada E, Lomena F, Font M, Pareto D, Gutierrez F, Simo M, Fernández-Egea E, Pavia J, Ros D, Bernardo M (2008) Fluordeoxyglucose-PET study in first-episode schizophrenic patients during the hallucinatory state, after remission and during linguistic-auditory activation. Nucl Med Commun 29:894–900

Peled A, Geva A, Kremen K, Blankfeld H, Esfandiarfard R, Nordahl T (2001) Functional connectivity and working memory in schizophrenia: an EEG study. Int J Neurosci 106:47–61

Peterson L, Peterson M (1959) Short-term retention of individual items. J Exp Psychol 58:193–198

Posner MI, Rossman E (1965) Effect of size and location of information transforms on short-term memory. J Exp Psychol 70:496–505

Potkin SG, Turner JA, Brown GG, McCarthy G, Greve DN, Glover GH, Manoach DS, Belger A, Diaz M, Wible CG, Ford JM, Mathalon DH, Gollub R, Lauriello J, O'Leary D, van Erp TG, Toga AW, Preda A, Lim KO, FBIRN (2009) Working memory and DLPFC inefficiency in schizophrenia: the FBIRN study. Schizophr Bull 35:19–31

Quintana J, Wong T, Ortiz-Portillo E, Kovalik E, Davidson T, Marder SR, Mazziotta JC (2003) Prefrontal-posterior parietal networks in schizophrenia: primary dysfunctions and secondary compensations. Biol Psychiatry 53:12–24

Raichle ME, MacLeod AM, Abraham Z, Snyder AZ, Powers WJ, Gusnard DA, Shulman GL (2001) A default mode of brain function. Proc Natl Acad Sci USA 98:676–682

Rao SM, Bandettini PA, Binder JR, Bobholz JA, Hammeke TA, Stein EA, Hyde JS (1996) Relationship between finger movement rate and functional magnetic resonance signal change in human primary motor cortex. J Cereb Blood Flow Metab 16:1250–1254

Rubinov M, Knock SA, Stam CJ, Micheloyannis S, Harris AWF, Williams LM, Breakspear M (2009) Small-world properties of nonlinear brain activity in schizophrenia. Hum Brain Mapp 30:403–416

Salvador R, Suckling J, Schwarzbauer C, Bullmore E (2005) Undirected graphs of frequency-dependent functional connectivity in whole brain networks. Philos Trans R Soc Lond B Biol Sci 360:937–946

Schloesser R, Gesierich T, Kaufmann B, Vucurevic G, Hunsche S, Gawehn J, Stoeter P (2003) Altered effective connectivity during working memory performance in schizophrenia: a study with fMRI and structural equation modeling. Neuroimage 10:751–763

Seal ML, Aleman A, McGuire PK (2004) Compelling imagery, unanticipated speech and deceptive memory: neurocognitive models of auditory verbal hallucinations. Cogn Neuropsychiatry 9:43–72

Shergill SS, Brammer MJ, Williams SC, Murray RM, McGuire PK (2000) Mapping auditory hallucinations in schizophrenia using functional magnetic resonance imaging. Arch Gen Psychiatry 57:1033–1038

Siegle GJ, Thompson WK, Carter CS, Steinhauer SR, Thase ME (2007) Increased amygdala and decreased dorsolateral prefrontal BOLD responses in unipolar depression: related and independent features. Biol Psychiatry 61:198–209

Silbersweig DA, Stern E, Frith C, Cahill C, Holmes A, Grootoonk S, Seaward J, McKenna P, Chua SE, Schnorr L, Jones T, Frackowiak RSJ (1995) A functional neuroanatomy of hallucinations in schizophrenia. Nature 378:176–179

Simons CJ, Tracy DK, Sanghera KK, O'Daly O, Gilleen J, Dominguez MD, Krabbendam L, Shergill SS (2010) Functional magnetic resonance imaging of inner speech in schizophrenia. Biol Psychiatry 67:232–237

Spence SA, Liddle PF, Stefan MD, Hellewell JS, Sharma T, Friston KJ, Hirsch SR, Frith CD, Murray RM, Deakin JFW, Grasby PM (2000) Functional anatomy of verbal fluency in people with schizophrenia and those at genetic risk. Br J Psychiatry 196:2–60

Spencer KM, Nestor PG, Niznikiewicz MA, Salisbury DF, Shenton ME, McCarley RW (2003) Abnormal neural synchrony in schizophrenia. J Neurosci 23:7407–7411

Sporns O, Zwi J (2004) The small world of the cerebral cortex. Neuroinformatics 2:145–162

Stephan KE, Friston KJ, Frith CD (2009) Dysconnection in schizophrenia: from abnormal synaptic plasticity to failures of self-monitoring. Schizophr Bull 35:509–527

Sternberg S (1969) Memory scanning: new findings and current controversies. In: Deutsch D, Deutsch JA (eds) Short-term memory. Academic, New York, pp 196–231

Sun FT, Miller LM, D'Esposito M (2004) Measuring interregional functional connectivity using coherence and partial coherence analyses of fMRI data. Neuroimage 21:647–658

Toga AW, Mazziotta JC (2002) Brain mapping: the methods. Academic, San Diego

Troiani V, Fernández-Seara MA, Wang Z, Detre JA, Ash S, Grossman M (2008) Narrative speech production: an fMRI study using continuous arterial spin labeling. Neuroimage 40:932–939

Turner JA, Smyth P, Macciardi F, Fallon JH, Kennedy JL, Potkin SG (2006) Imaging phenotypes and genotypes in schizophrenia. Neuroinformatics 4:21–49

van de Ven VG, Formisano E, Röder CH, Prvulovic D, Bittner RA, Dietz MG, Hubl D, Dierks T, Federspiel A, Esposito F, Di Salle F, Jansma B, Goebel R, Linden DE (2005) The spatiotemporal pattern of auditory cortical responses during verbal hallucinations. Neuroimage 27:644–655

Van der Werf YD, Witter MP, Groenewegen HJ (2002) The intralaminar and midline nuclei of the thalamus. Anatomical and functional evidence for participation in processes of arousal and awareness. Brain Res Brain Res Rev 39:107–140

Volkow ND, Wolf AP, Van Gelder P, Brodie JD, Overall JE, Cancro R, Gomez-Mont F (1987) Phenomenological correlates of metabolic activity in 18 patients with chronic schizophrenia. Am J Psychiatry 144:151–158

Watts DJ, Strogatz SH (1998) Collective dynamics of 'small-world' networks. Nature 393:409–410

Weinberger DR, Aloia MS, Goldberg TE, Berman KF (1994) The frontal lobes and schizophrenia. J Neuropsychiatry Clin Neurosci 6:419–427; Review. Erratum in: J Neuropsychiatry Clin Neurosci 1995;7:121

Weinberger DR, Berman KF (1996) Prefrontal function in schizophrenia: confounds and controversies. Philos Trans R Soc Lond B Biol Sci 351:1495–1503

Weinberger DR, Berman KF, Suddath R, Torrey EF (1992) Evidence of dysfunction of a prefrontal-limbic network in schizophrenia: a magnetic resonance imaging and regional cerebral blood flow study of discordant monozygotic twins. Am J Psychiatry 149:890–897

Wernicke C (1969) The symptom complex of aphasia: a psychological study on an anatomical basis. In: Co RS (ed) Boston studies in philosophy of science, vol IV. Reudel, Dordrecht, The Netherlands, pp 34–97 (Original work published 1874)

Whalley HC, Simonotto E, Marshall I, Owens DGC, Goddard NH, Johnstone EC, Lawrie SM (2005) Functional disconnectivity in subjects at high genetic risk of schizophrenia. Brain 128:2097–2108

Whalley HC, Steele JD, Mukherjee P, Romaniuk L, McIntosh AM, Hall J, Lawrie SM (2009) Connecting the brain and new drug targets for schizophrenia. Curr Pharm Des 15:2615–2631

Williamson P (2007) The final common pathway of schizophrenia. Schizophr Bull 33:953–954

Windemuth A, Calhoun VD, Pearlson GD, Kocherla M, Jagannathan K, Ruaño G (2008) Physiogenomic analysis of localized FMRI brain activity in schizophrenia. Ann Biomed Eng 36:877–888

Winterer G, Coppola R, Egan MF, Goldberg TE, Weinberger DR (2003) Functional and effective frontotemporal connectivity and genetic risk for schizophrenia. Biol Psychiatry 54:1181–1192

Wolf DH, Gur RC, Valdez JN, Loughead J, Elliot MA, Gur RE, Ragland D (2007) Alterations of front-temporal connectivity during word-encoding in schizophrenia. Psychiatry Res 154: 221–232

Wolkin A, Sanfilipo M, Wolf AP, Angrist B, Brodie JD, Rotrosen J (1992) Negative symptoms and hypofrontality in chronic schizophrenia. Arch Gen Psychiatry 49:959–965

Wood JH (1987) Cerebral blood flow: physiological and clinical aspects. McGraw-Hill, New York

Woodruff PW, Wright IC, Bullmore ET, Brammer M, Howard RJ, Williams SC, Shapleske J, Rossell S, David AS, McGuire PK, Murray RM (1997) Auditory hallucinations and the temporal cortical response to speech in schizophrenia: a functional magnetic resonance imaging study. Am J Psychiatry 154:1676–1682

Yurgelun-Todd DA, Renshaw PF, Cohen BM (1995) Functional MRI of schizophrenics and normal controls during word production. Schizophr Res 15:104

Zhou D, Thompson WK, Siegle GJ (2009) MATLAB toolbox for functional connectivity. Neuroimage 47:1590–1607

Neurochemical Imaging in Schizophrenia

Nina Urban and Anissa Abi-Dargham

Contents

1 Introduction 216
2 Brief Overview of Neurochemical Imaging Techniques 217
3 Imaging Neurotransmitter Systems 219
    3.1 Dopamine 219
    3.2 Serotonin 225
    3.3 Gamma-Aminobutyric Acid 227
    3.4 N-Methyl-D-Aspartic Acid and Glutamate 228
4 Occupancy Studies (Pharmacological Studies) 228
    4.1 DA Receptor Occupancy 229
    4.2 Serotonin Occupancy 231
5 Future Directions 232
References 233

Abstract Recent advances in the development and applications of neurochemical brain imaging methods have improved the ability to study the neurochemistry of the living brain in normal processes as well as psychiatric disorders. In particular, positron emission tomography (PET) and single photon emission computed tomography (SPECT) have been used to determine neurochemical substrates of schizophrenia and to uncover the mechanism of action of antipsychotic medications. The growing availability of radiotracers for monoaminergic neurotransmitter synthesis, transporters and receptors, has enabled the evaluation of hypotheses regarding neurotransmitter function in schizophrenia derived from preclinical and clinical observations.

N. Urban
Department of Psychiatry, New York State Psychiatric Institute, Columbia University, New York, NY 10032, USA

A. Abi-Dargham (✉)
Department of Radiology, New York State Psychiatric Institute, Columbia University, New York, NY 10032, USA
e-mail: aa324@columbia.edu

This chapter reviews the studies using neurochemical brain imaging methods for (1) detection of abnormalities in indices of dopamine and serotonin transmission in patients with schizophrenia compared to controls, (2) development of new tools to study other neurotransmitters systems, such as gamma-aminobutyric acid (GABA) and glutamate, and (3) characterization of target occupancy by antipsychotic drugs, as well as its relationship to efficacy and side effects.

As more imaging tools become available, this knowledge will expand and will lead to better detection of disease, as well as better therapeutic approaches.

Keywords Neurochemical imaging techniques · Dopamine · Serotonin · Gamma-amino butyric acid (GABA) · Glutamate · Occupancy

1 Introduction

The introduction of neuroimaging techniques in the 1960s has revolutionized the study of the biology of psychiatric disorders. Functional imaging techniques including neurochemical/molecular imaging such as single photon emission computerized tomography (SPECT), positron emission tomography (PET), and magnetic resonance spectroscopy (MRS) have advanced our understanding of the pathophysiology of schizophrenia and other psychiatric disorders. This chapter summarizes current insights gained from application of molecular imaging techniques to the study of schizophrenia.

Prior to the advent of in vivo imaging, postmortem studies were the mainstay for developing an understanding of the neurochemical alterations in schizophrenia (for review, see Benes 2000). Postmortem studies have limitations; in particular, they do not allow exploring the functional aspect of neurochemical transmission. On the other hand, PET and SPECT allow direct or indirect measurement of neurotransmitter systems in living patients and can be used to explore alterations in neurotransmitter systems suggested by postmortem studies, as well as examine their clinical correlates.

Neuroreceptor imaging research in schizophrenia can be largely divided into (1) studies of pathophysiology and (2) studies of pharmacology. Pathophysiology studies examine neuroreceptor binding under baseline conditions, competition between the radioligand and endogenous neurotransmitters at the binding site to assess neurotransmitter release, as well as activity of enzymes. Pharmacology studies explore the mechanism of action for the existing treatments utilized in this disorder, by measuring dose-dependent occupancy of the drug in question at different receptor-binding sites. Insights from pharmacological studies are invaluable for improving psychiatric management and recently PET has become an increasingly used tool in the development of new psychiatric medications. Here, we will consider studies of both pathophysiology and pharmacology with PET and

SPECT, and discuss the application of these techniques to drug development relevant to schizophrenia.

2 Brief Overview of Neurochemical Imaging Techniques

The objective of PET or SPECT neuroreceptor imaging is to obtain quantitative information regarding the distribution of the target molecules in the living human brain. These studies involve the injection of a radioactively labeled tracer (radioligand) that binds specifically to the protein of interest, usually specific neuroreceptors or transporters (Laruelle et al. 1994). Relative to SPECT, PET allows visualizing a larger number of candidate targets in the brain and produces higher quality images due to higher resolution (better "signal-to-noise ratio") and sensitivity of the scanner. PET is also more quantitatively informative because tissue attenuation can be more accurately measured with PET technology and the associated radioisotopes than with SPECT. On the other hand, SPECT uses longer acting isotopes, allowing shipments of radioisotopes bypassing the need for an on-site cyclotron. In addition, when near-equilibrium methods of analysis can be applied, SPECT can be used to assess a relatively large number of subjects (Laruelle et al. 1994).

A crucial step for PET and SPECT technology is the synthesis of radiotracers. To be successfully used for in vivo molecular imaging, the chemical properties of a radiotracer must fall within a narrow range of appropriate combinations of lipophilicity, receptor affinity and selectivity, specificity, reversibility, and toxicity. The most commonly used positron emitting sources for PET imaging are carbon-11 ($[^{11}C]$) and fluoride-18 ($[^{18}F]$) (for in-depth review, see Townsend 2004). Radiotracer production has been the rate-limiting step in terms of exploring new targets in the brain. There are tracers for dopaminergic and serotonergic sites, but very few for sites outside of these systems that are available for use in humans. Once produced successfully, the radiotracer is injected into a vein, travels throughout the body, crosses the blood–brain barrier (BBB), and binds to the receptor (referred to as *specific binding*). The radiotracer also binds to other nonreceptor proteins in the brain (termed *nonspecific binding*).

Several factors must be taken into account in order to form accurate conclusions about receptor parameters. The activity recorded by the scanner in areas of the brain represents a combination of specifically bound, nonspecifically bound, and unbound or free radioligand. The free and nonspecifically bound radioligand are referred to as nondisplaceable binding or compartment. The proportions represented by each of these parts are time varying and interdependent. Additionally, peripheral clearance, regional cerebral blood flow, and transport of the radiotracer across the BBB influence the radioactivity profile over time and can vary significantly from subject to subject. Analysis of neuroreceptor studies requires model-based methods that relate the observed time activity in the region of interest (ROI) to the plasma time activity curve over the time course of the scan through a defined mathematical model (i.e., receptor parameter estimation is based on fitting data to a model of the underlying kinetics of ligand uptake in the brain). A variety of model-based methods have been developed (for in-depth review, see Slifstein and Laruelle 2001). The outcome measure that can be derived in neuroreceptor imaging studies is called the binding potential (BP), a term introduced by Mintun et al. (1984). It is proportional to the ratio B_{max}/K_D. The constant of proportionality differs according to the method of analysis used. K_D (nM) is the radioligand equilibrium dissociation constant and B_{max} (nM; receptor density) is the number of binding sites. In vitro, derivation of the affinity, $1/K_D$, and B_{max} is possible by using a radioactively labeled tracer, and varying the concentration of unlabeled tracer to obtain a range of receptor occupancies. In vivo studies of this type are difficult to perform in humans, as they would require multiple scans and pharmacological concentrations of the radioligand. With the very small concentrations of radioligand used in human PET studies ("tracer dose"), K_D and B_{max} cannot be measured separately. The outcome measure most often reported for PET studies is the "specific to nonspecific partition coefficient," BP_{ND} (unitless) (Fig. 2):

$$BP_{ND} = f_{ND} B_{max}/K_D.$$

Here, the constant of proportionality f_{ND} is the fraction of freely dissolved and nonspecifically bound radioligand that is freely dissolved in brain tissue. Measurement of BP_{ND} does not require measurement of arterial plasma concentration of the radioligand. Two other forms of BP that do require arterial plasma to be measured are BP_P:

$$BP_P = f_P B_{max}/K_D,$$

where f_P is the fraction of unmetabolized radioligand in arterial plasma that is not protein bound, and BP_F:

$$BP_F = B_{max}/K_D.$$

BP_F can be derived from BP_P if f_P has been measured. The choice of outcome measure is partly dictated by the experimental design – whether or not arterial plasma samples are collected – but it is also important to recognize that the outcome measures refer to different free pools of radioligand: free radioligand in the brain in the case of BP_{ND}, and in the arterial plasma in the cases of BP_P and BP_F. Ideally, all three measures provide equivalent information, but there can be differences across study groups in either free pool, leading to BP differences due to the proportionality constant rather than the receptor-related quantities B_{max} and K_D. Thus, it is important that investigators rule out these confounds in the analysis of their studies. Also, prior to the publication of the consensus nomenclature in Innis et al. (2007), the term BP was used interchangeably for any of the three outcome measures; in older literature, the particular choice must be inferred by context if it has not been made explicit. In the remainder of this chapter, the term BP is used exclusively, but can refer to any of these outcome measures.

The major disadvantages of PET and SPECT are the radiation exposure involved, limiting the number of scans which a subject may have, and the dependence of the technique on the availability of appropriate radioligands to label molecules of interest.

3 Imaging Neurotransmitter Systems

Exploration of the basic pathophysiology of schizophrenia includes studies of receptor or transporter expression in schizophrenia compared to controls, activity of enzymatic processes, and in vivo measures of neurotransmitter release. The majority of neuroreceptor studies in schizophrenia research focus on the dopamine (DA) and the serotonin (5-hydroxytryptamine – 5-HT) system, owing to both the radiotracers available for use, and current theories on the etiology of schizophrenia (for review, see Abi-Dargham 2007; Guillin et al. 2007).

3.1 Dopamine

Striatal DA activity, largely via DA D2 receptors (D2R), regulates response inhibition, temporal organization of information, and motor performance, while cortical DA transmission via DA D1 receptors (D1R) is likely affecting the maintenance and representation of ongoing behavior (Cropley et al. 2006).

Here, mostly studies with combined D2/D3 antagonist radiotracers are considered, as the lack of D3-selective radioligands has limited exploration of the functional role of the D3 receptors in the brain. Similarly, studies examining D1Rs do not distinguish between D1 and D5 as the radiotracers are not selective for one versus the other (Missale et al. 1998). In this chapter, we use D2R to refer to both D2 and D3 receptors and D1R to refer to both D1 and D5, unless indicated otherwise.

3.1.1 Striatal DA Parameters

D2 Receptors

Baseline Striatal D2 Receptor Density

Initial SPECT and PET ligand studies in schizophrenia focused on determining the number of DA receptors at baseline compared to controls, as the apparent overactivity of the DA system in patients with schizophrenia could be explained by an increased numbers of striatal DA receptors. By now, there has been an abundance of studies investigating striatal D2Rs in patients with schizophrenia, both treated and medication naïve, with varying results: The findings of the first studies were inconsistent, with some reporting increased D2R binding in schizophrenia (Crawley et al. 1986; Pearlson et al. 1993; Tune et al. 1993, 1996; Wong et al. 1986) and others no difference from controls (Abi-Dargham et al. 1998; Breier et al. 1997; Farde et al. 1990; Hietala et al. 1994; Martinot et al. 1991). Studies of medication-naïve patients with schizophrenia, using the PET D2R ligands [^{11}C]raclopride (Breier et al. 1997; Farde et al. 1990; Hietala et al. 1994; Talvik et al. 2006), [^{11}C]N-methylspiperone ([^{11}C]NMSP) (Nordstrom et al. 1995b; Okubo et al. 1997; Wong et al. 1986), and [^{76}Br]lisuride (Martinot et al. 1991, 1994) and the SPECT ligands [^{123}I]iodobenzamide ([^{123}I]IBZM) (Knable et al. 1997b; Laruelle et al. 1996; Parellada et al. 2004; Yang et al. 2004) and [^{76}Br]bromospiperone (Blin et al. 1989; Martinot et al. 1990), did not yield consistent results. Farde et al. (1990) found no general difference between groups, but patients had significant hemispheric asymmetry in D2R densities in the putamen.

Seeman and Seeman (1988) proposed to explain the discrepant results among studies with the differences in the sensitivities of different ligands to the effects of endogenous DA. There is a significantly larger effect size of studies employing butyrophenone radiotracers compared to radiotracers from other chemical families (benzamides and lisuride) (Laruelle 1998). If endogenous DA competes with the radioligand for the receptor, higher levels of DA will reduce BP for ligands that are more readily displaced by DA (such as [^{11}C]raclopride), thereby reducing the estimate of total D2R numbers. An alternative explanation is that discrepancies could be due to small sample size leading to underpowered studies. Thus, meta-analyses were performed to attempt to derive conclusions. Several meta-analyses (Kestler et al. 2001; Laruelle 1998; Weinberger and Laurelle 2001; Zakzanis and Hansen 1998) showed an overall modest (10–20%) elevation in striatal D2R density in schizophrenia. This increase is independent of the effects of antipsychotic drugs, as it was observed in drug-naïve patients (e.g., Wong et al. 1986). It is also regionally specific, as these increases are not seen in the extrastriatal regions (Buchsbaum et al. 2006; Suhara et al. 2002; Takahashi et al. 2006; Talvik et al. 2006). This increase could be genetically determined, as one SPECT study in drug-naïve patients with schizophrenia using [^{123}I]epidepride suggested: No significant differences in BP values were observed between patients and controls, but a significant correlation between frontal D2R BP values and positive symptoms in male patients with schizophrenia was found, as well as higher frontal BP values in male compared to female patients (Glenthoj et al. 2006). A twin study of patients with schizophrenia and their unaffected twins showed that monozygotic co-twins had increased caudate D2R density compared with unaffected dizygotic co-twins and healthy controls, and D2R BP was associated with a poor performance on cognitive tasks involving corticostriatal pathways (Hirvonen et al. 2005).

DA Release: Pharmacological Challenge Studies

Elevated baseline D2R is too small in magnitude to present a sufficient explanation for the increased dopaminergic tone in schizophrenia. For this reason, it was suspected that transmitter release might be abnormal. Pharmacological manipulations that induce DA release from the presynaptic dopaminergic neuron (e.g., amphetamine) allow evaluation of DA presynaptic activity or storage capacity. The released DA competes with the radioligand at the receptor and leads to a reduction in D2R radiotracer binding upon repeated scanning. The difference between baseline and postchallenge BP is considered to be an indirect index of DA transmission (Laruelle 2000; Laruelle et al. 1996). These interactions are present in rodents, nonhuman primates, and humans (for review, see Laruelle 2000). Combined microdialysis and imaging experiments in primates demonstrated that the magnitude of the ligand displacement correlated with the magnitude of the increase in amphetamine-induced intrasynaptic DA (Laruelle et al. 1997), suggesting that challenge studies provide an appropriate measure of the changes in synaptic DA levels. Agonist-induced receptor internalization may contribute to this effect, but the exact contribution of competition versus internalization to the resulting change in BP in vivo is difficult to assess (Ginovart 2005).

Studies using this approach have found evidence of roughly doubled radioligand displacement in patients with schizophrenia compared with controls (Abi-Dargham et al. 1998; Kestler et al. 2001; Laruelle et al. 1996). This was independent of the radioligand or imaging modality employed (Abi-Dargham et al. 1998; Laruelle et al. 1996).

Among patients, elevated amphetamine-induced DA release was associated with transient exacerbation of positive psychotic symptoms (Laruelle et al. 1996). The increased amphetamine-induced DA release was observed both in first episode, drug-naïve patients and in those previously treated with antipsychotic drugs. First episode patients and those who were experiencing an episode of illness exacerbation at the time of the scan showed relatively larger amphetamine-induced DA release, while patients in remission appeared no different from controls, although numerically higher (Laruelle et al. 1999). Thus, patients with schizophrenia are on average more sensitive to the DA-releasing effects of amphetamine compared with controls, but this hyperdopaminergic state is malleable and may reflect either an acute illness phase or a risk for relapse (Laruelle et al. 1999). Older literature has demonstrated that patients whose symptoms worsened following amphetamine or methylphenidate administration were more prone to relapse than those whose symptoms did not worsen (Lieberman et al. 1994).

Baseline DA Release

Amphetamine challenge imaging studies have the disadvantage of measuring changes in synaptic DA transmission following a nonphysiological stimulus and do not provide any information about synaptic DA levels at baseline. In rodents, acute depletion of synaptic DA is associated with an acute increase in the in vivo binding of [^{11}C]raclopride or [^{123}I]IBZM to D2Rs (Laruelle 2000). As increased expression of the receptor was not observed in vitro, increased radiotracer binding cannot be due to receptor upregulation, but to removal of endogenous DA and unmasking of D2Rs. Our group (Abi-Dargham et al. 2000) used an acute DA depletion paradigm in humans involving administration of oral alpha-methyl-*para*-tyrosine (AMPT), a tyrosine hydroxylase inhibitor, over 48 h, to assess the degree of occupancy of D2Rs by endogenous DA at baseline. We administered AMPT prior to estimating D2R density in patients with schizophrenia and control subjects with the SPECT tracer [^{123}I]IBZM. D2R availability after AMPT administration increased significantly more in patients than in control subjects (19 vs. 9%), consistent with the hypothesis that these patients have higher baseline levels of intrasynaptic DA. As further evidence for a striatal hyperdopaminergic state, AMPT administration led to an acute reduction in positive symptoms, and a higher level of intrasynaptic DA at baseline was predictive of rapid clinical response of positive symptoms to treatment (Abi-Dargham et al. 2000). Another study employing the depletion paradigm found that compared to healthy controls, D2R availability to the ligand [^{123}I]epidepride after DA depletion in patients correlated significantly with dysphoric symptom scores on the positive and negative symptom scale (PANSS) (Fujita et al. 2000).

Striatal D1 Receptors

The majority of imaging studies (Abi-Dargham et al. 2002; Karlsson et al. 2002; Okubo et al. 1997) have reported unaltered levels of striatal D1R in unmedicated patients with schizophrenia. Only one PET study with the D1R radioligand [^{11}C] SCH 23390 found reduced striatal D1R density in patients when compared to healthy controls and unaffected co-twins of the patients (Hirvonen et al. 2006). Yet, the finding of unchanged striatal D1 levels in the majority of the in vivo studies is consistent with the results of most postmortem studies (Seeman et al. 1987). One source of discrepancy may be the presence of differences in volumes between groups. Volume loss results in a larger fraction of partial volume effects in smaller structures compared to larger structures, leading to erroneous decreases in receptor BP measures which can be corrected by using partial volume correction methods (Rousset et al. 1998, 2008).

Dopamine Transporters

All data reviewed so far are consistent with higher DA output in striatal regions of patients with schizophrenia, which could also be explained by increased density of DA terminals. As striatal dopamine transporter (DAT) is exclusively localized on DA terminals, measurement of this receptor provides an indirect assessment of their density. Several studies have addressed this question by measuring binding of DAT radiotracers in patients with schizophrenia (Hsiao et al. 2003; Laakso et al. 2000,

2001; Laruelle et al. 2000; Lavalaye et al. 2001; Schmitt et al. 2005; Yoder et al. 2004). No differences in DAT binding between groups were reported for studies with either medication-free or medication-naïve patients at the time of scanning. A later study, however, found a significant correlation between DAT and D2R availability in the patient but not in the healthy control group and an inverse correlation between DAT and D2R availability and the extent of positive symptoms (Schmitt et al. 2008). When performed with currently medicated patients, varying results were found: no difference (Yoder et al. 2004), decreased (Laakso et al. 2001; Mateos et al. 2007), and also increased DAT binding (Sjoholm et al. 2004). Yet, the majority of evidence does not support increased presynaptic DA output in schizophrenia as a function of higher DA terminal density at baseline, which is consistent with postmortem studies (Laruelle et al. 2000) and a study evaluating the vesicular monoamine transporter type 2 (VMAT2) in the striatum: No difference between schizophrenic patients and control groups was found using the PET ligand [^{11}C] dihydrotetrabenazine ([^{11}C]DTBZ) (Taylor et al. 2000).

DA Synthesis

Presynaptic striatal dopaminergic function can be assessed using radiolabeled levodopa (L-dopa) or fluorodopa (F-dopa), which is converted to DA and stored in striatal DA nerve terminals. This provides an index of the synthesis of DA in the presynaptic terminals of striatal dopaminergic neurons (Moore et al. 2003).

The majority of studies in patients with schizophrenia using this technique to date have reported elevated presynaptic DA synthesis capacity in schizophrenia (Hietala et al. 1995, 1999; Howes et al. 2009; Lindstrom et al. 1999; McGowan et al. 2004; Meyer-Lindenberg et al. 2002), with moderate-to-large effect sizes. Two other studies, both in chronic patients, reported either a small but not significant elevation (Dao-Castellana et al. 1997) or a small reduction in L-dopa levels (Elkashef et al. 2000). All the studies that investigated patients during episodes of acute psychosis at the time of PET scanning found elevated presynaptic striatal DA availability (Hietala et al. 1995, 1999; Howes et al. 2009; Lindstrom et al. 1999) with relevant effect size (Howes et al. 2007). This is one of the most widely replicated brain dopaminergic abnormalities in schizophrenia.

Between 22 and 31% of individuals meeting clinical criteria for a high risk of psychosis, the "prodromal stage," develop a psychotic illness, predominantly schizophrenia, within 12 months, and 35% do so after 2.5 years of follow-up (Cannon et al. 2008). Prodromal patients also show an increase in striatal [^{18}F] F-dopa accumulation, which is positively correlated with more severe symptoms as it approaches the levels seen in patients with schizophrenia (Howes et al. 2009). Striatal [^{18}F]F-dopa was elevated in patients with prodromal symptoms of schizophrenia to an intermediate degree compared with that in patients with schizophrenia, confirming that DA overactivity predates the onset of schizophrenia (Howes et al. 2009). Elevated presynaptic striatal dopaminergic function is found also in other schizophrenia spectrum patients, such as schizotypal personality disorder (Abi-Dargham et al. 2004) and in relatives of patients with schizophrenia (Huttunen et al. 2008).

3.1.2 Extrastriatal Dopamine

D2 Receptors

Until recently, research has been hampered by the lack of suitable radioligands for detection of the low-density D2R DA receptor populations in the limbic and cortical DA systems that may be implicated in the pathophysiology of schizophrenia. The first generation of D2R ligands ([^{11}C]NMSP, [^{11}C]raclopride, [^{123}I]IBZM) enabled imaging of only striatal D2Rs. The recent introduction of high-affinity D2R radiotracers ([^{123}I]epidepride, [^{18}F]fallypride, [^{11}C]FLB 457) made visualization of extrastriatal D2Rs possible. Because of the low density of extrastriatal D2Rs (Hall et al. 1994) and therefore the low signal-to-noise ratio, radiotracers with high affinity or low nonspecific binding are required, such as [^{18}F]fallypride or [^{11}C]FLB 457 (K_D = 0.20 nM) (Halldin et al. 1995; Slifstein et al. 2004).

Several PET or SPECT studies have examined extrastriatal D2R levels in schizophrenia with the new generation of tracers. All have reported a decrease or no change in radiotracer (Buchsbaum et al. 2006; Suhara et al. 2002; Talvik et al. 2003; Tuppurainen et al. 2003; Yasuno et al. 2004b) binding in schizophrenia compared to controls. The most recent evaluation with [^{18}F]fallypride, however, found increased D2R levels in the substantia nigra bilaterally and decreased levels were seen in the left medial thalamus (Kessler et al. 2009) (Fig. 1) illustrates D2R binding in striatal and extrastriatal regions with [^{18}F]fallypride.

Extrastriatal D1 Receptors

Dopaminergic transmission in the prefrontal cortex (PFC) is mainly mediated by D1R, and D1R dysfunction has been linked to cognitive impairment and negative symptoms in schizophrenia (see review Tamminga 2006). Imaging studies of D1R present conflicting results: reduced D1R density measured with PET and [^{11}C]SCH 23390 was found to be related to the severity of negative symptoms and cognitive impairment (Okubo et al. 1997). A correlation between negative symptoms and [^{11}C]SCH 23390 binding in the frontal cortex persisted even when no significant difference in D1R binding in medication-naïve patients with schizophrenia compared to controls was found (Karlsson et al. 2002). Our group used [^{11}C] NNC 112 to investigate cortical D1R and found increased levels (Abi-Dargham et al. 2002). The increase correlated with cognitive impairment, which may be interpreted as the result of a compensatory upregulation in D1R density to chronic low levels of DA stimulation (Abi-Dargham and Moore 2003).

These differences in findings may be explained by different properties of the radiotracers: Guo et al. (2003) evaluated the effect of acute and subchronic DA

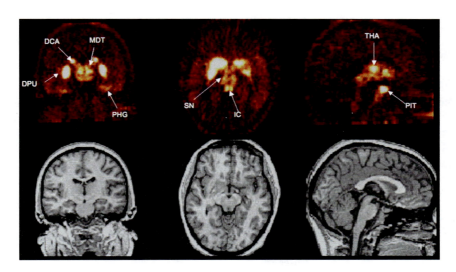

Fig. 1 PET scan in one subject under baseline conditions with coregistered structural MRI, showing activity distribution from 10 to 60 min following injection of the D2R antagonist [^{18}F] fallypride. Relatively more activity is seen in striatal compared to extrastriatal areas (dorsal caudate – DCA and dorsal putamen – DPU are shown), where concentration of D2R receptors is 10–20-fold higher. Extrastriatal binding is highest in mediodorsal thalamus (MDT), posterior hippocampal gyrus (PHG), substantia nigra (SN), inferior colliculus (IC), thalamus (THA), and pituitary gland (PIT)

depletion on the in vivo binding of [^{11}C]NNC 112 and [^{3}H]SCH 23390 in rats and found that in vivo binding of these radiotracers is differentially affected by changes in endogenous DA tone. Further studies in patients are required to clarify the exact role of D1R in schizophrenia, particularly because both tracers also bind to serotonergic 5-HT2A receptors (Ekelund et al. 2007) to a significant degree, resulting in lack of D1R selectivity in cortical areas (Slifstein et al. 2007).

3.2 Serotonin

The hypothesis that serotonin (or 5-HT) may be involved in the pathogenesis of schizophrenia was initially based on the observation of the structural similarity between serotonin and the hallucinogenic drug lysergic acid diethylamide (LSD) (Aghajanian and Marek 2000; Gaddum 1957). Postmortem studies have described alterations of 5-HT2A receptors (5-HT2AR), serotonin transporter (SERT), and above all, 5-HT1A receptors (5-HT1AR) (Gurevich and Joyce 1997; Sumiyoshi et al. 1996), and serotonin has been implicated in negative symptoms of schizophrenia (Meltzer and Sumiyoshi 2008). Cerebrospinal fluid studies and pharmacological challenges also suggest a deficit in serotonergic function in the cortex of patients with schizophrenia (Borg 2008; Sedvall 1990). 5-HT2AR antagonism appears to have beneficial effects on both positive and negative symptoms of the illness (Abi-Dargham 2007). 5-HT2AR antagonists are proposed to reduce negative symptoms in schizophrenia through activation of midbrain DA projections to the limbic system and cerebral cortex: activation of ventral tegmental area (VTA) neurons by 5-HT2AR antagonists might provide a basis for their effect on improvement of negative symptoms (Ugedo et al. 1989), as VTA DA neurons and their projections are involved in drive and reward (Bozarth 1986). Positive symptoms could be reduced by attenuation of DA phasic activity via blockade of 5-HT2A-stimulated tyrosine hydroxylase activity (Abi-Dargham et al. 1997; Bozarth 1986). Given the relatively recent development of radiotracers to study 5-HTAR, only a limited amount of data from imaging studies is available.

3.2.1 5-HT2A Receptors

The majority of postmortem studies have reported decreased 5-HT2AR in the PFC (Arora and Meltzer 1991; Burnet et al. 1996; Dean and Hayes 1996; Joyce et al. 1993; Laruelle et al. 1993). When comparing 5-HT2AR and serotonin uptake sites in PFC and occipital cortex of patients with schizophrenia, chronic schizoaffective disorder, nonpsychotic suicide victims, and controls, 5-HT2AR density was decreased in the PFC of chronic psychotic patients dying of natural causes, as opposed to the other groups (Laruelle et al. 1993). All but one (Ngan et al. 2000; using [^{18}F]setoperone) in vivo PET studies in drug-naïve or drug-free patients with schizophrenia reported normal cortical 5-HT2AR binding in the PFC, using various ligands: [^{11}C]N-methylspiperone (Okubo et al. 2000), [^{18}F]setoperone (Lewis et al. 1999; Trichard et al. 1998), and [^{18}F]altanserin (Erritzoe et al. 2008). Erritzoe et al. reported increased 5-HT2AR binding in the caudate nucleus (Erritzoe et al. 2008), a finding that requires further confirmation, especially in light of the low density of the 5-HT2AR in this brain region.

3.2.2 5-HT1A Receptors

The most consistent abnormality of 5-HT parameters reported in postmortem studies in schizophrenia is an increase in the density of 5-HT1AR in prefrontal cortical regions, including the dorsolateral prefrontal cortex (DLPFC), the anterior cingulate, and motor regions (Hashimoto et al. 1991; Joyce et al. 1993). More recently, three PET studies examined 5-HT1AR levels in vivo using [^{11}C]WAY 100635 in patients with schizophrenia and healthy controls (Frankle et al. 2006; Tauscher et al. 2002; Yasuno et al. 2004a). Frankle et al. (2006) did not detect differences in 5-HT1AR binding, whereas Tauscher et al. (2002) reported increased binding in the temporal lobe and Yasuno et al. (2004a) found decreased binding in the amygdala. A significant negative correlation was observed between BP in the amygdala and the negative and depression/anxiety symptom scores on a subscale of the PANSS (Yasuno et al. 2004a). The lack of consistency with postmortem findings may relate to the different resolutions of the different techniques. Some postmortem studies show more pronounced increase in 5-HT1AR density within superficial cortical layers (Gurevich and Joyce 1997), while others show no difference while exploring specific cellular locations within the PFC (Cruz et al. 2004). With the currently available PET technology, it is not possible to explore differences in receptor density specific to cortical layers.

3.2.3 Serotonin Transporters

The role of the SERT is the inactivation of 5-HT via uptake into the presynaptic nerve terminals. Initially, SPECT studies in humans using the SERT radiotracer [^{123}I]β-CIT were limited to imaging the midbrain. [^{11}C]McN 5652 was the first PET radiotracer successfully developed as a SERT-specific ligand, but lately a newer ligand with higher specific to nonspecific binding (Huang et al. 2002) has been in use to explore SERT density in schizophrenia: [^{11}C]DASB proved of advantage for measurement of SERT in regions with moderate density, such as the limbic regions (Frankle et al. 2004b). A study evaluating ten brain regions, including the striatum, midbrain, amygdala, hippocampus, and anterior cingulate cortex found no difference in SERT density between subjects with schizophrenia and controls (Frankle et al. 2005). A limitation of this tracer is the ability to reliably quantify SERT in cortical regions, where SERT density is low (Frankle et al. 2004b).

3.3 Gamma-Aminobutyric Acid

A significant body of preclinical findings suggests deficiency of GABAergic neurotransmission in the PFC in schizophrenia (for review, see Lewis 2000). Upregulation of GABA-A receptor-binding activity is observed throughout most subregions of the hippocampus (Benes et al. 1997). In vivo evaluation of GABAergic systems in schizophrenia was initially limited to assessment of benzodiazepine (BDZ) receptor densities with SPECT and [^{123}I]iomazenil, and none of the studies comparing patients with schizophrenia to controls reported significant regional differences (Abi-Dargham et al. 1999; Busatto and Pilowsky 1995; Busatto et al. 1997; Verhoeff et al. 1999), contrary to postmortem findings. However, alterations of GABAergic systems in schizophrenia might be restricted to certain cortical layers or classes of GABAergic cells that are beyond the resolution of current imaging techniques (Benes et al. 1997). A recently developed technique may allow to measure acute GABA fluctuations in patients with schizophrenia in vivo: The increase in the affinity of GABA-A receptors in the presence of increased GABA levels, as suggested by preclinical work, can be detected as an increase in the binding of GABA-A BDZ-receptor site-specific radioligands. Acute elevation in GABA levels was achieved through the blockade of the GABA membrane transporter (GAT1) in a PET study with [^{11}C]flumazenil. In healthy controls, this resulted in significant increases in [^{11}C]flumazenil BP compared to baseline in cortical regions (Frankle et al. 2009). This new technique may be applied in future studies to measure in vivo the responsivity of GABA transmission in patients with schizophrenia compared to controls.

3.4 N-Methyl-D-Aspartic Acid and Glutamate

The *N*-methyl-D-aspartic acid (NMDA) receptor hypofunction hypothesis of schizophrenia arose from the observation that subanesthetic doses of dissociative anesthetic drugs such as phencyclidine (PCP) and ketamine, both noncompetitive antagonists at the NMDA receptor, reliably induce the hallucinations, delusions, cognitive impairments, brain functional abnormalities, and, most notably, negative symptoms of schizophrenia in normal subjects (Krystal et al. 1994; Olney and Farber 1995; Vollenweider and Geyer 2001). The primary target may be NMDA receptors expressed on GABAergic interneurons in the thalamus and basal forebrain (Olney and Farber 1995) and action may be mediated through excess glutamate release and hyperactivity at non-NMDA glutamate receptors (Aghajanian and Marek 2000). Both drugs bind to an intrachannel site of the NMDA receptor and prevent calcium influx into the cell (Anis et al. 1983). The binding site's physical location complicates the development of suitable PET and SPECT radiotracers for this site (Waterhouse 2003). To date, one study used [^{123}I]CNS 1261, an intrachannel NMDA receptor SPECT ligand, and showed significant reductions in relative NMDA receptor binding in the left hippocampus in medication-free, but not antipsychotic-treated, patients with schizophrenia compared to healthy subjects (Pilowsky et al. 2006). Other studies have explored the effects of NMDA antagonists on indices of dopaminergic function and showed that S-ketamine increased DA release in the striatum, as evidenced by decreased [^{11}C]raclopride BP after ketamine administration (Vollenweider et al. 2000), and potentiated the effect of amphetamine-induced DA release in healthy volunteers to levels similar to those observed in patients with schizophrenia (van Berckel et al. 2006). On the other hand, chronic ketamine users had alterations in cortical D1R of similar direction and magnitude to one of the reports of D1R alterations observed in schizophrenia (Guo et al. 2003; Kapur et al. 1999). These findings from imaging studies suggest that the glutamate and DA dysregulations in schizophrenia may be interrelated and potentiate each other, forming a vicious circle.

4 Occupancy Studies (Pharmacological Studies)

The most widespread use of neuroreceptor imaging in schizophrenia over the last decade has been the assessment of receptor occupancy achieved by typical and atypical antipsychotic drugs. The main focus has been on D2R occupancy, particularly in the striatum, but 5-HT2AR, 5-HT1AR, and D1R occupancy have also been studied (Abi-Dargham and Laruelle 2005; Frankle 2007; Frankle and Laruelle 2002; Kapur et al. 1999; Nyberg et al. 1998). PET and SPECT imaging are used not only to assess the in vivo properties of currently marketed drugs, but also to help with the development of new medications for psychiatric conditions. Overall, many insights into the mechanisms of action of antipsychotics have been gained from brain imaging studies, as we will illustrate and summarize below.

4.1 DA Receptor Occupancy

4.1.1 D2 Receptor Occupancy

Implications for Treatment

Currently, most antipsychotics are administered empirically according to clinical dose-finding studies, in which arbitrarily selected doses are tested to find the most efficient dose range in a patient population. Imaging studies have consistently shown that all categories of antipsychotic drugs induced a marked occupancy of D2R at clinical doses and a threshold exists above which extrapyramidal side effects (EPS) are likely to occur ($\pm 80\%$). Farde et al. (1992) demonstrated that treatment of patients with schizophrenia with a range of antipsychotic medications resulted in blockade of 65–90% of striatal D2Rs, and that patients with acute EPS had higher levels of D2R occupancy. The minimum striatal D2 occupancy required for antipsychotic efficacy is less clearly defined, but occupancy above the EPS threshold does not appear to confer additional benefit: Clinical studies with haloperidol do not point to an advantage of doses exceeding 5 mg/day, which corresponds to $\pm 80\%$ occupancy.

Another question of clinical relevance is whether poor response to treatment can be explained by inadequate blockade of D2R. In this regard, "nonresponders" show little improvement despite high D2R occupancy rates (Kapur et al. 2000; Wolkin et al. 1989).

The onset of antipsychotic action appears to occur early (Leucht et al. 2005), but is dependent on the degree of striatal D2R occupancy (Agid et al. 2007). Low doses of selective D2R antagonists like haloperidol and raclopride require 50–60% occupancy for a rapid clinical response (Kapur et al. 2000; Nordstrom et al. 1993b), and D2 occupancy during the first 2 days predicts the nature of response over the next 2 weeks (Catafau et al. 2006).

PET studies helped to investigate the mechanisms that underlie clozapine's superior efficacy. Clozapine was the first "atypical antipsychotic" and three main hypotheses have been proposed: high 5-HT2/D2R binding ratio, loose binding to D2R, and regional specificity (Abi-Dargham and Laruelle 2005). In clozapine-treated patients with schizophrenia, the mean D2R occupancy was 47% (range 20–67%). It was also the first antipsychotic found to have a very high 5-HT2R occupancy in schizophrenia (range 84–94%) even at low doses (Farde et al. 1992;

Nordstrom et al. 1993a, 1995a). At high plasma concentrations, clozapine can induce high extrastriatal DA D2R occupancy, as found with [^{11}C]FLB 457 (Takano et al. 2006). Higher in vivo binding to cortical D2R than in the basal ganglia is suggested as an indicator of favorable profile (Xiberas et al. 2001) but has not been shown to relate to therapeutic effect, while striatal D2R occupancy has been related to therapeutic effects in more than one study with other atypical antipsychotics (Agid et al. 2007; Kegeles et al. 2008).

Studies evaluating the next two atypical antipsychotic medications, risperidone and olanzapine, found that therapeutically effective doses of risperidone blocked 46% to over 90% of striatal D2R (Knable et al. 1997a). In a study with [^{123}I]IBZM SPECT in patients treated with olanzapine, Raedler et al. observed a range of 33–81% D2 receptor blockade at 5 mg and 56–97% at 20 mg. No significant increase in EPS was found when patients' medication was increased from the lower to the higher olanzapine dose (Raedler et al. 1999). Sparing of substantia nigra and VTA D2R occupancy, demonstrated with [^{18}F] fallypride, was proposed to contribute to the low incidence of EPS in olanzapine-treated patients (Kessler et al. 2005) although many alternative explanations can be offered, the most parsimonious being low striatal D2R occupancy followed by anticholinergic blockade. Both risperidone and olanzapine in clinical dosages lead to levels of D2R occupancy that are comparable to those observed with low dosages of "typical" antipsychotics (Bressan et al. 2003; Knable et al. 1997a), but higher than with clozapine.

The striatal D2R occupancy of atypical antipsychotics ranges from 81% with risperidone to 30% with quetiapine (rank order: risperidone > olanzapine > clozapine > quetiapine) (Heinz et al. 1996). From the observed degree of elevation of synaptic DA levels in antipsychotic-naïve patients, Laruelle et al. (2005) have estimated that antipsychotic medications would need to occupy 48% of the D2R to normalize DA transmission. The fact that patients are generally treated with dosages of medication that result in 60–80% D2R blockade, and higher with standard dosing of typical antipsychotics, means that their DA tone is being maintained at slightly lower levels than those found in unmedicated healthy subjects (Frankle et al. 2004a). Ongoing D2R blockade and reduced DA tone may contribute to the dysphoria, secondary negative symptoms, and depression experienced by some patients even in the absence of EPS (Mizrahi et al. 2007).

More recently, ziprasidone has been demonstrated to result in levels of D2R occupancy similar to risperidone and olanzapine (Mamo et al. 2004; Vernaleken et al. 2008).

The advent of partial agonists may offer a way to achieve substantial D2R occupancy with a lesser impact on subjective well-being (Mamo et al. 2006). Aripiprazole at clinical doses occupies about 90% of its target receptor in the brain (Grunder et al. 2008). Occupancy levels are slightly higher in extrastriatal than striatal regions (Kegeles et al. 2008 see Fig. 2).

Quetiapine shows a transiently high D2R occupancy, which decreases to very low levels by the end of the dosing interval. Quetiapine's low D2R occupancy can explain its lack of EPS and some have speculated that its transient D2R occupancy may be sufficient for its antipsychotic effect (Kapur and Seeman 2001; Tauscher-Wisniewski et al. 2002).

Fig. 2 Parametric map of BP_{ND} of the dopamine D2R antagonist [^{18}F]fallypride, computed for nine unmedicated subjects (*baseline*), showing activity distribution in striatal regions summed over 240 min (*middle*). Coregistered structural MRI are on the *left*. On the *right*, parametric map for two subjects treated with 30 mg of aripiprazole daily. The image after [^{18}F] fallypride injection was summed over 120 min. Aripiprazole reached over 90% receptor occupancy at this dose

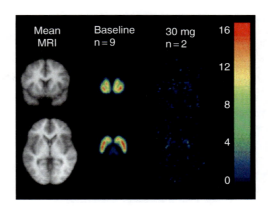

4.1.2 D1 Receptor Occupancy

Exploring the effects of antipsychotic medications on extrastriatal D1Rs has become of interest because of its implications for cognitive deficits in schizophrenia (Goldman-Rakic and Selemon 1997). Several small clinical trials with D1R antagonists have proved ineffective in reducing psychotic symptoms (Karlsson et al. 1995). Imaging studies of D1R occupancy have explored antipsychotic binding in the striatum of clozapine (33–59%), typical antipsychotics (0–44%), olanzapine (43%), quetiapine (12%), and risperidone (25%) (Farde et al. 1992; Nordstrom et al. 1995a). When comparing atypical antipsychotics for D1R occupancies studied with PET and [^{11}C]SCH 23390, mean striatal occupancies ranged from 55% with clozapine to 12% with quetiapine (rank order: clozapine > olanzapine > risperidone > quetiapine). The ratio of striatal D1/D2R occupancy was significantly higher for clozapine (0.88) relative to olanzapine (0.54), quetiapine (0.41), or risperidone (0.31) (Tauscher et al. 2004). D1R occupancy does not seem to contribute to the therapeutic effect of these drugs, as they are all D1R antagonists. Currently D1R agonists are under experimental testing.

4.2 Serotonin Occupancy

4.2.1 5-HT2A Receptor Occupancy

A fair number of recent imaging studies measured 5-HT2AR occupancy achieved by atypical medications in vivo, but information on the binding of typical antipsychotic medications to the 5-HT2AR in human subjects is very limited. Nordstrom et al. (1995b) used [^{11}C]NMSP to image the 5-HT2AR in patients with schizophrenia on clozapine. The receptor occupancy values were high despite the wide range of clozapine serum levels at the time of the scan (120–1,060 ng/ml); when compared to antipsychotic-naïve patients with schizophrenia, the 5-HT2AR occupancy ranged from 84 to 94%. Other studies have also revealed high levels of 5-HT2AR occupancy for risperidone. Greater than 80% occupancy was reported in a small sample of seven subjects taking 3 mg/day of risperidone (Nyberg et al. 1999). At 6 mg/day of risperidone, the 5-HT2AR occupancy was 95%. 5-HT2AR occupancy rates are associated with favorable treatment for depressive symptoms within schizophrenia and improvement of cognitive function (Kasper et al. 1999). It has been suggested that blockade of serotonin 5-HT2A/2C receptors may be responsible for the lower EPS observed when patients are treated with atypical antipsychotics (Heinz et al. 1996), but at high D2R occupancy levels, 5-HT2A/2C receptor antagonism is not sufficient to prevent EPS (Knable et al. 1997a), since the threshold of D2R occupancy associated with EPS is not markedly different between these drugs and the ones devoid of 5-HT2AR antagonism (Kapur et al. 1998; Knable et al. 1997a; Nyberg et al. 1998, 1999).

4.2.2 5-HT1A Receptor Occupancy

The atypical antipsychotic medications aripiprazole, clozapine, quetiapine, and ziprasidone all have a degree of agonist activity at the 5-HT1AR (Bantick et al. 2004b). Studies using [^{11}C]WAY 100635 to assess the degree of occupancy of clozapine and ziprasidone at that receptor in schizophrenia have been unsuccessful in detecting occupancy (Bantick et al. 2004a). Aripiprazole was found to have a mean occupancy of 16.4% in the temporal and frontal regions (range 0 to 39% and −2 to 43%, respectively) of patients. There was no correlation between dose (between 10 and 30 mg daily for 4 weeks) and 5-HT1AR occupancy (Mamo et al. 2007). Preclinical studies demonstrate that activation of this receptor can increase DA release in the PFC (for review, see Meltzer et al. 2003). Clinically, several small studies have shown that the addition of 5-HT1AR agonists, such as tandospirone and buspirone, to D2R antagonist medications improves negative and cognitive symptoms in schizophrenia (Sumiyoshi et al. 2001a, b). The presence of a pharmacological effect in the absence of detectable occupancy, or with minimal occupancy, is not unusual for full agonists, and has been described previously in other systems. The reasons may relate to the fact that an antagonist is used to label the sites, binding to both high- and low-affinity configurations of the receptor, while the agonist binds only to the fraction of those in the high-affinity state (Leff 1995).

5 Future Directions

The search for new and improved antipsychotic agents with more focus on neurotransmitters other than DA is an active area of research (for review, see Stone and Pilowsky 2007). Future drug development and research into the etiopathogenesis will focus on further identifying and manipulating the upstream factors that converge on the dopaminergic system (Howes and Kapur 2009). This process will unravel new mechanisms that can be used as therapeutic targets. PET can speed the process of drug discovery by (1) aiding in identifying these mechanisms of pathology, (2) providing rapid screening of new drugs with early fast decisions about which drugs are suitable to move into clinical testing, and (3) guiding dose selection. The availability of new tracers for transmitter systems that have not been studied to date will greatly facilitate this process.

Thus, neurochemical imaging has played a major role in advancing our knowledge of the pathology and treatment of schizophrenia and will continue to do so as technology improves and widens in scope.

References

Abi-Dargham A (2007) Alterations of serotonin transmission in schizophrenia. Int Rev Neurobiol 78:133–164
Abi-Dargham A, Laruelle M (2005) Mechanisms of action of second generation antipsychotic drugs in schizophrenia: insights from brain imaging studies. Eur Psychiatry 20:15–27
Abi-Dargham A, Moore H (2003) Prefrontal DA transmission at D1 receptors and the pathology of schizophrenia. Neuroscientist 5:404–416
Abi-Dargham A, Laruelle M, Aghajanian GK, Charney D, Krystal J (1997) The role of serotonin in the pathophysiology and treatment of schizophrenia. J Neuropsychiatry Clin Neurosci 9:1–17
Abi-Dargham A, Gil R, Krystal J, Baldwin RM, Seibyl JP, Bowers M, van Dyck CH, Charney DS, Innis RB, Laruelle M (1998) Increased striatal dopamine transmission in schizophrenia: confirmation in a second cohort. Am J Psychiatry 155:761–767
Abi-Dargham A, Laruelle M, Krystal J (1999) No evidence of altered in vivo benzodiazepine receptor binding in schizophrenia. Neuropsychopharmacology 20:650–661
Abi-Dargham A, Rodenhiser J, Printz D, Zea-Ponce Y, Gil R, Kegeles LS, Weiss R, Cooper TB, Mann JJ, Van Heertum RL, Gorman JM, Laruelle M (2000) Increased baseline occupancy of D2 receptors by dopamine in schizophrenia. Proc Natl Acad Sci USA 97:8104–8109
Abi-Dargham A, Mawlawi O, Lombardo I (2002) Prefrontal dopamine D1 receptors and working memory in schizophrenia. J Neurosci 22:3708–3719
Abi-Dargham A, Kegeles LS, Zea-Ponce Y, Mawlawi O, Martinez D, Mitropoulou V, O'Flynn K, Koenigsberg HW, Van Heertum R, Cooper T, Laruelle M, Siever LJ (2004) Striatal amphetamine-induced dopamine release in patients with schizotypal personality disorder studied with single photon emission computed tomography and [123I]iodobenzamide. Biol Psychiatry 55:1001–1006
Aghajanian GK, Marek GJ (2000) Serotonin model of schizophrenia: emerging role of glutamate mechanisms. Brain Res Brain Res Rev 31:302–312
Agid O, Mamo D, Ginovart N, Vitcu I, Wilson AA, Zipursky RB, Kapur S (2007) Striatal vs extrastriatal dopamine D2 receptors in antipsychotic response – a double-blind PET study in schizophrenia. Neuropsychopharmacology 32:1209–1215
Anis NA, Berry SC, Burton NR, Lodge D (1983) The dissociative anaesthetics, ketamine and phencyclidine, selectively reduce excitation of central mammalian neurones by N-methyl-aspartate. Br J Pharmacol 79:565–575
Arora RC, Meltzer HY (1991) Serotonin2 (5-HT2) receptor binding in the frontal cortex of schizophrenic patients. J Neural Transm Gen Sect 85:19–29
Bantick RA, Montgomery AJ, Bench CJ, Choudhry T, Malek N, McKenna PJ, Quested DJ, Deakin JF, Grasby PM (2004a) A positron emission tomography study of the 5-HT1A receptor in schizophrenia and during clozapine treatment. J Psychopharmacol 18:346–354

Bantick RA, Rabiner EA, Hirani E, de Vries MH, Hume SP, Grasby PM (2004b) Occupancy of agonist drugs at the 5-HT1A receptor. Neuropsychopharmacology 29:847–859

Benes FM (2000) Emerging principles of altered neural circuitry in schizophrenia. Brain Res Brain Res Rev 31:251–269

Benes FM, Wickramasinghe R, Vincent SL, Khan Y, Todtenkopf M (1997) Uncoupling of GABA (A) and benzodiazepine receptor binding activity in the hippocampal formation of schizophrenic brain. Brain Res 755:121–129

Blin J, Baron JC, Cambon H, Bonnet AM, Dubois B, Loc'h C, Maziere B, Agid Y (1989) Striatal dopamine D2 receptors in tardive dyskinesia: PET study. J Neurol Neurosurg Psychiatry 52:1248–1252

Borg J (2008) Molecular imaging of the 5-HT(1A) receptor in relation to human cognition. Behav Brain Res 195:103–111

Bozarth MA (1986) Neural basis of psychomotor stimulant and opiate reward: evidence suggesting the involvement of a common dopaminergic system. Behav Brain Res 22:107–116

Breier A, Su TP, Saunders R, Carson RE, Kolachana BS, de Bartolomeis A, Weinberger DR, Weisenfeld N, Malhotra AK, Eckelman WC, Pickar D (1997) Schizophrenia is associated with elevated amphetamine-induced synaptic dopamine concentrations: evidence from a novel positron emission tomography method. Proc Natl Acad Sci USA 94:2569–2574

Bressan RA, Erlandsson K, Jones HM, Mulligan RS, Ell PJ, Pilowsky LS (2003) Optimizing limbic selective D2/D3 receptor occupancy by risperidone: a [123I]-epidepride SPET study. J Clin Psychopharmacol 23:5–14

Buchsbaum MS, Christian BT, Lehrer DS, Narayanan TK, Shi B, Mantil J, Kemether E, Oakes TR, Mukherjee J (2006) D2/D3 dopamine receptor binding with [F-18]fallypride in thalamus and cortex of patients with schizophrenia. Schizophr Res 85:232–244

Burnet PW, Eastwood SL, Harrison PJ (1996) 5-HT1A and 5-HT2A receptor mRNAs and binding site densities are differentially altered in schizophrenia. Neuropsychopharmacology 15:442–455

Busatto GF, Pilowsky LS (1995) Neuroreceptor mapping with in-vivo imaging techniques: principles and applications. Br J Hosp Med 53:309–313

Busatto GF, Pilowsky LS, Costa DC, Ell PJ, David AS, Lucey JV, Kerwin RW (1997) Correlation between reduced in vivo benzodiazepine receptor binding and severity of psychotic symptoms in schizophrenia. Am J Psychiatry 154:56–63

Cannon TD, Cadenhead K, Cornblatt B, Woods SW, Addington J, Walker E, Seidman LJ, Perkins D, Tsuang M, McGlashan T, Heinssen R (2008) Prediction of psychosis in youth at high clinical risk: a multisite longitudinal study in North America. Arch Gen Psychiatry 65:28–37

Catafau AM, Corripio I, Perez V, Martin JC, Schotte A, Carrio I, Alvarez E (2006) Dopamine D2 receptor occupancy by risperidone: implications for the timing and magnitude of clinical response. Psychiatry Res 148:175–183

Crawley JC, Crow TJ, Johnstone EC, Oldland SR, Owen F, Owens DG, Smith T, Veall N, Zanelli GD (1986) Uptake of 77Br-spiperone in the striata of schizophrenic patients and controls. Nucl Med Commun 7:599–607

Cropley VL, Fujita M, Innis RB, Nathan PJ (2006) Molecular imaging of the dopaminergic system and its association with human cognitive function. Biol Psychiatry 59:898–907

Cruz DA, Eggan SM, Azmitia EC, Lewis DA (2004) Serotonin1A receptors at the axon initial segment of prefrontal pyramidal neurons in schizophrenia. Am J Psychiatry 161:739–742

Dao-Castellana MH, Paillere-Martinot ML, Hantraye P, Attar-Levy D, Remy P, Crouzel C, Artiges E, Feline A, Syrota A, Martinot JL (1997) Presynaptic dopaminergic function in the striatum of schizophrenic patients. Schizophr Res 23:167–174

Dean B, Hayes W (1996) Decreased frontal cortical serotonin2A receptors in schizophrenia. Schizophr Res 21:133–139

Ekelund J, Slifstein M, Narendran R, Guillin O, Belani H, Guo NN, Hwang Y, Hwang DR, Abi-Dargham A, Laruelle M (2007) In vivo DA D(1) receptor selectivity of NNC 112 and SCH 23390. Mol Imaging Biol 9:117–125

Elkashef AM, Doudet D, Bryant T, Cohen RM, Li SH, Wyatt RJ (2000) 6-(18)F-DOPA PET study in patients with schizophrenia. Positron emission tomography. Psychiatry Res 100:1–11

Erritzoe D, Rasmussen H, Kristiansen KT, Frokjaer VG, Haugbol S, Pinborg L, Baare W, Svarer C, Madsen J, Lublin H, Knudsen GM, Glenthoj BY (2008) Cortical and subcortical 5-HT2A receptor binding in neuroleptic-naive first-episode schizophrenic patients. Neuropsychopharmacology 33:2435–2441

Farde L, Wiesel FA, Stone-Elander S, Halldin C, Nordstrom AL, Hall H, Sedvall G (1990) D2 dopamine receptors in neuroleptic-naive schizophrenic patients. A positron emission tomography study with [11C]raclopride. Arch Gen Psychiatry 47:213–219

Farde L, Nordstrom AL, Wiesel FA, Pauli S, Halldin C, Sedvall G (1992) Positron emission tomographic analysis of central D1 and D2 dopamine receptor occupancy in patients treated with classical neuroleptics and clozapine. Relation to extrapyramidal side effects. Arch Gen Psychiatry 49:538–544

Frankle WG (2007) Neuroreceptor imaging studies in schizophrenia. Harv Rev Psychiatry 15:212–232

Frankle WG, Laruelle M (2002) Neuroreceptor imaging in psychiatric disorders. Ann Nucl Med 16:437–446

Frankle WG, Gil R, Hackett E, Mawlawi O, Zea-Ponce Y, Zhu Z, Kochan LD, Cangiano C, Slifstein M, Gorman JM, Laruelle M, Abi-Dargham A (2004a) Occupancy of dopamine D2 receptors by the atypical antipsychotic drugs risperidone and olanzapine: theoretical implications. Psychopharmacology (Berl) 175:473–480

Frankle WG, Huang Y, Hwang DR, Talbot PS, Slifstein M, Van Heertum R, Abi-Dargham A, Laruelle M (2004b) Comparative evaluation of serotonin transporter radioligands 11C-DASB and 11C-McN 5652 in healthy humans. J Nucl Med 45:682–694

Frankle WG, Narendran R, Huang Y, Hwang DR, Lombardo I, Cangiano C, Gil R, Laruelle M, Abi-Dargham A (2005) Serotonin transporter availability in patients with schizophrenia: a positron emission tomography imaging study with [11C]DASB. Biol Psychiatry 57: 1510–1516

Frankle WG, Lombardo I, Kegeles LS, Slifstein M, Martin JH, Huang Y, Hwang DR, Reich E, Cangiano C, Gil R, Laruelle M, Abi-Dargham A (2006) Serotonin 1A receptor availability in patients with schizophrenia and schizo-affective disorder: a positron emission tomography imaging study with [11C]WAY 100635. Psychopharmacology (Berl) 189:155–164

Frankle WG, Cho RY, Narendran R, Mason NS, Vora S, Litschge M, Price JC, Lewis DA, Mathis CA (2009) Tiagabine increases [11C]flumazenil binding in cortical brain regions in healthy control subjects. Neuropsychopharmacology 34:624–633

Fujita M, Verhoeff NP, Varrone A, Zoghbi SS, Baldwin RM, Jatlow PA, Anderson GM, Seibyl JP, Innis RB (2000) Imaging extrastriatal dopamine D(2) receptor occupancy by endogenous dopamine in healthy humans. Eur J Pharmacol 387:179–188

Gaddum JH (1957) Serotonin–LSD interactions. Ann NY Acad Sci 66:643–647, discussion 647–648

Ginovart N (2005) Imaging the dopamine system with in vivo [11C]raclopride displacement studies: understanding the true mechanism. Mol Imaging Biol 7:45–52

Glenthoj BY, Mackeprang T, Svarer C, Rasmussen H, Pinborg LH, Friberg L, Baare W, Hemmingsen R, Videbaek C (2006) Frontal dopamine D(2/3) receptor binding in drug-naive first-episode schizophrenic patients correlates with positive psychotic symptoms and gender. Biol Psychiatry 60:621–629

Goldman-Rakic PS, Selemon LD (1997) Functional and anatomical aspects of prefrontal pathology in schizophrenia. Schizophr Bull 23:437–458

Grunder G, Fellows C, Janouschek H, Veselinovic T, Boy C, Brocheler A, Kirschbaum KM, Hellmann S, Spreckelmeyer KM, Hiemke C, Rosch F, Schaefer WM, Vernaleken I (2008) Brain and plasma pharmacokinetics of aripiprazole in patients with schizophrenia: an [18F] fallypride PET study. Am J Psychiatry 165:988–995

Guillin O, Abi-Dargham A, Laruelle M (2007) Neurobiology of dopamine in schizophrenia. Int Rev Neurobiol 78:1–39

Guo N, Hwang DR, Lo ES, Huang YY, Laruelle M, Abi-Dargham A (2003) Dopamine depletion and in vivo binding of PET D1 receptor radioligands: implications for imaging studies in schizophrenia. Neuropsychopharmacology 28:1703–1711

Gurevich EV, Joyce JN (1997) Alterations in the cortical serotonergic system in schizophrenia: a postmortem study. Biol Psychiatry 42:529–545

Hall H, Sedvall G, Magnusson O, Kopp J, Halldin C, Farde L (1994) Distribution of D1- and D2- dopamine receptors, and dopamine and its metabolites in the human brain. Neuropsychopharmacology 11:245–256

Halldin C, Farde L, Hogberg T, Mohell N, Hall H, Suhara T, Karlsson P, Nakashima Y, Swahn CG (1995) Carbon-11-FLB 457: a radioligand for extrastriatal D2 dopamine receptors. J Nucl Med 36:1275–1281

Hashimoto T, Nishino N, Nakai H, Tanaka C (1991) Increase in serotonin 5-HT1A receptors in prefrontal and temporal cortices of brains from patients with chronic schizophrenia. Life Sci 48:355–363

Heinz A, Knable MB, Weinberger DR (1996) Dopamine D2 receptor imaging and neuroleptic drug response. J Clin Psychiatry 57(Suppl 11):84–88, discussion 89–93

Hietala J, Syvalahti E, Vuorio K, Nagren K, Lehikoinen P, Ruotsalainen U, Rakkolainen V, Lehtinen V, Wegelius U (1994) Striatal D2 dopamine receptor characteristics in neuroleptic-naive schizophrenic patients studied with positron emission tomography. Arch Gen Psychiatry 51:116–123

Hietala J, Syvalahti E, Vuorio K, Rakkolainen V, Bergman J, Haaparanta M, Solin O, Kuoppamaki M, Kirvela O, Ruotsalainen U, Salokangas RKR (1995) Presynaptic dopamine function in striatum of neuroleptic-naive schizophrenic patients. Lancet 346:1130–1131

Hietala J, Syvalahti E, Vilkman H, Vuorio K, Rakkolainen V, Bergman J, Haaparanta M, Solin O, Kuoppamaki M, Eronen E, Ruotsalainen U, Salokangas RK (1999) Depressive symptoms and presynaptic dopamine function in neuroleptic-naive schizophrenia. Schizophr Res 35:41–50

Hirvonen J, van Erp TG, Huttunen J, Aalto S, Någren K, Huttunen M, Lönnqvist J, Kaprio J, Hietala J, Cannon TD (2005) Increased caudate dopamine D2 receptor availability as a genetic marker for schizophrenia. Arch Gen Psychiatry 4:371–378

Hirvonen J, van Erp TG, Huttunen J, Aalto S, Nagren K, Huttunen M, Lonnqvist J, Kaprio J, Cannon TD, Hietala J (2006) Brain dopamine d1 receptors in twins discordant for schizophrenia. Am J Psychiatry 163:1747–1753

Howes OD, Kapur S (2009) The dopamine hypothesis of schizophrenia: version III – the final common pathway. Schizophr Bull 35:549–562

Howes OD, Montgomery AJ, Asselin MC, Murray RM, Grasby PM, McGuire PK (2007) Molecular imaging studies of the striatal dopaminergic system in psychosis and predictions for the prodromal phase of psychosis. Br J Psychiatry 51(Suppl):S13–S18

Howes OD, Montgomery AJ, Asselin MC, Murray RM, Valli I, Tabraham P, Bramon-Bosch E, Valmaggia L, Johns L, Broome M, McGuire PK, Grasby PM (2009) Elevated striatal dopamine function linked to prodromal signs of schizophrenia. Arch Gen Psychiatry 66:13–20

Hsiao MC, Lin KJ, Liu CY, Tzen KY, Yen TC (2003) Dopamine transporter change in drug-naive schizophrenia: an imaging study with 99mTc-TRODAT-1. Schizophr Res 65:39–46

Huang Y, Hwang DR, Narendran R, Sudo Y, Chatterjee R, Bae SA, Mawlawi O, Kegeles LS, Wilson AA, Kung HF, Laruelle M (2002) Comparative evaluation in nonhuman primates of five PET radiotracers for imaging the serotonin transporters: [11C]McN 5652, [11C]ADAM, [11C]DASB, [11C]DAPA, and [11C]AFM. J Cereb Blood Flow Metab 22:1377–1398

Huttunen J, Heinimaa M, Svirskis T, Nyman M, Kajander J, Forsback S, Solin O, Ilonen T, Korkeila J, Ristkari T, McGlashan T, Salokangas RK, Hietala J (2008) Striatal dopamine synthesis in first-degree relatives of patients with schizophrenia. Biol Psychiatry 63:114–117

Innis RB, Cunningham VJ, Delforge J, Fujita M, Gjedde A, Gunn RN, Holden J, Houle S, Huang SC, Ichise M, Iida H, Ito H, Kimura Y, Koeppe RA, Knudsen GM, Knuuti J,

Lammertsma AA, Laruelle M, Logan J, Maguire RP, Mintun MA, Morris ED, Parsey R, Price JC, Slifstein M, Sossi V, Suhara T, Votaw JR, Wong DF, Carson RE (2007) Consensus nomenclature for in vivo imaging of reversibly binding radioligands. J Cereb Blood Flow Metab 27:1533–1539

Joyce JN, Shane A, Lexow N, Winokur A, Casanova MF, Kleinman JE (1993) Serotonin uptake sites and serotonin receptors are altered in the limbic system of schizophrenics. Neuropsychopharmacology 8:315–336

Kapur S, Seeman P (2001) Does fast dissociation from the dopamine d(2) receptor explain the action of atypical antipsychotics? A new hypothesis. Am J Psychiatry 3:360–369

Kapur S, Zipursky RB, Remington G, Jones C, DaSilva J, Wilson AA, Houle S (1998) 5-HT2 and D2 receptor occupancy of olanzapine in schizophrenia: a PET investigation. Am J Psychiatry 155:921–928

Kapur S, Zipursky RB, Remington G (1999) Clinical and theoretical implications of 5-HT2 and D2 receptor occupancy of clozapine, risperidone, and olanzapine in schizophrenia. Am J Psychiatry 156:286–293

Kapur S, Zipursky R, Jones C, Remington G, Houle S (2000) Relationship between dopamine D(2) occupancy, clinical response, and side effects: a double-blind PET study of first-episode schizophrenia. Am J Psychiatry 4:514–520

Karlsson P, Smith L, Farde L, Harnryd C, Sedvall G, Wiesel FA (1995) Lack of apparent antipsychotic effect of the D1-dopamine receptor antagonist SCH39166 in acutely ill schizophrenic patients. Psychopharmacology (Berl) 121:309–316

Karlsson P, Farde L, Halldin C, Sedvall G (2002) PET study of D(1) dopamine receptor binding in neuroleptic-naive patients with schizophrenia. Am J Psychiatry 159:761–767

Kasper S, Tauscher J, Kufferle B, Barnas C, Pezawas L, Quiner S (1999) Dopamine- and serotonin-receptors in schizophrenia: results of imaging-studies and implications for pharmacotherapy in schizophrenia. Eur Arch Psychiatry Clin Neurosci 249(Suppl 4):83–89

Kegeles LS, Slifstein M, Frankle WG, Xu X, Hackett E, Bae SA, Gonzales R, Kim JH, Alvarez B, Gil R, Laruelle M, Abi-Dargham A (2008) Dose-occupancy study of striatal and extrastriatal dopamine D2 receptors by aripiprazole in schizophrenia with PET and [18F]fallypride. Neuropsychopharmacology 33:3111–3125

Kessler RM, Ansari MS, Riccardi P, Li R, Jayathilake K, Dawant B, Meltzer HY (2005) Occupancy of striatal and extrastriatal dopamine D2/D3 receptors by olanzapine and haloperidol. Neuropsychopharmacology 30:2283–2289

Kessler RM, Woodward ND, Riccardi P, Li R, Ansari MS, Anderson S, Dawant B, Zald D, Meltzer HY (2009) Dopamine D2 receptor levels in striatum, thalamus, substantia nigra, limbic regions, and cortex in schizophrenic subjects. Biol Psychiatry 65:1024–1031

Kestler LP, Walker E, Vega EM (2001) Dopamine receptors in the brains of schizophrenia patients: a meta-analysis of the findings. Behav Pharmacol 12:355–371

Knable MB, Heinz A, Raedler T, Weinberger DR (1997a) Extrapyramidal side effects with risperidone and haloperidol at comparable D2 receptor occupancy levels. Psychiatry Res 75:91–101

Knable MB, Egan MF, Heinz A, Gorey J, Lee KS, Coppola R, Weinberger DR (1997b) Altered dopaminergic function and negative symptoms in drug-free patients with schizophrenia. [123I]-iodobenzamide SPECT study. Br J Psychiatry 171:574–577

Krystal JH, Karper LP, Seibyl JP, Freeman GK, Delaney R, Bremner JD, Heninger GR, Bowers MB Jr, Charney DS (1994) Subanesthetic effects of the noncompetitive NMDA antagonist, ketamine, in humans. Psychotomimetic, perceptual, cognitive, and neuroendocrine responses. Arch Gen Psychiatry 51:199–214

Laakso A, Vilkman H, Alakare B, Haaparanta M, Bergman J, Solin O, Peurasaari J, Rakkolainen V, Syvalahti E, Hietala J (2000) Striatal dopamine transporter binding in neuroleptic-naive patients with schizophrenia studied with positron emission tomography. Am J Psychiatry 157:269–271

Laakso A, Bergman J, Haaparanta M, Vilkman H, Solin O, Syvalahti E, Hietala J (2001) Decreased striatal dopamine transporter binding in vivo in chronic schizophrenia. Schizophr Res 52:115–120
Laruelle M (1998) Imaging dopamine transmission in schizophrenia. A review and meta-analysis. Q J Nucl Med 42:211–221
Laruelle M (2000) Imaging synaptic neurotransmission with in vivo binding competition techniques: a critical review. J Cereb Blood Flow Metab 20:423–451
Laruelle M, Abi-Dargham A, Casanova MF, Toti R, Weinberger DR, Kleinman JE (1993) Selective abnormalities of prefrontal serotonergic receptors in schizophrenia. A postmortem study. Arch Gen Psychiatry 50:810–818
Laruelle M, Wallace E, Seibyl JP, Baldwin RM, Zea-Ponce Y, Zoghbi SS, Neumeyer JL, Charney DS, Hoffer PB, Innis RB (1994) Graphical, kinetic, and equilibrium analyses of in vivo [123I] beta-CIT binding to dopamine transporters in healthy human subjects. J Cereb Blood Flow Metab 14:982–994
Laruelle M, Abi-Dargham A, van Dyck CH, Gil R, D'Souza CD, Erdos J, McCance E, Rosenblatt W, Fingado C, Zoghbi SS, Baldwin RM, Seibyl JP, Krystal JH, Charney DS, Innis RB (1996) Single photon emission computerized tomography imaging of amphetamine-induced dopamine release in drug-free schizophrenic subjects. Proc Natl Acad Sci USA 93:9235–9240
Laruelle M, Iyer RN, al-Tikriti MS, Zea-Ponce Y, Malison R, Zoghbi SS, Baldwin RM, Kung HF, Charney DS, Hoffer PB, Innis RB, Bradberry CW (1997) Microdialysis and SPECT measurements of amphetamine-induced dopamine release in nonhuman primates. Synapse 25:1–14
Laruelle M, Abi-Dargham A, Gil R, Kegeles L, Innis R (1999) Increased dopamine transmission in schizophrenia: relationship to illness phases. Biol Psychiatry 46:56–72
Laruelle M, Abi-Dargham A, van Dyck C (2000) Dopamine and serotonin transporters in patients with schizophrenia: an imaging study with [(123)I]beta-CIT. Biol Psychiatry 47:371–379
Laruelle M, Frankle WG, Narendran R, Kegeles LS, Abi-Dargham A (2005) Mechanism of action of antipsychotic drugs: from dopamine D(2) receptor antagonism to glutamate NMDA facilitation. Clin Ther 27(Suppl A):S16–S24
Lavalaye J, Linszen DH, Booij J, Dingemans PM, Reneman L, Habraken JB, Gersons BP, van Royen EA (2001) Dopamine transporter density in young patients with schizophrenia assessed with [123]FP-CIT SPECT. Schizophr Res 47:59–67
Leff P (1995) The two-state model of receptor activation. Trends Pharmacol Sci 16:89–97
Leucht S, Busch R, Hamann J, Kissling W, Kane JM (2005) Early-onset hypothesis of antipsychotic drug action: a hypothesis tested, confirmed and extended. Biol Psychiatry 57: 1543–1549
Lewis DA (2000) GABAergic local circuit neurons and prefrontal cortical dysfunction in schizophrenia. Brain Res Brain Res Rev 31:270–276
Lewis R, Kapur S, Jones C, DaSilva J, Brown GM, Wilson AA, Houle S, Zipursky RB (1999) Serotonin 5-HT2 receptors in schizophrenia: a PET study using [18F]setoperone in neuroleptic-naive patients and normal subjects. Am J Psychiatry 156:72–78
Lieberman JA, Alvir J, Geisler S, Ramos-Lorenzi J, Woerner M, Novacenko H, Cooper T, Kane JM (1994) Methylphenidate response, psychopathology and tardive dyskinesia as predictors of relapse in schizophrenia. Neuropsychopharmacology 11:107–118
Lindstrom LH, Gefvert O, Hagberg G, Lundberg T, Bergstrom M, Hartvig P, Langstrom B (1999) Increased dopamine synthesis rate in medial prefrontal cortex and striatum in schizophrenia indicated by L-(beta-11C) DOPA and PET. Biol Psychiatry 46:681–688
Mamo D, Kapur S, Shammi CM, Papatheodorou G, Mann S, Therrien F, Remington G (2004) A PET study of dopamine D2 and serotonin 5-HT2 receptor occupancy in patients with schizophrenia treated with therapeutic doses of ziprasidone. Am J Psychiatry 161:818–825
Mamo D, Graff A, Romeyer F et al (2006) Central D2 and 5HT2A occupancy of aripiprazole in patients with schizophrenia. Schizophr Res 82:104–105

Mamo D, Graff A, Mizrahi R, Shammi CM, Romeyer F, Kapur S (2007) Differential effects of aripiprazole on D(2), 5-HT(2), and 5-HT(1A) receptor occupancy in patients with schizophrenia: a triple tracer PET study. Am J Psychiatry 164:1411–1417

Martinot JL, Peron-Magnan P, Huret JD, Mazoyer B, Baron JC, Boulenger JP, Loc'h C, Maziere B, Caillard V, Loo H, Syrota A (1990) Striatal D2 dopaminergic receptors assessed with positron emission tomography and [76Br]bromospiperone in untreated schizophrenic patients. Am J Psychiatry 147:44–50

Martinot JL, Paillere-Martinot ML, Loc'h C, Hardy P, Poirier MF, Mazoyer B, Beaufils B, Maziere B, Allilaire JF, Syrota A (1991) The estimated density of D2 striatal receptors in schizophrenia. A study with positron emission tomography and 76Br-bromolisuride. Br J Psychiatry 158:346–350

Martinot JL, Paillere-Martinot ML, Loc'h C, Lecrubier Y, Dao-Castellana MH, Aubin F, Allilaire JF, Mazoyer B, Maziere B, Syrota A (1994) Central D2 receptors and negative symptoms of schizophrenia. Br J Psychiatry 164:27–34

Mateos JJ, Lomena F, Parellada E, Mireia F, Fernandez-Egea E, Pavia J, Prats A, Pons F, Bernardo M (2007) Lower striatal dopamine transporter binding in neuroleptic-naive schizophrenic patients is not related to antipsychotic treatment but it suggests an illness trait. Psychopharmacology (Berl) 191:805–811

McGowan S, Lawrence AD, Sales T, Quested D, Grasby P (2004) Presynaptic dopaminergic dysfunction in schizophrenia: a positron emission tomographic [18F]fluorodopa study. Arch Gen Psychiatry 61:134–142

Meltzer HY, Sumiyoshi T (2008) Does stimulation of 5-HT(1A) receptors improve cognition in schizophrenia? Behav Brain Res 195:98–102

Meltzer HY, Li Z, Kaneda Y, Ichikawa J (2003) Serotonin receptors: their key role in drugs to treat schizophrenia. Prog Neuropsychopharmacol Biol Psychiatry 27:1159–1172

Meyer-Lindenberg A, Miletich RS, Kohn PD, Esposito G, Carson RE, Quarantelli M, Weinberger DR, Berman KF (2002) Reduced prefrontal activity predicts exaggerated striatal dopaminergic function in schizophrenia. Nat Neurosci 5:267–271

Mintun MA, Raichle ME, Kilbourn MR, Wooten GF, Welch MJ (1984) A quantitative model for the in vivo assessment of drug binding sites with positron emission tomography. Ann Neurol 15:217–227

Missale C, Nash SR, Robinson SW, Jaber M, Caron MG (1998) Dopamine receptors: from structure to function. Physiol Rev 78:189–225

Mizrahi R, Rusjan P, Agid O, Graff A, Mamo DC, Zipursky RB, Kapur S (2007) Adverse subjective experience with antipsychotics and its relationship to striatal and extrastriatal D2 receptors: a PET study in schizophrenia. Am J Psychiatry 164:543–546

Moore RY, Whone AL, McGowan S, Brooks DJ (2003) Monoamine neuron innervation of the normal human brain: an 18F-DOPA PET study. Brain Res 982:137–145

Ngan ET, Yatham LN, Ruth TJ, Liddle PF (2000) Decreased serotonin 2A receptor densities in neuroleptic-naive patients with schizophrenia: a PET study using [(18)F]setoperone. Am J Psychiatry 157:1016–1018

Nordstrom AL, Farde L, Halldin C (1993a) High 5-HT2 receptor occupancy in clozapine treated patients demonstrated by PET. Psychopharmacology (Berl) 110:365–367

Nordstrom AL, Farde L, Wiesel FA, Forslund K, Pauli S, Halldin C, Uppfeldt G (1993b) Central D2-dopamine receptor occupancy in relation to antipsychotic drug effects: a double-blind PET study of schizophrenic patients. Biol Psychiatry 33:227–235

Nordstrom AL, Farde L, Nyberg S, Karlsson P, Halldin C, Sedvall G (1995a) D1, D2, and 5-HT2 receptor occupancy in relation to clozapine serum concentration: a PET study of schizophrenic patients. Am J Psychiatry 152:1444–1449

Nordstrom FL, Eriksson L, Halldin C (1995b) No elevated D2 dopamine receptors in neuroleptic-naive schizophrenic patients revealed by positron emission tomography and [11C]N-methyl-spiperone. Psychiatry Res 61:67–83

Nyberg S, Nilsson U, Okubo Y, Halldin C, Farde L (1998) Implications of brain imaging for the management of schizophrenia. Int Clin Psychopharmacol 13(Suppl 3):S15–S20

Nyberg S, Eriksson B, Oxenstierna G, Halldin C, Farde L (1999) Suggested minimal effective dose of risperidone based on PET-measured D2 and 5-HT2A receptor occupancy in schizophrenic patients. Am J Psychiatry 156:869–875

Okubo Y, Suhara T, Suzuki K, Kobayashi K, Inoue O, Terasaki O, Someya Y, Sassa T, Sudo Y, Matsushima E, Iyo M, Tateno Y, Toru M (1997) Decreased prefrontal dopamine D1 receptors in schizophrenia revealed by PET. Nature 385:634–636

Okubo Y, Suhara T, Suzuki K, Kobayashi K, Inoue O, Terasaki O, Someya Y, Sassa T, Sudo Y, Matsushima E, Iyo M, Tateno Y, Toru M (2000) Serotonin 5-HT2 receptors in schizophrenic patients studied by positron emission tomography. Life Sci 66:2455–2464

Olney JW, Farber NB (1995) NMDA antagonists as neurotherapeutic drugs, psychotogens, neurotoxins, and research tools for studying schizophrenia. Neuropsychopharmacology 13:335–345

Parellada E, Lomena F, Catafau AM, Bernardo M, Font M, Fernandez-Egea E, Pavia J, Gutierrez F (2004) Lack of sex differences in striatal dopamine D2 receptor binding in drug-naive schizophrenic patients: an IBZM-SPECT study. Psychiatry Res 130:79–84

Pearlson GD, Tune LE, Wong DF, Aylward EH, Barta PE, Powers RE, Tien AY, Chase GA, Harris GJ, Rabins PV (1993) Quantitative D2 dopamine receptor PET and structural MRI changes in late-onset schizophrenia. Schizophr Bull 19:783–795

Pilowsky LS, Bressan RA, Stone JM, Erlandsson K, Mulligan RS, Krystal JH, Ell PJ (2006) First in vivo evidence of an NMDA receptor deficit in medication-free schizophrenic patients. Mol Psychiatry 11:118–119

Raedler TJ, Knable MB, Lafargue T, Urbina RA, Egan MF, Pickar D, Weinberger DR (1999) In vivo determination of striatal dopamine D2 receptor occupancy in patients treated with olanzapine. Psychiatry Res 90:81–90

Rousset OG, Ma Y, Evans AC (1998) Correction for partial volume effects in PET: principle and validation. J Nucl Med 39:904–911

Rousset OG, Collins DL, Rahmim A, Wong DF (2008) Design and implementation of an automated partial volume correction in PET: application to dopamine receptor quantification in the normal human striatum. J Nucl Med 49:1097–1106

Schmitt GJ, Meisenzahl EM, Frodl T, La Fougere C, Hahn K, Moller HJ, Dresel S (2005) The striatal dopamine transporter in first-episode, drug-naive schizophrenic patients: evaluation by the new SPECT-ligand[99mTc]TRODAT-1. J Psychopharmacol 19:488–493

Schmitt GJ, la Fougere C, Dresel S, Frodl T, Hahn K, Moller HJ, Meisenzahl EM (2008) Dual-isotope SPECT imaging of striatal dopamine: first episode, drug naive schizophrenic patients. Schizophr Res 101:133–141

Sedvall G (1990) Monoamines and schizophrenia. Acta Psychiatr Scand 358(Suppl):7–13

Seeman MV, Seeman P (1988) Psychosis and positron tomography. Can J Psychiatry 33:299–306

Seeman P, Bzowej NH, Guan HC, Bergeron C, Reynolds GP, Bird ED, Riederer P, Jellinger K, Tourtellotte WW (1987) Human brain D1 and D2 dopamine receptors in schizophrenia, Alzheimer's, Parkinson's, and Huntington's diseases. Neuropsychopharmacology 1:5–15

Sjoholm H, Bratlid T, Sundsfjord J (2004) 123I-beta-CIT SPECT demonstrates increased presynaptic dopamine transporter binding sites in basal ganglia in vivo in schizophrenia. Psychopharmacology (Berl) 173:27–31

Slifstein M, Laruelle M (2001) Models and methods for derivation of in vivo neuroreceptor parameters with PET and SPECT reversible radiotracers. Nucl Med Biol 28:595–608

Slifstein M, Hwang DR, Huang Y, Guo N, Sudo Y, Narendran R, Talbot P, Laruelle M (2004) In vivo affinity of [18F]fallypride for striatal and extrastriatal dopamine D2 receptors in nonhuman primates. Psychopharmacology (Berl) 175:274–286

Slifstein M, Kegeles LS, Gonzales R, Frankle WG, Xu X, Laruelle M, Abi-Dargham A (2007) [(11)C]NNC 112 selectivity for dopamine D(1) and serotonin 5-HT(2A) receptors: a PET study in healthy human subjects. J Cereb Blood Flow Metab 27:1733–1741

Stone JM, Pilowsky LS (2007) Novel targets for drugs in schizophrenia. CNS Neurol Disord Drug Targets 6:265–272

Suhara T, Okubo Y, Yasuno F, Sudo Y, Inoue M, Ichimiya T, Nakashima Y, Nakayama K, Tanada S, Suzuki K, Halldin C, Farde L (2002) Decreased dopamine D2 receptor binding in the anterior cingulate cortex in schizophrenia. Arch Gen Psychiatry 59:25–30

Sumiyoshi T, Stockmeier CA, Overholser JC, Dilley GE, Meltzer HY (1996) Serotonin1A receptors are increased in postmortem prefrontal cortex in schizophrenia. Brain Res 708:209–214

Sumiyoshi T, Matsui M, Nohara S, Yamashita I, Kurachi M, Sumiyoshi C, Jayathilake K, Meltzer HY (2001a) Enhancement of cognitive performance in schizophrenia by addition of tandospirone to neuroleptic treatment. Am J Psychiatry 158:1722–1725

Sumiyoshi T, Matsui M, Yamashita I, Nohara S, Kurachi M, Uehara T, Sumiyoshi S, Sumiyoshi C, Meltzer HY (2001b) The effect of tandospirone, a serotonin(1A) agonist, on memory function in schizophrenia. Biol Psychiatry 49:861–868

Takahashi H, Higuchi M, Suhara T (2006) The role of extrastriatal dopamine D2 receptors in schizophrenia. Biol Psychiatry 59:919–928

Takano A, Suhara T, Kusumi I, Takahashi Y, Asai Y, Yasuno F, Ichimiya T, Inoue M, Sudo Y, Koyama T (2006) Time course of dopamine D2 receptor occupancy by clozapine with medium and high plasma concentrations. Prog Neuropsychopharmacol Biol Psychiatry 30:75–81

Talvik M, Nordstrom AL, Olsson H, Halldin C, Farde L (2003) Decreased thalamic D2/D3 receptor binding in drug-naive patients with schizophrenia: a PET study with [11C]FLB 457. Int J Neuropsychopharmacol 6:361–370

Talvik M, Nordstrom AL, Okubo Y, Olsson H, Borg J, Halldin C, Farde L (2006) Dopamine D2 receptor binding in drug-naive patients with schizophrenia examined with raclopride-C11 and positron emission tomography. Psychiatry Res 148:165–173

Tamminga CA (2006) The neurobiology of cognition in schizophrenia. J Clin Psychiatry 67:e11

Tauscher J, Kapur S, Verhoeff NP, Hussey DF, Daskalakis ZJ, Tauscher-Wisniewski S, Wilson AA, Houle S, Kasper S, Zipursky RB (2002) Brain serotonin 5-HT1A receptor binding in schizophrenia measured by positron emission tomography and [11C]WAY-100635. Arch Gen Psychiatry 59:514–520

Tauscher J, Hussain T, Agid O, Verhoeff NP, Wilson AA, Houle S, Remington G, Zipursky RB, Kapur S (2004) Equivalent occupancy of dopamine D1 and D2 receptors with clozapine: differentiation from other atypical antipsychotics. Am J Psychiatry 161:1620–1625

Tauscher-Wisniewski S, Kapur S, Tauscher J, Jones C, Daskalakis ZJ, Papatheodorou G, Epstein I, Christensen BK, Zipursky RB (2002) Quetiapine: an effective antipsychotic in first-episode schizophrenia despite only transiently high dopamine-2 receptor blockade. J Clin Psychiatry 63:992–997

Taylor SF, Koeppe RA, Tandon R, Zubieta JK, Frey KA (2000) In vivo measurement of the vesicular monoamine transporter in schizophrenia. Neuropsychopharmacology 23:667–675

Townsend DW (2004) Physical principles and technology of clinical PET imaging. Ann Acad Med Singapore 33:133–145

Trichard C, Paillère-Martinot ML, Attar-Levy D, Blin J, Feline A, Martinot JL (1998) No serotonin 5-HT2A receptor density abnormality in the cortex of schizophrenic patients studied with PET. Schizophr Res 31:13–17

Tune LE, Wong DF, Pearlson G, Strauss M, Young T, Shaya EK, Dannals RF, Wilson AA, Ravert HT, Sapp J, Cooper T, Chase GA, Wagner HN Jr (1993) Dopamine D2 receptor density estimates in schizophrenia: a positron emission tomography study with 11C-N-methylspiperone. Psychiatry Res 49:219–237

Tune L, Barta P, Wong D, Powers RE, Pearlson G, Tien AY, Wagner HN (1996) Striatal dopamine D2 receptor quantification and superior temporal gyrus: volume determination in 14 chronic schizophrenic subjects. Psychiatry Res 67:155–158

Tuppurainen H, Kuikka J, Viinamaki H, Husso-Saastamoinen M, Bergstrom K, Tiihonen J (2003) Extrastriatal dopamine D 2/3 receptor density and distribution in drug-naive schizophrenic patients. Mol Psychiatry 8:453–455

Ugedo L, Grenhoff J, Svensson TH (1989) Ritanserin, a 5-HT2 receptor antagonist, activates midbrain dopamine neurons by blocking serotonergic inhibition. Psychopharmacology (Berl) 98:45–50

van Berckel BN, Kegeles LS, Waterhouse R, Guo N, Hwang DR, Huang Y, Narendran R, Van Heertum R, Laruelle M (2006) Modulation of amphetamine-induced dopamine release by group II metabotropic glutamate receptor agonist LY354740 in non-human primates studied with positron emission tomography. Neuropsychopharmacology 31:967–977

Verhoeff NP, Soares JC, D'Souza CD, Gil R, Degen K, Abi-Dargham A, Zoghbi SS, Fujita M, Rajeevan N, Seibyl JP, Krystal JH, van Dyck CH, Charney DS, Innis RB (1999) [123I] Iomazenil SPECT benzodiazepine receptor imaging in schizophrenia. Psychiatry Res 91:163–173

Vernaleken I, Fellows C, Janouschek H, Brocheler A, Veselinovic T, Landvogt C, Boy C, Buchholz HG, Spreckelmeyer K, Bartenstein P, Cumming P, Hiemke C, Rosch F, Schafer W, Wong DF, Grunder G (2008) Striatal and extrastriatal D2/D3-receptor-binding properties of ziprasidone: a positron emission tomography study with [18F]fallypride and [11C] raclopride (D2/D3-receptor occupancy of ziprasidone). J Clin Psychopharmacol 28:608–617

Vollenweider FX, Geyer MA (2001) A systems model of altered consciousness: integrating natural and drug-induced psychoses. Brain Res Bull 56:495–507

Vollenweider FX, Vontobel P, Oye I, Hell D, Leenders KL (2000) Effects of (S)-ketamine on striatal dopamine: a [11C]raclopride PET study of a model psychosis in humans. J Psychiatr Res 34:35–43

Waterhouse RN (2003) Imaging the PCP site of the NMDA ion channel. Nucl Med Biol 30:869–878

Weinberger DR, Laurelle M (2001) Neurochemical and neuropharmacological imaging in schizophrenia. In: Davis KL, Charney DS, Coyle JT, Nemeroff C (eds) Neuropharmacology – the fifth generation of progress. Lippincott Williams & Wilkins, Philadelphia, pp 833–856

Wolkin A, Barouche F, Wolf AP, Rotrosen J, Fowler JS, Shiue CY, Cooper TB, Brodie JD (1989) Dopamine blockade and clinical response: evidence for two biological subgroups of schizophrenia. Am J Psychiatry 146:905–908

Wong DF, Wagner HN Jr, Tune LE, Dannals RF, Pearlson GD, Links JM, Tamminga CA, Broussolle EP, Ravert HT, Wilson AA, Toung JK, Malat J, Williams JA, O'Tuama LA, Snyder SH, Kuhar MJ, Gjedde A (1986) Positron emission tomography reveals elevated D2 dopamine receptors in drug-naive schizophrenics. Science 234:1558–1563

Xiberas X, Martinot JL, Mallet L, Artiges E, Loc HC, Maziere B, Paillere-Martinot ML (2001) Extrastriatal and striatal D(2) dopamine receptor blockade with haloperidol or new antipsychotic drugs in patients with schizophrenia. Br J Psychiatry 179:503–508

Yang YK, Yu L, Yeh TL, Chiu NT, Chen PS, Lee IH (2004) Associated alterations of striatal dopamine D2/D3 receptor and transporter binding in drug-naive patients with schizophrenia: a dual-isotope SPECT study. Am J Psychiatry 161:1496–1498

Yasuno F, Suhara T, Ichimiya T, Takano A, Ando T, Okubo Y (2004a) Decreased 5-HT1A receptor binding in amygdala of schizophrenia. Biol Psychiatry 55:439–444

Yasuno F, Suhara T, Okubo Y, Sudo Y, Inoue M, Ichimiya T, Takano A, Nakayama K, Halldin C, Farde L (2004b) Low dopamine d(2) receptor binding in subregions of the thalamus in schizophrenia. Am J Psychiatry 161:1016–1022

Yoder KK, Hutchins GD, Morris ED, Brashear A, Wang C, Shekhar A (2004) Dopamine transporter density in schizophrenic subjects with and without tardive dyskinesia. Schizophr Res 71:371–375

Zakzanis KK, Hansen KT (1998) Dopamine D2 densities and the schizophrenic brain. Schizophr Res 32:201–206

A Selective Review of Volumetric and Morphometric Imaging in Schizophrenia

James J. Levitt, Laurel Bobrow, Diandra Lucia, and Padmapriya Srinivasan

Contents

1 Introduction 244
2 Limbic/Paralimbic Regions 245
3 Prefrontal Cortical Regions 256
4 Caudate Nucleus 264
5 Neocortical Temporal Lobe 268
6 Conclusion 276
References 277

Abstract Brain imaging studies have long supported that schizophrenia is a disorder of the brain, involving many discrete and widely spread regions. Generally, studies have shown decreases in cortical gray matter (GM) volume. Here, we selectively review recent papers studying GM volume changes in schizophrenia subjects, both first-episode (FE) and chronic, in an attempt to quantify and better understand differences between healthy and patient groups. We focused on the cortical GM of the prefrontal cortex, limbic and paralimbic structures, temporal lobe, and one subcortical structure (the caudate nucleus).

We performed a search of the electronic journal database PsycINFO using the keywords "schizophrenia" and "MRI," and selected for papers published between

---

J.J. Levitt (✉)
Department of Psychiatry at the VA Boston Healthcare System, Harvard Medical School, Brockton Campus, 116A4, 940 Belmont Street, Brockton, MA 02301, USA
Psychiatry Neuroimaging Laboratory, Department of Psychiatry at Brigham and Women's Hospital, Harvard Medical School, Boston, MA, USA
e-mail: james_levitt@hms.harvard.edu

L. Bobrow, D. Lucia, and P. Srinivasan
Psychiatry Neuroimaging Laboratory, Department of Psychiatry at Brigham and Women's Hospital, Harvard Medical School, Boston, MA, USA

This review represents original materials, it has not been previously published, and it is not under consideration for publication elsewhere.

2001 and 2008. We then screened for only those studies utilizing manual or manually edited tracing methodologies for determining regions of interest (ROIs). Each region of interest was indexed independently; thus, one paper might yield results for numerous brain regions.

Our review found that in almost all ROIs, cortical GM volume was decreased in the patient populations. The only exception was the caudate nucleus – most studies reviewed showed no change, while one study showed an increase in volume; this region, however, is particularly sensitive to medication effects. The reductions were seen in both FE and chronic schizophrenia.

These results clearly support that schizophrenia is an anatomical disorder of the brain, and specifically that schizophrenia patients tend to have decreased cortical GM in regions involved in higher cognition and emotional processing. That these reductions were found in both FE and chronic subjects supports that brain abnormalities are present at the onset of illness, and are not simply a consequence of chronicity. Additional studies assessing morphometry at different phases of the illness, including prodromal stages, together with longitudinal studies will elucidate further the role of progression in this disorder.

Keywords Anterior cingulate gyrus · Amygdala · Caudate nucleus 32 · Entorhinal cortex · Fusiform gyrus · Hippocampus · Heschl's gyrus · Inferior frontal gyrus · Inferior temporal gyrus · Middle frontal gyrus · Middle temporal gyrus · Orbitofrontal cortex · Parahippocampal cortex · Planum temporale · Prefrontal cortex · Straight gyrus · Superior frontal gyrus · Superior temporal gyrus · Temporal pole

1 Introduction

Schizophrenia is believed to be a disorder that affects many regions of the brain. One important source of evidence supporting this conclusion is the large body of imaging studies assessing brain volume and morphometry that describe abnormalities in multiple brain regions in this devastating disorder. In this paper, we review selective recent cortical GM findings reported between 2001 and today, dating from the important prior review by Shenton et al. (2001), a member of our group. Though abnormalities in white matter and cerebrospinal fluid have also been implicated in many studies, the primary focus of this review will be on cortical (GM) regions of interest (ROIs) that have been largely manually traced. Specifically, this review will focus on the cortical regions of prefrontal, limbic, paralimbic, and neocortical temporal lobe structures, and on the caudate nucleus subcortical structure. Though not inclusive of every region examined morphometrically in schizophrenia, we believe these regions cover a wide expanse of cortical territory and a key subcortical structure believed relevant to schizophrenia.

Although ROI manual tracing has certain difficulties – including that it is labor-intensive, that it makes it difficult to assess many ROIs simultaneously, and that rater bias can cause problems – this technique remains the gold standard for such studies. We acknowledge, however, that automatic parcellation methodologies, such as the FreeSurfer program or voxel based comparisons of GM concentration and volume (i.e., voxel based morphometry, VBM), are steadily gaining popularity. They permit more rapid assessments of the entire brain, avoid bias from a more hypothesis-driven approach of selecting specific brain regions for study, and are reliable. Yet these programs, as intriguing as they are, still have problems, especially in assessing regions of great sulcal variability due to problems registering brains to a standardized template, which is a serious issue for a number of the ROIs we will review. In the studies we will review, whole brain volumes (used to correct for head size) are nearly always obtained through automatic or semi-automatic methodology. We will restrict this review, however, to studies where the specific ROIs were primarily manually traced.

Additionally, we were interested in looking at different phases of the disorder. Unfortunately, there were too few prodromal studies to include in this review, but of the studies we did include, we divided results based on whether the study looked at first-episode (FE) or chronic patients. Our definition of "chronic" schizophrenia includes subjects who would not fall under the category of "first-episode," which is itself a potentially heterogeneous group due to differences including age of onset, length of FE, and the duration of the prodromal period. We also appreciate that the "chronic" grouping of subjects used in the literature represents a diverse population, due to wide variation in duration of illness, among other factors. This issue is discussed in more detail in the section on the amygdala.

We selected papers for this review by searching for the words "schizophrenia" and "MRI" in the psychology- and neuroscience-specific abstract database PsycINFO and then screening out studies that used voxels larger than 1.5 mm in any direction, or did not selectively measure GM volume, or that were based on automatic programs. We supplemented these papers by sometimes including papers found in the references of those we obtained from PsycINFO. We appreciate that there are omissions in our review of certain relevant papers, as this database is not infallible.

2 Limbic/Paralimbic Regions

There have been numerous GM morphometric studies of limbic/paralimbic regions in schizophrenia. For the purpose of this paper, we will focus upon the temporolimbic regions of the hippocampal gyrus and the amygdala (AMG), though in Tables 1 and 2, we will include studies relating to the parahippocampal gyrus (PHG) and the entorhinal cortex (EC). Furthermore, we will include in this section the paralimbic anterior cingulate gyrus (ACG), an important target of projections from the hippocampus, PHG, and EC. Our discussion of another principal prefrontal limbic/paralimbic region, the orbitofrontal cortex, will occur in the context of our review of frontal lobe findings.

Overall, as shown in Table 3, 32 papers on limbic/paralimbic regions in schizophrenia met our inclusion criteria. Of these 32 studies, 75% (24/32) found volume

Table 1 Parahippocampal gyrus and entorhinal cortex overview

Region	Total papers	Type of change	Laterality	Total FE papers	FE changes	Total chronic papers	Chronic changes
All limbic and paralimbic regions	32	24 (75%) − 0 (0%) + 8 (25%) n.c.	3 (12.5%) L 6 (25%) R	9[a]	6 (66.7%) − 0 (0%) + 3 (33.3%) n.c.	24[a]	20 (83.3%) − 0 (0%) + 4 (16.7%) n.c.
All prefrontal regions	11	8 (72.7%) − 0 (0%) + 3 (27.3%) n.c.	2 (25%) L 1 interaction	4	4 (100%) − 0 (0%) + 0 (0%) n.c.	7	4 (57.1%) − 0 (0%) + 3 (42.9%) n.c.
Caudate nucleus	6	0 (0%) − 1 (16.5%) + 5 (83.5%) n.c.	1 interaction	3	0 (0%) − 0 (0%) + 3 (100%) n.c.	3	0 (0%) − 1 (33.3%) + 2 (66.7%) n.c.
All neocortical temporal regions	12	10 (83.3%) − 0 (0%) + 2 (16.7%) n.c.	8 (80%) L	5	5 (100%) − 0 (0%) + 0 (0%) n.c.	7	5 (71.4%) − 0 (0%) + 2 (28.6%) n.c.

n.c. = no change in volume relative to controls
[a]One paper had both FE and chronic subjects, and hence is listed twice

Table 2 Parahippocampal gyrus and entorhinal cortex papers

Region overview	Total papers	Type of change	Laterality	Total FE papers	FE changes	Total chronic papers	Chronic changes
Parahippocampal gyrus and entorhinal cortex	7	3 (42.3%) − 0 (0%) + 4 (57.7%) n.c.	1 (33.3) R	3	2 (66.7%) − 0 (0%) + 1 (33.3%) n.c.	4	1 (25%) − 0 (0%) + 3 (75%) n.c.

n.c. = no change in volume relative to controls

decreases in schizophrenia subjects in at least one limbic structure, compared with controls; none found increases, and 25% (8/32) found no change. Of the studies finding volume change, 12.5% reported findings specifically in the left hemisphere, and 25% reported findings in the right.

In our database search of the above limbic regions, the region that had been most extensively studied in schizophrenia was the hippocampus (16 studies), followed by the ACG (10 studies), the amygdala (8 studies), and the parahippocampal and entorhinal cortices (7 studies).

As can be seen in Tables 4 and 5, summarizing the 16 studies that analyzed the hippocampus, 68.8% (11/16) of the studies reported hippocampal volume reductions and 31.2% (5/16) showed no change in volume. Of the 11 studies showing a decrease in volume, one showed changes lateralized exclusively to the left side and one showed changes exclusively on the right. Of five FE studies, three showed decreases and two showed no change. Similarly, of 12 chronic studies, 9 (75%) showed decreases and 3 (25%) showed no change.

The studies of the hippocampus in schizophrenia raise a number of issues that are relevant in general for regional volumetric studies in schizophrenia. One issue is how best to control for head size. The papers on the hippocampus use different approaches, which is one example of how lack of methodological uniformity makes

A Selective Review of Volumetric and Morphometric Imaging in Schizophrenia

Table 3 Total regional statistics

Type of change	Region paper	Population	Laterality	Method	Correlations, additional findings, and comments
Parahippocampal gyrus					
–	Prasad et al. (2004)	33 FE SZ (23m), 43 NC (21m)	R sig. if outliers removed	Intrinsic/ extrinsic	Negative correlation of parahippocampal volume and SAPS global positive symptom scores, severity of delusions, and formal thought disorder
None	Cahn et al. (2002)	20 FE SZ (16m), 20 NC (16m)	None	Intrinsic/ extrinsic	No correlations reported
None	Sim et al. (2006)	19(m) SZ, 19(m) NC	None	Intrinsic	No significant correlations found
None	Takahashi et al. (2006b)	65 SZ (35m), 72 NC (38m)	None	Intrinsic/ extrinsic	No significant correlations found
None	Yamasue et al. (2004)	27 chronic SZ (20m), 27 NC (20m)	None	Intrinsic/ extrinsic	No significant correlations found
Entorhinal cortex					
–	Joyal et al. (2002)	18 FE SZ (11m), 22 NC (14m)	Bilateral	Intrinsic/ extrinsic	No significant correlations found
–	Baiano et al. (2008)	70 SZ (45m), 77 NC (40m)	Bilateral	Intrinsic	No significant correlations found

m = male subjects

Table 4 Hippocampal overview

Region	Total papers	Type of change	Laterality	Total FE papers	FE changes	Total chronic papers	Chronic changes
Hippocampus	16	11 (68.8%) – 0 (0%) + 5 (31.2%) n.c.	1 (9.1%) L 1 (9.1%) R	5[a]	3 (60%) – 0 (0%) + 2 (40%) n.c.	12[a]	9 (75%) – 0 (0%) + 3 (25%) n.c.

n.c. = no change in volume relative to controls
[a]One paper had both FE and chronic subjects, and hence is listed twice

the comparison of studies from different research groups difficult. Since gender affects head size, this is particularly important if samples include both males and females. The literature addresses this issue in different ways, including using a single slice (e.g., Laakso et al. 2001), or finding the relative volume of a region of interest by comparing it to total intracranial contents (ICC; e.g., our group). Though ICC would seem to be the best means of controlling for head size, the way in which ICC is measured can be imprecise and creates possible inconsistencies. For example, many of the analyses of the hippocampus are based on carefully traced manual drawing, but then ICC will be based upon an automatic method (e.g., Expectation

Table 5 Hippocampal papers

Type of change	Paper	Population	Laterality	Method	Correlations, additional findings, and comments
–	Kuroki et al. (2006a)	24(m) chronic SZ, 31(m) NC	Bilateral	Intrinsic/extrinsic	Positive correlation of bilateral hippocampus and WMS-III general memory in SZs
–	Nestor et al. (2007)	44(m) SZ, 43(m) NC	Bilateral	Intrinsic	Positive correlation of R hippocampus and overall memory in SZs. Positive correlation of L hippocampus and verbal memory
–	Pegues et al. (2003)	27(m) SZ, 24(m) NC	Bilateral	Intrinsic/extrinsic	No correlations reported
–	Phillips et al. (2002)	32 FE SZ (25m), 139 NC (82m)	Bilateral	Intrinsic	No correlations reported
–	Rupp et al. (2005)	33(m) SZ, 40(m) NC	Bilateral	Intrinsic/extrinsic	Correlation between hippocampal volume and olfactory discrimination performance
–	Sim et al. (2006)	19(m) chronic SZ, 19(m) NC	Bilateral	Intrinsic	Positive correlation of R hippocampal volume and total PANSS score
–	Thoma et al. (2008)	22 chronic SZ (16m), 22 NC (17m)	Bilateral	Intrinsic	Negative correlation between anterior hippocampal volume and sensory gating ratio
–	Velakoulis et al. (2001)	45 chronic SZ (39m), 139 NC (82m)	Bilateral	Intrinsic	No correlations reported
–	Weiss et al. (2005)	25(m) chronic SZ, 25(m) NC	Bilateral	Intrinsic/extrinsic	No correlations reported
–	Whitworth et al. (2005)	21(m) FE SZ, 17(m) chronic SZ, 20(m) NC with 2–4 year follow-up	R sig. at follow-up L not sig.	Intrinsic/extrinsic	Greater annual L hippocampal decrease. No significant correlations found
–	Yamasue et al. (2004)	27 chronic SZ (20m), 27 NC (20m)	L post.	Intrinsic/extrinsic	No significant correlations found

A Selective Review of Volumetric and Morphometric Imaging in Schizophrenia 249

	Study	Subjects		Intrinsic/extrinsic	Correlations
None	Cahn et al. (2002)	20 FE SZ (16m), 20 NC (16m)	None	Intrinsic/extrinsic	No correlations reported
None	Kasai et al. (2003a)	13 FE SZ (10m), 14 NC (13m)	None	Intrinsic/extrinsic	No correlations reported
None	Laakso et al. (2001)	18 FE SZ (11m), 22 NC (14m)	None	Intrinsic	No correlations reported
None	Szendi et al. (2006)	13(m) SZ, 13(m) NC	None	Intrinsic	Positive correlation of R volume and severity of negative symptoms
None	Tanskanen et al. (2005)	56 SZ (33m), 104 NC (62m)	None	Intrinsic/extrinsic	Some patients had comorbid disorders

m = male subjects

Maximization or FreeSurfer), and though intraclass correlations are reported for ROIs, rarely is the reliability of ICC included.

Another issue is that of sample size and statistical power considerations. Studies which report negative findings of group differences but are based upon small samples raise the issue of a Type 2 error. Such studies include that by Szendi et al. (2006), which reported negative results but had only assessed 13 young, chronic male patients and 13 male controls, and Anderson et al. (2002), which also reported negative results for the hippocampus in chronic schizophrenia but only had a sample size of 16 male patients and 15 male normal controls.

A further issue applicable to many morphometric studies is that tracing criteria differ between studies, again complicating attempts at comparing results. Studies in the hippocampus, for example, have combined the amygdala and hippocampus using an extrinsic boundary such as the mammillary bodies to separate the two structures (e.g., Yamasue et al. 2004). Also, more subtly, some groups have included the subiculum in their definition (e.g., Shenton et al. 1992), whereas others have not. The subiculum is mesocortex, that is, transitional cortex between neocortex and allocortex (Nieuwenhuys et al. 2008), and its inclusion, or not, might affect results, since mesocortex might be differently involved in schizophrenia than allocortex. Alternatively, some groups have measured the entire hippocampus (Kuroki et al. 2006a), whereas other groups have attempted to measure subregions of the hippocampus. For example, Weiss et al. (2005) divided the hippocampus into anterior and posterior regions and raised the interesting issue of connectional patterns. For example, they cite work suggesting that the anterior hippocampus is more closely connected to medial PFC, whereas the posterior hippocampus is more closely connected to dorsolateral PFC, which implies different functions for these two subregions and might suggest selective (if only one subregion is abnormal) as opposed to diffuse involvement of the hippocampus in schizophrenia. Another factor is that chemoarchitecture may differ even within a given ROI. For example, Velakoulis et al. (2001) point out that D2 receptor concentration is not uniform across the entire hippocampus. This could lead to differential development or to differential effects of medication on the volume of the hippocampus (as has been suggested for the caudate; see below). Currently, the conclusion is still under debate; according to Weiss et al. (2005), the involvement of the hippocampus is diffuse. Conversely, other groups (e.g., Csernansky et al. 2002; Pegues et al. 2003) have suggested greater involvement of the anterior compared to the posterior hippocampus.

Shape analyses are particularly useful methods in trying to address the more focal nature of problems. For example, innervations from one region to another (e.g., from the PFC to the caudate) are local phenomena, and false negative findings are made possible by including too large a territory for analysis (Nugent et al. 2006). In an interesting study by Joyal et al. (2002), they examined the EC separate from the PHG. In this study of FE schizophrenia, they reported a significant bilateral reduction in EC volume, though in the same sample, they did not see any reduction in hippocampal volume. This suggests that the volume loss in the medial temporal lobe in FE schizophrenia may be regionally specific rather than diffuse. The EC is a critical junction for receiving input and sending output from the neocortex to the hippocampus, and is thought to be highly relevant for the cognitive operation of declarative memory (Joyal et al. 2002). Further, cytoarchitectonically, it is six layered cortex as opposed to the three layers seen in the allocortical hippocampus. This again suggests the importance of more focal analyses as nearby regions may not even share the same cytoarchitecture. A more focal approach is, thus, the best way to determine how regionally specific versus how diffuse morphometric abnormalities may be in schizophrenia.

A final issue raised by these papers on the hippocampus is that of the number of regions that will be assessed simultaneously. In the very labor-intensive paper of Yamasue et al. (2004), they looked at 30 subregions, 15 on each side, using manual tracing methodology. They compared results across all regions by looking at the effect size for each region. This informative approach allows one to try to answer the important question of whether abnormalities in schizophrenia are focal, multifocal, or diffuse. Another way of looking at multiple regions at once is by employing automatic methods, such as FreeSurfer, which parcellates many cortical and subcortical regions, and VBM, which performs voxel by voxel comparisons. But as stated earlier, it appears that validating the findings of more automatic procedures with manual tracing still remains important at this point.

As can be seen in Tables 6 and 7, summarizing the eight studies that analyzed the amygdala, six (75%) of the studies revealed amygdala volume reductions while two (25%) showed no change. Of the six studies showing reduced volume, four showed bilateral reductions and two showed changes lateralized to the right side. Of the two studies that included FE subjects, both showed decreases. Conversely, of seven chronic studies, five (71.4%) showed decreases and two (28.6%) showed no change.

The studies of the amygdala in schizophrenia raise further issues of general interest. A number of the studies report asymmetry differences between schizophrenia subjects and controls (e.g., Joyal et al. 2003). Asymmetry abnormalities in the literature have been used to suggest the role of neurodevelopmental factors in schizophrenia, either on the basis of environmental issues such as prenatal and perinatal birth complications, hormonal environment during neural development, or a genetic etiology (Niu et al. 2004; Yamasue et al. 2004). Methodologically, to be confident regarding asymmetry findings, studies need to control for handedness. Some studies do this by requiring that all subjects be right-handed (e.g., Yamasue et al. 2004), whereas other researchers use strategies such as group-matching for handedness (e.g., Joyal et al. 2003). Though considered acceptable, this latter approach raises the possibility of a confound for asymmetry findings. Other confounding issues for

Table 6 Amygdala overview

Region	Total papers	Type of change	Laterality	Total FE papers	FE changes	Total chronic papers	Chronic changes
Amygdala	8	6 (75%) − 0 (0%) + 2 (25%) n.c.	2 (33%) R	2[a]	2 (100%) − 0 (0%) + 0 (0%) n.c.	7[a]	5 (71.4%) − 0 (0%) + 2 (28.6%)

n.c. = no change in volume relative to controls
[a]One paper had both FE and chronic subjects, and hence is listed twice

Table 7 Amygdal papers

Type of change	Paper	Population	Laterality	Method	Correlations, additional findings, and comments
–	Exner et al. (2004)	16 SZ (11m), 16 NC (11m)	R sig; L not sig.	Intrinsic/ extrinsic	Positive correlation of R amygdala volume and emotional learning in both SZs and NCs
–	Joyal et al. (2003)	18 FE SZ (11m), 22 NC (14m)	Bilateral	Intrinsic	No significant correlations found
–	Namiki et al. (2007)	20 SZ (10m), 20 NC (10m)	Bilateral	Intrinsic/ extrinsic	Positive correlation of bilateral volume and recognition of facial happiness
–	Niu et al. (2004)	40 SZ (20m), 40 NC (20m)	Bilateral in males; R only in females	Intrinsic	Negative correlation of L volume and illness duration
–	Rupp et al. (2005)	33(m) SZ, 40(m) NC	Bilateral	Intrinsic/ extrinsic	No significant correlations found
–	Whitworth et al. (2005)	21(m) FE SZ, 17(m) chronic SZ, 20(m) NC	Bilateral	Intrinsic/ extrinsic	No significant correlations found
None	Tanskanen et al. (2005)	56 SZ (33m), 104 NC (62m)	None	Intrinsic/ extrinsic	Some patients had comorbid disorders
None	Yamasue et al. (2004)	27 chronic SZ (20m), 27 NC (20m)	None	Intrinsic/ extrinsic	No significant correlations found

m = male subjects

morphometric brain measures include gender and the effects of chronicity such as illness duration, chronic hospitalization, diet, and medication. Again, efforts to minimize these confounds include separate gender analyses and correlating results with clinical measures of duration or with medication dosages, using chlorpromazine milligram equivalences. Another complicating factor in explaining a neuroleptic drug effect on brain region measures is that dopamine and dopamine receptors are not uniformly distributed through the brain. The antagonism of dopamine receptors is one potential mechanism for changing brain region size (Levitt et al. 2002). Dopamine receptors are highly concentrated in certain brain structures, such as the striatum, and can be found in other structures including the amygdala, hippocampus, and cingulate, but have lower receptor densities in areas such as in the prefrontal cortex. Thus, this differential receptor distribution might variably affect the size of selective brain regions in subjects taking neuroleptics.

Not all studies of the amygdala have looked at gender differences, but whether regional sexual dimorphisms are present or not remains an important issue. A problem with gender analysis is that sample size can be compromised by subgrouping analyses into gender. For example, in a study by Exner et al. (2004), using a sample of 16 subjects with schizophrenia and 16 matched controls, a 13% reduction was found in total size of the amygdala, which on follow-up testing was significant for the right side. The size of the right amygdala, in turn, was positively related to emotional learning in both schizophrenia patients and controls. However, both patient and control groups had 11 males and 5 females; thus, it can be seen that larger samples would be necessary to clarify the role of gender (the authors did compare effect size for percentage volume differences between the sexes, and males had a larger effect size for percentage volume decrease). Another study, by Niu et al. (2004), which did have a sufficient sample to examine the role of gender, found significantly reduced amygdala volume bilaterally in males and unilaterally on the right in females, suggesting a more "diffuse" and pronounced abnormality in the amygdala in male patients. Finally, these authors reported a size asymmetry in male patients, showing a left less than right volume of the amygdala within the group. They pointed out that the left and right amygdala may differ functionally, citing another work showing that the left amygdala was especially involved in external stimuli processing as it relates to behaviors of withdrawal. These types of findings emphasize the importance of comparing left and right hemisphere structures independently, as there may be lateralized function in a number of brain structures besides, for example, the well-known lateralization of language function in the temporal/parietal lobe.

A final interesting point raised in the Niu et al. (2004) article is the negative correlation found between duration of illness and the volume of the left, but not right, amygdala, suggesting that the reduction in volume, at least in the left, may be a consequence of postonset effects. This again supports the importance of exploring brain–behavior relationships independently for structures in the left and right hemisphere. Also, as can be seen in their sample, the range of illness duration was relatively large (0.5–180 months) which clearly may have facilitated the finding of this correlation. A number of studies in the literature fail to show such correlations, but if there is little variability in duration of illness, say in a chronic sample ill for 20 years, this will not lend itself to such correlations. The issue of duration of illness is problematic as many studies, in an attempt to reduce heterogeneity, divide patients into FE or chronic samples. There remains heterogeneity within chronic samples, however, as subjects with a mean age in their 20s and ill for 5 years intuitively seem quite different from subjects in their 40s who have been ill for 20 years. Even FE subjects, for example, if defined by their first hospitalization (with varying age at their first break), can differ greatly if the durations of their prodromal periods were varied. The above confounds have led to the idea of assessing endophenotype markers, including MRI measures, in prodromal subjects and in adolescents at high genetic risk for schizophrenia. The important emerging MRI literature on these types of subjects, however, is beyond the scope of this review.

Another significant issue is the importance of assessing the amygdala separately from the hippocampus. Early studies (e.g., Shenton et al. 1992) combined the amygdala and hippocampus into one structure, the amygdala–hippocampal complex, and then divided the two regions based on an extrinsic anatomic boundary, the mammillary bodies. With the advent of software imaging programs that allow for the simultaneous display of three orthogonal views, more recent studies (using the sagittal view in particular) have been able to assess the hippocampus–amygdala boundary based on intrinsic anatomic criteria, for example, the alveus (Niu et al. 2004). Given the different functions subserved by the amygdala and the hippocampus and their different anatomic connectivity patterns (Nolte 2009), it seems clear that these structures should be measured as independently as possible.

As can be seen in Tables 8 and 9, summarizing the ten studies that analyzed the ACG, 70% (7/10) of the studies revealed a volumetric decrease in at least one subregion and 30% (3/10) showed no change. Of the seven studies showing reduced volume, two (28.6%) showed bilateral reductions while three (42.9%) and two (28.6%) showed changes lateralized to the right and left side, respectively, with the remaining study having a more complicated pattern. From Tables 8 and 9, one notes that the single FE study showed a decrease, and of ten chronic studies, seven (70%) showed decreases and three (30%) showed no change.

The anterior cingulate is believed to play important roles in multiple functions including emotional, cognitive, and attentional processes, and in social cognition, including empathy and Theory of Mind (Fujiwara et al. 2008; Koo et al. 2008). Though a number of studies have found decreased ACG volume in schizophrenia (see Tables 8 and 9), the measurement of the volume of this region is complicated by the nearby structures of the paracingulate sulcus (PCS) and gyrus. Controversy exists as regards how best to account for the variable presence of this sulcus/gyrus, which is estimated to be present in 30–60% of cases (Fornito et al. 2008). Fornito et al. (2008) argues that the presence of a PCS has an important impact on measures of both ACG and paracingulate gyrus volume. These authors argued that in order to assess the volume of the ACG, PCS morphology should be controlled for, taking into account its presence or absence in both hemispheres. When they assessed the ACG and PCS in this manner (using FreeSurfer software that assesses surface-based morphometry), they reported no group difference in volume for (FE) schizophrenia subjects versus controls. However, they reported finding bilaterally reduced thickness of paralimbic regions of the ACG in schizophrenia, together with an increase in both limbic and paralimbic ACG surface area. Conversely, a recent report by Koo et al. (2008), assessing the ACG using more traditional manual tracing methodology, argued that the paracingulate gyrus "was mainly BA 32," and hence cytoarchitectonically different from the rest of the ACG. Thus, they did not include it in their cingulate gyrus measurement. These authors studied FE schizophrenia and FE schizoaffective subjects cross-sectionally and longitudinally. They found group volume differences in specific subregions of the ACG, with FE

Table 8 Anterior cingulate overview

Region	Total papers	Type of change	Laterality	Total FE papers	FE changes	Total chronic papers	Chronic changes
Anterior cingulate gyrus	10	7 (70%) − 0 (0%) + 3 (30%) n.c.	2 (28.6%) L 3 (42.9%) R	1	1 (100%) − 0 (0%) + 0 (0%) n.c.	9	6 (66.7%) − 0 (0%) + 3 (33.3%) n.c.

n.c. = no change in volume relative to controls

Table 9 Anterior cingulate gyrus papers

Type of change	Paper	Population	Laterality	Method	Correlations, additional findings, and comments
–	Fujiwara et al. (2008)	24 SZ (12m), 20 NC (10m)	Females only: bilateral ventral; L dorsal	Intrinsic/Extrinsic	Negative correlation of L dorsal volume and personal distress scores in females
–	Haznedar et al. (2004)	27 SZ (20m), 32 NC (25m)	L only	Intrinsic/Extrinsic	No significant correlations found
–	Koo et al. (2008)	39 FE SZ (30m), 40 NC (31m). 1.5 year follow-up: 17 FE SZ (14m), 18 NCs (15m).	Bilateral	Intrinsic/Extrinsic	Negative correlation BPRS withdrawal with R affective volume and its decrease between scans. Positive correlation of R cognitive volume and WAIS digit span and information scores. Negative correlation of bilateral volume (especially affective) with interscan interval. Larger decrease in R affective in poor responders
–	Takahashi et al. (2002a)	40 SZ (20m), 40 NC (20m)	R only, in females only	Intrinsic/extrinsic	No significant correlations found
–	Takahashi et al. (2002b)	40 SZ (20m), 48 NC (24m)	R only, in females only	Intrinsic/extrinsic	No significant correlations found
–	Yamasue et al. (2004)	27 SZ (20m), 27 NC (20m)	Bilateral	Intrinsic/extrinsic	Negative correlation of bilateral volume and difficulty in abstract thinking. Negative correlation of R ACG with severity of motor retardation
–	Zhou et al. (2005)	59 SZ (31m), 58 NC (30m)	R only	Intrinsic/extrinsic	No significant correlations found
None	Coryell et al. (2005)	10 SZ (6m), 10 NC (6m)	None	Intrinsic/extrinsic	Studied subgenual region only
None	Riffkin et al. (2005)	18 SZ (8m), 18 NC (8m)	None	Intrinsic/extrinsic	No significant correlations found
None	Szendi et al. (2006)	13(m) SZ, 13(m) NC	None	Intrinsic	No significant correlations found

m = male subjects

SZs showing significantly smaller left subgenual, left and right affective, right cognitive, and right posterior cingulate GM subregions compared to controls. Furthermore, these same subjects were reported to have progressive longitudinal volume reductions in the GM of subgenual, affective and posterior cingulate subregions. The above discrepant findings suggest that this important area of the brain, with complex sulcal boundaries, requires further study and demonstrates the complexity of assessing discrepant results based on different methodologies.

Two further interesting studies of the ACG carried out by the same group (Takahashi et al. 2002a; Zhou et al. 2005) looked at GM asymmetry in contrast to sulcal pattern asymmetry and gender effects. While initially they reported gender effects (2002a), with a larger sample – and it was not clear whether this sample was an overlapping sample – these gender differences disappeared. In their later and larger sample study (Zhou et al. 2005), the authors reported a right-sided decrease in volume in the ACG and a bilateral decrease in GM volume in the posterior cingulate in both males and females. In these last two studies, the subjects were in their mid-20s and were ill for about 3–5 years, again raising the issue of whether the findings in a "chronic" sample such as this one might differ from those in chronic patients in their 40s and ill for 20 years or more.

Lastly, a study by Zetzsche et al. (2007) found that in chronic male schizophrenia patients, 30.0 ± 8.4 years old, ACG GM volume of the right precallosal subregion and right total ACG was significantly reduced, as was the left subgenual ACG GM, though there were no statistically significant differences between patients and controls in asymmetry quotients in the four subregions they examined. The idea of subdividing the ACG is of interest as function differs for the different subregions (Koo et al. 2008). Nonetheless, the subregions are defined geometrically and arbitrarily as well as based on extrinsic landmarks, such as the genu of the corpus collusum or the first emergence of the internal capsule, which Koo uses as the posterior boundary for the subgenual ACG. Thus, the assessment of this region would appear to benefit from employing local shape analytic methodology (Levitt et al. 2009). A point to keep in mind for the ACG, but applying as well to other brain ROIs, is that investigators use different landmarks to define similarly labeled regions. So, whereas Koo's posterior boundary for the affective subregion of the ACG is the vertical plane passing through the posterior surface of the genu of the corpus callosum, authors referring to this subregion of the ACG as "precallosal" use the anterior surface of the genu for this boundary. The absence of uniformity across studies in defining ROIs makes comparisons across such studies more difficult.

3 Prefrontal Cortical Regions

There have been fewer manually driven morphometric studies of prefrontal GM cortical regions compared with the number of such studies of limbic/paralimbic regions in schizophrenia. Assessing usable studies, using our criteria, we identified

11 such papers assessing the prefrontal cortex compared with 32 such papers assessing limbic/paralimbic cortex. This may be due to the greater difficulty in outlining these structures, or due perhaps to technical issues including greater variability in prefrontal sulcal patterns and increased segmentation of sulci – that is, discontinuous sulci, e.g., the superior and inferior frontal sulci – causing further difficulty in the task of manual tracing. For the purpose of this paper, we will focus our attention upon the neocortical GM prefrontal cortex, the lateral surface of the prefrontal cortex [which can be subdivided into three regions: the DLPFC/DMPFC (i.e., the superior frontal gyrus), the DLPFC (i.e., the middle frontal gyrus), and the inferior frontal gyrus], and the orbital surface of the prefrontal cortex including the orbitofrontal gyri, which itself can be divided into subregions, and the gyrus rectus. The medial frontal cortex was largely covered in the limbic/paralimbic section above with the primary focus being on the cingulate and paracingulate gyri.

Overall, as outlined in Table 3, 11 papers on prefrontal cortical regions in schizophrenia met our inclusion criteria. Of these studies, 72.7% (8/11) found decreases in volume in schizophrenia patients in at least one prefrontal cortical structure, 0% found an increase in at least one structure, and 27.3% (3/11) found no change in any prefrontal structure examined. Overall, none of the studies reported abnormal findings on the right, 25% (2/8) reported abnormal findings on the left or had some left but otherwise bilateral results; one study showed different hemispheric volume asymmetry patterns between patient and control groups, and the rest (62.5% or 5/8) reported bilateral findings.

As per our database search of the prefrontal cortical regions, the regions that have most extensively been studied in schizophrenia were the largely unparcellated PFC (8/11 studies), the orbitofrontal cortex (5/11 studies), and middle frontal gyrus (4/11 studies). Next followed the inferior frontal gyrus and the superior frontal gyrus with 3/11 studies each and then the straight gyrus with 2/11 studies.

Though the unparcellated prefrontal cortex was not always defined in the same way (i.e., some studies focused on only the lateral PFC, while others analyzed the entire PFC), we categorized these studies together in one group. As can be seen in Tables 10 and 11 which summarizes these eight studies, 75% (6/8) found volume reductions and 25% (2/8) showed no change in volume. Of four FE studies, all found a decrease. Of the four chronic subject studies, two (50%) reported decreases and two (50%) showed no change.

Further, as can be seen in Tables 12 and 13, summarizing the three studies of the superior frontal gyrus, 33.3% (1/3) found bilateral decreases in volume and 66.7% (2/3) found no change in volume. These three studies were performed using chronic subjects which, as indicated in Tables 14–21, was also the case for all of the following studies of other PFC subregions as there were no FE studies. Tables 14 and 15 summarize the four studies of the middle frontal gyrus; one (25%) found reduced volume and three (75%) found no change. In the study showing decreased volume, they found only left-sided changes. Of the three inferior frontal gyrus studies (Tables 16 and 17), two (66.7%) found decreases and one (33.3%) found no change in volume. For one of the studies showing a decrease in volume, it was also reported that there was a significant gender effect, with male patients showing

Table 10 Prefrontal cortex overview

Region	Total papers	Type of change	Laterality	Total FE papers	FE changes	Total chronic papers	Chronic changes
Total prefrontal cortex	8	6 (75%) − 0 (0%) + 2 (25%) n.c.	2 (33%) left	4	4 (100%) − 0 (0%) + 0 (0%) n.c.	4	2 (50%) − 0 (0%) + 2 (50%) n.c.

n.c. = no change in volume relative to controls

decreases on the left only, whereas in female subjects this finding was bilateral; the other study found bilateral findings and did not explore gender effects. Of the five studies that analyzed the orbitofrontal cortex (Tables 18 and 19), 20% (1/5) found reduced volume and 80% (4/5) found no change in volume. In the study showing decreased volume, they found bilateral changes. Lastly, as can be seen in Tables 20 and 21, of the two straight gyrus studies, one found decreases and the other found no change.

A number of important further issues emerged in our review of the prefrontal cortex studies, including some involving brain–behavior relationships. Nakamura et al. (2007) assessed what they called neocortical GM. They reported a reduction in neocortical GM in FE schizophrenia compared to healthy controls, but did not report a group difference for the frontal lobe. Nonetheless, subjects were rescanned about 1.5 years later and the authors found significant group differences in progressive decreases in volume in both frontal and temporal regions. This paper, which argues for the value of parcellation, shows how complex the analysis of multiple subregions can become, especially when combined with a longitudinal design. Another issue raised by this paper was the potential effect of medication on MRI volume analyses. FE affective subjects were scanned longitudinally, and over time the GM volumes in these subjects increased, raising the possibility that mood stabilizers (such as lithium and valproic acid) were causing an increase in cortical GM volume. The authors cautioned that this may prove an important confound in MRI studies, particularly in those involving bipolar subjects.

Another issue that emerges from these studies is the importance of sample size. The prefrontal cortex covers a wide expanse of territory over the brain and the manual tracing of these structures, along sulci that are difficult to identify, can be extremely labor-intensive. This predisposes to small samples and to Type 2 errors. A study by Szendi et al. (2006), which looked at a number of brain regions together with relevant clinical and neuropsychological functions and neurological signs, had only 13 chronic SZ subjects and 13 controls. They did not find regional differences in volumes, but did report a significant difference in the straight gyrus with regard to laterality, with controls having a tendency toward dominance on the left while patients had a right-sided dominance. If sample sizes were larger, however, significant volume differences might have emerged. For instance, in a study by Suzuki et al. (2005), a number of prefrontal and temporolimbic ROIs were manually traced in a sample of 25 patients with schizotypal disorder, 53 patients with schizophrenia, and 59 healthy controls. These investigators were able to show that schizophrenia

Table 11 Prefrontal cortex papers

Type of change	Paper	Population	Laterality	Method	Correlations, additional findings, and comments
–	Dickey et al. (2004)	12 FE (9m), 15 NC (14m) 1.5 year follow-up	Bilateral	Intrinsic/ extrinsic[a]	No correlations reported Significant decrease at time 1; no further change over time
–	Hirayasu et al. (2001)	17 FE, 17 NC (15m each)	L sig; R not sig	Intrinsic/ extrinsic[a]	No correlations reported
–	Nakamura et al. (2007)	29 FE (24m), 36 NC (31m); follow-up: 17 FE (14m) 26 NC (22m)	Bilateral	Intrinsic/ extrinsic[a]	Negative correlation of PFC and BPRS score increases in total measure, thought disturbance factor, and anxiety-depression factor No change at Time 1; difference appeared at 18 month follow-up
–	Suzuki et al. (2005)	53 SZ (32m), 59 NC (35m)	Bilateral	Intrinsic/ extrinsic	No significant correlations
–	Wiegand et al. (2004)	17 FE, 17 NC (15m each)	L sig.; R not sig.	Intrinsic/ extrinsic[a]	Negative correlation of total absolute thickness (and R absolute thickness) with age, and with age at first medication No significant group differences for thickness or complexity
–	Zhou et al. (2005)	59 SZ (31m), 58 NC (30m)	Bilateral	Intrinsic/ extrinsic	No significant correlations
None	Sapara et al. (2007)	28 chronic SZ (24m), 20 NC (17m)	None	Intrinsic	Negative correlations of illness duration with total and L PFC. Positive correlation of total PFC with total insight and insight into illness
None	Wible et al. (2001)	17(m) chronic SZ, 17(m) NC	None	Intrinsic/ extrinsic[a]	Positive correlation of R PFC with amygdala–hippocampus volume. Negative correlation of L PFC and total SANS

[a]Used EMSeg for GM segmentation, but used manual boundaries for ROI
m = male subjects

Table 12 Superior frontal gyrus overview

Region	Total papers	Type of change	Laterality	Total FE papers	FE changes	Total chronic papers	Chronic changes
Superior frontal gyrus	3	1 (33.3%) − 0 (0%) + 2 (66.7%) n.c.	All bilateral	0	0 (0%) − 0 (0%) + 0 (0%) n.c.	3	1 (33.3%) − 0 (0%) + 2 (66.7%) n.c.

n.c. = no change in volume relative to controls

Table 13 Superior frontal gyrus papers

Type of change	Paper	Population	Laterality	Method	Correlations, additional findings, and comments
−	Suzuki et al. (2005)	53 SZ (32m), 59 NC (35m)	Bilateral	Mostly intrinsic	No significant correlations found
None	Sapara et al. (2007)	28 chronic SZ (24m), 20 NC (17m)	None	Intrinsic	Negative correlation of total L SFG and illness duration. Positive correlation of R SFG and insight into illness; males have correlation for total SFG
None	Yamasue et al. (2004)	27 chronic SZ (20m), 27 NC (20m)	None	Intrinsic/ extrinsic	No significant correlations found

m = male subjects

Table 14 Middle frontal gyrus overview

Region	Total papers	Type of change	Laterality	Total FE papers	FE changes	Total chronic papers	Chronic changes
Middle frontal gyrus	4	1 (25%) − 0 (0%) + 3 (75%) n.c.	1 (100%) left	0	0 (0%) − 0 (0%) + 0 (0%) n.c.	4	1 (25%) − 0 (0%) + 3 (75%) n.c.

n.c. = no change in volume relative to controls

patients had bilaterally reduced total PFC GM volumes compared to schizotypal subjects and to controls, and that patients with schizophrenia had reduced GM volumes relative to schizotypal patients in their "right superior frontal gyrus, left middle frontal gyrus, bilateral inferior frontal gyrus and bilateral straight gyrus." A study by Zhou et al. (2005) also showed significant volume differences. These investigators studied 59 patients with schizophrenia and 58 controls, and found that patients had cortical volume reductions in bilateral whole frontal lobes, the prefrontal area (which they defined as "the frontal lobe other than the PCG and cingulate gyrus"), precentral gyrus, the right ACG, and the posterior cingulate gyrus bilaterally. (As an intriguing aside, with regard to the abnormality of the precentral gyrus, the authors speculated that the mirror-neuron system, which they indicate is in the primary motor area (MI) among other locations, could be

Table 15 Middle frontal gyrus papers

Type of change	Paper	Population	Laterality	Method	Correlations, additional findings, and comments
–	Suzuki et al. (2005)	53 SZ (32m), 59 NC (35m)	L sig; R not sig	Mostly intrinsic	No significant correlations found
None	Sapara et al. (2007)	28 chronic SZ (24m), 20 NC (17m)	None	Intrinsic	Positive correlation of L MFG and insight into need for treatment (not significant for males)
None	Szendi et al. (2006)	13(m) SZ, 13(m) NC	None	Mostly intrinsic	Negative correlation of R MFG and age of onset. Positive correlation of L MFG and backward digit span
None	Yamasue et al. (2004)	27 chronic SZ (20m), 27 NC (20m)	None	Intrinsic/ extrinsic	No significant correlations found

m = male subjects

Table 16 Inferior frontal gyrus overview

Region	Total papers	Type of change	Laterality	Total FE papers	FE changes	Total chronic papers	Chronic changes
Inferior frontal gyrus	3	2 (66.7%) – 0 (0%) + 1 (33.3%) n.c.	All bilateral	0	0 (0%) – 0 (0%) + 0 (0%) n.c.	3	2 (66.7%) – 0 (0%) + 1 (33.3%) n.c.

n.c. = no change in volume relative to controls

Table 17 Inferior frontal gyrus papers

Type of change	Paper	Population	Laterality	Method	Correlations, additional findings, and comments
–	Suzuki et al. (2005)	53 SZ (32m), 59 NC (35m)	Bilateral	Mostly intrinsic	No significant correlations found Decreased IFG SZ (significant gender interaction – males only have decrease on L; females bilateral)
–	Yamasue et al. (2004)	27 chronic SZ (20m), 27 NC (20m)	Bilateral	Intrinsic/ extrinsic	R IFG negatively correlated with total negative symptoms and stereotyped thinking
None	Sapara et al. (2007)	28 chronic SZ (24m), 20 NC (17m)	None	Intrinsic	Positive correlation of total IFG with total insight and insight into illness; gender interaction – males only have correlation on L

m = male subjects

Table 18 Orbitofrontal gyrus overview

Region	Total papers	Type of change	Laterality	Total FE papers	FE changes	Total chronic papers	Chronic changes
Orbitofrontal gyrus	5	1 (20%) − 0 (0%) + 4 (80%) n.c.	All bilateral	0	0 (0%) − 0 (0%) + 0 (0%) n.c.	5	1 (20%) − 0 (0%) + 4 (80%) n.c.

n.c. = no change in volume relative to controls

Table 19 Orbitofrontal gyrus papers

Type of change	Paper	Population	Laterality	Method	Correlations, additional findings, and comments
−	Nakamura et al. (2008)	24(m) chronic, 25 NC (19m)	Bilateral	Mostly intrinsic	Negative correlation of R MidOFG and illness duration. Negative correlation of L MidOFG and SAPS formal thought disorder
None	Sapara et al. (2007)	28 chronic SZ (24m) 20 NC (17m)	None	Intrinsic	Negative correlation with illness duration. Positive correlation with total insight and insight into illness; R OFG positive correlation with insight into symptoms
None	Suzuki et al. (2005)	53 SZ (32m), 59 NC (35m)	None	Mostly intrinsic	No significant correlations found
None	Szendi et al. (2006)	13(m) SZ, 13(m) NC	None	Mostly intrinsic	No significant correlations found
None	Yamasue et al. (2004)	27 chronic SZ (20m), 27 NC (20m)	None	Intrinsic/ extrinsic	No significant correlations found

m = male subjects

Table 20 Straight gyrus/gyrus rectus overview

Region	Total papers	Type of change	Laterality	Total FE papers	FE changes	Total chronic papers	Chronic changes
Straight gyrus	2	1 (50%) − 0 (0%) + 1 (50%) n.c.	1 interaction	0	0 (0%) − 0 (0%) + 0 (0%) n.c.	2	1 (50%) − 0 (0%) + 1 (50%) n.c.

n.c. = no change in volume relative to controls

compromised and that this might have a negative impact on "the cognitive–motor interface" in subjects with schizophrenia.) The use of reporting effect sizes is one way that investigators have attempted to address the recurring problem of small sample size.

Table 21 Straight gyrus/gyrus rectus papers

Type of change	Paper	Population	Laterality	Method	Correlations, additional findings, and comments
–	Suzuki et al. (2005)	53 SZ (32m), 59 NC (35m)	Bilateral	Mostly intrinsic	No significant correlations found
None	Szendi et al. (2006)	13(m) SZ, 13(m) NC	Interaction: NC L > R; SZ R > L	Mostly intrinsic	Negative correlation of SG volume and Visual Patterns Test. Negative correlation of L SG volume and anhedonia

m = male subjects

The frontal lobe mediates a number of important processes which may impact executive functions, working memory, abstract reasoning, social behavior, self-monitoring, and impulse control in schizophrenia. Additional higher-order functions which have been examined in relation to the PFC include insight (Sapara et al. 2007) and empathy (Fujiwara et al. 2008). Sapara et al. (2007) carried out a study in 28 chronic, stable outpatient schizophrenia subjects and 20 controls. Despite not being able to show group differences in PFC regions, they reported that smaller total PFC GM volume in patients was associated with a lower level of insight. This was true for both hemispheres, but it seems that the left IFG volume was the "strongest contributor, after controlling for the duration of illness, to the observed total PFC-insight into the mental illness relationship;" additionally, this type of insight showed the most consistent association with prefrontal volumes. Typically, patients with schizophrenia are thought to deny that they are ill, so this study provides a potential cortical substrate for this common symptom. In a study of the ACG by Fujiwara et al. (2008), a structure which can be thought of as both part of the paralimbic system and of the prefrontal cortex, they note that the ACG "is one of the critical structures for empathy processing, the pathology of this structure might be a major source of social dysfunction, including interpersonal miscommunication in schizophrenia." These authors showed that in female schizophrenia patients, left dorsal ACG volume inversely correlated with "personal distress subscale scores," suggesting that specific ACG subdivision pathology might affect "specific empathic disabilities in schizophrenia, with potential gender specificity." In sum, though difficult to measure morphometrically, PFC subregions are involved in processes so central to more sophisticated human functioning that they are common targets for researchers interested in human nature.

The OFC is another interesting region in terms of the potential behavioral functions it subserves. Work by our group (Nakamura et al. 2008) showed a bilateral reduction in volume in a subregion of the OFC – what the authors labeled the middle orbital gyrus (MiOG) – together with an altered sulcal pattern of the MiOG, using a sample of 24 chronic SZ subjects and 25 matched controls. We also found that longer duration of illness and more severe positive formal thought disorder was associated with a reduction in volume of the MiOG, suggesting that this ROI has an effect on clinical state. In addition, we reported that there was a different sulcal pattern distribution in the so-called H-shaped sulci in the OFC in patients compared with controls, with patients having a twofold increase of the most rare sulcal pattern seen in controls (25% in SZ vs. 12% in HC). Furthermore, in schizophrenia, worse SES and lower scores in verbal comprehension were associated with this more rare sulcal pattern, again suggesting that the morphometric characteristics of OFC sulcal patterns have clinical relevance.

Another useful approach to understanding the prefrontal cortex is to examine not only its volume, but to attempt to assess its thickness and surface area as an index of its cortical complexity or degree of folding. In such an attempt, our group (Wiegand et al. 2005; Wiegand et al. 2004) measured prefrontal cortical thickness in age-matched FE schizophrenia, affective psychosis, and control subjects in a sample in which we had previously assessed prefrontal cortical volume and found a left lateralized reduction in PFC volume in FE schizophrenia patients (Hirayasu et al. 2001). In this study, which measured both cortical thickness and cortical complexity, the latter was measured by counting the number of voxels at the GM CSF boundary, with an increased number of counted voxels indicative of a larger surface area and, hence, a higher degree of gyrification. Although we did not find a group difference in cortical thickness, we did find that it negatively correlated with age only in the schizophrenia group. Furthermore, there was a group difference in asymmetry coefficients for complexity, with FE schizophrenia subjects showing a loss of a normal left-greater-than-right asymmetry pattern. Overall, in this study we found that schizophrenia patients had decreased prefrontal cortical volume with a loss of the normal left-greater-than-right cortical complexity asymmetry, and a clinical correlation of thickness with age in schizophrenia, suggesting the usefulness of combining thickness and surface area measures as well as volume measures in morphometric studies of the brain.

4 Caudate Nucleus

There has been increasing interest in the study of subcortical structures in the context of schizophrenia. The caudate nucleus has been a particular focus of research, as this basal ganglia nucleus is thought to be an especially relevant component of subcortical nuclei for cognitive function. An influential paper by Alexander et al. (1986) described multiple parallel but segregated frontal subcortical circuits in the brain, and emphasized that the frontal output from the basal ganglia was not restricted to motor regions, but also projected to nonmotor prefrontal cortical regions subserving higher-order cognitive and limbic functions. Thus, on anatomic grounds, it was appreciated that the basal ganglia could modulate these nonmotor higher-order brain functions. Furthermore, fMRI studies in both primates and humans have shown the involvement of the caudate nucleus in working memory, a core higher cognitive function which appears to be compromised in schizophrenia. Thus, for the purpose of this review, we will restrict our review of subcortical structures to the caudate nucleus.

There have been a number of manually driven morphometric studies of the caudate nucleus in schizophrenia. The biggest issue affecting volumetric assessment in these studies is the potential confound of the use of neuroleptic medications. For example, as described below, neuroleptic medications – typical neuroleptics in particular – have been reported to increase the volume of the caudate nucleus (Chakos et al. 1994; Keshavan et al. 1994), and thus it is not clear if the reported abnormal findings of the caudate are intrinsic to the disease or are medication effects, or if the reported absence of abnormality may be due to a masking effect of neuroleptic medication. The volume of the caudate nucleus, we believe, is deceptively challenging to measure as there are a number of edges where a good deal of partial voluming effects are present, and thus we have chosen to review only those papers that fully manually edited this structure.

As can be seen in Tables 22 and 23, summarizing the six studies that analyzed the caudate, no study showed a decrease in volume; 83.5% (5/6) of the studies revealed no change and 16.5% (1/6) revealed a bilateral increase in volume. Of three FE studies, all showed no change in caudate volume, and of the three chronic studies, one (33.3%) showed an increase in volume, while the other two (66.7%) showed no change. The remaining two papers listed in Table 23 have not been included in the overall account of caudate nucleus findings due to their lack of control groups.

The confounding issue of medication, as noted above, is of particular relevance to the assessment of the caudate nucleus. One technique to address this is to study subjects with "minimal or no prior" treatment with antipsychotic medications. Only three of the eight papers cited in Tables 22 and 23 performed such studies. Of these, as briefly reviewed here, all three found no group differences in caudate volume. In a study by Cahn et al. (2002), in 20 antipsychotic-naïve patients with FE schizophrenia and 20 healthy comparisons, no difference in volume was found for the caudate. In a longitudinal study by Tauscher-Wisniewski et al. (2005), they found no group difference in caudate volume at baseline for 37 unmedicated FE psychotic patients and 37 healthy controls. These authors, however, did not report relative volumes. They indicated that patients were matched to controls for age and sex and reported that ANCOVA analysis found significant effects for both age and gender, but no significant group difference. This study also reported that ten patients, who had received 12 weeks of quetiapine with a "mean daily dose of 494 mg" were rescanned in follow-up, and they did not find a significant change in caudate volume, although a Type 2 error here is certainly possible given the small sample size.

Table 22 Caudate nucleus overview

Region	Total papers	Type of change	Laterality	Total FE papers	FE changes	Total chronic papers	Chronic changes
Caudate nucleus	6	0 (0%) – 1 (16.5%) + 5 (83.5%) n.c.	1 interaction	4	0 (0%) – 0 (0%) + 4 (100%) n.c.	2	0 (0%) – 1 (50%) + 1 (50%) n.c.

n.c. = no change in volume relative to controls

Table 23 Caudate nucleus papers

Type of change	Paper	Population	Laterality	Method	Medication status/comments
+	Lawyer et al. (2006)	45 chronic SZ, 27 NC (genders not given)	Bilateral	Unclear	Most appear to have been on antipsychotic medication
None	Cahn et al. (2002)	20 FE SZ (16m), 20 NC (16m)	None	Intrinsic/extrinsic	Antipsychotic naïve
None	Glenthoj et al. (2007)	19 FE SZ (14m), 19 NC (11m)	NCs L > R; SZs R > L	Intrinsic/extrinsic	16 drug naïve, 3 minimally medicated
None	Riffkin et al. (2005)	18 chronic SZ (8m), 18 NC (8m)	None	Intrinsic/extrinsic	All SZs on antipsychotics, 6 on antidepressants (2 TCAs, 4 SSRIs)
None	Tauscher-Wisniewski et al. (2002)	15 SZ (10m), 10 NC (7m)	None	Intrinsic/extrinsic	On antipsychotics, mixture of subjects on typical or atypical neuroleptics
None	Tauscher-Wisniewski et al. (2005)	37 FE SZ (22m), 37 NC (22m)	None	Intrinsic/extrinsic	Unmedicated
N/A (no NC group)	Scheepers et al. (2001a)	26 SZ (18m), 24 week follow-up. No NCs	None	Intrinsic/extrinsic	Longitudinal; 24 week follow-up on patients before and after being switched to clozapine from typical neuroleptics
N/A (no NC group)	Scheepers et al. (2001b)	22 SZ (gender not specified), 52 week follow-up. No NCs	None	Intrinsic/extrinsic	Longitudinal; 52 week follow-up on patients before and after being switched to clozapine from typical neuroleptics

m = male subjects

As the control group was not rescanned, this data is difficult to interpret. Lastly, in a longitudinal study by Glenthoj et al. (2007), they compared 16 antipsychotic-naïve and three minimally medicated FE SZ subjects with 19 matched controls. It appeared that the three minimally medicated patients were scanned on a 1.0 T scanner instead of the 1.5 T scanner on which the rest of the subjects were scanned. This study did not find a group difference in caudate volume. They did, however, report a significant hemisphere by group interaction ($p = 0.052$), with controls having left larger than right caudate nuclei, and patients having "marginally" right larger than left caudate nuclei – an effect they reported was stronger when the three minimally medicated patients were excluded from the analyses ($p = 0.030$). The authors also did not find a significant increase in caudate volume after 3 month follow-up either with risperidone ($N = 11$) or zuclopenthixol ($N = 8$), a typical neuroleptic. This study's results are difficult to interpret, however, due to the small sample and because they did not rescan controls.

The other five studies were all in medicated subjects. The study by Lawyer et al. (2006) used a sample consisting of 71 chronic SZ subjects, 64 of whom were on antipsychotic medication, and 65 healthy controls. They measured the caudate in 45 of the SZ subjects and in 27 of the controls, and found that the caudate was enlarged in the patients. From the paper, however, it is not clear how many of these 45 SZ subjects were on medication. Furthermore, the authors did not control for head size and reported caudate absolute volumes, and within the subset of subjects whose caudates were measured, it was not clear if the patients and controls were matched for age and gender. Moreover, the statistics employed were t-tests for 16 brain structures and multiple comparisons were corrected for by applying FDR to the p-values, which is not the typical analytic approach in morphometric studies. In a study by Riffkin et al. (2005), they assessed the caudate nucleus in three groups (OCD, chronic schizophrenia, and healthy controls) with 18 subjects each. As the primary goal of this paper was to compare OCD with the other two groups, statistics directly comparing SZ and healthy controls were not presented. Nonetheless, across the three groups, no group difference was detected in caudate volume. Despite the SZ patients all being on antipsychotic medication, the caudate was bilaterally smaller in patients compared with HCs (no p-value included). The paper, however, did not indicate what type of antipsychotic medication (i.e., typical or atypical or the dosage) the SZ patients were taking.

In the longitudinal study by Tauscher-Wisniewski et al. (2002), they found at baseline no group difference between 15 FE psychosis subjects compared with 10 normal controls. At the time of the first scan, eight patients were neuroleptic-naïve and seven were not. Furthermore, on follow-up approximately 5 years later, the authors reported that the caudate volume decreased 9% in both groups. A possible confound, both in light of the results of this study and in light of their other study cited above (Tauscher-Wisniewski et al. (2005), in which they showed that caudate volume inversely correlated with age for unmedicated SZs and for controls) is that patients were significantly younger at the time of the first scan (23.0 ± 6.2 vs. 29.4 ± 8.6 years). During the follow-up period patients were treated with a variety of typical and atypical neuroleptic medications. These authors note that the dosing with typical neuroleptics in this study was low and that most of the patients during the follow-up period were on atypical neuroleptics. This study points out the potential problems of mixing typical and atypical medications in analyzing results, as well as having small sample sizes further complicating any type of subgroup analysis. In addition to neuroleptic type, this paper also raises the issue of neuroleptic dosing as another factor that potentially influences neuroplastic drug effects. The last two studies that assessed the effect of neuroleptic medication on caudate volume were both by Scheepers et al. (2001a, 2001b). These studies did not assess caudate volume in controls, but rather longitudinally assessed caudate volume in schizophrenia patients who did not respond to typical neuroleptics "before discontinuing typical antipsychotics and after 24 weeks of treatment with clozapine" (Scheepers et al. 2001a) and after 52 weeks of treatment with clozapine (Scheepers et al. 2001b). In the 24-week study, the authors found a significant decrease in caudate volume between baseline and follow-up. In addition, in the 24-week follow-up study there was no treatment by side interaction. Furthermore, "no difference in caudate volume changes were found between responders and nonresponders." In the 52-week study, an extension of the previous study using the same sample but with additional drop-outs (four subjects), the authors observed a significant decrease in the left caudate but not in the right. In addition, they found that the left caudate volume change was significant in responders, but not in nonresponders. Moreover, at 52 weeks the authors reported a significant correlation between left caudate volume change and a reduction in scores on the PANSS positive and general symptoms, but not negative symptoms scales. This latter finding addresses the issue of the clinical implication of volume change in caudate for SZ patients, but as the authors note, this study was performed in those not responding to typical neuroleptics and their results may not generalize to all schizophrenia subjects.

5 Neocortical Temporal Lobe

Given the characteristic symptoms of auditory hallucinations and formal thought disorder in schizophrenia, it has been believed that the disturbance in brain function involves abnormal auditory perceptual processing and abnormal language processing. The neocortical temporal lobe has been a region of particular interest in schizophrenia research, as language and auditory processing is initially perceived in the STG in Heschl's gyrus, and further auditory processing, including language comprehension, occurs in the nearby planum temporale (PT), particularly in the left in right-handed individuals. The neocortical temporal lobe has been parcellated into a number of subregions including the STG, the PT, Heschl's gyrus, temporal pole (which is actually paralimbic, not neocortical, cortex), MTG, ITG, and the fusiform gyrus. For the purpose of this review, we will focus on findings in the STG, the PT, and Heschl's gyrus, although findings in other subregions are noted in Tables 24–31.

Overall, we reviewed 12 papers on neocortical temporal lobe regions in schizophrenia (see Table 3). Of these 12 studies, 83.3% (10/12) found volume decreases

A Selective Review of Volumetric and Morphometric Imaging in Schizophrenia   269

Table 24 Temporal pole overview

Region	Total papers	Type of change	Laterality	Total FE papers	FE changes	Total chronic papers	Chronic changes
Temporal pole	2	1 (50%) − 0 (0%) + 1 (50%) n.c.	1 (100%) left	1	1 (100%) − 0 (0%) + 0 (0%) n.c.	1	0 (0%) − 0 (0%) + 1 (100%) n.c.

n.c. = no change in volume relative to controls

Table 25 Temporal pole papers

Type of change	Paper	Population	Laterality	Method	Correlations, additional findings, and comments
−	Kasai et al. (2003c)	27 FE (23m), 29 NC (24m)	L sig.; R not sig.	Intrinsic/ extrinsic	No significant correlations found Lack of normal L > R asymmetry
None	Takahashi et al. (2006a)	53 SZ (32m), 59 NC (35m)	None	Mostly intrinsic	No significant correlations found

m = male subjects

Table 26 Middle temporal gyrus overview

Region	Total papers	Type of change	Laterality	Total FE papers	FE changes	Total chronic papers	Chronic changes
Middle temporal gyrus	3	2 (66.7%) − 0 (0%) + 1 (33.3%) n.c.	1 (50%) left	1	1 (100%) − 0 (0%) + 0 (0%) n.c.	2	1 (50%) − 0 (0%) + 1 (50%) n.c.

n.c. = no change in volume relative to controls

Table 27 Middle temporal gyrus papers

Type of change	Paper	Population	Laterality	Method	Correlations, additional findings, and comments
−	Kuroki et al. (2006b)	20 FE (16m), 23 NC (20m)	Bilateral	Intrinsic/ extrinsic	Negative correlation with age at first medication in bilateral MTG
−	Onitsuka et al. (2004)	23(m) SZ, 28(m) NC	L sig. R not sig.	Intrinsic/ extrinsic	Negative correlation with and global hallucination rating and L MTG. Hallucinators ($n = 13$) had a more significantly decreased L MTG than nonhallucinators ($n = 6$)
None	Takahashi et al. (2006b)	65 SZ (35m), 72 NC (38m)	None	Intrinsic/ extrinsic	No significant correlations found

m = male subjects

Table 28 Inferior temporal gyrus overview

Region	Total papers	Type of change	Laterality	Total FE papers	FE changes	Total chronic papers	Chronic changes
Inferior temporal gyrus	3	2 (66.7%) − 0 (0%) + 1 (33.3%) n.c	All bilateral	1	1 (100%) − 0 (0%) + 0 (0%) n.c.	2	1 (50%) − 0 (0%) + 1 (50%) n.c.

n.c. = no change in volume relative to controls

Table 29 Inferior temporal gyrus papers

Type of change	Paper	Population	Laterality	Method	Correlations, additional findings, and comments
−	Kuroki et al. (2006b)	20 FE (16m), 23 NC (20m)	Bilateral	Intrinsic/extrinsic	Negative correlation of age at first medication and L ITG. Negative correlation of BPRS hostility/suspiciousness and R MidITG
−	Onitsuka et al. (2004)	23(m) SZ, 28(m) NC	Bilateral	Intrinsic/extrinsic	No significant correlations found
None	Takahashi et al. (2006b)	65 SZ (35m), 72 NC (38m)	None	Intrinsic/extrinsic	No significant correlations found

m = male subjects

Table 30 Fusiform gyrus overview

Region	Total papers	Type of change	Laterality	Total FE papers	FE changes	Total chronic papers	Chronic changes
Fusiform gyrus	3	3 (100%) − 0 (0%) + 0 (0%) n.c.	1 (33.3%) left	1	1 (100%) − 0 (0%) + 0 (0%) n.c.	2	2 (100%) − 0 (0%) + 0 (0%) n.c.

n.c. = no change in volume relative to controls

Table 31 Fusiform gyrus papers

Type of change	Paper	Population	Laterality	Method	Correlations, additional findings, and comments
−	Lee et al. (2002)	22 FE SZ (17m), 24 NC (21m)	L sig.; R not sig.	Mostly intrinsic	No significant correlations found
−	Onitsuka et al. (2004)	23(m) SZ, 28(m) NC	Bilateral	Intrinsic/extrinsic	No significant correlations found
−	Takahashi et al. (2006b)	65 SZ (35m), 72 NC (38m)	Bilateral	Mostly intrinsic	Negative correlation of R fusiform with SANS affective flattening or blunting

m = male subjects

in schizophrenia in at least one neocortical temporal lobe structure, none found an increase in any structure, and 16.7% (2/12) found no change in volume. Of studies reporting volume differences, 80% (8/10) of studies reported stronger differences on the left side. Of five FE studies, all (100%) showed decreases. Of the seven chronic studies, five (71.4%) showed decreases and two (28.6%) showed no change (Fig. 1).

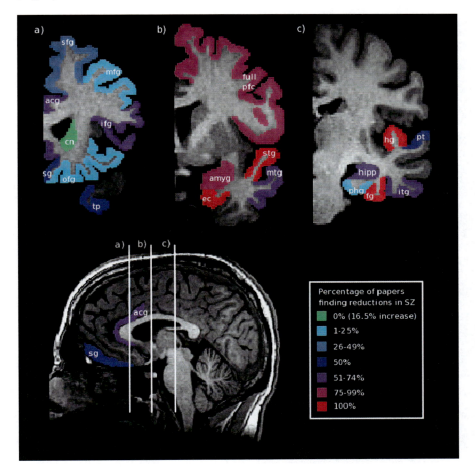

Fig. 1 Percentages of papers finding volume reductions in patients with schizophrenia in the following regions: (**a**) anterior cingulate gyrus (acg), superior frontal gyrus (sfg), middle frontal gyrus (mfg), inferior frontal gyrus (ifg), orbitofrontal gyrus (ofg), straight gyrus (sg), caudate nucleus (cn), and temporal pole (tp); (**b**) full prefrontal cortex (pfc), superior temporal gyrus (stg), middle temporal gyrus (mtg), entorhinal cortex (ec), and amygdala (amyg); (**c**) Heschl's gyrus (hg), planum temporale (pt), inferior temporal gyrus (itg), fusiform gyrus (fg), parahippocampal gyrus (phg), and hippocampus (hipp). A sagittal slice is provided for reference for the location of the three coronal slices. In order to facilitate the display of maximal regions in coronal slice (**b**), the full PFC is shown on a slice slightly posterior to where it would actually be drawn. The parietal, occipital, and insular cortices did not have a sufficient number of publications to be included in our review

As can be seen in Tables 32 and 33, summarizing the six studies that analyzed the STG, all (100%) of the studies revealed STG volume reductions, as no studies showed either an increase or no change in volume. Four (66.7%) of these studies showed stronger differences on the left side. Two of these studies were in FE subjects and four were in chronic subjects.

Table 32 Superior temporal gyrus overview

Region	Total papers	Type of change	Laterality	Total FE papers	FE changes	Total chronic papers	Chronic changes
Superior temporal gyrus	6	6 (100%) − 0 (0%) + 0 (0%) n.c.	4 (80%) left	2	2 (100%) − 0 (0%) + 0 (0%) n.c.	4	4 (100%) − 0 (0%) + 0 (0%) n.c.

n.c. = no change in volume relative to controls

Table 33 Superior temporal gyrus papers

Type of change	Paper	Population	Laterality	Method	Correlations, additional findings, and comments
−	Anderson et al. (2002)	16(m) chronic SZ, 15(m) NC	L sig. R trend	Intrinsic/ extrinsic	Negative correlation of PANSS poor attention with R post. STG. Negative correlation of TDI and L STG, and R post. STG (11 SZs included)
−	Kasai et al. (2003a)	13 FE (10m), 13 NC (12m); also 1.5 year follow-up	L sig. R not sig.	Intrinsic/ extrinsic	No significant correlations found. Faster L post. STG decrease in volume between initial scan and follow-up in SZs.
−	Kuroki et al. (2006b)	20 FE (16m), 23 NC (20m)	L sig. R not sig.	Intrinsic/ extrinsic	No significant correlations found
−	Onitsuka et al. (2004)	23(m) SZ, 28(m) NC	L sig. R not sig.	Intrinsic/ extrinsic	Negative correlation of L STG with total SAPS and global hallucination rating. Hallucinators ($n = 13$) had a significantly greater decrease in L STG than nonhallucinators ($n = 6$)
−	Takahashi et al. (2006a)	53 SZ (32m), 59 NC (35m)	Bilateral	Intrinsic/ extrinsic	Negative correlation of L rostral STG and SAPS delusions, and hallucinations in the 20 FE subjects
−	Yamasue et al. (2004)	27 chronic SZ (20m), 27 NC (20m)	Bilateral	Intrinsic/ extrinsic	No significant correlations found. Findings in posterior STG only.

m = male subjects

A Selective Review of Volumetric and Morphometric Imaging in Schizophrenia   273

Table 34 Planum temporale overview

Region	Total papers	Type of change	Laterality	Total FE papers	FE changes	Total chronic papers	Chronic changes
Planum temporale	4	2 (50%) − 0 (0%) + 2 (50%) n.c.	2 (100%) left	1	1 (100%) − 0 (0%) + 0 (0%) n.c.	3	1 (33.3%) − 0 (0%) + 2 (66.7%) n.c.

n.c. = no change in volume relative to controls

Table 35 Planum temporale papers

Type of change	Paper	Population	Laterality	Method	Correlations, additional findings, and comments
−	Kasai et al. (2003b)	13 FE (10m), 13 NC (12m); also 1.5 year follow-up	L sig. R not sig.	Intrinsic/ extrinsic	Negative correlation of L PT and BPRS mean conceptual disorganization. L PT decrease between initial scan and follow-up only in SZs
−	Takahashi et al. (2006a)	53 SZ (32m), 59 NC (35m)	Bilateral, but larger difference on left	Mostly intrinsic	No significant correlations found
None	Meisenzahl et al. (2002)	30(m) SZ, 30(m) NC	None	Intrinsic/ extrinsic	No significant correlations found
None	Shapleske et al. (2001)	74(m) SZ, 32(m) NC	None	Intrinsic/ extrinsic	No significant correlations found

m = male subjects

As can be seen in Tables 34 and 35, summarizing the four studies that analyzed the PT, 50% (2/4) of the studies revealed PT volume reductions, none showed an increase, and 50% (2/4) showed no change. Of the two studies showing a reduction, both (100%) showed a stronger difference on the left. The one FE study showed a decrease, whereas of three chronic studies, one (33.3%) showed a decrease and two (66.7%) showed no change in volume.

Lastly, as can be seen in Tables 36 and 37, summarizing the two papers that analyzed Heschl's gyrus (also known as the transverse temporal gyrus), 100% (2/2) of the studies revealed HG volume reductions. The studies were evenly divided between one FE study, with left lateralized findings, and one chronic study with bilateral findings.

Studies of the neocortical temporal lobe particularly highlight issues regarding sulcal complexity inherent in the morphometric assessment of brain regions in schizophrenia. One issue is that of basing volumetric measurements on tracing sulci. Clearly, this is the approach on which manual tracing is based. The neocortical

Table 36 Heschl's/transverse temporal gyrus overview

Region	Total papers	Type of change	Laterality	Total FE papers	FE changes	Total chronic papers	Chronic changes
Heschl's/ transverse temporal gyrus	2	2 (100%) − 0 (0%) + 0 (0%) n.c.	1 (50%) left	1	1 (100%) − 0 (0%) + 0 (0%) n.c.	1	1 (100%) − 0 (0%) + 0 (0%) n.c.

n.c. = no change in volume relative to controls

Table 37 Heschl's/transverse temporal gyrus papers

Type of change	Paper	Population	Laterality	Method	Correlations, additional findings, and comments
−	Kasai et al. (2003b)	13 FE (10m), 13 NC (12m); also 1.5 year follow-up	L sig. at follow-up, but not R	Intrinsic/ extrinsic	Negative correlation of HG volume decrease and BPRS mean conceptual disorganization, mean suspiciousness, baseline somatic concern, and baseline anxiety-depression. L HG decreased between initial scan and follow-up only in SZs
−	Takahashi et al. (2006a)	53 SZ (32m), 59 NC (35m)	Bilateral	Mostly intrinsic	No significant correlations found

m = male subjects

temporal lobe, especially the STG, has relatively deep and unbroken sulci defining its superior (the Sylvian fissure) and inferior borders (the STG sulcus) (Kasai et al. 2003a, b, c; Takahashi et al. 2006a). This is an advantage for tracing in the temporal lobe, as sulci that are more frequently noncontinuous (like frontal sulci such as the superior and inferior frontal sulci) complicate the task of manual tracing due to issues of how best to bridge such discontinuities. This advantage for neocortical temporal lobe versus PFC may be one reason that volumetric group differences have been more readily documented in this brain region compared to the PFC. A further complication is that not only may the sulcal patterns be more complex in one region over another, for example, the PFC versus the lateral superior temporal lobe, but there may also be group differences in sulcal pattern. For example, Nakamura et al. (2008) showed distinct group differences between patients and controls in the H-shaped sulcus that defines orbitofrontal subregions in the PFC, thus further complicating straightforward interpretations of subregional volume measurements between groups. Another issue is that the anterior and posterior boundaries are not anatomic, and thus one needs to select boundary criteria, which at a minimum must be reliable. In the literature, this has often resulted in the selection of extrinsic boundaries, which of necessity are somewhat arbitrary. For example, a common division chosen between anterior and posterior STG is the extrinsic structure of the mammillary bodies. The anterior boundary utilized has often been the temporal stem and the posterior boundary, the crus of the fornix. Again, these latter two structures are not intrinsic to the STG (Anderson et al. 2002; Kasai et al. 2003a). Subregional analyses also suffer in that cytoarchitectonically, and hence probably functionally, the brain typically changes in a gradient fashion rather than abruptly (Mesulam 2000) and thus distinct dividing lines between subregions, such as between anterior and posterior STG, or even between HG and PT, do not capture this. Nonetheless, a number of interesting findings have been reported in the STG in schizophrenia in the six studies we reviewed, as briefly described below.

In two papers by Kasai et al. (2003a, b), the researchers reported a progressive GM decrease in FE subjects in STG, HG, and PT lateralized to the left side. The first study (Kasai et al. 2003a) included 13 patients with FE SZ, 15 patients with FE affective psychosis, mainly manic, and 14 healthy comparisons, and scanned them again approximately 1.5 years later. They reported a regionally selective decrease in left temporal STG, with posterior STG attaining significance in SZ subjects compared with both affective psychosis and healthy controls. Conversely, for medial temporal lobe structures, including what they labeled the anterior and posterior amygdala–hippocampal complex, they did not find this progressive decrease in volume. In their second study (Kasai et al. 2003b), using an overlapping sample of subjects (13 schizophrenia, 15 affective psychosis, mainly manic, and 22 healthy controls), they found a left, but not right, progressive volume reduction in the HG and PT GM in FE SZ subjects. Both of these studies required their subjects to be right-handed. It was thus of interest that left STG, HG and PT showed progressive change consistent with this brain region, at least with regard to the PT, which subserves language function in right-handed individuals. From this same group, Anderson et al. (2002) looked at STG and other temporal lobe ROIs in 16 chronic schizophrenia patients and 15 controls in a cross-sectional study. Though the repeated measures ANOVA only showed a trend main effect for an interaction of ROI and group ($p < 0.087$), the findings were highly suggestive of a decrease in STG volume in schizophrenia with somewhat of a leftward bias, as the reported effect size for left STG (0.91) was large and for right STG (0.53) was moderate. Takahashi et al. (2006a) looked at STG, HG and PT, and other temporal lobe subregions in a considerably larger sample of 65 schizophrenia patients (mean duration of illness of 4.0 ± 4.1 years), 39 SPD patients, and 72 healthy controls. They reported diminished volume of the whole STG bilaterally, bilateral HG reductions, "left lateralized PT volume reduction," and "bilateral volume reduction of the caudal STG" in schizophrenia compared with controls. In this larger sample, it would appear that the PT was the structure which primarily demonstrated a left lateralized pattern of reduction. Conversely, in another large sample study including 74 male chronic schizophrenia patients (30 of whom did not have a history of auditory verbal hallucinations and 44 of whom did) and 32 matched controls, they found that PT volume and surface area, as well as Sylvian fissure length, were larger in the left hemisphere in all three groups, but there were no significant differences in the degree of leftward asymmetry among the three groups or between the two patient subgroups. Furthermore, this study did not show group differences in the surface area or volume of the PT or in the length of the Sylvian Fissure (Shapleske et al. 2001). This study, however, was not restricted to right-handed subjects, although it did covary for handedness and found a trend ($p = 0.09$) for left-handers having "slightly bigger PTs." Lastly, a study by Meisenzahl et al. (2002) used three definitions of the PT border and, using all definitions, also failed to show volume or asymmetry coefficient group differences between 30 right-handed male schizophrenia patients and 30 right-handed controls. Thus, findings to date regarding the volume and asymmetry differences for the PT, in particular, for schizophrenia are not all in agreement.

6 Conclusion

The volumetric and morphometric GM MRI literature in schizophrenia thus far supports the idea that schizophrenia is an expression of a disordered brain. Most brain regions, certainly the frontotemporal cortex in particular, are reported to have volumetric reductions. Our review suggests that such reductions are generally found both in FE and in chronic schizophrenia subjects, and not predominantly in chronic subjects as might be expected if such brain abnormalities occur exclusively as a progressive component of the illness. Magnetic resonance volumetric studies in adult schizophrenia thus support the notion that schizophrenia is a developmental disorder, as suggested by findings that abnormal brain morphometry exists at the onset of the illness and is not simply a consequence of having schizophrenia. This statement, of course, does not preclude the possibility that there is some degree of progressive deterioration in brain structure with aging, and that this may well be accelerated in patients with schizophrenia compared to healthy controls (which is suggested by a number of longitudinal studies mentioned in our review). Though beyond the scope of this chapter, assessing subjects at an earlier age (e.g., a prodromal phase of the illness) and assessing endophenotypic markers, in spectrum conditions and in relatives, are clearly important approaches to help determine how early brain abnormalities begin, and whether such abnormalities can exist in the absence of full-blown symptomatology, that is, represent causes rather than consequences of the illness.

We have restricted our review to manual tracing studies, as we believe this remains the gold standard for volumetric studies. We acknowledge, however, the importance of large sample sizes, a problem exacerbated by the importance of performing subgroup comparisons based on variables such as gender, medication status or sulcal pattern, which is a challenge for the highly labor-intensive manual tracing approach. Thus we also acknowledge that as automatic methodologies and computation expertise improve, allowing for larger sample studies, automatic studies will increase in importance; but for a time, at least, we believe that employing manual editing to images rendered by automatic programs may yield the best balance between validity and efficiency. In reviewing our tables, for example, we are struck by the absence of recent studies measuring PFC subregions in FE subjects (see Tables 12–21) despite the obvious importance of the PFC in higher cognitive functions. The variability and discontinuities of sulcal patterns pose a challenge for morphometry with any current approach, whether manual, automatic, or combined. To some extent, we wonder whether the higher success rates of finding brain abnormalities in certain subregions over other subregions (e.g., the STG, or the hippocampus) may be due to the less varied and less interrupted patterns of more universally uniform sulci, such as the Sylvian fissure, compared to more variable sulci, such as the superior and inferior frontal sulci in the PFC. Thus, multimodal studies – including techniques such as fMRI, electrophysiolgic measures, diffusion tensor imaging, and postmortem studies – are needed in schizophrenia brain research to clarify the findings suggested by MRI morphometry studies.

Acknowledgment This work was supported by a VA Merit Award (JJL).

References

Alexander GE, DeLong MR, Strick PL (1986) Parallel organization of functionally segregated circuits linking basal ganglia and cortex. Annu Rev Neurosci 9:357–381

Anderson JE, Wible CG, McCarley RW, Jakab M, Kasai K, Shenton ME (2002) An MRI study of temporal lobe abnormalities and negative symptoms in chronic schizophrenia. Schizophr Res 58:123–134

Baiano M, Perlini C, Rambaldelli G, Cerini R, Dusi N, Bellani M, Spezzapria G, Versace A, Balestrieri M, Mucelli RP, Tansella M, Brambilla P (2008) Decreased entorhinal cortex volumes in schizophrenia. Schizophr Res 102:171–180

Cahn W, Hulshoff Pol HE, Bongers M, Schnack HG, Mandl RC, Van Haren NE, Durston S, Koning H, Van Der Linden JA, Kahn RS (2002) Brain morphology in antipsychotic-naive schizophrenia: a study of multiple brain structures. Br J Psychiatry (Suppl 43):S66–S72

Chakos MH, Lieberman JA, Bilder RM, Borenstein M, Lerner G, Bogerts B, Wu H, Kinon B, Ashtari M (1994) Increase in caudate nuclei volumes of first-episode schizophrenic patients taking antipsychotic drugs. Am J Psychiatry 151:1430–1436

Coryell W, Nopoulos P, Drevets W, Wilson T, Andreasen NC (2005) Subgenual prefrontal cortex volumes in major depressive disorder and schizophrenia: diagnostic specificity and prognostic implications. Am J Psychiatry 162:1706–1712

Csernansky JG, Wang L, Jones D, Rastogi-Cruz D, Posener JA, Heydebrand G, Miller JP, Miller MI (2002) Hippocampal deformities in schizophrenia characterized by high dimensional brain mapping. Am J Psychiatry 159:2000–2006

Dickey CC, Salisbury DF, Nagy AI, Hirayasu Y, Lee CU, McCarley RW, Shenton ME (2004) Follow-up MRI study of prefrontal volumes in first-episode psychotic patients. Schizophr Res 71:349–351

Exner C, Boucsein K, Degner D, Irle E, Weniger G (2004) Impaired emotional learning and reduced amygdala size in schizophrenia: a 3-month follow-up. Schizophr Res 71:493–503

Fornito A, Yucel M, Wood SJ, Adamson C, Velakoulis D, Saling MM, McGorry PD, Pantelis C (2008) Surface-based morphometry of the anterior cingulate cortex in first episode schizophrenia. Hum Brain Mapp 29:478–489

Fujiwara H, Shimizu M, Hirao K, Miyata J, Namiki C, Sawamoto N, Fukuyama H, Hayashi T, Murai T (2008) Female specific anterior cingulate abnormality and its association with empathic disability in schizophrenia. Prog Neuropsychopharmacol Biol Psychiatry 32: 1728–1734

Glenthoj A, Glenthoj BY, Mackeprang T, Pagsberg AK, Hemmingsen RP, Jernigan TL, Baare WF (2007) Basal ganglia volumes in drug-naive first-episode schizophrenia patients before and after short-term treatment with either a typical or an atypical antipsychotic drug. Psychiatry Res 154:199–208

Haznedar MM, Buchsbaum MS, Hazlett EA, Shihabuddin L, New A, Siever LJ (2004) Cingulate gyrus volume and metabolism in the schizophrenia spectrum. Schizophr Res 71:249–262

Hirayasu Y, Tanaka S, Shenton ME, Salisbury DF, DeSantis MA, Levitt JJ, Wible C, Yurgelun-Todd D, Kikinis R, Jolesz FA, McCarley RW (2001) Prefrontal gray matter volume reduction in first episode schizophrenia. Cereb Cortex 11:374–381

Joyal CC, Laakso MP, Tiihonen J, Syvalahti E, Vilkman H, Laakso A, Alakare B, Räkköläinen V, Salokangas RK, Hietala J (2002) A volumetric MRI study of the entorhinal cortex in first episode neuroleptic-naive schizophrenia. Biol Psychiatry 51:1005–1007

Joyal CC, Laakso MP, Tiihonen J, Syvalahti E, Vilkman H, Laakso A, Alakare B, Räkköläinen V, Salokangas RK, Hietala J (2003) The amygdala and schizophrenia: a volumetric magnetic resonance imaging study in first-episode, neuroleptic-naive patients. Biol Psychiatry 54: 1302–1304

Kasai K, Shenton ME, Salisbury DF, Hirayasu Y, Lee CU, Ciszewski AA, Yurgelun-Todd D, Kikinis R, Jolesz FA, McCarley RW (2003a) Progressive decrease of left superior temporal gyrus gray matter volume in patients with first-episode schizophrenia. Am J Psychiatry 160:156–164

Kasai K, Shenton ME, Salisbury DF, Hirayasu Y, Onitsuka T, Spencer MH, Yurgelun-Todd DA, Kikinis R, Jolesz FA, McCarley RW (2003b) Progressive decrease of left Heschl gyrus and planum temporale gray matter volume in first-episode schizophrenia: a longitudinal magnetic resonance imaging study. Arch Gen Psychiatry 60:766–775

Kasai K, Shenton ME, Salisbury DF, Onitsuka T, Toner SK, Yurgelun-Todd D, Kikinis R, Jolesz FA, McCarley RW (2003c) Differences and similarities in insular and temporal pole MRI gray matter volume abnormalities in first-episode schizophrenia and affective psychosis. Arch Gen Psychiatry 60:1069–1077

Keshavan MS, Bagwell WW, Haas GL, Sweeney JA. Schooler NR, Pettegrew JW (1994) Changes in caudate volume with neuroleptic treatment. Lancet 344:1434

Koo MS, Levitt JJ, Salisbury DF, Nakamura M, Shenton ME, McCarley RW (2008) A cross-sectional and longitudinal magnetic resonance imaging study of cingulate gyrus gray matter volume abnormalities in first-episode schizophrenia and first-episode affective psychosis. Arch Gen Psychiatry 65:746–760

Kuroki N, Kubicki M, Nestor PG, Salisbury DF, Park HJ, Levitt JJ, Woolston S, Frumin M, Niznikiewicz M, Westin CF, Maier SE, McCarley RW, Shenton ME (2006a) Fornix integrity and hippocampal volume in male schizophrenic patients. Biol Psychiatry 60:22–31

Kuroki N, Shenton ME, Salisbury DF, Hirayasu Y, Onitsuka T, Ersner-Hershfield H, Yurgelun-Todd D, Kikinis R, Jolesz FA, McCarley RW (2006b) Middle and inferior temporal gyrus gray matter volume abnormalities in first-episode schizophrenia: an MRI study. Am J Psychiatry 163:2103–2110

Laakso MP, Tiihonen J, Syvalahti E, Vilkman H, Laakso A, Alakare B, Räkköläinen V, Salokangas RK, Koivisto E, Hietala J (2001) A morphometric MRI study of the hippocampus in first-episode, neuroleptic-naive schizophrenia. Schizophr Res 50:3–7

Lawyer G, Nyman H, Agartz I, Arnborg S, Jonsson EG, Sedvall GC, Hall H (2006) Morphological correlates to cognitive dysfunction in schizophrenia as studied with Bayesian regression. BMC Psychiatry 6:31

Lee CU, Shenton ME, Salisbury DF, Kasai K, Onitsuka T, Dickey CC, Yurgelun-Todd D, Kikinis R, Jolesz FA, McCarley RW (2002) Fusiform gyrus volume reduction in first-episode schizophrenia: a magnetic resonance imaging study. Arch Gen Psychiatry 59:775–781

Levitt JJ, McCarley RW, Dickey CC, Voglmaier MM, Niznikiewicz MA, Seidman LJ, Hirayasu Y, Ciszewski AA, Kikinis R, Jolesz FA, Shenton ME (2002) MRI study of caudate nucleus volume and its cognitive correlates in neuroleptic-naive patients with schizotypal personality disorder. Am J Psychiatry 159:1190–1197

Levitt JJ, Styner M, Niethammer M, Bouix S, Koo MS, Voglmaier MM, Dickey CC, Niznikiewicz MA, Kikinis R, McCarley RW, Shenton ME (2009) Shape abnormalities of caudate nucleus in schizotypal personality disorder. Schizophr Res 110:127–139

Meisenzahl EM, Zetzsche T, Preuss U, Frodl T, Leinsinger G, Moller HJ (2002) Does the definition of borders of the planum temporale influence the results in schizophrenia? Am J Psychiatry 159:1198–1200

Mesulam M (2000) Brain, mind, and the evolution of connectivity. Brain Cogn 42:4–6

Nakamura M, Salisbury DF, Hirayasu Y, Bouix S, Pohl KM, Yoshida T, Koo MS, Shenton ME, McCarley RW (2007) Neocortical gray matter volume in first-episode schizophrenia and first-episode affective psychosis: a cross-sectional and longitudinal MRI study. Biol Psychiatry 62:773–783

Nakamura M, Nestor PG, Levitt JJ, Cohen AS, Kawashima T, Shenton ME, McCarley RW (2008) Orbitofrontal volume deficit in schizophrenia and thought disorder. Brain 131:180–195

Namiki C, Hirao K, Yamada M, Hanakawa T, Fukuyama H, Hayashi T, Murai T (2007) Impaired facial emotion recognition and reduced amygdalar volume in schizophrenia. Psychiatry Res 156:23–32

Nestor PG, Kubicki M, Kuroki N, Gurrera RJ, Niznikiewicz M, Shenton ME, McCarley RW (2007) Episodic memory and neuroimaging of hippocampus and fornix in chronic schizophrenia. Psychiatry Res 155:21–28

Nieuwenhuys R, Voogd J, Cv H (2008) The human central nervous system. Springer, New York

Niu L, Matsui M, Zhou SY, Hagino H, Takahashi T, Yoneyama E, Kawasaki Y, Suzuki M, Seto H, Ono T, Kurachi M (2004) Volume reduction of the amygdala in patients with schizophrenia: a magnetic resonance imaging study. Psychiatry Res 132:41–51

Nolte J (2009) The human brain: an introduction to its functional anatomy. Mosby/Elsevier, Philadelphia, PA

Nugent AC, Milham MP, Bain EE, Mah L, Cannon DM, Marrett S, Zarate CA, Pine DS, Price JL, Drevets WC (2006) Cortical abnormalities in bipolar disorder investigated with MRI and voxel-based morphometry. Neuroimage 30:485–497

Onitsuka T, Shenton ME, Salisbury DF, Dickey CC, Kasai K, Toner SK, Frumin M, Kikinis R, Jolesz FA, McCarley RW (2004) Middle and inferior temporal gyrus gray matter volume abnormalities in chronic schizophrenia: an MRI study. Am J Psychiatry 161:1603–1611

Pegues MP, Rogers LJ, Amend D, Vinogradov S, Deicken RF (2003) Anterior hippocampal volume reduction in male patients with schizophrenia. Schizophr Res 60:105–115

Phillips LJ, Velakoulis D, Pantelis C, Wood S, Yuen HP, Yung AR, Desmond P, Brewer W, McGorry PD (2002) Non-reduction in hippocampal volume is associated with higher risk of psychosis. Schizophr Res 58:145–158

Prasad KM, Rohm BR, Keshavan MS (2004) Parahippocampal gyrus in first episode psychotic disorders: a structural magnetic resonance imaging study. Prog Neuropsychopharmacol Biol Psychiatry 28:651–658

Riffkin J, Yucel M, Maruff P, Wood SJ, Soulsby B, Olver J, Kyrios M, Velakoulis D, Pantelis C (2005) A manual and automated MRI study of anterior cingulate and orbito-frontal cortices, and caudate nucleus in obsessive-compulsive disorder: comparison with healthy controls and patients with schizophrenia. Psychiatry Res 138:99–113

Rupp CI, Fleischhacker WW, Kemmler G, Kremser C, Bilder RM, Mechtcheriakov S, Szeszko PR, Walch T, Scholtz AW, Klimbacher M, Maier C, Albrecht G, Lechner-Schoner T, Felber S, Hinterhuber H (2005) Olfactory functions and volumetric measures of orbitofrontal and limbic regions in schizophrenia. Schizophr Res 74:149–161

Sapara A, Cooke M, Fannon D, Francis A, Buchanan RW, Anilkumar AP, Barkataki I, Aasen I, Kuipers E, Kumari V (2007) Prefrontal cortex and insight in schizophrenia: a volumetric MRI study. Schizophr Res 89:22–34

Scheepers FE, de Wied CC, Hulshoff Pol HE, van de Flier W, van der Linden JA, Kahn RS (2001a) The effect of clozapine on caudate nucleus volume in schizophrenic patients previously treated with typical antipsychotics. Neuropsychopharmacology 24:47–54

Scheepers FE, Gispen de Wied CC, Hulshoff Pol HE, Kahn RS (2001b) Effect of clozapine on caudate nucleus volume in relation to symptoms of schizophrenia. Am J Psychiatry 158: 644–646

Shapleske J, Rossell SL, Simmons A, David AS, Woodruff PW (2001) Are auditory hallucinations the consequence of abnormal cerebral lateralization? A morphometric MRI study of the sylvian fissure and planum temporale. Biol Psychiatry 49:685–693

Shenton ME, Kikinis R, Jolesz FA, Pollak SD, LeMay M, Wible CG, Hokama H, Martin J, Metcalf D, Coleman M et al (1992) Abnormalities of the left temporal lobe and thought disorder in schizophrenia. A quantitative magnetic resonance imaging study. N Engl J Med 327:604–612

Shenton ME, Dickey CC, Frumin M, McCarley RW (2001) A review of MRI findings in schizophrenia. Schizophr Res 49:1–52

Sim K, DeWitt I, Ditman T, Zalesak M, Greenhouse I, Goff D, Weiss AP, Heckers S (2006) Hippocampal and parahippocampal volumes in schizophrenia: a structural MRI study. Schizophr Bull 32:332–340

Suzuki M, Zhou SY, Takahashi T, Hagino H, Kawasaki Y, Niu L, Matsui M, Seto H, Kurachi M (2005) Differential contributions of prefrontal and temporolimbic pathology to mechanisms of psychosis. Brain 128:2109–2122

Szendi I, Kiss M, Racsmany M, Boda K, Cimmer C, Voros E, Kovács ZA, Szekeres G, Galsi G, Pléh C, Csernay L, Janka Z (2006) Correlations between clinical symptoms, working memory functions and structural brain abnormalities in men with schizophrenia. Psychiatry Res 147:47–55

Takahashi T, Kawasaki Y, Kurokawa K, Hagino H, Nohara S, Yamashita I, Nakamura K, Murata M, Matsui M, Suzuki M, Seto H, Kurachi M (2002a) Lack of normal structural asymmetry of the anterior cingulate gyrus in female patients with schizophrenia: a volumetric magnetic resonance imaging study. Schizophr Res 55:69–81

Takahashi T, Suzuki M, Kawasaki Y, Kurokawa K, Hagino H, Yamashita I, Zhou SY, Nohara S, Nakamura K, Seto H, Kurachi M (2002b) Volumetric magnetic resonance imaging study of the anterior cingulate gyrus in schizotypal disorder. Eur Arch Psychiatry Clin Neurosci 252: 268–277

Takahashi T, Suzuki M, Zhou SY, Tanino R, Hagino H, Kawasaki Y, Matsui M, Seto H, Kurachi M (2006a) Morphologic alterations of the parcellated superior temporal gyrus in schizophrenia spectrum. Schizophr Res 83:131–143

Takahashi T, Suzuki M, Zhou SY, Tanino R, Hagino H, Niu L, Kawasaki Y, Seto H, Kurachi M (2006b) Temporal lobe gray matter in schizophrenia spectrum: a volumetric MRI study of the fusiform gyrus, parahippocampal gyrus, and middle and inferior temporal gyri. Schizophr Res 87:116–126

Tanskanen P, Veijola JM, Piippo UK, Haapea M, Miettunen JA, Pyhtinen J, Bullmore ET, Jones PB, Isohanni MK (2005) Hippocampus and amygdala volumes in schizophrenia and other psychoses in the Northern Finland 1966 birth cohort. Schizophr Res 75:283–294

Tauscher-Wisniewski S, Tauscher J, Logan J, Christensen BK, Mikulis DJ, Zipursky RB (2002) Caudate volume changes in first episode psychosis parallel the effects of normal aging: a 5-year follow-up study. Schizophr Res 58:185–188

Tauscher-Wisniewski S, Tauscher J, Christensen BK, Mikulis DJ, Zipursky RB (2005) Volumetric MRI measurement of caudate nuclei in antipsychotic-naive patients suffering from a first episode of psychosis. J Psychiatr Res 39:365–370

Thoma RJ, Hanlon FM, Petropoulos H, Miller GA, Moses SN, Smith A, Parks L, Lundy SL, Sanchez NM, Jones A, Huang M, Weisend MP, Cañive JM (2008) Schizophrenia diagnosis and anterior hippocampal volume make separate contributions to sensory gating. Psychophysiology 45:926–935

Velakoulis D, Stuart GW, Wood SJ, Smith DJ, Brewer WJ, Desmond P, Singh B, Copolov D, Pantelis C (2001) Selective bilateral hippocampal volume loss in chronic schizophrenia. Biol Psychiatry 50:531–539

Weiss AP, Dewitt I, Goff D, Ditman T, Heckers S (2005) Anterior and posterior hippocampal volumes in schizophrenia. Schizophr Res 73:103–112

Whitworth AB, Kemmler G, Honeder M, Kremser C, Felber S, Hausmann A, Walch T, Wanko C, Weiss EM, Stuppaeck CH, Fleischhacker WW (2005) Longitudinal volumetric MRI study in first- and multiple-episode male schizophrenia patients. Psychiatry Res 140:225–237

Wible CG, Anderson J, Shenton ME, Kricun A, Hirayasu Y, Tanaka S, Levitt JJ, O'Donnell BF, Kikinis R, Jolesz FA, McCarley RW (2001) Prefrontal cortex, negative symptoms, and schizophrenia: an MRI study. Psychiatry Res 108:65–78

Wiegand LC, Warfield SK, Levitt JJ, Hirayasu Y, Salisbury DF, Heckers S, Dickey CC, Kikinis R, Jolesz FA, McCarley RW, Shenton ME (2004) Prefrontal cortical thickness in first-episode psychosis: a magnetic resonance imaging study. Biol Psychiatry 55:131–140

Wiegand LC, Warfield SK, Levitt JJ, Hirayasu Y, Salisbury DF, Heckers S, Bouix S, Schwartz D, Spencer M, Dickey CC, Kikinis R, Jolesz FA, McCarley RW, Shenton ME (2005) An *in vivo* MRI study of prefrontal cortical complexity in first-episode psychosis. Am J Psychiatry 162:65–70

Yamasue H, Iwanami A, Hirayasu Y, Yamada H, Abe O, Kuroki N, Fukuda R, Tsujii K, Aoki S, Ohtomo K, Kato N, Kasai K (2004) Localized volume reduction in prefrontal, temporolimbic, and paralimbic regions in schizophrenia: an MRI parcellation study. Psychiatry Res 131:195–207

Zetzsche T, Preuss U, Frodl T, Watz D, Schmitt G, Koutsouleris N, Born C, Reiser M, Möller HJ, Meisenzahl EM (2007) *In-vivo* topography of structural alterations of the anterior cingulate in patients with schizophrenia: new findings and comparison with the literature. Schizophr Res 96:34–45

Zhou SY, Suzuki M, Hagino H, Takahashi T, Kawasaki Y, Matsui M, Seto H, Kurachi M (2005) Volumetric analysis of sulci/gyri-defined *in vivo* frontal lobe regions in schizophrenia: precentral gyrus, cingulate gyrus, and prefrontal region. Psychiatry Res 139:127–139

Neurophysiological Measures of Sensory Registration, Stimulus Discrimination, and Selection in Schizophrenia Patients

Anthony J. Rissling and Gregory A. Light

Contents

1 Introduction 284
   1.1 Automatic and Attention Dependent Processes 285
   1.2 Event-Related Potentials 286
   1.3 The Oddball Paradigm 287
   1.4 Basic Processes 289
2 N1 ERP 290
   2.1 N1 Deficits in Schizophrenia 291
   2.2 N1 Stability, Reliability, and Heritability 292
3 MMN ERP 293
   3.1 MMN Deficits in Schizophrenia 294
   3.2 MMN Stability, Reliability, and Heritability 296
4 P300 ERP 297
   4.1 P300 Deficits in Schizophrenia 298
   4.2 P3 Stability, Reliability, and Heritability 299
5 Discussion 300
References 301

Abstract Cortical Neurophysiological event related potentials (ERPs) are multi-dimensional measures of information processing that are well suited to efficiently parse automatic and controlled components of cognition that span the range of deficits exhibited in schizophrenia patients. Components following a stimulus reflect the sequence of neural processes triggered by the stimulus, beginning with early automatic sensory processes and proceeding through controlled decision and response related processes. Previous studies employing ERP paradigms have reported deficits of information processing in schizophrenia across automatic through attention dependent processes including sensory registration (N1), automatic change detection (MMN), the orienting or covert shift of attention towards novel or

---

A.J. Rissling and G.A. Light (✉)
Department of Psychiatry, University of California, San Diego, CA, USA
e-mail: glight@ucsd.edu

infrequent stimuli (P3a), and attentional allocation following successful target detection processes (P3b). These automatic and attention dependent information components are beginning to be recognized as valid targets for intervention in the context of novel treatment development for schizophrenia and related neuropsychiatric disorders. In this review, we describe three extensively studied ERP components (N1, mismatch negativity, P300) that are consistently deficient in schizophrenia patients and may serve as genetic endophenotypes and as quantitative biological markers of response outcome.

Keywords Event Related Potentials (ERPs) · Endophenotypes · Attention · Mismatch Negativity (MMN) · N1 · P300

1 Introduction

Understanding the basic neural processes that underlie complex higher-order cognitive operations and functional domains is a fundamental goal of cognitive neuroscience. Event-related potentials (ERPs) allow investigators to probe sensory, perceptual, and cognitive processing with millisecond precision. This high temporal resolution lends itself to the study of the earliest stages of information processing and the subsequent transitions from sensory-based perceptual processing to the higher cognitive operations that are necessary to successfully navigate through the complex stimulus-laden environments of everyday life. As ERPs are objective indices of human information processing that reliably evaluate core neurophysiological functions that span across a wide spectrum of cognitive operations, they have proven useful for the identification of critical abnormalities that are evident in neuropsychiatric patient populations (Coull 1998; O'Donnell et al. 1999). In schizophrenia, ERP studies have revealed abnormalities across nearly all stages of information processing, including deficits at the most basic levels of sensory registration, automatic sensory discrimination, novelty and target detection, and cognitive resource allocation as well as in higher-order cognitive operations (e.g., language processes Kiang et al. 2007, 2008). These automatic and attention dependent information processes are regarded as key cognitive measures and are beginning to be recognized as valid targets for developing novel treatments of schizophrenia and related neuropsychiatric disorders (e.g., Lavoie et al. 2008). On a macro level, these core functions may relate to important demographic, clinical, neuropsychological, and functional outcome variables (Braff and Light 2004; Braff et al. 1999; Green 1996; Green et al. 2000; Light and Braff 2005a) and as well provide a useful platform for exploring the underlying neural substrates and genetic architecture that supports these functions (Braff and Freedman 2002; Braff and Light 2004; Turetsky et al. 2007) (see Fig. 1). In this review, we will describe three extensively studied ERP components (N1, mismatch negativity or MMN, P300) that are consistently deficient in schizophrenia patients and may serve as endophenotypes and targets for measuring response outcome.

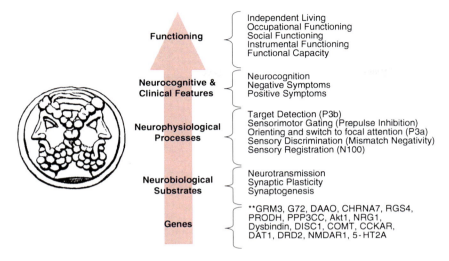

Fig. 1 Multilayered analysis of deficits in schizophrenia. As metaphorically suggested by the two faces of Janus, probes of neurophysiological processes provide investigators a unique vantage point and the ability to look simultaneously in two directions: down to the underlying neural and genetic underpinnings of the automatic and attention dependent processes and up to the associated clinical, neurocognitive, and functional features of the illness

1.1 Automatic and Attention Dependent Processes

Successful processing of sensory inputs requires the ability to screen out or inhibit intrinsic responses to redundant or irrelevant inputs and, reciprocally, to enhance or facilitate responses to deviant, novel, or salient stimuli. Schizophrenia patients evidence deficits in both of these discrete but related processes. The deficits may be due to excess resource allocation to task-irrelevant stimuli (distraction), low resources, or an inability to mobilize and allocate resources to salient stimuli (Braff 1993). ERPs allow investigators to disentangle these interacting but dissociable processes.

Sensory and decision making processes in humans are governed by two qualitatively different types of processes that modulate neural responses to sensory stimuli (Hillyard and Anllo-Vento 1998; Luck et al. 2000; Umbricht et al. 2006). The first, automatic or preattentive processes are responsible for filtering out redundant or irrelevant environmental changes. These automatic sensory processes are presumed to be elicited automatically, and require little or no effort, engagement, or awareness on the part of the subject (Braff and Light 2004; Näätänen 1992). If deviant, novel, or salient changes in the ongoing stream of stimuli occur, a cascade of neural processes is inhibited whereby, the change is automatically detected. Depending on the contextual demands, these physiological processes are then apportioned to facilitate or appropriate responses. Top-down processes bias stimulus selection on the basis of salience or relevance to a goal directed task. While often viewed as discrete or dichotomous, automatic and controlled processes represent a "spectrum" from automatic/preattentive to controlled/attention dependent processes (Braff and Light 2004). There is evidence that at least some higher-order attention dependent processes are "governed" by automatic involuntary sensory processes in a bottom-up manner (Tiitinen et al. 1994). Conversely, some automatic processes can be influenced by top-down attention dependent processes (e.g., Trejo et al. 1995; Woldorff et al. 1991).

1.2 Event-Related Potentials

Electroencephalography (EEG) is a noninvasive and relatively inexpensive method for assessing neurophysiological function. EEG measures the electrical activity of large, synchronously firing populations of neurons in the brain with electrodes placed on the scalp (Light et al. 2010). In an ERP experimental design, a large number of time-locked experimental trials are averaged together. ERPs provide a functional measure of neuroelectric brain activity that occurs time locked to a significant event, reflecting successive stages of information processing (Pfefferbaum et al. 1995) which allow investigators to probe sensory, perceptual, and cognitive processing with millisecond precision (see Fig. 2). As noted above, this high

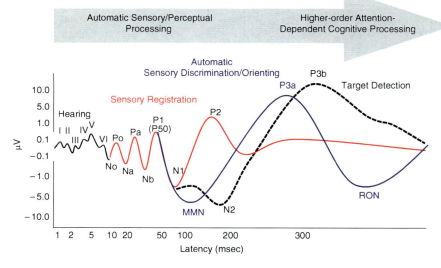

Fig. 2 "Idealized" event-related potential waveforms representing stages of information processing representing shifts from automatic to higher-order controlled processes. These waveforms cannot be elicited from a single paradigm and are thus schematic representations

temporal resolution allows for the study of the earliest stages of information processing and the subsequent transitions from automatic sensory-based perceptual processing to higher-order and integrative cognitive operations. Specifically, the amplitude and latency of the successive peaks can be used to quantify the level or amount of processing resources and the time course of cognitive processing which may vary due to mental state, attentional demand, or distraction. The distribution of voltage over the scalp can be used alone or in conjunction with other imaging techniques to estimate the neuroanatomical loci of these processes.

ERPs have many strengths for psychiatric neuroscience applications, including the exploration of schizophrenia, its treatment, and the contribution of lower-level sensory processes to more complex cognitive, clinical, and functional characteristics. Typically, ERP components are assessed via waveform averaging of ongoing EEG responses to sensory stimuli recorded over a recording session. More recent advances in EEG processing analyses now provide many advantages over traditional ERP averaging, allowing for the decomposition of event-related time frequency dynamics that occur over time that may not be well represented in standard response averages (Makeig et al. 2004a). In addition, independent component analysis (ICA) is a novel approach for empirically defining dissociable response components and can enhance the ability to identify the sources of these responses with substantially greater spatial resolution (Makeig et al. 2002, 2004a, b). ICA allows for the separate filtering and analysis of activities and scalp projections of up to dozens of concurrently active and temporally distinct EEG sources, while retaining all the temporal resolution of traditionally measured composite scalp recordings.

1.3 The Oddball Paradigm

The prototypical experimental paradigm employed to elicit the N1, MMN, and P300 ERP responses has been the oddball task in which a sequence of repetitive standard sounds are presented with a high probability (80–90%) and are interrupted infrequently (5–20%) by physiologically deviant, "oddball" stimuli (e.g., stimuli that differ in duration, pitch, or intensity) (see Fig. 3). Although the N1, MMN, and P300 can be elicited during the oddball task by stimuli in any sensory modality, it is the auditory modality that has been most widely studied in schizophrenia patients and arguably demonstrates the strongest effects (Pfefferbaum et al. 1989). The oddball paradigm has proven a very versatile tool in testing both normal and deficient information processing. This is due to the fact that different ERP components are elicited to the standard and deviant stimuli that can be differentiated by their distinct relationship to the experimental conditions of the oddball paradigm employed including stimulus probability, stimulus onset asynchrony (SOA), and the contextual salience of the stimulus (target or distracter) (Fig. 4). Based on the experimental conditions of the oddball paradigm employed, automatic or attention dependent processes may be tested that result in ERP components elicited, that vary

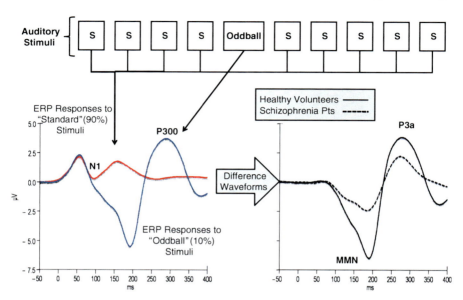

Fig. 3 Auditory "oddball" paradigm. Tones are presented to subjects. "Standard" tones (S) are frequently presented (i.e., usually 80–90%). Occasionally (i.e., ~10%), tones that differ in some physical characteristic such as pitch, duration, or intensity are presented to subjects (referred to as "oddball" tones). EEG responses to the standards and oddball trials are averaged separately. The averaged standard response is then subtracted from the ERP average of the oddball trials to generate a difference wave for assessing MMN/P3a components. P300 subcomponents (both P3a and P3b) are typically measured in ERP response to the oddball trials rather than using difference waveforms

in latency, scalp distribution, and underlying neural networks responsible for component generation.

To measure controlled attention dependent processes during an oddball paradigm, an active task is employed where a participant is instructed to pay attention to the stimulus stream and respond covertly or overtly to one stimulus while ignoring other standard or distracter stimuli. The participant is required to discriminate the infrequent target stimuli from the frequent standard stimuli by noting the occurrence of the target, typically by pressing a button or mentally counting. In contrast to the active oddball task, the passive version does not require a behavioral response from the subject. In this context, the subject's attention is often directed away from the sequence of standard and deviant tones toward another, moderately demanding task, usually in a different modality (e.g., subject is instructed to read, watch a silent video, or even perform a difficult visual continuous performance task).

In the single stimulus task, the target is presented infrequently in time with no other stimuli. In the traditional two-stimulus oddball, an infrequent deviant oddball occurs in a background of frequent standard stimuli. In the three-stimulus variant of the oddball paradigm, an additional infrequent-nontarget stimulus is inserted into a sequence of infrequent target and frequent standard stimuli. The three-stimulus

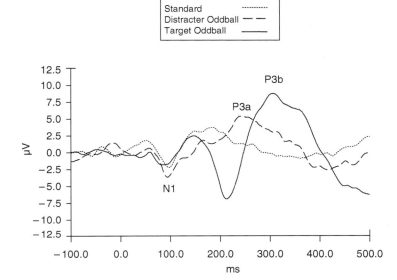

Fig. 4 Grand average ERP components for both a target oddball stimulus and a nontarget distracter from healthy volunteers. The figure depicts the variation in ERP amplitude across the N1, P3a, and P3b components in response to the two stimuli (target and distracter) that vary in contextual salience during a three stimulus oddball paradigm

oddball task allows for the separate testing of the effects of nontarget and distracter stimuli on behavioral performance (during active tasks) and may help to elucidate the source of the neural generated effect (scalp distribution).

1.4 Basic Processes

At least two cerebral mechanisms operate within the auditory modality preattentively, to screen relevant changes in incoming auditory stimuli. One mechanism is based on a neurophysiological reaction to transient increments or decrements in stimulus intensity and has been associated with the auditory changes in the amplitude of N1 (Näätänen and Picton 1987). A second mechanism is responsible for the detection of any stimulus change that does not fit in with a neural trace of an established model of regularity in the acoustic environment. This mechanism of change detection is associated with the generation of the MMN (Näätänen 1990; Näätänen et al. 2007). The N1 and MMN are elicited when the automatic or preattentive system detects stimuli which carry previously unavailable information that may require conscious attention dependent processing. When a significant change in the background stimuli is detected, a transition to further more focal or attention dependent processing occurs. The subcomponents of the P3 reflect the transition to focal attention or orienting to the stimulus (P3a) and the elicitation of attention dependent processes (P3b).

The N1, MMN, and P3 ERP components index different processes that span sensory through decision making processes, with varying neural substrates involved in their elicitation. The three components are not strongly coupled (e.g., P3a can be elicited without concurrent N1 or MMN elicitation; Horváth et al. 2008). Separately they index specific and discrete or dissociable processes and indicate deficits with varying underlying neural substrates and potential bottom-up relationship to clinical variables, such as neuropsychological performance, symptoms, and functional outcome. However, when viewed together from one difference wave comparing an oddball stimulus to the standard stimuli, the breakdown in the flow of processing and the potential effects of distracters may be differentiated (Cortiñas et al. 2008; Escera et al. 2002).

2 N1 ERP

The N1 is a mid-latency auditory ERP response arising primarily from the auditory cortex, approximately 100 ms after the presentation of an auditory stimulus. The N1 reflects the early phase of stimulus registration and processing (Boutros et al. 2004; McCarley et al. 1991; Parasuraman and Beatty 1980). The N1 may be elicited by abrupt changes in spectral sound intensity (Näätänen and Picton 1987) from background whether in or out of attentional focus and therefore shares features of both automatic and attention dependent components. N1 is sensitive to the physical characteristics of the stimulus (e.g., duration, intensity, rise time) but less sensitive to contextual effects than later evoked potentials.

N1 sums activation from several brain areas, including the auditory cortex (Levänen et al. 1993; Vaughan and Ritter 1970). Therefore, decreases in N1 amplitude indicate impairment in early auditory sensory processing localized to primary and secondary auditory cortex. Although a large part of the N1 response is not stimulus specific (Näätänen 1990), the location and parameters of some N1 subcomponents are sensitive to primary acoustic features, such as frequency and intensity (Roberts and Poeppel 1996).

During the oddball paradigm, the frequently presented standard stimuli elicit a small N1 amplitude, whereas a passive or distracter oddball stimulus or active "target" oddball stimulus will often produce a larger amplitude N1 response. According to Näätänen's (1990) influential model, the N1 may reflect a process involved in redirecting the focus of attention to the onset of new especially salient stimuli. This suggestion is supported by evidence that large N1 amplitude responses are followed by (1) a large P3a amplitude response to passive or distracter oddball stimuli indicating the initiation of an orienting response and (2) large P3b amplitude responses to target oddball stimuli indicating a shift from automatic to attention dependent processes.

2.1 N1 Deficits in Schizophrenia

Reduced amplitude of the auditory N100 evoked potential is a robust physiological abnormality in schizophrenia (Laurent et al. 1999; Strik et al. 1992; Turetsky et al. 2008). Because the N1 may be elicited by an abrupt change in sound energy from background, several approaches have been employed to test N1 deficits in schizophrenia including presenting tones at varying interstimulus intervals (Ebmeier 1991; Roth et al. 1980) as well as the oddball and S1–S2 paradigms in both humans and rodents (Connolly et al. 2004; Maxwell et al. 2004).

Previous studies have focused mainly on the demonstrated N1 amplitude deficit to the frequent standard tones in patients with schizophrenia compared to healthy controls during an oddball paradigm (Iwanami et al. 1994; Pfefferbaum et al. 1989; Roth et al. 1991). The amplitude deficit in the N1 during the standard stimulus suggests a deficit in early sensory processing of auditory stimuli or potentially lower resources allocated to the processing of basic change in auditory stimuli from background. One advantage in testing the N1 during the oddball paradigm is the ability to test the subsequent transition to more focal or attention dependent processes reflected by the P3a or P3b (Fig. 4) and the high signal to noise ratio components that are generated in response to hundreds or even thousands of stimuli. A growing literature in support of the decreased N1 amplitude deficit in schizophrenia exists from studies testing the N1 ERP during a task originally developed to test P50 ERP suppression. During the basic S1–S2 paired-click paradigm, the first of a pair of stimuli (S1) is followed shortly (usually 500 ms) by an identical stimulus (S2) (Fig. 5). During the task, the subject is administered a pair of brief

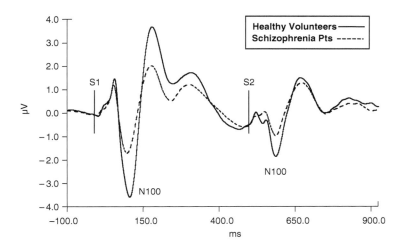

Fig. 5 Mean N100 amplitude of schizophrenia patients and healthy volunteers during a basic S1–S2 paired-click paradigm designed to test P50 and N100 amplitude suppression

auditory stimuli (clicks) and brain evoked responses are recorded. The click stimuli elicit a distinct complex of ERP components including P50 and N1.When auditory click stimuli are presented in pairs with inter-click interval of less than 1 s, the P50 and N1 responses to the second of the two identical clicks are suppressed, or gated out. According to the sensory gating hypothesis, the first stimulus elicits the initial excitatory response of the neuronal population giving rise to the P50 response and also activates inhibitory pathways attenuating the response to the second stimulus (Adler et al. 1999; Freedman et al. 1983).

Because of several factors including an increased signal to noise ratio and the ease of measurement during the same S1–S2 paradigm, several studies have tested the potential deficits in N1 amplitude and N1 amplitude suppression in schizophrenia and other disorders. In support of our discussion of deficits in the N1, several studies have reported decrements in N1 ERP amplitude to the S1 in schizophrenia patients when compared to controls (Boutros et al. 2004; Clementz and Blumenfeld 2001; Nagamoto et al. 1989). This paradigm and homologous N1 ERP responses (i.e., N40) have also been examined in rats (Swerdlow et al. 2005) and mice (Maxwell et al. 2004, 2006).

2.2 N1 Stability, Reliability, and Heritability

The N1 ERP amplitude response deficit meets several criteria for a genetic endophenotype for schizophrenia. First, the N1 ERP amplitude response is reliable. Rentzsch et al. (2008) testing the reliability of N100 amplitude in 41 healthy subjects in two sessions, 4 weeks apart, using intraclass correlation reported good reliability for the S1 (0.71) and S2 (0.34) during the S1–S2 paradigm. Second, reduced N1 amplitude is a stable deficit in schizophrenia. It is reported in both recent onset (Sumich et al. 2006) and medication-free patients (Ogura et al. 1991) and persists after clinical stabilization and medication withdrawal (Laurent et al. 1999). Finally, data suggests reduced N1 amplitude is a heritable vulnerability factor and that the magnitude of the decrement correlates with the degree of genetic risk. Ahveninen et al. (2006) reported reduced N1 amplitude in both schizophrenia patients and their unaffected co-twins, relative to controls. Moreover, the magnitude of this amplitude decrement correlated with the degree of genetic resemblance to the schizophrenia patients. Turetsky et al. (2007) reported heritability estimates of 0.40 and 0.29 for N1 amplitudes during the S1 and S2 click paradigm in schizophrenia patients and their unaffected family members. However, contrary evidence has also been reported. Four previous studies that contrasted first-degree relatives of schizophrenia patients with healthy comparison subjects failed to find any significant differences in N1 amplitude (Frangou et al. 1997; Karoumi et al. 2000; Waldo et al. 1988; Winterer et al. 2001) when the S1–S2 paradigm was employed.

3 MMN ERP

Following the sensory registration exhibited by the elicitation of the N1, the MMN probes the next stage of stimulus processing during the oddball paradigm. Measurement of MMN is carried out on the difference waveform constructed by subtracting the auditory evoked potential response to the standard tone from that of the deviant during an oddball paradigm. The MMN appears as a prominent negative potential on this difference waveform. The MMN occurs rapidly beginning 50 ms following the onset of the oddball and peak amplitude after an additional 100–150 ms. Auditory MMN latency is typically between 150 and 200 ms poststimulus, with maximal MMN amplitude responses evident at frontocentral scalp recording sites.

The MMN amplitude has been shown to increase while the peak latency decreases with increasing deviance in tone frequency between the standard and the deviant sounds (Amenedo and Escera 2000; Berti et al. 2004; Näätänen et al. 1982; Pakarinen et al. 2007; Tiitinen et al. 1994), which suggests that the amplitude of MMN is a function of the magnitude of the physical difference between the standard and deviant auditory stimuli.

MMN may be regarded as a probe of sensory filtering/discrimination mechanisms that become activated only when a sound carries new information that may require further focal or attention dependent processing (see Näätänen and Alho 1997 for reviews; Näätänen and Winkler 1999). In support of this view, MMN is elicited when the standard repeating stimulus in an oddball paradigm is interrupted infrequently by an oddball deviant (e.g., pitch, intensity, duration) stimulus. It has been shown that the process generating the MMN is initiated by a mismatch between the incoming sound and a memory record representing the regularities of the immediate history of auditory stimulation (Näätänen 1992; Winkler and Czigler 1998; Winkler et al. 1996). It has been suggested to reflect deviation of the incoming stimulus from the memory representation of the standard sound (Näätänen 1990) or from the sound input predicted by a neural model encoding the detected regularities (Winkler et al. 1996). Physiologically, the MMN is the first measurable brain response component that differentiates between usual (standard) and unusual (deviant) auditory stimuli (Näätänen et al. 1989b). This is supported by the observation that no MMN is elicited by the repetitive standard stimulus (Näätänen et al. 1989a).

The MMN may also be elicited during active oddball tasks with long SOAs employed in order to elicit the N1 and P300 ERPs (Fig. 2). However, the prototypical paradigm used for MMN elicitation utilizes a passive oddball task with a brief interstimulus interval (e.g., 300–500 ms) (Fig. 4). The relatively rapid stimulus presentation ensures that the echoic memory trace of the preceding stimulus is still active when the subsequent stimulus is presented. The mismatch response appears to reflect a predominantly automatic and preattentive process that is not under direct subject control: MMN elicitation requires no behavioral response from subjects; participants do not need to be engaged in a cognitive task; and it can be elicited while subjects perform other mental activities in parallel without apparent interaction or interference (Alho 1992; Rinne et al. 2001; Sussman et al. 2003). Well defined MMN waveforms can be obtained from fetuses using magnetoencephalography (Draganova et al. 2005), as well as from sleeping infants (Huotilainen et al. 2003) and adults (Nashida et al. 2000), patients with severe brain injuries (Kaipio et al. 2001), and even comatose individuals who ultimately regain consciousness (Fischer et al. 2000; Kane et al. 1996; Morlet et al. 2000). While some studies suggest that even preattentive processes can be influenced by top-down factors under specific circumstances for elicitation (Trejo et al. 1995; Woldorff et al. 1991), other studies do not find MMN to be substantially affected by task-related top-down processes (Alain and Woods 1997; Ritter et al. 1999). The automatic MMN process may have an important role in initiating involuntary switching of attention to an auditory stimulus change occurring outside the focus of attention (Giard et al. 1990; Lyytinen et al. 1992; Näätänen 1990, 1992; Näätänen and Michie 1979) as evidenced by a P3a component that usually follows the MMN response, indicating a sign of attention switching (Squires et al. 1975).

Magnetoencephalographic studies and dipole modeling of surface ERP data strongly indicate that the auditory MMN is generated within the primary and secondary auditory cortices (Alho et al. 1996; Rinne et al. 2000; Waberski et al. 2001) and gets a contribution from at least two intracranial processes: (1) a bilateral supratemporal process generating the supratemporal MMN subcomponent and the polarity-reversed "MMN" in nose-referenced mastoid recordings (Giard et al. 1990) and (2) a predominantly right-hemispheric frontal process, generating the frontal MMN subcomponent (Giard et al. 1990; Näätänen et al. 1978; Rinne et al. 2000). The supratemporal component is presumed associated with automatic change detection, whereas the frontal component appears to be related to an attention switch (Escera et al. 1998; Giard et al. 1990; Näätänen et al. 1978; Näätänen and Michie 1979; Rinne et al. 2000; Schröger 1997) and the subsequent elicitation of the P3a response.

3.1 MMN Deficits in Schizophrenia

Deficits in MMN generation represent a remarkably robust finding in schizophrenia (Fig. 6). Several published reports of reduced MMN in schizophrenia patients have been reported utilizing various stimulation parameters (e.g., pitch, duration, and intensity stimulus manipulations) and conditions (Alain et al. 1998; Hirayasu et al. 1998a; Javitt 2000; Javitt et al. 2000a; Jessen et al. 2001; Kiang et al. 2009; Michie et al. 2000; O'Donnell et al. 1994; Oades et al. 1997; Schall et al. 1999; Todd et al. 2000; Umbricht et al. 2003). In a recent meta-analysis performed by Umbricht and Krljes (2005), the mean effect size for the schizophrenia deficit was ~ 1.0 – a large deficit according to common standards. As with other cognitive measures, MMN and P3a reductions also occur with increased age (Kiang et al. 2009). Importantly, our recent study of 250 schizophrenia patients and 150 nonpsychiatric comparison

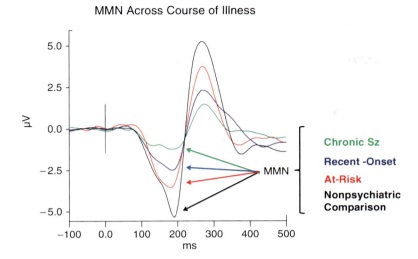

Fig. 6 Mismatch negativity and P3a amplitude from four groups: individuals at risk for schizophrenia, recent onset schizophrenia patients, chronic schizophrenia patients, and healthy volunteers. Note that although the data is cross-sectional, the MMN and P3a amplitudes decrease with the progression of the disease

subjects demonstrated that schizophrenia patients still exhibit approximately 1.0 SD effect size deficit across adult age ranges. In contrast to most other physiological indices, MMN deficits appear to be relatively specific to schizophrenia. Bipolar, major depressive (Umbricht and Krljes 2005), and obsessive-compulsive disorder patients (Oades et al. 1997) all have normal MMNs, though there are reports of MMN deficits among chronic alcoholics (van der Stelt et al. 1997).

Several studies have reported, in schizophrenia patients, deficits in MMN to be highly associated with patients' impairments in daily functioning, level of independence in their community living situation, and functional outcome (Light and Braff 2005a, b). Across the studies, schizophrenia patients with more severe functional impairments had relatively smaller (i.e., less negative) MMN amplitudes than higher functioning patients. Light and Braff (2005a) tested schizophrenia patients and healthy nonpsychiatric comparison subjects in an MMN ERP paradigm to determine if a schizophrenia-linked deficit in MMN is associated with impairments in everyday functional status, level of independence in living situation, and the ability to perform tasks routinely encountered in everyday situations. The authors reported that (1) schizophrenia patients had significantly reduced MMN, (2) greater levels of MMN impairment were associated with lower Global Assessment of Functioning Scale ratings, and (3) patients with greater MMN impairments were more likely to live in highly structured versus independent settings. In a follow-up study, Light and Braff (2005b) confirmed that the MMN amplitude in schizophrenia patients was highly stable (ICCs = 0.90) over a 15 month test–retest interval and that the MMN/function relationship was also very stable (rs = -0.63 to rs = -0.68).

These findings of a relationship between MMN and indicators of functional status have been recently replicated and extended by others (Rasser et al. 2009; Kawakubo et al. 2007; Wynn et al. 2010). Overall, the data suggest an important bottom-up relationship of the automatic sensory-based cognitive processes that elicit the MMN response to functional outcome.

3.2 MMN Stability, Reliability, and Heritability

The MMN ERP amplitude response meets several criteria for a genetic endophenotype in schizophrenia. First, there is substantial evidence to indicate that MMN has good test–retest reliability (Pekkonen et al. 1995). In a study of 15 healthy individuals tested on two separate occasions 1–27 days apart, Tervaniemi et al. (1999) examined the reliability of MMN for deviant stimuli that varied in duration, pitch, or intensity. Reliability was greatest for the duration deviant ($r = 0.78$) and lowest for the pitch deviant ($r = 0.53$). Kathmann et al. (1999) reported very similar estimates from a study of 45 subjects tested 2–4 weeks apart. Test–retest reliability, in this case, was > 0.8 for a duration deviant and ∼0.5 for a pitch deviant. Escera et al. (2000) observed reliabilities of 0.72 and 0.80 for a duration-deviant MMN, depending upon whether the peak or the mean within a defined time interval was used to measure MMN amplitude. Kujala et al. (2001) noted test–retest correlations of 0.60–0.75, depending on the degree of deviance of the infrequent stimulus. Only one small study ($n = 14$) (Joutsiniemi et al. 1998) reported correlation coefficients that were described as unacceptable. As noted above, the one study reporting the results of repeat testing in schizophrenia patients over a 15 month interval (Light and Braff 2005b) found intraclass correlations of ∼0.90 after 1 year.

Second, MMN deficits in schizophrenia show evidence of heritability. Clinically unaffected family members of schizophrenia patients (Jessen et al. 2001; Michie et al. 2002), children at risk for developing schizophrenia (Bar-Haim et al. 2003; Schreiber et al. 1992), and recent onset patients (Javitt et al. 2000b; Umbricht et al. 2006) have all been reported to have reduced MMN amplitudes. However, there have also been reports of normal MMNs in unaffected family members (Bramon et al. 2004). Moreover, in contrast to the virtually universal finding of abnormal MMNs among chronic schizophrenia patients, MMNs have been reported to be normal in first-episode patients (Salisbury et al. 2002; Umbricht et al. 2006). Salisbury et al. (2002) reported specifically that the pitch-deviant MMN reductions reported present in patients with chronic schizophrenia are not present at a first hospitalization. They suggested that the sensory and memory functions indexed by MMN seem unaffected early in the schizophrenia disease process, and that reductions in MMN amplitude may develop over time and index the progression of the disorder.

Studies conducted in our laboratory and in collaboration with our colleagues Drs. Jahchan, Cadenhead, Malaguti, and Braff using duration-deviant stimuli for MMN elicitation suggest that MMN reductions may be less early in the course of illness or perhaps even in individuals who are at risk for developing schizophrenia. Figure 6 depicts cross-sectional duration MMN and P3a amplitude in individuals at risk for schizophrenia, recent onset schizophrenia patients, chronic schizophrenia patients, and healthy volunteers. Clearly, additional studies are necessary to delineate the nature of the MMN abnormality, its prevalence among prodromal and/or first-episode schizophrenia subjects, and its utility for predicting conversion to psychosis in individuals at genetically high risk for developing the disorder (Brockhaus-Dumke et al. 2005).

4 P300 ERP

When a significant change in the background stimuli is detected following the N1 and MMN, a transition to further processing exhibited by elicitation of the P300 (also referred to as "P3") occurs. The P3 is a change detection index that reflects a variety of cognitive processes including attention and memory. There is general consensus that P3 is not a unitary brain potential but represents the summation of activity from various widely distributed areas in the brain. However, a distinction can be made between the automatic and passively elicited P3a and the attention dependent P3b as they may be evoked by different stimulus or task contexts, have distinct topographic amplitude distributions, and result from functionally distinct but overlapping neural activation patterns (Gaeta et al. 2003; Polich and Criado 2006; Spencer et al. 2001; Turetsky et al. 1998a).

The P3a subcomponent is thought to reflect an orienting response and the covert shifting of focal attention that occurs when novel or infrequent oddball stimuli are presented. Passive oddball stimuli or distracters in a three-stimulus oddball paradigm elicit a P3a with a central/parietal scalp distribution and a short peak latency (Katayama and Polich 1998) (Fig. 4). However, novel nontarget "distracter" stimuli that are presented with target and standard stimuli during an active oddball paradigm will elicit a "novelty P3" that has a frontal/central maximum amplitude distribution and short peak latency and that habituates rapidly (Almasy and Blangero 2001; Iacono et al. 2000). As nontarget distracter stimuli do not require an overt response, the topographic scalp distribution differences among the types of P3a potentials reflect stimulus driven attributes that can be manipulated by discrimination task demands.

The P3b is an attention dependent response that is elicited to correctly identified target oddball stimuli during an active oddball paradigm (Fig. 4) with a latency of approximately 300–400 ms and a parietal scalp distribution. Most P3a and P3b tasks have employed easy stimulus discrimination tasks to facilitate target detection. In each case, the target elicits a large positive going potential that increases in amplitude from the frontal to parietal electrodes and has a peak latency of about 300 ms for auditory stimuli (Johnson 1993; Mertens and Polich 1997). When the perceptual discrimination between the target and standard stimulus is difficult, increased frontal/central amplitude for the nontarget distracter P3a and a parietal maximum for the target P3b are obtained.

The P3b is thought to reflect contextual updating and memory processes as well as target stimulus classification (Donchin and Coles 1988), its amplitude reflecting the allocation of and amount of attentional resources (Kok 2001; Wronka et al. 2007), and its peak latency related to stimulus evaluation time (Kutas et al. 1977). Intracranial electrophysiological monitoring and fMRI studies have similarly discerned multiple sources of P300 ERP activity, including the hippocampus, thalamus inferior parietal lobe, superior temporal gyrus, and frontal lobe (e.g., Kiehl et al. 2001; Smith et al. 1990). Intracortical recordings indicate that the P3a component is generated in a widespread area including the dorsolateral prefrontal cortex, the anterior cingulate lobe as well as the posterior cingulate lobe, the supramarginal gyrus, and several parts of the temporal lobe. Depth-recordings and lesion studies in humans and monkeys have found evidence for multiple generators of the P3b in the mediotemporal lobe, the inferior temporal lobe, the midtemporal lobe, superior temporal gyrus, the frontal lobe, and the parietal cortex (Smith et al. 1990). Convergent evidence from ERP source localization and fMRI activation studies suggests that P3b scalp activity arises primarily from the inferior parietal cortex, particularly the supramarginal gyrus, while the P3a reflects the activity of lateral prefrontal and superior temporal areas (Linden 2005).

4.1 P300 Deficits in Schizophrenia

Chronic schizophrenia patients show a reduction of P300 amplitude of approximately one standard deviation relative to healthy subjects (Price et al. 2006) and include P3a as well as P3b components (Hall et al. 2006; Iacono et al. 2000). The P3a has not been studied as extensively in schizophrenia patients as the P3b. Several studies, however, have found its amplitude to be decreased in schizophrenia patients, in response to infrequent nontarget or distracter stimuli during a target detection task (Grillon et al. 1990; Turetsky et al. 1998a) and following the MMN in the context of a passive auditory paradigm (Fig. 6).

Reduction of the auditory P3b amplitude is one of the most robust biological findings in schizophrenia. Most previous studies have demonstrated a reduction in the P3b amplitude in schizophrenic patients compared with those of normal controls (Braff 1993; Duncan 1988; Michie et al. 1990; Turetsky et al. 1998a). P300 latency deficits are not as robust as the reported amplitude differences. Longer P300 latencies in schizophrenia patients have been observed in some studies (Blackwood et al. 1987; O'Donnell et al. 1995; Pfefferbaum et al. 1989; Romani et al. 1987; Souza et al. 1995) while other studies have reported no differences (Pfefferbaum 1984). Several comprehensive reviews on P300 in the assessment of schizophrenia are available elsewhere (Duncan 1988; Ford et al. 1992; McCarley et al. 1991).

4.2 P3 Stability, Reliability, and Heritability

The P3b ERP amplitude response meets several criteria for a genetic endophenotype in schizophrenia. First, measures of P3b amplitude in both schizophrenia and nonpsychiatric samples exhibit very good test–retest reliability (Kileny and Kripal 1987; Kinoshita et al. 1996; Mathalon et al. 2000; Pollock and Schneider 1992; Segalowitz and Barnes 1993; Sinha et al. 1992). P3 studies of healthy control subjects with repeated oddball testing produce good test–retest correlation coefficients for both amplitude (0.50–0.91) and latency (0.40–0.84) measures (Karniski and Blair 1989; Kileny and Kripal 1987; Sandman and Patterson 2000; Segalowitz and Barnes 1993). In a study of both schizophrenia patients and controls tested twice over the course of 1–3 years, Turetsky et al. (1998b) reported reliabilities of 0.86 for controls and 0.61 for patients.

The P3b amplitude deficit in schizophrenia reflects a stable trait factor but shows state influences. Auditory P3b amplitude reduction has been shown to be a stable biological feature of schizophrenia that is relatively insensitive to clinical status or medication after the onset of the illness. P3b deficits persist after typical or atypical antipsychotic medication treatment and in patients withdrawn from medication (Faux et al. 1993; Pfefferbaum et al. 1989), in drug-naive schizophrenia patients (Hirayasu et al. 1998a) after symptoms improve (Blackwood et al. 1987; Coburn et al. 1998; Pass et al. 1980; Turetsky et al. 1998b) or are largely remitted (Rao et al. 1995), as well as in first-episode patients (Hirayasu et al. 1998b), which suggests trait characteristics of P3 amplitude deficits.

However, the P3b amplitude reduction in schizophrenia has been reported to be associated with both negative (Egan et al. 1994; Hirayasu et al. 1998a; Pfefferbaum et al. 1989; Strik et al. 1993) and positive (Egan et al. 1994; Iwanami et al. 2000; Kawasaki et al. 1997; McCarley et al. 1989) symptom severity suggesting a state marker of the disease. The association to both negative and positive symptoms may be due to the dependence on attentional processes for the elicitation of the P3b.

The P3b amplitude reduction is not exclusive to schizophrenia. It has been reported in other psychiatric disorders, such as depression and alcoholism (Blackwood et al. 1987; Pfefferbaum 1984).

Heritability of P3b amplitude is high (70–80%) (Hall et al. 2006), suggesting that P300 may serve as a risk or trait marker of the genetic risk for schizophrenia (Winterer et al. 2001). Several studies have reported associations of P3b abnormalities with genetic variations that are thought to contribute to this genetic risk (Blackwood and Muir 2004; Gallinat et al. 2003). Similar, though less severe P3b amplitude reductions have been reported in unaffected family members of schizophrenia patients including discordant and concordant twins (Blackwood et al. 1991; Frangou et al. 1997; Karoumi et al. 2000; Saitoh et al. 1984; Schreiber et al. 1992; Turetsky et al. 2000; Weisbrod et al. 1999) and healthy siblings (Blackwood et al. 1991; Frangou et al. 1997; Kidogami et al. 1992; Roxborough et al. 1993; Saitoh et al. 1984). These data suggest that P3b amplitude reduction may be a marker of genetic vulnerability to schizophrenia. However, other studies did not replicate this finding (Friedman et al. 1988; Karoumi et al. 2000).

5 Discussion

Schizophrenia is considered to be a heritable disorder with a complex genetic architecture interacting with environmental factors. With the goal of reducing the genetic complexity of schizophrenia, the field has increasingly turned to the study of biological or behavioral traits as intermediate phenotypes termed "endophenotypes". These endophenotypes are commonly defined as laboratory-based measures that assess relatively simple and quantifiable biobehavioral characteristics that segregate with the illness. As these characteristics tend to be simpler than the complex phenotype of schizophrenia, the hope is that endophenotypes will sharpen the search for primary susceptibility genes. It is generally agreed that an ideal neurophysiological endophenotype should (1) exhibit heritability by indicating a robust deficit in both patients and unaffected family members, (2) be stable across symptom state and reflect a trait characteristic of the disorder, (3) be easily and rapidly measured with minimal subject demands, (4) demonstrate excellent test–retest reliability, and (5) reflect a discrete neurobiological mechanism that is informative for the pathophysiology of the disorder (Bearden and Freimer 2006; Gottesman and Gould 2003; Turetsky et al. 2007; Braff et al. 2008; Braff and Light 2005).

Decreased N1, MMN, and P300 amplitudes are heritable and robust deficits present in both schizophrenia patients and their clinically unaffected family members. The N1 and MMN are stable across symptom state while the P300 indicates both trait and state characteristics that may be due to the attention dependent properties required for its elicitation. The three ERPs are easily and rapidly measured with minimal subject demands and demonstrate excellent test–retest reliability. N1, MMN, and P300 are good candidates for genetic association studies aimed at the identification of functionally relevant susceptibility genes.

With the employment of the oddball paradigm, early perceptual, stimulus discrimination, and target detection processes may be probed by the elicitation and measurement of the N1, MMN, and P300 ERP components. These three components span the spectrum from automatic to controlled attention dependent processes and are beginning to be recognized as valid targets for testing novel treatments of schizophrenia and related disorders.

These measures also demonstrate an intriguing relationship of basic sensory processing with higher-order measures of learning and memory and psychosocial functioning in healthy individuals. It is conceivable that the neural substrates that regulate automatic monitoring and detection of environmental changes trigger a cascade of higher-order processing. The outcome of this cascade might be a determination of whether environmental cues are salient and in need of further processing vs. trivial, allowing for suppression and inattention in accordance with classic information processing theories (Johnston and Heinz 1978; Norman 1968; Treisman 1960). Efficiency at such elementary neurophysiological levels can free up attention dependent, controlled cognitive resources for the successful encoding, retrieval, and discrimination of task-relevant information which in turn facilitates the iterative and responsive processing necessary for adaptive cognitive and social functioning (Näätänen et al. 2001).

References

Adler L, Freedman R, Ross R, Waldo M (1999) Elementary phenotypes in the neurobiological and genetic study of schizophrenia. Biol Psychiatry 4:8–18
Ahveninen J, Jääskeläinen I, Osipova D, Huttunen M, Ilmoniemi RJ, Kaprio J, Lönnqvist J, Manninen M, Pakarinen S, Therman S, Näätänen R, Cannon TD (2006) Inherited auditory-cortical dysfunction in twin pairs discordant for schizophrenia. Biol Psychiatry 60:612–620
Alain C, Woods D (1997) Attention modulates auditory pattern memory as indexed by event-related brain potentials. Psychophysiology 34:534–546
Alain C, Hargrave R, Woods D (1998) Processing of auditory stimuli during visual attention in patients with schizophrenia. Biol Psychiatry 44:1151–1159
Alho K (1992) Selective attention in auditory processing as reflected by event-related brain potentials. Psychophysiology 29:247–263
Alho K, Tervaniemi M, Huotilainen M, Lavikainen J, Tiitinen H, Ilmoniemi R, Knuutila J, Näätänen R (1996) Processing of complex sounds in the human auditory cortex as revealed by magnetic brain responses. Psychophysiology 33:369–375
Almasy L, Blangero J (2001) Endophenotypes as quantitative risk factors for psychiatric disease: rationale and study design. Am J Med Genet 105:42–44
Amenedo E, Escera C (2000) The accuracy of sound duration representation in the human brain determines the accuracy of behavioural perception. Eur J Neurosci 12:2570–2574
Bar-Haim Y, Marshall P, Fox N, Schorr E, Gordon-Salant S (2003) Mismatch negativity in socially withdrawn children. Biol Psychiatry 54:17–24
Bearden CE, Freimer NB (2006) Endophenotypes for psychiatric disorders: ready for primetime? Trends Genet 22:306–313
Berti S, Roeber U, Schröger E (2004) Bottom-up influences on working memory: behavioral and electrophysiological distraction varies with distractor strength. Exp Psychol 51:249–257
Blackwood D, Muir W (2004) Clinical phenotypes associated with DISC1, a candidate gene for schizophrenia. Neurotox Res 6:35–41
Blackwood D, Whalley L, Christie J, Blackburn I, St Clair D, McInnes A (1987) Changes in auditory P3 event-related potential in schizophrenia and depression. Br J Psychiatry 150:154–160
Blackwood D, St Clair D, Muir W, Duffy J (1991) Auditory P300 and eye tracking dysfunction in schizophrenic pedigrees. Arch Gen Psychiatry 48:899–909
Boutros N, Korzyukov O, Jansen B, Feingold A, Bell M (2004) Sensory gating deficits during the mid-latency phase of information processing in medicated schizophrenia patients. Psychiatry Res 126:203–215
Braff D (1993) Information processing and attention dysfunctions in schizophrenia. Schizophr Bull 19:233–259
Braff D, Freedman R (2002) Endophenotypes in studies of the genetics of schizophrenia. In: Davis KL, Charney DS, Coyle JT, Nemeroff C (eds) Neuropsychopharmacology: the fifth generation of progress. Linppincott Williams & Wilkins, Philadelphia, pp 703–716

Braff D, Light G (2004) Preattentional and attentional cognitive deficits as targets for treating schizophrenia. Psychopharmacology (Berl) 174:75–85

Braff D, Light G (2005) The use of neurophysiological endophenotypes to understand the genetic basis of schizophrenia. Dialogues Clin Neurosci 7:125–135

Braff D, Swerdlow N, Geyer M (1999) Symptom correlates of prepulse inhibition deficits in male schizophrenic patients. Am J Psychiatry 156:596–602

Braff D, Swerdlow NR, Light GA, Greenwood TA, Schork NJ, the COGS investigators (2008) Advances in endophenotyping schizophrenia: the consortium on the genetics of schizophrenia (COGS). World Psychiatry 7:11–18

Bramon E, Croft R, McDonald C, Virdi GK, Gruzelier JG, Baldeweg T, Sham PC, Frangou S, Murray RM (2004) Mismatch negativity in schizophrenia: a family study. Schizophr Res 67:1–10

Brockhaus-Dumke A, Tendolkar I, Pukrop R, Schultze-Lutter F, Klosterkotter J, Ruhrmann S (2005) Impaired mismatch negativity generation in prodromal subjects and patients with schizophrenia. Schizophr Res 73:297–310

Clementz BA, Blumenfeld LD (2001) Multichannel electroencephalographic assessment of auditory evoked response suppression in schizophrenia. Exp Brain Res 139:377–390

Coburn K, Shillcutt S, Tucker K, Estes K, Brin F, Merai P, Moore N (1998) P300 delay and attenuation in schizophrenia: reversal by neuroleptic medication. Biol Psychiatry 44:466–474

Connolly P, Maxwell C, Liang Y, Kahn J, Kanes S, Abel T, Gur R, Turetsky B, Siegel S (2004) The effects of ketamine vary among inbred mouse strains and mimic schizophrenia for the P80, but not P20 or N40 auditory ERP components. Neurochem Res 29:1179–1188

Cortiñas M, Corral M, Garrido G, Garolera M, Pajares M, Escera C (2008) Reduced novelty-P3 associated with increased behavioral distractibility in schizophrenia. Biol Psychol 78:253–260

Coull J (1998) Neural correlates of attention and arousal: insights from electrophysiology, functional neuroimaging and psychopharmacology. Prog Neurobiol 55:343–361

Donchin E, Coles M (1988) Is the P300 component a manifestation of context updating? Behav Brain Sci 11:357–374

Draganova R, Eswaran H, Murphy P, Huotilainen M, Lowery C, Preissl H (2005) Sound frequency change detection in fetuses and newborns, a magnetoencephalographic study. Neuroimage 28:354–361

Duncan CC (1988) Event-related brain potentials: a window on information processing in schizophrenia. Schizophr Bull 14:199–203

Ebmeier K (1991) Effect sizes of P300 in schizophrenics. Biol Psychiatry 29:1159–1160

Egan MF, Duncan CC, Suddath RL, Kirch DG, Mirsky AF, Wyatt RJ (1994) Event-related potential abnormalities correlate with structural brain alterations and clinical features in patients with chronic schizophrenia. Schizophr Res 11:259–271

Escera C, Alho K, Winkler I, Näätänen R (1998) Neural mechanisms of involuntary attention to acoustic novelty and change. J Cogn Neurosci 10:590–604

Escera C, Yago E, Polo M, Grau C (2000) The individual replicability of mismatch negativity at short and long interstimulus intervals. Clin Neurophysiol 111:546–551

Escera C, Corral M, Yago E (2002) An electrophysiological and behavioral investigation of involuntary attention towards auditory frequency, duration and intensity changes. Cogn Brain Res 14:325–332

Faux S, Shenton M, McCarley R, Nestor P, Marcy B, Ludwig A (1990) Preservation of P300 event-related potential topographic asymmetries in schizophrenia with use of either linked-ear or nose reference sites. Electroencephalogr Clin Neurophysiol 75:378–391

Faux SF, McCarley RW, Nestor PG, Shenton ME, Pollak SD, Penhune V, Mondrow E, Marcy B, Peterson A, Horvath T et al (1993) P300 topographic asymmetries are present in unmedicated schizophrenics. Electroencephalogr Clin Neurophysiol 88(1):32–41

Fischer C, Morlet D, Giard M (2000) Mismatch negativity and N100 in comatose patients. Audiol Neurootol 5:192–197

Ford J, Pfefferbaum A, Roth W (1992) P3 and schizophrenia. Ann N Y Acad Sci 658:146–162

Frangou S, Sharma T, Alarcon G, Sigmudsson T, Takei N, Binnie C, Murray RM (1997) The Maudsley family study, II: Endogenous event-related potentials in familial schizophrenia. Schizophr Res 23:45–53

Freedman R, Adler LE, Waldo MC, Pachtman E, Franks RD (1983) Neurophysiological evidence for a defect in inhibitory pathways in schizophrenia: comparison of medicated and drug-free patients. Biol Psychiatry 18:537–551

Friedman D, Cornblatt B, Vaughan H, Erlenmeyer-Kimling L (1988) Auditory event-related potentials in children at risk for schizophrenia: the complete initial sample. Psychiatry Res 26:203–221

Gaeta H, Friedman D, Hunt G (2003) Stimulus characteristics and task category dissociate the anterior and posterior aspects of novelty P3. Psychophysiology 40:198–208

Gallinat J, Senkowski D, Wernicke C, Juckel G, Becker I, Sander T, Smolka M, Hegerl U, Rommelspacher H, Winterer G (2003) Allelic variants of the functional promoter polymorphism of the human serotonin transporter gene is associated with auditory cortical stimulus processing. Neuropsychopharmacology 28:530–532

Giard M, Perrin F, Pernier J, Bouchet P (1990) Brain generators implicated in the processing of auditory stimulus deviance: a topographic event-related potential study. Psychophysiology 27:627–640

Gottesman II, Gould TD (2003) The endophenotype concept in psychiatry: etymology and strategic intentions. Am J Psychiatry 160:636–645

Green MF (1996) What are the functional consequences of neurocognitive deficits in schizophrenia? Am J Psychiatry 153:321–330

Green M, Kern R, Braff D, Mintz J (2000) Neurocognitive deficits and functional outcome in schizophrenia: are we measuring the "right stuff"? Schizophr Bull 26:119–136

Grillon C, Courchesne E, Ameli R, Geyer M, Braff D (1990) Increased distractibility in schizophrenic patients. Electrophysiologic and behavioral evidence. Arch Gen Psychiatry 47:171–179

Hall MH, Schulze K, Rijsdijk F, Picchioni M, Ettinger U, Bramon E, Freedman R, Murray RM, Sham P (2006) Heritability and reliability of P300, P50 and duration mismatch negativity. Behav Genet 36:845–857

Hillyard SA, Anllo-Vento L (1998) Event-related brain potentials in the study of visual selective attention. Proc Natl Acad Sci USA 95:781–787

Hirayasu Y, Asato N, Ohta H, Hokama H, Arakaki H, Ogura C (1998a) Abnormalities of auditory event-related potentials in schizophrenia prior to treatment. Biol Psychiatry 43:244–253

Hirayasu Y, Shenton ME, Salisbury DF, Dickey CC, Fischer IA, Mazzoni P, Kisler T, Arakaki H, Kwon JS, Anderson JE, Yurgelun-Todd D, Tohen M, McCarley RW (1998b) Lower left temporal lobe MRI volumes in patients with first-episode schizophrenia compared with psychotic patients with first-episode affective disorder and normal subjects. Am J Psychiatry 155:1384–1391

Horváth J, Winkler I, Bendixen A (2008) Do N1/MMN, P3a, and RON form a strongly coupled chain reflecting the three stages of auditory distraction? Biol Psychol 79:139–147

Huotilainen M, Kujala A, Hotakainen M, Shestakova A, Kushnerenko E, Parkkonen L, Fellman V, Näätänen R (2003) Auditory magnetic responses of healthy newborns. NeuroReport 14: 1871–1875

Iacono WG, Carlson SR, Malone SM (2000) Identifying a multivariate endophenotype for substance use disorders using psychophysiological measures. Int J Psychophysiol 38:81–96

Iwanami A, Suga I, Kanamori R (1994) N1 component derived from the temporal region during an auditory passive event-related potential paradigm in schizophrenics. Clin Electroencephalogr 25:50–53

Iwanami A, Okajima Y, Kuwakado D, Isono H, Kasai K, Hata A, Nakagome K, Fukuda M, Kamijima K (2000) Event-related potentials and thought disorder in schizophrenia. Schizophr Res 42:187–191

Javitt DC (2000) Intracortical mechanisms of mismatch negativity dysfunction in schizophrenia. Audiol Neurootol 5:207–215

Javitt DC, Shelley A, Ritter W (2000a) Associated deficits in mismatch negativity generation and tone matching in schizophrenia. Clin Neurophysiol 111:1733–1737

Javitt DC, Shelley AM, Silipo G, Lieberman JA (2000b) Deficits in auditory and visual context-dependent processing in schizophrenia: defining the pattern. Arch Gen Psychiatry 57: 1131–1137

Jessen F, Fries T, Kucharski C, Nishimura T, Hoenig K, Maier W, Falkai P, Heun R (2001) Amplitude reduction of the mismatch negativity in first-degree relatives of patients with schizophrenia. Neurosci Lett 309:185–188

Johnson R (1993) On the neural generators of the P300 component of the event related potential. Psychophysiology 30:90–97

Johnston WA, Heinz SP (1978) Flexibility and capacity demands of attention. J Exp Psychol 107:420–435

Joutsiniemi S, Ilvonen T, Sinkkonen J, Huotilainen M, Tervaniemi M, Lehtokoski A, Rinne T, Näätänen R (1998) The mismatch negativity for duration decrement of auditory stimuli in healthy subjects. Electroencephalogr Clin Neurophysiol 108:154–159

Kaipio M, Novitski N, Tervaniemi M, Alho K, Öhman J, Salonen O, Näätänen R (2001) Fast vigilance decrement in closed head injury patients as reflected by the mismatch negativity (MMN). NeuroReport 12:1517

Kane NM, Curry SH, Rowlands CA, Manara AR, Lewis T, Moss T, Cummins BH, Butler SR (1996) Event-related potentials–neurophysiological tools for predicting emergence and early outcome from traumatic coma. Intensive Care Med 22:39–46

Karniski W, Blair R (1989) Topographical and temporal stability of the P300. Electroencephalogr Clin Neurophysiol 72:373

Karoumi B, Laurent A, Rosenfeld F, Rochet T, Brunon A, Dalery J, d'Amato T, Saoud M (2000) Alteration of event related potentials in siblings discordant for schizophrenia. Schizophr Res 41:325–334

Katayama J, Polich J (1998) Stimulus context determines P3a and P3b. Psychophysiology 35: 23–33

Kathmann N, Frodl-Bauch T, Hegerl U (1999) Stability of the mismatch negativity under different stimulus and attention conditions. Clin Neurophysiol 110:317–323

Kawakubo Y, Kamio S, Nose T, Iwanami A, Nakagome K, Fukuda M, Kato N, Rogers MA, Kasai K (2007) Phonetic mismatch negativity predicts social skills acquisition in schizophrenia. Psychiatry Res 152:261–265

Kawasaki Y, Maeda Y, Higashima M, Nagasawa T, Koshino Y, Suzuki M, Ide Y (1997) Reduced auditory P300 amplitude, medial temporal volume reduction and psychopathology in schizophrenia. Schizophr Res 26:107–115

Kiang M, Kutas M, Light G, Braff D (2007) Electrophysiological insights into conceptual disorganization in schizophrenia. Schizophr Res 92:235–236

Kiang M, Kutas M, Light G, Braff D (2008) An event-related brain potential study of direct and indirect semantic priming in schizophrenia. Am J Psychiatry 165:74–81

Kiang M, Braff DL, Sprock J, Light GA (2009) The relationship between preattentive sensory processing deficits and age in schizophrenia patients. Clin Neurophysiol 120:1949–1957

Kidogami Y, Yoneda H, Asaba H, Sakai T (1992) P300 in first degree relatives of schizophrenics. Schizophr Res 6:9–13

Kiehl K, Laurens K, Duty T, Forster B, Liddle P (2001) Neural sources involved in auditory target detection and novelty processing: an event-related fMRI study. Psychophysiology 38:133–142

Kileny PR, Kripal JP (1987) Test-retest variability of auditory event-related potentials. Ear Hear 8:110–114

Kinoshita S, Inoue M, Maeda H, Nakamura J, Morita K (1996) Long-term patterns of change in ERPs across repeated measurements. Physiol Behav 60:1087–1092

Kok A (2001) On the utility of P3 amplitude as a measure of processing capacity. Psychophysiology 38:557–577

Kujala T, Kallio J, Tervaniemi M, Näätänen R (2001) The mismatch negativity as an index of temporal processing in audition. Clin Neurophysiol 112:1712–1719

Kutas M, McCarthy G, Donchin E (1977) Augmenting mental chronometry: P300 as a measure of stimulus evaluation time. Science 197:792–795

Laurent A, Garcia-Larrea L, d'Amato T, Bosson J, Saoud M, Marie-Cardine M, Maugière F, Dalery J (1999) Auditory event-related potentials and clinical scores in unmedicated schizophrenic patients. Psychiatry Res 86:229–238

Lavoie S, Murray M, Deppen P, Knyazeva M, Berk M, Boulat O, Bovet P, Bush A, Conus P, Copolov D, Fornari E, Meuli R, Solida A, Vianin P, Cuénod M, Buclin T, Do K (2008) Glutathione precursor, N-acetyl-cysteine, improves mismatch negativity in schizophrenia patients. Neuropsychopharmacology 33:2187–2199

Levänen S, Hari R, McEvoy L, Sams M (1993) Responses of the human auditory cortex to changes in one versus two stimulus features. Exp Brain Res 97:177–183

Light G, Braff D (2005a) Mismatch negativity deficits are associated with poor functioning in schizophrenia patients. Arch Gen Psychiatry 62:127–136

Light G, Braff D (2005b) Stability of mismatch negativity deficits and their relationship to functional impairments in chronic schizophrenia. Am J Psychiatry 162:1741–1743

Light G, Williams L, Sprock J, Swerdlow N, Braff D (2010) Electroencephalography (EEG) and event-related potentials (ERP's) with human participants. Curr Protoc Neurosci (in press)

Linden D (2005) The P300: where in the brain is it produced and what does it tell us? Neuroscientist 11:563–576

Luck SJ, Woodman GF, Vogel EK (2000) Event-related potential studies of attention. Trends Cogn Sci 4:432–440

Lyytinen H, Blomberg A, Näätänen R (1992) Event-related potentials and autonomic responses to a change in unattended auditory stimuli. Psychophysiology 29:523–534

Makeig S, Westerfield M, Jung TP, Enghoff S, Townsend J, Courchesne E, Sejnowski TJ (2002) Dynamic brain sources of visual evoked responses. Science 295:690–694

Makeig S, Debener S, Onton J, Delorme A (2004a) Mining event-related brain dynamics. Trends Cogn Sci 8:204–210

Makeig S, Delorme A, Westerfield M, Jung T, Townsend J, Courchesne E, Sejnowski T (2004b) Electroencephalographic brain dynamics following manually responded visual targets. PLOS Biol 2:747–762

Mathalon D, Ford J, Pfefferbaum A (2000) Trait and state aspects of P300 amplitude reduction in schizophrenia: a retrospective longitudinal study. Biol Psychiatry 47:434–449

Maxwell C, Liang Y, Weightman B, Kanes S, Abel T, Gur R, Turetsky B, Bilker W, Lenox R, Siegel S (2004) Effects of chronic olanzapine and haloperidol differ on the mouse N1 auditory evoked potential. Neuropsychopharmacology 29:739–746

Maxwell C, Ehrlichman R, Liang Y, Gettes D, Evans D, Kanes S, Abel T, Karp J, Siegel S (2006) Corticosterone modulates auditory gating in mouse. Neuropsychopharmacology 31:897–903

McCarley R, Faux S, Shenton M, LeMay M, Cane M, Ballinger R, Duffy F (1989) CT abnormalities in schizophrenia. A preliminary study of their correlations with P300/P200 electrophysiological features and positive/negative symptoms. Arch Gen Psychiatry 46:698–708

McCarley R, Faux S, Shenton M, Nestor P, Adams J (1991) Event-related potentials in schizophrenia: their biological and clinical correlates and a new model of schizophrenic pathophysiology. Schizophr Res 4:209–231

Mertens R, Polich J (1997) P300 from a single-stimulus paradigm: passive versus active tasks and stimulus modality. Electroencephalogr Clin Neurophysiol 104:488–497

Michie P, Fox A, Ward P, Catts S, McConaghy N (1990) Event-related potential indices of selective attention and cortical lateralization in schizophrenia. Psychophysiology 27:209–227

Michie PT, Budd TW, Todd J, Rock D, Wichmann H, Box J, Jablensky AV (2000) Duration and frequency mismatch negativity in schizophrenia. Clin Neurophysiol 111:1054–1065

Michie P, Innes-Brown H, Todd J, Jablensky A (2002) Duration mismatch negativity in biological relatives of patients with schizophrenia spectrum disorders. Biol Psychiatry 52:749–758

Morlet D, Bouchet P, Fischer C (2000) Mismatch negativity and N100 monitoring: potential clinical value and methodological advances. Audiol Neurootol 5:198–206

Näätänen R (1990) The role of attention in auditory information processing as revealed by event-related potentials and other brain measures of cognitive function. Behav Brain Sci 13:201–288

Näätänen R (1992) Attention and brain function. Erlbaum, Hillsdale, NJ

Näätänen R, Alho K (1997) Mismatch negativity – the measure for central sound representation accuracy. Audiol Neurootol 2:341–353

Näätänen R, Michie P (1979) Early selective-attention effects on the evoked potential: a critical review and reinterpretation. Biol Psychol 8:81–135

Näätänen R, Picton T (1987) The N1 wave of the human electric and magnetic response to sound: a review and an analysis of the component structure. Psychophysiology 24:375–425

Näätänen R, Winkler I (1999) The concept of auditory stimulus representation in cognitive neuroscience. Psychol Bull 125:826–859

Näätänen R, Gaillard A, Mantysalo S (1978) Early selective-attention effect on evoked potential reinterpreted. Acta Psychol (Amst) 42:313–329

Näätänen R, Simpson M, Loveless N (1982) Stimulus deviance and evoked potentials. Biol Psychol 14:53–98

Näätänen R, Paavilainen P, Alho K, Reinikainen K, Sams M (1989a) Do event-related potentials reveal the mechanism of the auditory sensory memory in the human brain? Neurosci Lett 98:217–221

Näätänen R, Paavilainen P, Reinikainen K (1989b) Do event-related potentials to infrequent decrements in duration of auditory stimuli demonstrate a memory trace in man? Neurosci Lett 107:347–352

Näätänen R, Tervaniemi M, Sussman E, Paavilainen P, Winkler I (2001) "Primitive intelligence" in the auditory cortex. Trends Neurosci 24:283–288

Näätänen R, Paavilainen P, Rinne T, Alho K (2007) The mismatch negativity (MMN) in basic research of central auditory processing: a review. Clin Neurophysiol 118:2544–2590

Nagamoto HT, Adler LE, Waldo MC, Freedman R (1989) Sensory gating in schizophrenics and normal controls: effects of changing stimulation interval. Biol Psychiatry 25:549–561

Nashida T, Yabe H, Sato Y, Hiruma T, Sutoh T, Shinozaki N, Kaneko S (2000) Automatic auditory information processing in sleep. Sleep 23:821–828

Norman DA (1968) Toward a theory of memory and attention. Psychol Rev 75:522–536

O'Donnell BF, Hokama H, McCarley RW, Smith RS, Salisbury DF, Mondrow E, Nestor PG, Shenton ME (1994) Auditory ERPs to non-target stimuli in schizophrenia: relationship to probability, task-demands, and target ERPs. Int J Psychophysiol 17:219–231

O'Donnell BF, Faux SF, McCarley RW, Kimble MO, Salisbury DF, Nestor PG, Kikinis R, Jolesz FA, Shenton ME (1995) Increased rate of P300 latency prolongation with age in schizophrenia. Electrophysiological evidence for a neurodegenerative process. Arch Gen Psychiatry 52:544–549

O'Donnell B, McCarley R, Potts G, Salisbury D, Nestor P, Hirayasu Y, Niznikiewicz M, Barnard J, Shen Z, Winstein D, Bookstein DL, Shenton M (1999) Identification of neural circuits underlying P300 abnormalities in schizophrenia Psychophysiology 36:388–398

Oades R, Dittmann-Balcar A, Zerbin D, Grzella I (1997) Impaired attention-dependent augmentation of MMN in nonparanoid vs paranoid schizophrenic patients: a comparison with obsessive-compulsive disorder and healthy subjects. Biol Psychiatry 41:1196–1210

Ogura C, Nageishi Y, Matsubayashi M, Omura F, Kishimoto A, Shimokochi M (1991) Abnormalities in event-related potentials, N100, P200, P300 and slow wave in schizophrenia. Jpn J Psychiatry Neurol 45:57–65

Pakarinen S, Takegata R, Rinne T, Huotilainen M, Näätänen R (2007) Measurement of extensive auditory discrimination profiles using the mismatch negativity (MMN) of the auditory event-related potential (ERP). Clin Neurophysiol 118:177–185

Parasuraman R, Beatty J (1980) Brain events underlying detection and recognition of weak sensory signals. Science 210:80–83
Pass H, Klorman R, Salzman L, Klein R, Kaskey G (1980) The late positive component of the evoked response in acute schizophrenics during a test of sustained attention. Biol Psychiatry 15:9–20
Pekkonen E, Rinne T, Näätänen R (1995) Variability and replicability of the mismatch negativity. Electroencephalogr Clin Neurophysiol 96:546–554
Pfefferbaum A (1984) Clinical application of the P3 component of event-related potentials: II. Dementia, depression and schizophrenia. Electroencephalogr Clin Neurophysiol 59:104–124
Pfefferbaum A, Ford J, White P, Roth W (1989) P3 in schizophrenia is affected by stimulus modality, response requirements, medication status, and negative symptoms. Arch Gen Psychiatry 46:1035–1044
Pfefferbaum A, Roth W, Ford J (1995) Event-related potentials in the study of psychiatric disorders. Arch Gen Psychiatry Res 52:559–563
Polich J, Criado J (2006) Neuropsychology and neuropharmacology of P3a and P3b. Int J Psychophysiol 60:172–185
Pollock VE, Schneider LS (1992) Reliability of late positive component activity (P3) in healthy elderly adults. J Gerontol 47:M88–M92
Price GW, Michie PT, Johnston J, Innes-Brown H, Kent A, Clissa P, Jablensky AV (2006) A multivariate electrophysiological endophenotype, from a unitary cohort, shows greater research utility than any single feature in the Western Australian family study of schizophrenia. Biol Psychiatry 60:1–10
Rao K, Ananthnarayanan C, Gangadhar B, Janakiramaiah N (1995) Smaller auditory P300 amplitude in schizophrenics in remission. Neuropsychobiology 32:171–174
Rasser PE, Schall U, Todd J, Michie PT, Ward PB, Johnston P, Helmbold K, Case V, Søyland A, Tooney PA, Thompson PM (2009) Gray matter deficits, mismatch negativity, and outcomes in schizophrenia. Schizophr Bull (Epub ahead of print, June 26)
Rentzsch J, Jockers-Scherubl MC, Boutros NN, Gallinat J (2008) Test-retest reliability of P50, N100 and P200 auditory sensory gating in healthy subjects. Int J Psychophysiol 67:81–90
Rinne T, Alho K, Ilmoniemi R, Virtanen J, Näätänen R (2000) Separate time behaviors of the temporal and frontal mismatch negativity sources. Neuroimage 12:14–19
Rinne T, Antila S, Winkler I (2001) Mismatch negativity is unaffected by top-down predictive information. NeuroReport 12:2209–2213
Ritter W, Sussman E, Deacon D, Cowan N, Vaughan HG Jr (1999) Two cognitive systems simultaneously prepared for opposite events. Psychophysiology 36:835–838
Roberts T, Poeppel D (1996) Latency of auditory evoked M100 as a function of tone frequency. NeuroReport 7:1138–1140
Romani A, Merello S, Gozzoli L, Zerbi F, Grassi M, Cosi V (1987) P300 and CT scan in patients with chronic schizophrenia. Br J Psychiatry 151:506–513
Roth W, Horvath T, Pfefferbaum A, Kopell B (1980) Event-related potentials in schizophrenics. Electroencephalogr Clin Neurophysiol 48:127–139
Roth W, Goodale J, Pfefferbaum A (1991) Auditory event-related potentials and electrodermal activity in medicated and unmedicated schizophrenics. Biol Psychiatry 29:585–599
Roxborough H, Muir W, Blackwood D, Walker M, Blackburn I (1993) Neuropsychological and P300 abnormalities in schizophrenics and their relatives. Psychol Med 23:305–314
Saitoh O, Niwa S, Hiramatsu K, Kameyama T, Rymar K, Itoh K (1984) Abnormalities in late positive components of event-related potentials may reflect a genetic predisposition to schizophrenia. Biol Psychiatry 19:293–303
Salisbury D, Shenton M, Griggs C, Bonner-Jackson A, McCarley R (2002) Mismatch negativity in chronic schizophrenia and first-episode schizophrenia. Arch Gen Psychiatry 59:686–694
Sandman C, Patterson J (2000) The auditory event-related potential is a stable and reliable measure in elderly subjects over a 3 year period. Clin Neurophysiol 111:1427–1437
Schall U, Catts SV, Karayanidis F, Ward PB (1999) Auditory event-related potential indices of fronto-temporal information processing in schizophrenia syndromes: valid outcome prediction of clozapine therapy in a three-year follow-up. Int J Neuropsychopharmacol 2:83–93

Schreiber H, Stolz-Born G, Kornhuber H, Born J (1992) Event related potential correlates of impaired selective attention in children at high risk for schizophrenia. Biol Psychiatry 32:634–651

Schröger E (1997) On the detection of auditory deviations: a pre-attentive activation model. Psychophysiology 34:245–257

Segalowitz SJ, Barnes KL (1993) The reliability of ERP components in the auditory oddball paradigm. Psychophysiology 30:451–459

Sinha R, Bernardy N, Parsons OA (1992) Long-term test-retest reliability of event-related potentials in normals and alcoholics. Biol Psychiatry 32:992–1003

Smith M, Halgren E, Sokolik M, Baudena P, Musolino A, Liegeois-Chauvel C, Chauvel P (1990) The intracranial topography of the P3 event-related potential elicited during auditory oddball. Electroencephalogr Clin Neurophysiol 76:235–248

Souza VBN, Muir WJ, Walker MT, Glabus MF, Roxborough HM, Sharp CW, Dunan JR, Blackwood DHR (1995) Auditory P300 event-related potentials and neuropsychological performance in schizophrenia and bipolar affective disorder. Biol Psychiatry 37:300–310

Spencer K, Dien J, Donchin E (2001) Spatiotemporal analysis of the late ERP responses to deviant stimuli. Psychophysiology 38:343–358

Squires NK, Squires KC, Hillyard SA (1975) Two varieties of long-latency positive waves evoked by unpredictable auditory stimuli in man. Electroencephalogr Clin Neurophysiol 38:387–340

Strik W, Dierks T, Boning J, Osterheider M, Caspari A, Korber J (1992) Disorders of smooth pursuit eye movement and auditory N100 in schizophrenic patients. Psychiatry Res 41:227–235

Strik W, Dierks T, Maurer K (1993) Amplitudes of auditory P300 in remitted and residual schizophrenics: correlations with clinical features. Neuropsychobiology 27:54–60

Sumich A, Harris A, Flynn G, Whitford T, Tunstall N, Kumari V, Brammer M, Gordon E, Williams LM (2006) Event-related potential correlates of depression, insight and negative symptoms in males with recent-onset psychosis. Clin Neurophysiol 117:1715–1727

Sussman E, Winkler I, Wang W (2003) MMN and attention: competition for deviance detection. Psychophysiology 40:430–435

Swerdlow NR, Geyer MA, Shoemaker JM, Light GA, Braff DL, Stevens KE, Sharp R, Breier M, Neary A, Auerbach PP (2005) Convergence and divergence in the neurochemical regulation of prepulse inhibition of startle and N40 suppression in rats. Neuropsychopharmacology 31:506–515

Tervaniemi M, Lehtokoski A, Sinkkonen J, Virtanen J, Ilmoniemi R, Näätänen R (1999) Test-retest reliability of mismatch negativity for duration, frequency and intensity changes. Clin Neurophysiol 110:1388–1393

Tiitinen H, May P, Reinikainen K, Näätänen R (1994) Attentive novelty detection in humans is governed by pre-attentive sensory memory. Nature 372:90–92

Todd J, Michie PT, Budd TW, Rock D, Jablensky AV (2000) Auditory sensory memory in schizophrenia: inadequate trace formation? Psychiatry Res 96:99–115

Treisman A (1960) Contextual cues in selective listening. Q J Exp Psychol 12:242–248

Trejo L, Ryan-Jones D, Kramer A (1995) Attentional modulation of the mismatch negativity elicited by frequency differences between binaurally presented tone bursts. Psychophysiology 32:319–328

Turetsky B, Colbath E, Gur R (1998a) P300 subcomponent abnormalities in schizophrenia: I. Physiological evidence for gender and subtype specific differences in regional pathology. Biol Psychiatry 43:84–96

Turetsky B, Colbath E, Gur R (1998b) P300 subcomponent abnormalities in schizophrenia: longitudinal stability and relationship to symptom change. Biol Psychiatry 43:31–39

Turetsky B, Colbath E, Gur R (2000) P300 subcomponent abnormalities in schizophrenia: III. Deficits in unaffected siblings of schizophrenic probands. Biol Psychiatry 47:380–390

Turetsky B, Calkins M, Light G, Olincy A, Radant A, Swerdlow N (2007) Neurophysiological endophenotypes of schizophrenia: the viability of selected candidate measures. Schizophr Bull 33:69–94

Turetsky B, Greenwood T, Olincy A, Radant A, Braff D, Cadenhead K, Dobie D, Freedman R, Green M, Gur R, Gur R, Light G, Mintz J, Nuechterlein K, Schork N, Seidman L, Siever L, Silverman J, Stone W, Swerdlow N, Tsuang D, Tsuang M, Calkins M (2008) Abnormal auditory N100 amplitude: a heritable endophenotype in first-degree relatives of schizophrenia probands. Biol Psychiatry 64:1051–1059

Umbricht D, Krljes S (2005) Mismatch negativity in schizophrenia: a meta-analysis. Schizophr Res 76:1–23

Umbricht D, Koller R, Schmid L, Skrabo A, Grübel C, Huber T, Stassen H (2003) How specific are deficits in mismatch negativity generation to schizophrenia? Biol Psychiatry 53:1120–1131

Umbricht D, Bates J, Lieberman J, Kane J, Javitt D (2006) Electrophysiological indices of automatic and controlled auditory information processing in first-episode, recent-onset and chronic schizophrenia. Biol Psychiatry 59:762–772

van der Stelt O, Gunning W, Snel J, Kok A (1997) No electrocortical evidence of automatic mismatch dysfunction in children of alcoholics. Alcohol Clin Exp Res 21:569–575

Vaughan H, Ritter W (1970) The sources of auditory evoked responses recorded from the human scalp. Electroencephalogr Clin Neurophysiol 28:360–367

Waberski TD, Kreitschmann-Andermahr I, Kawohl W, Darvas F, Ryang Y, Gobbele R, Buchner H (2001) Spatio-temporal source imaging reveals subcomponents of the human auditory mismatch negativity in the cingulum and right inferior temporal gyrus. Neurosci Lett 308:107–110

Waldo M, Adler LE, Freedman R (1988) Defects in auditory sensory gating and their apparent compensation in relatives of schizophrenics. Schizophr Res 1:19–24

Weisbrod M, Hill H, Niethammer R, Sauer H (1999) Genetic influence on auditory information processing in schizophrenia: P300 in monozygotic twins. Biol Psychiatry 46:721–725

Winkler I, Czigler I (1998) Mismatch negativity: deviance detection or the maintenance of the standard. NeuroReport 9:3809

Winkler I, Karmos G, Näätänen R (1996) Adaptive modeling of the unattended acoustic environment reflected in the mismatch negativity event-related potential. Brain Res 742:239–252

Winterer G, Egan M, Radler T, Coppola R, Weinberger D (2001) Event-related potentials and genetic risk for schizophrenia. Biol Psychiatry 50:407–417

Woldorff MG, Hillyard SA (1991) Modulation of early auditory processing during selective listening to rapidly presented tones. Electroencephalogr Clin Neurophysiol 79:170–91

Wronka E, Kuniecki M, Kaiser J, Coenen A (2007) The P3 produced by auditory stimuli presented in a passive and active condition: modulation by visual stimuli. Acta Neurobiol Exp 67:155–164

Wynn J, Sugar C, Horan W, Kern R, Green M (2010) Mismatch negativity, social cognition, and functioning in schizophrenia patients. Biol Psychiatry (Epub ahead of print, Jan 13)

Eye Tracking Dysfunction in Schizophrenia: Characterization and Pathophysiology

Deborah L. Levy, Anne B. Sereno, Diane C. Gooding, and Gilllian A. O'Driscoll

Contents

1 Introduction 312
2 Components of the Smooth Pursuit Eye Tracking Response 315
3 Characterization of ETD 318
4 Pathophysiology of ETD 322
    4.1 Behavioral Evaluations of the Contribution of Motion Processing to ETD 322
    4.2 Extraretinal Processes in Pursuit 331
    4.3 Neuroimaging of Pursuit and Component Processes 334
5 Association Between Genetic Polymorphisms and ETD 336
6 Summary 337
References 337

Abstract Eye tracking dysfunction (ETD) is one of the most widely replicated behavioral deficits in schizophrenia and is over-represented in clinically unaffected first-degree relatives of schizophrenia patients. Here, we provide an overview of research relevant to the characterization and pathophysiology of this impairment. Deficits are most robust in the maintenance phase of pursuit, particularly during the tracking of predictable target movement. Impairments are also found in pursuit initiation and correlate with performance on tests of motion processing, implicating early sensory processing of motion signals. Taken together, the evidence suggests

---

D.L. Levy (✉)
Psychology Research Laboratory, McLean Hospital, 115 Mill Street, Belmont, MA 02478, USA
e-mail: dlevy@mclean.harvard.edu

A.B. Sereno
Department of Neurobiology and Anatomy, University of Texas Medical School at Houston, Houston, TX, USA

D.C. Gooding
Department of Psychology, University of Wisconsin-Madison, Madison, WI, USA

G.A. O'Driscoll
Department of Psychology, McGill University, Montreal, QC, Canada

that ETD involves higher-order structures, including the frontal eye fields, which adjust the gain of the pursuit response to visual and anticipated target movement, as well as early parts of the pursuit pathway, including motion areas (the middle temporal area and the adjacent medial superior temporal area). Broader application of localizing behavioral paradigms in patient and family studies would be advantageous for refining the eye tracking phenotype for genetic studies.

Keywords Eye tracking dysfunction · Smooth pursuit eye movements · Motion processing · Extraretinal processes · Schizophrenia · Genetics · Endophenotypes

1 Introduction

In 1908, Allen Diefendorf, a psychiatrist, and Raymond Dodge, an experimental psychologist, collaborated on the first study of ocular motor function in psychiatric patients (Diefendorf and Dodge 1908). Dodge's development of a method for photographically recording eye movements (e.g., the photochronograph) allowed objective quantification of certain eye movement metrics and made experimental studies feasible. They reasoned that because eye movements were a ubiquitous aspect of everyday functioning, patients and controls would have comparable degrees of acquired proficiency. Diefendorf and Dodge chose to study smooth pursuit and reflexive saccades in order to capitalize on over-learned visual behaviors and to avoid the confounding effects of tasks that were "too complicated" or had "too unusual demands" for chronically ill patients to perform. In this way, any deficits found would suggest disease-related dysfunction in a potentially informative neural system. Thus, from both the scientific and methodological vantage points, Diefendorf and Dodge's landmark study of eye movements in psychiatric patients laid the foundation for investigations that continue to this day.

That first empirical study compared patients with dementia praecox (now schizophrenia), manic-depressive psychosis (now bipolar disorder), and various organic conditions (e.g., epilepsy, neurosyphilis) with controls on simple pursuit and saccade tasks. They found such a strong and selective association between impaired smooth pursuit eye movements and dementia praecox that they described it as "praecox pursuit." Surprisingly, the finding of a specific psychophysiological abnormality that differentiated one major psychosis from other functional and organic psychotic conditions was pursued only twice in the ensuing six decades;[1]

---

[1] Two studies explicitly followed up the Diefendorf and Dodge report (Couch and Fox 1934; White 1938). Both studies replicated the finding of impaired pursuit in schizophrenic patients, but questioned its specificity and independence from clinical state, especially in manic-depressive patients. Modern psychotropic drugs were not yet in use, but barbiturates were commonly used to control agitation. Impaired pursuit was found during periods of clinical exacerbation,

it was after this long fallow period that the modern era of research on ocular motor function in schizophrenia began.

The independent rediscovery of smooth pursuit eye movement impairment, otherwise known as eye tracking dysfunction (ETD), by Holzman and colleagues (Holzman et al. 1973, 1974a) was a serendipitous byproduct of an empirical study designed to assess the integrity of the vestibular system in schizophrenia. A consistent finding in schizophrenia at that time was vestibular hyporeactivity (Holzman 1969). Tests of vestibular function routinely include vestibularly induced eye movements (e.g., nystagmus, a slow eye movement in one direction followed by a fast eye movement in the opposite direction) as well as smooth pursuit and saccadic eye movements (Baloh and Honrubia 1990). Unexpectedly, the vestibulo-ocular reflex of schizophrenia patients was found to function normally (Levy et al. 1978).[2] However, smooth pursuit eye movements (or "eye tracking patterns") were abnormal, not only in patients but also in their clinically unaffected first-degree biological relatives (Holzman et al. 1973, 1974a). Unbeknownst to Holzman and colleagues, they had replicated and extended the findings of Diefendorf and Dodge from six and a half decades earlier (Stevens 1974; Holzman et al. 1974b).

Within 20 years of Holzman and colleagues' first two eye tracking papers (Holzman et al. 1973, 1974a), over 80 replications of the finding of ETD in schizophrenia patients were published. Issues of specificity, psychotropic medication effects, stage of illness, temporal stability, and effects of clinical state and attention were addressed by independent groups all over the world. Multiple replications of the familial aggregation of ETD in relatives of schizophrenia patients also followed, suggesting that it might be heritable. Studies of twins discordant for schizophrenia as well as healthy twins supported the idea that eye tracking performance was under genetic control (Holzman et al. 1977, 1988; Iacono and Lykken 1979; Bell et al. 1994; Katsanis et al. 2000). The elevated rate of ETD in clinically unaffected relatives and in clinically discordant co-twins provided evidence that ETD could not be attributed to treatment, hospitalization, or other confounding factors. Rather, it raised the possibility that ETD might be an alternative manifestation of genetic liability for schizophrenia. The significantly higher rate of ETD than recurrence risk for schizophrenia in first-degree relatives of schizophrenia patients suggested that ETD might be a more penetrant, pleiotropic expression of the same genes that were risk factors for the clinical disorder (Holzman et al. 1988; Holzman and

---

corresponding to periods of barbiturate treatment, whereas pursuit normalized during periods of remission, corresponding to barbiturate discontinuation. Only later were barbiturates discovered to impair pursuit (Rashbass and Russell 1961; Schalen et al. 1988), suggesting that what appeared at the time to be an association between clinical state and pursuit performance was actually a drug-induced epiphenomenon.

[2] A discussion of possible reasons for the difference between these results and those of earlier investigators as well as a critical review of the literature on vestibular function in psychopathological conditions can be found elsewhere (Levy et al. 1983). The status of visual–vestibular interaction remains unclear, with some data supporting normal responses in schizophrenic patients (Levy et al. 1978) and other data supporting abnormal responses (Jones and Pivik 1983; Yee et al. 1987; Warren and Ross 1998).

Matthysse 1990; Matthysse and Parnas 1992). This research also demonstrated the value of studying clinically unaffected relatives of patients, a once neglected resource that is now widely utilized in psychopathology research to unravel the pattern of genetic transmission of a schizophrenia-endophenotype complex.

The dedication of an entire recent issue of *Brain & Cognition* [volume 68(3), 2008] to eye movement research in psychiatry, coinciding with the 100th anniversary of Diefendorf and Dodge's seminal paper, attests to the importance of eye movement research in psychopathology research. Although schizophrenia has tended to be the primary focus of this research, ocular motor function has been studied in many other psychiatric conditions as well – bipolar, major depressive and obsessive-compulsive disorders, anorexia nervosa, schizophrenia-related personality disorders, substance use (including nicotine effects), schizotypal traits, and childhood and adolescent-onset disorders [e.g., (Iacono et al. 1982; Clementz et al. 1996; Jacobsen et al. 1996; Pallanti et al. 1996, 1998; Thaker et al. 1996a; Bauer 1997; Farber et al. 1997; O'Driscoll et al. 1998; Sweeney et al. 1998b; Gooding et al. 2000; Larrison et al. 2000, 2004; Ross et al. 2000; Kumra et al. 2001; Depatie et al. 2002; Kathmann et al. 2003; Ceballos and Bauer 2004; Lenzenweger and O'Driscoll 2006; Sereno et al. 2009)]. Further, oculomotor control in psychiatric populations has now been studied with a range of tasks much broader than the standard pursuit and reflexive saccade paradigms. Researchers have employed tasks that include smooth pursuit during sudden changes in predictable target motion (Allen et al. 1990; Clementz et al. 1996; Thaker et al. 1998, 1999; Trillenberg et al. 1998; Hong et al. 2005a; Avila et al. 2006) and pursuit on textured backgrounds (Yee et al. 1987; Schlenker et al. 1994; Arolt et al. 1998; Hutton et al. 2000). In addition, several different voluntary saccade paradigms have been used, including saccades to predictable targets (Levin et al. 1982; Abel et al. 1992; Clementz et al. 1994; Crawford et al. 1995a, b; Karoumi et al. 1998; Hutton et al. 2001; Krebs et al. 2001; O'Driscoll et al. 2005; Spengler et al. 2006; Sailer et al. 2007) [see also review by (Gooding and Basso 2008); saccades away from targets (antisaccades) (Thaker et al. 1989; Fukushima et al. 1990; Clementz et al. 1994; Sereno and Holzman 1995; Katsanis et al. 1997; McDowell and Clementz 1997; Rosenberg et al. 1997; Hutton et al. 1998; Maruff et al. 1998; O'Driscoll et al. 1998; Gooding 1999; Castellanos et al. 2000; Curtis et al. 2001; Gooding and Tallent 2001; Mostofsky et al. 2001; Barton et al. 2002; Sweeney et al. 2002; Brownstein et al. 2003; Calkins et al. 2003; Munoz et al. 2003; Ettinger et al. 2004; Levy et al. 2004; Radant et al. 2007; Barton et al. 2008); saccades to remembered or attended targets (Park and Holzman 1992; Ross et al. 1994; Park et al. 1995; Everling et al. 1996; McDowell and Clementz 1996; Sweeney et al. 1998a; Muller et al. 1999; Larrison-Faucher et al. 2002; Winograd-Gurvich et al. 2006)]; and saccades to target sequences (Biscaldi et al. 1998; LeVasseur et al. 2001; Ram-Tsur et al. 2006). Fixation (Amador et al. 1991; Gooding et al. 2000; Munoz et al. 2003; Smyrnis et al. 2004; Barton et al. 2008), the oculocephalic reflex (Lipton et al. 1980), and optokinetic and vestibular responses (Levy et al. 1978, 1983; Latham et al. 1981; Jones and Pivik 1983; Yee et al. 1987; Cooper and Pivik 1991; Warren and Ross 1998) have been studied as well.

The rationale for Diefendorf and Dodge's study implicitly acknowledged a fundamental connection between schizophrenia and brain dysfunction that might be elucidated by the investigation of eye movements. Much of the work by modern investigators is based on the same assumption. Indeed, one reason that the study of eye movements has become so widely adopted in psychopathology laboratories is that they can be mapped to specific neural structures [for overviews see (Thier and Ilg 2005; Leigh and Zee 2006)]. Investigations of the pathophysiology of ocular motor dysfunction using neurologically informative behavioral paradigms hold the potential to clarify aspects of normal and disrupted brain circuitry in schizophrenia. In this chapter, we present an overview of selected topics relevant to the characterization and pathophysiology of smooth pursuit ETD in schizophrenia.

2 Components of the Smooth Pursuit Eye Tracking Response

Smooth pursuit eye movements are slow movements of the eye (less than about 100 deg/s) that function to keep a small moving target on the fovea (the retinal area that has the greatest visual acuity) by matching eye velocity to target velocity (Lisberger et al. 1987). Saccadic eye movements, on the other hand, rapidly shift gaze (up to 900 deg/s) to bring a new target onto the fovea. In general, pursuit begins first (latency around 100–150 ms) and is interrupted by an initial catch-up saccade (CUS) (latency around 200–250 ms) that brings the target onto the fovea (Sereno et al. 2009), after which the two systems work together to maintain it there.

Pursuit has been divided into two phases, an initiation phase and a maintenance phase, which differ in terms of the principal processes driving pursuit. When the pursuit system is initially stimulated by the perception of motion across the retina, the eye begins to accelerate after a latency of about 100 ms (Lisberger and Westbrook 1985; Barnes et al. 1987). The first 100 ms of the pursuit response is called pursuit initiation or "open-loop pursuit." It is driven primarily by the perception of a target moving slowly across the retina and reflects an initial estimate of the target speed. In this first 100 ms, no feedback from the retina influences the motor response, as the delay of information from the retina to the brainstem is approximately 100 ms (Krauzlis and Lisberger 1994). However, after 100 ms of pursuit, the relevant structures receive feedback from the retina regarding residual velocity and position error; at this point, the loop is closed, and the maintenance phase of pursuit begins. Pursuit maintenance uses velocity and position information from the retina as well as extraretinal information, such as corollary discharge from the motor system to sensory regions regarding the pursuit commands being issued, information about the position of the eyes in the head and the head in space, and accumulating experience with the target.

To study the smooth pursuit response in the initiation phase without the contribution of an orienting saccade that brings the target on to the fovea, researchers often use the "Rashbass" paradigm (Rashbass 1961). In the Rashbass paradigm (illustrated in Fig. 1), the central target steps off the fovea and then ramps

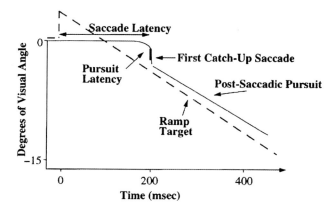

Fig. 1 Schematic presentation of a foveopetal (Rashbass type) step-ramp task used to assess pursuit initiation and pursuit gain. Reprinted with permission from Sweeney et al. (1998a)

(i.e., slides) back toward the fovea at a speed that returns it to center in less than 200 ms. Since the latency of a saccade is about 200 ms, and the target is back on the fovea at this point, pursuit begins without being interrupted by a saccade. Thus, by using the Rashbass paradigm, it is possible to isolate the smooth component of pursuit initiation. The integrity of pursuit initiation is quantified using measures of eye velocity or acceleration during the first 100 ms of pursuit as well as pursuit latency.

The adequacy of the pursuit response during the maintenance phase is often quantified by "pursuit gain" (the ratio of eye velocity to target velocity). The closer pursuit gain is to 1.0, the greater is the correspondence between the eye velocity and target velocity, and the more stable the target is on the fovea.[3] When pursuit gain is less than 1.0, the eyes are moving slower than the target, and compensatory CUSs can be used to reposition the eyes on the target (see Fig. 2, top tracing). Conversely, when gain is greater than 1.0, the eyes are moving faster than the target, and compensatory back-up saccades bring the eyes back to the target. For predictable target trajectories, such as sinusoidal waveforms (e.g., Figs. 2 and 4) and constant velocity ramps (e.g., Fig. 3), the match between eye velocity and target velocity can be quantified either as average gain across the trace or, in the case of sinusoidal targets, "peak gain" (gain during a brief period when target velocity is highest).

Saccades that occur during pursuit can be classified as compensatory or intrusive. Compensatory saccades include catch-up and back-up saccades that reposition the eyes on the target and thus reduce position error. Intrusive saccades, in contrast, disrupt the correspondence between the eye and target position and increase position error. Three types of intrusive saccades have been included in the quantitative

---

[3]This function of gain was discovered by the same Dodge who collaborated with Diefendorf in the first study of oculomotor function in schizophrenia (Dodge 1903).

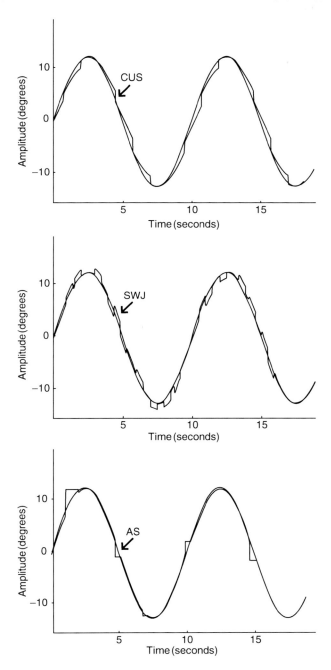

Fig. 2 A 0.1 Hz sinusoidal target (*lighter gray*) and simulations of low gain pursuit and catch-up saccades (CUS) (*top*), square wave jerks (SWJ) (*middle*), and anticipatory saccades (AS) (*bottom*). Adapted with permission from Abel and Ziegler (1988)

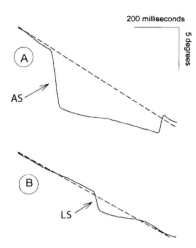

Fig. 3 Two segments of eye movement tracing. *Dotted lines* represent target motion as it moves from right (*top*) to left (*bottom*) at 16.7 deg/s. Seven hundred milliseconds are presented in each tracing. *Arrows* identify anticipatory saccades. Panel A: A large anticipatory saccade with an amplitude of 6.9°, followed by 312 ms of slowed smooth pursuit at 6 deg/s, then 110 ms of slowed smooth pursuit at 8 deg/s, followed by a saccade to return gaze to target location. Panel B: A small anticipatory saccade (or leading saccade, LS) with an amplitude of 2.7°, followed by 210 ms of slowed smooth pursuit at 7 deg/s. Reprinted with permission from Ross et al. (1999)

characterization of ETD in psychiatric populations. Square wave jerks (SWJ) consist of oppositely directed pairs of small (1–5°) saccades in which the first saccade takes the eyes off the target and the second saccade returns the eyes to the target. The intersaccadic interval is ~130–450 ms, during which pursuit continues (Fig. 2, middle tracing). Anticipatory saccades (AS) are large amplitude (>4–5°) saccades that move the eyes ahead of the target and are followed by periods of low gain pursuit (Fig. 2, bottom tracing; Fig. 3a) (Abel and Ziegler 1988; Leigh and Zee 2006). Leading saccades are saccades that take the eyes ahead of the target but have no minimum amplitude criterion, and are generally in the 1–4° range (Fig. 3b) (Ross et al. 1999). Other types of saccadic intrusions are found in certain neurological populations, but have not been studied in psychiatric populations (e.g., macro-SWJ, macrosaccadic oscillations, ocular flutter, and opsoclonus) (Leigh and Zee 2006).

3 Characterization of ETD

The early years of modern studies of ETD used global ratings that were either qualitative or quantitative. Qualitative ratings were judgments of how closely the eye position trace corresponded to the target position trace, either by dichotomizing

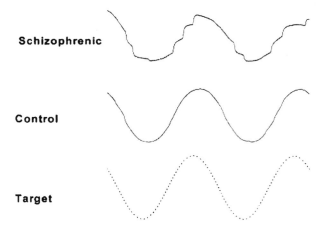

Fig. 4 Illustrative tracings of smooth pursuit eye movements of a schizophrenia patient (*top panel*) and of a normal control (*middle panel*). The target is a 0.4 Hz sine wave (*bottom panel*, *dotted line*). The record of the schizophrenia patient shows many irregularities that suggest low gain pursuit with frequent catch-up saccades. The record of the normal control shows an occasional small catch-up saccade. Reprinted with permission from Holzman (2000)

the degree of correspondence as "normal" or "abnormal" (Fig. 4), or by using an ordinal scale to reflect varying degrees of deviation from the position trace. Quantitative measures included frequency of velocity arrests, the natural logarithm of the signal-to-noise ratio, root mean square error, and total saccade frequency, among others [for a review see (Levy et al. 1993)]. These measures consistently established the presence of an eye tracking abnormality in schizophrenia patients and their relatives. Indeed, in two recent meta-analyses, global measures such as these had among the largest effect sizes (Calkins et al. 2008; O'Driscoll and Callahan 2008).

Although global measures are effective in identifying deviance, a disadvantage of these measures is that they cannot specify *what* is abnormal about the eye tracking. As Abel and Ziegler pointed out, global measures do not distinguish between "abnormalities *of* pursuit" and "abnormalities *during* pursuit" (Abel and Ziegler 1988). Specifically, global measures could not distinguish among abnormalities of the smooth pursuit system, disinhibition of the saccadic system, or some combination (Levin 1984). Thus, they cannot provide insight into the processes or physiological substrates of eye tracking deviance.

Specific measures of pursuit, however, can help to clarify the nature of the deficit. For example, saccadic intrusions in the context of normal gain suggest disinhibition of the saccadic system. Reduced gain in the context of increased CUS implicates a disturbance in the pursuit system for which CUS are compensating. Decreased gain with no increase in CUS suggests a pursuit disturbance as well as increased tolerance for position error. The converse, normal gain in the context of increased compensatory saccades, indicates reduced tolerance for position error (Levy et al. 1993). As these various scenarios make clear, parsing ETD into its specific components is an essential step both toward identifying the specific processes that underlie ETDs and identifying the pathophysiological substrates of the deficits.

A recent meta-analysis of ETD in schizophrenia quantified the results of studies that used global and specific measures (O'Driscoll and Callahan 2008). The analysis included studies comparing pursuit in schizophrenia patients and controls published subsequent to a 1993 review (Levy et al. 1993). Fifty-nine studies met criteria for inclusion and involved 2,107 schizophrenia patients and 1,965 controls. A summary of mean effect sizes and 95% confidence intervals for different eye tracking measures is shown in Fig. 5 (from O'Driscoll and Callahan (2008) with permission). The analysis confirmed strong differences between schizophrenia patients and controls in eye tracking performance for global and certain specific measures. The effect sizes (Cohen's d) for global variables were large; indeed, the largest effect size was obtained for qualitative ratings ($d = 1.55$). The latter finding is consistent with several reports indicating that qualitative ratings discriminate

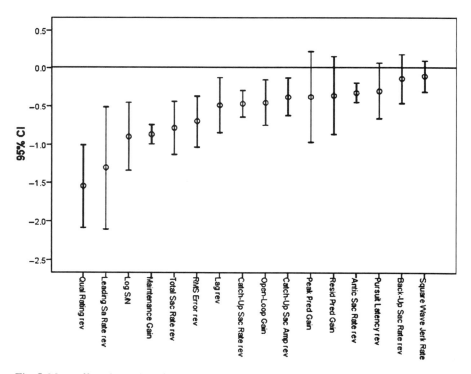

Fig. 5 Mean effect size and confidence intervals for patient-control differences in 16 measures of eye tracking performance. To allow a visual comparison of the magnitude of the effects, all ds have been made negative. Positive ds that have been reversed for the figure have "rev" appended to the variable name. The actual sign of the d based on the formula meanSz-MEANCOntrol/(Pooled SD) is shown in Table 3 of the published paper. Reprinted with permission from O'Driscoll and Callahan (2008)

patients and relatives from controls better than specific quantitative measures [e.g., (Friedman et al. 1995; Keefe et al. 1997; Levy et al. 2000)]. Two of the specific indices, maintenance gain and leading saccade rate (i.e., anticipatory saccades with no minimum amplitude criterion) had large effect sizes ($d = -0.87$ and $d = 1.31$, respectively)[4] as well as the smallest and largest 95% confidence intervals, respectively. The effect size for total saccade rate was also large. Effect sizes in the medium range were found for CUS, open-loop gain, and predictive gain measures (the latter variables are discussed below). O'Driscoll and Callahan concluded that the results did "not yield a clear-cut distinction between involvement of the pursuit or saccade system in the eye tracking deficit in schizophrenia; both pursuit and intrusive saccade measures yield at least one large effect size. It is also clear... that global measures generally yield larger effect sizes than specific measures" (p. 366). These findings notwithstanding, the authors correctly recognized that "in terms of neurophysiological informativeness, specific measures... allow precise hypotheses to be generated... in relation to areas in the pursuit pathway" (p. 366). They also noted several important caveats in interpreting the results of the meta-analysis. First, the amount of the recording on which a dependent measure is based seemed to be positively correlated with effect size. Qualitative ratings and maintenance gain, for example, are based on a larger proportion of the record than variables that, of necessity, are based on smaller segments (e.g., open-loop gain, predictive gain). As the reliability of a variable increases with the amount of data used to measure it, variables that are measured for longer periods of time may produce stronger results because of their enhanced statistical properties. Second, effect sizes for maintenance gain and CUS varied as a function of matching for sex in patients and controls, with larger effect sizes when the groups were matched than when they were not matched. This finding reflects a minor tendency for men to have higher maintenance gain than women (Lenzenweger and O'Driscoll 2006) and for men to be over-represented in patient samples.

In a recent complementary meta-analysis of studies on first-degree relatives of schizophrenia patients, Calkins and colleagues reported very similar results to those of O'Driscoll and Callahan. They found the largest effect sizes for global measures and for the specific measures, maintenance gain, and anticipatory saccades (a subset of leading saccades) (Calkins et al. 2008).

One possible reason for the apparent superiority of global ratings in terms of differentiating patients from controls is that global measures sum across different types of deficits in much the same way that in a depression questionnaire, the global question "Have you been been feeling down, depressed or hopeless?" will identify more individuals who subsequently meet criteria for depression than specific items like "Do you have trouble sleeping?" Global ratings average across different kinds of deviance that express or present in different severities in different individuals, while specific measures do not have this flexibility.

---

[4]Positive and negative values for effect sizes correspond to whether patients had higher or lower mean scores than controls, respectively.

Thus, an advantage of global measures of ETD, in addition to their greater sensitivity to between-group differences, is that they can be used to take into account the within-group heterogeneity in ways that specific measures often do not or cannot [see (Gibbons et al. 1984; Levy et al. 1993) for detailed discussions of the use of mixture analysis to resolve within-group heterogeneity; see (Levy et al. 2000) for an example of how global and specific measures can be used in tandem to clarify the nature of within-group heterogeneity; see (Buchsbaum and Rieder 1979) for a discussion of the impact of heterogeneity on traditional between-group comparisons].

In both the above meta-analyses, it is important to note that the amount of research devoted to different specific measures varied widely (e.g., from five studies of schizophrenia for predictive gain to 42 for maintenance gain, and generally fewer for each variable in relatives). Thus, for some of the newest measures where there are not enough data currently to draw firm conclusions, there should be some caution in interpretation.

4 Pathophysiology of ETD

Below we discuss several different approaches to identifying the neural substrates of ETD, each of which draws heavily on the effects of spontaneously occurring lesions in humans and experimental lesions and single-cell recordings in nonhuman primates. We begin with investigations of motion processing, a sensory function mediated in extrastriatal regions, and proceed to investigations of higher-order cognitive contributions that implicate regions later in the pursuit pathway.

4.1 Behavioral Evaluations of the Contribution of Motion Processing to ETD

A key component of the pursuit response is the processing of target velocity. This component contributes more to pursuit initiation, or "open-loop" pursuit, than to pursuit maintenance (Lisberger et al. 1987). This is because, generally, the stimulus for pursuit initiation is the movement of a novel target across the retina, the velocity of which must initially be estimated entirely perceptually. Once the maintenance phase of pursuit begins, other components of the pursuit response – predictions regarding target movement based on velocity memory, corollary discharge of the motor command to sensory areas regarding movement of the eyes in the head and the head in space, etc. – begin to contribute; at the same time, motion changes on the retina (i.e., retinal slip) decrease as the eye and target are now moving at approximately the same speed in the same direction.

Two regions of the extrastriate cortex, the middle temporal (MT) area and adjacent medial superior temporal (MST) area (in humans V5/V5a), play a critical role in the processing of visual motion. These regions respond to the passive perception of moving stimuli during smooth pursuit (Zeki 1974; Van Essen and Maunsell 1983). When these motion-sensitive regions of the brain are damaged, initial pursuit eye velocity is reduced, pursuit latency is increased, and motion perception is temporarily impaired (Wurtz et al. 1990).

Psychophysical studies investigating the potential contribution of motion processing deficits to ETD have taken several approaches. The first approach requires participants to make judgments about the velocity or direction of a motion stimulus (e.g., Fig. 6). The second approach requires participants to generate saccades to moving targets based on their velocity and direction (Figs. 1 and 7). Both approaches have been shown to index the integrity of extrastriate motion areas in nonhuman primates and in neurological populations. Nonhuman primates with lesions of MT (but not with lesions of the frontal eye fields) generate saccades that underestimate target speed, suggesting that the accuracy of saccades to moving targets is sensitive and somewhat specific to the integrity of extrastriate motion areas (Newsome et al. 1985; Thurston et al. 1988). The third approach involves evaluating the integrity of open-loop pursuit vs. closed-loop pursuit with the expectation that open-loop would be more compromised than closed-loop if motion processing were the major contributor to tracking deficits. The reason is that prediction is the predominant driver of closed-loop pursuit (Vandenberg 1988), while motion perception is the predominant driver of open-loop pursuit (Lisberger et al. 1987). In the two oculomotor approaches, the contribution of prediction to performance (which can compensate for motion perception deficits) can be controlled by varying target velocity, direction, and timing on a trial-by-trial basis (see Figs. 1 and 7).

4.1.1 Psychophysical Judgment Studies of Motion Perception

Using a standard motion perception task, one early study addressed the question of whether motion perception contributed to ETD in schizophrenia (Stuve et al. 1997). This study used a direction discrimination paradigm to assess motion perception in patients with schizophrenia and controls. In this task, participants watch a screen in which hundreds of dots move in random directions (illustrated in Fig. 8). The proportion of dots that move in a fixed direction (i.e., "motion coherence") is varied, and the level of coherence that is needed to correctly identify the direction is the individual's motion perception threshold (Newsome and Pare 1988). This task has been extensively used in single-unit recordings from nonhuman primates and has also been used in studies of neurological populations with lesions to MT/MST. Neuronal firing in this region significantly predicts the direction the monkey will choose on a trial-by-trial basis (Britten et al. 1996); stimulation of neurons in MT biases the monkey's judgment in the preferred direction of the stimulated neurons (Salzman et al. 1992). Lesions to MT/MST significantly increase direction discrimination thresholds in nonhuman primates (Newsome and Pare 1988) and in a patient with a V5 (MT) lesion (Baker et al. 1991). Stuve and colleagues found that

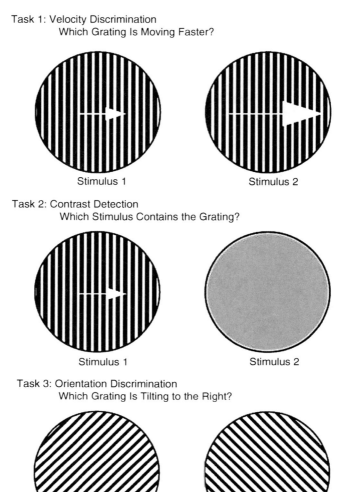

Fig. 6 A schematic representation of the stimuli used for the velocity discrimination, contrast detection, and orientation discrimination tasks. Reprinted with permission from Chen et al. (1999c)

patients with schizophrenia had significantly elevated motion thresholds that were correlated with pursuit deficits but not with performance on a sustained attention task. Accumulating research has provided consistent evidence that schizophrenia patients have a higher threshold for detecting the direction of coherent motion than controls (Wertheim et al. 1985; Stuve et al. 1997; Li 2002; Chen et al. 2003;

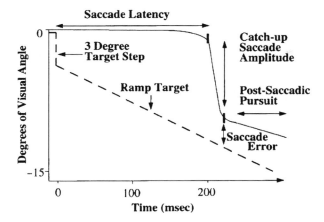

Fig. 7 Schematic presentation of a foveofugal step-ramp task used to assess the use of motion information by the pursuit and saccadic eye movement systems. Reprinted with permission from Sweeney et al. (1998a)

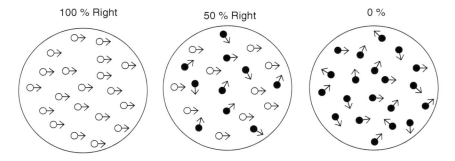

Fig. 8 Schematic representation of coherent motion at 100, 50, and 0% movement in a rightward direction. In the actual stimulus display, the dots moving coherently and those moving at random (i.e., noise) are the same color. Reprinted with permission from Slaghuis et al. (2007b)

Slaghuis et al. 2005, 2007a; Kim et al. 2006) and three of these studies found that the magnitude of the deficit correlated with closed-loop gain (Stuve et al. 1997; Slaghuis et al. 2005, 2007b).

Another method of assessing the functional integrity of the motion processing system is to measure the amount of contrast necessary to perform a velocity discrimination task. When the processing of visual signals is impaired, higher levels of contrast are necessary (Plant and Nakayama 1993; Pasternak and Merrigan 1994). Thus, measuring contrast sensitivity during velocity discrimination can index the integrity of the motion processing system. Contrast sensitivity measured independent of movement, such as in pure contrast detection tasks or in orientation discrimination tasks (shown in Fig. 6), provide valuable control conditions for movement. Chen and colleagues used this approach to establish a selective deficit in motion processing in schizophrenia that correlated with pursuit performance. They found that non-hospitalized schizophrenia patients needed higher amounts of contrast than controls to detect small differences in velocity (11 vs. 9 deg/s), but not to detect large differences in velocity (15 vs. 5 deg/s) (Fig. 9, top). The groups did not differ in detecting contrast or orientation (Fig. 9, bottom) (Chen et al. 1999c). Another study showed that the deficits were found in patients (Fig. 10) and in their clinically unaffected relatives (Fig. 11) at intermediate velocities (e.g., 10 deg/s), but not at slow (3.8 deg/s) and fast (26.2 deg/s)

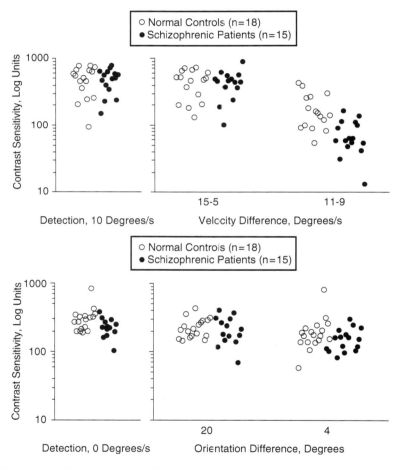

Fig. 9 *Top panel*: Contrast sensitivity for contrast detection (*left panel*) and for velocity discrimination (*right panel*). The groups differed significantly only on velocity discriminations of 11 vs. 9 deg/s. *Bottom panel*: Contrast sensitivity for detection (*left panel*) and for orientation discrimination (*right panel*). Patients and normal controls performed similarly. Reprinted with permission from Chen et al. (1999c)

velocities (Chen et al. 1999b). At slow and fast velocities, non-velocity cues can be used to help make velocity discriminations – position information at slow velocities (McKee 1981; Nakayama and Tyler 1981) and contrast differences at fast velocities (Pantle 1978). Manipulations to remove these non-velocity cues raised the velocity thresholds of both patients and relatives, indicating that the deficit was velocity-specific and could be partially compensated for by reliance on non-velocity cues (Chen et al. 1999b).

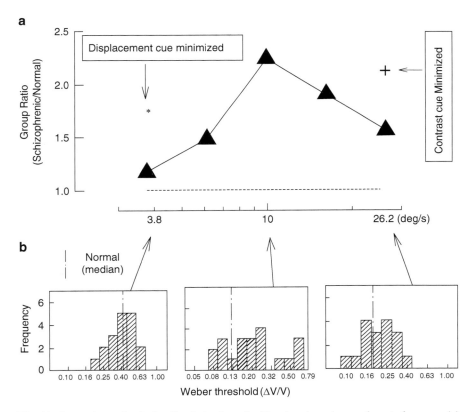

Fig. 10 Comparison of velocity discrimination of schizophrenia and normal control groups. (a) Group ratios (schizophrenia/normal control) of Weber thresholds plotted as a function of base velocity. The Weber fraction ($\Delta V/V$) is the just-noticeable differences between the velocities of the targets being compared. A ratio of unity, shown in the *dotted horizontal line*, indicates equivalent performance by the two groups. The larger the ratio is, the higher the velocity discrimination threshold of the patients relative to the normal controls. The *asterisk* and *cross sign* represent the group ratios after exposure time for the 3.8 deg/s target (*asterisk*), and the amount of contrast for the 26.2 deg/s target (*cross sign*) was randomized. (b) Histograms in the three panels (from left to right) represent distributions of individual patients' thresholds at the slowest (3.8 deg/s), middle (10 deg/s), and fastest (26.2 deg/s) base velocities. The *vertical line* in each panel indicates the median threshold of the normal control group. Reprinted with permission from Chen et al. (1999b)

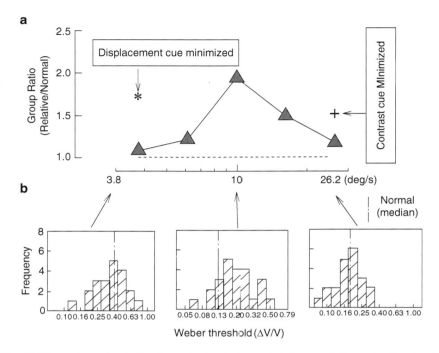

Fig. 11 Comparison of velocity discrimination between first-degree relatives of schizophrenia patients and normal controls. (**a**) Group ratio (as in Fig. 10, but here for relatives/normal controls) of Weber fraction thresholds plotted as a function of base velocity. The *asterisk* and *cross sign* represent group ratios after exposure time and amounts of contrast of the two velocity comparison targets were randomized. (**b**) Histograms in the three panels represent, from left to right, the distributions of individual relatives' thresholds at the slowest, middle, and fastest velocities. Other details are similar to those in Fig. 10. Reprinted with permission from Chen et al. (1999b)

A subsequent study isolated the motion deficit to later stages of visual processing (Chen et al. 2004). However, studies done by other laboratories have suggested deficits in early visual processing as well (Schwartz et al. 1987; Slaghius 1998; Butler et al. 2001; Green et al. 2003; Coleman et al. 2009; also see Slaghuis et al. 2007a).

We could find only one study that examined the relationship between open-loop gain (Fig. 12) and motion perception measures (Chen et al. 1999a). These authors found an association between both open- and closed-loop gain and reduced sensitivity to velocity information, supporting a connection between impaired motion processing and deficits in both the initiation and maintenance of pursuit (Chen et al. 1999a). The stronger association with open-loop gain ($r = 0.70$, $p < 0.01$, $n = 15$; Fig. 13), which depends on sensory input without feedback about target position, than for closed-loop gain ($r = 0.53$, $p < 0.05$, $n = 15$) is expected, given the primacy of motion processing in driving pursuit in the open-loop phase.

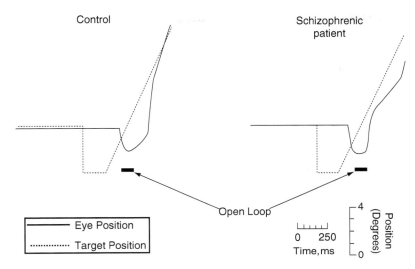

Fig. 12 Step-ramp pursuit of a normal control (*left*) and a schizophrenia patient (*right*). The target (*dotted line*) steps abruptly to the left and remains stationary for 200 ms before beginning a 20 deg/s ramp trajectory to the right. The open-loop period, denoted by the *black horizontal bars*, begins 130 ms after the target starts its ramp and continues for 100 ms. In response, at about 150 ms after the start of the ramp, the normal control begins a smooth eye movement that accelerates at a rate that is discernibly faster than that of the schizophrenia patient, whose initial eye movement barely accelerates. Reprinted with permission from Chen et al. (1999a)

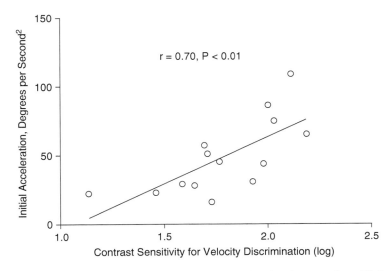

Fig. 13 Scatter diagram of the relationship within the schizophrenia group ($n = 15$) between open-loop acceleration for the 10 deg/s target and velocity discrimination between two targets (11 deg/s vs. 9 deg/s). Reprinted with permission from Chen et al. (1999a)

4.1.2 Saccadic Studies of Motion Perception

Several groups have assessed motion processing in schizophrenia by evaluating the accuracy of saccades to moving targets (Clementz 1996; Thaker et al. 1996b; Sweeney et al. 1998a, 1999; Lencer et al. 2004). This paradigm originated in the nonhuman primate literature and involves targets that step off the fovea and then ramp either away from the fovea (foveofugal) or toward the fovea (foveopetal) at different speeds (Newsome et al. 1985) (Figs. 1 and 7, respectively). MT lesions increase saccade latency and reduce the sensitivity of saccade amplitude to differences in ramp speed and ramp direction (i.e., foveofugal vs. foveopetal) (Newsome et al. 1985). All studies of schizophrenia have found that patients adjust saccadic amplitude according to ramp speed and direction to the same extent as controls and have normal saccade latencies (Clementz 1996; Thaker et al. 1996b; Sweeney et al. 1998a, 1999; Lencer et al. 2004) regardless of medication status and chronicity (Sweeney et al. 1998a, 1999). These studies suggest that saccadic motion estimates are unaffected in schizophrenia (Sweeney et al. 1998a, 1999), a conclusion that is inconsistent with patients' performance on motion perception tests. One possible explanation for this inconsistency is that motion perception studies have found impairments in fine velocity discriminations (e.g., 9 vs. 11 deg/s target speeds) but not in gross velocity discriminations (e.g., 5 vs. 15 deg/s) (Chen et al. 1999c). Studies that used saccades-to-moving-target paradigms in schizophrenia have generally used ramp speeds that differ widely (e.g., 8 vs. 16 deg/s, and even 8 vs. 24 deg/s, 9 vs. 27 deg/s), partly because saccadic endpoints to moving targets have some scatter, and larger differences in target speeds allow clearer distinctions between endpoints. However, the large differences in target speeds may reduce the difficulty of the motion component of the task and allow non-velocity cues (for example, changes in contrast and position) to aid saccade targeting.

4.1.3 Pursuit Initiation Studies

Several studies have used pursuit initiation in schizophrenia to examine the contribution of motion processing to pursuit deficits. Larger deficits in pursuit initiation (open-loop pursuit) than in pursuit maintenance (closed-loop pursuit) would be consistent with an impairment in motion processing. Deficits similar in magnitude in the two phases, or larger in the pursuit maintenance phase, suggest deficits in other functions (prediction, corollary discharge) that play a greater role in closed-loop pursuit (see Sect. 2, Components of the Smooth Pursuit Eye Tracking Response). Pursuit initiation has been studied both subsequent to the initial saccade (Feil 1997; Sweeney et al. 1999; Chen et al. 1999a; Sherr et al. 2002; Lencer et al. 2004; Avila et al. 2006) and without an initial saccade using the Rashbass paradigm (Clementz 1996; Ross et al. 1996; Farber et al. 1997; Radant et al. 1997; Hong et al. 2003). The schizophrenia-control difference in average effect size for studies that eliminate the saccade ($d = -0.54 \pm 0.28$) vs. those that do not ($d = -0.36 \pm 0.62$) is modest, and the average effect size across studies of open-loop pursuit is medium (see Fig. 5). Eight studies measured open- and closed-loop pursuit in the same patients (Clementz and McDowell 1994; Farber et al. 1997; Feil 1997; Radant et al. 1997; Sweeney et al. 1999; Chen et al. 1999a; Sherr et al. 2002; Lencer et al. 2004). Five of these studies found larger effects for open-loop than for closed-loop pursuit (Clementz and McDowell 1994; Radant et al. 1997; Sweeney et al. 1999; Chen et al. 1999a; Lencer et al. 2004),[5] two studies found larger effects for closed-loop than for open-loop pursuit (Sherr et al. 2002; Hong et al. 2003), and one study found no deficits in closed-loop pursuit or in pursuit acceleration during the first 100 ms (Farber et al. 1997).[6] However, across all studies published since 1993 (which include all open-loop studies and a large subset of closed-loop studies), open-loop pursuit measures have yielded a medium effect size, d of -0.45 (± 0.47, $n = 12$), whereas closed-loop pursuit gain has yielded a large effect size, d, of -0.87 (± 0.42, $n = 42$). For measures of both open- and closed-loop pursuit, deficits have been found even in neuroleptic naïve and unmedicated patients (Hutton et al. 1998; Sweeney et al. 1998a, 1999; Thaker et al. 1999; Lencer et al. 2008). These findings suggest that if motion processing deficits contribute to ETD, higher-order processes that would normally compensate for motion processing deficits are affected as well. In the studies by Sweeney and colleagues (Sweeney et al. 1998a, 1999), schizophrenia patients had delayed pursuit initiation and decreased closed-loop gain, normal CUS latency and amplitude, and reduced gain of postsaccadic pursuit compared with controls. The authors concluded that the pattern of deficits was consistent with involvement of the frontal eye fields (FEFs) (Sharpe and Morrow 1991; Keating 1993). The pattern seen after MT lesions – which is similar but includes dysmetric saccades to moving targets (Newsome et al. 1985; Thurston et al. 1988) – was not observed and seemed to militate against a motion processing explanation of pursuit deficits (but see caveat in Sect. 4.1.2).

4.2 Extraretinal Processes in Pursuit

The robust deficits in maintenance pursuit in schizophrenia [see (O'Driscoll and Callahan 2008)] could reflect impairments in extraretinal processes, rather than or as well as deficits in motion processing. Recent studies have focused on whether the predictive component of pursuit is impaired in schizophrenia as prediction of target movement is critical to high-gain closed-loop pursuit (Vandenberg 1988). An early psychophysical study addressed this question by having patients and controls watch a smoothly moving target disappear behind a screen and press a button at the moment they expected the target to reappear (Hooker and Park 2000). Patients had larger timing errors than controls, consistent with a deficit in motion prediction

---

[5]Larger for 10 deg/s targets, no difference for 20 deg/s targets.
[6]Differences were found in the last 40 ms of pursuit initiation, but not in the first 60 ms. Other investigators averaged across these epochs.

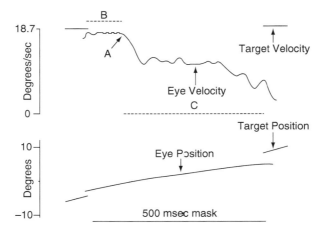

Fig. 14 The *top panel* shows eye and target velocity data, and the *bottom panel* shows corresponding position data from a 500-ms mask occurring during a ramp. Eye velocity remained unchanged for about 95 ms after the target was extinguished (B), presumably still influenced by the prior closed-loop response. After this initial period, the eye velocity stabilized to a lower level (58% of the closed-loop response) (C), arguably the response based on extraretinal motion signals. Residual predictive gain was calculated by dividing average eye velocity during C by expected target velocity. The transition point from closed-loop to extraretinal response (A) was identified by an algorithm. The program searched for the time point within the mask when the eye velocity first decreased by 50% of the premask value. From this point backwards, the algorithm searches for the local minimum or maximum value (depending on target direction) by analyzing the smoothed first (velocity) and second (acceleration) derivatives of position. This is identified as the transition point. Reprinted with permission from Thaker et al. (2003)

and the finding could not be attributed to motor slowing. Other studies of prediction have analyzed the speed of pursuit during brief periods when the target disappears. Figure 14 shows an example of a paradigm used to evaluate the predictive component of pursuit. Masking the trajectory of the pursuit target for short periods (i.e., 500 ms) eliminates retinal feedback and requires that extraretinal information, such as corollary discharge, velocity memory, and predictions regarding the target movement, drive pursuit (Lisberger et al. 1987; Newsome et al. 1988). The ratio of eye velocity to target velocity during epochs when the target is masked (i.e., predictive gain) indexes the efficacy of extraretinal signals in sustaining pursuit. A few studies have reported that schizophrenia patients (Thaker et al. 1999; Hong et al. 2003, 2005a), as well as their clinically unaffected relatives (Thaker et al. 1998, 2003; Hong et al. 2008), have lower predictive gain than controls.

A decrease in eye velocity during target blanking could reflect a reduction of motion signals in memory or a reduction in the gain of the signals driving the smooth pursuit system (Orban de Xivry et al. 2008). The effect sizes for this deficit are in the medium range. However, as larger effect sizes are found for measures of closed-loop pursuit (Fig. 5) that combine prediction and retinal information (i.e., gain and leading saccades), ETD likely reflects impairments in both motion processing and in prediction, implicating motion areas and FEFs, or possibly other areas in which both motion signals and predictive signals are represented [e.g., MST (Newsome et al. 1988); ventral intraparietal area (Schlack et al. 2003)].

The FEF contribution to pursuit has been studied in both nonhuman primates and in neurological populations. The characteristic features of pursuit after damage to the FEFs in nonhuman primates and in neurological populations include low initial and maintenance gain[7] (Keating 1991; MacAvoy et al. 1991; Rivaud et al. 1994; Morrow and Sharpe 1995; Heide et al. 1996; Lekwuwa and Barnes 1996; Shi et al. 1998) and impaired predictive pursuit (pursuit during target blanking) (Keating 1991, 1993; MacAvoy et al. 1991). In FEFs, the smooth velocity of the eye is rate-coded, such that increased eye velocity is associated with increased firing (Gottlieb et al. 1994). Microstimulation of FEF neurons increases smooth eye velocity (Gottlieb et al. 1993). Predictive pursuit, or pursuit during target blanking, is thought to depend on a neural representation of target motion. Neural correlates of internal representations of target motion, even changing target motion, have been found in FEFs, with neural activity coding target motion estimates during target blanking (Tanaka and Fukushima 1998; Barborica and Ferrera 2003, 2004; Xiao et al. 2003). Such a representation might be reconstructed from an efference copy of the pursuit motor command combined with retinal slip when the target is visible. The FEFs are also thought to play a critical role in controlling the "gain" of the signals driving pursuit (Tanaka and Lisberger 2001, 2002a, b). This notion of "gain" is distinct from pursuit gain, and describes the amplification of the pursuit response to visual or predictive signals driving pursuit. Tanaka and Lisberger showed that microstimulation of the pursuit area of the FEFs increases the gain of the pursuit system, that is, increases the magnitude of the pursuit response to retinal slip (Tanaka and Lisberger 2002c). In humans, transcranial magnetic stimulation of the FEFs also increases the magnitude of the pursuit response to predicted target motion (Gagnon et al. 2006).

Neurons in MST are sensitive to velocity and direction signals on the retina (Newsome et al. 1985), and also code extraretinal information, in that neurons in MST continue to fire during pursuit of a target that has briefly disappeared (Newsome et al. 1988; Bremmer et al. 1997). The extraretinal firing may code corollary discharge from motor areas (Newsome and Pare 1988; Komatsu and Wurtz 1989) or a representation of target movement in space (Thier and Erickson 1992). In nonhuman primates, lesions to MST do not affect saccades to moving targets (Fig. 1), but lesions to MST do reduce closed-loop pursuit gain (postsaccadic pursuit in Figs. 1 and 7) (Dursteler and Wurtz 1988) and reduce eye acceleration during pursuit initiation (Fig. 12). Lesions to the lateral portion of MST reduce sensitivity to retinal slip during ongoing pursuit (Komatsu and Wurtz 1989).

---

[7]If lesion is unilateral, deficits may be for ipsiversive pursuit only (Morrow and Sharpe 1995) or may affect pursuit in both directions (Lekwuwa and Barnes 1996).

4.3 Neuroimaging of Pursuit and Component Processes

Several neuroimaging studies have investigated the neural substrates of ETD in schizophrenia patients and in their first-degree relatives. Paradigms used have included closed-loop smooth pursuit and predictive pursuit as well as tasks tapping motion perception.

An early imaging study relating neural activation to ETD found that reduced FEF activation during an attentional task was correlated with measures of pursuit quality outside the scanner (Ross et al. 1995). Subsequent studies of ETD in patients have compared the activation observed during smooth pursuit in schizophrenia patients with that seen in controls. Results are somewhat difficult to summarize across studies because coordinates differ by up to 4 cm across studies for both putative MT/MST and for FEF. Setting these anatomical discrepancies aside, a few studies have reported lower activation in schizophrenia patients than in controls in MT/MST (Lencer et al. 2005; Keedy et al. 2006) and an adjacent anterior temporal region (Hong et al. 2005b), as well as in FEFs (Tregellas et al. 2004; Hong et al. 2005b; Keedy et al. 2006), supplementary eye fields (Hong et al. 2005b), parietal cortex (Keedy et al. 2006), and cingulate (Hong et al. 2005b; Keedy et al. 2006). Differences have also been found outside the traditional pursuit pathway, with replications of increased activity in patients in hippocampus (Tregellas et al. 2004; Tanabe et al. 2006), thalamus (Tregellas et al. 2004; Nagel et al. 2007), and right fusiform gyrus (Tregellas et al. 2004; Tanabe et al. 2006). The scatter in coordinates for canonical regions does not occur in comparing pursuit to fixation, but in comparing the pursuit-related activation in schizophrenia to pursuit-related activation in controls. These outlying activations, which fall in the periphery of a region of interest, could result from a comparison of two different size peaks (in controls vs. patients) centered on the same location. Higher peaks have wider peripheries (due to spatial smoothing), so two activations in the same location may yield maximal statistical differences in the periphery of the peaks where standard deviations for the group with the small peak will be very low.

There are several limitations in the interpretation of these studies. First, for most studies, differences in activations between groups may not be due to ETD, but rather to other factors associated with the diagnosis (e.g., medication, institutionalization) that could affect brain function. To minimize these differences, Keedy and colleagues (2006) included only first-episode, neuroleptic-naive patients; their study found extensive deficits in pursuit activation, and the authors concluded that there was a "system-wide" involvement of cortical oculomotor areas. Another limitation of most of the studies is that schizophrenia patients with pursuit deficits are compared with controls with no pursuit deficits. Since the groups differ in eye tracking performance, activation differences between the groups may simply reflect group differences in engagement in the task. Hong and colleagues attempted to minimize this problem by comparing patients and controls who were matched for average pursuit performance. Group differences in visual processing areas (increased activation), and in FEFs and supplementary eye fields (decreases in schizophrenia), were still found (Hong et al. 2005b). However, if there are no group differences in average pursuit performance, the extent to which the differences in activation are attributable to pursuit rather than to diagnosis remains unclear.

A more compelling design might involve comparing poor tracking and good tracking patients with each other and with controls [see (Levy et al. 2000)]. Such a comparison has the advantage of clarifying the neural correlates of ETD unconfounded by neural differences that are specific to the diagnosis rather than to tracking. A study that used this type of approach to examine ETD in unaffected first-degree relatives of schizophrenia patients made a strong case for FEF dysfunction as a substrate of low gain pursuit (O'Driscoll et al. 1999). Controls and relatives with normal pursuit both significantly activated FEFs during smooth pursuit, whereas demographically similar relatives with ETD as a group did not ($p > 0.9$). A correlational analysis relating regional neural activation to pursuit gain in the relatives found the highest correlation to be in the FEFs ($r = 0.74$). The peak correlation was located only 3 mm from the site of maximum FEF activation in controls. No group differences in activation were found in motion perception areas.

The extraretinal component of pursuit was examined in one imaging study of schizophrenia (Nagel et al. 2007). Patients and controls were examined during predictive tracking of a target that was periodically blanked. There were no significant performance differences between groups during target blanking, although gain values during blanking dropped to the 0.2 range, suggesting that neither group was able to sustain predictive pursuit. The schizophrenia group was found to have reduced activation in cerebellum during predictive tracking compared with controls, and increased activation in right anterior cingulate and in an area referred to as FEFs, although the very posterior location, $y = -20$, suggests that this may correspond to motor strip eye field, [see (Tehovnik et al. 2000)], an area that has been implicated in oculomotor prediction (Gagnon et al. 2002).

The integrity of motion processing areas supporting pursuit has been assessed in several imaging studies. One study had schizophrenia patients and controls make speed discriminations and contrast discriminations in the scanner (Chen et al. 2008). Controls showed strong activation (BOLD signal changes) in area MT/MST during motion tasks, consistent with the known role of this region in sensory processing of motion stimuli. Schizophrenia patients showed significantly less activation than controls in MT/MST. The groups did not differ in activation patterns while processing nonmotion stimuli. During motion processing, patients activated the inferior convexity of the prefrontal cortex more than controls did, suggesting that cognitive processing may have been used to help compensate for deficient sensory processing. Another study compared activation in first-episode neuroleptic-naïve schizophrenia patients and controls during passive viewing of motion stimuli compared with fixation. Patients had widespread reductions in activation, including in lateral and medial geniculate nuclei of right thalamus, a ventral region of FEF, as well as in occipital cortex, temporal lobe, and inferior parietal lobe (Braus et al. 2002). Widespread abnormalities were also found in schizophrenia in a study investigating the integrity of magnocellular vs. parvocellular pathways (Martinez et al. 2008). Magnocellular pathways are preferentially involved in motion processing, and some studies have suggested that schizophrenia patients are selectively impaired on tasks that tap magnocellular function as opposed to parvocellular function [(Kéri et al. 2004; Delord et al. 2006), but see also (Skottun and Skoyles 2007)]. Patients and controls viewed sinusoidal gratings biased to preferentially activate magnocellular (low spatial frequency and low contrast) or parvocellular (high spatial frequency) pathways. Differences between groups emerged only in the magnocellular condition. Reduced activation was found throughout the magnocellular system, including visual cortex, temporal cortex, and the dorsal parietal pathway (Martinez et al. 2008).

In sum, neuroimaging studies of maintenance pursuit have reported reduced activation of FEFs and motion processing areas in schizophrenia, with some studies finding that the reductions are more widespread and others finding as well greater activation in some areas outside the traditional pursuit pathway. Studies of motion processing are similarly divided between findings of focal reduction in motion processing areas and in generalized reductions that include thalamus, visual cortex, parietal cortex, and other regions in the dorsal stream, with some evidence of compensatory activations outside the motion pathway. To date, studies comparing patients with and without pursuit deficits or with and without motion processing deficits have not been conducted.

5 Association Between Genetic Polymorphisms and ETD

When an endophenotype is a more penetrant, pleiotropic expression of the same genes that are risk factors for schizophrenia, it can increase power to detect linkage for schizophrenia susceptibility genes compared with the clinical disorder alone (Lander 1988; Holzman and Matthysse 1990; Matthysse and Parnas 1992; Holzman 1994; Freedman et al. 1999). Indeed, this is the primary rationale for incorporating endophenotypes (Gottesman and Gould 2003) into linkage studies of complex diseases. The reason for this improvement in power is that the endophenotype (in this case, ETD) would improve accurate identification of non-penetrant gene carriers (Matthysse and Parnas 1992; Botstein and Risch 2003).

The first effort to examine the usefulness of ETD measures in linkage studies was conducted by Arolt and colleagues (Arolt et al. 1996, 1999). Using a gain score dichotomized into normal or abnormal pursuit, they calculated two point linkage analyses between ETD and 16 microsatellite markers on chromosome 6p21–23. A maximum LOD (logarithm of the odds to the base 10) score of 3.51 was obtained for marker D6S271 ($\theta = 0.0$); marker D6S282 yielded a maximum LOD score of 3.44 at $\theta = 0.05$ (Arolt et al. 1996). The results were quite similar when the analyses were repeated on a slightly larger sample using additional markers in the same region. Independent support for these results was found in other studies that combined qualitative ratings of ETD and schizophrenia as part of a latent trait model (Matthysse and Holzman 1987; Holzman et al. 1988); a LOD score of 2.05 was found for a marker within 3 cm of the positive markers studied by Arolt and colleagues (Matthysse et al. 2004).

Several studies have examined the relation between the COMT (catechol-O-methyltransferase) genotype and ETD. Rybakowski and colleagues reported that the Met/Met genotype was significantly associated with *better* closed-loop gain in male schizophrenia patients (Rybakowski et al. 2002). A similar association between this genotype and predictive gain was found in controls in a study by Thaker and colleagues (Thaker et al. 2004). However, in that study, patients with this genotype did not differ in maintenance gain and had *worse* predictive gain than patients with the Val/Val or Val/Met genotypes. Haraldsson and colleagues recently reported no association between the rs4680 val^{158}met COMT polymorphism and either schizophrenia or steady-state pursuit gain and saccade frequency (Haraldsson et al. 2009). Further studies are needed to clarify this assortment of different findings with respect to COMT. Polymorphisms in other genes have also been examined in several samples, with reported but unconfirmed associations between pursuit performance and genotype (Rybakowski et al. 2001; Bogacki et al. 2005).

6 Summary

ETD is a robust finding associated with schizophrenia and shows significant co-familiality. Using well-characterized paradigms that were developed in nonhuman primate single-unit work, researchers have attempted to link specific component processes of pursuit to specific neural substrates. Despite variability in quantitative measures and behavioral paradigms, there is general agreement that ETD seems to involve impairments in motion processing and in higher-order processes such as prediction and gain control of signals driving pursuit. Motion-sensitive regions (MT/MST) and the FEF have been implicated as neural substrates of ETD, although some neuroimaging studies suggest a more system-wide pattern of dysfunction in the dorsal stream. Genetic associations with ETD have not yet conclusively implicated any one chromosomal region or specific genes.

Acknowledgments This work was supported in part by NIMH grants R01 MH071523 and MH31340, the Sidney R. Baer, Jr. Foundation, the Essel Foundation, the National Association for Research on Schizophrenia and Depression (NARSAD), an Essel Investigator NARSAD and NSF grant 0924636, a grant from the Canadian Institute of Health Research, a William Dawson Scholar Award, and a Stairs Memorial Foundation grant. The authors thank Dr. Larry Abel for making the original material for Fig. 2 available for adaptation and Joshua Ritz for formatting the figures.

References

Abel L, Ziegler A (1988) Smooth pursuit eye movements in schizophrenics – what constitutes quantitative assessment? Biol Psychiatry 24:747–761

Abel LA, Levin S, Holzman PS (1992) Abnormalities of smooth pursuit and saccadic control in schizophrenia and affective disorders. Vision Res 32:1009–1014

Allen JS, Matsunaga K, Hacisalihzade S, Stark L (1990) The smooth pursuit eye movements of normal and schizophrenic subjects tracking an unpredictable target. Biol Psychiatry 28:795–720

Amador XF, Sackheim HA, Mukherjee S, Halperin R, Neeley P, Maclin E, Schnur D (1991) Specificity of smooth pursuit eye movement and visual fixation abnormalities in schizophrenia: comparison to mania and normal controls. Schizophr Res 5:135–144

Arolt V, Lencer R, Nolte A, Muller-Myhsok B, Purmann S, Schurmann M, Leutelt J, Pinnow M, Schwinger E (1996) Eye tracking dysfunction is a putative phenotypic susceptibility marker of schizophrenia and maps to a locus on chromosome 6p in families with multiple occurrence of the disease. Am J Med Genet 67:564–579

Arolt V, Teichert H-M, Steege D, Lencer R, Heide W (1998) Distinguishing schizophrenic patients from healthy controls by quantitative measurement of eye movement parameters. Biol Psychiatry 44:448–458

Arolt V, Lencer R, Purmann S, Schurmann M, Muller-Myhsok B, Krecker K, Schwinger E (1999) Testing for linkage of eye tracking dysfunction to markers on chromosomes 6, 8, 9, 20, 22 in families multiply affected with schizophrenia. Am J Med Genet 88:603–606

Avila MT, Hong LE, Moates A, Turano KA, Thaker GK (2006) Role of anticipation in schizophrenia-related pursuit initiation deficits. J Neurophysiol 95:593–601

Baker CL, Hess RF, Zihl J (1991) Residual motion perception in a "motion blind" patient assessed with limited-lifetime random dot stimuli. J Neurosci 11:451–461

Baloh RW, Honrubia V (1990) Clinical neurophysiology of the vestibular system. F.A. Davis Company, Philadelphia

Barborica A, Ferrera VP (2003) Estimating invisible target speed from neuronal activity in monkey frontal eye field. Nat Neurosci 6:66–74

Barborica A, Ferrera VP (2004) Modification of saccades evoked by stimulation of frontal eye field during invisible target tracking. J Neurosci 24:3260–3267

Barnes GR, Donnelly SF, Eason RD (1987) Predictive velocity estimation in the pursuit reflex response to pseudo-random and step displacement stimuli in man. J Physiol 389:111–136

Barton JJS, Cherkasova M, Lindgren K, Goff DC, Intriligator JM, Manoach DS (2002) Antisaccades and task-switching: studies of control processes in saccadic function in normal subjects and schizophrenic patients. Neurobiology of eye movements: from molecules to behavior. Ann N Y Acad Sci 956:250–263

Barton JJS, Pandita M, Thakkar K, Goff DC, Manoach DS (2008) The relation between antisaccade errors, fixation stability and prosaccade errors in schizophrenia. Exp Brain Res 186:273–282

Bauer LO (1997) Smooth pursuit eye movement dysfunction in abstinent cocaine abusers. Clin Exp Res 21:910–915

Bell BB, Abel LA, Li W, Christian JC, Yee RD (1994) Concordance of smooth pursuit and saccadic measures in normal monozygotic twins. Biol Psychiatry 36:522–526

Biscaldi M, Gezeck S, Stuhr V (1998) Poor saccadic control correlates with dyslexia. Neuropsychologia 36:1189–1202

Bogacki PA, Borkowska A, Wojanowska-Bogacka M, Rybakowski JK (2005) Relationship between class I and II HLA antigens in schizophrenia and eye movement disturbances: a preliminary study. Neuropsychobiology 51:204–210

Botstein D, Risch N (2003) Discovering genotypes underlying human phenotypes: past successes for Mendelian disease, future approaches for complex disease. Nat Genet 33(Suppl):228–237

Braus DF, Weber-Fahr W, Tost H, Ruf M, Henn FA (2002) Sensory information processing in neuroleptic-naive first-episode schizophrenic patients. Arch Gen Psychiatry 59:696–701

Bremmer F, Ilg UJ, Thiele A, Distler C, Hoffmann KP (1997) Eye position effects in monkey cortex. I. Visual and pursuit-related activity in extrastriate areas MT and MST. J Neurophsyiol 77:944–961

Britten KH, Newsome WT, Shadlen MN, Celebrini S, Movshon JA (1996) A relationship between behavioral choice and the visual response of neurons in macaque MT. Vis Neurosci 13:87–100

Brownstein J, Krastoshevsky O, McCollum C, Kundamal S, Matthysse S, Holzman PS, Mendell NR, Levy DL (2003) Antisaccade performance is abnormal in schizophrenia patients but not in their biological relatives. Schizophr Res 63:13–25

Buchsbaum MS, Rieder RO (1979) Biological heterogeneity and psychiatric research. Arch Gen Psychiatry 36:1163–1169

Butler PD, Schechter I, Zemon V, Schwartz SG, Greenstein VC, Gordon J, Schroeder CE, Javitt DC (2001) Dysfunction in early-stage visual processing in schizophrenia. Am J Psychiatry 158:1126–1133

Calkins ME, Iacono WG, Curtis CE (2003) Smooth pursuit and antisaccade performance evidence trait stability in schizophrenia patients and their relatives. Int J Psychophysiol 49:139–146

Calkins ME, Iacono WG, Ones DS (2008) Eye movement dysfunction in first-degree relatives of patients with schizophrenia: a meta-analytic evaluation of candidate endophenotypes. Brain Cogn 68:436–461

Castellanos FX, Marvasti FF, Ducharme JL, Walter JM, Israel ME, Krain A, Pavlovsky C, Hommer DW (2000) Executive function oculomotor tasks in girls with ADHD. J Am Acad Child Adolesc Psychiatry 39:644–650

Ceballos NA, Bauer LO (2004) Effects of antisocial personality, cocaine and opioid dependence, and gender on eye movement control. Psychol Rep 95:551–563

Chen Y, Levy DL, Nakayama K, Matthysse S, Holzman PS (1999a) Dependence of impaired eye tracking on deficient velocity discrimination of schizophrenia. Arch Gen Psychiatry 56:155–161

Chen Y, Nakayama K, Levy DL, Matthysse S, Holzman PS (1999b) Psychophysical isolation of a motion-processing deficit in schizophrenics and their relatives and its association with impaired smooth pursuit. Proc Natl Acad Sci USA 96:4724–4729

Chen Y, Palafox G, Nakayama K, Levy DL, Matthysse S, Holzman PS (1999c) Motion perception in schizophrenia. Arch Gen Psychiatry 56:149–154

Chen Y, Nakayama K, Levy DL, Matthysse S, Holzman PS (2003) Processing of global, but not local, motion direction is deficient in schizophrenia. Schizophr Res 61:215–227

Chen Y, Levy DL, Sheremata S, Holzman PS (2004) Compromised late-stage motion processing in schizophrenia. Biol Psychiatry 55:834–841

Chen Y, Grossman ED, Bidwell LC, Yurgelun-Todd D, Gruber SA, Levy DL, Nakayama K, Holzman PS (2008) Altered posterior and prefrontal cortical activation during visual motion processing in schizophrenia. Cogn Affect Behav Neurosci 8:293–303

Clementz BA (1996) Saccades to moving targets in schizophrenia: evidence for normal posterior cortex functioning. Psychophysiology 33:650–654

Clementz BA, McDowell JE (1994) Smooth-pursuit in schizophrenia: abnormalities of open-loop and closed-loop responses. Psychophysiology 31:79–86

Clementz BA, McDowell JE, Zisook S (1994) Saccadic system functioning among schizophrenia patients and their first-degree biological relatives. J Abnorm Psychol 103:277–287

Clementz BA, Farber RH, Lam MN, Swerdlow NR (1996) Ocular motor responses to unpredictable and predictable smooth pursuit stimuli among patients with obsessive-compulsive disorder. J Psychiatry Neurosci 21:21–28

Coleman MJ, Cestnick L, Krastoshevsky O, Krause V, Huang Z, Mendell NR, Levy DL (2009) Schizophrenia patients show deficits in shifts of attention to different levels of global–local stimuli: evidence for magnocellular dysfunction. Schizophr Bull 35:1108–1116

Cooper PM, Pivik RT (1991) Abnormal visual–vestibular interaction and smooth pursuit tracking in psychosis. J Psychiatry Neurosci 16:30–40

Couch FH, Fox JC (1934) Photographic study of ocular movements in mental disease. Arch Neurol Psychiatry 34:556–578

Crawford TJ, Haegar B, Kennard C, Reveley MA, Henderson L (1995a) Saccadic abnormalities in psychotic patients. I. Neuroleptic-free psychotic patients. Psychol Med 25:461–471

Crawford TJ, Haegar B, Kennard C, Reveley MA, Henderson L (1995b) Saccadic abnormalities in psychotic patients. II. The role of neuroleptic treatment. Psychol Med 25:473–483

Curtis CE, Calkins ME, Grove WM, Feil KJ, Iacono WG (2001) Saccadic disinhibition in patients with acute and remitted schizophrenia and their first-degree biological relatives. Am J Psychiatry 158:100–106

Delord S, Ducato MG, Pins D, Devinck F, Thomas P, Boucart M, Knoblauch K (2006) Psychophysical assessment of magno- and parvocellular function in schizophrenia. Vis Neurosci 23:645–650

Depatie L, O'Driscoll GA, Holahan A-LV, Atkinson V, Thavundayil JX, Kin NY, Lal S (2002) Nicotine and behavioral markers of risk for schizophrenia: a double-blind, placebo-controlled, cross-over study. Neuropsychopharmacology 27:1056–1070

Diefendorf AR, Dodge R (1908) An experimental study of the ocular reactions of the insane from photographic records. Brain 31:451–489

Dodge R (1903) Five types of eye movement in the horizontal meridian plane in the field of regard. Am J Physiol 8:307–329

Dursteler MR, Wurtz RH (1988) Pursuit and optokinetic deficits following chemical lesions of cortical areas MT and MST. J Neurophsyiol 60:940–965

Ettinger U, Kumari V, Crawford TJ, Corr PJ, Das M. Zachariah E, Hughes C, Sumich AL, Rabe-Hesketh S, Sharma T (2004) Smooth pursuit and antisaccade eye movements in siblings discordant for schizophrenia. J Psychiatr Res 38:177–184

Everling S, Krappman P, Pruess S, Brand A, Flohr H (1996) Hypometric primary saccades of schizophrenics in a delayed response task. Exp Brain Res 111:289–295

Farber RH, Clementz BA, Swerdlow NR (1997) Characteristics of open- and closed-loop smooth pursuit responses among obsessive-compulsive disorder, schizophrenia, and nonpsychiatric individuals. Psychophysiology 34:157–162

Feil K (1997) Smooth pursuit eye movement dysfunctions in remitted and acute schizophrenia: a failure to replicate. Department of Psychology, University of Minnesota, Minneapolis, MN

Freedman R, Adler LE, Leonard S (1999) Alternative phenotypes for the complex genetics of schizophrenia. Biol Psychiatry 45:551–558

Friedman L, Jesberger J, Siever L, Thompson P, Mohs R, Meltzer H (1995) Smooth pursuit performance in patients with affective disorders or schizophrenia and normal controls: analysis with specific oculomotor measures, RMS error and qualitative ratings. Psychol Med 25:387–403

Fukushima J, Morita N, Fukushima K, Chiba T, Tanaka S, Yamashita I (1990) Voluntary control of saccadic eye movements in patients with schizophrenic and affective disorders. J Psychiatr Res 24:9–24

Gagnon D, O'Driscoll GA, Petrides M, Pike GB (2002) The effect of spatial and temporal information on saccades and neural activity in oculomotor structures. Brain 125:123–139

Gagnon D, Paus T, Grosbras M-H, Pike GB, O'Driscoll GA (2006) Transcranial magnetic stimulation of frontal oculomotor regions during smooth pursuit. J Neurosci 26:458–466

Gibbons RD, Dorus E, Ostrow D, Pandey GN, Davis JM, Levy DL (1984) Mixture distributions in psychiatric research. Biol Psychiatry 19:935–961

Gooding DC (1999) Antisaccade task performance in questionnaire-identified schizotypes. Schizophr Res 35:157–166

Gooding DC, Basso MA (2008) The tell-tale tasks: a review of saccadic research in psychiatric patient populations. Brain Cogn 68:371–390

Gooding DC, Tallent KA (2001) The association between antisaccade task and working memory task performance in schizophrenia and bipolar disorder. J Nerv Ment Dis 189:8–16

Gooding DC, Miller MD, Kwapil TR (2000) Smooth pursuit eye tracking and visual fixation in psychosis-prone individuals. Psychiatry Res 93:41–54

Gottesman II, Gould TD (2003) The endophenotype concept in psychiatry: etymology and strategic intentions. Am J Psychiatry 130:636–645

Gottlieb JP, Bruce CJ, MacAvoy MG (1993) Smooth pursuit eye movements elicited by microstimulation of the primate frontal eye field. J Neurophysiol 69:786–799

Gottlieb JP, MacAvoy MG, Bruce CJ (1994) Neural responses related to smooth pursuit eye movements and their correspondence with electrically elicited smooth eye movements in the primate frontal eye field. J Neurophysiol 72:1634–1653

Green MF, Mintz J, Salveson D, Nuechterlein KH, Breitmeyer B, Light GA, Braff DL (2003) Visual masking as a probe for abnormal gamma range activity in schizophrenia. Biol Psychiatry 53:113–1119

Haraldsson HM, Ettinger U, Magnusdottir BB, Sigmundsson T, Sigurdsson E, Ingason A, Petursson H (2009) COMT val^{158}met genotype and smooth pursuit eye movements in schizophrenia. Psychiatry Res 169:173–175

Heide W, Kurzidim K, Kompf D (1996) Deficits of smooth pursuit eye movements after frontal and parietal lesions. Brain 119:1951–1969

Holzman PS (1969) Perceptual aspects of psychopathology. In: Zubin J, Shagass C (eds) Neurobiological aspects of psychopathology. Grune & Stratton, New York, pp 144–178

Holzman PS (1994) The role of psychological probes in genetic studies of schizophrenia. Schizophr Res 13:1–9

Holzman PS (2000) Eye movements and the search for the essence of schizophrenia. Brain Res Rev 31:350–356

Holzman PS, Matthysse S (1990) The genetics of schizophrenia: a review. Psychol Sci 1:279–286

Holzman PS, Proctor LR, Hughes DW (1973) Eye-tracking patterns in schizophrenia. Science 181:179–181

Holzman PS, Proctor LR, Levy DL, Yasillo NJ, Meltzer HY, Hurt SW (1974a) Eye tracking dysfunctions in schizophrenic patients and their relatives. Arch Gen Psychiatry 31:143–151

Holzman PS, Proctor LR, Yasillo NJ, Levy DL (1974b) Reply to Stevens. Science 184:1203–1204

Holzman PS, Kringlen E, Levy DL, Proctor LR, Haberman S, Yasillo NJ (1977) Abnormal-pursuit eye movements in schizophrenia: evidence for a genetic indicator. Arch Gen Psychiatry 34:802–805

Holzman PS, Kringlen E, Matthysse S, Flanagan SD, Lipton RB, Cramer G, Levin S, Lange K, Levy DL (1988) A single dominant gene can account for eye tracking dysfunctions and schizophrenia in offspring of discordant twins. Arch Gen Psychiatry 45:641–647

Hong LE, Avila MT, Adami H, Elliot A, Thaker GK (2003) Components of the smooth pursuit function in deficit and nondeficit schizophrenia. Schizophr Res 63:39–48

Hong LE, Avila MT, Thaker GK (2005a) Response to unexpected target changes during sustained visual tracking in schizophrenic patients. Exp Brain Res 165:125–131

Hong LE, Tagamets M, Avila MT, Wonodi I, Holcomb H, Thaker GK (2005b) Specific motion processing pathway deficit during eye tracking in schizophrenia: a performance-matched functional magnetic resonance imaging study. Biol Psychiatry 57:726–732

Hong LE, Turano KA, O'Neill J, Hao LIW, McMahon RP, Elliott A, Thaker GK (2008) Refining the predictive pursuit endophenotype in schizophrenia. Biol Psychiatry 63:458–464

Hooker C, Park S (2000) Trajectory estimation in schizophrenia. Schizophr Res 45:83–92

Hutton SB, Crawford TJ, Puri BK, Duncan LJ, Chapman M, Kennard C, Barnes TRE, Joyce EM (1998) Smooth pursuit and saccadic abnormalities in first-episode schizophrenia. Psychol Med 28:685–692

Hutton SB, Crawford TJ, Kennard C, Barnes TRE, Joyce EM (2000) Smooth pursuit tracking over a structured background in first-episode schizophrenic patients. Eur Arch Psychiatry Clin Neurosci 250:221–225

Hutton S, Cuthbert I, Crawford TJ, Kennard C, Barnes TRE, Joyce EM (2001) Saccadic hypometria in drug-naïve and drug-treated schizophrenic patients: a working memory deficit? Psychophysiology 38:125–132

Iacono W, Lykken D (1979) Electro-oculographic recording and scoring of smooth pursuit and saccadic eye tracking: a parametric study using monozygotic twins. Psychophysiology 16:94–107

Iacono W, Peloquin LJ, Lumry AE, Valentine RH, Tuason VB (1982) Eye tracking in patients with unipolar and bipolar affective disorders in remission. J Abnorm Psychol 91:35–44

Jacobsen LK, Hong WL, Hommer DW, Hamburger SD, Castellanos FX, Frazier JA, Giedd JN, Gordon CT, Karp BI, McKenna K, Rapoport JL (1996) Smooth pursuit eye movements in child onset schizophrenia: comparison with attention-deficit hyperactivity disorder and normal controls. Biol Psychiatry 40:1144–1154

Jones AM, Pivik RT (1983) Abnormal visual–vestibular interactions in psychosis. Biol Psychiatry 18:45–61

Karoumi B, Ventre-Dominey J, Dalery J (1998) Predictive saccade behavior is enhanced in schizophrenia. Cognition 68:B81–B91

Kathmann N, Hochrein A, Uwer R, Bondy B (2003) Deficits in gain of smooth pursuit eye movements in schizophrenia and affective disorder patients and their unaffected relatives. Am J Psychiatry 160:696–702

Katsanis J, Kortenkamp S, Iacono WG, Grove WM (1997) Antisaccade performance in patients with schizophrenia and affective disorder. J Abnorm Psychol 106:468–472

Katsanis J, Taylor J, Iacono WG, Hammer MA (2000) Heritability of different measures of smooth pursuit eye tracking dysfunction: a study of normal twins. Psychophysiology 37:724–730

Keating EG (1991) Frontal eye field lesions impair predictive and visually-guided pursuit eye movements. Exp Brain Res 86:311–323

Keating EG (1993) Lesions of the frontal eye field impair pursuit eye movements, but preserve the predictions driving them. Behav Brain Res 53:91–104

Keedy SK, Ebens CL, Keshavan MS, Sweeney JA (2006) Functional magnetic resonance imaging studies of eye movements in first episode schizophrenia: smooth pursuit, visually guided saccades and the oculomotor delayed response task. Psychiatry Res 146:199–211

Keefe R, Silverman J, Mohs R, Siever L, Harvey P, Friedman L, Roitman S, DuPre R, Smith C, Schmeider J, Davis K (1997) Eye tracking, attention, and schizotypal symptoms in nonpsychotic relatives of patients with schizophrenia. Arch Gen Psychiatry 54:169–176

Kéri S, Kelemen O, Benedek G, Janka Z (2004) Vernier threshold in patients with schizophrenia and in their unaffected siblings. Neuropsychology 18:537–542

Kim D, Wylie G, Pasternak R, Butler PD, Javitt DC (2006) Magnocellular contributions to impaired motion processing in schizophrenia. Schizophr Res 82:1–8

Komatsu H, Wurtz RH (1989) Modulation of pursuit eye movements by stimulation of cortical areas MT and MST. J Neurophysiol 62:31–47

Krauzlis RJ, Lisberger SG (1994) Temporal properties of visual motion signals for the initiation of smooth-pursuit eye-movements in monkeys. J Neurophysiol 72:150–162

Krebs MO, Gut-Fayand A, Amado I, Daban C, Bourdel M, Poirier MF, Berthoz A (2001) Impairment of predictive saccades in schizophrenia. Neuroreport 12:465–469

Kumra S, Sporn A, Hommer DW, Nicolson R, Thaker G, Israel E, Lenane M, Bedwell J, Jacobsen LK, Gochman P, Rapoport JL (2001) Smooth pursuit eye-tracking impairment in childhood-onset psychotic disorders. Am J Psychiatry 158:1291–1298

Lander E (1988) Splitting schizophrenia. Nature 336:105–106

Larrison AL, Ferrante CF, Briand KA, Sereno AB (2000) Schizotypal traits, attention and eye movements. Prog Neuropsychopharmacol Biol Psychiatry 24:357–372

Larrison AL, Briand KA, Sereno AB (2004) Nicotine improves antisaccade task performance without affecting prosaccades. Hum Psychopharmacol 19:409–419

Larrison-Faucher A, Briand KA, Sereno AB (2002) Delayed onset of inhibition of return in schizophrenia. Prog Neuropsychopharmacol Biol Psychiatry 26:505–512

Latham C, Holzman PS, Manschrek TC, Tole J (1981) Optokinetic nystagmus and pursuit eye movements in schizophrenia. Arch Gen Psychiatry 38:997–1003

Leigh RJ, Zee DS (2006) The neurology of eye movements. Oxford University Press, New York

Lekwuwa GU, Barnes GR (1996) Cerebral control of eye movements: the relationship between cerebral lesions sites and smooth pursuit deficits. Brain 119:473–490

Lencer R, Trillenberg P, Trillenberg-Krecker K, Junghanns K, Kordon A, Broocks A, Hohagen F, Heide W, Arolt V (2004) Smooth pursuit deficits in schizophrenia, affective disorder and obsessive-compulsive disorder. Psychol Med 34:451–460

Lencer R, Nagel M, Sprenger A, Heide W, Binkofski F (2005) Reduced neuronal activity in the V5 complex underlies smooth-pursuit deficit in schizophrenia: evidence from an fMRI study. Neuroimage 24:1256–1259

Lencer R, Sprenger A, Harris MSH, Reilly JL, Keshavan MS, Sweeney JA (2008) Effects of second-generation antipsychotic medication on smooth pursuit performance in antipsychotic-naive schizophrenia. Arch Gen Psychiatry 65:1146–1154

Lenzenweger MF, O'Driscoll GA (2006) Smooth pursuit eye movement and schizotypy in the community. J Abnorm Psychol 115:779–786

LeVasseur AL, Flanagan JR, Riopelle RJ, Munoz DP (2001) Control of volitional and reflexive saccades in Tourette's syndrome. Brain 124:2045–2058

Levin S (1984) Frontal lobe dysfunction in schizophrenia. I. Eye movement impairments. J Psychiatr Res 18:27–55

Levin S, Jones A, Stark L, Merrin EL, Holzman PS (1982) Saccadic eye movements of schizophrenic patients measured by reflected light technique. Biol Psychiatry 17:1277–1287

Levy DL, Holzman PS, Proctor LR (1978) Vestibular responses in schizophrenia. Arch Gen Psychiatry 35:972–981

Levy DL, Holzman PS, Proctor LR (1983) Vestibular dysfunction and psychopathology. Schizophr Bull 9:383–483

Levy DL, Holzman PS, Matthysse S, Mendell NR (1993) Eye tracking dysfunction and schizophrenia: a critical perspective. Schizophr Bull 19:461–536

Levy DL, Lajonchere CM, Dorogusker B, Min D, Lee S, Tartaglini A, Lieberman JA, Mendell NR (2000) Quantitative characterization of eye tracking dysfunction in schizophrenia. Schizophr Res 42:171–185

Levy DL, O'Driscoll G, Cook SR, Matthysse S, Holzman PS, Mendell NR (2004) Antisaccade performance in biological relatives of schizophrenia patients: a meta-analysis. Schizophr Res 71:113–125

Li CR (2002) Impaired detection of visual motion in schizophrenic patients. Prog Neuropsychopharmacol Biol Psychiatry 26:929–934

Lipton RB, Levin S, Holzman PS (1980) Horizontal and vertical pursuit eye movements, the oculocephalic reflex and the functional psychoses. Psychiatr Res 3:193–203

Lisberger SG, Westbrook LE (1985) Properties of visual inputs that initiate horizontal smooth-pursuit eye movements in monkeys. J Neurosci 5:1662–1673

Lisberger SG, Morris EJ, Tyschen L (1987) Visual motion processing and sensory-motor integration for smooth pursuit eye movements. Annu Rev Neurosci 10:97–129

MacAvoy MG, Gottlieb JP, Bruce CJ (1991) Smooth pursuit eye movement representation in the primate frontal eye field. Cereb Cortex 1:95–102

Martinez A, Hillyard SA, Dias EC, Hagler DJ Jr, Butler PD, Guilfoyle DN, Jalbrzikowski M, Silipo G, Javitt DC (2008) Magnocellular pathway impairment in schizophrenia: evidence from functional magnetic resonance. J Neurosci 28:7492–7500

Maruff P, Danckert J, Pantelis C, Currie J (1998) Saccadic and attentional abnormalities in patients with schizophrenia. Psychol Med 28:1091–1100

Matthysse S, Holzman PS (1987) Genetic latent structure models: implications for research on schizophrenia. Psychol Med 17:271–274

Matthysse S, Parnas J (1992) Extending the phenotype of schizophrenia: implications for linkage analysis. J Psychiatr Res 26:329–344

Matthysse S, Holzman PS, Gusella J, Levy DL, Harte C, Jorgensen A, Moller L, Parnas J (2004) Linkage of eye movement dysfunction to chromosome 6p in schizophrenia: A confirmation. Am J Hum Genet (Neuropsychiatric Genetics) 128B:30–36

McDowell JE, Clementz BA (1996) Oculomotor delayed response task performance among schizophrenia patients. Neuropsychobiology 34:67–71

McDowell JE, Clementz BA (1997) The effect of fixation condition manipulations on antisaccade performance in schizophrenia: studies of diagnostic specificity. Exp Brain Res 115:333–344

McKee SP (1981) A local mechanism for differential velocity detection. Vision Res 21:491–500

Morrow MJ, Sharpe JA (1995) Deficits of smooth-pursuit eye movement after unilateral frontal lobe lesions. Ann Neurol 37:443–451

Mostofsky SH, Lasker AG, Cutting LE, Denckla MB, Zee DS (2001) Oculomotor abnormalities in attention deficit hyperactivity disorder: a preliminary study. Neurology 57:423–430

Muller N, Riedel M, Eggert T, Straube A (1999) Internally and externally guided voluntary saccades in unmedicated and medicated schizophrenics. II. Saccadic latency, gain, and fixation suppression errors. Eur Arch Psychiatry Clin Neurosci 249:7–14

Munoz DP, Armstrong IT, Hampton KA, Moore KD (2003) Altered control of visual fixation and saccadic eye movements in attention-deficit hyperactivity disorder. J Neurophysiol 90:503–514

Nagel M, Sprenger A, Nitschke M, Zapf S, Heide W, Binkofksi F, Lencer R (2007) Different extraretinal neuronal mechanisms of smooth pursuit eye movements in schizophrenia: an fMRI study. Neuroimage 34:300–309

Nakayama K, Tyler CW (1981) Psychophysical isolation of movement sensitivity by removal of familiar position cues. Vision Res 21:427–433

Newsome WT, Pare EB (1988) A selective impairment of motion perception following lesions of the middle temporal visual area (MT). J Neurosci 8:2201–2211

Newsome WT, Wurtz RH, Dursteler MR, Mikami A (1985) Deficits in visual motion processing following ibotenic acid lesions of the middle temporal visual area of the macaque monkey. J Neurosci 5:825–840

Newsome WT, Wurtz RH, Komatsu H (1988) Relation of cortical areas MT and MST to pursuit eye movements. II. Differentiation of retinal from extraretinal inputs. J Neurophysiol 60:604–620

O'Driscoll G, Callahan BL (2008) Smooth pursuit in schizophrenia: a meta-analytic review of research since 1993. Brain Cogn 68:359–370

O'Driscoll G, Lenzenweger MF, Holzman PS (1998) Antisaccades and smooth pursuit eye movements and schizotypy. Arch Gen Psychiatry 55:837–843

O'Driscoll GA, Benkelfat C, Florencio PS, Wolff AL, Joober R, Lal S, Evans AC (1999) Neural correlates of eye tracking deficits in first-degree relatives of schizophrenic patients: a positron emission tomography study. Arch Gen Psychiatry 56:1127–1134

O'Driscoll GA, Depatie L, Holahan ALV, Savion-Lemieux T, Barr RG, Jolicoeur C, Douglas VI (2005) Executive functions and methylphenidate response in subtypes of attention-deficit/hyperactivity disorder. Biol Psychiatry 57:1452–1460

Orban de Xivry JJ, Missal M, Lefèvre P (2008) A dynamic representation of target motion drives predictive smooth pursuit during target blanking. J Vis 8(6):1–13

Pallanti S, Grecu LM, Gangemi PF, Massi S, Parigi A, Arnetoli G, Quercioli L, Zaccara G (1996) Smooth-pursuit eye movement and saccadic intrusions in obsessive-compulsive disorder. Biol Psychiatry 40:1164–1172

Pallanti S, Quercioli L, Zaccara G, Ramacciotti AB, Arnetoli G (1998) Eye movement abnormalities in anorexia nervosa. Psychiatry Res 78:59–70

Pantle A (1978) Temporal frequency response characteristic of motion channels measured with three different psychophysical techniques. Percept Psychophys 24:285–294

Park S, Holzman PS (1992) Schizophrenics show spatial working memory deficits. Arch Gen Psychiatry 49:975–982

Park S, Holzman PS, Lenzenweger MF (1995) Individual differences in spatial working memory in relation to schizotypy. J Abnorm Psychol 104:355–363

Pasternak T, Merrigan WH (1994) Motion perception following lesions of the superior temporal sulcus in the monkey. Cereb Cortex 4:247–259

Plant GT, Nakayama K (1993) The characteristics of residual motion perception in the hemifield contralateral to lateral occipital lesions in humans. Brain 116:1337–1353

Radant AD, Claypoole K, Wingerson DK, Cowley DS, Roy-Byrne P (1997) Relationship between neuropsychological and oculomotor measures in schizophrenia patients and normal controls. Biol Psychiatry 42:797–805

Radant A, Dobie DJ, Calkins ME, Olincy A, Braff DL, Cadenhead KS, Freedman R, Green MF, Greenwood TA, Gur RE, Light GA, Meichle SP, Mintz J, Nuechterlein KH, Schork NJ, Seidman LJ, Siever LJ, Silverman JM, Stone WS, Swerdlow NR, Tsuang MT, Turetsky BI,

Tsuang DW (2007) Successful multi-site measurement of antisaccade performance deficits in schizophrenia. Schizophr Res 89:320–329

Ram-Tsur R, Faust M, Caspi A, Gordon CR, Zivotofsky AZ (2006) Evidence for oculomotor deficits in developmental dyslexia: application of the double-step paradigm. Invest Ophthalmol Vis Sci 47:4401–4409

Rashbass C (1961) The relationship between saccadic and smooth tracking eye movements. J Physiol 159:326–338

Rashbass C, Russell GFM (1961) Action of a barbiturate drug (amylobarbitone sodium) on the vestibulo-ocular reflex. Brain 84:329–335

Rivaud S, Muri RM, Gaymard B, Vermersch AI, Pierrot-Deseilligny C (1994) Eye movement disorders after frontal lobe lesions in humans. Exp Brain Res 102:110–120

Rosenberg DR, Dick EL, O'Hearn KM, Sweeney JA (1997) Response inhibition deficits in obsessive compulsive disorder: an indicator of dysfunction of frontostriatal circuits. J Psychiatry Neurosci 22:29–38

Ross RG, Hommer D, Breiger D, Varley C, Radant AD (1994) Eye movement task related to frontal lobe functioning in children with attention deficit hyperactivity disorder. J Am Acad Child Adolesc Psychiatry 33:869–874

Ross DE, Thaker GK, Holcomb HH, Cascella NG, Medoff DR, Tamminga CA (1995) Abnormal smooth pursuit eye movements in schizophrenic patients are associated with cerebral glucose metabolism in oculomotor regions. Psychiatry Res 58:53–67

Ross DE, Thaker GK, Buchanan RW, Lahti AC, Medoff D, Bartko JJ, Moran M, Thaker GK (1996) Association of abnormal smooth pursuit eye movements with the deficit syndrome in schizophrenic patients. Am J Psychiatry 153:1158–1165

Ross RG, Olincy A, Radant A (1999) Amplitude criteria and anticipatory saccades during smooth pursuit eye movements in schizophrenia. Psychophysiology 36:464–468

Ross RG, Olincy A, Harris JG, Sullivan B, Radant A (2000) Smooth pursuit eye movements in schizophrenia and attentional dysfunction: adults with schizophrenia, ADHD, and a normal comparison group. Biol Psychiatry 48:197–203

Rybakowski JK, Borkowska A, Czerski PM, Hauser J (2001) Dopamine D3 receptor (DRD3) gene polymorphism is associated with the intensity of eye movement disturbances in schizophrenic patients and healthy subjects. Mol Psychiatry 6:718–724

Rybakowski JK, Borkowska A, Czerski PM, Hauser J (2002) Eye movement disturbances in schizophrenia and a polymorphism of catechol-O-methyltransferase gene. Psychiatry Res 113:49–57

Sailer U, Eggert T, Strassnig M, Reidel M, Straube A (2007) Predictive eye and hand movements are differentially affected by schizophrenia. Eur Arch Psychiatry Clin Neurosci 257:413–422

Salzman CD, Murasugi CM, Britten KH, Newsome WT (1992) Microstimulation in visual area MT: effects on direction discrimination performance. J Neurosci 12:2331–2355

Schalen L, Pyykko I, Korttila K, Magnusson M, Enbom H (1988) Effects of intravenously given barbiturate and diazepam on eye motor performance in man. Adv Otolaryngol 42:260–264

Schlack A, Hoffmann KP, Bremmer F (2003) Selectivity of macaque ventral intraparietal area (area VIP) for smooth pursuit eye movements. J Physiology 551:551–556

Schlenker R, Cohen R, Berg P, Hubman W, Mohr F, Watzl H, Werther P (1994) Smooth-pursuit eye movement dysfunction in schizophrenia: the role of attention and general psychomotor dysfunctions. Eur Arch Psychiatry Clin Neurosci 244:153–160

Schwartz BD, McGinn T, Winstead DK (1987) Disordered spatiotemporal processing in schizophrenics. Biol Psychiatry 22:688–698

Sereno AB, Holzman PS (1995) Antisaccades and smooth pursuit eye movements in schizophrenia. Biol Psychiatry 37:394–401

Sereno AB, Babin SL, Hood AJ, Jeter CB (2009) Executive functions: eye movements and neuropsychiatric disorders. In: Squire LR (ed) Encyclopedia of neuroscience, vol 4, pp 117–122

Sharpe JA, Morrow MJ (1991) Cerebral hemispheric smooth pursuit disorders. Acta Neurol Belg 91:81–96

Sherr JD, Myers CS, Avila MT, Elliott A, Blaxton TA, Thaker GK (2002) The effects of nicotine on specific eye tracking measures in schizophrenia. Biol Psychiatry 52:721–728

Shi DX, Friedman PR, Bruce CJ (1998) Deficits in smooth pursuit eye movements after muscimol inactivation within the primate's frontal eye field. Neurophysiology 80:458–464

Skottun BC, Skoyles JR (2007) Contrast sensitivity and magnocellular function in schizophrenia. Vision Res 47:2923–2933

Slaghius WL (1998) Contrast sensitivity for stationary and drifting frequency gratings in positive- and negative-symptom schizophrenia. J Abnorm Psychol 107:49–62

Slaghuis WL, Bowling AC, French RV (2005) Smooth pursuit eye movement and direction motion contrast sensitivity in schizophrenia. Exp Brain Res 166:89–101

Slaghuis WL, Hawkes A, Holthouse T, Bruno R (2007a) Eye movement and visual motion perception in schizophrenia I: apparent motion evoked smooth pursuit eye movement reveals a hidden dysfunction in smooth pursuit eye movement in schizophrenia. Exp Brain Res 182:399–413

Slaghuis WL, Holthouse T, Hawkes A, Bruno R (2007b) Eye movement and visual motion perception in schizophrenia II: global coherent motion as a function of target velocity and stimulus density. Exp Brain Res 182:415–426

Smyrnis N, Kattoulas E, Evdokomidis I, Stefanis NC, Avramopoulos D, Pantes G, Theleritis C, Stefanis CN (2004) Active fixation performance in 940 young men: effects of IQ, schizotypy, anxiety and depression. Exp Brain Res 156:1–10

Spengler D, Trillenberg P, Sprenger A, Nagel M, Kordon A, Junhanns K, Heide W, Arolt V, Hohagen F, Lencer R (2006) Evidence from increased anticipation of predictive saccades for a dysfunction of front-striatal circuits in obsessive compulsive disorder. Psychiatry Res 143:77–88

Stevens J (1974) Letter in response to eye tracking patterns in schizophrenia. Science 184:1201–1202

Stuve TA, Friedman L, Jesberger JA, Gilmore GC, Strauss ME, Meltzer HY (1997) The relationship between smooth pursuit performance, motion perception and sustained visual attention in patients with schizophrenia and normal controls. Psychol Med 27:143–152

Sweeney JA, Luna B, Srinivasagam NM, Keshavan MS, Schooler NR, Haas GL, Carl JR (1998a) Eye tracking abnormalities in schizophrenia: evidence for dysfunction in the frontal eye fields. Biol Psychiatry 44:698–708

Sweeney JA, Strojwas MH, Mann JJ, Thase ME (1998b) Prefrontal and cerebellar abnormalities in major depression: evidence from oculomotor studies. Biol Psychiatry 43:584–594

Sweeney JA, Luna B, Haas GL, Keshavan MS, Mann JJ, Thase ME (1999) Pursuit tracking impairments in schizophrenia and mood disorders: step-ramp studies with unmedicated patients. Biol Psychiatry 46:671–680

Sweeney JA, Levy DL, Harris MSH (2002) Eye movement research with clinical populations. In: Hyona J, Munoz DP, Heide W, Radach R (eds) Progress in brain research. Brain's eyes: neurobiological and clinical aspects of oculomotor research, vol 140. Elsevier Science, Amsterdam, pp 507–522

Tanabe JL, Tregellas JR, Martin LF, Freedman R (2006) Effects of nicotine on hippocampal and cingulate activity during smooth pursuit eye movement in schizophrenia. Biol Psychiatry 59:754–761

Tanaka M, Fukushima K (1998) Neuronal responses related to smooth pursuit eye movements in the periarcuate cortical area of monkeys. J Neurophysiol 80:28–47

Tanaka M, Lisberger SG (2001) Regulation of the gain of visually guided smooth-pursuit eye movements by frontal cortex. Nature 409:191–194

Tanaka M, Lisberger SG (2002a) Role of arcuate frontal cortex of monkeys in smooth pursuit eye movements. II. Relation to vector averaging pursuit. J Neurophysiol 87:2700–2714

Tanaka M, Lisberger SG (2002b) Role of arcuate frontal cortex of monkeys in smooth pursuit eye movements. I. Basic response properties to retinal image motion and position. J Neurophsyiol 87:2684–2699

Tanaka M, Lisberger SG (2002c) Enhancement of multiple components of pursuit eye movement by microstimulation in the arcuate frontal pursuit area in monkeys. J Neurophysiol 87:802–818

Tehovnik EJ, Sommer MA, Chou IH, Slocum WM, Schiller PH (2000) Eye fields in the frontal lobes of primates. Brain Res Brain Res Rev 32:413–448

Thaker GK, Nguyen JA, Tamminga CA (1989) Increased saccadic distractibility in tardive dyskinesia: functional evidence for subcortical GABA dysfunction. Biol Psychiatry 25:49–59

Thaker G, Cassady S, Adami H, Moran M, Ross D (1996a) Eye movements in spectrum personality disorders: comparison of community subjects and relatives of schizophrenic patients. Am J Psychiatry 153:362–368

Thaker GK, Ross DE, Buchanan RW, Moran M, Lahti AC, Kim CE (1996b) Does pursuit abnormality in schizophrenia represent a deficit in the predictive mechanism? Psychiatry Res 59:221–237

Thaker GK, Ross DE, Cassady SL, Adami HM, LaPorte D, Medoff DR, Lahti AC (1998) Smooth pursuit eye movements to extraretinal motion signals: deficits in relatives of patients with schizophrenia. Arch Gen Psychiatry 55:830–836

Thaker GK, Ross DE, Buchanan RW, Adami HM, Medoff DR (1999) Smooth pursuit eye movements to extraretinal motion signals: deficits in patients with schizophrenia. Psychiatry Res 88:209–219

Thaker GK, Avila MT, Hong LE, Medoff DR, Ross DE, Adami HM (2003) A model of smooth pursuit eye movement deficit associated with the schizophrenia phenotype. Psychophysiology 40:277–284

Thaker GK, Wonodi I, Avila MT, Hong LE, Stine OC (2004) Catechol O-methyltransferase polymorphism and eye tracking in schizophrenia: a preliminary report. Am J Psychiatry 161:2320–2322

Thier P, Erickson RG (1992) Responses of visual tracking neurons from cortical area MST-1 to visual, eye and head motion. Eur J Neurosci 4:539–553

Thier P, Ilg UJ (2005) The neural basis of smooth-pursuit eye movements. Curr Opin Neurobiol 15:645–652

Thurston SE, Leigh RJ, Crawford TJ, Thompson A, Kennard C (1988) Two distinct deficits of visual tracking caused by unilateral lesions of cerebral cortex in humans. Ann Neurol 23:266–273

Tregellas JR, Tanabe JL, Miller DE, Ross RG, Olincy A, Freedman R (2004) Neurobiology of smooth pursuit eye movement deficits in schizophrenia: an fMRI study. Am J Psychiatry 161:315–321

Trillenberg P, Heide W, Junghanns B, Blankenburg M, Arolt V, Kompf D (1998) Target anticipation and impairment of smooth pursuit eye movements in schizophrenia. Exp Brain Res 120:316–324

Van Essen DC, Maunsell JHR (1983) Hierarchical organization and the functional streams in the visual system. Trends Neurosci 6:370–375

Vandenberg AV (1988) Human smooth pursuit during transient perturbations of predictable and unpredictable target movement. Exp Brain Res 72:95–108

Warren S, Ross RG (1998) Deficient cancellation of the vestibular ocular reflex in schizophrenia. Schizophr Res 34:187–193

Wertheim AH, Van Gelder P, Peselow LE, Cohen N (1985) High thresholds for movement perception in schizophrenia may indicate abnormal extraneous noise levels of central vestibular activity. Biol Psychiatry 20:1197–1210

White HR (1938) Ocular pursuit in normal and psychopathological subjects. J Exp Psychol 22:17–31

Winograd-Gurvich C, Georgiou-Karistianis N, Fitzgerald PB, Millist L, White OB (2006) Oculomotor differences between melancholic and non-melancholic depression. J Affect Disord 93:193–203

Wurtz RH, Komatsu H, Yamasaki DSG, Dursteler MR (1990) Cortical visual processing for oculomotor control. In: Cohen B, Bodis-Wollner I (eds) Vision and brain. Raven Press, New York, pp 211–231

Xiao Q, Barborica A, Ferrera VP (2003) Modulation of visual responses in macaque frontal eye field during covert tracking of invisible targets. Cereb Cortex 17:918–928

Yee R, Baloh R, Marder S, Levy DL, Sakala S, Honrubia V (1987) Eye movements in schizophrenia. Invest Ophthalmol Vis Sci 28:366–374

Zeki SM (1974) Functional organization of a visual area in the posterior bank of the superior temporal sulcus of the rhesus monkey. J Physiol 236:549–573

Prepulse Inhibition of the Startle Reflex: A Window on the Brain in Schizophrenia

David L. Braff

Contents

1 Background of PPI Studies 350
2 Prepulse Inhibition Deficits in Schizophrenia Spectrum (and Other) Patients 353
3 Sex, Symptoms, Cognitive, and Functional Correlates of PPI Deficits
  in Schizophrenia Patients 355
4 Pharmacological Studies of PPI in Human Subjects Relevant to Schizophrenia 355
   4.1 Dopamine 356
   4.2 Nicotine 357
5 Antipsychotic Medications in Schizophrenia Patients 358
6 Genomic Influences on PPI in Schizophrenia 359
7 Summary and Future Directions 363
References 365

Abstract Prepulse inhibition (PPI) of the startle response is an important measure of information processing deficits and inhibitory failure in schizophrenia patients. PPI is especially useful because it occurs in the same lawful manner in all mammals, from humans to rodents, making it an ideal candidate for cross-species translational research. PPI deficits occur across the "schizophrenia spectrum" from schizophrenia patients to their clinically unaffected relatives. Parallel animal model and human brain imaging studies have demonstrated that PPI is modulated by cortico-striato-pallido-thalamic (and pontine) circuitry. This circuitry is also implicated in schizophrenia neuropathology and neurophysiology. The finding of PPI deficits in schizophrenia patients has been replicated by many groups, and these deficits correlate with measures of thought disorder and appear to be "normalized" by second generation antipsychotic (SGA) medications. Consistent pharmacological effects on PPI have been demonstrated; among these, dopamine agonists induce

D.L. Braff
Department of Psychiatry, University of California, San Diego (UCSD), 9500 Gilman Drive, La Jolla, CA 92093-0804, USA
e-mail: dbraff@ucsd.edu

PPI deficits and (in animal models) these are reversed by first and SGA medications. PPI is also significantly heritable in humans and animals and can be used as a powerful endophenotype in studies of families of schizophrenia patients. Genomic regions, including the NRGL-ERBB4 complex with its glutamatergic influences, are strongly implicated in PPI deficits in schizophrenia. PPI continues to hold promise as an exciting translational cross-species measure that can be used to understand the pathophysiology and treatment of the schizophrenias via pharmacological, anatomic, and genetic studies.

Keywords Schizophrenia · Prepulse inhibition · Information processing · Endophenotype · Animal models · Dopamine

1 Background of PPI Studies

Prepulse inhibition (PPI) of the startle reflex is a particularly informative measure of inhibitory function and information processing deficits in schizophrenia. The mammalian startle reflex is an automatic or reflexive contraction of the skeletal and facial muscles in response to a sudden, intense stimulus in one of several modalities (e.g., visual, auditory, or tactile). This reflex is usually classified as a special subtype of a defensive (as opposed to an orienting) response (Turpin 1986). A major value of the startle reflex paradigm in schizophrenia research is that startle shows plasticity and also is amenable to cross species pharmacological, neural substrate, genomic, and translational research (Bellesi et al. 2009; Braff and Geyer 1990; Braff et al. 1978, 2001b, 2007a, b, 2008; Csomor et al. 2009; Geyer and Braff 1987; Geyer et al. 2001; Greenwood et al. 2007; Li et al. 2009; Powell et al. 2009; Shilling et al. 2008; Swerdlow et al. 2001b, c, d, 2006; Wolf et al. 2007; Wynn et al. 2007). In human research, the eyeblink component of whole body startle is measured using electromyography (EMG) of the orbicularis oculi muscle (Graham 1975). For rodents, in cross species translational research, stabilimeter chambers assess whole-body flinch elicited by startling stimuli. Perhaps most importantly, the startle response exhibits plasticity across species, including PPI (Graham 1975), habituation (Geyer and Braff 1987; Hoffman and Searle 1968), and fear potentiation (Brown et al. 1951). Many of these forms of plasticity are modulated by the forebrain and related structures, such as cortico-striato-pallido-thalamic circuitry (Swerdlow and Koob 1987) implicated in schizophrenia neuropathology and neurophysiological deficits. Thus, startle plasticity, especially in PPI, is an excellent candidate for cross species translational research (cf. Braff et al. 2001b; Geyer et al. 2001; Swerdlow et al. 2001b, c, d). In 1978, Braff et al. in Enoch Callaways's lab at UCSF first reported PPI deficits in schizophrenia patients using potentiometers kindly supplied by Francis Graham. Since that time, there has been a virtual avalanche of papers involving PPI in schizophrenia patients (see Fig. 1). A recent

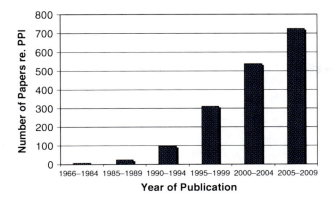

Fig. 1 Dramatic increase in PubMed PPI citations over time

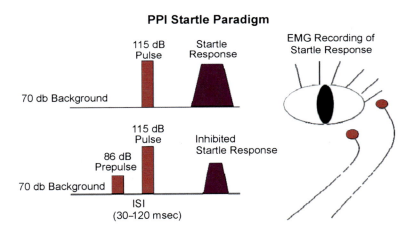

Fig. 2 The basic, neutral, uninstructed PPI paradigm

PubMed search found over 550 papers for the key words "PPI + schizophrenia" in the past 8 years, reflecting the importance of this measure in schizophrenia research.

In the PPI paradigm, the first weak sensory event (the prepulse) inhibits or "gates" the motor response to a startling stimulus; hence, we use the term "sensorimotor gating" to describe the construct measured by PPI (as opposed to a purely sensory paradigm such as P50 or N100 suppression) (see Fig. 2).

PPI is a common, lawfully mediated, and robust mammalian phenomenon that occurs when the prepulse and startling stimuli are in the same or different sensory modalities (Blumenthal and Gescheider 1987; Graham 1980; Hoffman and Ison 1980). It occurs in all mammals tested to date, and it is not a form of conditioning, because it occurs on the first exposure to the combination of weak prepulse and strong startling stimuli (Blumenthal et al. 1996). PPI (unlike startle itself) does not exhibit habituation or extinction over multiple trials and is a relatively stable neurobiological marker: PPI measurements repeated at 1-month intervals for 3 months yielded intraclass correlations of 0.94 (Cadenhead et al. 1999). Neurobiologically, the prepulse has inhibitory influences that can be regulated by connections between limbic, cortical, basal ganglia, and pontine circuitry (cf. Lee et al. 1996; Swerdlow and Koob 1987; Swerdlow et al. 1992). Thus, PPI reflects the activation of behavioral gating processes that are regulated by forebrain neural circuitry. This circuitry exerts a "down-stream tonic" regulatory or modulating influence, but the signal of the prepulse need not traverse or be transmitted across forebrain circuitry in order to produce PPI (Davis and Gendelman 1977). The forebrain regulation of PPI has been the target of many mammalian neurobiological and psychopathology studies (Swerdlow et al. 2001d).

In order to elicit PPI, a weak prepulse stimulus activates brain-based processes that blunt responsivity to sensory events during a subsequent relatively brief temporal window. The temporal limits of the time period of "gating" attributed to the prepulse are empirically determined to be approximately 30–500 ms in duration, in both humans and rodents (Graham 1975; Hoffman and Searle 1968). Prepulse-to-pulse intervals of 30–240 ms are typically utilized in human PPI experiments, with maximal amplitude inhibition generally occurring with intervals of approximately 120 ms. Compared to amplitude inhibition, startle latency reduction ("latency facilitation") typically occurs at briefer prepulse-to-pulse intervals (e.g., 30 ms). The period of reduced reflex responsivity after the prepulse presentation has been hypothesized to transiently "protect" information contained in the weak stimulus, so that maximal information can be processed from the prepulse stimulus, without interference from subsequent strong or disrupting startling stimuli (Blumenthal et al. 1996; Braff et al. 1978, 2001a, b; Swerdlow et al. 1999).

PPI is studied in the laboratory, but under "natural" circumstances, the process of sensorimotor gating is conceptualized as helping the organism to regulate environmental inputs in order to navigate successfully in a stimulus-laden world full of supernumerary and nonsalient stimuli, and to selectively allocate attentional resources to salient stimuli (Braff et al. 1978, 1992, 1999; Grillon et al. 1992). Typically, an individual's gating processes are viewed as showing behavioral plasticity and to have both state and trait determinants influenced by a combination of environmental and genetic factors. Human PPI is viewed as being substantially trait or genetically determined (i.e., heritable) on the basis of data from many sources including the Consortium on the Genetics of Schizophrenia (COGS) (Greenwood et al. 2007). Mammalian PPI is also sensitive to some neurodevelopmentally timed environmental factors in rodents, such as neonatal hippocampal insult (Lipska et al. 1995) or social isolation rearing (Geyer et al. 1993), and to perturbations as well as toxic environmental influences of the neurochemical and hormonal milieu of the nervous system (cf. Swerdlow et al. 2000). Of course, these "environmental" vectors can result in methylation or other epigenetic changes to the genome and to abnormal gene expression.

2 Prepulse Inhibition Deficits in Schizophrenia Spectrum (and Other) Patients

Interest in PPI as a measure of sensorimotor gating in the commonly used neutral uninstructed paradigm has been stimulated by findings that disorders with known dysfunction in brain substrates that regulate PPI are accompanied by evidence of impaired cognitive, motor, or sensorimotor inhibition. Deficient PPI has been reported repeatedly in patients with schizophrenia since Braff et al.'s (1978) initial original report (e.g., Bolino et al. 1994; Braff et al. 1992, 1999; Kumari et al. 1999, 2000; Kunugi et al. 2007; Takahashi et al. 2008; Weike et al. 2000; see Braff et al. 2001b for a review). Nonmedicated, nonpsychotic schizotypal patients (Cadenhead et al. 1993) and clinically unaffected relatives of schizophrenia patients (Cadenhead et al. 2000) also show PPI deficits. But PPI deficits are not specific to schizophrenia and are observed in patients with disorders that might be viewed clinically as a "family of gating disorders," including obsessive compulsive disorder (OCD) (Swerdlow et al. 1993; Hoenig et al. 2005), Huntington's disease (HD) (Swerdlow et al. 1995b; Valls-Solé et al. 2004), nocturnal enuresis and attention deficit disorder (Ornitz et al. 1992), Tourette Syndrome (Castellanos et al. 1996; Swerdlow et al. 2001a), blepharospasm (Gomez-Wong et al. 1998), nonepileptic seizures (Pouretemad et al. 1998), and, perhaps, post-traumatic stress disorder (PTSD) (Grillon et al. 1996). Strikingly, these disorders are all characterized by a loss of gating in sensory, motor, or cognitive domains, and in some cases, by known or proposed abnormalities in the cortico-striato-pallido-pontine (CSPP) circuitry that modulates PPI. Thus, via downward influences through the pons on the primary startle circuit, PPI deficits are not unique to a single form of psychopathology, but fit into a rational picture of CSPP dysfunction across multiple disorders (e.g., Swerdlow et al. 1992).

The neurobiology of schizophrenia has been intensely studied using PPI, because of the consistent deficits seen in schizophrenia patients (see above) and the striking relevance of the PPI "anatomy" to the pathophysiology of this disorder (Swerdlow et al. 2008), and because deficits in gating of cognitive and sensory information are clinically important features of schizophrenia. Thus schizophrenia patients are impaired in measures of sustained or "voluntary" attention, but clinically and experimentally they also demonstrate "automatic or involuntary" (Callaway and Naghdi 1982) information processing deficits, including an inability to automatically filter or "gate" irrelevant thoughts and sensory stimuli from intruding into conscious awareness (Braff et al. 1978; McGhie and Chapman 1961; Venables 1960). It is this automatic processing, modeled operationally in the neutral, uninstructed PPI paradigm, that forms a bridge from human to animal model studies of PPI (cf. Braff et al. 2001a, b; Geyer et al. 2001; Swerdlow et al. 2001b, c, d) (Fig. 3).

PPI deficits and correlated cognitive and symptomatic abnormalities have been reported in schizophrenia patients and individuals across the "schizophrenia spectrum" of disorders from clinically unaffected relatives to schizotypal and prodromal schizophrenia patients (Cadenhead et al. 1993, 2000; Bolino et al. 1994; Braff et al.

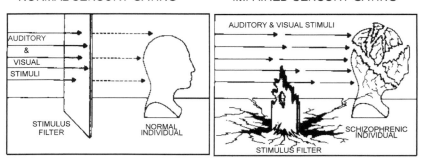

Fig. 3 An illustration of how a loss of "gating" of environmental stimuli is associated with cognitive fragmentation. This "gating" model is a useful heuristic for cognitive and clinical symptoms in schizophrenia. Laboratory measures like PPI are used as operational models to study processes believed to contribute to this loss of "gating," but it important to emphasize that clinical symptoms are not *caused by* laboratory-based deficits (e.g., low PPI): rather, a working and testable hypothesis is that these two phenomena (symptoms and PPI deficits) may reflect overlapping neurobiological and genetic processes

1978, 1992, 1999, 2001a, 2007a; Grillon et al. 1992; Karper et al. 1996; Kumari et al. 1999, 2000; Parwani et al. 2000; Perry and Braff 1996; Perry et al. 1998; Quednow et al. 2008; Weike et al. 2000). Studies also confirm that schizophrenia patients exhibit reduced PPI across auditory, tactile, and electrocutaneous stimulus modalities (Bolino et al. 1994; Braff et al. 1978, 1992, 2001a, b; Grillon et al. 1992; Kumari et al. 1999, 2000; Weike et al. 2000). Both intramodal and cross-modal PPIs have been found to be reduced in schizophrenia patients (Braff et al. 1992). The finding that PPI is reduced in the prodrome of schizophrenia (Quednow et al. 2008) and in clinically unaffected relatives of schizophrenia probands compared to normal controls (Cadenhead et al. 2000), makes PPI a robust and valid candidate as an endophenotypic marker in genetic studies (Braff and Freedman 2002; Braff et al. 2007a, b; Gottesman and Gould 2003; Greenwood et al. 2007).

Might these finding be attributable to the effects of antipsychotic medications rather than schizophrenia per se? Not likely. Studies consistently demonstrate that antipsychotic medications either have no effect, or increase PPI (i.e., potentially oppose PPI deficits) in schizophrenia patients and laboratory animals (cf. Braff et al. 2001b; Geyer et al. 2001; Swerdlow et al. 2001b, c, d). The validity of PPI as a familial and heritable schizophrenia endophenotype (Anokhin et al. 2003) has recently been strongly reinforced in a large multisite study from the COGS, with significant heritability of PPI in families of schizophrenia patients (Greenwood et al. 2007) further strengthening the argument that PPI deficits in schizophrenia are not due to medication effects. Furthermore, PPI deficits found in unmedicated schizophrenia patients (e.g., Swerdlow et al. 2006), and clinically unaffected relatives of schizophrenia patients (see above) "seal the deal" for schizophrenia patients' PPI: deficits are not significantly or primarily attributable to antipsychotic medication effects.

3 Sex, Symptoms, Cognitive, and Functional Correlates of PPI Deficits in Schizophrenia Patients

Many informative studies have investigated the relationship between PPI and other important variables. With commonly used stimulus parameters, PPI levels appear to be higher in normal men than in normal women, so sex must be accounted for in human PPI studies (e.g., Swerdlow et al. 2006). Some (but not all) studies report correlations of PPI deficits with specific clinical variables such as positive symptoms in schizophrenia patients (Braff et al. 1999). Perhaps most importantly, PPI deficits are associated with thought disorder (Perry and Braff 1994), even more highly and strikingly ($R = -0.78$) when both PPI and thought disorder are assessed in a single test session (Perry et al. 1999). Karper et al. (1996) reported that lower PPI is associated with the important schizophrenia related feature of greater distractibility (Grillon et al. 1992). While PPI levels are positively associated with neurocognitive measures in normal comparison subjects (e.g., Bitsios and Giakoumaki 2005; Giakoumaki et al. 2006), the correlation of PPI deficits with a large number of neuropsychological tests in schizophrenia patients has proven less than fully compelling (Swerdlow et al. 2006). Other demographic and symptom measures associated with reduced PPI in schizophrenia include early age of onset (Kumari et al. 2000) and greater positive or negative symptoms (Braff et al. 1999). In both cases, however, there are conflicting reports in the literature (Braff et al. 1999; Kumari et al. 1999; Perry and Braff 1994), suggesting that some of these observed PPI relationships may reflect characteristics peculiar to specific patient cohort factors such as restricted age or symptom range, stage of illness, or experimental parameters (cf. Swerdlow et al. 2006).

Despite the overwhelming evidence supporting the occurrence and importance of PPI deficits in schizophrenia patients, there are reports to the contrary. However, these reports typically utilized complex paradigms where factors such as "directed attentional" forms of PPI and patient cohort effects may have influenced the results (Dawson et al. 1993; Ford et al. 1999). A large definitive study by Swerdlow et al. (2006) confirmed the finding of PPI deficits in schizophrenia. Additionally, the influence of medications, sex (see above), and an association between smoking and higher (more normal) PPI levels in patients was observed. Lastly, via quartile analyses of PPI levels, Swerdlow et al. (2006) determined that impaired (lower) PPI is associated with impaired daily functioning in schizophrenia patients.

4 Pharmacological Studies of PPI in Human Subjects Relevant to Schizophrenia

As an outgrowth of our knowledge of PPI deficits in patients with various neuropsychiatric disorders and in parallel with translational animal model studies (cf. Geyer et al. 2001; Swerdlow et al. 2001b, c, d), investigators have increasingly utilized psychopharmacological strategies in normal humans with a number of carefully selected compounds. This section of the chapter is organized to review findings from a few compounds of importance in schizophrenia research (dopamine agonists, nicotine, antipsychotic medications) out of the many psychopharmacological agents that have been studied in conjunction with PPI (cf. Braff et al. 2001b; Geyer et al. 2001; Swerdlow et al. 1994, 2001b, c, d).

4.1 Dopamine

The dopamine (DA) hypothesis of schizophrenia has been the pre-eminent neurotransmitter based explanation of schizophrenia, although glutamatergic and many other neurotransmitter "contributors" to schizophrenia have been identified by Coyle, Javitt, and others (e.g., Lisman et al. 2008). Across many studies, the activation of dopamine receptors by either indirect DA agonists, such as amphetamine, or direct agonists, such as apomorphine, leads to robust "schizophrenia-like" deficits in PPI in humans and in rats. In human volunteers, Abduljawad et al. (1998) found that the D2 dopamine agonist bromocriptine significantly reduced PPI in human subjects. This result was found in a within subjects study utilizing 12 men in four sessions where subjects received oral doses of placebo, the D2 receptor agonist (and probable D1 receptor partial agonist) bromocriptine (1.25 mg), the D2 receptor antagonist haloperidol (3 mg), and a combined treatment with the cited doses of bromocriptine and haloperidol in a balanced double-blind protocol. Haloperidol was effective in antagonizing the PPI-disruptive effects of bromocriptine, confirming the similar very robust results found reported using other direct DA agonists in rodents (Geyer et al. 2001; Swerdlow et al. 2001b, c, d).

Studies of the effects of dopamine agonists on PPI in humans have utilized a number of other drugs including the indirect dopamine agonist amphetamine, which reliably reduces PPI in rodents. Hutchison and Swift (1999) reported that a 20-mg oral dose of D-amphetamine disrupted PPI in normal (nonsmoking) human subjects. Both male and female ($N = 18$ each) subjects were tested at 60, 90, and 120 min after drug ingestion in a brief repeatable PPI test using a single prepulse condition. A 58 dB(A) tone against a background of 54 dB(A) white noise (for a prepulse of 4 dB(A)) was used vs. the more commonly used, "stronger" 15 dB(A) prepulses (e.g., 85 dB(A) over a 70 dB(A) background). As predicted from rodent studies, D-amphetamine significantly reduced PPI, in the 90-min post-drug test. In contrast, subjective measures of the effects of amphetamine were affected at all time-points and heart rate was elevated at both the 90- and 120-min time-points. No gender differences were observed in these low levels of PPI, and gender did not interact with the effect of amphetamine on PPI.

Subsequent to these reports by Abduljawad et al. (1998) and Hutchison and Swift (1999), several groups have reported on the effects on PPI of both the indirect DA agonists amphetamine (Swerdlow et al. 2003; Talledo et al. 2009) and amantadine (Swerdlow et al. 2002a, b; Bitsios et al. 2005), as well as the direct DA agonists pergolide (Bitsios et al. 2005), ropinirole (Giakoumaki et al. 2007), pramipexole

(Swerdlow et al. 2009a), and apomorphine (Schellekens et al. 2010). Despite the fact that these drugs act through a wide range of different mechanisms and DA receptor subtypes, two consistent patterns have emerged from these elegant studies. First, in clinically normal humans, as in laboratory rats (e.g., Swerdlow et al. 2001b, c, d, 2004), DA agonists can either decrease or increase PPI, depending on the prepulse stimulus characteristics, particularly prepulse intensity and interval. Second, DA agonist effects on PPI in normal humans depend strongly on the baseline level of PPI; this relationship is also seen in the comparisons of several rodent strains (e.g., Talledo et al. 2009). Increasing evidence suggests that this relationship between baseline PPI and DA agonist effects in both humans and rodents may reflect the association of baseline PPI with specific genetic polymorphisms (discussed below). More generally, DA agonist effects on PPI appear to exhibit inter-individual differences that are associated predictably with specific physiological and perhaps genetic characteristics. Clearly, this ability to study DA agonist effects on PPI across species will enable the field to better understand the potential role of brain DA activity in schizophrenia-related PPI deficits.

Further studies with additional compounds, doses, and stimulus parameters will be useful to clarify the relationships between changes in PPI and other effects of dopamine agonists and antagonists, as well as the specific conditions and populations. Currently, the results of these studies provide evidence for the involvement of D2 (and probably other) dopamine receptors in the modulation of PPI in humans. This is an excellent example of a "translational" approach, because it had already been shown that PPI in rats is reliably disrupted by both direct and indirect dopamine agonists, and this disruption can be reversed by antipsychotic medications (e.g., Swerdlow et al. 1994). Dopamine's role in regulating PPI is also discussed in the genomics section of this chapter (below).

4.2 Nicotine

A number of studies report the effect of nicotine on PPI (Hong et al. 2008; Swerdlow et al. 2006; Woznica et al. 2009). A major observationally based reason for examining nicotine effects is that purely uncontrolled anecdotal reports and more theoretical formulations have been proffered that link the high level of smoking in schizophrenia patients to a potential "naturalistic" attempt to normalize sensorimotor (and perhaps other) neurocognitive deficits (Dalack et al. 1998; Forchuk et al. 1997; Markou et al. 1998). Part of the motivation for assessing the effects of nicotine on PPI in human subjects is the rather extensive literature on nicotinic effects in humans on a different measure of inhibition: the suppression of the sensory P50 event-related-potential (i.e., sensory gating). In addition, some studies in rats indicate that PPI is increased by nicotine administration (Acri et al. 1994; Curzon et al. 1994). In this context, Kumari et al. (1996) assessed the effect of cigarette smoking on PPI in healthy male smokers who were deprived of cigarettes overnight. Cigarette smoking in this group increased PPI. Similarly, DellaCasa et al. (1998)

compared nonsmokers, deprived smokers, and smokers who smoked during the test session after deprivation or after ad lib smoking. They also found that smoking during the startle session increased PPI but did not affect startle magnitude or habituation. In this study, significant gender effects were noted: in men, smoking resulted in increased PPI compared to nonsmokers while, in women, nicotine deprivation appeared to decrease PPI and smoking restored PPI to the level of nonsmokers. Kumari et al. (1997) systematically assessed the effect of acute nicotine (6 and 12 μg/kg delivered subcutaneously) on PPI in healthy male nonsmokers. The higher dose of nicotine increased PPI significantly when percent scores were used, but not when absolute difference scores were used. Thus, although nicotine had no significant effect on startle magnitude, the disparity between the percent and difference scores prompts some caution in drawing firm conclusions from these data. Cumulatively, these results were interpreted to provide some support for the previous findings indicating that PPI is increased by cigarette smoking in overnight- nicotine deprived smokers, and more generally, for the notion that nicotinic receptor stimulation increases sensorimotor gating as it does for sensory P50 gating. Because of the rapid desensitization and sensitization of nicotinic receptors that are associated with nicotine administration and withdrawal, the physiological bases of these results are quite complicated and difficult to interpret. Kumari et al. (1997) concluded that these findings indicate that previously observed effects of smoking on percent PPI in smoking-deprived subjects were not attributable to the reversal of a deficit that had been induced by smoking withdrawal, but instead represent a direct pharmacological action of nicotine itself. As noted above, Swerdlow et al. (2006) reported a definitive association between smoking and higher PPI in a large cohort of schizophrenia patients.

Overall, the results of these studies in humans appear to be consistent with related reports of increased PPI induced by acutely administered nicotine in animal experiments. Because the modulation of PPI by nicotine may be crucially influenced via dopaminergic, cholinergic, or other mechanisms, Swerdlow et al. (2006) concluded that their study could not be definitively interpreted mechanistically. At a "functional" level, they suggested that the effect of nicotine in increasing PPI might relate to its attention/cognitive facilitating properties (Jones et al. 1992). These properties have been repeatedly observed in human subjects and might be related to the fact that PPI deficits correlate with cognitive and attentional abnormalities (Braff et al. 1992; Karper et al. 1995; Perry and Braff 1994; Swerdlow et al. 1995a).

5 Antipsychotic Medications in Schizophrenia Patients

Several studies have reported that second generation antipsychotic (SGA) medication "normalizes" PPI in schizophrenia patients (Kumari et al. 2000; Weike et al. 2000). It is difficult to interpret the full implications of these "positive" reports, because they were based on between-subject comparisons. Clearly, a definitive study using PPI as a biomarker in longitudinal clinical trials designs is still needed. Prodromal (Quednow et al. 2008) and unmedicated schizophrenia patients (Weike et al. 2000) and even schizophrenia patients treated with "typical" antipsychotics (Braff et al. 1978; Kumari et al. 2000; Swerdlow et al. 2006; Mackeprang et al. 2002; Perry et al. 2002) exhibit PPI deficits compared to normal controls. Findings of PPI deficits in clinically unaffected relatives of schizophrenia patients (Kumari et al. 2005; Cadenhead et al. 2000), and schizotypal patients (Cadenhead et al. 1993) also support the idea that PPI deficits occur across the schizophrenia spectrum in the absence of SGAs. In contrast, in some studies, PPI in schizophrenia patients medicated with atypical antipsychotics (Kumari et al. 2000) or a mixture of typical and atypical antipsychotics (Swerdlow et al. 2006; Weike et al. 2000) did not differ significantly from PPI in normal controls. A number of well-designed studies (e.g., Mackeprang et al. 2002; Minassian et al. 2007; Wynn et al. 2007; Csomor et al. 2009) also implicate antipsychotics as "normalizing" PPI deficits in schizophrenia patients. Thus, it appears that antipsychotic medications, especially SGAs, are associated with more normal PPI values in schizophrenia. In clinical practice, the pathway by which a given patient (or group of patients) is treated with atypical antipsychotics is also determined by many "nonbiological" factors such as insurance company rules, institutional policies and practices, drug prices, etc. lending to all sorts of complex ascertainment biases. A between subject experimental design also does not control for the likely neurobiological differences between groups of patients who are treated with SGAs antipsychotic medications, may be more likely to respond clinically, and have lower levels of acute neurological side effects vs. groups of patients who are treated with first generation antipsychotic medications and may be more likely to be clinically unresponsive or have higher neurological side effect profiles from high potency dopamine receptor blockade. In any case, without the necessary well powered longitudinal and controlled crossover studies, it is still a bit premature to definitively conclude that atypical antipsychotics actually fully "normalize" PPI in schizophrenia patients, although some degree of "normalization" seems clear. In a cross species context, the utility of PPI in predicting first and SGA medication is vividly illustrated by Fig. 4.

6 Genomic Influences on PPI in Schizophrenia

Evidence regarding the profound genetic influences on PPI comes from both human and animal model studies. Heritability, strain relatedness, and other specific genetic effects abound in the human (Greenwood et al. 2007) and rodent literature (cf. Powell et al. 2009). In human studies, a first step in understanding the genetics of schizophrenia and its endophenotypes (Braff et al. 2007a, b) is to demonstrate familial transmission and significant heritability, which has been done with PPI (Anokhin et al. 2003; Greenwood et al. 2007). A second step is identifying, via linkage and association studies, genetic loci and specific alleles or other genomic abnormalities influencing PPI. One paradox pervades this area of inquiry: I term it

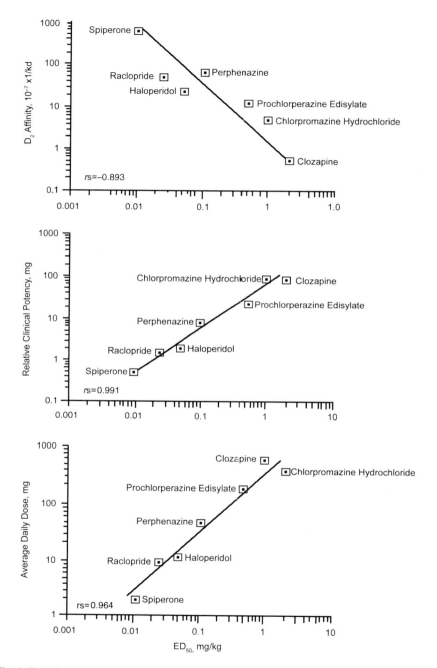

Fig. 4 From Swerdlow et al. (1994). This figure illustrates the predictive validity of the PPI model, in which the PPI-disruptive effects of DA agonists are prevented by antipsychotic medications according to their clinical potency

"Large N, Small Effect Size Conundrum". For Genome Wide Association Studies (GWAS), tens of thousands of human subjects might be studied to identify the genetic basis of a highly heritable trait such as height. A problem arises in these atheoretical nonbiologically informed "agnostic" studies of the genetic basis of the highly heritable traits such as height in humans: large N's (30,000 or more samples) may be needed to find "significant" results implicating 10 or so alleles that account for only 3% of height (Weedon and Frayling 2008). Applying this methodology to a heterogeneous disorder such as schizophrenia would seem to be "a bridge too far". In advancing the field of schizophrenia genetics, already replete with failures to replicate on the basis of varying ascertainment strategies and other confounds (e.g., Sanders et al. 2008), the use of a neurobiologically "valid" endophenotype such as PPI vs. the qualitative "fuzzy" heterogeneous diagnosis of schizophrenia offers profound advantages (Braff et al. 2007a, b, 2008; Thaker 2007; Walters and Owen 2007).

Endopheno- or intermediate phenotypes are gene influenced laboratory measures (such as PPI) that show deficits in schizophrenia patients and their clinically unaffected relatives (Gottesman and Gould 2003). The driving force behind the use of intermediate phenotypes in many disorders from schizophrenia (Gottesman and Gould 2003; Braff et al. 2007a, b) to hypertension (O'Connor et al. 2000) and Type 2 diabetes (Scott et al. 2007) is that the biological (intermediate) phenotype has a closer biological link to a gene associated with a significant trait of a human disorder. The endophenotype is "closer" to the protein gene product and is simpler to identify than a complex qualitative clinical disorder. Thus an alternative to "agnostic" GWAS in PPI/schizophrenia studies is to use existing GWAS, candidate gene, and human and animal model neurobiological information and, via strong inference (Platt 1964), create robust a priori hypotheses about the association of specific genes related to schizophrenia endophenotypes. This could be termed a "Functionally Selective X Gene Interrogation" method and avoids the GWAS-related "Large N, Small Effect Size Conundrum" where it would take 1,000's of (heterogeneous) schizophrenia patients and control subjects to achieve statistical significance but the importance of a finding is only marginal due to the small genetic contributions of loci harbored across many genes. In contrast, a small number of genes that have been found to influence schizophrenia endophenotypes, such as PPI, can be interrogated with a strong inferential basis for expecting specific genetic influences (cf. Stefansson et al. 2002; see below).

PPI is also highly heritable in rodents, with striking strain-related differences. One interesting experiment from Thomas Insel's laboratory (Francis et al. 2003) illustrates the power of endophenotypes such as PPI: Francis et al. used C57BL/6J (B6) and BALB/cJ (BALB) mouse strains that differ in PPI levels in what might be termed a "nature vs. nurture" design that parsed genetic (G) and environmental (E) factors on a number of commonly used rodent behavioral phenotypes. The results showed the profound genetic influence on PPI. Adult mice were tested in several behavioral measures after developing from either within − or across-strain embryonic transplantation or within − or across-strain fostering. Commonly used cross-species behavioral paradigms such as open field activity and water maze tests were highly influenced by the embryonic or rearing environment. In contrast, PPI phenotypes were refractory to these environment effects. Thus, PPI was seen as being highly "genetically determined," and levels of PPI remained linked to embryonic genotype rather than to pre- or postnatal manipulations which significantly influenced other behaviors. While heritability/phenotype status established the powerful genetic regulation of PPI, the actual genes involved remain unclear despite the known strain-related genetic differences of the B6 and BALB mice.

Another good example of a "neurobiologically informed" genetic approach is a study by Stefansson et al. (2002). This group found that in their Icelandic isolate schizophrenia population the genes for neuregulin (NRG1) and its receptor ERBB4 were associated with schizophrenia. NRG1 and ERBB4 are profoundly implicated in glutamate signaling which is disordered in schizophrenia as per the work of Coyle and Javitt and others (e.g., Lisman et al. 2008). So, Steffanssen et al. did two things: First, they found evidence that NRG1 and ERBB4 genes are significantly associated with schizophrenia in humans in their kindreds. Then, they "knocked out" NRG1 and ERBB4 and saw profound "schizophrenia-like" PPI deficits in the hypomorphs vs. wild-type mice. These studies make it clear that a search for genetic associations of PPI deficits in schizophrenia may not be best resolved by an "agnostic" (GWAS) genomics approach using extremely large N's which yield significant but small effect size results but rather by evidence based, strong inference guided (Platt 1964) specific genetic studies. The COGS has confirmed the heritability of PPI (Greenwood et al. 2007) in schizophrenia kindreds and the data linking PPI deficits and NRGL and ERBB4 have been presented at meetings and have now been submitted for peer review.

So, the search for specific genetic influences on PPI continues. Since Steffanson et al.'s study in 2002, there has been a proliferation of publications identifying the genetic substrates of PPI in mammalian neurobiology and PPI in schizophrenia. If one were to look at various QTL's, loci, associations studies, knock out and knock down rodent studies, and the many genetic loci modeling PPI (Powell et al. 2009), you might throw up your hands in despair and say "PPI is associated with too many genes". But extensive neural circuits (e.g., CSPT) modulating PPI rely on widespread neural, autonomic, and neurotransmitter substrates and many corresponding genomic loci. An expanding literature is identifying the genomic basis of PPI. For example, COMT rs4818 C/G, risk haplotype PRODH, COMT Val158Met, and 5HT2a genes all have been implicated in PPI modulation and/or drug sensitivity in normal individuals, or in PPI levels in schizophrenia patients (Giakoumaki et al. 2008; Maier et al. 2008; Roussos et al. 2008, 2009; Quednow et al. 2008). Many of these loci rationally fit into the PPI modulation by distributed neural circuitry identified in the mammalian literature from rodent lesion to human brain imaging studies (see above, also cf. Swerdlow et al. 2001b, c, d; Swerdlow and Koob 1987). Of course, because of complex gene–gene interactions, epigenetic, copy number variations (CNVs), de novo mutations, and gene x environment interactions, the genetic control of the PPI endophenotype is bound to be a very complex but ultimately rational story involving many genes expressed in the brain (Insel and Collins 2003).

Both stable, familial mutations and de novo CNV mutations occur in "far flung" regions of the genome in schizophrenia (Sanders et al. 2008; Walsh et al. 2008). The heritable, familial mutations support the "common disease, common variation" hypothesis for schizophrenia. The de novo CNV mutations support the "common disease rare variation" hypothesis for at least some schizophrenia patients. But, these de novo mutations occur across many genes that contribute to key neurodevelopmental pathways (Walsh et al. 2008) and are far from random. This pattern of CNV abnormalities is paralleled by widespread PPI neural substrate abnormalities in schizophrenia. Ideally, the Walsh et al. (2008) study of CNVs in schizophrenia needs to be fully "mapped onto" crucial and disparate neurodevelopmental pathways in schizophrenia. Thus, the hope is that genetic abnormalities influencing mammalian PPI can ultimately be "mapped onto" the distributed neural circuitry modulating PPI. Recent studies in rodents (Shilling et al. 2008) have identified brain regional differences in gene expression between rat strains that also differ in their sensitivity to the PPI-disruptive effects of dopamine agonists. When we merge the vast neural circuit basis of an endophenotype like PPI with our knowledge of the distributed neural circuit/genomic basis of schizophrenia, a more rational, understandable picture emerges. It is the output of these complex circuits that is of importance in mammalian biology. Ultimately one (very ambitious) goal of researchers is to find genetically mediated risk markers such as PPI deficits that identify ultra high risk children and adolescents, and determine whether they benefit from psychosocial and perhaps medication prophylaxis. The benefits might be reversal of PPI deficits and ultimately a reduction of rates of conversion to psychosis and normalization of PPI with its relationship to real world function (e.g., Swerdlow et al. 2006). This strategy offers a possible robust biomarker-defined path for identifying and tracking disease onset, its treatment, and new antipsychotic medications and even efficacious nonmedication therapies.

7 Summary and Future Directions

PPI is a critically important measure that may be used to understand schizophrenia across multiple domains: from neural and genomic substrates to early identification to treatment and outcome. Since publication of our last 2001 reviews related to PPI in human studies (Braff et al. 2001b), its preclinical psychopharmacology (Geyer et al. 2001), and its neural circuitry (Swerdlow et al. 2001c), a PubMed search reveals over 550 articles with the key words "schizophrenia" and "PPI". These PPI studies have been central to efforts to understand the neurobiology of schizophrenia. However, as recently discussed by Swerdlow et al. (2008), many issues linking reduced PPI to schizophrenia need resolution.

Some key points raised in this chapter are summarized as follows:

First, in human research, PPI represents one of several measures of inhibitory failure associated with primary "core" Bleulerian features of cognitive fragmentation and symptoms of schizophrenia.

Second, PPI deficits extend across the schizophrenia spectrum, from schizophrenia patients to clinically unaffected relatives of schizophrenia patients, schizotypal patients, and prodromal schizophrenia subjects.

Third, PPI has robust face, construct, and predictive validity. It probably has been the single most productive and informative cross species translational paradigm in schizophrenia research based on multiple PubMed and ISI searches.

Fourth, the neural circuitry that modulates PPI and may therefore be responsible for reduced PPI in pathological states "maps onto" the known neural circuit dysfunction of schizophrenia in many crucial forebrain areas.

Fifth, mammalian PPI is significantly heritable in humans and rodents and is amenable to cross species translational genomic investigations involving single nucleotide polymorphisms, CNVs, epigenetics, and a host of other informative genomic strategies, which are now underway.

Sixth, in schizophrenia patients, SGA treatment is associated with "more normal" PPI levels. A definitive longitudinal study has yet to confirm these compelling between groups data.

Seventh, in cross species translational research, PPI deficits are induced by NMDA antagonists and DA agonists in rats and are reversed by efficacious antipsychotic medications according to their clinical potency (see Fig. 4). This is useful information for new antipsychotic medication development.

Future Directions: Antipsychotic medication effects on PPI present a challenging set of issues for future PPI studies (cf. Swerdlow et al. 2008); the use of controlled longitudinal, crossover designs, especially in initially nonmedicated patients to allow within-subject comparisons may be particularly critical for understanding the impact of antipsychotic medications on PPI in schizophrenia. If such studies continue to confirm a "normalizing" impact of atypical antipsychotics on PPI (and an associated normalizing of social and vocational function) in schizophrenia, this finding will be of even more importance. In this case, creative strategies for studying never medicated schizophrenia spectrum patients, in prodromal or "first-break" designs, as well as family studies, will be increasingly important in assessing the impact of antipsychotic medications on symptoms and function in schizophrenia patients. Perhaps most importantly, the genomic "signatures" of PPI deficits in schizophrenia patients may point the way for new strong inference based molecular (Platt 1964) and psychosocial therapy targets for the differential, "personalized medicine" treatment of patients from the group of schizophrenias. It is quite possible that we will identify which genetic "signatures" are associated with poor outcome, side effects and treatment response, as has been seen in breast cancer treatment and anticoagulant dosing studies (cf. Braff and Freedman 2008). PPI offers a salient window on the brain of the neural and genomic abnormalities of schizophrenia patients from multigenerational familial to de novo mutations, and from gene expression to epigenetic factors, across the landscape of modern genomic science.

Acknowledgments This work was supported in part by grants from the National Institute of Mental Health (MH42228, MH065571), a NARSAD Distinguished Investigator Award and the Department of Veteran Affairs, VISN 22 MIRECC (Mental Illness Research, Education, and

Clinical Center) on Schizophrenia and Psychosis. The author thanks his colleagues Neal Swerdlow and Mark Geyer who authored two of the three cited 2001 Psychopharmacology review papers. This chapter is an update and extension of Braff et al. (2001). The author also would like to thank Dr. Swerdlow for his very helpful comments on this manuscript.

References

Abduljawad KAJ, Langley RW, Bradshaw CM, Szabadi E (1998) Effects of bromocriptine and haloperidol on prepulse inhibition of the acoustic startle response in man. J Psychopharmacol 12:239–245

Acri JB, David EM, Popke EJ, Grunberg NE (1994) Nicotine increases sensory gating measured as inhibition of the acoustic startle reflex in rats. Psychopharmacology 114:369–374

Anokhin AP, Heath AC, Myers E, Ralano A, Wood S (2003) Genetic influences on prepulse inhibition of startle reflex in humans. Neurosci Lett 353:45–48

Bellesi M, Melone M, Gubbini A, Battistacci S, Conti F (2009) GLT-1 upregulation impairs prepulse inhibition of the startle reflex in adult rats. Glia 57:703–713

Bitsios P, Giakoumaki SG (2005) Relationship of prepulse inhibition of the startle reflex to attentional and executive mechanisms in man. Int J Psychophysiol 55:229–241

Bitsios P, Giakoumaki SG, Frangou S (2005) The effects of dopamine agonists on prepulse inhibition in healthy men depend on baseline PPI values. Psychopharmacology 182(1):144–152

Blumenthal TD, Gescheider GA (1987) Modification of the acoustic startle response by a tactile prepulse: effects of stimulus onset asynchrony and prepulse intensity. Psychophysiology 24:320–327

Blumenthal TD, Schicatano EJ, Chapman JG, Norris CM, Ergenzinger ER (1996) Prepulse effects on magnitude estimation of startle-eliciting stimuli and startle responses. Percept Psychophys 58:73–80

Bolino F, DiMichele V, DiCicco L, Manna V, Daneluzzo E, Casacchia M (1994) Sensorimotor gating and habituation evoked by electro-cutaneous stimulation in schizophrenia. Biol Psychiatry 36:670–679

Braff DL, Freedman R (2002) Endophenotypes in studies of the genetics of schizophrenia. In: Davis KL, Charney D, Coyle JT, Nemeroff C (eds) Neuropsychopharmacology: the fifth generation of progress. Lippincott Williams & Wilkins, Philadelphia, pp 703–716

Braff DL, Freedman R (2008) Clinically responsible genetic testing in neuropsychiatric patients: a bridge too far and too soon. Am Psychiatry 165:952–955

Braff DL, Geyer MA (1990) Sensorimotor gating and schizophrenia: human and animal model studies. Arch Gen Psychiatry 47:181–188

Braff DL, Stone C, Callaway E, Geyer MA, Glick I, Bali L (1978) Prestimulus effects on human startle reflex in normals and schizophrenics. Psychophysiology 15:339–343

Braff DL, Grillon C, Geyer MA (1992) Gating and habituation of the startle reflex in schizophrenic patients. Arch Gen Psychiatry 49:206–215

Braff DL, Swerdlow NR, Geyer MA (1999) Symptom correlates of prepulse inhibition deficits in male schizophrenic patients. Am J Psychiatry 156:596–602

Braff DL, Geyer MA, Light GA, Sprock J, Perry W, Cadenhead KS, Swerdlow NR (2001a) Impact of prepulse characteristics on the detection of sensorimotor gating deficits in schizophrenia. Schizophrenia Res 49:173–180

Braff DL, Geyer MA, Swerdlow NR (2001b) Human studies of prepulse inhibition of startle: normal subjects, patient groups, and pharmacological studies. Psychopharmacology 156:234–258

Braff DL, Light GA, Swerdlow NR (2007a) Prepulse inhibition and P50 suppression are both deficient but not correlated in schizophrenia patients. Biol Psychiatry 61:1204–1207

Braff DL, Freedman R, Schork NJ, Gottesman II (2007b) Deconstructing schizophrenia: an overview of the use of endophenotypes in order to understand a complex disorder. Schizophr Bull 33:21–32

Braff DL, Greenwood TA, Swerdlow NR, Light GA. Schork NJ (2008) The investigators of the Consortium on the Genetics of Schizophrenia. Advances in endophenotyping schizophrenia. World Psychiatry 7:11–18

Brown JS, Kalish HI, Farber IE (1951) Conditional fear as revealed by magnitude of startle response to an auditory stimulus. J Exp Psychol 41:317–328

Cadenhead KS, Geyer MA, Braff DL (1993) Impaired startle prepulse inhibition and habituation in patients with schizotypal personality disorder. Am J Psychiatry 150:1862–1867

Cadenhead KS, Carasso BS, Swerdlow NR, Geyer MA, Braff DL (1999) Prepulse inhibition and habituation of the startle response are stable neurobiological measures in a normal male population. Biol Psychiatry 45:360–364

Cadenhead KS, Swerdlow NR, Shafer KM, Diaz M. Braff DL (2000) Modulation of the startle response and startle laterality in relatives of schizophrenia patients and schizotypal personality disordered subjects: evidence of inhibitory deficits. Am J Psychiatry 157:1660–1668

Callaway E, Naghdi S (1982) An information processing model for schizophrenia. Arch Gen Psychiatry 39:339–347

Castellanos FX, Fine EJ, Kaysen D, Marsh WL, Rapoport JL, Hallet M (1996) Sensorimotor gating in boys with Tourette's syndrome and ADHD: preliminary results. Biol Psychiatry 39:33–41

Csomor PA, Yee BK, Feldon J, Theodoridou A, Studerus E, Vollenweider FX (2009) Impaired prepulse inhibition and prepulse-elicited reactivity but intact reflex circuit excitability in unmedicated schizophrenia patients: a comparison with healthy subjects and medicated schizophrenia patients. Schizophr Bull 35:244–255

Curzon P, Kim DJ, Decker MW (1994) Effects of nicotine, lobeline and mecamylamine on sensory gating in the rat. Pharmacol Biochem Behav 49:877–882

Dalack GW, Healy DJ, Meador-Woodruff JH (1998) Nicotine dependence in schizophrenia: clinical phenomena and laboratory findings. Am J Psychiatry 155:1490–1501

Davis M, Gendelman PM (1977) Plasticity of the acoustic startle response in the acutely decerebrate rat. J Comp Physiol Psychol 91:549–563

Dawson ME, Hazlett EA, Filion DL, Nuechterlein KH, Schell AM (1993) Attention and schizophrenia: impaired modulation of the startle reflex. J Abnorm Psychol 104:633–641

DellaCasa V, Hofer I, Weiner I, Feldon J (1998) The effects of smoking on acoustic prepulse inhibition in healthy men and women. Psychopharmacology 137:362–368

Forchuk C, Norman R, Malla A, Vos S, Martin ML (1997) Smoking and schizophrenia. J Psychiatr Ment Health Nurs 4:355–359

Ford JM, Roth WT, Menon V, Pfefferbaum A (1999) Failures of automatic and strategic processing in schizophrenia: comparisons of event-related brain potential and startle blink modification. Schizophr Res 37:149–163

Francis DD, Szegda K, Campbell G, Martin WD, Insel TR (2003) Epigenetic sources of behavioral differences in mice. Nat Neurosci 6:445–446

Geyer MA, Braff DL (1987) Startle habituation and sensorimotor gating in schizophrenia and related animal models. Schizophr Bull 13:643–668

Geyer MA, Wilkinson LS, Humby T, Robbins TW (1993) Isolation rearing of rats produces a deficit in prepulse inhibition of acoustic startle similar to that in schizophrenia. Biol Psychiatry 34:361–372

Geyer MA, Krebs-Thomson K, Braff DL, Swerdlow NR (2001) Pharmacological studies of prepulse inhibition models of sensorimotor gating deficits in schizophrenia: a decade in review. Psychopharmacology 156:117–154

Giakoumaki SG, Bitsios P, Frangou S (2006) The level of prepulse inhibition in healthy individuals may index cortical modulation of early information processing. Brain Res 1078:168–170

Giakoumaki SG, Roussos P, Frangou S, Bitsios P (2007) Disruption of prepulse inhibition of the startle reflex by the preferential D(3) agonist ropinirole in healthy males. Psychopharmacology 194:289–295

Giakoumaki SG, Roussos P, Bitsios P (2008) Improvement of prepulse inhibition and executive function by the COMT inhibitor tolcapone depends on COMT Val158Met polymorphism. Neuropsychopharmacology 33:3058–3068

Gomez-Wong E, Marti MJ, Tolosa E, Valls-Sole J (1998) Sensory modulation of the blink reflex in patients with blepharospasm. Arch Neurol 55:1233–1237

Gottesman II, Gould TD (2003) The endophenotype concept in psychiatry: etymology and strategic intentions. Am J Psychiatry 160:636–645

Graham F (1975) The more or less startling effects of weak prestimuli. Psychophysiology 12:238–248

Graham FK (1980) Control of reflex blink excitability. In: Thompson RF, Hicks LH, Shryrkov VB (eds) Neural mechanisms of goal-directed behavior and learning. Academic, New York, pp 511–519

Greenwood TA, Braff DL, Light GA, Cadenhead KS, Calkins ME, Dobie DJ, Freedman R, Green MF, Gur RE, Gur RC, Mintz J, Nuechterlein KH, Olincy A, Radant AD, Seidman LJ, Siever LJ, Silverman JM, Stone WS, Swerdlow NR, Tsuang DW, Tsuang MT, Turetsky BI, Schork NJ (2007) Initial heritability analyses of endophenotypic measures for schizophrenia: the Consortium on the Genetics of Schizophrenia. Arch Gen Psychiatry 64:1242–1250

Grillon C, Ameli R, Charney DS, Krystal J, Braff DL (1992) Startle gating deficits occur across prepulse intensities in schizophrenic patients. Biol Psychiatry 32:939–943

Grillon C, Morgan CA, Southwick SM, Davis M, Charney DS (1996) Baseline startle amplitude and prepulse inhibition in Vietnam veterans with posttraumatic stress disorder. Psychiatry Res 64:169–178

Hoenig K, Hochrein A, Quednow BB, Maier W, Wagner M (2005) Impaired prepulse inhibition of acoustic startle in obsessive-compulsive disorder. Biol Psychiatry 57:1153–1158

Hoffman HS, Ison JR (1980) Reflex modification in the domain of startle. I: Some empirical findings and their implications for how the nervous system processes sensory input. Psychol Rev 2:175–189

Hoffman HS, Searle JL (1968) Acoustic and temporal factors in the evocation of startle. J Acoust Soc Am 43:269–282

Hong LE, Wonodi I, Lewis J, Thaker GK (2008) Nicotine effect on prepulse inhibition and prepulse facilitation in schizophrenia patients. Neuropsychopharmacology 33:2167–2174

Hutchison KE, Swift R (1999) Effect of d-amphetamine on prepulse inhibition of the startle reflex in humans. Psychopharmacology 143:394–400

Insel TR, Collins FS (2003) Psychiatry in the genomics era. Am J Psychiatry 160:616–620

Jones GM, Sahakian BJ, Levy R, Warburton DM, Gray JA (1992) Effects of acute subcutaneous nicotine on attention, information processing and short-term memory in Alzheimer's disease. Psychopharmacology 108:458–494

Karper LP, Grillon C, Abi-Saab D, Morgan CA, Charney DS, Krystal JH (1995) The effect of ketamine on prepulse inhibition and attention. Schizophr Res 15:180

Karper LP, Freeman GK, Grillon C, Morgan CA III, Charney DS, Krystal HJ (1996) Preliminary evidence of an association between sensorimotor gating and distractibility in psychosis. J Neuropsychiatry Clin Neurosci 8:60–66

Kumari V, Checkley SA, Gray JA (1996) Effect of cigarette smoking on prepulse inhibition of the acoustic startle reflex in healthy male smokers. Psychopharmacology 128:54–60

Kumari V, Cotter PA, Checkley SA, Gray JA (1997) Effect of acute subcutaneous nicotine on prepulse inhibition of the acoustic startle reflex in healthy male nonsmokers. Psychopharmacology 132:389–395

Kumari V, Soni W, Sharma T (1999) Normalization of information processing deficits in schizophrenia with clozapine. Am J Psychiatry 156:1046–1051

Kumari V, Soni W, Mathew VM, Sharma T (2000) Prepulse inhibition of the startle response in men with schizophrenia: effects of age of onset of illness, symptoms, and medication. Arch Gen Psychiatry 57:609–614

Kumari V, Das M, Zachariah E, Ettinger U, Sharma T (2005) Reduced prepulse inhibition in unaffected siblings of schizophrenia patients. Psychophysiology 42:588–594

Kunugi H, Tanaka M, Hori H, Hashimoto R, Saitoh O, Hironaka N (2007) Prepulse inhibition of acoustic startle in Japanese patients with chronic schizophrenia. Neurosci Res 59:23–28

Lee Y, Lopez DE, Meloni EG, Davis M (1996) A primary acoustic startle pathway: obligatory role of cochlear root neurons and the nucleus reticularis pontis caudalis. J Neurosci 16: 3775–3789

Li L, Du Y, Li N, Wu X, Wu Y (2009) Top-down modulation of prepulse inhibition of the startle reflex in humans and rats. Neurosci Biobehav Rev 33:1157–1167

Lipska B, Swerdlow NR, Geyer MA, Jaskiw GE, Braff DL, Weinberger DR (1995) Neonatal excitoxic hippocampal damage in rats causes post-pubertal changes in prepulse inhibition of startle and its disruption by apomorphine. Psychopharmacology 122:35–43

Lisman JE, Coyle JT, Green RW, Javitt DC, Benes FM, Heckers S, Grace AA (2008) Circuit-based framework for understanding neurotransmitter and risk gene interactions in schizophrenia. Trends Neurosci 31:234–242

Mackeprang T, Kristiansen KT, Glenthoj BY (2002) Effects of antipsychotics on prepulse inhibition of the startle response in drug-naïve schizophrenic patients. Biol Psychiatry 52:863–873

Maier W, Mössner R, Quednow BB, Wagner M, Hurlemann R (2008) From genes to psychoses and back: the role of the 5HT2alpha-receptor and prepulse inhibition in schizophrenia. Eur Arch Psychiatry Clin Neurosci 258(Suppl 5):40–43

Markou A, Kosten TR, Koob GF (1998) Neurobiological similarities in depression and drug dependence: a self-medication hypothesis. Neuropsychopharmacology 18:135–174

McGhie A, Chapman J (1961) Disorders of attention and perception in early schizophrenia. Br J Med Psychol 34:103–116

Minassian A, Feifel D, Perry W (2007) The relationship between sensorimotor gating and clinical improvement in acutely ill schizophrenia patients. Schizophr Res 89:225–231

O'Connor DT, Insel PA, Ziegler MG, Hook VY, Smith DW, Hamilton BA, Taylor PW, Parmer RJ (2000) Heredity and the autonomic nervous system in human hypertension. Curr Hypertens Rep 2:16–22

Ornitz EM, Hanna GL, de Traversay Y (1992) Prestimulation-induced startle modulation in attention-deficit hyperactivity disorder and nocturnal enuresis. Psychophysiology 29:437–451

Parwani A, Duncan EJ, Bartlett E, Madonick SH, Efferen TR, Rajan R, Sanfilipo M, Chappell PB, Chakravorty S, Gonzenbach S, Ko GN, Rotrosen JP (2000) Impaired prepulse inhibition of acoustic startle in schizophrenia. Biol Psychiatry 47:662–669

Perry W, Braff DL (1994) Information processing deficits and thought disorder in schizophrenia. Am J Psychiatry 151:363–367

Perry W, Braff DL (1996) Disturbed thought and information processing deficits in schizophrenia. Biol Psychiatry 39:549

Perry W, Swerdlow NR, McDowell JE, Braff DL (1998) Schizophrenia and frontal lobe functioning: evidence from neuropsychology, cognitive neuroscience, and psychophysiology. In: Cummings J, Miller B (eds) The frontal lobes. Guilford, New York, pp 509–521

Perry W, Geyer MA, Braff DL (1999) Sensorimotor gating and thought disturbance measured in close temporal proximity in schizophrenic patients. Arch Gen Psychiatry 56:277–281

Perry W, Feifel D, Minassian A, Bhattacharjie I, Braff DL (2002) Information processing deficits in acutely psychotic schizophrenia patients medicated and unmedicated at the time of admission. Am J Psychiatry 159:1375–1381

Platt JR (1964) Strong inference. Science 146:347–353

Pouretemad HR, Thompson PJ, Fenwick PBC (1998) Impaired sensorimotor gating in patients with non-epileptic seizures. Epilepsy Res 31:1–12

Powell SB, Zhou X, Geyer MA (2009) Prepulse inhibition and genetic mouse models of schizophrenia. Behav Brain Res 204:282–294

Quednow BB, Frommann I, Berning J, Kühn KU, Maier W, Wagner M (2008) Impaired sensorimotor gating of the acoustic startle response in the prodrome of schizophrenia. Biol Psychiatry 64:766–773

Roussos P, Giakoumaki SG, Bitsios P (2008) The dopamine D(3) receptor Ser9Gly polymorphism modulates prepulse inhibition of the acoustic startle reflex. Biol Psychiatry 64:235–240

Roussos P, Giakoumaki SG, Bitsios P (2009) Tolcapone effects on gating, working memory, and mood interact with the synonymous catechol-O-methyltransferase rs4818c/g polymorphism. Biol Psychiatry 66:997–1004

Sanders AR, Duan J, Levinson DF, Shi J, He D, Hou C, Burrell GJ, Rice JP, Nertney DA, Olincy A, Rozic P, Vinogradov S, Buccola NG, Mowry BJ, Freedman R, Amin F, Black DW, Silverman JM, Byerley WF, Crowe RR, Cloninger CR, Martinez M, Gejman PV (2008) No significant association of 14 candidate genes with schizophrenia in a large European ancestry sample: implications for psychiatric genetics. Am J Psychiatry 165:497–506

Schellekens AF, Grootens KP, Neef C, Movig KL, Buitelaar JK, Ellenbroek B, Verkes RJ (2010) Effect of apomorphine on cognitive performance and sensorimotor gating in humans. Psychopharmacology 207:559–569

Scott LJ, Mohlke KL, Bonnycastle LL, Willer CJ, Li Y, Duren WL, Erdos MR, Stringham HM, Chines PS, Jackson AU, Prokunina-Olsson L, Ding CJ, Swift AJ, Narisu N, Hu T, Pruim R, Xiao R, Li XY, Conneely KN, Riebow NL, Sprau AG, Tong M, White PP, Hetrick KN, Barnhart MW, Bark CW, Goldstein JL, Watkins L, Xiang F, Saramies J, Buchanan TA, Watanabe RM, Valle TT, Kinnunen L, Abecasis GR, Pugh EW, Doheny KF, Bergman RN, Tuomilehto J, Collins FS, Boehnke M (2007) A genome-wide association study of type 2 diabetes in Finns detects multiple susceptibility variants. Science 316:1341–1345

Shilling PD, Saint Marie RL, Shoemaker JM, Swerdlow NR (2008) Strain differences in the gating-disruptive effects of apomorphine: relationship to gene expression in nucleus accumbens signaling pathways. Biol Psychiatry 63:748–758

Stefansson H, Sigurdsson E, Steinthorsdottir V, Bjornsdottir S, Sigmundsson T, Ghosh S, Brynjolfsson J, Gunnarsdottir S, Ivarsson O, Chou TT, Hjaltason O, Birgisdottir B, Jonsson H, Gudnadottir VG, Gudmundsdottir E, Bjornsson A, Ingvarsson B, Ingason A, Sigfusson S, Hardardottir H, Harvey RP, Lai D, Zhou M, Brunner D, Mutel V, Gonzalo A, Lemke G, Sainz J, Johannesson G, Andresson T, Gudbjartsson D, Manolescu A, Frigge ML, Gurney ME, Kong A, Gulcher JR, Petursson H, Stefansson K (2002) Neuregulin1 and susceptibility to schizophrenia. Am J Hum Genet 71:877–892

Swerdlow NR, Koob GF (1987) Dopamine, schizophrenia, mania, and depression: toward a unified hypothesis of cortico-striato-pallido-thalamic function. Behav Brain Sci 10:197–245

Swerdlow NR, Caine SB, Braff DL, Geyer MA (1992) The neural substrates of sensorimotor gating of the startle reflex: a review. J Psychopharmacol 6:176–190

Swerdlow NR, Benbow CH, Zisook S, Geyer MA, Braff DL (1993) A preliminary assessment of sensorimotor gating in patients with obsessive compulsive disorder. Biol Psychiatry 33:298–301

Swerdlow NR, Braff DL, Taaid N, Geyer MA (1994) Assessing the validity of an animal model of deficient sensorimotor gating in schizophrenic patients. Arch Gen Psychiatry 51:139–154

Swerdlow NR, Filion D, Geyer MA, Braff DL (1995a) "Normal" personality correlates of sensorimotor, cognitive, and visuospatial gating. Biol Psychiatry 37:286–299

Swerdlow NR, Paulsen J, Braff DL, Butters N, Geyer MA, Swenson MR (1995b) Impaired prepulse inhibition of acoustic and tactile startle response in patients with Huntington's disease. J Neurol Neurosurg Psychiatry 58:192–200

Swerdlow NR, Geyer MA, Blumenthal TD, Hartman PL (1999) Effects of discrete acoustic prestimuli on perceived intensity and behavioral responses to startling acoustic and tactile stimuli. Psychobiology 27:547–556

Swerdlow NR, Eastvold A, Gerbranda T, Uyan KM, Hartman P, Doan Q, Auerbach P (2000) Effects of caffeine on sensorimotor gating of the startle reflex in normal control subjects: impact of caffeine intake and withdrawal. Psychopharmacology 151:368–378

Swerdlow NR, Karban B, Ploum Y, Sharp R, Geyer MA, Eastvold A (2001a) Tactile prepuff inhibition of startle in children with Tourette's syndrome: in search of an "fMRI-friendly" startle paradigm. Biol Psychiatry 50:578–585

Swerdlow NR, Eastvold A, Yuan KM, Ploum Y, Cadenhead K (2001b) Matching strategies for drug studies of prepulse inhibition in humans. Behav Pharmacol 12:45–52

Swerdlow NR, Geyer MA, Braff DL (2001c) Neural circuit regulation of prepulse inhibition of startle in the rat: current knowledge and future challenges. Psychopharmacology 156:194–215

Swerdlow NR, Platten A, Shoemaker J, Pitcher L, Auerbach P (2001d) Effects of pergolide on sensorimotor gating of the startle reflex in rats. Psychopharmacology 158:230–240

Swerdlow NR, Eastvold A, Karban B, Ploum Y, Stephany N, Geyer MA, Cadenhead K, Auerbach PP (2002a) Dopamine agonist effects on startle and sensorimotor gating in normal male subjects: time course studies. Biol Psychiatry 161:189–201

Swerdlow NR, Stephany N, Shoemaker JM, Ross L, Wasserman LC, Talledo J, Auerbach PP (2002b) Effects of amantadine and bromocriptine on startle and sensorimotor gating: parametric studies and cross-species comparisons. Psychopharmacology 164:82–92

Swerdlow NR, Stephany N, Wasserman LC, Talledo J, Shoemaker J, Auerbach PP (2003) Amphetamine effects on prepulse inhibition across species: replication and parametric extension. Neuropsychopharmacology 28:640–650

Swerdlow NR, Shoemaker JM, Auerbach PP, Pitcher L, Goins J, Platten A (2004) Heritable differences in the dopaminergic regulation of sensorimotor gating: II. Temporal, pharmacologic and generational analyses of apomorphine effects on prepulse inhibition. Psychopharmacology 174:452–462

Swerdlow NR, Light GA, Cadenhead KS, Sprock J, Hsieh MH, Braff DL (2006) Startle gating deficits in a large cohort of patients with schizophrenia: relationship to medications, symptoms, neurocognition, and level of function. Arch Gen Psychiatry 63:1325–1335

Swerdlow NR, Weber M, Qu Y, Light GA, Braff DL (2008) Realistic expectations of prepulse inhibition in translational models for schizophrenia research. Psychopharmacology 199:331–388

Swerdlow NR, Lelham SA, Sutherland Owens AN, Chang WL, Sassen SD, Talledo JA (2009a) Pramipexole effects on startle gating in rats and normal men. Psychopharmacology 205:689–698

Swerdlow NR, van Bergeijk DP, Bergsma F, Weber E, Talledo J (2009b) The effects of memantine on prepulse inhibition. Neuropsychopharmacology 34(7):1854–1864

Takahashi H, Iwase M, Ishii R, Ohi K, Fukumoto M, Azechi M, Ikezawa K, Kurimoto R, Canuet L, Nakahachi T, Iike N, Tagami S, Morihara T, Okochi M, Tanaka T, Kazui H, Yoshida T, Tanimukai H, Yasuda Y, Kudo T, Hashimoto R, Takeda M (2008) Impaired prepulse inhibition and habituation of acoustic startle response in Japanese patients with schizophrenia. Neurosci Res 62:187–194

Talledo JA, Sutherland Owens AN, Schortinghuis T, Swerdlow NR (2009) Amphetamine effects on startle gating in normal women and female rats. Psychopharmacology 204:165–175

Thaker GK (2007) Schizophrenia endophenotypes as treatment targets. Expert Opin Ther Targets 11:1189–1206

Turpin G (1986) Effects of stimulus intensity on autonomic responding: the problem of differentiating orienting and defense reflexes. Psychophysiology 23:1–14

Valls-Solé J, Muñoz JE, Valldeoriola F (2004) Abnormalities of prepulse inhibition do not depend on blink reflex excitability: a study in Parkinson's disease and Huntington's disease. Clin Neurophysiol 115(7):1527–1536

Venables PH (1960) The effect of auditory and visual stimulation on the skin potential responses of schizophrenics. Brain 83:77–92

Walsh T, McClellan JM, McCarthy SE, Addington AM, Pierce SB, Cooper GM, Nord AS, Kusenda M, Malhotra D, Bhandari A, Stray SM, Rippey CF, Roccanova P, Makarov V,

Lakshmi B, Findling RL, Sikich L, Stromberg T, Merriman B, Gogtay N, Butler P, Eckstrand K, Noory L, Gochman P, Long R, Chen Z, Davis S, Baker C, Eichler EE, Meltzer PS, Nelson SF, Singleton AB, Lee MK, Rapoport JL, King MC, Sebat J (2008) Rare structural variants disrupt multiple genes in neurodevelopmental pathways in schizophrenia. Science 320:539–543

Walters JT, Owen MJ (2007) Endophenotypes in psychiatric genetics. Mol Psychiatry 12:886–890

Weedon MN, Frayling TM (2008) Reaching new heights: insights into the genetics of human stature. Trends Genet 24:595–603

Weike AI, Bauer U, Hamm AO (2000) Effective neuroleptic medication removes prepulse inhibition deficits in schizophrenia patients. Biol Psychiatry 47:61–70

Wolf R, Paelchen K, Matzke K, Dobrowolny H, Bogerts B, Schwegler H (2007) Acute or subchronic clozapine treatment does not ameliorate prepulse inhibition (PPI) deficits in CPB-K mice with low levels of hippocampal NMDA receptor density. Psychopharmacology 194:93–102

Woznica AA, Sacco KA, George TP (2009) Prepulse inhibition deficits in schizophrenia are modified by smoking status. Schizophr Res 112:86–90

Wynn JK, Green MF, Sprock J, Light GA, Widmark C, Reist C, Erhart S, Marder SR, Mintz J, Braff DL (2007) Effects of olanzapine, risperidone and haloperidol on prepulse inhibition in schizophrenia patients: a double-blind, randomized controlled trial. Schizophr Res 95:134–142

Neurocognition in Schizophrenia

Solomon Kalkstein, Irene Hurford, and Ruben C. Gur

Contents

1 Introduction 374
2 General Intellectual Functioning 375
3 Attention 376
4 Processing Speed 377
5 Executive Functioning 377
6 Learning and Memory 379
7 Language 380
8 Visual Perceptual/Constructional Skills 381
9 Fine Motor Skills 381
10 Social Cognition 382
11 Deficits Among Populations at Risk and Endophenotypes 383
12 Longitudinal Studies of Neuropsychological Deficits 384
13 Future Directions of Research 385
References 386

Abstract Neuropsychological deficits among schizophrenia patients have been consistently documented in research over the past 20 years and are reviewed in this chapter. Discussion of general abilities is presented as a background and is followed by analysis of functioning in specific cognitive domains. Overall intellectual deficits are indicated by results from both general intelligence tests and composite test battery scores. Within specific cognitive domains, effect size

---

S. Kalkstein (✉), I. Hurford, and R.C. Gur
Schizophrenia Research Center, Neuropsychiatry Section, Department of Psychiatry, University of Pennsylvania School of Medicine, 10th Floor, Gates Building, 3400 Spruce Street, Philadelphia, PA 19104, USA
Philadelphia Veterans Affairs Medical Center, University of Pennsylvania School of Medicine, Philadelphia, PA 19104-6021, USA
e-mail: solomon.kalkstein@va.gov

differences are noted in numerous areas, including attention, with indications that working memory is affected more severely than simple attention, likely due to inclusion of an executive component in such tasks. There is also evidence of slowed processing speed among schizophrenia patients, likely contributing to deficits in other domains which rely on rapid and efficient assimilation of information. Executive impairments have been found on tests assessing set-shifting abilities, selective attention, and inhibition of inappropriate responses. Learning and memory deficits have been demonstrated extensively, with some evidence that recall of verbal material is more affected than recall of visual information, and that recognition abilities are comparatively less impaired than recall for both modalities. Receptive and expressive language abilities are compromised in schizophrenia patients, as well as visual perceptual, constructional, and fine motor skills. Social cognition is an area of particular importance due to its relevance to functional outcome. Deficits in expression and recognition of facial and prosodic affect have been demonstrated, although subjective experience of emotion appears to be relatively well preserved. Neuropsychological deficits described in this review appear to generally remain stable throughout adulthood, supporting neurodevelopmental, rather than neurodegenerative, models of the illness. Finally, cognitive deficits are increasingly used as endophenotypes, which is likely an important direction of future research.

Keywords Cognition · Endophenotypes · Memory · Neuropsychology · Schizophrenia

1 Introduction

Schizophrenia is a chronic and severe mental disorder characterized by positive (i.e., delusions, hallucinations, disorganized speech, bizarre behavior) and negative (i.e., flat or inappropriate affect, alogia, avolition, anhedonia) symptoms and occurs worldwide in approximately 1% of the population (American Psychiatric Association 2000). In the past, the prevailing belief within the neuropsychology community was that cognitive deficits among schizophrenia patients were minimal or attributable to the effects of symptoms or medication. However, during the last two decades, significant and extensive deficits across neuropsychological domains have been consistently documented (Saykin et al. 1991; Reichenberg and Harvey 2007). Moreover, these deficits exist in patients at first presentation (Saykin et al. 1994) and even prior to their first manifestations and diagnosis of the disease (Davidson et al. 1999). Furthermore, deficits have also been identified in unaffected relatives (Sitskoorn et al. 2004). Therefore, specific neuropsychological deficits have been suggested as representing valuable endophenotypes with predictive usefulness in premorbid populations (Gur et al. 2007b). In this chapter, we discuss findings with regard to deficits among schizophrenia patients across neuropsychological domains. Also discussed are deficits among populations at risk and in the prodromal stage of the illness, as well as longitudinal studies of cognition. Finally, we comment on implications of this research for competing etiological models of schizophrenia.

2 General Intellectual Functioning

Spearman (1927) introduced the concept of a general intelligence factor, which he termed "g." Derived from factor analysis of multiple mental tests, the g-factor was proposed as the underlying construct that accounts for performance on all measures of cognitive ability. This concept of general intellectual functioning has remained influential. Two approaches to assessing overall intellectual functioning have been employed in schizophrenia research using different measures of performance: general IQ scores derived from intelligence tests and composite scores from comprehensive neuropsychological test batteries. Bratti and Bilder (2006) have noted that these two methods can yield different results, as domains known to be impaired among schizophrenic patients are not measured by conventional intelligence tests. Most prominently, verbal learning and memory, one of the most severely impaired areas of cognition in schizophrenia, is not included in subtests of the Wechsler Adult Intelligence Scale (WAIS; The Psychological Corporation 1997). Reichenberg and Harvey (2007) list several meta-analytic studies, showing that both methods of measuring general intelligence yield significant impairments for schizophrenia patients as compared to controls, with moderate to large effect sizes according to Cohen's (1988) conventions. This consistency suggests that specific neuropsychological deficits may occur against a background of compromised general intellectual functioning.

Related to Spearman's g-factor of intelligence is the controversial question among schizophrenia researchers as to whether cognitive impairment in the disease is better conceptualized as broad and generalized, or as domain specific with distinct patterns of better-preserved and more impaired cognitive functions. Based upon factor analytic results, some researchers have posited a multifactorial model of cognitive dysfunction, in which neuropsychological tests measure relatively independent processes (Egan et al. 2001). The measurement and treatment research to improve cognition in schizophrenia (MATRICS) consensus-developed battery (Nuechterlein et al. 2008), for example, which was broadly supported among schizophrenia researchers, presumes separable, discrete cognitive domains that can be differentially affected by treatment. Recently, however, an accumulating body of evidence has demonstrated a large degree of shared variance among standard cognitive measures providing evidence for a generalized cognitive deficit as a core feature of schizophrenia (Dickinson et al. 2004, 2006, 2008), with specific areas of greater relative impairment, such as executive functioning, depending on the severity of global deficit (Bilder et al. 2000). The etiology of neural dysfunction producing such pervasive downstream effects is uncertain. Widespread reductions in gray matter, diminished white matter density and coherence, and poor neuronal signal integration have all been suggested as neurobiological mechanisms underlying these global deficits (Dickinson et al. 2008).

3 Attention

Attention refers generally to one's ability to detect the signal in complex incoming sensory information. However, as Lezak (2004) has pointed out, definitions of attention have been widely divergent and investigators have created numerous subcategories to describe nuances in the attentional process. The schizophrenia literature has employed several of these descriptors. Some tasks operate on the premise that the attentional system is limited in capacity such that only a finite amount of information can be processed within a given duration of time. A particular individual's storage capacity is a simple measure of available space within the attentional reserve. Schizophrenia patients have been shown to possess a compromised storage capacity using digit span forward, a test in which increasingly lengthy number strings must be maintained and repeated verbatim (Aleman et al. 1999; Dickinson et al. 2007).

More complex attentional tasks, which require not only the maintenance of information but also its manipulation, are referred to as *working memory tasks*. For example, digit span backward involves the repetition of a string of numbers in reverse order, requiring that information be held and reorganized according to particular rules. The other tasks which comprise the working memory index of the WAIS involve similar procedures: rearranging jumbled letter–number strings into a numeric–alphabetical order, and completing mathematical problems mentally, without the use of pen and paper. On tasks of working memory, meta-analytic studies have found large effect sizes separating schizophrenia patients from controls (Dickinson et al. 2007; Reichenberg and Harvey 2007). One of these meta-analyses includes 124 working memory studies, including the tasks listed above in addition to others which incorporate a very brief delay, suggesting that this deficit is consistent across different tasks and paradigms (Lee and Park 2005).

Meta-analytic studies of test performance among schizophrenia patients and healthy controls show that effect size differences are typically smaller on tasks of simple attention, such as digits forward, as compared to working memory and executive functioning tasks (Dickinson et al. 2007; Reichenberg and Harvey 2007). Similarly, research has shown that schizophrenia patients have greater deficits on tasks requiring allocation of attention to relevant objects, compared to tasks that require the identification of an object once attention has been focused on it (Fuller et al. 2006). A higher-level failure of executive-type processes in regulating the control of attention among schizophrenia patients represents one explanation for these findings and illustrates the involvement of executive functions in the application of attention.

An additional domain of attention is vigilance, or the ability to sustain one's attention over a prolonged period of time. This capacity is typically assessed using a continuous performance test paradigm that requires focus and concentration for a period of 5–25 min. The computerized Conners' continuous performance test (CPT; Conners 2000), for example, flashes letters on a computer screen and requires that the subject taps the space bar whenever the letter "X" appears. This tedious task continues for 14 min and generates error rates (false positives and false negatives) which provide a useful measure of sustained attention. The small number of errors on typical (search "X") CPTs limit their ability to measure individual differences. Nuechterlein (1983) addressed this problem by reducing the visibility of the signal, while Kurtz et al. (2001) addressed it by increasing the number of valid responses (press when you see any digit among seven-segment displays). Schizophrenia patients are found to have large effect size differences when compared to healthy controls on standard and adapted versions of sustained attention tests (Reichenberg and Harvey 2007).

4 Processing Speed

Schizophrenia patients exhibit deficits in their ability to process new information rapidly and efficiently. Processing speed impairment is of particular concern because many other cognitive operations, such as encoding and retrieval, will be affected by slowing in this domain. Processing speed has been measured using several tests, which typically allow 60–120 s for cognitive operations to be executed. The speed at which the subject reads words can be measured using the Stroop test (Golden 1978). The processing speed index on the WAIS is composed of two timed subtests, requiring rapid motor responses after integrating information during a brief visual scan. In the digit symbol coding subtest, a key must be consulted and symbols must be drawn under their corresponding numbers. In the second subtest, symbol search, target shapes must be searched for within an adjacent set and the subject must indicate whether they are present in the set or not. The trail making test A (Reitan and Wolfson 1985), which is also timed, contains consecutively numbered circles which must be connected by lines drawn as quickly as possible. On these tasks, meta-analysis has generally yielded large effects for patients as compared to controls (Reichenberg and Harvey 2007). The digit symbol coding subtest of the WAIS-III has been the subject of its own meta-analysis (Dickinson et al. 2007) and it was found to represent the largest single cognitive domain impairment in schizophrenia.

5 Executive Functioning

Among the neuropsychological deficits implicated in schizophrenia, executive processing has been identified as an area of particular vulnerability. Executive functions are higher-order cognitive abilities that control decision making and deal with the "how" and "whether" aspects of behavior, rather than the simpler "what" and "how much" functions addressed by other cognitive domains (Lezak 2004). Numerous such skills have been subsumed under the rubric of executive functioning and they include volition, motivation, self-awareness, planning, initiating purposeful behavior, inhibiting inappropriate responses, abstract reasoning, and mental flexibility. Executive functions have been localized to the frontal lobes, specifically, the prefrontal cortex (Lezak 2004).

Various measures have been developed to assess cognitive features within this domain and impaired performance in schizophrenia has been found on several of them. Many studies of executive functioning in schizophrenia have employed the Wisconsin Card Sorting Test (WCST; Heaton 1981), which assesses abstraction skills, inhibition of inappropriate responses, and the ability to shift cognitive strategies in response to changing environmental circumstances (Strauss et al. 2006). Heinrichs and Zakzanis' (1998) meta-analysis, based on 104 WCST studies, found a large effect that was moderated by IQ, suggesting that impaired WCST scores reflected low general intellectual abilities, a conclusion also reached by Laws (1999) in his meta-analysis. Bilder et al. (2000) have found that executive deficits are even more pronounced than global deficits in patients with relatively low general cognitive functioning. Other meta-analyses (Dickinson et al. 2007; Reichenberg and Harvey 2007) have also found severe WCST impairments in schizophrenia patients as compared to controls, as well as compared to other psychiatric groups (Johnson-Selfridge and Zalewski 2001). One meta-analysis reported that both positive and negative symptoms were correlated with executive functioning deficits (Johnson-Selfridge and Zalewski 2001). However, another meta-analysis found small to modest correlations only between negative and disorganization symptoms and WCST perseveration rates (Nieuwenstein et al. CitationRef 2001). The authors cite evidence that negative and disorganization symptoms are known to be associated with dysfunction in the frontal cortex, and correlations with perseverative errors are explained in terms of abnormalities in the frontal region, which cause disturbances to both areas of functioning.

The Stroop test (Golden 1978) is another measure of executive functioning that requires selective attention and the inhibition of habitual responses. Meta-analytic reviews have consistently documented large effect sizes, indicating impaired performance in schizophrenia patients as compared to controls (Dickinson et al. 2007; Reichenberg and Harvey 2007). In the trail making test B (Reitan and Wolfson 1985), subjects are asked to draw lines connecting circles in an alternating sequence, switching back and forth between those with numbers and those with letters. Meta-analytic studies indicate that schizophrenia patients also perform poorly on this test, with large effect sizes evident (Reichenberg and Harvey 2007). Finally, in addition to assessing language abilities, phonemic fluency tests, such as the Controlled Oral Word Association (COWA; Benton et al. 1994) test, are also thought to provide a measure of executive functioning since the subject must search the semantic store selectively, choosing only words that begin with the letter of interest (Strauss et al. 2006). Schizophrenia patients also exhibit large effect size impairments on these tasks (Reichenberg and Harvey 2007).

Computer paradigms have also been used to assess mental flexibility (Wylie et al. 2006). Similar to the WCST, they assess a skill called *task-switching* (sometimes referred to as *set-shifting*), which involves the processing of incoming sensory information according to computational rule sets. In these research paradigms, task-switching is typically assessed by asking subjects to initially classify contextual stimuli according to certain rules, and then, upon receiving a cue, to change that rule-set and to then categorize the same or similar stimuli according to a new set of criteria. Typically, in such paradigms, performance is poorer on the first trial after the switch for patients as well as controls, due to the application of a new interpretive cognitive rule. However, schizophrenia studies have yielded the surprising result that switch costs among patients are comparable to those of healthy comparisons, leading to the conclusion that set maintenance and task-switching reflect distinct information processing systems, with the latter system generally intact in schizophrenia (Kieffaber et al. 2006; Wylie et al. 2008).

6 Learning and Memory

In his early description of the disease, Bleuler (1952) erroneously postulated that schizophrenia does not affect memory functioning as such, but through affective symptoms, alters the patient's ability to recall successfully encoded memories. Instead, numerous tests assessing verbal and nonverbal, auditory and visual learning and memory in schizophrenia patients have demonstrated significant impairments in encoding as well as in recall and recognition (Saykin et al. 1991, 1994; Heinrichs and Zakzanis 1998). These impairments are of particular concern because verbal memory deficits are among the strongest predictors of functional outcome in this population (Green 1996).

When measures of memory are combined for the purposes of meta-analysis, large effect sizes have been demonstrated (Fioravanti et al. 2005). Tasks of learning and memory have also been surveyed separately by meta-analysis. On all studies examining list-learning ability, both immediate and delayed recall for schizophrenia patients was approximately one standard deviation or worse below the mean of healthy controls (Dickinson et al. 2007). On narrative memory tasks in which subjects were asked to remember a story of approximately one paragraph length, as well as on tasks in which words were paired arbitrarily with other words and subjects were cued with one word and asked to remember its match, large effect sizes were also found between patients and controls (Dickinson et al. 2007). Effect sizes are also significant for visual memory tasks (Heinrichs and Zakzanis 1998; Reichenberg and Harvey 2007). For example, on a task requiring subjects to draw a series of five designs after a brief exposure period, schizophrenia patients showed a large effect size difference when compared to controls (Dickinson et al. 2007).

Meta-analyses examining verbal and nonverbal recall impairments have yielded some evidence of larger effect size differences for the former as compared to the latter (Dickinson et al. 2007). Additionally, longitudinal studies have documented comparatively worse verbal, as compared to visual memory deficits as it was found in one study to be the most profoundly impaired cognitive domain at both intake and 19-month follow-up (Censits et al. 1997), and in another to be the only cognitive domain that did not improve over a 5-year follow-up period (Hoff et al. 1999). With regard to recognition memory, the reverse trend was found. A meta-analysis which focused on 84 studies found that visual recognition – which included faces, pictures, and designs – was significantly more impaired than verbal recognition (Pelletier et al. 2005), a trend that was found in an early meta-analysis, as well (Aleman et al. 1999). Recognition memory overall has been shown to be less impaired than recall for both verbal and nonverbal material, indicating poorer consolidation of information and retrieval deficits once information has been encoded (Aleman et al. 1999).

7 Language

Studies assessing language abilities in schizophrenia patients have examined both expressive and receptive language skills using a variety of measures. Reading tests require the recognition and correct pronunciation of letters and words, and vocabulary tests require specific word definitions, reflecting crystallized knowledge of verbal information. Word generation tests impose a time-limited interval within which words are generated according to semantic (i.e., animals, supermarket items) or phonemic (letter) categories. Confrontational naming tasks, such as the Boston naming test, present line drawings or pictures in increasing levels of difficulty and require the verbal identification of items. These methods require the communication of verbal material. Receptive language abilities require verbal or behavioral responses as indicators of comprehension of verbal material.

One meta-analysis found moderate to large effect size differences between schizophrenia and control subjects in studies using vocabulary tests and substantially larger differences where word generation tests were administered (Heinrichs and Zakzanis 1998). They concluded that language tests are "fairly powerful and moderately reliable discriminators of schizophrenia and control populations." A more recent meta-analytic study found a moderate effect for reading tests and large effect sizes for vocabulary as well as for both semantic and phonemic fluency tasks (Dickinson et al. 2007). Reichenberg and Harvey (2007) cite some studies reporting relatively milder impairments in schizophrenia patients on vocabulary and confrontational naming tests. Nevertheless, most studies indicate significant impairment in each area of language functioning and, when studies of language ability are combined, large effect sizes clearly result (Fioravanti et al. 2005). It appears from most of the research in this area that receptive and expressive language abilities are significantly compromised in schizophrenia patients.

Semantic category tests are considered to be less challenging than phonemic letter tests, as more words are consistently produced by healthy adults in response to categories (Lezak 2004). Perhaps, connections in the lexicon based upon semantic properties are stronger because the meanings of words are reinforced by personally meaningful experiences, unlike connections in the lexicon that are based on letters (Bokat and Goldberg 2003). Therefore, the meta-analytic finding that semantic fluency is relatively more impaired than letter fluency among schizophrenia patients is surprising (Dickinson et al. 2007; Henry and Crawford, 2005; Bokat and Goldberg 2003), especially considering that the semantic store of schizophrenia patients is less degraded than that of other patient groups, such as Alzheimer's disease (Doughty et al. 2008). While the underlying mechanism for semantic fluency deficits is unknown, the finding that schizophrenia patients have greater semantic fluency impairment may suggest important differences in neuroanatomical substrates.

8 Visual Perceptual/Constructional Skills

To assess visual perceptual ability, studies in the schizophrenia literature have typically used the judgment of line orientation test (Benton et al. 1983), which requires the matching of angled line segments to their equivalents on a multiple-choice card. Meta-analytic reviews have demonstrated moderate to large effect size differences between schizophrenia patients and healthy controls on this task (Heinrichs and Zakzanis 1998; Dickinson et al. 2007). These reviews have also indicated moderate to large effect size differences on the block design subtest of the WAIS. The block design subtest provides another measure of visuospatial processing, with an added constructional component requiring that the subject manually arranges blocks so that they resemble designs modeled on stimulus cards.

9 Fine Motor Skills

Two tests of manual dexterity and speed have been widely used in schizophrenia research. The finger-tapping test (Reitan and Wolfson 1985) involves rapidly pressing the index finger on a key with a device that records the total number of taps made within 10 s. As in other neuropsychological tests, the average tap rate is compared to healthy individuals of the same age. The Purdue pegboard (Purdue Research Foundation 1948) and grooved pegboard (Trites 1977) tests are also timed and require the placement of round or grooved pegs into slotted holes. On these measures, meta-analysis has demonstrated moderate to large effect size differences for schizophrenia patients as compared to controls for both the dominant and nondominant hands (Heinrichs and Zakzanis 1998; Dickinson et al. 2007).

10 Social Cognition

A relatively new but promising area of interest in schizophrenia research is social cognition, or the process of perceiving and interpreting social information. Its importance in understanding the illness is suggested by the decision of the MATRICS consensus battery panel to include the managing emotions component of the Mayer–Salovey–Caruso emotional intelligence test as part of the battery and to add social cognition as its seventh cognitive domain (Nuechterlein et al. 2008). Although social cognition is a construct not yet fully defined (Green and Leitman 2008), a National Institute of Mental Health (NIMH)-sponsored consensus meeting led to agreement that within schizophrenia research, it encompasses five general domains, which are acknowledged to overlap and are not mutually independent. These domains are emotional processing, social perception, social knowledge, theory of mind, and attributional bias (Green et al. 2008). Research on social cognition in schizophrenia reported domain-specific deficits and also demonstrated that several of these domains are related to measures of functional outcome such as social behavior in treatment settings, social/interactional skills, independent living skills, and everyday social problem-solving abilities (Couture et al. 2006).

Regarding perception of emotion, facial affect and prosody recognition have been studied extensively among schizophrenia patients (Heimberg et al. 1992; Kohler et al. 2003). A comprehensive review of these studies was performed by Edwards et al. (2002) and they noted that although deficits in both areas were generally reported for patients, significant interstudy variability made it difficult to draw conclusions concerning the nature of those deficits. An earlier review of facial expression studies (Mandal et al. 1998) revealed a general deficit in affective perception and expression, and particular difficulty in recognizing the negative emotions of fear and anger. Another recent meta-analysis (Hoekert et al. 2007) revealed a large impairment in the perception of emotional prosody in schizophrenia patients. The cause for this impairment, however, may not be a genuine affective processing deficit. Rather, basic early auditory dysfunction, as measured by tone-matching and melodic distortion tasks, predict difficulties in prosody processing, and may be responsible for schizophrenia patients' compromised ability to process emotional prosody (Leitman et al. 2005).

Expression of emotion has also been identified as an area of significant deficit in schizophrenia patients as compared both to healthy controls and to other psychiatric and neurological patient populations. Schizophrenia patients, both on and off medication, display fewer positive and negative facial expressions in response to a variety of evocative stimuli including film clips, foods, and social interactions (Kring and Moran 2008). Although some studies reported that they display subtle facial microexpressions of emotion, their relatively weak emotional signs are likely to be missed by observers, leading to generally negative social consequences. Schizophrenia patients also exhibit deficits in the expression of emotional prosody, with meta-analytic comparisons to healthy controls yielding an effect size greater than one (Hoekert et al. 2007).

In contrast to their compromised ability to express emotion, schizophrenia patients report subjectively experiencing positive and negative emotions in a manner appropriate to the stimulus or situation that is presented to them (Herbener et al. 2008). Given the lack of insight that is a core deficit in the illness, various reliability measures have been administered to address the question of whether these patients possess the ability to accurately report emotional states and results reflect high consistency within and across tests (Kring and Moran 2008). Moreover, when constructing narratives about life events, the content expressed by schizophrenia patients is generally similar to controls as they typically use words that are consistent with the emotional valence of the experiences they are describing (Gruber and Kring 2008). Thus, in their review, Kring and Moran (2008) conclude that the deficits associated with emotion processing in schizophrenia can be attributed to emotional expression, rather than to the experience of emotion, which is relatively well preserved.

Theory of mind, or the ability to infer and represent the mental states of others, is also impaired in schizophrenia (Brune 2005; Pickup and Frith 2001) and deficits in this area have been hypothesized as a strong contributing factor in the development and exacerbation of psychotic symptoms. Cognitive misrepresentations of others' intentions can lead to faulty attributional styles and this can exacerbate persecutory delusions and paranoia symptoms (Penn et al. 2008; Bentall et al. 2001). Thus, social cognition domains may interact with each other in compounding social difficulties as patients misinterpret behavioral cues, make false negative attributions to others, and experience difficulty in expressing their emotions effectively. These deficits may partially explain the finding that schizophrenia patients have poor insight into the nature and extent of their cognitive deficits, as they are less likely to benefit from social feedback. It has therefore been suggested that psychoeducation not only include discussions of psychotic symptoms but also address cognitive impairments so that insight and motivation for treatment can be improved (Medalia and Thysen 2008).

11 Deficits Among Populations at Risk and Endophenotypes

Endophenotypes are identifiable and measurable characteristics of a disorder, which are presumed to have a simpler inheritance pattern than the full phenotype of the illness and which may serve as markers, signaling risk for development of a psychiatric illness (Bratti and Bilder 2008; Gottesman and Gould 2003). Several neurocognitive deficits have been proposed as endophenotypes since they are present in individuals at genetic risk for schizophrenia (Cannon et al. 1994). Indeed, intermediate neurocognitive deficits on measures of vigilance, processing speed, and executive functioning, which place individuals between population norms and first-episode impairments, are present during the prodromal stage of schizophrenia (Hawkins et al. 2004; Keefe et al. 2006). Unaffected relatives of schizophrenia patients also have a milder form of the deficits characteristic of the illness (Cannon et al. 1994). A study in which cognitive tasks were administered to 193 unaffected siblings of schizophrenia patients revealed poorer performance among siblings as compared to healthy controls on tasks of executive functioning and verbal memory (Egan et al. 2001). Additionally, cognitive deficits in a particular domain did not strongly predict deficits in other domains, indicating that impairments were distinct and represented independent markers of risk for the development of the illness.

Another study utilized a computerized neuropsychological testing paradigm, the Penn Computerized Neuropsychological Battery (CNB; Gur et al. 2001a). The CNB allows for analysis of both accuracy and speed and has been used to demonstrate pervasive neurocognitive impairment in schizophrenia (Gur et al. 2001b). In a multiplex, multigenerational cohort, unaffected relatives of schizophrenia patients performed with significantly poorer accuracy than controls on measures of spatial processing, abstraction, and mental flexibility, and their speed was significantly slower on tests of attention, face memory, spatial processing, and sensorimotor dexterity (Gur et al. 2007a). In a different computerized study of individuals at high genetic risk for schizophrenia, compromised performance was found on a measure of spatial working memory and slowed processing speed was evident on a test of spatial planning (O'Connor et al. 2009). Several meta-analyses of studies on unaffected relatives have been performed recently and they have revealed moderate impairments on tasks of attention (Trandafir et al. 2006; Snitz et al. 2006), verbal memory (Sitskoorn et al. 2004; Trandafir et al. 2006; Snitz et al. 2006), and executive functioning (Sitskoorn et al. 2004; Snitz et al. 2006) among other domains. An NIMH-funded, multisite collaboration identified as the Consortium on the Genetics of Schizophrenia (COGS), investigating cognitive endophenotypes related to schizophrenia, has identified three neurocognitive areas of deficit as candidates; attention (sustained focused attention), verbal declarative memory, and working memory, and proposed two additional domains, face recognition memory, and emotion processing, based on promising results of computerized testing (Gur et al. 2007b). These areas of compromised performance among unaffected relatives of schizophrenia patients represent potential endophenotypes that may be useful in marking risk for development of the disease.

12 Longitudinal Studies of Neuropsychological Deficits

Neuropsychological dysfunction is clearly a core feature of schizophrenia. Deficits are already evident at the time of first episode of psychosis and have been documented as early as age 6 or 7 (Bratti and Bilder 2006). Most of the meta-analyses cited above included a large majority of studies in which schizophrenia patients were middle-aged adults and were age-matched to controls, demonstrating pervasive neurocognitive impairments in adulthood (Aleman et al. 1999; Heinrichs and Zakzanis 1998; Johnson-Selfridge and Zalewski 2001; Pelletier et al. 2005; Dickinson et al. 2007). Cross-sectional studies have also been conducted among elderly patients and have indicated the presence of global deficits and compromised performance in numerous specific cognitive domains (Rajji and Mulsant 2008). Thus, neuropsychological impairment in schizophrenia remains present throughout the lifespan.

Results of most studies employing neuropsychological measures at baseline and follow-up to assess for longitudinal change have not yielded significant differences among adults. One study (Russell et al. 1997) obtained the childhood IQ scores of 34 schizophrenia patients using records from a childhood psychiatry service that they had attended on average 20 years earlier. While they found significant impairment compared to general population norms, no significant differences between childhood and adulthood IQs were found, suggesting a stable deficit in general intellectual function beginning in childhood and lasting over two decades. However, Bratti and Bilder (2006) cite several studies which indicate that there is a significant cognitive decline between the ages of 12 and 17. Rund (1998) reviewed 15 longitudinal studies involving mostly adult patients with a follow-up period of at least 1 year and found no evidence of cognitive decline. A more recent meta-analysis of 53 studies involving adult patients (Szoke et al. 2008) not only failed to find evidence of cognitive decline over time, but also indicated that improvement took place among schizophrenia patients, likely a result either of cognitive remediation treatments or neuropsychological test practice effects. Kurtz's (2005) review of ten studies yielded similar findings of stable neuropsychological test results among outpatient, community-dwelling adults, although he noted gross mental status declines among elderly, institutionalized patients. Additionally, a recent meta-analysis was conducted on studies examining cognitive performance in late-life schizophrenia (Irani et al. 2009). While global deficits were evident when this population was compared to age-matched controls, no unique effect beyond expected age-associated decline was found longitudinally, from baseline to follow-up.

Numerous studies therefore appear to show that while cognitive decline takes place during early-to-middle teenage years, neuropsychological deficits remain stable as adult schizophrenia patients' age. Neurodevelopmental models of schizophrenia, which emphasize disruptions during critical periods of brain development in explaining the etiology of the disease, are most consistent with a static encephalopathy process, while neurodegenerative models would predict significant decline as the result of an ongoing degenerative progression. Although Kraepelin (1919) emphasized the deterioration of cognitive abilities beyond the initial stages of the illness, the majority of longitudinal studies of schizophrenia patients appear to support a neurodevelopmental, rather than a neurodegenerative, model of schizophrenia (Censits et al. 1997; Malaspina 2006; Rapoport et al. 2005; Woods 1998).

13 Future Directions of Research

Although pervasive cognitive deficits have been demonstrated in schizophrenia over the past two decades, it was only recently that a consensus neuropsychological battery was developed for research purposes. The NIMH-sponsored MATRICS initiative employed a panel of experts and after surveying more than 90 tests, developed a brief consensus battery which can be used in evaluating both clinical drug trials and cognitive remediation programs whose aim is to improve cognition in schizophrenia (Kern et al. 2004; Nuechterlein et al. 2008). Doubts have been raised about the ability of the MATRICS, or indeed any neuropsychological battery, to reveal domain-specific treatment effects due to overlap among the dimensions of neuropsychological test performance (Dickinson and Gold 2008). Nevertheless, a recently published study indicated that the MATRICS battery is effective in differentiating between an early-onset schizophrenia adolescent group and healthy controls on all domains with the exception of social cognition (Holmen et al. 2009). This result suggests that the MATRICS battery in its current incarnation is insensitive to deficits in social cognition, because several domains of social cognition have demonstrated large effect sizes for differences between patients with schizophrenia and controls (see also Kohler et al. 2009). Furthermore, the MATRICS relies on old, and sometimes outdated paper-and-pencil tests that are cumbersome to administer, require extensive training in administration and scoring, and produce a long paper trail requiring clumsy procedures for data entry and analysis. New computerized batteries are now available and plans are underway to replace the MATRICS battery with a neuroscience-based computerized format (Carter et al. 2009). It is hoped that such consensus batteries will improve the quality of outcome research by becoming the standard assessment tool used in future studies of cognition in schizophrenia.

Outcome measurement, for which such batteries will be helpful, is especially important in light of promising studies on improvement of functioning with treatment interventions. A meta-analytic review indicated that cognitive remediation moderately improves functioning in overall cognition as well as across almost all domains assessed, with social cognition showing the largest effect size (McGurk et al. 2007). Additionally, treatment with atypical antipsychotic medications has produced significant improvements in social skills performance for schizophrenia patients, perhaps resulting from a concurrent enhancement of cognitive abilities (Harvey et al. 2008). Novel therapeutic interventions with "cognitive enhancers" may further test the hypothesis that outcome in schizophrenia can be ameliorated by targeting the cognitive deficits. Longitudinal studies of individuals at risk for psychosis have also demonstrated that improvement in neurocognitive symptoms reliably predicts better functional outcome (Niendam et al. 2007). Future analysis of the nature of neurocognitive deficits in schizophrenia and the effectiveness of treatment interventions in improving deficit areas will remain an important component of research into this debilitating and complex disease.

References

Aleman A, Hijman R, de Haan EHF, Kahn RS (1999) Memory impairment in schizophrenia: a meta-analysis. Am J Psychiatry 156:1358–1366
American Psychiatric Association (2000) Diagnostic and statistical manual of mental disorders, revised 4th edn. American Psychiatric Association, Washington, DC

Bentall RP, Corcoran R, Howard R, Blackwood N, Kinderman P (2001) Persecutory delusions: a review and theoretical integration. Clin Psychol Rev 21:1143–1192

Benton AL, Sivan AB, Hamsher KdeS, Varney NR, Spreen O (1983) Contributions to neuropsychological assessment. Psychological Assessment Resources, Orlando, FL

Benton AL, Hamsher KdeS, Sivan AB (1994) Multilingual aphasia examination, 3rd edn. Psychological Corporation, San Antonio, TX

Bilder RM, Goldman RS, Robinson D, Reiter G, Bell L, Bates JA, Pappadopulos E, Willson DF, Alvir JM, Woerner MG, Geisler S, Kane JM, Lieberman JA (2000) Neuropsychology of first-episode schizophrenia: initial characterization and clinical correlates. Am J Psychiatry 157:549–559

Bleuler E (1952) Dementia praecox; or the group of schizophrenias. International Universities Press, New York, NY

Bokat CE, Goldberg TE (2003) Letter and category fluency in schizophrenic patients: a meta-analysis. Schizophr Res 64:73–78

Bratti I, Bilder R (2006) Neurocognitive deficits and first-episode schizophrenia: characterization and course. In: Sharma T, Harvey PD (eds) The early course of schizophrenia. Oxford University Press, Oxford, pp 87–110

Bratti I, Bilder R (2008) Cognitive functioning in schizophrenia: cognitive impairments as clues for treatment development. Dir Psychiatry 28:33–49

Brune M (2005) "Theory of mind" in schizophrenia: a review of the literature. Schizophr Bull 31:21–42

Cannon TD, Zorrilla LE, Shtasel D, Gur RE, Gur RC, Marco EJ, Moberg P, Price A (1994) Neuropsychological functioning in siblings discordant for schizophrenia and healthy volunteers. Arch Gen Psychiatry 51:651–661

Carter CS, Barch DM, Gur R, Gur R, Pinkham A, Ochsner K (2009) CNTRICS final task selection: social cognitive and affective neuroscience-based measures. Schizophr Bull 35:153–162

Censits DM, Ragland JD, Gur RC, Gur RE (1997) Neuropsychological evidence supporting a neurodevelopmental model of schizophrenia: a longitudinal study. Schizophr Res 24:289–298

Cohen J (1988) Statistical power analysis for the behavioral sciences. Lawrence Erlbaum Associates, Hillsdale, NJ

Conners CK (2000) Conners' continuous performance test II. Multi-Health Systems, Toronto, ON

Couture SM, Penn DL, Roberts DL (2006) The functional significance of social cognition in schizophrenia: a review. Schizophr Bull 32:S44–S63

Davidson M, Reichenberg A, Rabinowitz J, Weiser M, Kaplan Z, Mark M (1999) Behavioral and intellectual markers for schizophrenia in apparently healthy male adolescents. Am J Psychiatry 156:1328–1335

Dickinson D, Gold JM (2008) Less unique variance than meets the eye: overlap among traditional neuropsychological dimensions in schizophrenia. Schizophr Bull 34:423–434

Dickinson D, Iannone VN, Wilk CM, Gold JM (2004) General and specific cognitive deficits in schizophrenia. Biol Psychiatry 55:826–833

Dickinson D, Ragland JD, Calkins ME, Gold JM, Gur RC (2006) A comparison of cognitive structure in schizophrenia patients and healthy controls using confirmatory factor analysis. Schizophr Res 85:20–29

Dickinson D, Ramsey ME, Gold JM (2007) Overlooking the obvious: a meta-analytic comparison of digit symbol coding tasks and other cognitive measures in schizophrenia. Arch Gen Psychiatry 64:532–542

Dickinson D, Ragland JD, Gold JM, Gur RC (2008) General and specific cognitive deficits in schizophrenia: Goliath defeats David. Biol Psychiatry 64:823–827

Doughty OJ, Done DJ, Lawrence VA, Al-Mousawi A, Ashaye K (2008) Semantic memory impairment in schizophrenia – deficit in storage or access of knowledge? Schizophr Res 105:40–48

Edwards J, Jackson HJ, Pattison, PE (2002) Emotion recognition via facial expression and affective prosody in schizophrenia: a methodological review. Clinical Psychology Review 22:789–832

Egan MF, Goldberg TE, Gscheidle T, Weirich M, Rawlings R, Hyde TM, Bigelow L, Weinberger DR (2001) Relative risk for cognitive impairments in siblings of patients with schizophrenia. Biol Psychiatry 50:98–107

Fioravanti M, Carlone O, Vitale B, Cinti ME, Clare L (2005) A meta-analysis of cognitive deficits in adults with a diagnosis of schizophrenia. Neuropsychol Rev 15:73–95

Fuller RL, Luck SJ, Braun EL, Robinson BM, McMahon RP, Gold JM (2006) Impaired control of visual attention in schizophrenia. J Abnorm Psychol 115:266–275

Golden CJ (1978) Stroop color and word test manual. Stoelting, Wood Dale, IL

Gottesman II, Gould TD (2003) The endophenotype concept in psychiatry: etymology and strategic intentions. Am J Psychiatry 160:636–645

Green MF (1996) What are the functional consequences of neurocognitive deficits in schizophrenia? Am J Psychiatry 153:321–330

Green MF, Leitman DI (2008) Social cognition in schizophrenia. Schizophr Bull 34:670–672

Green MF, Penn DL, Bentall R, Carpenter WT, Gaebel W, Gur RC, Kring AM, Park S, Silverstein SM, Heinssen R (2008) Social cognition in schizophrenia: a NIMH workshop on definitions, assessment and research opportunities. Schizophr Bull 34:1211–1220

Gruber J, Kring AM (2008) Narrating emotional events in schizophrenia. J Abnorm Psychol 117:520–533

Gur RC, Ragland JD, Moberg PJ, Turner TB, Bilker WB, Kohler C, Siegel SJ, Gur RE (2001a) Computerized neurocognitive scanning: I. Methodology and validation in healthy people. Neuropsychopharmacology 25:766–776

Gur RC, Ragland JD, Moberg PJ, Bilker WB, Kohler C, Siegel SJ, Gur RE (2001b) Computerized neurocognitive scanning. II. The profile of schizophrenia. Neuropsychopharmacology 25:777–788

Gur RE, Nimgaonkar VL, Almasy L, Calkins ME, Ragland JD, Pogue-Geile MF, Kanes S, Blangero J, Gur RC (2007a) Neurocognitive endophenotypes in a multiplex multigenerational family study of schizophrenia. Am J Psychiatry 164:813–819

Gur RE, Calkins ME, Gur RC, Horan WP, Nuechterlein KH, Seidman LJ, Stone WS (2007b) The consortium on the genetics of schizophrenia: neurocognitive endophenotypes. Schizophr Bull 33:49–68

Harvey PD, Patterson TL, Potter LS, Zhong K, Brecher M (2008) Improvement in social competence with short-term atypical anti-psychotic treatment: a randomized, double-blind comparison of quetiapine versus risperidone for social competence, social cognition and neuropsychological functioning. Am J Psychiatry 163:1918–1925

Hawkins KA, Addington J, Keefe RSE, Christensen B, Perkins DO, Zipursky R, Woods SW, Miller TJ, Marquez E, Breier A, McGlashan TH (2004) Neuropsychological status of subjects at high risk for a first episode of psychosis. Schizophr Res 67:115–122

Heaton RK (1981) Wisconsin Card Sorting Test manual. Psychological Assessment Resources, Odessa, FL

Heimberg C, Gur RE, Erwin RJ, Shtasel DL, Gur RC (1992) Facial emotion discrimination. III. Behavioral findings in schizophrenia. Psychiatry Res 42:253–265

Heinrichs RW, Zakzanis KK (1998) Neurocognitive deficit in schizophrenia: a quantitative review of the evidence. Neuropsychology 12:426–445

Henry JD, Crawford JR (2005) A meta-analytic review of verbal fluency deficits in schizophrenia relative to other neurocognitive deficits. Cogn Neuropsychiatry 10:1–33

Herbener ES, Song W, Khine TT, Sweeney JA (2008) What aspects of emotional functioning are impaired in schizophrenia? Schizophr Res 98:239–246

Hoekert M, Kahn RS, Pijnenborg M, Aleman A (2007) Impaired recognition and expression of emotional prosody in schizophrenia: review and meta-analysis. Schizophr Res 96:135–145

Hoff AL, Sakuma M, Wieneke M, Horon R, Kushner M, DeLisi LE (1999) Longitudinal neuropsychological follow-up study of patients with first-episode schizophrenia. Am J Psychiatry 156:1336–1341

Holmen A, Juul-Langseth M, Thormodsen R, Melle I, Rund BR (2009) Neuropsychological profile in early-onset schizophrenia-spectrum disorders: measured with the MATRICS battery. Schizophr Bull. doi:10.1093/schbul/sbn174

Irani F, Kalkstein S, Moberg E, Moberg PJ (2009) Neuropsychological deficits in elderly patients with schizophrenia: a meta-analytic review. Poster presented at American Psychological Association convention, Toronto, Canada

Johnson-Selfridge M, Zalewski C (2001) Moderator variables of executive functioning in schizophrenia: meta-analytic findings. Schizophr Bull 27:305–316

Keefe RSE, Perkins DO, Gu H, Zipursky RB, Christensen BK, Lieberman JA (2006) A longitudinal study of neurocognitive function in individuals at-risk for psychosis. Schizophr Res 88:26–35

Kern RS, Green MF, Neuchterlein KH, Deng BH (2004) NIMH-MATRICS survey on assessment of neurocognition in schizophrenia. Schizophr Res 72:11–19

Kieffaber PD, Kappenman ES, Bodkins M, Shekhar A, O'Donnell BF, Hetrick WP (2006) Switch and maintenance of task set in schizophrenia. Schizophr Res 84:345–358

Kohler CG, Turner TH, Bilker WB, Brensinger CM, Siegel SJ, Kanes SJ, Gur RE, Gur RC (2003) Facial emotion recognition in schizophrenia: intensity effects and error pattern. Am J Psychiatry 160:1768–1774

Kohler CG, Walker JB, Martin EA, Healey KM, Moberg PJ (2009) Facial emotion perception in schizophrenia: a meta-analytic review. Schizophr Bull. doi:10.1093/schbul/sbn192

Kraepelin E (1919) Dementia praecox and paraphrenia. E & S Livingstone, Edinburgh

Kring AM, Moran EK (2008) Emotional response deficits in schizophrenia: insights from affective science. Schizophr Bull 34:819–834

Kurtz MM (2005) Neurocognitive impairment across the lifespan in schizophrenia: an update. Schizophr Res 74:15–26

Kurtz MM, Ragland JD, Bilker W, Gur RC, Gur RE (2001) Comparison of the continuous performance test with and without working memory demands in healthy controls and patients with schizophrenia. Schizophr Res 48:307–316

Laws KR (1999) A meta-analytic review of Wisconsin Card Sort studies in schizophrenia: general intellectual deficit in disguise? Cogn Neuropsychiatry 4:1–35

Lee J, Park S (2005) Working memory impairments in schizophrenia: a meta-analysis. J Abnorm Psychol 114:599–611

Leitman DI, Foxe JJ, Butler PD, Saperstein A, Revheim N, Javitt DC (2005) Sensory contributions to impaired prosodic processing in schizophrenia. Biol Psychiatry 58:56–61

Lezak MD (2004) Neuropsychological assessment, 4th edn. Oxford University Press, New York, NY

Malaspina D (2006) Schizophrenia: a neurodevelopmental or a neurodegenerative disorder. J Clin Psychiatry 67:e07

Mandal MK, Pandey R, Prasad AB (1998) Facial expression of emotions and schizophrenia: a review. Schizophr Bull 24:399–412

McGurk SR, Twamley EW, Sitzer DI, McHugo GJ, Mueser KT (2007) A meta-analysis of cognitive remediation in schizophrenia. Am J Psychiatry 164:1791–1802

Medalia A, Thysen J (2008) Insight into neurocognitive dysfunction in schizophrenia. Schizophr Bull 34:1221–1230

Niendam TA, Bearden CE, Zinberg J, Johnson JK, O'Brien M, Cannon TD (2007) The course of neurocognition and social functioning in individuals at ultra high risk for psychosis. Schizophr Bull 33:772–781

Nieuwenstein, MR, Aleman, A, de Haan, EHF (2001) Relationship between symptom dimensions and neurocognitive functioning in schizophrenia: a meta-analysis of WCST and CPT studies. Journal of Psychiatric Research 35:119–125

Nuechterlein KH (1983) Signal detection in vigilance tasks and behavioral attributes among offspring of schizophrenic mothers and among hyperactive children. J Abnorm Psychol 92:4–28

Nuechterlein KH, Green MF, Kern RS, Baade LE, Barch DM, Cohen JD, Essock S, Fenton WS, Frese FJ, Gold JM, Goldberg T, Heaton RK, Keefe RSE, Kraemer H, Mesholam-Gately R,

Seidman LJ, Stover E, Weinberger DR, Young AS, Zalcman S, Marder SR (2008) The MATRICS consensus cognitive battery, part 1: test selection, reliability, and validity. Am J Psychiatry 165:203–213

O'Connor M, Harris JM, McIntosh AM, Owens DGC, Lawrie SM, Johnstone EC (2009) Specific cognitive deficits in a group at genetic high risk of schizophrenia. Psychol Med 39:1649–1655. doi:10.1017/S0033291709005303

Pelletier M, Achim AM, Montoya A, Lal S, Lepage M (2005) Cognitive and clinical moderators of recognition memory in schizophrenia: a meta-analysis. Schizophr Res 74:233–252

Penn DL, Sanna LJ, Roberts DL (2008) Social cognition in schizophrenia: an overview. Schizophr Bull 34:408–411

Pickup GJ, Frith CD (2001) Theory of mind impairments in schizophrenia: symptomatology, severity and specificity. Psychol Med 31:207–220

Purdue Research Foundation (1948) Purdue Pegboard Test. Lafayette Instrument, Lafayette, IN

Rajji TK, Mulsant BH (2008) Nature and course of cognitive function in late-life schizophrenia: a systematic review. Schizophr Res 102:122–140. doi:10.1016/j.schres.2008.03.015

Rapoport JL, Addington AM, Frangou S, Psych MRC (2005) The neurodevelopmental model of schizophrenia: update 2005. Mol Psychiatry 10:434–449

Reichenberg A, Harvey PD (2007) Neuropsychological impairments in schizophrenia: integration of performance-based and brain imaging findings. Psychol Bull 133:833–858

Reitan RM, Wolfson D (1985) The Halstead–Reitan neuropsychological test battery. Neuropsychology Press, Tucson, AZ

Rund BR (1998) A review of longitudinal studies of cognitive functions in schizophrenia patients. Schizophr Bull 24:425–435

Russell AJ, Munro JC, Jones PB, Hemsley DR, Murray RM (1997) Schizophrenia and the myth of intellectual decline. Am J Psychiatry 154:635–639

Saykin AJ, Gur RC, Gur RE, Mozley PD, Mozley LH, Resnick SM, Kester DB, Stafiniak P (1991) Neuropsychological function in schizophrenia: selective impairment in memory and learning. Arch Gen Psychiatry 48:618–624

Saykin AJ, Shtasel DL, Gur RE, Kester DB, Mozley LH, Stafiniak P, Gur RC (1994) Neuropsychological deficits in neuroleptic naïve patients with first-episode schizophrenia. Arch Gen Psychiatry 51:124–131

Sitskoorn MM, Aleman A, Ebisch SJH, Appels MCM, Kahn RS (2004) Cognitive deficits in relatives of patients with schizophrenia: a meta-analysis. Schizophr Res 71:285–295

Snitz BE, MacDonald AW III, Carter CS (2006) Cognitive deficits in unaffected first-degree relatives of schizophrenia patients: a meta-analytic review of putative endophenotypes. Schizophr Bull 32:179–194

Spearman C (1927) The abilities of man: their nature and measurement. Macmillan, New York, NY

Strauss E, Sherman EMS, Spreen O (2006) A compendium of neuropsychological tests: administration, norms, and commentary, 3rd edn. Oxford University Press, New York, NY

Szoke A, Trandafir A, Dupont ME, Meary A, Schurhoff F, Leboyer M (2008) Longitudinal studies of cognition in schizophrenia: meta-analysis. Br J Psychiatry 192:248–257

The Psychological Corporation (1997) Wechsler Adult Intelligence Scale test manual, 3rd edn. The Psychological Corporation, San Antonio, TX

Trandafir A, Meary A, Schurhoff F, Leboyer M, Szoke A (2006) Memory tests in first-degree adult relatives of schizophrenic patients: a meta-analysis. Schizophr Res 81:217–226

Trites RL (1977) Neuropsychological test manual. Royal Ottawa Hospital, Ottawa, ON

Woods BT (1998) Is schizophrenia a progressive neurodevelopmental disorder? Toward a unitary pathogenetic mechanism. Am J Psychiatry 155:1661–1670

Wylie GR, Javitt DC, Foxe JJ (2006) Jumping the gun: is effective preparation contingent upon anticipatory activation in task relevant neural circuitry? Cereb Cortex 16:394–404

Wylie GR, Clark EA, Butler PD, Javitt DC (2008) Schizophrenia patients show task switching deficits consistent with N-methyl-D-aspartate system dysfunction but not global executive deficits: implications for pathophysiology of executive dysfunction in schizophrenia. Schizophr Bull. doi:10.1093/schbul/sbn119 [Epub ahead of print, Oct 3]

Animal Models of Schizophrenia

Jared W. Young, Xianjin Zhou, and Mark A. Geyer

Contents

1	Introduction	392
2	Criteria Used to Validate Animal Models	395
	2.1 Reliability	395
	2.2 Face Validity	396
	2.3 Predictive Validity	396
	2.4 Construct Validity	396
	2.5 Etiological Validity	397
3	Modeling Schizophrenia in Animals	397
4	Behavioral Measures by Symptom Group	398
	4.1 Positive Symptoms	398
	4.2 Negative Symptoms	399
	4.3 Cognitive Symptoms	400
	4.4 Sensorimotor Gating Paradigms	403
	4.5 Latent Inhibition	404
5	Experimental Manipulations for Animal Models of Schizophrenia	405
	5.1 Dopaminergic Agonist Models	405
	5.2 Glutamatergic Antagonist Models	407
	5.3 Serotonergic Agonist Models	409
	5.4 Cholinergic Antagonist Models	410
	5.5 Lesion Models	411
	5.6 Genetic Models for Schizophrenia	412
	5.7 Candidate Genes from Animal Models	415
	5.8 Transgenic Mouse Models	416
6	Conclusions	419
References		420

J.W. Young, X. Zhou, and M.A. Geyer (✉)
Department of Psychiatry, University of California San Diego, 9500 Gilman Drive MC 0804, La Jolla, CA, 92093-0804, USA
e-mail: mgeyer@ucsd.edu

Abstract Schizophrenia may well represent one of the most heterogenous mental disorders in human history. This heterogeneity encompasses (1) etiology; where numerous putative genetic and environmental factors may contribute to disease manifestation, (2) symptomatology; with symptoms characterized by group; positive – behaviors not normally present in healthy subjects (e.g. hallucinations), negative – reduced expression of normal behaviors (e.g. reduced joy), and cognitive – reduced cognitive capabilities separable from negative symptoms (e.g. impaired attention), and (3) individual response variation to treatment. The complexity of this uniquely human disorder has complicated the development of suitable animal models with which to assay putative therapeutics. Moreover, the development of animal models is further limited by a lack of positive controls because currently approved therapeutics only addresses psychotic symptoms, with minor negative symptom treatment.

Despite these complexities however, many animal models of schizophrenia have been developed mainly focusing on modeling individual symptoms. Validation criteria have been established to assay the utility of these models, determining the (1) face, (2) predictive, (3) construct, and (4) etiological validities, as well as (5) reproducibility of each model. Many of these models have been created following the development of major hypotheses of schizophrenia, including the dopaminergic, glutamatergic, and neurodevelopmental hypotheses. The former two models have largely consisted of manipulating these neurotransmitter systems to produce behavioral abnormalities with some relevance to symptoms or putative etiology of schizophrenia. Given the serotonergic link to hallucinations and cholinergic link to attention, other models have manipulated these systems also. Finally, there has also been a drive toward creating mouse models of schizophrenia utilizing transgenic technology. Thus, there are opportunities to combine both environmental and genetic factors to create more suitable models of schizophrenia. More sophisticated animal tasks are also being created with which to ascertain whether these models produce behavioral abnormalities consistent with patients with schizophrenia.

While animal models of schizophrenia continue to be developed, we must be cognizant that (1) validating these models are limited to the degree by which Clinicians can provide relevant information on the behavior of these patients, and (2) any putative treatments that are developed are also likely to be given with concurrent antipsychotic treatment. While our knowledge of this devastating disorder increases and our animal models and tasks with which to measure their behaviors become more sophisticated, caution must still be taken when validating these models to limit complications when introducing putative therapeutics to human trials.

1 Introduction

Kraepelin (1896) first described as dementia praecox the disorder now known as schizophrenia (cited in Hirsch and Weinberger 1985). It carries a lifetime risk of 1% (Cannon and Jones 1996), and has both a genetic and an environmental etiology lifetime risk for a dizygotic twin of a patient with schizophrenia is ~40%, (Klaning 1999; van Os and McGuffin 2003). Traditional diagnosis of schizophrenia has been based on characterized positive and negative symptoms (Andreasen et al. 1997; Pearlson 2000). Positive symptoms are generally defined as features or behaviors not normally present but are so due to the disease process. These include hallucinations, delusions, and bizarre behavior. Negative symptoms are features or behaviors normally present but are reduced due to the disease process, including alogia, affective flattening, anhedonia, and avolition (Ellenbroek and Cools 2000). The cognitive symptoms of these patients first recognized by Kraepelin are, however, once again recognized as core to the disorder and correlate most closely with functional outcome (Green 1996, 2006a). Although our understanding of this disease has evolved and changed over time, knowledge of the relevant clinical literature of schizophrenia remains vital when attempting to develop appropriate animal models of the disorder.

An animal model is defined as any experimental preparation developed to study a particular human condition or phenomenon in animals – in our case the psychopathology associated with the group of schizophrenia disorders. While developing and assessing an animal model, it is important to specify the purpose intended for the model, because the intended purpose determines the criteria that the model must satisfy to establish its validity. At one extreme, one can attempt to develop an animal model that mimics the schizophrenia syndrome in its entirety. In the early years of psychopharmacology, the term "animal model" often denoted such an attempt to reproduce a psychiatric disorder in a laboratory animal. Unfortunately, the group of schizophrenia disorders is characterized by considerable heterogeneity and a complex clinical course that reflects many factors that cannot be reproduced readily in animals. Thus, the frequent attempts to model the syndromes of schizophrenia in animals have usually met with failure, thereby prompting skepticism regarding this entire approach.

At the other extreme, a more limited use of an animal model related to schizophrenia is to systematically study the effects of antipsychotic treatments. Here, the behavior of the model is only intended to reflect the efficacy of known therapeutic agents, and thus lead to the discovery of related pharmacotherapies. Because the explicit purpose of the model is to predict treatment efficacy, the principle guiding this approach has been termed "pharmacological isomorphism" (Matthysse 1986). The fact that such models are developed and validated by reference to the effects of known therapeutic drugs frequently limits their ability to identify new drugs with novel mechanisms of action. Similarly, an important limitation that is inherent in this approach is that it is not designed to identify new antipsychotics that might better treat the symptoms of schizophrenia that are refractory to current treatments, such as cognitive dysfunction.

Because of the complexity of schizophrenia, another approach for the development of relevant animal models relies on focusing only on specific signs or symptoms associated with schizophrenia, rather than mimicking the entire syndrome. In such cases, specific symptoms in schizophrenia patients provide a focus for study in experimental animals. The particular behavior being studied may or may not be pathognomonic for or even symptomatic of schizophrenia, but must be defined objectively and observed reliably. It is important to emphasize that the reliance of such a model on specific observables minimizes a fundamental problem plaguing animal models of the syndrome of schizophrenia. Specifically, the difficulties inherent in conducting experimental studies of patients with schizophrenia have limited the number of definitive clinical findings with which one can validate an animal model of schizophrenia. As mentioned earlier, the validation of any animal model can be only as sound as the information available in the relevant clinical literature (Segal and Geyer 1985). By focusing on specific signs or symptoms rather than syndromes, one can increase the confidence in the cross-species validity of the model. The narrow focus of this approach generally leads to pragmatic advantages in the conduct of mechanistic studies addressing the neurobiological substrates of the behavior in question. By contrast, in models intended to reproduce the entire syndrome of schizophrenia, the need for multiple simultaneous endpoints makes it relatively difficult to apply the invasive experimental manipulations required to establish underlying mechanisms. Furthermore, the study of putatively homologous behaviors in both human and nonhuman subjects effectively addresses and bypasses the nonconstructive criticism that complex mental disorders cannot possibly be modeled in nonhuman animals.

Another approach for the development of animal models is based more theoretically upon psychological constructs believed to be affected in schizophrenia. Such an identification of underlying psychological processes or behavioral dimensions (Matthysse 1986; Segal and Geyer 1985) involves the definition of a hypothetical construct and subsequent establishment of operational definitions suitable for the experimental testing of the validity of the construct. Constructs, such as vigilance, perseveration, sensorimotor gating, or working memory, have been used in this manner in schizophrenia research. This approach is most fruitful when conceptually and/or procedurally related experiments are undertaken both in the relevant patient population and in the putative animal model. That is, studies of appropriate patients are needed to establish the operational definitions of the hypothetical construct, and the construct's relevance to schizophrenia. In concert, parallel studies of the potentially homologous construct, process, or dimension are required to determine the similarity of the animal model to the human phenomena. An important and advantageous aspect of this approach is that the validation of the hypothetical construct, and its cross-species homology can be established by studies of normal humans and animals, in addition to schizophrenia patients or experimentally manipulated animals. This approach, therefore, benefits from the existing literature relevant to the hypothetical construct upon which the model is based. In a sense, this approach explicitly recognizes that the experimental study of schizophrenia in humans involves as much of a modeling process as does the study of the disorder in animals. The Measurement and Treatment Research to Improve Cognition in Schizophrenia (MATRICS) group attempted to identify such constructs that are consistently deficient in schizophrenia patients. Seven such constructs were identified including attention, working memory, speed of processing, reasoning and problem solving, social cognition, visual learning and memory, and verbal learning and memory. The identification and operational definitions of these constructs provide a clear opportunity for cross-species validation of putative animal models of schizophrenia (Young et al. 2009b).

2 Criteria Used to Validate Animal Models

Validation of a model refers to the extent to which a model is useful for a given purpose. Thus, different types of validity are relevant depending on the desired purpose of the test one wishes to validate. Furthermore, in considering the validity of a model, both the independent variable (i.e., inducing manipulation) and the dependent measures (i.e., behavioral outcome) need to be evaluated. The reliability and predictive validity of the model system are relevant to both the independent variable and the dependent measures and are the most important criteria to satisfy (for further discussion of the various types of validity, see Braff and Geyer 1990; Geyer and Braff 1987; Geyer and Markou 1995). Additional criteria relevant to the independent variable include etiological, construct, and face validity, with etiological validity being the most relevant. The criteria relevant to the dependent variable include construct, convergent, discriminant, and face validity. Clearly, the more types of validity a model satisfies its value, utility, and relevance to the human condition is likely to be greater. Some forms of validity are more important than others, however, and although many investigators and most nonscientists place considerable emphasis on face validity, it has been argued that predictive validity and reliability are the only necessary and sufficient criteria for the initial evaluation of any animal model (Geyer and Markou 1995).

2.1 Reliability

Reliability of a model is relevant to both the independent and the dependent variables and refers to the consistency and stability with which the variables of interest are observed (Geyer and Markou 1995; Segal and Geyer 1985). Consistency should be evident at the following levels: (a) ability to manipulate the independent variable with a high degree of precision; (b) ability to measure the dependent variable objectively; (c) small within-subject variability of the dependent variable; (d) small between-subject variability of the dependent variable; (e) reproducibility of the phenomenon under maximally similar conditions; and (f) reproducibility of the effects of manipulations (Gao et al. 2000).

2.2 Face Validity

Face validity refers to the phenomenological similarity between the dependent variable (behavior of the animal) and the specific symptoms of the human condition (Lieberman et al. 1997). Although face validity is an intuitively appealing criterion with which to validate models, appearing desirable (Lipska and Weinberger 2000; Weiner et al. 1996), it (a) is not actually necessary, (b) can be misleading, and (c) is difficult to defend rigorously. The latter proves most difficult as they almost invariably involve subjective arbitrary arguments that are not necessarily accepted by all investigators in the field (see Lipska and Weinberger 1995). Thus, although face validity may provide a heuristic starting point for the development of an animal model, it cannot be used to establish the validity of the model.

2.3 Predictive Validity

The ability of a model to make correct predictions about the human phenomenon of interest is referred to as predictive validity (Geyer and Braff 1987; Geyer and Markou 1995). The term predictive validity is often used in the narrow sense of the model's ability to identify drugs having therapeutic value in humans (i.e. pharmacological isomorphism, Matthysse 1986). The use of the term in this way is limited because it ignores other important ways in which a model can be validated by making successful predictions (Ellenbroek and Cools 2000). For example, predictive validation of the experimental preparation is also observed whereby variables have similar influences in both the model and the modeled phenomenon and can enhance one's understanding of the phenomenon.

2.4 Construct Validity

Construct validity is most commonly defined as the accuracy with which the test measures that which it is intended to measure (Geyer et al. 1999). Although considered by investigators in a variety of fields as the most important property of a test or measure (Geyer et al. 1999; Lipska et al. 1995), the establishment of construct validity is rare. The process of construct validation of a test is not different in any essential way from the general scientific procedure for developing and testing theories (Geyer et al. 1999), and thus for developing animal models. Further problems arise as conceptions of what a test is supposed to measure or a model is supposed to mimic are constantly changing as scientific theories and theoretical constructs are modified. Thus, a model's usefulness, and hence its overall validity, cannot be determined simply by the degree of construct validation that it has. Nevertheless, the process of construct validation is valuable in the never-ending process of further development and refinement of the model. As new experimental and observational evidence accrues from both the animal model and the clinical conditions, the model is refined and therefore enables more accurate predictions.

A model or a test has convergent (or concurrent) and discriminant validity only in relation to other models, tests, or measures. Convergent validity is the degree to which a test correlates with other tests that attempt to measure the same construct (Taiminen et al. 2000). Discriminant validity is the degree to which a test measures aspects of a phenomenon that are different from other aspects of the phenomenon that other tests assess (Taiminen et al. 2000).

2.5 Etiological Validity

The concept of etiological validity is closely related to the concept of construct validity. A model has etiological validity if the etiologies of the phenomenon in the animal model and the human condition are identical. As such, assessing etiological validity involves an evaluation of the inducing conditions, and the implicit or explicit hypotheses about the etiology of the disease. Accordingly, like construct validation, the process of etiological validation is a fundamental component of scientific investigation. When etiological validity can be established, the model can become extremely useful in the development of treatments. The limitations of treatment-oriented models based on pharmacological isomorphism, alluded to above, can be overcome if an etiologically based model is found. Unfortunately, the etiologies of psychiatric disorders are seldom known. Hence, etiological validity in this context is generally limited to hypotheses regarding a possible etiology. Indeed, the purpose for the development of animal models is often to enable the identification of or to test a hypothesis about the etiology of the disease.

3 Modeling Schizophrenia in Animals

Schizophrenia is a uniquely human disorder, and may in fact confer humanity upon us (Horrobin 1998). Thus, modeling the disease in animals may prove problematic. Further complications on developing animal models of this disease arise whereby patients do not exhibit a uniform set of symptoms. Bleuler, who coined the term "schizophrenia" following Kraepelin's observations, referred to the disease as the "group of schizophrenias," indicative of the recognized heterogeneity of the disorder. Not only is the heterogeneity of the disease problematic, but it is also difficult to identify animal traits consistent with the DSM-IV criteria. As discussed earlier, greater understanding of the disorder may arise from modeling specific signs and symptoms, as opposed to the entire syndrome. Many studies therefore focus on examining behavioral measures in tasks related to symptoms in schizophrenia, or that are predictive of antipsychotic activity. For example, prepulse inhibition (PPI) is a cross-species measurement of sensorimotor gating – deficient in schizophrenia patients (Braff et al. 1978) – that can also be assessed in rodents. Thus, PPI can be disrupted in animals and these experimentally induced PPI deficits can be used to identify potential therapeutics, which may increase PPI in patients (Powell et al. 2009). Other behavioral measurements, such as drug-induced increases in horizontal locomotion, do not necessarily correspond to symptoms of patients with schizophrenia as they are not more active in an exploratory setting (Perry et al. 2009). These measurements have been primarily used to provide a functional measure for the antidopaminergic activity of neuroleptics.

In addition to behavioral assessments, cellular and molecular markers that are based on described changes in human postmortem and imaging studies are potentially useful measures for establishing the validity of animal models. Furthermore, because of the inherent limitations of modeling in laboratory animals, some of the most prominent behavioral abnormalities of schizophrenia, such as delusions and hallucinations, cellular and molecular characterizations can complement behavioral investigations in evaluating developmental and genetic models of schizophrenia.

Developing a model of schizophrenia includes not only an appropriate measure such as behavior or molecular markers, but also a manipulation, which attempts to cause the disease-like phenomenon. Thus while PPI constitutes the measure, the model is created by administering a drug, e.g. the noncompetitive NMDA receptor antagonist phencyclidine (PCP), to animals and assessing its effects on PPI. Although there are numerous ways in which animal models could be subdivided (Ellenbroek and Cools 2000), in the remaining chapter, we will attempt to review the measures by which the effects of experimental manipulations are assessed and then review those manipulations and their utility. The construct validity of these tasks, etiological validity of the manipulations, and predictive validity of the observed effects will be discussed.

4 Behavioral Measures by Symptom Group

4.1 Positive Symptoms

Hallucinations, delusions, and thought disorders cannot be measured in animals, especially not in tasks with construct validity (Jones et al. 2008). It has been postulated, however, that treating animals with drugs that induce psychotic symptoms in humans and exacerbate symptoms in schizophrenia patients (psychotomimetics), then measuring the behavioral effects may proffer a means by which to experimentally investigate psychosis. Such postulations have led to the development of novel antipsychotics and will be reviewed under pharmacological manipulations. Tasks that can assess any positive symptoms with cross-species translational validity are rare, however. In a reverse-translational approach, a task has been developed with which to assess the exploratory behavior of schizophrenia patients, termed the human behavioral pattern monitor (BPM). This approach is termed "reverse-translational" as the task was based on a rodent version of the BPM, already used for rats (Geyer et al. 1986), and mice (Risbrough et al. 2006; Young et al. 2009a), leading to the creation of a human version (Perry et al. 2009; Risbrough et al. 2006; Young et al. 2009a, 2007c). Given that schizophrenia patients exhibit a unique behavioral profile among patient groups with psychosis tested to date (Perry et al. 2009), this task may be suitable for characterizing the unique exploratory patterns that accompany psychomotor activation in these patients (Jones et al. 2008). Other avenues are being explored to measure and quantify psychomotor agitation utilizing activity meters (Farrow et al. 2006; Minassian et al. 2009).

4.2 Negative Symptoms

As with positive symptoms, some negative symptoms of schizophrenia are virtually impossible to model in laboratory animals, including alogia and affective flattening, with avolition and apathy also proving very difficult (Ellenbroek and Cools 2000). Anhedonia, which refers to a reduced ability to experience pleasure, may, however, lend itself to animal modeling. To measure anhedonia, however, we must first be able to assess hedonic behavior in animals. While such behavior may prove difficult to assess, behaviors that are voluntarily initiated and repeated – although not compulsively such as grooming or stereotypy – are considered to be rewarding/hedonic. Several examples of this behavior exist in the literature, ranging from sucrose vs. water preference, increasing responses to obtain the same value of reward (progressive ratio breakpoint), as well as intracranial self-stimulation (ICSS).

4.2.1 Progressive Ratio Breakpoint Studies

Rodents can be trained to lever press or nose poke to obtain a food reward. Increasing the number of responses required to obtain the same food reward (progressive ratio) provides a measure of how willing the animal is to work for a single reward – the point at which the animal stops responding is considered to be its breakpoint. Thus, the breakpoint is interpreted as the value the animal puts on the reward. Treatments that reduce the breakpoint without concomitant effects on activity levels may prove useful as models of anhedonia (Ellenbroek and Cools 2000). This task also proves useful, however, when examining the effects of experimental manipulations on the performance of appetitively motivated tasks. Thus if an experimental manipulation affects performance on an appetitively motivated task such as the 5-choice serial reaction-time (5-CSR) task (see below), but also lowers breakpoint, it would be difficult to interpret the effects on 5-CSR task performance as solely attentional in nature (Young et al. 2009b).

4.2.2 Intracranial Self-Stimulation

ICSS was developed by Olds and Milner (1954) and involves implanting an electrode into specific brain regions such as the septal region or medial forebrain bundle, then allowing the rat to voluntarily self-administer small electric currents following lever presses. Consistent with the progressive ratio breakpoint study, if an experimental manipulation reduces the number of responses during ICSS without affecting activity levels, it could prove useful as a model of anhedonia (Ellenbroek and Cools 2000).

4.3 Cognitive Symptoms

Previously, much of the focus of schizophrenia research was to develop novel antipsychotic medication to treat positive symptomatology. It is becoming apparent, however, that cognitive deficits are core symptoms of the disorder. These deficits often precede the manifestation of psychosis (Cornblatt et al. 1998, 1997; Erlenmeyer-Kimling 2000), and are orthogonal to positive and negative symptoms (Goldberg and Weinberger 1995; Nieuwenstein et al. 2001). More importantly is that cognitive performance of patients most closely correlates with functional outcome (Green 1996, 2006b), thus there has been a major drive toward understanding and developing treatments for the cognitive symptoms experienced by schizophrenia patients (Floresco et al. 2005; see Barch chapter this text). In response to the lack of effective cognitive treatments, the National Institute of Mental Health (NIMH) sponsored the Measurement and Treatment Research to Improve Cognition in Schizophrenia (MATRICS) initiative (Marder and Fenton 2004) to bring together academic, industrial, and governmental bodies to address this great "unmet therapeutic need." This MATRICS group comprised mostly of clinical psychologists and identified seven cognitive domains as commonly deficient in schizophrenia patients: attention/vigilance; working memory; reasoning and problem solving; processing speed; visual learning and memory; verbal learning and memory; and social cognition (Nuechterlein et al. 2004). From this list, MATRICS identified a standard paper-and-pen test battery with which to assess the cognitive performance of schizophrenia patients. The Treatment Units for Research on Neurocognition and Schizophrenia (TURNS) was then devised to test putative therapeutics in patients using the MATRICS test battery (Buchanan et al. 2007). This group also had a preclinical subcommittee which began to identify tasks which may map onto this test battery (Young et al. 2007b), elaborating on previous suggestions (Hagan and Jones 2005). The Cognitive Neuroscience Treatment Research to Improve Cognition in Schizophrenia (CNTRICS) group was then created to assess cognition in schizophrenia from using specific cognitive constructs with recognition for the need of cross-species translational validity for animal testing (Carter and Barch 2007).

Thus, there is now extensive research which focuses on identifying the specific cognitive deficits apparent in schizophrenia patients. Suitable tasks, their cross-species translational validity, and the effects of experimental manipulations as models of schizophrenia have also been discussed (Young et al. 2009b). Construct validity for the appropriate cognitive domains are discussed elsewhere (Young et al. 2009b), here a brief summary of tasks will be provided.

4.3.1 Attentional Dysfunction

5-Choice Serial Reaction-Time Task

The 5-choice serial reaction-time (5-CSR) task requires rodents to attend to an array of five holes and nose poke in any hole that is illuminated. Although initially developed in Cambridge (Carli et al. 1983) to assess rat models of attention deficit hyperactivity disorder, the task is now also used in mice (Humby et al. 1999), world-wide, and extensively for schizophrenia research (Chudasama and Robbins 2004). While originally designed as analogous to Leonard's 5-choice reaction-time paradigm for humans (Carli et al. 1983), the 5-CSR task has been used extensively to investigate distinct neuroanatomical mechanisms of attention. Further development of the task has recently been described which may confer further cross-species translational validity for the continuous performance test (CPT) – a common test used in humans and chosen by MATRICS – coined the 5-choice CPT (Young et al. 2009b).

Sustained Attention Task

The sustained attention task (SAT) was developed by Bushnell et al. (1994), with construct validity as an attentional task provided by McGaughy and Sarter (1995). The SAT is performed in a two-lever operant chamber where the rat is required to respond to one lever following the presentation of a stimulus, or to another lever if the rat perceived that no stimulus was presented. Although limited availability to test mice, a human version of the task has been developed and the distractor condition was chosen by CNTRICS to assess top down attentional control (Nuechterlein et al. 2009).

4.3.2 Executive Function

Attentional Set-Shifting Task

The Wisconsin Card Sorting Task (WCST) is a traditional "pen-and-paper" task used to assess executive functioning in humans. A more formalized version of the task was developed by Owen et al. (1991), coined the intradimensional/

extradimensional (ID/ED) task, as part of the Cambridge Neuropsychological Test Automated Battery (CANTAB). The ID/ED task, which can also be tested in monkeys, retains key features of the WCST but allows greater analyses of the pattern of performance. Schizophrenia patients initially exhibit selective deficits in ED shifting, but gradually also exhibit poorer reversal learning compared to control subjects (Pantelis et al. 1999). The ID/ED task was also chosen by CNTRICS as a mean to assay executive control (Barch et al. 2009). The attentional set-shifting task (ASST) was developed as a rodent version of the ID/ED task and to date has been tested in rats (Barense et al. 2002; Birrell and Brown 2000), and in mice (Bissonette et al. 2008; Young et al. 2009b). This task requires rodents to utilize stimuli from at least two dimensions (e.g. odor and digging medium) to repeatedly solve the location of a buried food reward with no training given prior to testing. Following acquisition of the rule (e.g. reward indicated by the relevant dimension odor and the relevant stimulus paprika, with acquisition demonstrated by six consecutively correct responses), the rule can be changed to examine reversal (the previously irrelevant stimulus in the relevant dimension now dictates reward location), intradimensional (a novel stimulus in the relevant dimension dictates reward location), and extradimensional (a novel stimulus in the previously irrelevant dimension dictates reward location), learning in a manner consistent with monkeys and humans.

4.3.3 Working Memory

Radial Arm Maze

The radial arm maze (RAM) represents one of the oldest tests of higher cognitive functioning in rodents (Olton and Werz 1978). The RAM traditionally uses an octagonal starting hub that has eight arms leading from it and simply requires the rodent to visit each arm to obtain a food reward, adopting a win-shift strategy to avoid reentering arms in which they have already obtained a reward. The task is designed such that the rodent utilizes extra-maze spatial cues to identify arms not yet entered and the number of arms visited prior to reentering an arm is taken as the spatial span capacity of the animal. Given that there is a recognition component to this task, it may not assess working memory consistent with that defined by Baddeley (1986), however, (Dudchenko 2004). Moreover, care must be taken when interpreting data from this task as methodological differences could result in the measurement of a different cognitive construct than that of working memory span capacity (Young et al. 2009b).

Odor Span Task

Consistent with the RAM, the odor span task (OST) requires rodents to remember increasing number of items of information to select the novel item (Dudchenko et al. 2000). In the case of the OST, increasing numbers of nonspatial odor cues are presented, providing a putative measure of nonspatial span capacity. Again, however, this task has a recognition component and so data interpreted with reference to working memory should be viewed with caution.

4.3.4 Visual Learning and Memory

Morris Water Maze

The water maze was developed by Morris (1981) and it utilizes the natural preference of rodents to escape a marine environment. A circular, high-sided pool filled with warm water and a submerged platform is used. The water is opaque so that for the rodents to locate the platform they must use extra-maze cues. Traditionally, rodents are trained to locate a stationary platform for several sessions a day over several days, with latency and path lengths to locate the submerged platform taken as a measure of learning and memory. Although the focus of using this task has been to investigate memory encoding and hippocampal activity, many have used this task in schizophrenia research. Perhaps, most exciting is the reverse translation of this task for use in humans. While a pool and opaque water is not used, subjects don a virtual reality mask and are required to navigate an environment utilizing spatial cues to locate certain objects. Initial research assessing the performance of schizophrenia patients suggests profound learning and memory deficits, which can then be back-translated for further development of animal models of schizophrenia.

Novel Object Recognition Task

The novel object recognition task (NORT) is another task commonly used in research to investigate memory performance. This task utilizes the innate preference of animals (also observed in humans) to investigate novel objects. The rodent is presented with two identical novel objects and given time to explore these objects (sample phase). The animal is then removed for a predetermined period of time (delay period – studies range from 3 s to 24 h), following which they are presented with two objects, one identical to the previously presented objects, and the other one novel (choice phase). The preference of the animal to explore the novel over familiar object is taken as the memory capacity of that animal for that given delay period.

4.4 Sensorimotor Gating Paradigms

4.4.1 Prepulse Inhibition

The prepulse inhibition (PPI) paradigm is based on the fact that a weak prestimulus presented 30–500 ms prior to a startling stimulus reduces (gates) the amplitude of the startle response. The cross-species translational validity and reliability of this robust phenomenon is clear: PPI is observed in many species including humans, monkeys, rats, and mice; PPI is evident both within and between multiple sensory modalities using a variety of stimulus parameters; and PPI does not require learning or instruction comprehension. Virtually all the evidence available supports the belief that PPI is homologous from rodents to humans. As reviewed elsewhere (Geyer et al. 1999; Swerdlow and Geyer 1998), several laboratories have reported significant deficits in PPI in patients with schizophrenia, schizotypy, and presumably psychosis-prone subjects using a variety of testing procedures and stimulus parameters. Although PPI deficits are not unique to patients diagnosed with schizophrenia, there is evidence that PPI is useful for probing the neurobiology and genetics of sensorimotor gating deficits in schizophrenia (Powell et al. 2009; Swerdlow et al. 2008).

4.4.2 Auditory Gating

The auditory gating paradigm assesses the P50 evoked potential elicited when two acoustic clicks are presented in rapid succession, usually 500 ms apart. In normal individuals, the P50 ERP to the second click is reduced or gated relative to the ERP to the first click. Patients with schizophrenia and their first-degree relatives exhibit less of this sensorimotor gating (Freedman et al. 1994). While researchers have used the difference between hippocampal P20 and N40 ERPs as being analogous to the human P50 (Stevens et al. 1996), this has been questioned (Connolly et al. 2004; Umbricht et al. 2004), with data suggesting the P20 ERP alone in rodents may be analogous to the human P50 (Metzger et al. 2007).

4.5 *Latent Inhibition*

It has been suggested that latent inhibition (LI) may prove a useful cross-species translational test to assay the neurobiology underpinning schizophrenia (see Gray and Snowden 2005; Moser et al. 2000). LI is a measure by which subjects are inhibited from learning a classic CS–US association due to earlier exposure of the CS, which did not predict the US, and although the underlying mechanism is unknown, it has been proposed that this phenomenon has been linked to attentional and/or switching strategy deficits. Abnormal LI has been observed in schizophrenia patients (Baruch et al. 1988), but more recent data suggest no deficit in patients or in fact increased LI in medicated patients (Swerdlow et al. 1996). Certainly, these data limit the interpretation of LI being linked to cognitive performance, but moreover limit its utility in schizophrenia research until a better understanding of the phenomenon is achieved (Swerdlow 2010).

5 Experimental Manipulations for Animal Models of Schizophrenia

The most common approach for developing animal models of schizophrenia has been to exploit pharmacological treatments or "drug-induced states" that produce schizophrenia-like symptoms in nonpatient humans. These models generally have some predictive and/or construct validity, and have been instrumental in establishing three of the most prominent theories of schizophrenia, the dopamine, the glutamate, and the serotonin (or serotonin–dopamine) hypotheses. Recently, there has also been increased interest in the role of cholinergic receptors in schizophrenia given the cognitive disruption observed in these patients. Finally, lesion and genetic models have also been developed. For pharmacological models, both acute and repeated administration techniques are used. Acute models often assess a putative therapeutic while a drug is within the animals system. Repeated administration models assess behavioral performance when no drug is on board and the animals are not exhibiting symptoms of withdrawal. Any abnormalities observed following a repeated administration technique would suggest that there has been an underlying alteration in the neuroanatomy subserving the behavior being measured, thus avoiding complication of drug model through treatment interactions which may confound acute administration models. Three techniques for repeated administration are commonly utilized, subchronic (approximately 5 days), chronic (approximately 30 days), and intermittent/escalating administration. For both subchronic and chronic administration techniques, the day and time of administration are often kept consistent. An intermittent/escalating dosing regimen utilizes the unpredictability of dosing both time and day as well as increasing the dose, in the belief that this unpredictability and drug severity may contribute to the resulting abnormal behavioral profile. Each of these techniques will be covered where appropriate below.

5.1 Dopaminergic Agonist Models

Dopamine remains the neurotransmitter that is most closely linked with schizophrenia. The dopamine hypothesis of schizophrenia was proposed as early as 1967 suggesting that the disorder was as a result of hyperdopaminergic activity in the striatum. Support for this theory is observed whereby antipsychotic treatment for schizophrenia act via D2 receptor blockade (Seeman et al. 1976). Moreover, chronic psychostimulant abuse, such as amphetamine, can lead to psychotic episodes, while low dose amphetamine administration in schizophrenia patients worsens their symptoms (Laruelle et al. 1999). More recently, the dopamine hypothesis has evolved to include a putative hypofrontal dopaminergic system that may contribute to negative and cognitive symptoms experienced by schizophrenia patients (Davis et al. 1991).

5.1.1 Acute Models

The validity of acute amphetamine administration on behavior in a variety of tasks has been assessed as an animal model of schizophrenia. For example, amphetamine-induced hyperactivity in an open field paradigm has often been used in drug discovery as a screen for antipsychotic efficacy. While antipsychotic-induced reversal of hyperactivity is often observed, this model may limit the development of novel therapeutics. This limitation stems from the putative interpretation whereby the antipsychotic drugs tested often have dopamine D2 antagonist properties, thereby what is truly observed may simply reflect dopamine agonist-induced hyperactivity and dopamine antagonist-induced attenuation. This direct receptor interaction is often referred to as receptor tautology (Geyer 2006), and amphetamine-induced hyperactivity may simply be a screen for drugs with D2 antagonist properties, thus maintaining the status quo for current antipsychotics. Furthermore, the validity of amphetamine-induced hyperactivity in an animal model of schizophrenia has been questioned by the observation that schizophrenia patients exhibit a behavioral exploratory pattern distinct from amphetamine administration (Perry et al. 2009). While such findings are preliminary, caution must be taken in the amphetamine-induced hyperactivity model of schizophrenia. Similar concerns are raised when utilizing amphetamine-or apomorphine-induced disruption of sensorimotor gating to identify drugs with antipsychotic properties. For example, antipsychotic-induced restoration of PPI in apomorphine-treated rats correlates highly with their clinical potency (Swerdlow et al. 1994). Thus, the receptor mechanism of the agonist used to induce the schizophrenia-like abnormality predicts the antagonists the behavioral test will identify. Acute dopaminergic models have rarely been utilized as models of negative symptoms or cognitive disruption for schizophrenia.

5.1.2 Repeat Administration Models

The most common form of repeated administration with dopaminergic agonists is the intermittent/escalating dosing regimen. The behavior of rats following such a dosing regimen has been investigated both in cognitive and in motivational tasks. Several studies in the SAT have demonstrated that these rats exhibit impaired sustained attention, which may result from cholinergic abnormalities (Deller and Sarter 1998; Martinez and Sarter 2008). These deficits were reversed through low doses of antipsychotic treatment (Martinez and Sarter 2008). Given that antipsychotic treatment does not reverse cognitive deficits in schizophrenia patients however, the validity of these results for schizophrenia research has been questioned. Sarter et al. (2009) describe the use of doses, which only result in 50% D2 receptor occupancy, a level rarely used in clinical studies, which has some support for antipsychotic-induced cognitive improvements in schizophrenia patients (Green et al. 2002). These improvements were, however, very modest, not the full reversal observed in the study of Martinez and Sarter (2008). Thus, debate remains on the utility of this mode of schizophrenia. This intermittent/escalating amphetamine model does appear to result in consistent cognitive deficits, however. Impaired sustained attention was also observed when using this model in the 5-CSR task, which was subsequently reversed through intraprefrontal infusions of the dopamine D1 agonist SKF38393 (Fletcher et al. 2007). Selective set-shifting deficits have also been observed using this model, which were also reversed following intra-PFC infusion of the D1 agonist SKF38393 (Fletcher et al. 2005). Cognitive deficits resulting from this model may, therefore, be sensitive to dopaminergic-manipulation-induced reversal. This treatment regimen does not appear to affect global cognitive performance, however, for hippocampal spatial learning and delayed memory as measured by the water maze and delayed-non-match-to-sample tasks were unaffected (Russig et al. 2003; Shoblock et al. 2003; Stefani and Moghaddam 2002). If this intermittent-escalating amphetamine regimen does selectively result in aberrant frontal-mediated responses, it is interesting to note that this model does not reproduce negative symptoms despite speculation that these symptoms originate from the frontal cortex. In fact, this model results in a hypersensitivity to reward (Wyvell and Berridge 2001). Therefore, these rats exhibit hedonic-like as opposed anhedonic-like behaviors, in contrast with schizophrenia, but consistent with bipolar disorder. The only negative symptom-like behavior that is observed from this model is during periods of withdrawal which detracts from the utility of the model with behavioral affects to be separate from withdrawal (Barr and Phillips 1999; Barr et al. 2002).

5.2 Glutamatergic Antagonist Models

Noncompetitive glutamatergic antagonists can act as psychotomimetics (Luby et al. 1959), reproducing positive, negative, as well as cognitive symptoms in normal humans (Krystal et al. 1994; Malhotra et al. 1996; Tamminga 1998). Moreover, such drugs may also exacerbate symptoms in schizophrenia patients (Malhotra et al. 1997). Furthermore, and in contrast with serotonergic agonists (see below), prolonged exposure to these drugs can still exert psychotomimetic effects and may, in fact, produce a more persistent schizophrenia-like symptomatology including cognitive deficiencies (Cosgrove and Newell 1991). Thus, there has been an increased interest in developing glutamatergic antagonist models of (Jentsch et al. 1997), as well as glutamatergic treatments for (Javitt et al. 1999; Moghaddam and Adams 1998; Moreno et al. 2009) schizophrenia.

Glutamatergic antagonist models vary in administration regimen with both acute and repeated dosing models observed (Jentsch and Roth 1999). These models will M.A. Geyer (✉), be discussed in greater detail in the following section. Some glutamatergic animal models of schizophrenia also utilize prenatal or prepubertal, acute and subchronic administration regimens, in an attempt to mimic the neurodevelopmental aspect of schizophrenia (Anastasio and Johnson 2008; Lu et al. 2009). As these animals or not behaviorally assessed until postpubertal, these models benefit from not having drug manipulations within the animals system during the time of testing. Moreover, several have demonstrated consistency with neurodevelopmental alterations in schizophrenia patients. These models are covered in greater detail, however, in Powell (2010).

5.2.1 Acute Models

In animals, glutamatergic antagonists can increase activity and reduce PPI consistent with dopaminergic drugs (see above), but appear to do so through a nondopaminergic pathway (Adams and Moghaddam 1998). Typical antipsychotics, which typically act on D2 receptors alone, do not reverse, for example, PCP-induced hyperactivity or disruption in PPI, yet can be reversed through atypical antipsychotic administration (Swerdlow and Geyer 1998). These effects may differ, however, as for example ketamine-induced disruption in PPI can be reversed through both chlorpromazine and clozapine administration (Swerdlow et al. 1998). Despite these reports however, and some suggestions PCP may act through a D2 mechanism (Seeman and Guan 2008), PCP-induced disruption in behaviors has been used as a means to assess putative antipsychotics without a dopaminergic – thus putative receptor tautological – confound. As mentioned earlier, however, glutamatergic antagonist administration can alter a variety of behaviors that are consistent with schizophrenia symptomatology. Acute PCP administration can reduce social behaviors in the social interaction test (Sams-Dodd 1999), interpreted as relevance to the negative symptoms of schizophrenia. Moreover, acute PCP can reduce voluntary sucrose consumption in rats 20 h postadministration exacerbated by acute clozapine and haloperidol, but not by subchronic clozapine (Turgeon and Hulick 2007). The effect of acute PCP on the progressive ratio task has yet to be assessed, however, although perinatal PCP in this task (Wiley and Compton 2004) and repeated PCP in ICSS (Amitai et al. 2009) had no effect, or increasing responding, respectively.

Cognitive deficits are apparent in normal subjects treated acutely with PCP or ketamine (Krystal et al. 1994). Some cognitive deficits have been replicated in animal studies including impaired attention as measured in the 5-CSR task (Greco et al. 2005; Le Pen et al. 2003), although this effect may have been as a result of reduced responding in rats (Amitai et al. 2007), and hyperactivity in mice (Greco et al. 2005). Impaired extradimensional set-shifting has been observed in rats performing the ASST 22 h postadministration (Egerton et al. 2005), with normal performance in other domains indicating the effect was unlikely to be due to putative anhedonic effects described earlier (Wiley and Compton 2004).

5.2.2 Repeated Dosing Models

Repeated administration of PCP can impair NORT performance in mice (Kunitachi et al. 2009). An affect ameliorated by the AChEI donepezil but not physostigmine, which the authors suggest may be sigma-1 mediated (Kunitachi et al. 2009). Repeated administration of phencyclidine actually produced a hedonic response in rats as measured by ICSS (Amitai et al. 2009). As this effect was reversed in clozapine-treated rats, questions must be raised regarding the ability of PCP to produce negative symptoms in animal models, as well as the ability of clozapine as a positive control in this model. Moreover, repeated PCP administration increased social novelty preference in mice, questioning whether such treatment may be a viable model for negative symptoms of schizophrenia (Brigman et al. 2009). The effects of PCP on repeated dosing in social behaviors in rats have produced conflicting results, however. While reduced social behaviors have been observed (Snigdha and Neill 2008), others have observed no differences (Sams-Dodd 2004).

Repeated PCP administration has been utilized similarly to repeat amphetamine where cognitive performance can be assessed several days after the administration regimen has ceased so that no PCP by treatment interactions are likely to occur, and after withdrawal effects wear off. While repeated PCP administration impaired spatial learning in the water maze, the relevance to schizophrenia may be questioned given that atypical antipsychotic treatment reversed this deficit (Didriksen et al. 2007) but do not treat cognitive disruption in schizophrenia. This theme of repeated PCP administration-induced deficiency and atypical antipsychotic-induced reversal is observed in numerous other cognitive tasks modeling cognitive disruption in schizophrenia. For example, in a reversal learning paradigm (Abdul-Monim et al. 2006), or in the ASST of problem solving (Goetghebeur and Dias 2009; Nikiforuk et al. 2010), short-term memory in the RAM (He et al. 2006), and NORT (Grayson et al. 2007), atypical antipsychotics such as clozapine, olanzapine, quetiapine, sertindole, or risperidone reversed the PCP- and ketamine-induced deficits, whereas administration of typical antipsychotics, such as haloperidol and chlorpromazine did not. These studies have often described their findings of positive effects of atypical but not typical antipsychotics in their models as providing predictive validity of the model and as "positive controls" (Abdul-Monim et al. 2007; Nikiforuk et al. 2010), despite clinical acknowledgement that antipsychotics do not improve cognitive in schizophrenia and certainly do not reverse cognitive disruption. Thus questions on the utility of the glutamatergic antagonist-induced disruption in cognitive performance has been brought into question and further studies are clearly required.

5.3 Serotonergic Agonist Models

Interest in manipulating the serotonergic system to create a model for schizophrenia began following the identification of putative psychotomimetic properties of serotonergic agonists. The behavioral effects of lysergic acid diethylamide (LSD) appeared consistent with a review of behavioral reports of early stage psychoses patients, including those with schizophrenia (Bowers and Freedman 1966). These similarities were later empirically studied with data providing support for the psychotomimetic properties of LSD (Gouzoulis-Mayfrank et al. 1998). The discovery that LSD acts as a serotonin agonist led to the serotonin hypothesis of schizophrenia that (a) there was an abnormal regulation of serotonin in patients with schizophrenia, (b) serotonergic agonists may model the disease, and (c) serotonergic antagonists may prove to be a viable treatment [reviewed in Geyer and Vollenweider (2008)].

The serotonin hypothesis has been questioned, however, given the chronic nature of schizophrenia, yet people develop a tolerance to the hallucinogenic effects of LSD. Moreover, LSD often produces visual hallucinations as opposed to auditory hallucinations associated with schizophrenia. Evidence suggests, however, that in early psychotic episodes in patients with schizophrenia there is a predominance of visual compared to auditory hallucinations (Freedman and Chapman 1973). These data combined with reports of feelings of initial exhilaration in early episode schizophrenia patients consistent with LSD suggests that serotonergic effects may correlate with early stages of schizophrenia. As LSD is a 5-HT2A agonist, this hypothesis is further supported by findings of reduced 5-HT2A receptor densities in the prefrontal cortex of subjects at risk of schizophrenia and drug-naïve patients with schizophrenia (Hurlemann et al. 2008; Ngan et al. 2000). Finally, while the importance of D2 antagonism in antipsychotic treatments has been discussed and is well established, atypical antipsychotics also act as antagonists at 5-HT2A receptors. Thus, given the evidence, it is likely that the serotonergic system plays a role in the initial psychotic even in the lives of patients with schizophrenia. Therefore, 5-HT2A agonism may prove a viable model for identifying putative novel antipsychotics. The utility of 5-HT2A agonists as a model for schizophrenia beyond this stage as well as in negative and cognitive symptoms has yet to be established, however. It appears that 5-HT2A agonism may affect time perception and temporal control in humans, however, which may contribute to resulting working memory deficits (Wittmann et al. 2007). These findings may explain why 5-HT2A agonism in rats result in increased premature responding in the 5-CSR task, yet do not affect accuracy when an appropriately timed response is made (Koskinen et al. 2000). Further studies are required to identify whether this model can be used in other aspects of schizophrenia symptomatology.

5.4 Cholinergic Antagonist Models

For some time, researchers have been aware of cognitive disruption following cholinergic manipulation. Cognitive decline in Alzheimer's disease was correlated with reduced choline acetyltransferase binding in patients, resulting in the development of acetylcholinesterase inhibitor treatment for Alzheimer's patients. While the putative benefits of cholinergic treatment have been investigated for schizophrenia (see Stip et al. 2007; Scarr and Dean 2008; Martin et al. 2004; Ferreri et al. 2006), little effort has focused on using cholinergic models as cognitive disruption in schizophrenia (Barak 2009). It is important to note, however, that blockade muscarinic- or nicotinic-acetylcholine receptors (mAChR and nAChR respectively) via scopolamine or mecamylamine (respectively) can impair cognitive performance in rodents in a number of cognitive domains relevant to schizophrenia. These domains include attention as measured by the 5-CSR task (Grottick and Higgins 2000; Hodges et al. 2009; Pattij et al. 2007; Waters et al. 2005), problem solving in the ASST (Chen et al. 2004a; Christakou et al. 2001), short-term memory in the NORT (Rutten et al. 2008; Sambeth et al. 2007), temporal working memory-like behavior in the radial arm maze (Hodges et al. 2009; Levin et al. 2005). Scopolamine affects behavior in other domains, however. Scopolamine-induced hyperactivity is observed, which was reversed through antipsychotic treatment albeit at doses that affected activity levels alone (Shannon and Peters 1990). Scopolamine administration can also reduce PPI, an effect attenuated by haloperidol treatment (Jones et al. 2005). Beyond pharmacological challenges, cholinergic lesions have also been used to probe putative therapeutics for the treatment of schizophrenia (detailed below). One of the major concerns of utilizing cholinergic manipulations as a model of cognitive disruption in schizophrenia is, however, that many of the induced deficits are reversible with acetylcholinesterase inhibitor treatment and emerging data suggest that these drugs do not ameliorate cognitive disruption in schizophrenia patients (Akhondzadeh et al. 2008; Dyer et al. 2008; Fagerlund et al. 2007; Kohler et al. 2007). Thus the validity of these manipulations as viable models for developing therapeutics for schizophrenia has been questioned. While this may be true, one important tenet of a pro-cognitive therapeutic for the treatment of schizophrenia is that it would likely be used as an adjunctive therapy with antipsychotic treatment. To date it appears that little if any acetylcholinesterase inhibitors have been tested alleviating a cholinergic model-induced disruption in cognition in conjunction with antipsychotic treatment. Thus future research may still include a cholinergic component to drug discovery in schizophrenia research.

5.5 Lesion Models

The neonatal ventral hippocampal lesion model of schizophrenia is probably the most widely used lesion model of schizophrenia (Lipska et al. 1995). Given that the lesion is performed prepubertal it is deemed that this model represents a neurodevelopmental model of schizophrenia. Neurodevelopmental models of schizophrenia are reviewed elsewhere however (see Powell 2010) and so will not be covered here. Here lesions are discussed in terms of brain regions which may subserve behavioral abnormalities in schizophrenia. For example, basal forebrain lesions impairs 5-CSR task (Muir et al. 1992; Robbins et al. 1989) and SAT performance (McGaughy et al. 1996). Given that basal forebrain lesion-induced deficits in attention can be reversed through acetylcholinesterase inhibitor treatment (Muir et al. 1992), which does not improve attention in schizophrenia patients, the validity/use of this lesion as a model of schizophrenia must be questioned. Lesion of the medial prefrontal cortex (mPFC) in rats leads to impaired cognitive flexibility (Birrell and Brown 2000; Tait et al. 2009), and although deficits have been observed in attentional tasks, these have been described as executive dysfunction in nature

(Miner et al. 1997). Both cognitive flexibility and executive functions require online monitoring of performance and lesions of the mPFC appears to also impair rats performing the RAM but only when distracters are presented (Gisquet-Verrier and Delatour 2006). As an animal model of schizophrenia, mPFC lesions actually increase PPI and rats are desensitized to the stimulant effects of amphetamine (Lacroix et al. 1998). Lesion models have not been used extensively and focus has been placed on other models.

5.6 Genetic Models for Schizophrenia

5.6.1 Schizophrenia as a Genetic Disease

Despite the complexity and ill-defined nature of schizophrenia, it has been known that schizophrenia runs in families (Gottesman and Erlenmeyer-Kimling 2001; Sullivan 2008). People carry significantly higher risks of developing schizophrenia if their close relatives have the disease; however, the transmission of the disease does not follow the classic Mendelian genetics pattern of inheritance (Gottesman and Erlenmeyer-Kimling 2001). To further reduce genetic heterogeneity to assess genetic contributions to schizophrenia, monozygotic twin studies provide compelling evidence that schizophrenia etiology is at least, in part, genetic. Several identical twin studies have concluded that one twin would have 50% chance to develop schizophrenia (1% chance for general population) if the cotwin had the disease (Tsuang 2000). While the studies suggested strong genetic components for schizophrenia, they also highlighted significant environmental effects and interactions between genetics and environment (Kato et al. 2005; Tsuang 2000). In both family and twin studies, nongenetic factors, such as family culture, behavior, diet and environmental insults, could contribute to the pathogenesis of the disease. Adoption studies, which remove factors of familial environment, indicate that adopted children had approximately 10% chances to develop schizophrenia if their biological parents suffered from schizophrenia, indicating a tenfold risk factor over the general population (Kendler and Diehl 1993; Sullivan 2008). Therefore, classic genetics and genetic epidemiology have concluded that schizophrenia is a heritable disease; however, it has been challenging to identify susceptibility genes for schizophrenia (Eisener et al. 2007; Kendler and Diehl 1993; Sullivan 2008).

5.6.2 Candidate Susceptibility Genes for Schizophrenia

There are several different approaches to identify candidate susceptibility genes for schizophrenia. Almost all of the candidate genes have been so far identified first from human genetic studies, and subsequently studied in animal models for relevant endophenotypes. However, the reverse approach that starts from genes involved in the modulation of endophenotypes to subsequent genetic verification in human population studies could also be fruitful. In this review, we will not attempt to include all candidate susceptibility genes reported in the literature, but focus on a few representatives identified from human genetic linkage and association studies, the classic cytogenetic studies on schizophrenia family, biological studies on candidate genes coming from animal models, and recent studies on copy number variation in human genome.

Human Genetic Linkage and Association

With the completion of the human genome sequence, it becomes feasible to conduct genome wide linkage and association studies to identify susceptibility loci for schizophrenia and other psychiatric disorders. Both families and case–control (unrelated individuals) studies have been conducted to map genetic susceptibility loci for schizophrenia. Many putative candidate genes have therefore been suggested for schizophrenia (Norton et al. 2006). Only a few, however, have been extensively examined in replication studies.

Neuregulin-1

Several genome-wide scans indicated that there was a susceptibility locus on chromosome 8p22–p11, which was also supported by meta-analysis of published linkage data (Green et al. 2005; Petryshen et al. 2005; Stefansson et al. 2002). Subsequent fine mapping with both microsatellite markers and single nucleotide polymorphisms (SNPs) identified three risk haplotypes around the first exon of neuregulin 1 gene (NRG1) in Icelandic case–control samples (Stefansson et al. 2002). Due to its involvement in neurodevelopment, glutamatergic neurotransmission and synaptic plasticity, neuregulin 1 (NRG1) becomes a plausible susceptibility gene (Corfas et al. 2004; Harrison and Law 2006). However, extensive replication of the association between NRG1 and schizophrenia has been inconsistent, and no functional genetic mutations have been identified (Harrison and Law 2006; Norton et al. 2006).

Dysbindin

Similar to the finding of NRG1 as a susceptibility gene for schizophrenia, linkage analysis of the Irish Study of High-Density Schizophrenia Families first detected a linkage peak on chromosome 6p24–22 (Straub et al. 2002). Evidence for linkage in this region was replicated in several other studies (Schwab et al. 2003). Subsequent fine mapping in the Irish sample identified several SNPs within DTNBP1 gene strongly associated with schizophrenia (Straub et al. 2002). Significant associations were also found in multiple haplotypes. However, replication of the association between DTNBP1 and schizophrenia has been mixed, and no functional mutations have been identified either (Mutsuddi et al. 2006; Norton et al. 2006; Peters et al. 2008).

Cytogenetic Studies

Disrupted in Schizophrenia 1

In contrast to the identification of NRG1 and DTNBP1 from genome-wide scans for susceptibility genes, Blackwood et al. reported a large Scottish schizophrenia family carrying a balanced chromosome translocation between chromosomes 1 and 11 (Blackwood et al. 2001; Millar et al. 2001). In the family, about half people suffer from schizophrenia, major recurrent depression, bipolar disorders, and other behavior disorders. The chromosome translocation demonstrated more than 70% penetrance for mental illness (Blackwood et al. 2001). Subsequent molecular cloning identified two novel genes DISC1 (Disrupted-in-schizophrenia 1) and DISC2 (Disrupted-in-schizophrenia 2) physically disrupted at chromosome 1q42 (Millar et al. 2000). The breakpoint was localized within the intron 8 of the DISC1 gene (Millar et al. 2001). The balanced chromosome translocation did not have any DNA deletion or any other DNA rearrangements, suggesting that no surrounding genes were physically disrupted (Millar et al. 2000). Therefore, DISC1 was regarded as the most promising candidate susceptibility gene for schizophrenia and other psychiatric disorders (Carter 2006; Muir et al. 2008; Porteous et al. 2006; Sawamura and Sawa 2006). However, another gene, named Boymaw, was recently found to be disrupted by the translocation on chromosome 11 (Zhou et al. 2008). The translocation could cause the generation of two fusion transcripts between DISC1 and Boymaw genes.

Many genetic association studies have been conducted to confirm the association of DISC1 and psychiatric disorders in different ethnic populations (Callicott et al. 2005; Cannon et al. 2005; Hennah et al. 2009, 2005; Hodgkinson et al. 2004; Thompson et al. 2005). However, replication of the association between DISC1 and schizophrenia has been inconsistent, and most importantly no functional mutations have been identified (Sanders et al. 2008). Nevertheless, recent molecular studies of DISC1 surprisingly found that DISC1 become insoluble in the postmortem human brains in approximately 20% of sporadic patients with schizophrenia, major depression, and bipolar disorder (Leliveld et al. 2008). This finding suggested a more general role for DISC1 in the pathogenesis of these major mental illnesses.

COMT

It has been known that approximately 25–30% of DiGeorge syndrome patients fulfill DSM-IV criteria for schizophrenia (Arinami 2006; Murphy et al. 1999). These patients typically carry a large DNA deletion encompassing approximately 30 individual genes on chromosome 22q11. The gene encoding catechol O-Methyltransferase (COMT), a critical enzyme in monoamine (including dopamine) metabolism, is localized in the deleted region (Karayiorgou et al. 1998; Shashi et al. 2006). Psychopharmacological studies have demonstrated that dopamine neurotransmission was disrupted in schizophrenia patients. These data, therefore, strongly suggest that COMT could be a plausible susceptibility gene for schizophrenia. Extensive genetic studies have been conducted to confirm the association between COMT and schizophrenia (Diaz-Asper et al. 2008; Lewandowski 2007; Slifstein et al. 2008). However, it remains controversial whether COMT represents a susceptibility gene (Sanders et al. 2008; Williams et al. 2007). In contrast to neuregulin-1 and dysbindin, there is a potential functional variant (Val158Met polymorphism) identified in the coding sequence of the human COMT gene. The polymorphism Val158Met significantly altered the human COMT enzymatic activity (Chen et al. 2004b).

NPAS3

NPAS3 gene, a member of bHLH-PAS domain family of transcription factors, is expressed during embryogenesis and exclusively in adult mouse brain (Brunskill et al. 1999). Two schizophrenia patients were found to carry a truncated NPAS3 gene in the same family by a balanced chromosome translocation at t(9;14)(q34; q13) (Kamnasaran et al. 2003; Pickard et al. 2005). Subsequent human genetic association studies provided further support for NPAS3 as a susceptibility gene for schizophrenia (Lavedan et al. 2009; Pickard et al. 2009).

5.7 *Candidate Genes from Animal Models*

5.7.1 Sp4

Sp4, a member of Sp-family transcription factors, is restrictively expressed in neuronal cells in adult mouse brain. The absence of Sp4 gene results in impaired postnatal development of hippocampal dentate gyrus (Zhou et al. 2007). The reduced expression of Sp4 gene causes hippocampal vacuolization and deficits in sensorimotor gating and memory (Zhou et al. 2005), putative endophenotypes for schizophrenia, and other psychiatric disorders. The Sp4 gene exhibits haploinsufficiency in the modulation of sensorimotor gating, and the deficient sensorimotor gating can be partially reversed through the administration of dopamine antagonists. Subsequent human genetic studies found several SNPs associated with bipolar disorder in both European Caucasians and Chinese Han population (Zhou et al. 2009). The same alleles from the same SNPs displayed significant associations with bipolar disorder in two different ethic groups, suggesting Sp4 as a candidate susceptibility gene for bipolar disorder.

5.7.2 Copy Number Variation

Genomic abnormalities, such as chromosome 22q11 deletion, have been demonstrated to associate with schizophrenia. However, genome-wide scale examination of structural alteration of human genome became feasible only after the completion of human genome and the advance of microarray technology. Recent studies found that most copy number variations (CNV) (from 1 kb to several Mb) are rare (less than 1% in population), and many of them are de novo mutation (Bassett et al. 2008; Williams et al. 2009). Although schizophrenia patients have significantly higher number of rare CNVs, it remains unclear which individual CNV carries the risk for the disease. However, several rare large DNA deletions (more than 500 kb to several Mb) have been found to have a high penetrance for schizophrenia.

5.7.3 Chromosome 22q11 Deletion

DiGeorge syndrome patients typically carry a large DNA deletion encompassing approximately 30 individual genes on chromosome 22q11, and 25–30% of them fulfill DSM-IV criteria for schizophrenia (Arinami 2006; Murphy et al. 1999). Thirteen large deletions on chromosome 22q11 were identified from more than 3,000 patients, none was found in the same number of controls (Bassett et al. 2008). In another independent study, eight large deletions on chromosome 22q11 were found in 3,833 patients, none was found in more than 39,000 controls (Stefansson et al. 2008). The deletion contributes to 0.3% cases of schizophrenia population. The candidate susceptibility gene COMT is within the deletion.

5.7.4 Chromosome 15q13.3 Deletion

Genome-wide examination of CNVs uncovered a novel locus on chromosome 15q13.3 having a high penetrance for schizophrenia. The deletion was found with 0.17% frequency in schizophrenia population and 0.02% in the controls (Stefansson et al. 2008), and a similar finding was reported in another study (Bassett et al. 2008). Interestingly, the alpha-7 nicotinic receptor gene (CHRNA7), a candidate susceptibility gene for schizophrenia, was localized within the deletion.

5.8 Transgenic Mouse Models

The mouse genetic models are invaluable for both behavioral and psychopharmacological studies not only to validate susceptibility genes but also to provide a cross-species, translational model for identification of efficacious new treatments for patients with schizophrenia.

5.8.1 Neuregulin

Despite the association of NRG1 with schizophrenia, there were no functional variants identified that causes the disease. Furthermore, there are at least 15 different NRG1 isoforms produced from the NRG1 genomic locus (Falls 2003). Different NRG1 isoforms were deleted in mouse, and homozygous mutants were lethal (Stefansson et al. 2002; Wolpowitz et al. 2000). In NRG1 transmembrane-domain–knockout mice, hyperactivity, deficient sensorimotor gating, and reduced expression of NMDA receptors were observed. The administration of clozapine reversed the hyperactivity of the NRG1 hypomorphs (Stefansson et al. 2002). Disruption of social novelty was also observed in the heterozygous NRG1 mice. However, the spatial memory appeared intact in the heterozygous mice (O'Tuathaigh et al. 2008). In Type III NRG1 heterozygous mice, enlarged lateral ventricles, deficient sensorimotor gating, decreased dendritic spine density, and impaired short-term memory were reported, while nicotine administration reversed the PPI deficit (Chen et al. 2008).

5.8.2 Dysbindin

The null mutation of dysbindin was first described in sandy (sdy) mouse, a model for human Hermansky–Pudlak syndrome (Li et al. 2003). Because the finding that dysbindin may function as a susceptibility gene for schizophrenia, there has been much interest in examining behavioral abnormalities in the mutant mice. The first behavioral studies were reported from the mutant mice in DBA/2J background where the spontaneous sdy mutation occurred. Lower locomotor activities, social interaction deficit, and higher dopamine turnover were found in the sdy mice than their wild-type controls in open field test; however, no difference was found in PPI (Hattori et al. 2008; Takao and Miyakawa 2008). No spatial memory deficit was found in the sdy mice, although reduced long-term memory retention was reported (Takao and Miyakawa 2008). The effects of sdy mutation were also examined in C57BL/6J background. The sdy mutant mice instead displayed hyperactivities in open field test, impaired spatial memory in water maze (Cox et al. 2009). Inbred mouse strains often carry many genetic mutations (Cox et al. 2009). It is unclear whether the opposite behavioral abnormalities observed come from different inbred strains. Therefore, it could be helpful to assess the sdy functions in hybrid F1 generation from two different mouse strains to minimize effects from other genetic mutations.

5.8.3 DISC1

The first mouse model came from the finding that there was a natural deletion of 25-bp coding sequence in exon 6 of the DISC1 gene in 129S6/SvEv mouse strain (Koike et al. 2006). The deletion caused a frameshift in the translation of DISC1 proteins, which resulted in a premature termination in exon 7. However, no truncated DISC1 proteins were detected from the mouse brain. No gross behavioral and anatomical abnormalities were found except for subtle deficits in hippocampally mediated short-term memory when the mutation was transferred to C57BL/6J genetic background (Koike et al. 2006). It should be noted that the site of the mouse DISC1 gene mutation was different from the breakpoint in the Scottish schizophrenia family. Moreover, it remains controversial whether the deletion abolished the production of all DISC1 proteins instead of a specific DISC1 isoform in 129S6/SvEv mouse strain (Ishizuka et al. 2007). So far, three transgenic mouse lines have been generated to express the truncated DISC1 proteins in mouse brain. Two of them contained a transgene expressing C-terminal truncated DISC1 proteins encoded from the first 8 exons of human DISC1 gene (the human translocation breakpoint localized in the intron 8) with either a heterologous CAMKII promoter or an inducible Tet-off artificial promoter (Hikida et al. 2007; Pletnikov et al. 2008). Again, no gross anatomical abnormalities were found. Subtle behavioral abnormalities and reduction of hippocampal volume were reported in transgenic mice. Li et al. (2007) reported the third transgenic mouse line expressing an inducible C-terminal portion of human DISC1 protein. Besides subtle reduced hippocampal volume, transgenic mice displayed a deficit in hippocampally mediated short-term memory, reduced sociability, and reduced dendritic complexity in the dentate gyrus. The overexpressed C-terminal DISC1 proteins were very different from the truncated DISC1 proteins proposed for the Scottish schizophrenia family, however.

Two different DISC1 mutant mouse lines were also created from ENU (*N*-ethyl *N*-nitrosourea) mutagenesis. The mice carrying different missense mutations in the mouse DISC1 gene displayed sensorimotor gating deficits, a putative endophenotype for schizophrenia, which could be reversed through the treatment of antipsychotics (Clapcote et al. 2007). Different behavioral abnormalities were observed in mice with different missense mutations, however, while neither missense mutation has been found in human schizophrenia patients yet. Recently, a human-like mouse DISC1 model was created to produce truncated mouse DISC1 proteins by inserting a poly(A) site in intron 8 of mouse DISC1 gene (Kvajo et al. 2008). In contrast to all other mouse DISC1 models, the truncated DISC1 proteins were produced from the endogenous mouse DISC1 locus. No gross anatomical abnormalities were found in the brain of homozygous mouse mutants, with only a subtle volume reduction in prefrontal cortex was reported.

Taken together, several mouse lines expressing varied truncated DISC1 proteins have been generated. No consistent abnormalities have been found, however. Such inconsistency may indicate a subtle effect of the DISC1 truncation, suggesting that it may contribute only, in part, to the high risk for neuropsychiatric disorders in the Scottish schizophrenia family. Further examination of the utility of these models would prove useful however, in tasks with great cross-species translational validity (Young et al. 2009b).

5.8.4 COMT

Dysfunction of dopamine (DA) neurotransmission has been demonstrated in a variety of human psychiatric disorders and drug addiction. A single nucleotide polymorphism, causing a missense mutation of 158 Valine to Methionine which decreases the enzymatic activity from 40% to threefold in different literatures (Chen et al. 2004b), was found in human COMT to associate with human cognitive function by neuroimaging studies (Egan et al. 2001), and possibly contribute to the pathogenesis of several psychiatric disorders in genetic association studies (Qian et al. 2003). There are no mouse models available yet for humanized COMT to directly examine the different in vivo effects of the mutation, although there are COMT null mutant mice as well as transgenic mice expressing human COMT isoforms (Paterlini et al. 2005; Stark et al. 2009; Suzuki et al. 2009).

5.8.5 CHRNA7

Genetic linkage studies have implicated chromosome 15q13–14 in the etiology of schizophrenia (Freedman et al. 1997; Liu et al. 2001). Specifically, the alpha 7 nAChR gene (CHRNA7) may be downregulated in the brains of schizophrenia patients and linked to neuropsychological deficits in schizophrenia (Freedman et al. 1997). To investigate the role of this receptor in the mammalian brain Orr-Urtreger et al. (1997) created a null mutation for the alpha 7 subunit in mice. These mice while apparently normal have been assessed in a number of behavioral and cognitive domains relevant to schizophrenia. These mice exhibited normal cue-conditioning learning, spatial learning in the water maze, activity levels, and PPI levels (Paylor et al. 1998). More complex cognitive functioning has also been assessed however, and these alpha-7 null mutants exhibit impaired sustained attention as measured by the 5-CSR task, as measured by two different laboratories (Hoyle et al. 2006; Young et al. 2007a, 2004). Moreover, these mice exhibit impaired odor span task performance, which may stem from their attentional deficits (Young et al. 2007a). Moreover, while classical conditioning appears intact in alpha-7 null mutants instrumental learning may be impaired (Keller et al. 2005; Young et al. 2004). Finally, these mice also exhibit impaired spatial learning in the radial arm maze (Levin et al. 2009). Thus, although these mice do not appear to exhibit abnormal performance in paradigms relating to positive or negative symptoms to date, they may prove to be a viable model for the cognitive disruption observed in patients with schizophrenia.

6 Conclusions

There are arguably fewer neurodegenerative diseases with such heterogeneity of patient symptoms than schizophrenia. Since being described by Bleuler, it was understood that the disease we term "schizophrenia" can also be referred to as a group of diseases. This observation is evident in the variation of symptoms that are observed, the group of positive, negative, and now cognitive symptoms experienced by these patients, which not only vary within individuals both in symptoms and in degree, but also alter over time and following current treatment. Attempting to generate animal models for this disease could obviously prove daunting. Our increased understanding of these symptoms, however, the progression, the underlying neurobiology, and the genetic contribution to the disease, increases our opportunity to model aspects of schizophrenia. Within this chapter, we have covered many of the animal models of schizophrenia that have been used to date, for positive, negative, and now cognitive symptoms. The variation in manipulation to create these models and in the tasks used to measure their validity highlight the breadth and depth of the field. Our increased knowledge of the genetic contribution to schizophrenia and whether these genes contribute to specific symptoms will begin to play a greater role in animal modeling in the future. Furthermore, the likelihood that antipsychotic treatment will remain a part of schizophrenia medication in the future would suggest that future models should also examine the impact of coantipsychotic treatment. Ultimately, the knowledge generated by clinicians can only enhance future models of this devastating disorder, leading to novel therapeutics.

References

Abdul-Monim Z, Reynolds GP, Neill JC (2006) The effect of atypical and classical antipsychotics on sub-chronic PCP-induced cognitive deficits in a reversal-learning paradigm. Behav Brain Res 169:263–273

Abdul-Monim Z, Neill JC, Reynolds GP (2007) Sub-chronic psychotomimetic phencyclidine induces deficits in reversal learning and alterations in parvalbumin-immunoreactive expression in the rat. J Psychopharmacol 21:198–205

Adams B, Moghaddam B (1998) Corticolimbic dopamine neurotransmission is temporally dissociated from the cognitive and locomotor effects of phencyclidine. J Neurosci 18:5545–5554

Akhondzadeh S, Gerami M, Noroozian M, Karamghadiri N, Ghoreishi A, Abbasi SH, Rezazadeh SA (2008) A 12-week, double-blind, placebo-controlled trial of donepezil adjunctive treatment to risperidone in chronic and stable schizophrenia. Prog Neuropsychopharmacol Biol Psychiatry 32:1810–1815

Amitai N, Semenova S, Markou A (2007) Cognitive-disruptive effects of the psychotomimetic phencyclidine and attenuation by atypical antipsychotic medications in rats. Psychopharmacology (Berl) 193:521–537

Amitai N, Semenova S, Markou A (2009) Clozapine attenuates disruptions in response inhibition and task efficiency induced by repeated phencyclidine administration in the intracranial self-stimulation procedure. Eur J Pharmacol 602:78–84

Anastasio NC, Johnson KM (2008) Differential regulation of the NMDA receptor by acute and sub-chronic phencyclidine administration in the developing rat. J Neurochem 104:1210–1218

Andreasen NC, Flaum M, Schultz S, Duzyurek S, Miller D (1997) Diagnosis, methodology and subtypes of schizophrenia. Neuropsychobiology 35:61–63

Arinami T (2006) Analyses of the associations between the genes of 22q11 deletion syndrome and schizophrenia. J Hum Genet 51:1037–1045

Baddeley AD (1986) Working memory. Oxford University Press, Oxford

Barak S (2009) Modeling cholinergic aspects of schizophrenia: focus on the antimuscarinic syndrome. Behav Brain Res 204:335–451

Barch DM, Carter CS, Arnsten A, Buchanan RW, Cohen JD, Geyer M, Green MF, Krystal JH, Nuechterlein K, Robbins T, Silverstein S, Smith EE, Strauss M, Wykes T, Heinssen R (2009) Selecting paradigms from cognitive neuroscience for translation into use in clinical trials: proceedings of the third CNTRICS meeting. Schizophr Bull 35:109–214

Barense MD, Fox MT, Baxter MG (2002) Aged rats are impaired on an attentional set-shifting task sensitive to medial frontal cortex damage in young rats. Learn Mem 9:191–201

Barr AM, Phillips AG (1999) Withdrawal following repeated exposure to d-amphetamine decreases responding for a sucrose solution as measured by a progressive ratio schedule of reinforcement. Psychopharmacology (Berl) 141:99–106

Barr AM, Zis AP, Phillips AG (2002) Repeated electroconvulsive shock attenuates the depressive-like effects of d-amphetamine withdrawal on brain reward function in rats. Psychopharmacology (Berl) 159:196–202

Baruch I, Hemsley DR, Gray JA (1988) Differential performance of acute and chronic schizophrenics in a latent inhibition task. J Nerv Ment Dis 176:598–606

Bassett AS, Marshall CR, Lionel AC, Chow EW, Scherer SW (2008) Copy number variations and risk for schizophrenia in 22q11.2 deletion syndrome. Hum Mol Genet 17:4045–4053

Birrell JM, Brown VJ (2000) Medial frontal cortex mediates perceptual attentional set shifting in the rat. J Neurosci 20:4320–4324

Bissonette GB, Martins GJ, Franz TM, Harper ES, Schoenbaum G, Powell EM (2008) Double dissociation of the effects of medial and orbital prefrontal cortical lesions on attentional and affective shifts in mice. J Neurosci 28:1124–1130

Blackwood DH, Fordyce A, Walker MT, St Clair DM, Porteous DJ, Muir WJ (2001) Schizophrenia and affective disorders–cosegregation with a translocation at chromosome 1q42 that directly disrupts brain-expressed genes: clinical and P300 findings in a family. Am J Hum Genet 69:428–433

Braff DL, Geyer MA (1990) Sensorimotor gating and schizophrenia. Human and animal model studies. Arch Gen Psychiatry 47:181–188

Braff D, Stone C, Callaway E, Geyer M, Glick I, Bali L (1978) Prestimulus effects on human startle reflex in normals and schizophrenics. Psychophysiology 15:339–343

Brigman JL, Ihne J, Saksida LM, Bussey TJ, Holmes A (2009) Effects of subchronic phencyclidine (PCP) treatment on social behaviors, and operant discrimination and reversal learning in C57BL/6J mice. Front Behav Neurosci 3:2

Brunskill EW, Witte DP, Shreiner AB, Potter SS (1999) Characterization of npas3, a novel basic helix-loop-helix PAS gene expressed in the developing mouse nervous system. Mech Dev 88:237–241

Buchanan RW, Freedman R, Javitt DC, Abi-Dargham A, Lieberman JA (2007) Recent advances in the development of novel pharmacological agents for the treatment of cognitive impairments in schizophrenia. Schizophr Bull 33:1120–1130

Bushnell PJ, Kelly KL, Crofton KM (1994) Effects of toluene inhalation on detection of auditory signals in rats. Neurotoxicol Teratol 16:149–160

Callicott JH, Straub RE, Pezawas L, Egan MF, Mattay VS, Hariri AR, Verchinski BA, Meyer-Lindenberg A, Balkissoon R, Kolachana B, Goldberg TE, Weinberger DR (2005) Variation in DISC1 affects hippocampal structure and function and increases risk for schizophrenia. Proc Natl Acad Sci USA 102:8627–8632

Cannon M, Jones P (1996) Schizophrenia. J Neurol Neurosurg Psychiatry 60:604–613

Cannon TD, Hennah W, van Erp TG, Thompson PM, Lonnqvist J, Huttunen M, Gasperoni T, Tuulio-Henriksson A, Pirkola T, Toga AW, Kaprio J, Mazziotta J, Peltonen L (2005) Association of DISC1/TRAX haplotypes with schizophrenia, reduced prefrontal gray matter, and impaired short- and long-term memory. Arch Gen Psychiatry 62:1205–1213

Carli M, Robbins TW, Evenden JL, Everitt BJ (1983) Effects of lesions to ascending noradrenergic neurones on performance of a 5-choice serial reaction task in rats; implications for theories of dorsal noradrenergic bundle function based on selective attention and arousal. Behav Brain Res 9:361–380

Carter CJ (2006) Schizophrenia susceptibility genes converge on interlinked pathways related to glutamatergic transmission and long-term potentiation, oxidative stress and oligodendrocyte viability. Schizophr Res 86:1–14

Carter CS, Barch DM (2007) Cognitive neuroscience-based approaches to measuring and improving treatment effects on cognition in schizophrenia: the CNTRICS initiative. Schizophr Bull 33:1131–1137

Chen KC, Baxter MG, Rodefer JS (2004a) Central blockade of muscarinic cholinergic receptors disrupts affective and attentional set-shifting. Eur J Neurosci 20:1081–1088

Chen X, Wang X, O'Neill AF, Walsh D, Kendler KS (2004b) Variants in the catechol-o-methyltransferase (COMT) gene are associated with schizophrenia in Irish high-density families. Mol Psychiatry 9:962–967

Chen YJ, Johnson MA, Lieberman MD, Goodchild RE, Schobel S, Lewandowski N, Rosoklija G, Liu RC, Gingrich JA, Small S, Moore H, Dwork AJ, Talmage DA, Role LW (2008) Type III neuregulin-1 is required for normal sensorimotor gating, memory-related behaviors, and corticostriatal circuit components. J Neurosci 28:6872–6883

Christakou A, Robbins TW, Everitt BJ (2001) Functional disconnection of a prefrontal cortical-dorsal striatal system disrupts choice reaction time performance: implications for attentional function. Behav Neurosci 115:812–825

Chudasama Y, Robbins TW (2004) Psychopharmacological approaches to modulating attention in the five-choice serial reaction time task: implications for schizophrenia. Psychopharmacology (Berl) 174:86–98

Clapcote SJ, Lipina TV, Millar JK, Mackie S, Christie S, Ogawa F, Lerch JP, Trimble K, Uchiyama M, Sakuraba Y, Kaneda H, Shiroishi T, Houslay MD, Henkelman RM, Sled JG, Gondo Y, Porteous DJ, Roder JC (2007) Behavioral phenotypes of Disc1 missense mutations in mice. Neuron 54:387–402

Connolly PM, Maxwell C, Liang Y, Kahn JB, Kanes SJ, Abel T, Gur RE, Turetsky BI, Siegel SJ (2004) The effects of ketamine vary among inbred mouse strains and mimic schizophrenia for the P80, but not P20 or N40 auditory ERP components. Neurochem Res 29:1179–1188

Corfas G, Roy K, Buxbaum JD (2004) Neuregulin 1-erbB signaling and the molecular/cellular basis of schizophrenia. Nat Neurosci 7:575–580

Cornblatt B, Obuchowski M, Schnur DB, O'Brien JD (1997) Attention and clinical symptoms in schizophrenia. Psychiatr Q 68:343–359

Cornblatt B, Obuchowski M, Schnur D, O'Brien JD (1998) Hillside study of risk and early detection in schizophrenia. Br J Psychiatry Suppl 172:26–32

Cosgrove J, Newell TG (1991) Recovery of neuropsychological functions during reduction in use of phencyclidine. J Clin Psychol 47:159–169

Cox MM, Tucker AM, Tang J, Talbot K, Richer DC, Yeh L, Arnold SE (2009) Neurobehavioral abnormalities in the dysbindin-1 mutant, sandy, on a C57BL/6J genetic background. Genes Brain Behav 8:390–397

Davis KL, Kahn RS, Ko G, Davidson M (1991) Dopamine in schizophrenia: a review and reconceptualization. Am J Psychiatry 148:1474–1486

Deller T, Sarter M (1998) Effects of repeated administration of amphetamine on behavioral vigilance: evidence for "sensitized" attentional impairments. Psychopharmacology (Berl) 137:410–414

Diaz-Asper CM, Goldberg TE, Kolachana BS, Straub RE, Egan MF, Weinberger DR (2008) Genetic variation in catechol-O-methyltransferase: effects on working memory in schizophrenic patients, their siblings, and healthy controls. Biol Psychiatry 63:72–79

Didriksen M, Skarsfeldt T, Arnt J (2007) Reversal of PCP-induced learning and memory deficits in the Morris' water maze by sertindole and other antipsychotics. Psychopharmacology (Berl) 193:225–233

Dudchenko PA (2004) An overview of the tasks used to test working memory in rodents. Neurosci Biobehav Rev 28:699–709

Dudchenko PA, Wood ER, Eichenbaum H (2000) Neurotoxic hippocampal lesions have no effect on odor span and little effect on odor recognition memory but produce significant impairments on spatial span, recognition, and alternation. J Neurosci 20:2964–2977

Dyer MA, Freudenreich O, Culhane MA, Pachas GN, Deckersbach T, Murphy E, Goff DC, Evins AE (2008) High-dose galantamine augmentation inferior to placebo on attention, inhibitory control and working memory performance in nonsmokers with schizophrenia. Schizophr Res 102:88–95

Egan MF, Goldberg TE, Kolachana BS, Callicott JH, Mazzanti CM, Straub RE, Goldman D, Weinberger DR (2001) Effect of COMT Val108/158 Met genotype on frontal lobe function and risk for schizophrenia. Proc Natl Acad Sci USA 98:6917–6922

Egerton A, Reid L, McKerchar CE, Morris BJ, Pratt JA (2005) Impairment in perceptual attentional set-shifting following PCP administration: a rodent model of set-shifting deficits in schizophrenia. Psychopharmacology (Berl) 179:77–84

Eisener A, Pato MT, Medeiros H, Carvalho C, Pato CN (2007) Genetics of schizophrenia: recent advances. Psychopharmacol Bull 40:168–177

Ellenbroek BA, Cools AR (2000) Animal models for the negative symptoms of schizophrenia. Behav Pharmacol 11:223–233

Erlenmeyer-Kimling L (2000) Neurobehavioral deficits in offspring of schizophrenic parents: liability indicators and predictors of illness. Am J Med Genet 97:65–71

Fagerlund B, Soholm B, Fink-Jensen A, Lublin H, Glenthoj BY (2007) Effects of donepezil adjunctive treatment to ziprasidone on cognitive deficits in schizophrenia: a double-blind, placebo-controlled study. Clin Neuropharmacol 30:3–12

Falls DL (2003) Neuregulins: functions, forms, and signaling strategies. Exp Cell Res 284:14–30

Farrow TF, Hunter MD, Haque R, Spence SA (2006) Modafinil and unconstrained motor activity in schizophrenia: double-blind crossover placebo-controlled trial. Br J Psychiatry 189:461–462

Ferreri F, Agbokou C, Gauthier S (2006) Cognitive dysfunctions in schizophrenia: potential benefits of cholinesterase inhibitor adjunctive therapy. J Psychiatry Neurosci 31:369–376

Fletcher PJ, Tenn CC, Rizos Z, Lovic V, Kapur S (2005) Sensitization to amphetamine, but not PCP, impairs attentional set shifting: reversal by a D1 receptor agonist injected into the medial prefrontal cortex. Psychopharmacology (Berl) 183:190–200

Fletcher PJ, Tenn CC, Sinyard J, Rizos Z, Kapur S (2007) A sensitizing regimen of amphetamine impairs visual attention in the 5-choice serial reaction time test: reversal by a D1 receptor agonist injected into the medial prefrontal cortex. Neuropsychopharmacology 32:1122–1132

Floresco SB, Geyer MA, Gold LH, Grace AA (2005) Developing predictive animal models and establishing a preclinical trials network for assessing treatment effects on cognition in schizophrenia. Schizophr Bull 31:888–894

Freedman B, Chapman LJ (1973) Early subjective experience in schizophrenic episodes. J Abnorm Psychol 82:46–54

Freedman R, Adler LE, Bickford P, Byerley W, Coon H, Cullum CM, Griffith JM, Harris JG, Leonard S, Miller C et al (1994) Schizophrenia and nicotinic receptors. Harv Rev Psychiatry 2:179–192

Freedman R, Coon H, Myles-Worsley M, Orr-Urtreger A, Olincy A, Davis A, Polymeropoulos M, Holik J, Hopkins J, Hoff M, Rosenthal J, Waldo MC, Reimherr F, Wender P, Yaw J, Young DA, Breese CR, Adams C, Patterson D, Adler LE, Kruglyak L, Leonard S, Byerley W (1997) Linkage of a neurophysiological deficit in schizophrenia to a chromosome 15 locus. Proc Natl Acad Sci USA 94:587–592

Gao XM, Sakai K, Roberts RC, Conley RR, Dean B, Tamminga CA (2000) Ionotropic glutamate receptors and expression of N-methyl-D-aspartate receptor subunits in subregions of human hippocampus: effects of schizophrenia. Am J Psychiatry 157:1141–1149

Geyer MA (2006) The family of sensorimotor gating disorders: comorbitities or diagnostic overlaps. Neurotox Res 10:211–220

Geyer MA, Braff DL (1987) Startle habituation and sensorimotor gating in schizophrenia and related animal models. Schizophr Bull 13:643–668

Geyer MA, Markou A (1995) Animal models of psychiatric disorders. In: Bloom FE, Kupfer D (eds) Psychopharmacology: the fourth generation of progress. Raven Press, New York, pp 787–798

Geyer MA, Vollenweider FX (2008) Serotonin research: contributions to understanding psychoses. Trends Pharmacol Sci 29:445–453

Geyer MA, Russo PV, Masten VL (1986) Multivariate assessment of locomotor behavior: pharmacological and behavioral analyses. Pharmacol Biochem Behav 25:277–288

Geyer MA, Braff D, Swerdlow NR (1999) Startle-response measures of information processing in animals. In: Haug M, Whalen RE (eds) Animal models of human emotion & cognition. APA Books, Washington, DC, pp 103–116

Gisquet-Verrier P, Delatour B (2006) The role of the rat prelimbic/infralimbic cortex in working memory: not involved in the short-term maintenance but in monitoring and processing functions. Neuroscience 141:585–596

Goetghebeur P, Dias R (2009) Comparison of haloperidol, risperidone, sertindole, and modafinil to reverse an attentional set-shifting impairment following subchronic PCP administration in the rat-a back translational study. Psychopharmacology (Berl) 202:287–293

Goldberg TE, Weinberger DR (1995) Thought disorder, working memory and attention: interrelationships and the effects of neuroleptic medications. Int Clin Psychopharmacol 10(Suppl 3):99–104

Gottesman II, Erlenmeyer-Kimling L (2001) Family and twin strategies as a head start in defining prodromes and endophenotypes for hypothetical early-interventions in schizophrenia. Schizophr Res 51:93–102

Gouzoulis-Mayfrank E, Habermeyer E, Hermle L, Steinmeyer A, Kunert H, Sass H (1998) Hallucinogenic drug induced states resemble acute endogenous psychoses: results of an empirical study. Eur Psychiatry 13:399–406

Gray NS, Snowden RJ (2005) The relevance of irrelevance to schizophrenia. Neurosci Biobehav Rev 29:989–999

Grayson B, Idris NF, Neill JC (2007) Atypical antipsychotics attenuate a sub-chronic PCP-induced cognitive deficit in the novel object recognition task in the rat. Behav Brain Res 184:31–38

Greco B, Invernizzi RW, Carli M (2005) Phencyclidine-induced impairment in attention and response control depends on the background genotype of mice: reversal by the mGLU(2/3) receptor agonist LY379268. Psychopharmacology (Berl) 179:68–76

Green MF (1996) What are the functional consequences of neurocognitive deficits in schizophrenia? Am J Psychiatry 153:321–330

Green MF (2006a) Cognitive impairment and functional outcome in schizophrenia and bipolar disorder. J Clin Psychiatry 67(Suppl 9):3–8, discussion 36–42

Green MF (2006b) Cognitive impairment and functional outcome in schizophrenia and bipolar disorder. J Clin Psychiatry 67:e12

Green MF, Marder SR, Glynn SM, McGurk SR, Wirshing WC, Wirshing DA, Liberman RP, Mintz J (2002) The neurocognitive effects of low-dose haloperidol: a two-year comparison with risperidone. Biol Psychiatry 51:972–978

Green EK, Raybould R, Macgregor S, Gordon-Smith K, Heron J, Hyde S, Grozeva D, Hamshere M, Williams N, Owen MJ, O'Donovan MC, Jones L, Jones I, Kirov G, Craddock N (2005) Operation of the schizophrenia susceptibility gene, neuregulin 1, across traditional diagnostic boundaries to increase risk for bipolar disorder. Arch Gen Psychiatry 62:642–648

Grottick AJ, Higgins GA (2000) Effect of subtype selective nicotinic compounds on attention as assessed by the five-choice serial reaction time task. Behav Brain Res 117:197–208

Hagan JJ, Jones DN (2005) Predicting drug efficacy for cognitive deficits in schizophrenia. Schizophr Bull 31:830–853

Harrison PJ, Law AJ (2006) Neuregulin 1 and schizophrenia: genetics, gene expression, and neurobiology. Biol Psychiatry 60:132–140

Hattori S, Murotani T, Matsuzaki S, Ishizuka T, Kumamoto N, Takeda M, Tohyama M, Yamatodani A, Kunugi H, Hashimoto R (2008) Behavioral abnormalities and dopamine reductions in sdy mutant mice with a deletion in Dtnbp1, a susceptibility gene for schizophrenia. Biochem Biophys Res Commun 373:298–302

He J, Xu H, Yang Y, Rajakumar D, Li X, Li XM (2006) The effects of chronic administration of quetiapine on the phencyclidine-induced reference memory impairment and decrease of Bcl-XL/Bax ratio in the posterior cingulate cortex in rats. Behav Brain Res 168:236–242

Hennah W, Tuulio-Henriksson A, Paunio T, Ekelund J, Varilo T, Partonen T, Cannon TD, Lonnqvist J, Peltonen L (2005) A haplotype within the DISC1 gene is associated with visual memory functions in families with a high density of schizophrenia. Mol Psychiatry 10:1097–1103

Hennah W, Thomson P, McQuillin A, Bass N, Loukola A, Anjorin A, Blackwood D, Curtis D, Deary IJ, Harris SE, Isometsa ET, Lawrence J, Lonnqvist J, Muir W, Palotie A, Partonen T, Paunio T, Pylkko E, Robinson M, Soronen P, Suominen K, Suvisaari J, Thirumalai S, St Clair D, Gurling H, Peltonen L, Porteous D (2009) DISC1 association, heterogeneity and interplay in schizophrenia and bipolar disorder. Mol Psychiatry 14:865–873

Hikida T, Jaaro-Peled H, Seshadri S, Oishi K, Hookway C, Kong S, Wu D, Xue R, Andrade M, Tankou S, Mori S, Gallagher M, Ishizuka K, Pletnikov M, Kida S, Sawa A (2007) Dominant-negative DISC1 transgenic mice display schizophrenia-associated phenotypes detected by measures translatable to humans. Proc Natl Acad Sci USA 104:14501–14506

Hodges DB Jr, Lindner MD, Hogan JB, Jones KM, Markus EJ (2009) Scopolamine induced deficits in a battery of rat cognitive tests: comparisons of sensitivity and specificity. Behav Pharmacol 20:237–251

Hodgkinson CA, Goldman D, Jaeger J, Persaud S, Kane JM, Lipsky RH, Malhotra AK (2004) Disrupted in schizophrenia 1 (DISC1): association with schizophrenia, schizoaffective disorder, and bipolar disorder. Am J Hum Genet 75:862–872

Horrobin DF (1998) Schizophrenia: the illness that made us human. Med Hypotheses 50:269–288

Hoyle E, Genn RF, Fernandes C, Stolerman IP (2006) Impaired performance of alpha7 nicotinic receptor knockout mice in the five-choice serial reaction time task. Psychopharmacology (Berl) 189:211–223

Humby T, Laird FM, Davies W, Wilkinson LS (1999) Visuospatial attentional functioning in mice: interactions between cholinergic manipulations and genotype. Eur J Neurosci 11:2813–2823

Hurlemann R, Matusch A, Kuhn KU, Berning J, Elmenhorst D, Winz O, Kolsch H, Zilles K, Wagner M, Maier W, Bauer A (2008) 5-HT2A receptor density is decreased in the at-risk mental state. Psychopharmacology (Berl) 195:579–590

Ishizuka K, Chen J, Taya S, Li W, Millar JK, Xu Y, Clapcote SJ, Hookway C, Morita M, Kamiya A, Tomoda T, Lipska BK, Roder JC, Pletnikov M, Porteous D, Silva AJ, Cannon TD, Kaibuchi K, Brandon NJ, Weinberger DR, Sawa A (2007) Evidence that many of the DISC1 isoforms in C57BL/6J mice are also expressed in 129S6/SvEv mice. Mol Psychiatry 12:897–899

Javitt DC, Liederman E, Cienfuegos A, Shelley AM (1999) Panmodal processing imprecision as a basis for dysfunction of transient memory storage systems in schizophrenia. Schizophr Bull 25:763–775

Jentsch JD, Roth RH (1999) The neuropsychopharmacology of phencyclidine: from NMDA receptor hypofunction to the dopamine hypothesis of schizophrenia. Neuropsychopharmacology 20:201–225

Jentsch JD, Redmond DE Jr, Elsworth JD, Taylor JR, Youngren KD, Roth RH (1997) Enduring cognitive deficits and cortical dopamine dysfunction in monkeys after long-term administration of phencyclidine. Science 277:953–955

Jones CK, Eberle EL, Shaw DB, McKinzie DL, Shannon HE (2005) Pharmacologic interactions between the muscarinic cholinergic and dopaminergic systems in the modulation of prepulse inhibition in rats. J Pharmacol Exp Ther 312:1055–1063

Jones DNC, Garlton JE, Minassian A, Perry W, Geyer MA (2008) Developing new drigs for schizophrenia: From animals to the clinic. In: McArthur R, Borsini F (eds) Animal and translational models for CNS drug discovery: psychiatric disorders. Elsevier Inc., New York, pp 199–262

Kamnasaran D, Muir WJ, Ferguson-Smith MA, Cox DW (2003) Disruption of the neuronal PAS3 gene in a family affected with schizophrenia. J Med Genet 40:325–332

Karayiorgou M, Gogos JA, Galke BL, Wolyniec PS, Nestadt G, Antonarakis SE, Kazazian HH, Housman DE, Pulver AE (1998) Identification of sequence variants and analysis of the role of the catechol-O-methyl-transferase gene in schizophrenia susceptibility. Biol Psychiatry 43:425–431

Kato T, Iwamoto K, Kakiuchi C, Kuratomi G, Okazaki Y (2005) Genetic or epigenetic difference causing discordance between monozygotic twins as a clue to molecular basis of mental disorders. Mol Psychiatry 10:622–630

Keller JJ, Keller AB, Bowers BJ, Wehner JM (2005) Performance of alpha7 nicotinic receptor null mutants is impaired in appetitive learning measured in a signaled nose poke task. Behav Brain Res 162:143–152

Kendler KS, Diehl SR (1993) The genetics of schizophrenia: a current, genetic-epidemiologic perspective. Schizophr Bull 19:261–285

Klaning U (1999) Greater occurrence of schizophrenia in dizygotic but not monozygotic twins. Register-based study. Br J Psychiatry 175:407–409

Kohler CG, Martin EA, Kujawski E, Bilker W, Gur RE, Gur RC (2007) No effect of donepezil on neurocognition and social cognition in young persons with stable schizophrenia. Cogn Neuropsychiatry 12:412–421

Koike H, Arguello PA, Kvajo M, Karayiorgou M, Gogos JA (2006) Disc1 is mutated in the 129S6/SvEv strain and modulates working memory in mice. Proc Natl Acad Sci USA 103:3693–3697

Koskinen T, Ruotsalainen S, Puumala T, Lappalainen R, Koivisto E, Mannisto PT, Sirvio J (2000) Activation of 5-HT2A receptors impairs response control of rats in a five-choice serial reaction time task. Neuropharmacology 39:471–481

Krystal JH, Karper LP, Seibyl JP, Freeman GK, Delaney R, Bremner JD, Heninger GR, Bowers MB Jr, Charney DS (1994) Subanesthetic effects of the noncompetitive NMDA antagonist, ketamine, in humans. Psychotomimetic, perceptual, cognitive, and neuroendocrine responses. Arch Gen Psychiatry 51:199–214

Kunitachi S, Fujita Y, Ishima T, Kohno M, Horio M, Tanibuchi Y, Shirayama Y, Iyo M, Hashimoto K (2009) Phencyclidine-induced cognitive deficits in mice are ameliorated by subsequent subchronic administration of donepezil: role of sigma-1 receptors. Brain Res 1279:189–196

Kvajo M, McKellar H, Arguello PA, Drew LJ, Moore H, MacDermott AB, Karayiorgou M, Gogos JA (2008) A mutation in mouse Disc1 that models a schizophrenia risk allele leads to specific alterations in neuronal architecture and cognition. Proc Natl Acad Sci USA 105:7076–7081

Lacroix L, Broersen LM, Weiner I, Feldon J (1998) The effects of excitotoxic lesion of the medial prefrontal cortex on latent inhibition, prepulse inhibition, food hoarding, elevated plus maze, active avoidance and locomotor activity in the rat. Neuroscience 84:431–442

Laruelle M, Abi-Dargham A, Gil R, Kegeles L, Innis R (1999) Increased dopamine transmission in schizophrenia: relationship to illness phases. Biol Psychiatry 46:56–72

Lavedan C, Licamele L, Volpi S, Hamilton J, Heaton C, Mack K, Lannan R, Thompson A, Wolfgang CD, Polymeropoulos MH (2009) Association of the NPAS3 gene and five other loci with response to the antipsychotic iloperidone identified in a whole genome association study. Mol Psychiatry 14:804–819

Le Pen G, Grottick AJ, Higgins GA, Moreau JL (2003) Phencyclidine exacerbates attentional deficits in a neurodevelopmental rat model of schizophrenia. Neuropsychopharmacology 28:1799–1809

Leliveld SR, Bader V, Hendriks P, Prikulis I, Sajnani G, Requena JR, Korth C (2008) Insolubility of disrupted-in-schizophrenia 1 disrupts oligomer-dependent interactions with nuclear distribution element 1 and is associated with sporadic mental disease. J Neurosci 28:3839–3845

Levin ED, Petro A, Beatty A (2005) Olanzapine interactions with nicotine and mecamylamine in rats: effects on memory function. Neurotoxicol Teratol 27:459–464

Levin ED, Petro A, Rezvani AH, Pollard N, Christopher NC, Strauss M, Avery J, Nicholson J, Rose JE (2009) Nicotinic alpha7- or beta2-containing receptor knockout: effects on radial-arm maze learning and long-term nicotine consumption in mice. Behav Brain Res 196:207–213

Lewandowski KE (2007) Relationship of catechol-O-methyltransferase to schizophrenia and its correlates: evidence for associations and complex interactions. Harv Rev Psychiatry 15:233–244

Li W, Zhang Q, Oiso N, Novak EK, Gautam R, O'Brien EP, Tinsley CL, Blake DJ, Spritz RA, Copeland NG, Jenkins NA, Amato D, Roe BA, Starcevic M, Dell'Angelica EC, Elliott RW, Mishra V, Kingsmore SF, Paylor RE, Swank RT (2003) Hermansky-Pudlak syndrome type 7 (HPS-7) results from mutant dysbindin, a member of the biogenesis of lysosome-related organelles complex 1 (BLOC-1). Nat Genet 35:84–89

Li W, Zhou Y, Jentsch JD, Brown RA, Tian X, Ehninger D, Hennah W, Peltonen L, Lonnqvist J, Huttunen MO, Kaprio J, Trachtenberg JT, Silva AJ, Cannon TD (2007) Specific developmental disruption of disrupted-in-schizophrenia-1 function results in schizophrenia-related phenotypes in mice. Proc Natl Acad Sci USA 104:18280–18285

Lieberman JA, Sheitman BB, Kinon BJ (1997) Neurochemical sensitization in the pathophysiology of schizophrenia: deficits and dysfunction in neuronal regulation and plasticity. Neuropsychopharmacology 17:205–229

Lipska BK, Weinberger DR (1995) Genetic variation in vulnerability to the behavioral effects of neonatal hippocampal damage in rats. Proc Natl Acad Sci USA 92:8906–8910

Lipska BK, Weinberger DR (2000) To model a psychiatric disorder in animals: schizophrenia as a reality test. Neuropsychopharmacology 23:223–239

Lipska BK, Swerdlow NR, Geyer MA, Jaskiw GE, Braff DL, Weinberger DR (1995) Neonatal excitotoxic hippocampal damage in rats causes post-pubertal changes in prepulse inhibition of startle and its disruption by apomorphine. Psychopharmacology (Berl) 122:35–43

Liu CM, Hwu HG, Lin MW, Ou-Yang WC, Lee SF, Fann CS, Wong SH, Hsieh SH (2001) Suggestive evidence for linkage of schizophrenia to markers at chromosome 15q13–14 in Taiwanese families. Am J Med Genet 105:658–661

Lu L, Mamiya T, Lu P, Toriumi K, Mouri A, Hiramatsu M, Kim HC, Zou LB, Nagai T, Nabeshima T (2009) Prenatal exposure to phencyclidine produces abnormal behaviour and NMDA receptor expression in postpubertal mice. Int J Neuropsychopharmacol 19:1–13

Luby ED, Cohen BD, Rosenbaum G, Gottlieb JS, Kelley R (1959) Study of a new schizophrenomimetic drug; sernyl. AMA Arch Neurol Psychiatry 81:363–369

Malhotra AK, Pinals DA, Weingartner H, Sirocco K, Missar CD, Pickar D, Breier A (1996) NMDA receptor function and human cognition: the effects of ketamine in healthy volunteers. Neuropsychopharmacology 14:301–307

Malhotra AK, Pinals DA, Adler CM, Elman I, Clifton A, Pickar D, Breier A (1997) Ketamine-induced exacerbation of psychotic symptoms and cognitive impairment in neuroleptic-free schizophrenics. Neuropsychopharmacology 17:141–150

Marder SR, Fenton W (2004) Measurement and treatment research to improve cognition in schizophrenia: NIMH MATRICS initiative to support the development of agents for improving cognition in schizophrenia. Schizophr Res 72:5–9

Martin LF, Kem WR, Freedman R (2004) Alpha-7 nicotinic receptor agonists: potential new candidates for the treatment of schizophrenia. Psychopharmacology (Berl) 174:54–64

Martinez V, Sarter M (2008) Detection of the moderately beneficial cognitive effects of low-dose treatment with haloperidol or clozapine in an animal model of the attentional impairments of schizophrenia. Neuropsychopharmacology 33:2635–2647

Matthysse S (1986) Animal models in psychiatric research. Prog Brain Res 65:259–270

McGaughy J, Sarter M (1995) Behavioral vigilance in rats: task validation and effects of age, amphetamine, and benzodiazepine receptor ligands. Psychopharmacology (Berl) 117:340–357

McGaughy J, Kaiser T, Sarter M (1996) Behavioral vigilance following infusions of 192 IgG-saporin into the basal forebrain: selectivity of the behavioral impairment and relation to cortical AChE-positive fiber density. Behav Neurosci 110:247–265

Metzger KL, Maxwell CR, Liang Y, Siegel SJ (2007) Effects of nicotine vary across two auditory evoked potentials in the mouse. Biol Psychiatry 61:23–30

Millar JK, Wilson-Annan JC, Anderson S, Christie S, Taylor MS, Semple CA, Devon RS, St Clair DM, Muir WJ, Blackwood DH, Porteous DJ (2000) Disruption of two novel genes by a translocation co-segregating with schizophrenia. Hum Mol Genet 9:1415–1423

Millar JK, Christie S, Anderson S, Lawson D, Hsiao-Wei Loh D, Devon RS, Arveiler B, Muir WJ, Blackwood DH, Porteous DJ (2001) Genomic structure and localisation within a linkage hotspot of disrupted in schizophrenia 1, a gene disrupted by a translocation segregating with schizophrenia. Mol Psychiatry 6:173–178

Minassian A, Henry BL, Geyer MA, Paulus MP, Young JW, Perry W (2009) The quantitative assessment of motor activity in mania and schizophrenia. J Affect Disord 120:200–206

Miner LA, Ostrander M, Sarter M (1997) Effects of ibotenic acid-induced loss of neurons in the medial prefrontal cortex of rats on behavioral vigilance: evidence for executive dysfunction. J Psychopharmacol 11:169–178

Moghaddam B, Adams BW (1998) Reversal of phencyclidine effects by a group II metabotropic glutamate receptor agonist in rats. Science 281:1349–1352

Moreno JL, Sealfon SC, Gonzalez-Maeso J (2009) Group II metabotropic glutamate receptors and schizophrenia. Cell Mol Life Sci 66:3777–3785

Morris RGM (1981) Spatial localization does not require the presence of local cues. Learn Motiv 12:239

Moser PC, Hitchcock JM, Lister S, Moran PM (2000) The pharmacology of latent inhibition as an animal model of schizophrenia. Brain Res Brain Res Rev 33:275–307

Muir JL, Dunnett SB, Robbins TW, Everitt BJ (1992) Attentional functions of the forebrain cholinergic systems: effects of intraventricular hemicholinium, physostigmine, basal forebrain lesions and intracortical grafts on a multiple-choice serial reaction time task. Exp Brain Res 89:611–622

Muir WJ, Pickard BS, Blackwood DH (2008) Disrupted-in-schizophrenia-1. Curr Psychiatry Rep 10:140–147

Murphy KC, Jones LA, Owen MJ (1999) High rates of schizophrenia in adults with velo-cardio-facial syndrome. Arch Gen Psychiatry 56:940–945

Mutsuddi M, Morris DW, Waggoner SG, Daly MJ, Scolnick EM, Sklar P (2006) Analysis of high-resolution HapMap of DTNBP1 (Dysbindin) suggests no consistency between reported common variant associations and schizophrenia. Am J Hum Genet 79:903–909

Ngan ET, Yatham LN, Ruth TJ, Liddle PF (2000) Decreased serotonin 2A receptor densities in neuroleptic-naive patients with schizophrenia: A PET study using [(18)F]setoperone. Am J Psychiatry 157:1016–1018

Nieuwenstein MR, Aleman A, de Haan EH (2001) Relationship between symptom dimensions and neurocognitive functioning in schizophrenia: a meta-analysis of WCST and CPT studies. Wisconsin Card Sorting Test. Continuous Performance Test. J Psychiatr Res 35:119–125

Nikiforuk A, Golembiowska K, Popik P (2010) Mazindol attenuates ketamine-induced cognitive deficit in the attentional set shifting task in rats. Eur Neuropsychopharmacol 20:37–48

Norton N, Williams HJ, Owen MJ (2006) An update on the genetics of schizophrenia. Curr Opin Psychiatry 19:158–164

Nuechterlein KH, Barch DM, Gold JM, Goldberg TE, Green MF, Heaton RK (2004) Identification of separable cognitive factors in schizophrenia. Schizophr Res 72:29–39

Nuechterlein KH, Luck SJ, Lustig C, Sarter M (2009) CNTRICS final task selection: control of attention. Schizophr Bull 35:182–196

O'Tuathaigh CM, O'Connor AM, O'Sullivan GJ, Lai D, Harvey R, Croke DT, Waddington JL (2008) Disruption to social dyadic interactions but not emotional/anxiety-related behaviour in mice with heterozygous 'knockout' of the schizophrenia risk gene neuregulin-1. Prog Neuropsychopharmacol Biol Psychiatry 32:462–466

Olds J, Milner P (1954) Positive reinforcement produced by electrical stimulation of septal area and other region of rat brain. J Comp Physiol Psychol 47:419–427

Olton DS, Werz MA (1978) Hippocampal function and behavior: spatial discrimination and response inhibition. Physiol Behav 20:597–605

Orr-Urtreger A, Goldner FM, Saeki M, Lorenzo I, Goldberg L, De Biasi M, Dani JA, Patrick JW, Beaudet AL (1997) Mice deficient in the alpha7 neuronal nicotinic acetylcholine receptor lack alpha-bungarotoxin binding sites and hippocampal fast nicotinic currents. J Neurosci 17:9165–9171

Owen AM, Roberts AC, Polkey CE, Sahakian BJ, Robbins TW (1991) Extra-dimensional versus intra-dimensional set shifting performance following frontal lobe excisions, temporal lobe excisions or amygdalo-hippocampectomy in man. Neuropsychologia 29:993–1006

Pantelis C, Barber FZ, Barnes TR, Nelson HE, Owen AM, Robbins TW (1999) Comparison of set-shifting ability in patients with chronic schizophrenia and frontal lobe damage. Schizophr Res 37:251–270

Paterlini M, Zakharenko SS, Lai WS, Qin J, Zhang H, Mukai J, Westphal KG, Olivier B, Sulzer D, Pavlidis P, Siegelbaum SA, Karayiorgou M, Gogos JA (2005) Transcriptional and behavioral interaction between 22q11.2 orthologs modulates schizophrenia-related phenotypes in mice. Nat Neurosci 8:1586–1594

Pattij T, Janssen MC, Loos M, Smit AB, Schoffelmeer AN, van Gaalen MM (2007) Strain specificity and cholinergic modulation of visuospatial attention in three inbred mouse strains. Genes Brain Behav 6:579–587

Paylor R, Nguyen M, Crawley JN, Patrick J, Beaudet A, Orr-Urtreger A (1998) Alpha7 nicotinic receptor subunits are not necessary for hippocampal-dependent learning or sensorimotor gating: a behavioral characterization of Acra7-deficient mice. Learn Mem 5:302–316

Pearlson GD (2000) Neurobiology of schizophrenia. Ann Neurol 48:556–566

Perry W, Minassian A, Paulus MP, Young JW, Kincaid MJ, Ferguson EJ, Henry BL, Masten VL, Sharp RF, Geyer MA (2009) A reverse-translational study of dysfunctional exploration in psychiatric disorders: from mice to men. Arch Gen Psychiatry 60:1072–1080

Peters K, Wiltshire S, Henders AK, Dragovic M, Badcock JC, Chandler D, Howell S, Ellis C, Bouwer S, Montgomery GW, Palmer LJ, Kalaydjieva L, Jablensky A (2008) Comprehensive analysis of tagging sequence variants in DTNBP1 shows no association with schizophrenia or with its composite neurocognitive endophenotypes. Am J Med Genet B Neuropsychiatr Genet 147B:1159–1166

Petryshen TL, Middleton FA, Kirby A, Aldinger KA, Purcell S, Tahl AR, Morley CP, McGann L, Gentile KL, Rockwell GN, Medeiros HM, Carvalho C, Macedo A, Dourado A, Valente J, Ferreira CP, Patterson NJ, Azevedo MH, Daly MJ, Pato CN, Pato MT, Sklar P (2005) Support for involvement of neuregulin 1 in schizophrenia pathophysiology. Mol Psychiatry 10(366–374):328

Pickard BS, Malloy MP, Porteous DJ, Blackwood DH, Muir WJ (2005) Disruption of a brain transcription factor, NPAS3, is associated with schizophrenia and learning disability. Am J Med Genet B Neuropsychiatr Genet 136B:26–32

Pickard BS, Christoforou A, Thomson PA, Fawkes A, Evans KL, Morris SW, Porteous DJ, Blackwood DH, Muir WJ (2009) Interacting haplotypes at the NPAS3 locus alter risk of schizophrenia and bipolar disorder. Mol Psychiatry 14:874–884

Pletnikov MV, Ayhan Y, Nikolskaia O, Xu Y, Ovanesov MV, Huang H, Mori S, Moran TH, Ross CA (2008) Inducible expression of mutant human DISC1 in mice is associated with brain and behavioral abnormalities reminiscent of schizophrenia. Mol Psychiatry 13(173–186):115

Porteous DJ, Thomson P, Brandon NJ, Millar JK (2006) The genetics and biology of DISC1 – an emerging role in psychosis and cognition. Biol Psychiatry 60:123–131

Powell SB (2010) Models of neurodevelopmental abnormalities in schizophrenia. Curr Top Behav Neurosci. doi: 10.1007/7854_2010_57

Powell SB, Zhou X, Geyer MA (2009) Prepulse inhibition and genetic mouse models of schizophrenia. Behav Brain Res 204:282–294

Qian Q, Wang Y, Zhou R, Li J, Wang B, Glatt S, Faraone SV (2003) Family-based and case-control association studies of catechol-O-methyltransferase in attention deficit hyperactivity disorder suggest genetic sexual dimorphism. Am J Med Genet B Neuropsychiatr Genet 118B:103–109

Risbrough VB, Masten VL, Caldwell S, Paulus MP, Low MJ, Geyer MA (2006) Differential contributions of dopamine D(1), D(2), and D(3) receptors to MDMA-induced effects on locomotor behavior patterns in mice. Neuropsychopharmacology 31:2349–2358

Robbins TW, Everitt BJ, Marston HM, Wilkinson J, Jones GH, Page KJ (1989) Comparative effects of ibotenic acid- and quisqualic acid-induced lesions of the substantia innominata on attentional function in the rat: further implications for the role of the cholinergic neurons of the nucleus basalis in cognitive processes. Behav Brain Res 35:221–240

Russig H, Durrer A, Yee BK, Murphy CA, Feldon J (2003) The acquisition, retention and reversal of spatial learning in the morris water maze task following withdrawal from an escalating dosage schedule of amphetamine in wistar rats. Neuroscience 119:167–179

Rutten K, Reneerkens OA, Hamers H, Sik A, McGregor IS, Prickaerts J, Blokland A (2008) Automated scoring of novel object recognition in rats. J Neurosci Methods 171:72–77

Sambeth A, Riedel WJ, Smits LT, Blokland A (2007) Cholinergic drugs affect novel object recognition in rats: relation with hippocampal EEG? Eur J Pharmacol 572:151–159

Sams-Dodd F (1999) Phencyclidine in the social interaction test: an animal model of schizophrenia with face and predictive validity. Rev Neurosci 10:59–90

Sams-Dodd F (2004) (+) MK-801 and phencyclidine induced neurotoxicity do not cause enduring behaviours resembling the positive and negative symptoms of schizophrenia in the rat. Basic Clin Pharmacol Toxicol 95:241–246

Sanders AR, Duan J, Levinson DF, Shi J, He D, Hou C, Burrell GJ, Rice JP, Nertney DA, Olincy A, Rozic P, Vinogradov S, Buccola NG, Mowry BJ, Freedman R, Amin F, Black DW, Silverman JM, Byerley WF, Crowe RR, Cloninger CR, Martinez M, Gejman PV (2008) No significant association of 14 candidate genes with schizophrenia in a large European ancestry sample: implications for psychiatric genetics. Am J Psychiatry 165:497–506

Sarter M, Martinez V, Kozak R (2009) A neurocognitive animal model dissociating between acute illness and remission periods of schizophrenia. Psychopharmacology (Berl) 202:237–258

Sawamura N, Sawa A (2006) Disrupted-in-schizophrenia-1 (DISC1): a key susceptibility factor for major mental illnesses. Ann N Y Acad Sci 1086:126–133

Scarr E, Dean B (2008) Muscarinic receptors: do they have a role in the pathology and treatment of schizophrenia? J Neurochem 107:1188–1195

Schwab SG, Knapp M, Mondabon S, Hallmayer J, Borrmann-Hassenbach M, Albus M, Lerer B, Rietschel M, Trixler M, Maier W, Wildenauer DB (2003) Support for association of schizophrenia with genetic variation in the 6p22.3 gene, dysbindin, in sib-pair families with linkage and in an additional sample of triad families. Am J Hum Genet 72:185–190

Seeman P, Guan HC (2008) Phencyclidine and glutamate agonist LY379268 stimulate dopamine D2High receptors: D2 basis for schizophrenia. Synapse 62:819–828

Seeman P, Lee T, Chau-Wong M, Wong K (1976) Antipsychotic drug doses and neuroleptic/dopamine receptors. Nature 261:717–719

Segal DS, Geyer MA (1985) Animal models of psychopathology. In: Judd LL, Groves PM (eds) Psychobiological foundations of clinical psychiatry. J.B. Lippincott Co., Philadelphia

Shannon HE, Peters SC (1990) A comparison of the effects of cholinergic and dopaminergic agents on scopolamine-induced hyperactivity in mice. J Pharmacol Exp Ther 255:549–553

Shashi V, Keshavan MS, Howard TD, Berry MN, Basehore MJ, Lewandowski E, Kwapil TR (2006) Cognitive correlates of a functional COMT polymorphism in children with 22q11.2 deletion syndrome. Clin Genet 69:234–238

Shoblock JR, Maisonneuve IM, Glick SD (2003) Differences between d-methamphetamine and d-amphetamine in rats: working memory, tolerance, and extinction. Psychopharmacology (Berl) 170:150–156

Slifstein M, Kolachana B, Simpson EH, Tabares P, Cheng B, Duvall M, Gordon Frankle W, Weinberger DR, Laruelle M, Abi-Dargham A (2008) COMT genotype predicts cortical-limbic D1 receptor availability measured with [(11)C]NNC112 and PET. Mol Psychiatry 13:821–827

Snigdha S, Neill JC (2008) Efficacy of antipsychotics to reverse phencyclidine-induced social interaction deficits in female rats – a preliminary investigation. Behav Brain Res 187:489–494

Stark KL, Burt RA, Gogos JA, Karayiorgou M (2009) Analysis of prepulse inhibition in mouse lines overexpressing 22q11.2 orthologues. Int J Neuropsychopharmacol 11:1–7

Stefani MR, Moghaddam B (2002) Effects of repeated treatment with amphetamine or phencyclidine on working memory in the rat. Behav Brain Res 134:267–274

Stefansson H, Sigurdsson E, Steinthorsdottir V, Bjornsdottir S, Sigmundsson T, Ghosh S, Brynjolfsson J, Gunnarsdottir S, Ivarsson O, Chou TT, Hjaltason O, Birgisdottir B, Jonsson H, Gudnadottir VG, Gudmundsdottir E, Bjornsson A, Ingvarsson B, Ingason A, Sigfusson S, Hardardottir H, Harvey RP, Lai D, Zhou M, Brunner D, Mutel V, Gonzalo A, Lemke G, Sainz J, Johannesson G, Andresson T, Gudbjartsson D, Manolescu A, Frigge ML, Gurney ME, Kong A, Gulcher JR, Petursson H, Stefansson K (2002) Neuregulin 1 and susceptibility to schizophrenia. Am J Hum Genet 71:877–892

Stefansson H, Rujescu D, Cichon S, Pietilainen OP, Ingason A, Steinberg S, Fossdal R, Sigurdsson E, Sigmundsson T, Buizer-Voskamp JE, Hansen T, Jakobsen KD, Muglia P, Francks C, Matthews PM, Gylfason A, Halldorsson BV, Gudbjartsson D, Thorgeirsson TE, Sigurdsson A, Jonasdottir A, Bjornsson A, Mattiasdottir S, Blondal T, Haraldsson M, Magnusdottir BB, Giegling I, Moller HJ, Hartmann A, Shianna KV, Ge D, Need AC, Crombie C, Fraser G, Walker N, Lonnqvist J, Suvisaari J, Tuulio-Henriksson A, Paunio T, Toulopoulou T, Bramon E, Di Forti M, Murray R, Ruggeri M, Vassos E, Tosato S, Walshe M, Li T, Vasilescu C, Muhleisen TW, Wang AG, Ullum H, Djurovic S, Melle I, Olesen J, Kiemeney LA, Franke B, Sabatti C, Freimer NB, Gulcher JR, Thorsteinsdottir U, Kong A, Andreassen OA, Ophoff RA, Georgi A, Rietschel M, Werge T, Petursson H, Goldstein DB, Nothen MM, Peltonen L, Collier DA, St Clair D, Stefansson K (2008) Large recurrent microdeletions associated with schizophrenia. Nature 455:232–236

Stevens KE, Freedman R, Collins AC, Hall M, Leonard S, Marks MJ, Rose GM (1996) Genetic correlation of inhibitory gating of hippocampal auditory evoked response and alpha-bungarotoxin-binding nicotinic cholinergic receptors in inbred mouse strains. Neuropsychopharmacology 15:152–162

Stip E, Sepehry AA, Chouinard S (2007) Add-on therapy with acetylcholinesterase inhibitors for memory dysfunction in schizophrenia: a systematic quantitative review, part 2. Clin Neuropharmacol 30:218–229

Straub RE, Jiang Y, MacLean CJ, Ma Y, Webb BT, Myakishev MV, Harris-Kerr C, Wormley B, Sadek H, Kadambi B, Cesare AJ, Gibberman A, Wang X, O'Neill FA, Walsh D, Kendler KS (2002) Genetic variation in the 6p22.3 gene DTNBP1, the human ortholog of the mouse dysbindin gene, is associated with schizophrenia. Am J Hum Genet 71:337–348

Sullivan PF (2008) The dice are rolling for schizophrenia genetics. Psychol Med 38:1693–1696, discussion 1818–1820

Suzuki G, Harper KM, Hiramoto T, Funke B, Lee M, Kang G, Buell M, Geyer MA, Kucherlapati R, Morrow B, Mannisto PT, Agatsuma S, Hiroi N (2009) Over-expression of a human chromosome 22q11.2 segment including TXNRD2, COMT and ARVCF developmentally affects incentive learning and working memory in mice. Hum Mol Genet 18:3914–3925

Swerdlow NR (2010) A cautionary note about latent inhibition in schizophrenia: Are we ignoring relevant information. In: Lubow RE, Weiner I (eds) Latent inhibition: cognition, neuroscience, and applications to schizophrenia. Cambridge University Press, Cambridge, pp 448–456

Swerdlow NR, Geyer MA (1998) Using an animal model of deficient sensorimotor gating to study the pathophysiology and new treatments of schizophrenia. Schizophr Bull 24:285–301

Swerdlow NR, Braff DL, Taaid N, Geyer MA (1994) Assessing the validity of an animal model of deficient sensorimotor gating in schizophrenic patients. Arch Gen Psychiatry 51:139–154

Swerdlow NR, Braff DL, Hartston H, Perry W, Geyer MA (1996) Latent inhibition in schizophrenia. Schizophr Res 20:91–103

Swerdlow NR, Bakshi V, Waikar M, Taaid N, Geyer MA (1998) Seroquel, clozapine and chlorpromazine restore sensorimotor gating in ketamine-treated rats. Psychopharmacology (Berl) 140:75–80

Swerdlow NR, Weber M, Qu Y, Light GA, Braff DL (2008) Realistic expectations of prepulse inhibition in translational models for schizophrenia research. Psychopharmacology (Berl) 199:331–388

Taiminen T, Jaaskelainen S, Ilonen T, Meyer H, Karlsson H, Lauerma H, Leinonen KM, Wallenius E, Kaljonen A, Salokangas RK (2000) Habituation of the blink reflex in first-episode schizophrenia, psychotic depression and non-psychotic depression. Schizophr Res 44:69–79

Tait DS, Marston HM, Shahid M, Brown VJ (2009) Asenapine restores cognitive flexibility in rats with medial prefrontal cortex lesions. Psychopharmacology (Berl) 202:295–306

Takao K, Miyakawa T (2008) Investigating genes-to-behavior pathways in psychiatric diseases: an approach using a comprehensive behavioral test battery on genetically engineered mice. Nihon Shinkei Seishin Yakurigaku Zasshi 28:135–142

Tamminga CA (1998) Schizophrenia and glutamatergic transmission. Crit Rev Neurobiol 12:21–36

Thompson JL, Watson JR, Steinhauer SR, Goldstein G, Pogue-Geile MF (2005) Indicators of genetic liability to schizophrenia: a sibling study of neuropsychological performance. Schizophr Bull 31:85–96

Tsuang M (2000) Schizophrenia: genes and environment. Biol Psychiatry 47:210–220

Turgeon SM, Hulick VC (2007) Differential effects of acute and subchronic clozapine and haloperidol on phencyclidine-induced decreases in voluntary sucrose consumption in rats. Pharmacol Biochem Behav 86:524–530

Umbricht D, Vyssotky D, Latanov A, Nitsch R, Brambilla R, D'Adamo P, Lipp HP (2004) Midlatency auditory event-related potentials in mice: comparison to midlatency auditory ERPs in humans. Brain Res 1019:189–200

van Os J, McGuffin P (2003) Can the social environment cause schizophrenia? Br J Psychiatry 182:291–292

Waters KA, Burnham KE, O'Connor D, Dawson GR, Dias R (2005) Assessment of modafinil on attentional processes in a five-choice serial reaction time test in the rat. J Psychopharmacol 19:149–158

Weiner I, Gal G, Rawlins JN, Feldon J (1996) Differential involvement of the shell and core subterritories of the nucleus accumbens in latent inhibition and amphetamine-induced activity. Behav Brain Res 81:123–133

Wiley JL, Compton AD (2004) Progressive ratio performance following challenge with antipsychotics, amphetamine, or NMDA antagonists in adult rats treated perinatally with phencyclidine. Psychopharmacology (Berl) 177:170–177

Williams HJ, Owen MJ, O'Donovan MC (2007) Is COMT a susceptibility gene for schizophrenia? Schizophr Bull 33:635–641

Williams HJ, Owen MJ, O'Donovan MC (2009) Schizophrenia genetics: new insights from new approaches. Br Med Bull 91:61–74

Wittmann M, Carter O, Hasler F, Cahn BR, Grimberg U, Spring P, Hell D, Flohr H, Vollenweider FX (2007) Effects of psilocybin on time perception and temporal control of behaviour in humans. J Psychopharmacol 21:50–64

Wolpowitz D, Mason TB, Dietrich P, Mendelsohn M, Talmage DA, Role LW (2000) Cysteine-rich domain isoforms of the neuregulin-1 gene are required for maintenance of peripheral synapses. Neuron 25:79–91

Wyvell CL, Berridge KC (2001) Incentive sensitization by previous amphetamine exposure: increased cue-triggered "wanting" for sucrose reward. J Neurosci 21:7831–7840

Young JW, Finlayson K, Spratt C, Marston HM, Crawford N, Kelly JS, Sharkey J (2004) Nicotine improves sustained attention in mice: evidence for involvement of the alpha7 nicotinic acetylcholine receptor. Neuropsychopharmacology 29:891–900

Young JW, Crawford N, Kelly JS, Kerr LE, Marston HM, Spratt C, Finlayson K, Sharkey J (2007a) Impaired attention is central to the cognitive deficits observed in alpha 7 deficient mice. Eur Neuropsychopharmacol 17:145–155

Young JW, Geyer MA, subcommittee p (2007b) Report on putative preclinical models of the MATRICS battery. http://www.turns.ucla.edu/preclinical-TURNS-report-2006b.pdf

Young JW, Minassian A, Paulus MP, Geyer MA, Perry W (2007c) A reverse-translational approach to bipolar disorder: rodent and human studies in the Behavioral Pattern Monitor. Neurosci Biobehav Rev 31:882–896

Young JW, Goey AK, Minassian A, Perry W, Paulus MP, Geyer MA (2009a) GBR 12909 administration as a mouse model of bipolar disorder mania: mimicking quantitative assessment of manic behavior. Psychopharmacology (Berl) 208:443s

Young JW, Powell SB, Risbrough V, Marston HM, Geyer MA (2009b) Using the MATRICS to guide development of a preclinical cognitive test battery for research in schizophrenia. Pharmacol Ther 122:150–202

Zhou X, Long JM, Geyer MA, Masliah E, Kelsoe JR, Wynshaw-Boris A, Chien KR (2005) Reduced expression of the Sp4 gene in mice causes deficits in sensorimotor gating and memory associated with hippocampal vacuolization. Mol Psychiatry 10:393–406

Zhou X, Qyang Y, Kelsoe JR, Masliah E, Geyer MA (2007) Impaired postnatal development of hippocampal dentate gyrus in Sp4 null mutant mice. Genes Brain Behav 6:269–276

Zhou X, Geyer MA, Kelsoe JR (2008) Does disrupted-in-schizophrenia (DISC1) generate fusion transcripts? Mol Psychiatry 13:361–363

Zhou X, Tang W, Greenwood TA, Guo S, He L, Geyer MA, Kelsoe JR (2009) Transcription factor SP4 is a susceptibility gene for bipolar disorder. PLoS ONE 4:e5196

Models of Neurodevelopmental Abnormalities in Schizophrenia

Susan B. Powell

Contents

1 Neurodevelopmental Models of Schizophrenia 437
    1.1 Developmental Theory of Schizophrenia 437
    1.2 Animal Models of Developmental Hypothesis 438
2 Behavioral Measures 439
    2.1 Spontaneous and Drug-Induced Locomotor Activity 440
    2.2 Gating Deficits 440
    2.3 Attention 441
    2.4 Cognitive Deficits 441
    2.5 Social Interaction 442
3 Epidemiologic-Based Developmental Manipulations 442
    3.1 Viral and Immune-Activating Models of Schizophrenia 443
    3.2 Maternal Malnutrition 448
    3.3 Obstetric Complications 450
    3.4 Prenatal/Postnatal Stress 452
    3.5 Postweaning Social Isolation 454
4 Heuristic Neurodevelopmental Models 458
    4.1 Neonatal Ventral Hippocampal Lesion Model 458
    4.2 Prenatal Toxin 461
    4.3 Postnatal/Neonatal NMDA Antagonists 461
5 Discussion 462
References 464

Abstract The neurodevelopmental hypothesis of schizophrenia asserts that the underlying pathology of schizophrenia has its roots in brain development and that these brain abnormalities do not manifest themselves until adolescence or early adulthood. Animal models based on developmental manipulations have provided

---

S.B. Powell
University of California, San Diego, 9500 Gilman Dr., La Jolla, CA 92093-0804, USA
e-mail: sbpowell@ucsd.edu

insight into the vulnerability of the developing fetus and the importance of the early environment for normal maturation. These models have provided a wide range of validated approaches to answer questions regarding environmental influences on both neural and behavioral development. In an effort to better understand the developmental hypothesis of schizophrenia, animal models have been developed, which seek to model the etiology and/or the pathophysiology of schizophrenia or specific behaviors associated with the disease. Developmental models specific to schizophrenia have focused on epidemiological risk factors (e.g., prenatal viral insult, birth complications) or more heuristic models aimed at understanding the developmental neuropathology of the disease (e.g., ventral hippocampal lesions). The combined approach of behavioral and neuroanatomical evaluation of these models strengthens their utility in improving our understanding of the pathophysiology of schizophrenia and developing new treatment strategies.

Keywords Development · Immune · Social isolation · Stress · Neonatal ventral hippocampal lesion · Protein deprivation · Prenatal · Neonatal · Postnatal · Animal model · Schizophrenia · Toxin · Obstetric complications · Behavior

Abbreviations

ASST	Attentional set-shifting task
DA	Dopamine
GABA	Gamma-aminobutyric acid
IL-1α	Interleukin-1alpha
IL-6	Interleukin-6
LI	Latent inhibition
LPS	Lipopolysaccharide
NMDA	N-methyl-D-aspartate
NORT	Novel object recognition
Nox2	NADPH oxidase 2
nVH	Neonatal ventral hippocampal lesion
PCP	Phencyclidine
PFC	Prefrontal cortex
PND	Postnatal day
pNM	Perinatal NMDA antagonist
PolyI:C	Polyriboinosinic–polyribocytidilic acid
PPI	Prepulse inhibition
PV	Parvalbumin
TH	Tyrosine hydroxylase

1 Neurodevelopmental Models of Schizophrenia

1.1 Developmental Theory of Schizophrenia

Over the past two decades, development of the central nervous system has become critical in understanding the neurobiology of schizophrenia (Fatemi and Folsom 2009; Lewis and Levitt 2002; Murray and Lewis 1988; Weinberger 1987). The neurodevelopmental hypothesis of schizophrenia asserts that the underlying pathology of schizophrenia has its roots in brain development and that these brain abnormalities do not manifest themselves until adolescence or early adulthood (Fatemi and Folsom 2009; Rapoport et al. 2005). In addition to the course of illness, support for a neurodevelopmental etiology comes from neuroanatomical and cytoarchitectural abnormalities in the brains of patients with schizophrenia. For example, ventricular enlargement and decreased cortical, hippocampal, and amygdalar volumes are present without any evidence of gliosis (i.e., trauma or neurodegeneration, Arnold et al. 1997; Fatemi and Folsom 2009; Weinberger 1987). Additionally, misplaced and clustered neurons, particularly in the entorhinal cortex, indicate problems of neuronal migration and suggest an early developmental anomaly (Arnold et al. 1991; Falkai et al. 2000; Jakob and Beckmann 1986). Pyramidal neurons in the hippocampus and neocortex have smaller cell bodies and fewer dendritic spines and dendritic arborizations (reviewed in Harrison and Weinberger 2005). Additionally, decreased presynaptic proteins such as synaptophysin, SNAP-25, and complexin II have been observed in schizophrenia brains as well as decreased density of interneurons (e.g., parvalbumin-immunoreactive cells) (Harrison and Weinberger 2005). There are also reports of decreases in cell numbers in the thalamus and a decreased number of oligodendrocytes. Neuroimaging data and postmortem studies have shown that N-acetylaspartate (NAA), a marker of neuronal integrity, is decreased in first episode and in never-medicated patients (Bertolino and Weinberger 1999; Nudmamud et al. 2003). On the basis of these neuropathological changes, investigators have conceptualized schizophrenia as a disease of functional "dysconnectivity" (Friston and Frith 1995; McGlashan and Hoffman 2000; Weinberger et al. 1992) or a "disorder of the synapse" affecting the machinery of the synapse (Frankle et al. 2003; reviewed in Harrison and Weinberger 2005; Mirnics et al. 2001). Recent evidence from MRI studies of reduced white matter supports the disconnection model of schizophrenia (Fatemi and Folsom 2009).

Epidemiological studies support the notion that environmental factors contribute to the incidence of schizophrenia (Cannon et al. 2003; Rapoport et al. 2005). For example, season of birth is a risk factor, with late winter/early spring births associated with an increased risk of schizophrenia (Boyd et al. 1986; Machon et al. 1983; Mino and Oshima 2006; Torrey et al. 1997). Recent studies showed that social factors such as urbanicity, immigrant status, and social isolation are associated with an increased risk for schizophrenia (Cannon et al. 2008; Dean et al. 2003; Marcelis et al. 1998). Hence, the developmental hypothesis of schizophrenia has led to the examination of environmental and epigenetic factors associated with schizophrenia, asserting that an early environmental insult, such as a viral exposure to the developing fetus (Brown and Susser 2002; Mednick et al. 1988; O'Callaghan et al. 1991; Takei et al. 1996) or obstetric complications (Owen and Lewis 1988), causes dysfunction in neural systems that normally reach maturity in late adolescence and early adulthood. Thus, the symptoms of schizophrenia would not express themselves until the point in development in which these brain areas (e.g., dorsal prefrontal cortex) mature (Weinberger 1987).

1.2 *Animal Models of Developmental Hypothesis*

Several animal models are being used to understand neurobiological processes relevant to the developmental hypothesis of schizophrenia (Fatemi and Folsom 2009; Lipska and Weinberger 2000; Meyer and Feldon 2009a, b; Powell and Geyer 2002). Although recreating a uniquely human condition such as schizophrenia is not feasible in animals, animal models have been useful in aiding our understanding of the pathophysiology of the disease (Geyer and Markou 2002; Powell and Geyer 2007; Swerdlow et al. 1994) (Young et al. 2010). These animal models are evaluated based on their face, construct, and predictive validity, and involve both the manipulations (e.g., pharmacological, developmental insult) and measures (e.g., prepulse inhibition, cognitive flexibility). For a more detailed description of the differences in various forms of validity, see Young et al. this text, Geyer and Markou (2002), Swerdlow et al. (1994). When evaluating the relevance of the model to schizophrenia, one should consider both the manipulations and the measures.

Developmental models specific to schizophrenia have focused on the intrauterine environment [e.g., viral insult, exposure to neurotoxins, prenatal maternal stress; (Fatemi et al. 2005; Meyer et al. 2009)], birth complications [e.g., cesarean section, hypoxia; (Boksa and El-Khodor 2003; Vaillancourt and Boksa 2000; Wakuda et al. 2008)], perinatal insult [e.g., ventral hippocampal lesions (Tseng et al. 2009)], prenatal stress (Koenig et al. 2002), and postnatal maternal and/or social deprivation (Ellenbroek et al. 1998; Fone and Porkess 2008; Powell and Geyer 2002). Additionally, genetic models, particularly those targeting developmental genes or those that display age-dependent emergence of a phenotype, can also address neurodevelopmental aspects of schizophrenia (Powell et al. 2009). The combined approach of behavioral and neuroanatomical evaluation of these models strengthens their utility in improving our understanding of the pathophysiology of schizophrenia and developing new treatment strategies (Meyer and Feldon 2009a) including prophylaxis (Meyer et al. 2008d; Powell et al. 2003).

There are several specific experimental factors that need to be considered when conducting developmental studies such as comparisons of the time course of CNS maturation across species, timing of the environmental manipulation, litter effects, and cross-fostering. For a more detailed review of these experimental considerations, see Meyer and Feldon (2009a). Rats and mice differ greatly from humans in the timing of brain development, with brain development in rodents occurring at a much faster pace than that in humans. Considering the proportion of time relative to the lifespan of the organism and the timing of specific neuronal processes, late gestation in humans most likely corresponds to the early postnatal period in rats and mice (Clancy et al. 2001, 2007). Comparisons in brain development between species can be estimated using an algorithm, originally described by (Finlay and Darlington 1995), which is now accessible through a web site (http://translatingtime.net/) (Clancy et al. 2007). Thus, the nature of the schizophrenia-relevant risk factor or neuropathology being modeled and the timing of insult are important factors to consider when evaluating neurodevelopmental models of schizophrenia.

Other important experimental considerations to take into account are in utero environment and maternal behavior. In prenatal challenge models, cross-fostering involves transferring pups from one dam to another lactating surrogate dam to account for any effects the challenge had on mother–pup interactions. For example, in order to confirm significant effects of your manipulation (e.g., prenatal immune activation) it is important to rule out effects of maternal behavior on the observed experimental results. An additional experimental consideration in developmental experiments is litter effects. In multiparous species such as rats and mice, there are often anywhere from 6 to 12 pups born at the same time. This fecundity, while one of the main reasons that rodents are the preferred laboratory species, presents problems to experimental design and statistical analyses, particularly in developmental studies (reviewed in Zorrilla 1997). Owing to shared genes, intrauterine environment, and common postnatal environment, littermates are more similar to each other than nonlittermates and are thus not independent observations. This interdependence complicates statistical analyses and, when each littermate is treated as an independent sample in the ANOVA, inflates the sample size and increases the likelihood of observing a false positive or a false negative (Zorrilla 1997). There are several ways to handle litter effects in the experimental design and statistical analysis, which are discussed in the final section of this chapter and in Zorrilla (1997).

2 Behavioral Measures

Several behavioral measures with certain degrees of validity have been used to assess neurodevelopmental manipulations of relevance to schizophrenia. In this chapter, we will focus primarily on the developmental manipulations themselves and report on the subsequent behavioral and neuronal abnormalities produced by the manipulation. For a more complete review of the "measures" or behavioral tasks used in animal models of schizophrenia, the reader is referred to Jones et al. (2008), Powell and Geyer (2007), Young et al. (2009), Young et al. this text. Briefly, developmental models have been evaluated across several behaviors of relevance to schizophrenia. Generally, these measures fall into four categories: locomotor activity (e.g., spontaneous and drug-induced), gating (e.g., prepulse inhibition of startle, auditory gating), cognitive (e.g., learning and memory, behavioral flexibility), and social (e.g., social interaction, social recognition).

2.1 Spontaneous and Drug-Induced Locomotor Activity

Measures of locomotion and stereotypy in animals have been useful in the identification of drugs that treat the positive symptoms of schizophrenia (e.g., dopamine D2 receptor antagonists). Typically, these experiments involve administering a psychostimulant such as amphetamine to a rat or mouse and observing both the quantity and quality of motor activity produced (reviewed in Segal and Geyer 1985; Segal et al. 1981; Swerdlow et al. 1986). In the absence of a pharmacological manipulation, as is the case with developmental models, spontaneous locomotor activity is often used to evaluate exploratory behavior and unconditioned anxiety produced by a developmental insult. Many of the developmental models reviewed here have also been evaluated for drug-induced locomotor activity (e.g., response to amphetamine, phencyclidine) to probe the functional integrity of the dopamine and glutamate systems, respectively. The measurement of amphetamine-induced locomotor activity is based on the more general dopamine hyperactivity hypothesis of schizophrenia and more specifically, the finding that patients with schizophrenia show exaggerated dopamine release and an exacerbation of symptoms in response to amphetamine (Laruelle et al. 1996).

2.2 Gating Deficits

Deficient gating of sensory input or intrusive thoughts in schizophrenia patients has been recognized for a number of years (Kietzman et al. 1985). In experimental animal studies, gating deficits are evaluated using three primary measures: prepulse inhibition (PPI) of startle, auditory gating, and latent inhibition (LI). Perhaps, the most common behavioral measure assessed in animal models of schizophrenia is PPI of the startle response, an operational measure of sensorimotor gating [reviewed in (Swerdlow and Geyer 1998; Swerdlow et al. 2008)]. PPI is disrupted in schizophrenia patients (Braff et al. 2001) (See Braff 2010) and in pharmacological, developmental, and genetic animal models of schizophrenia (Geyer et al. 2001; Powell and Geyer 2002; Powell et al. 2009). PPI is reliable, can be tested repeatedly in the same animal and has demonstrated face, construct, and predictive validity in animal models of schizophrenia (Geyer and Moghaddam 2002; Swerdlow et al. 1994, 2008). The ability to test the same behavior in the same animals repeatedly is a particularly attractive feature for developmental models that involve the assessment of behaviors both pre- and postpuberty. Together with amphetamine stimulated locomotor activity, sensorimotor gating, as measured by PPI, is often the "gold standard" behavioral endpoint in neurodevelopmental models of schizophrenia research.

The auditory "sensory gating" paradigm in animal studies is based on a similar paired-stimulus paradigm in humans in which the P50 event-related potential (ERP) elicited by the second of two audible clicks is normally reduced relative to the ERP elicited by the first click (Freedman et al. 1999). Because schizophrenia patients do now show the normal reduction in ERP to the second click, a rodent version based on the N40 ERP generated from the hippocampus has been evaluated in animal models of schizophrenia (Freedman et al. 1999; Stevens et al. 1997). LI is conceptually related to the gating theories of schizophrenia disorders and refers to the observation that repeated exposures to a sensory stimulus (i.e., habituation) retards the rate at which a subject will subsequently acquire a stimulus–response association based on this stimulus (Weiner and Arad 2009; Weiner et al. 1988). Meyer et al. (2005) hypothesized that decreases in LI may reflect increased distraction by irrelevant stimuli.

2.3 Attention

Attentional problems in schizophrenia are among the core features of the disease (Addington et al. 1997). In laboratory tests, schizophrenia patients show deficits in the continuous performance task (CPT), which measures sustained attention (Nestor and O'Donnell 1998; Orzack and Kornetsky 1966). Rodent models of attention include the 5-choice serial reaction time task (5-CSRTT) developed by Robbins and colleagues (Chudasama and Robbins 2004; Robbins 2002). Other rodent tasks of attention include the sustained attention task pioneered by Sarter and colleagues (Sarter et al. 2001). For a more complete review of attentional tasks in animal models of schizophrenia, see Young et al. (2009).

2.4 Cognitive Deficits

Because cognitive deficits in schizophrenia are part of the core features of the illness, are associated with poor quality of life, and are relatively resistant to current treatments, there is a renewed focus on defining and treating cognitive deficits (Green et al. 2004; Nuechterlein et al. 2004). The cognitive deficits in schizophrenia include impairments in working memory as well as problem solving, social cognition, and learning and memory (Cannon et al. 2005; Hagan and Jones 2005; Nuechterlein et al. 2004). Animal models mapping onto the specific cognitive domains deficient in schizophrenia are extensively reviewed elsewhere (Young et al. 2009). Briefly, rodent models have focused primarily on assessments of learning and memory (e.g., novel object recognition, Morris water maze; fear conditioning), working memory (e.g., delayed alternation in T-maze), and cognitive flexibility (e.g., set shifting, reversal learning; (Floresco et al. 2009). These tasks are outlined in the previous section (Young et al. this text) and in a recent review paper (Young et al. 2009). Important to developmental models of schizophrenia is the observation that cognitive deficits (e.g., processing speed, working memory, executive function, verbal memory) often predate the onset of psychotic symptoms (Eastvold et al. 2007). This early emergence of cognitive deficits in the prodromal phase should be considered in relation to the postpubertal emergence criteria adopted by many neurodevelopmental animal models. Thus, the emergence of early deficits in cognitive function in an animal model could strengthen its usefulness, particularly in relation to early intervention studies aimed at modeling prodromal treatments.

2.5 Social Interaction

Social withdrawal is included among the negative symptoms of schizophrenia and is often one of the earliest symptoms to occur (Johnstone et al. 2005; McClellan et al. 2003; Miller et al. 2002). Animal models of social impairments fall into three primary categories: social interaction allowing contact between the animal, social approach without contact, and social recognition/social novelty. Social interaction models involve exposing a rat to a nonaggressive conspecific and scoring the amount and type of social interaction [e.g., rough and tumble play, allogrooming; (Sams-Dodd 1996, 1998)]. Rodent social interaction tests such as these have shown their usefulness as a screen for putative antipsychotic medications (e.g., Bruins Slot et al. 2005). A simple test of social approach and novelty was established recently by Crawley, Moy, and colleagues. In this paradigm, test mice are placed in a three-chambered arena, in one chamber, a "stranger" mouse is placed under a wire cup, and in the opposite chamber there is an empty wire container (Crawley 2007; Moy et al. 2004). Exploratory behavior of the test mouse is quantified for 10 min, with social approach measured by comparing the number of contacts and time spent at the container with the "stranger" mouse compared to the empty container. After 10 min, social novelty is tested by putting a new mouse into the previously empty container, and comparing the exploration of the stranger mouse explored in the approach test and the new mouse. This measure of "social novelty" is very similar to other social recognition tests, which involve assessing the time spent investigating a novel, unfamiliar conspecific in the presence of a familiar conspecific (Engelmann et al. 1995; Ferguson et al. 2002; Thor and Holloway 1981; Winslow and Camacho 1995).

3 Epidemiologic-Based Developmental Manipulations

Historically, the manipulations used to produce animal models involved assessment of drug-induced changes in behavior, particularly in response to psychotomimetic drugs (e.g., amphetamine-induced hyperactivity, phencyclidine-induced PPI disruptions). While pharmacological models have proven useful in the evaluation of antipsychotics, developmental manipulations offer the unique ability to probe etiological factors associated with schizophrenia and provide models for the assessment of novel therapeutics (Meyer and Feldon 2009a). For example, assessment of drugs in amphetamine-induced hyperactivity model will tend to screen for compounds predominantly on dopamine antagonist properties similar to currently available antipsychotics. Neurodevelopmental models can be either epidemiologic and focus on the specific risk factors in the human populations (e.g., prenatal infection) (for review, see Meyer and Feldon 2009a) or be heuristic models of developmental neuropathology observed in schizophrenia (e.g., neonatal ventral hippocampal lesion) (for review, see Tseng et al. 2009). Hence, for the purposes of this discussion, the relevant neurodevelopmental animal models of schizophrenia are divided into two categories: epidemiologic and heuristic models (Table 1). Of course, there is some overlap between the two categories and it can be debated how well each model represents the epidemiological risk factor and/or neuropathology it attempts to address. Nevertheless, this distinction is used to guide the discussion of the developmental models reviewed in this chapter.

3.1 Viral and Immune-Activating Models of Schizophrenia

Epidemiological studies have linked prenatal exposure to viral and bacterial infections during early to mid-gestation with an increased risk for schizophrenia (reviewed in Brown and Susser 2002; Fatemi and Folsom 2009; Patterson 2009; but see also Selten et al. 1999a). Early studies focused on the link between influenza and schizophrenia (Mednick et al. 1988; O'Callaghan et al. 1991), but other infectious agents such as toxoplasmosis (Brown et al. 2005) and bacterial infections (Sorensen et al. 2009) have also been associated with the disease. Epidemiological findings have been corroborated by serologic evidence of gestational influenza infection during early to mid-pregnancy increasing the risk of schizophrenia threefold (Brown et al. 2004a). In addition to influenza, there is also serologic evidence of increased maternal levels of cytokines such as TNF-alpha (Buka et al. 2001) and IL-8 (Brown et al. 2004b) during pregnancy in mothers of patients with schizophrenia. Additional evidence for alterations in immune function in schizophrenia comes from the observation that higher levels of antibodies and alterations in other measures of immune function are reported in schizophrenia patients (reviewed in Patterson 2009; Schwarz et al. 2001). To examine and identify the causal relationship between the neural and behavioral consequences of prenatal exposure and immune challenges, the effects of maternal challenges with influenza virus (Shi et al. 2003), as well as other viruses [e.g., borna disease virus, lymphocytic choriomeningitis, cytomegalovirus; (Lipska and Weinberger 2000)], and immune activating agents have been investigated in animal models (for more thorough reviews, see Meyer and Feldon 2009a, b; Patterson 2009). When given at the appropriate time during gestation, these viruses or

Table 1 Overview of neurodevelopmental models

Risk factor/pathology	Rodent model	References
Epidemiological models		
Maternal infection	Prenatal viral infection	Shi et al. (2003), Fatemi et al. (1998, 2002a, b, 2004)
	Prenatal PolyI:C	Shi et al. (2003), Meyer et al. (2009)[a], Meyer and Feldon (2009b)[a]
	Prenatal LPS	Borrell et al. (2002), Romero et al. (2007, 2008), Fortier et al. (2004)
	Prenatal cytokine	Smith et al. (2007)
Neonatal infection	Neonatal viral infection	Asp et al. (2009), Rothschild et al. (1999)
	Neonatal cytokine exposure	Tohmi et al. (2004), Tsuda et al. (2006), Watanabe et al. (2004)
	Neonatal PolyI:C	Ibi et al. (2009)
	Neonatal LPS	Jenkins et al. (2009)
Prenatal stress/maternal deprivation	Prenatal restraint stress	Gue et al. (2004), Kapoor et al. (2009), Lemaire et al. (2000), Son et al. (2006), Szuran et al. (2000), Wu et al. (2007), Yang et al. (2006)
	Prenatal repeated variable stress	Koenig (2006)[a], Koenig et al. (2005), Lee et al. (2007)
	Postnatal maternal deprivation	Ellenbroek et al. (1998), Ellenbroek and Cools (2000)
Obstetric complications	Cesarean section	Brake et al. (1997, 2000), El-Khodor and Boksa (1998, 2000), Juarez et al. (2005), Vaillancourt and Boksa (2000)
	Hypoxia/anoxia	El-Khodor and Boksa (1997), Huang et al. (2009), Pereira et al. (2007), Sandager-Nielsen et al. (2004), Fendt et al. (2008)
	Placental insufficiency	Dieni and Rees (2003, 2005), Mallard et al. (1999), Rehn et al. (2004)
Nutritional deficiency	Prenatal protein deprivation	Palmer et al. (2004, 2008), Tonkiss et al. (1998), Ranade et al. (2008)
	Vitamin D deficiency	Eyles et al. (2009)[a], Burne et al. (2004a, b), O'Loan et al. (2007)
Social isolation	Postweaning social isolation rearing	Fone and Porkess (2008)[a], McLean et al. (2010), Powell and Geyer (2002)[a], Schiavone et al. (2009)
Heuristic models		
Developmental hippocampal pathology	Neonatal ventral hippocampal lesion	Bertrand et al. (2010), Lipska and Weinberger (2000)[a], Tseng et al. (2009)[a]
Disruption in neuronal migration	Prenatal MAM	Le Pen et al. (2006), Lodge and Grace (2009)[a], Moore et al. (2006)
Disruption in perinatal brain development	Neonatal NMDA antagonist	du Bois et al. (2008), Mouri et al. (2007)[a], Nakatani-Pawlak et al. (2009), Wang et al. (2008)

[a]Indicates review article; refer to the text for a more complete overview of the models

immune-activating agents can have rather selective effects on neuronal development and behavior. These prenatal immune paradigms, conducted in both rats and mice, have emerged over the last decade as some of the most important neurodevelopmental models of schizophrenia. These animal models involve exposure of pregnant rats or mice to an immune challenge with either influenza, the bacterial endotoxin lipopolysaccharide (LPS), or the viral mimic polyriboinosinic–polyribocytidilic acid (PolyI:C) during gestation and corresponding assessment of brain and behavioral effects in the offspring.

3.1.1 Prenatal Viral Exposure

Exposure of mice to influenza virus on gestation day 9 results in behavioral and brain abnormalities reminiscent of schizophrenia (Fatemi et al. 1998). Specifically, influenza-exposed mice showed deficits in PPI, decreased exploratory behavior, and decreased social interaction (Shi et al. 2003). Neuroanatomical abnormalities associated with in utero exposure to influenza include pyramidal cell atrophy and macrocephaly, increased glial fibrillary acidic protein (GFAP) immunoreactivity, increased glutamic acid decarboxylase (GAD) 65 and 67 proteins (Fatemi et al. 2002a, b, 2004). Prenatal exposure to influenza also results in decreased size of the lateral ventricles, disrupted corticogenesis, and reduced Reelin immunoreactivity in the frontal cortex and hippocampus (Fatemi et al. 1999).

3.1.2 Prenatal PolyI:C Exposure

Interestingly, the behavioral impairments (specifically PPI) in influenza-exposed offspring appeared to be associated with the *maternal* immune response and not with the viral infection per se because similar alterations in behavior were observed when the pregnant dam was treated with PolyI:C, which elicits an immune response in the mother similar to that observed with influenza (Shi et al. 2003). PolyI:C has been extensively studied in both rats and mice with varying outcomes based on the timing of exposure (Meyer and Feldon 2009b; Meyer et al. 2006b). Additional behavioral, neuropathological, and neurochemical studies further supported the prenatal PolyI:C model as a valid model of schizophrenia. Specifically, behavioral impairments in PPI, LI, reversal learning, novel object recognition, and working memory in addition to an increased sensitivity to dopamine agonists and glutamate antagonists are all observed in the offspring of mice and rats exposed to gestational PolyI:C (see also Fortier et al. 2007; Meyer et al. 2006a; Ozawa et al. 2006; Shi et al. 2003; Smith et al. 2007; Wolff and Bilkey 2008; Zuckerman et al. 2003; Zuckerman and Weiner 2003, 2005). Using MRI, a recent study showed increased lateral ventricle size and PPI deficits in mice exposed to PolyI:C in utero (Li et al. 2009). Several neuropathological studies revealed alterations in dopamine, gamma-aminobutyric acid

(GABA), and glutamate systems and decreased Reelin expression in the PFC and hippocampus (reviewed in Meyer and Feldon 2009b). Alterations in dopamine neurocircuitry in the PolyI:C model include increased tyrosine hydroxylase (TH) immunoreactivity in the nucleus accumbens (NAC), decreased dopamine receptors in the prefrontal cortex (PFC), and alterations in basal and stimulated dopamine release depending on the timing of insult (Meyer et al. 2008b, c; Winter et al. 2009). Similar to the observation of decreased calcium-binding protein parvalbumin (PV) immunoreactivity in schizophrenia brain (Beasley et al. 2002; Reynolds et al. 2004), mice exposed to PolyI:C in utero have decreased PV staining in hippocampus and PFC (Meyer et al. 2008c). Increased limbic GABA-A receptor immunoreactivity in brains of PolyI:C-exposed mice further support a role for GABA in the neuropathology of the immune insult (Nyffeler et al. 2006). Prenatal PolyI:C also leads to alterations in the glutamate system, as evidenced by decreased expression of the N-methyl-D-aspartate (NMDA) receptor subunit 1 (NR1) (Meyer et al. 2008c).

3.1.3 Prenatal LPS Exposure

Administration of the bacterial endotoxin LPS to mammalian species mimics the innate immune response that is typically seen after infection with gram-negative bacteria. Hence, neurodevelopmental animal models of schizophrenia have also utilized LPS as an infectious agent during gestation. Initial studies with prenatal LPS conducted by Borrell and Romero and colleagues administered LPS every other day throughout pregnancy (Borrell et al. 2002; Romero et al. 2007). Similar to prenatal viral exposure, when pregnant rats were exposed to LPS, their offspring exhibited PPI deficits that emerged postpuberty and were reversed by administration of antipsychotics (Borrell et al. 2002; Romero et al. 2007, 2008). The offspring also showed increased TH immunoreactivity and basal dopamine levels in the NAC as well as decreased DARP-32 in frontal cortex and increased synaptophysin in hippocampus and cortex (Borrell et al. 2002; Romero et al. 2007, 2008). Subsequent studies revealed that discrete administration of LPS during gestation (embryonic days 15–16 or 18–19) disrupts PPI and increases amphetamine-induced locomotor activity in the offspring (Fortier et al. 2004, 2007). Additional CNS abnormalities in adult offspring include morphological changes in pyramidal neurons of the hippocampus and PFC (Baharnoori et al. 2009; Nolan et al. 2003).

LPS administered to pregnant rats has been shown to increase cytokines in the amniotic fluid (Urakubo et al. 2001) and in fetal plasma (Ashdown et al. 2006). Borrell et al. (Borrell et al. 2002) showed that serum levels of the cytokines IL-6 and IL-2 were significantly higher in adult offspring of LPS-infected dams. These authors suggest that elevated cytokine levels in the adult offspring may contribute to the PPI deficits observed following prenatal LPS. These data are in line with reports describing elevated levels of cytokines in the CSF and plasma of schizophrenia patients (Licinio et al. 1993; Mittleman et al. 1997).

3.1.4 Role of Cytokines in Prenatal Immune Models

Studies with PolyI:C indicated that the maternal immune response was responsible for the schizophrenia-like behavioral and neuronal effects of prenatal immune challenge but the specific elements of the immune response contributing to these impairments was not known. Subsequent studies went on to examine the role of specific cytokines in the prenatal immune models. PolyI:C administration increases both proinflammatory (IL-6, IL-1β, TNF alpha) and anti-inflammatory (IL-10) cytokines. Both IL-6 and IL-10 play an important role in mediating the effects of prenatal PolyI:C administration. For example, many of the behavioral and neuropathological effects of prenatal PolyI:C are mimicked by gestational administration of IL-6 and blocked in IL-6 knockout mice (Smith et al. 2007). Conversely, overexpression of the anti-inflammatory cytokine IL-10 blocked the emergence of behavioral abnormalities in the offspring exposed to PolyI:C in utero but lead to some behavioral abnormalities when overexpressed on its own in the absence of any immunogenic agent (Meyer et al. 2008a).

3.1.5 Neonatal Immune Activation

In addition to prenatal immune activation, other immune-based developmental models have focused on neonatal exposure of rat or mouse pups to a viral or immunogenic agent (Nawa et al. 2000). These neonatal infections are thought to model infection during the late second/early third trimester. Earlier work showed that rats with neonatal exposure to cytomegalovirus show an increased sensitivity to the PPI-disruptive effects of apomorphine (Rothschild et al. 1999). Neonatal influenza administered on postnatal day 3 or 4 resulted in PPI deficits in adult *Tap1*-/- (transporter associated with antigen processing 1) mice expressing reduced levels of MHC Class I (Asp et al. 2009). The PPI deficits were accompanied by increases in transcripts encoding indoleamine-pyrrole 2,3-dioxygenase (IDO) and transient increases in other enzymes in the kynurenine pathway of tryptophan metabolism and kynurenic acid (KYNA) (Asp et al. 2009; Holtze et al. 2008). These studies raise the possibility that elevations in KYNA, an endogenous NMDA antagonist and nicotinic acetylcholine (nAch) alpha-7 antagonist, may mediate in part the behavioral effects of neonatal immune activation. Neonatal exposure to PolyI:C also results in behavioral and neurochemical abnormalities later in life (Ibi et al. 2009). Again, the effects of these immune-activating agents may be mediated through increases in cytokines. Administration of cytokines (e.g., IL-1α, leukemia inhibitory factor [LIF]) to neonatal rat or mouse pups also results in locomotor hyperactivity, decreased PPI, impaired social interaction, and neuroanatomical abnormalities such as increased TH and dopamine metabolism (Tohmi et al. 2004; Tsuda et al. 2006; Watanabe et al. 2004). Similar to effects of prenatal PolyI:C treatment in mice, administration of LPS on postnatal day 7 and 9 to rat pups also produced decreased PV immunoreactivity in the hippocampus (Jenkins et al. 2009).

3.1.6 Discussion of Immune Models

Like all the studies of epidemiological risk factors, one important aspect to consider is that the risk factor of prenatal or neonatal viral infection only confers a modest increased risk for developing the disorder. Obviously, not all individuals exposed to an infection in utero subsequently develop schizophrenia. One advantage to the environmental risk factors outlined here is the possibility to study gene–environment interactions. Inducing an effect of immune activation in genetically compromised or genetically susceptible animal but not in the wildtype animal would be extremely compelling and useful to our understanding of gene–environment interactions in schizophrenia. These gene–environment interactions are beginning to be explored in relation to immune challenge models (reviewed in Ayhan et al. 2009). Examples of these gene–environment interactions include the neonatal exposure to influenza in TAP1-/- mice described above (Asp et al. 2009) and recent studies examining the effects of neonatal PolyI:C administration in DISC1 dominant-negative (DISC1 DN) mutant mice (Ibi et al. 2010). Specifically, DISC1 DN mice displayed a more pronounced response to the behavioral effects of neonatal PolyI:C compared to wildtype mice (Ibi et al. 2009). Hence, immune models are ripe for studies of gene–environment interactions and may prove useful in our understanding of the dynamic interplay between susceptibility genes and environmental risk factors.

As mentioned previously, the role of maternal behavior in the effects of prenatal immune activation need to be taken into account. In studies, in which cross-fostering was employed, mother–pup behavioral interactions accounted for some of the observed effects (Meyer et al. 2006c, 2008b). One additional consideration in studies of prenatal immune activation is that the immune challenge may produce a nutritional deficiency in the mom, which can have significant effects on the offspring (see Sect. 3.2). Dams do tend to lose weight in response to the immune-activating agent.

3.2 *Maternal Malnutrition*

Epidemiological studies suggest that prenatal nutritional deficiency increases the risk of developing schizophrenia (Brown and Susser 2008; Susser et al. 1996). Perhaps the most robust examination of this relationship comes from the two periods of famine, one during 1944–1945 in The Netherlands termed "The Dutch Hunger Winter" and the other during the "Chinese Famine" that took place during 1959–1961. Those offspring exposed to the famine during early gestation had a twofold increase of developing schizophrenia as adults (Susser et al. 1996; Xu et al. 2009). Candidate micronutrients that may be responsible for these abnormalities include folate, vitamin D, essential fatty acids, retinoids, and iron, all of which play a role in normal fetal brain development (reviewed in Brown and Susser 2008). In order to determine the mechanism responsible for altered brain development in response to early gestational malnutrition, several animal models have been developed: prenatal protein deficiency (or protein–calorie malnutrition) and prenatal vitamin D deficiency (reviewed in Meyer and Feldon 2009a).

3.2.1 Prenatal Protein Deficiency

Prenatal protein deficiency typically involves depriving the dam of protein prior to and during pregnancy and comparing the behavioral effects in the offspring to that of control rats that received normal levels of protein during gestation (e.g., low casein diets (6%) or adequate casein diets (25%)). Protein deprivation in rats leads to many alterations in brain development consistent with an animal model of schizophrenia including structural differences in the hippocampus, alterations in dopamine and serotonin release, and changes in glutamate receptor binding (Meyer and Feldon 2009a for review). As far as behavioral effects of prenatal protein deficiency, several alterations have been reported. Namely, female rats that underwent prenatal protein deprivation display a postpubertal emergence of PPI deficits (Palmer et al. 2004), an increased responsiveness to dopamine agonists and NMDA receptor antagonists (Palmer et al. 2008; Tonkiss et al. 1998). Impairments in working memory measured in a radial arm maze have been reported in rats exposed to prenatal protein deprivation (Ranade et al. 2008); whereas other studies have shown no difference in working memory as measured in a T-maze alternation task and in an operant delayed alternation task (Tonkiss and Galler 1990). There have yet to be any report showing reversal of the functional impairments in prenatal protein deficiency with drug treatment. Such future studies would clarify the predictive validity of the prenatal protein deficiency model of schizophrenia.

3.2.2 Prenatal Vitamin D Deficiency

Maternal vitamin D deficiency has also been examined in animal models for its role in the development of schizophrenia (Eyles et al. 2009). McGrath (1999) argues that Vitamin D deficiency may explain several risk factors for schizophrenia including maternal malnutrition, increased winter births in schizophrenia, urbanicity, and dark-skinned immigrants in cold climates. The hypothesis of vitamin D deficiency is intriguing because vitamin D is important for normal fetal brain development and deficiencies (e.g., rickets) can have profound impact on health (Eyles et al. 2009). The epidemiological evidence for this link is mixed. While maternal vitamin D supplementation reduced the risk of schizophrenia in an examination of data from a Finnish Birth Cohort (McGrath et al. 2004), other studies have supported the link only weakly (McGrath et al. 2003) or not at all (Kendell and Adams 2002; Özer et al. 2004). Nevertheless, animal models of maternal vitamin D deficiency have supported a link between the deficiency and brain and behavioral abnormalities related to schizophrenia (reviewed in Eyles et al. 2009). For example, prenatal vitamin D deficiency in rats is associated with enlarged lateral ventricles and smaller neocortical width (Eyles et al. 2003), decreased neurotrophin levels (Eyles et al. 2003; Feron et al. 2005), altered neurogenesis (Cui et al. 2007), and increased long-term potentiation (LTP) (Grecksch et al. 2009). Maternal vitamin D deficiency also produces changes in PFC, hippocampal, and NAC gene and protein expression in pathways involved in oxidative stress, synaptic plasticity, calcium homeostasis, and neurotransmission (Almeras et al. 2007; Eyles et al. 2007; McGrath et al. 2008).

Maternal vitamin D deficiency is associated with several behavioral impairments with validity for schizophrenia. For example, rats exposed to prenatal vitamin D deficiency demonstrate heightened locomotor activity in a novel environment (Burne et al. 2004a, 2006) and increased sensitivity to the NMDA antagonist MK-801 and the dopamine D2 receptor antagonist haloperidol (Kesby et al. 2006). Deficits in PPI are observed when the prenatal vitamin D deficiency is continued through the tenth postnatal week, but not when exposed in utero only (Burne et al. 2004b). Regarding cognitive tasks, rats exposed to prenatal vitamin D deficiency show impaired habituation in the hole board but no differences in spatial learning in the radial arm maze or active avoidance learning in the shuttle box (Becker et al. 2005). The normal learning abilities are not surprising, considering the reported increase in LTP (Grecksch et al. 2009). Although most studies produce vitamin D deficiency throughout gestation, there is some evidence that the detrimental effects of maternal vitamin D deficiency may be more pronounced when the deficiency occurs during the late gestational period (O'Loan et al. 2007). Initial studies of maternal vitamin D deficiency in mice suggest that while adult offspring displayed increased locomotor activity, vitamin D-deprived and control mice did not differ in PPI or social behavior (Harms et al. 2008).

3.3 Obstetric Complications

Another set of prenatal and perinatal risk factors that have been well documented are obstetric complications. As Rapoport et al. (2005) point out, the relative risk for schizophrenia-associated obstetric complications is low, with an odds ratio for the exposure to obstetric complications increasing the risk of schizophrenia estimated at 2.0 (Rapoport et al. 2005). Nevertheless, obstetric complications have been well documented and linked to schizophrenia in several independent studies. Specifically, birth complications such as pre-eclampsia, cesarean section, and perinatal hypoxia are associated with an increased risk of schizophrenia (Cannon et al. 2002; Hultman et al. 1997; Zornberg et al. 2000). These birth complications have all been modeled in animals (reviewed in Boksa 2004; Meyer and Feldon 2009a).

3.3.1 Cesarean Section

For Cesarean section (C-section), the experimental litter is removed from the uterus and kept warm until being placed with a foster mom (El-Khodor and Boksa 1997).

In rats and the more precocious species, guinea pigs, C-section is associated with heightened sensitivity to the locomotor-activating effects of amphetamine and stress, deficits in PPI, and decreased and increased dopamine in the PFC and NAC, respectively (Brake et al. 1997, 2000; El-Khodor and Boksa 1998, 2000; Juarez et al. 2005; Vaillancourt and Boksa 2000). Additionally, rats born by C-section demonstrate a postpubertal increase in dopamine D1 receptor binding and increased functional response to D1 agonists (Boksa et al. 2002).

3.3.2 Perinatal Hypoxia

Both intrauterine and neonatal hypoxia have been tested in rats, mice, and guinea pigs. Both manipulations lead to widespread effects on brain morphology and neurochemistry, specifically with decreased hippocampal volume or neuronal cell loss coupled with reduced dendritic spine density and/or elongation (reviewed in Meyer and Feldon 2009a). Intrauterine hypoxia typically involves removing the intact uterus from the dam and placing it in a 37°C water bath for a certain period of time (typically around 15 min) (El-Khodor and Boksa 1997). Hence, perinatal hypoxia manipulations in practice also involve C-section. Neonatal hypoxia involves placing a pup in a chamber without oxygen or with very low levels of oxygen (e.g., 8%) for a specified period of time (Fendt et al. 2008; Nadri et al. 2007). Some methods combine the low-oxygen environment with occlusion of the carotid artery (Rice et al. 1981). The behavioral effects of intrauterine and neonatal hypoxia are mixed (reviewed in Meyer and Feldon 2009a). Whereas the near-term intrauterine hypoxia produced deficits in PPI and working memory in guinea pigs (Becker and Donnell 1952; Vaillancourt and Boksa 2000) and impairments in spatial learning in the water maze in rats(Boksa et al. 1995), no impairments in working memory as measured by spontaneous alternation in the T maze were reported in rats (Boksa et al. 1995). Rats exposed to intrauterine hypoxia did display decreased social and exploratory behavior and increased response to stress and dopamine agonists (Brake et al. 1997; reviewed in Meyer and Feldon 2009a).

Neonatal hypoxia is associated with reference and working memory impairments in the water maze and decreased hippocampal volume (Huang et al. 2009; Pereira et al. 2007). Interestingly, environmental enrichment blocked the memory impairments and hippocampal volume reduction produced by neonatal hypoxia (Pereira et al. 2007). Depending on the timing and severity of postnatal hypoxia, differing effects on PPI have been reported. Although hypoxia at postnatal day 9 altered mesolimbic dopamine neurochemistry, it did not produce differences in PPI (Sandager-Nielsen et al. 2004). Subsequent studies using subchronic exposure to hypoxia during postnatal day 4–8 did result in PPI deficits in the adult rat (Fendt et al. 2008). These animal studies of hypoxia are useful for our understanding of perinatal complications as a risk factor as they have shown many neuroanatomical abnormalities consistent with schizophrenia. The behavioral abnormalities, on the other hand, are less consistent thus making the model less useful for pharmacological intervention studies.

3.3.3 Placental Insufficiency

Placental insufficiency is another prenatal risk factor for schizophrenia (Cannon et al. 2002) which involves loss of blood flow to the developing fetus. Placental insufficiency is achieved experimentally in guinea pigs by ligation of the uterine artery and results in decreased PPI, enlargement of the lateral ventricles, reduced volume of the basal ganglia and septum, and reduced hippocampal BDNF (Dieni and Rees 2003, 2005; Mallard et al. 1999, 2000; Rehn et al. 2004). The neuroanatomical abnormalities, without any evidence of gliosis, together with the observation of deficient PPI, suggest that placental insufficiency may have face validity for schizophrenia. However, further behavioral tests and pharmacological interventions are warranted in this model (Meyer and Feldon 2009a).

3.4 Prenatal/Postnatal Stress

There is some evidence that psychological stressors during pregnancy increase the risk for schizophrenia in offspring. This association comes from the studies examining the occurrence of schizophrenia in offspring whose mothers were exposed to a stressful experience during pregnancy (reviewed in Koenig 2006; Koenig et al. 2002). Because assessing psychological stress retrospectively is very difficult, studies have focused on discrete events for the individual (e.g., death of a relative) or periods of "stress" for an entire community (e.g., war, flood). In a retrospective study of the 5-day Nazi invasion of the Netherlands in 1940, schizophrenia risk was increased when exposed to the traumatic event during the first trimester (van Os and Selten 1998). Additionally, the risk of schizophrenia was increased in offsprings whose mothers experienced the death of a relative during the first trimester pregnancy (Khashan et al. 2008). Other studies, however, have failed to report a relationship between prenatal stress and the later development of schizophrenia. For example, there was no relationship between a deadly 1953 flood in Holland and the later development of schizophrenia in exposed offspring (Selten et al. 1999b).

3.4.1 Prenatal Stress

On the basis of the association between prenatal stress and risk for schizophrenia and the observation that maternal stress may alter the programming of the fetal brain (Weinstock 2008), consequences of prenatal stress have been evaluated for their effects on behavior and neurochemical alterations associated with schizophrenia. The effects of prenatal stress on schizophrenia-related behaviors have been mixed. Rats exposed to prenatal stress show hyperactivity in a novel environment (Son et al. 2007) and an increased sensitivity to amphetamine (reviewed in Meyer and Feldon 2009a). Pups born to dams that had been restrained three times a day for 30 min during gestational days 15–22 did not exhibit deficits in PPI or LI when tested as adults (Lehmann et al. 2000). These rats actually showed slight increases in PPI, which was normalized by the combination of prenatal stress and maternal separation. Subsequent studies, however, did show impaired LI in the offspring of dams exposed to more severe stress of repeated electric foot shock. Prenatal repeated variable stress paradigm, on the other hand, produce behavioral alterations relevant to schizophrenia in the exposed offspring (Koenig 2006). Repeated variable stress described in these studies involved exposing the pregnant rat to several different stressors including 60 min restraint stress, cold exposure (4°C) for 6 h, overnight food deprivation, overcrowding during dark phase of cycle, swim stress, and lights on for 24 h. Rats are exposed to 2–3 of these stressors each day from gestation day 14 until parturition, i.e., the third week of pregnancy. Rats exposed to repeated variable stress showed impaired social interaction (Lee et al. 2007), PPI deficits, and increased sensitivity to amphetamine challenge postpuberty (Koenig et al. 2005). Interestingly, prenatally stressed rats also showed a trend toward deficits in N40 gating (Koenig et al. 2005). Thus, repeated variable stress appears to exert more robust effects on gating measures than does prenatal restraint stress.

Cognitive deficits, on the other hand, have been well documented in offspring of dams exposed to prenatal stress. Specifically, prenatal restraint stress leads to impairments in spatial learning in the water maze in rats (Lemaire et al. 2000; Szuran et al. 2000; Wu et al. 2007; Yang et al. 2006) and guinea pigs (Kapoor et al. 2009). Prenatally stressed rats also show impaired reversal learning in the water maze (Szuran et al. 2000). Additional impairments in working memory in the radial arm maze in mice (Son et al. 2006) and in the T maze delayed alternation test in rats have been associated with prenatal stress (Gue et al. 2004). Importantly, the working memory deficits in the T maze delayed alternation task were observed in prepubertal rats. The early emergence of working memory impairments is interesting to schizophrenia models because of the observation of early development of cognitive deficits in prodromal patients (Eastvold et al. 2007). Many sex differences in the behavioral effects of prenatal stress have been reported, with prenatal stress conferring more profound cognitive disruptions in males and more affective or anxiety disturbances in females (reviewed in Meyer and Feldon 2009a; Weinstock 2008). Few pharmacological studies have been conducted in the prenatal stress model, with the exception of oxytocin reversing impairments in social interaction in the model (Lee et al. 2007). Thus, further pharmacological studies should be conducted to evaluate the predictive validity of the model.

In addition to the behavioral abnormalities, prenatal stress alters brain development. Many studies report increased HPA axis activity and corresponding decreases in glucocorticoid and mineralcorticoid receptors (Szuran et al. 2000). Hippocampal abnormalities are consistently reported in mice exposed to prenatal stress. These abnormalities include decreased neurogenesis and decreased density of granule cells in the hippocampus (Lemaire et al. 2000), decreased LTP and enhanced LTD in hippocampal CA1 region (Son et al. 2006; Yang et al. 2006), and reduced NR2A and NR2B subunits of NMDA receptor in hippocampus (Son et al. 2006). Alterations in dopamine function including increased DA turnover and alterations in DA receptors and DA transporter (reviewed in Meyer and Feldon 2009a; Son et al. 2007).

In conclusion, although the prenatal stress model was developed primarily to examine factors contributing to anxiety and depressive disorders (Weinstock 2008), many of the behavioral and neurochemical abnormalities suggest that the model may prove useful for schizophrenia research (Koenig 2006).

3.4.2 Maternal Deprivation

Stress manipulations in the early postnatal period have also been assessed for their effects on schizophrenia-related behaviors. These experiments typically involve removing the pups from the dam during the first few weeks of life for various lengths of time, with shorter periods of time (3–15 min; early handling) resulting in *decreased* HPA axis response and longer periods of time (3 h or 24 h; maternal separation, MS) resulting in *increased* HPA axis activity (Meaney et al. 1991, 1993). Although early handling (i.e., brief separations) affects neuroendocrine and anxiety-related behavior, it does not appear to affect PPI (Pryce et al. 2001). Early handling increases LI when compared to nonhandled rats (Feldon et al. 1990; Weiner et al. 1985). PPI following more prolonged periods of maternal separation in rats has also been assessed (for review see Weiss and Feldon 2001). Ellenbroek et al. (1998) showed that separation from the dam for 24 h at PND 3, 6, and 9 produced deficits in PPI on PND 69 in male and female rats of the Wistar strain (Ellenbroek and Cools 2000). Maternal separation for shorter periods of time (e.g., 1–4 h/day) did not affect PPI in rats (Finamore and Port 2000; Weiss et al. 2001) or mice (Millstein et al. 2006), but did impair acoustic startle habituation (Finamore and Port 2000). These data indicate that longer periods of maternal separation (e.g., 24 h) at one point during the preweanling period may have a greater effect on PPI than shorter, repeated separations from the mother, suggesting that a certain amount of nutritional deprivation may be necessary to observe the effects of MS on PPI. Maternal separation does, however, result in cognitive deficits in adulthood. Specifically, MS rats display learning impairments in the water maze and the NORT (Aisa et al. 2007, 2008). These behavioral impairments are consistent with the observed alterations in hippocampal development (Huot et al. 2002) among other neurochemical and neuroanatomical differences (Holmes et al. 2005). In conclusion, MS produces cognitive impairments but does not consistently alter PPI. This observation, together with the lack of a clear epidemiological link between postnatal stress and schizophrenia, suggests that the MS model may not be a particularly useful model for schizophrenia research.

3.5 *Postweaning Social Isolation*

Social withdrawal and isolation are common features of schizophrenia that have received recent attention because of the role social factors play in the risk for schizophrenia and conversion to psychosis in prodromal patients (Addington et al.

2008). Indeed, social functioning, among other factors, predicts conversion to psychosis in patients at a high risk of developing psychosis (Cannon et al. 2008). Because of this observation, coupled with social factors contributing to the etiology of schizophrenia, we have categorized social isolation rearing as an epidemiological model in this review. When evaluating animal models of schizophrenia, one should consider both proximate and distal risk factors to the development of the disease. Thus far, we have discussed distal risk factors for the development of schizophrenia. Postweaning social isolation can be considered a model of a more proximal risk factor – social isolation.

Social isolation rearing of rodents is a developmental model relevant to schizophrenia that involves more subtle environmental manipulations leading to profound effects on behavior and brain development. Social isolation rearing of young rodents provides a nonpharmacologic method of inducing long-term alterations reminiscent of several symptoms seen in schizophrenia patients (Geyer et al. 1993; Powell and Geyer 2002). Rearing animals in social isolation is particularly consequential for species that rely on social contact after being weaned from the mother. Specifically, isolation rearing deprives rodents of social interactions during a developmental period in which play behavior emerges (Einon and Morgan 1977). Thus, as a consequence of social isolation, animals are deprived of stimuli critical to behavioral and neurobiological development (reviewed in Hall 1998). The lack of early social contact provides a model of the social isolation and social withdrawal which occurs early in the course of schizophrenia and predicts conversion to psychosis in patients at a high risk of developing psychosis (Cannon et al. 2008). Behavioral and neurochemical changes after isolation rearing in rats provide a nonlesion and nonpharmacological model to enhance our understanding of the developmentally linked emergence of neural and behavioral abnormalities in schizophrenia patients (Geyer et al. 1993; Powell and Geyer 2002).

3.5.1 Isolation Rearing: Neuroanatomical Abnormalities

Rats reared in social isolation exhibit profound abnormalities in behavior, drug responses, and neurochemistry compared to rats reared in social groups (Fone and Porkess 2008; Hall 1998; Powell and Geyer 2002). The most well documented set of studies are those that support isolation-reared rats as a model for dopamine hyper-reactivity associated with schizophrenia, such as (1) increased behavioral sensitivity to dopamine agonists (Bowling and Bardo 1994; Jones et al. 1990, 1992; Sahakian et al. 1975), (2) reduced responsivity to dopamine antagonists (Sahakian et al. 1977), (3) elevated basal and amphetamine-stimulated dopamine release in the NAC (Hall et al. 1998; Jones et al. 1992), and (4) elevated dopamine concentrations (Jones et al. 1992) and altered dopamine turnover (Blanc et al. 1980) in the frontal cortex. In addition to alterations in dopamine function, isolation-reared rats display abnormalities in the hippocampus and frontal cortex. Isolation-reared rats have increased density of 5-HT_{1A} receptors in the hippocampus (Del-Bel et al. 2002;

Preece et al. 2004). Synaptophysin is a synapse-specific protein involved in neurotransmitter release and its expression is reduced within certain hippocampal subfields in schizophrenia (Eastwood and Harrison 1995). Varty et al. (1999) also reported reduced synaptophysin immunoreactivity in the dentate gyrus of isolation-reared rats. There is also evidence of reduced BDNF in the hippocampus (Scaccianoce et al. 2006) and decreased spine density in isolation-reared rats (Silva-Gomez et al. 2003b). More recent studies have pointed to further alterations in the hippocampus of isolation-reared rats. Loss of PV-positive GABA interneurons observed in isolation-reared rats (Harte et al. 2007; Schiavone et al. 2009) is very similar to that reported in the hippocampus and frontal cortex of schizophrenia patients (Reynolds et al. 2004; Reynolds and Beasley 2001). Isolation-reared rats show abnormalities in the PFC including (1) abnormal firing of pyramidal cells in the PFC upon dopamine stimulation from VTA neurons (Peters and O'Donnell 2005), (2) decreased volume of PFC (Day-Wilson et al. 2006; Schubert et al. 2009; Silva-Gomez et al. 2003b), and (3) decreased dendritic arborization in the PFC (Pascual et al. 2006; Silva-Gomez et al. 2003b).

3.5.2 Isolation Rearing: Behavioral Abnormalities

Rats reared in social isolation show deficits in PPI (Cilia et al. 2001, 2005; Geyer et al. 1993; Varty and Geyer 1998; Varty and Higgins 1995) and slow rates of startle habituation (for reviews, see Geyer et al. 1993, 2001; Powell and Geyer 2002; Weiss and Feldon 2001). More recent studies have also shown that several different strains of mice (e.g., ddY, 129T2, C57BL/6) exhibit deficits in PPI when reared in social isolation from weaning (Dai et al. 2004; Sakaue et al. 2003; Varty et al. 2006; see also Pietropaolo et al. 2008). Deficits in PPI produced by isolation rearing are developmentally specific in that they only appear when social isolation occurs early, during the postnatal period, and not in rats isolated as adults (Wilkinson et al. 1994). PPI deficits in isolation-reared rats can be reversed with both typical (Geyer et al. 1993; Varty and Higgins 1995) and atypical (Bakshi et al. 1998; Cilia et al. 2001; Varty and Higgins 1995; but see Barr et al. 2006) antipsychotic drugs. Thus, several investigators have shown predictive validity of the PPI deficits in the isolation-rearing model (summarized in Geyer et al. 2001). In addition to PPI deficits, isolated rats also exhibit abnormalities in motor activity. When tested in novel environments, isolated rats show elevated levels and slowed habituation of locomotor activity (Hall 1998; Jones et al. 1989, 1990; Lapiz et al. 2000; Paulus et al. 1998; Sahakian et al. 1975; Varty et al. 2000), increased investigatory behavior (e.g., rearings, holepokes; Lapiz et al. 2000; Paulus et al. 1998), and an increased preference for a novel environment (Hall et al. 1997). Additionally, isolation-reared rats and mice show increased anxiety-like behavior (Da Silva et al. 1996; Wright et al. 1991), deficits in fear learning (Voikar et al. 2005; Weiss et al. 2004), impaired recognition memory (e.g., novel object recognition; Bianchi et al. 2006; McLean et al. 2010; Voikar et al. 2005), reduced spatial memory (Ibi et al. 2008), and cognitive inflexibility as demonstrated by deficits in reversal learning (Krech et al. 1962; Schrijver et al. 2004) and extradimensional set-shifting tasks (McLean et al. 2010; Schrijver and Wurbel 2001). Thus, isolation rearing of rats and mice is associated with impaired sensorimotor gating, cognitive inflexibility, reductions in PFC volume and hippocampal synaptic plasticity, hyperfunction of mesolimbic dopaminergic systems, and hypofunction of mesocortical dopamine, strikingly similar behavioral and neuroanatomical abnormalities as those observed in schizophrenia. Taken together, these results point towards the usefulness of the social isolation model in mimicking some behavioral, neurochemical, and neuropathological phenomena characteristic of schizophrenia.

Few studies have directly tested the mechanism by which isolation rearing exerts its effects on brain and behavioral development. Recently, a clear role for nicotinamide adenosine dinucleotide phosphate (NADPH) oxidase 2 (Nox2)-dependent oxidative mechanisms in the loss of PV interneurons and development of schizophrenia-like behavior in the isolation-rearing model was demonstrated (Schiavone et al. 2009). Corroborating our earlier work (Harte et al. 2007), Schiavone et al. (2009) found decreased PV immunoreactivity in the brains of rats reared in social isolation. This loss of PV interneurons was associated with elevations in Nox2, and the decrease in PV-staining and deficits in novel object recognition were blocked by treatment with the Nox2 inhibitor apocynin (Schiavone et al. 2009). Recent studies in mice have shown reductions in the expression of two developmental genes, Nurr1 and Npas4, in mice reared in social isolation (Ibi et al. 2008).

3.5.3 Discussion of Isolation-Rearing Model

There are several advantages to the isolation-rearing model that make it an appealing preclinical model of schizophrenia. First, the isolation-rearing model has shown a high degree of predictive validity for antipsychotic drugs. Second, there is a wealth of data on the relevant behavioral and neurochemical/neuroanatomical differences associated with postweaning social isolation, and many of these findings have been reported in both rats and mice. Finally, isolation rearing is a relatively easy procedure to conduct but does require ample housing space for individual cages. There are also several disadvantages to the model. Since the insult is ongoing, it is difficult to determine the precise timing of the effects since age is confounded with duration of isolation. Thus, there is a need for more discrete manipulations of the duration of isolation (e.g., early postnatal, pubertal). Relative to most of the other models reviewed here, isolation rearing occurs rather late in development and may thus only be relevant to early childhood and pubertal insults. Like the nVH lesion model discussed below, it can also be considered as a heuristic model to guide future studies into the pathophysiology of and treatments for schizophrenia.

4 Heuristic Neurodevelopmental Models

4.1 Neonatal Ventral Hippocampal Lesion Model

One of the first and most widely studied neurodevelopmental models of schizophrenia is the neonatal ventral hippocampal lesion (nVH) model. On the basis of the observation of developmental abnormalities in the hippocampus of schizophrenia patients, attempts to model the developmental perturbation and delayed behavioral symptomatology, similar to that of schizophrenia, have been undertaken (Lipska and Weinberger 2000; Tseng et al. 2009).

4.1.1 Neonatal Ventral Hippocampal Lesion Model: Behavioral Studies

Rats with nVH lesions show increased responsiveness (e.g., hyperlocomotion, increased stereotypy) to dopamine agonists (Lipska et al. 1993; Lipska and Weinberger 1993, 1994a; Sams-Dodd et al. 1997) and NMDA antagonists (Al-Amin et al. 2000, 2001), deficits in PPI (Francois et al. 2009a; Le Pen et al. 2003; Le Pen and Moreau 2002; Lipska et al. 1995), and more recently alterations in N40 gating (Swerdlow et al. 2007; Vohs et al. 2009). The locomotor sensitivity to stimulants and PPI deficits in rats with nVH lesions exhibit a delayed temporal pattern and do not appear until postpuberty (Lipska et al. 1993, 1995; Lipska and Weinberger 1993; but see Swerdlow et al. 2007). Additionally, nVH lesions produce social and cognitive impairments, strengthening the relationship to this model and schizophrenia symptomatology (Lipska 2004). Rats with nVH lesions display decreased social interactions, which emerge prepuberty (Becker et al. 1999; Flores et al. 2005; Sams-Dodd et al. 1997), and impaired social recognition memory (Becker and Grecksch 2000).

Cognitive deficits associated with nVH lesions encompass many of the cognitive domains deficient in schizophrenia. For example, rats sustaining nVH lesions display deficits in spatial learning in the water maze and radial arm maze (Chambers et al. 1996; Le Pen et al. 2000; Silva-Gomez et al. 2003a) as well as impaired avoidance learning (Le Pen et al. 2000) and novel object recognition . Continuous spatial delayed alternation task and discrete paired-trial variable-delay task were impaired in rats that sustained nVH lesions but not in rats sustaining adult VH lesions, lending further support to the neurodevelopmental construct validity of the nVH lesion model (Lipska et al. 2002a). Subsequent studies aimed at dissecting the relative contribution of the hippocampus and the PFC to the impairments on working memory tasks showed that nVH lesions lead to impairments in T maze delayed alternation and discrimination learning (Marquis et al. 2006). Impairments in T maze delayed alternation (Marquis et al. 2006) and in the radial arm maze (Chambers et al. 1996) were apparent at the juvenile age, suggesting early cognitive impairments in this model. Marquis et al. (2006) argue that the distinction of whether a given task is delay-dependent or not is critical to dissecting the relative contribution of mPFC and the hippocampus to working memory impairment. For example, lesions of the PFC impair performance when no delay is involved; whereas dorsal hippocampal lesions impair performance only when there are delays introduced (Winocur 1991). In several of the working memory tasks reviewed here, nVH lesions produced deficits independent of delay suggesting impaired PFC function. Additional probes of PFC function are set-shifting tasks, which probe cognitive flexibility or problem solving (Young et al. 2009). nVH lesions were associated with set-shifting deficits in the ASST (Marquis et al. 2008) and T maze set shifting (Brady 2009), with set-shifting impairments in the ASST occurring prepuberty.

Transient inactivation of the hippocampus during a critical developmental period produces many of the behavioral alterations observed in the nVH lesion model (Lipska 2004). Tetrodotoxin (TTX), which blocks voltage-gated sodium channels, infused into the VH on PND7 produced increased sensitivity to amphetamine and MK-801-induced locomotor activity in adulthood (Lipska et al. 2002b). These data suggest that the neonatal blockade during a critical period alters the development of hippocampal and related neurocircuits.

Important for the predictive validity of the nVH lesion neurodevelopmental model, some of the behavioral effects are reversed with antipsychotic drugs and putative antipsychotics. Antipsychotics blocked stress- and drug-induced hyperactivity (Al-Amin et al. 2000; Lipska and Weinberger 1994b; Rueter et al. 2004; Sams-Dodd et al. 1997) and PPI deficits (Le Pen and Moreau 2002; Rueter et al. 2004) but failed to block social impairments (Rueter et al. 2004; Sams-Dodd et al. 1997; but see Becker and Grecksch 2003) produced by nVH lesions. The nVH lesion model has also been used to test the efficacy of putative antipsychotics. For example, glycine and the glycine transporter inhibitor ORG24598 reverse PPI deficits associated with nVH lesions (Le Pen et al. 2003), and the AMPA antagonist LY293558 blocked MK-801-induced hyperactivity (Al-Amin et al. 2000). Thus, the nVH lesion model has shown predictive validity for schizophrenia pharmacotherapy.

4.1.2 Neonatal Ventral Hippocampal Lesion Model: Neuropathological Studies

In the outset of the model, the ventral hippocampus and subiculum were targeted because of the consistent alterations in hippocampus in schizophrenia patients and because the ventral hippocampus has important connections with the PFC and NAC (Lipska 2004). Indeed, lesions of the ventral hippocampus on PND7 result in many neuropathological changes that mirror many of the brain alterations observed in schizophrenia, particularly alterations in the PFC. O'Donnell, Tseng and colleagues have suggested that reorganization occurs within the PFC following nVH lesions (reviewed in O'Donnell et al. 2002; Tseng et al. 2009). These PFC changes are indicated by a postpubertal emergence of altered dopamine–glutamate interactions

(Tseng et al. 2007; Tseng and O'Donnell 2007). The inhibitory GABA interneuron system is dysregulated in response to nVH lesion. For example, several studies have shown decreased expression of GAD67 and PV in the PFC (Francois et al. 2009a; Lipska et al. 2003). Other studies, however, did not report changes in GAD67 or PV mRNA but did report abnormal responses to D2 stimulation in these interneurons (Tseng et al. 2008). While the development of a "noisy" circuit in the PFC may occur postpuberty, there is evidence of decreased dendritic spine densities in the PFC as early as PND36 (Marquis et al. 2008) and increased glucose metabolism as early as PND21 (Francois et al. 2009b). These changes in interneurons are supported by regional changes in $GABA_A$ receptor expression (Endo et al. 2007). Other neuroanatomical alterations include compromised neuronal function as evidenced by reduced NAA and glycogen synthase kinase-3β (GSK-3β), as well as reduced expression of BDNF mRNA, glutamate receptor GluR3, and glutamate transporter EAAC1 (reviewed in Lipska 2004).

4.1.3 Conclusions for nVH Lesion Model

While not modeling an epidemiological risk factor per se, the nVH model has proven extremely useful in preclinical animal studies of schizophrenia. Perhaps because it was one of the early neurodevelopmental models, there is a wealth of neuroanatomical and behavioral data on the model. While the cardinal measure of the nVH lesion model has been the postpubertal emergence of an increased sensitivity to amphetamine-induced locomotor activation, many other behavioral deficits are apparent including some that emerge prepuberty (e.g., social impairments, cognitive deficits). This prepubertal emergence of behavioral abnormalities may be very relevant to schizophrenia as early social and cognitive deficits continue to be recognized as symptoms occurring early in the progression of the disease. As reviewed here, the nVH lesion model has been used in preclinical drug development with some success. Additionally, changes in gene expression in PFC in the nVH lesion model may reveal new targets for schizophrenia genetic studies, thus offering a hypothesis-generating model for schizophrenia genetics (Wong et al. 2005). While there are many strengths to the nVH lesion model, there are also a few weaknesses. The procedure itself is technically challenging and often requires extra animals in the lesion group because some have to be excluded postmortem for improper placement or lack of a significant lesion. Many laboratories have, of course, successfully implemented the nVH lesion procedure as evidenced by the large number of publications on the topic. Additionally, recent advancements in MRI technology offer the ability to detect lesion size and regional extent *in vivo*, saving time and animals in a procedure that typically requires postmortem confirmation of lesion size and location (Bertrand et al. 2010). Thus, the nVH lesion model has shown face, predictive, and construct validity as an animal model of schizophrenia and is proving to be a useful tool in preclinical schizophrenia research.

4.2 Prenatal Toxin

Evidence of cytoarchitectural abnormalities in the brains of individuals with schizophrenia has led to the creation of models focused on disrupted neurogenesis (for a more complete review, see Lipska and Weinberger 2000; Lodge and Grace 2009). For example, rats exposed to the mitotic toxin methylazoxymethanol acetate (MAM) during gestation exhibit morphological abnormalities in brain regions implicated in schizophrenia (e.g., hippocampus, frontal, and entorhinal cortices; reviewed in Lodge and Grace 2009; Talamini et al. 1998, 1999) and behavioral abnormalities including deficits in PPI (Le Pen et al. 2006; Moore et al. 2006; with other reports of only modest effects, Talamini et al. 2000) and LI (Flagstad et al. 2005; Lodge et al. 2009). MAM-treated rats also show increased locomotor response to dopamine and glutamate psychostimulants (Lena et al. 2007; Moore et al. 2006) and impaired social interaction (Le Pen et al. 2006). The increased locomotor activity and PPI deficits emerged postpuberty (Le Pen et al. 2006; Moore et al. 2006). Gestational MAM also produces several cognitive deficits including impaired spatial recognition memory (Le Pen et al. 2006) and reversal learning (Flagstad et al. 2005; Moore et al. 2006). These behavioral differences produced by gestational MAM are associated with alterations in mesolimbic and mesocortical dopamine systems (Lodge and Grace 2007; Flagstad et al. 2004) and decreased PV interneuron number in the dorsolateral PFC and the hippocampus (Lodge et al. 2009; Penschuck et al. 2006). Experiments have administered MAM either during mid-gestation (E9–12) or late gestation (E17–18), with the more consistent behavioral effects occurring at the later gestational time point.

Other disruptors of neurogenesis have been evaluated for their effects on brain and behavioral development. For example, adult-onset PPI deficits are observed from disturbing neurogenesis with the antimitotic cytosine arabinocide (Ara-C) at embryonic days 19.5 and 20.5 (Elmer et al. 2004). Additionally, neonatal exposure of rats to the NOS inhibitor (L-nitroarginine) induces locomotor hypersensitivity to amphetamine and deficits in PPI (Black et al. 1999). Thus, gestational exposure to neurotoxins can have profound effects on brain development. Regarding its use as a neurodevelopmental model of schizophrenia, the MAM model has shown a certain degree of face and construct validity. Few studies, however, have tested the ability of antipsychotics or other compounds to reverse the behavioral effects. Thus, the pharmacological predictive validity of this model is yet to be determined.

4.3 Postnatal/Neonatal NMDA Antagonists

Accumulating evidence shows that perinatal NMDA-R antagonist exposures (pNM) can produce persistent behavioral and neurochemical deficits and the loss of PV interneurons (Andersen and Pouzet 2004; du Bois et al. 2008, 2009; Nakatani-Pawlak et al. 2009; Sircar and Rudy 1998; Stefani and Moghaddam 2005;

Wang et al. 2008; Wiley et al. 2003). Blockade of NMDA receptors in the postnatal period leads to a range of behavioral abnormalities relevant to schizophrenia from enhancement of exploration to impaired working memory in the delayed alternation task (reviewed in Mouri et al. 2007). Perinatal NMDA receptor antagonist exposure also leads to impairments in sensorimotor gating, spatial memory, social interaction behavior, and cognitive flexibility in adulthood (Boctor and Ferguson 2009; Broberg et al. 2008, 2009; Lei et al. 2009; Mouri et al. 2007; Secher et al. 2009; Wang et al. 2003). In addition to cognitive deficits, typical of schizophrenia, rats treated postnatally with NMDA receptor antagonists also showed higher level of fear exhibited in the elevated plus maze (Wedzony et al. 2008) and impairments in conditioned fear (Hunt 2006). A decrease in the number of PV-positive cells and spine density in the frontal cortex, NAC and hippocampus was also shown in both rats (Wang et al. 2008) and mice (Nakatani-Pawlak et al. 2009) when analyzed in adulthood. Oxidative mechanisms in this model were suggested by results showing that antioxidants can prevent the appearance of behavioral disruptions in adulthood (Wang et al. 2003).

5 Discussion

Over the last two decades, several neurodevelopmental animal models of schizophrenia have emerged to assess the pathophysiology associated with schizophrenia risk factors, the consequence of early brain insult, and the efficacy of putative antipsychotics. Many of these models reviewed here have convincingly shown face, construct, and predictive validity for schizophrenia. These models support the developmental hypothesis of schizophrenia by demonstrating similar behavioral and neuropathological abnormalities to those observed in the clinical condition. Many of these models converge on several key behavioral and neuropathological abnormalities. The two most common and consistent behavioral phenotypes are deficits in sensorimotor gating as measured by PPI and increased sensitivity to the locomotor-activating effects of amphetamine. The question is then whether these behavioral abnormalities emerge out of a common pathway or are they merely the most sensitive to multiple neurocircuit abnormalities? Increased mesolimbic dopamine activity is common to most of the models which exhibit PPI deficits and increased sensitivity to amphetamine. In addition to overactive mesolimbic dopamine, one of the most consistent findings in the models is that of decreased PV immunoreactivity in either the PFC or hippocampus, indicating abnormalities in the GABAergic inhibitory circuits that are critical to normal neuronal activity. Indeed, Behrens and Sejnowski argue that these GABA interneurons are slow to develop and are particularly sensitive to early environmental insult, particularly oxidative stress (Behrens and Sejnowski 2009).

The recent progress in developing and characterizing neurodevelopmental models, combined with the progress in genetic models of schizophrenia (reviewed by Young et al. this text), offer a unique opportunity to study gene–environment interactions. Additionally, neurodevelopmental models also offer the ability to explore the "2-hit" model described by Keshavan et al. (Keshavan and Hogarty 1999) in which maldevelopment during two critical time periods, early brain development and then adolescence, may lead to the development of schizophrenia. While animal models offer us the ability to probe in more depth the underlying pathology associated with an early environmental manipulation, *in vivo* animal imaging techniques such as MRI are emerging as a useful tool to evaluate structural changes in brain in the developing animal. Initial studies using MRI have been applied to social isolation rearing (Schubert et al. 2009), nVHs (Bertrand et al. 2010), and prenatal PolyI:C (Li et al. 2009; Piontkewitz et al. 2009). Future neurodevelopmental studies may benefit from this approach which allows for the determination of the ontogeny of structural abnormalities produced by the developmental manipulations.

Findings from neurodevelopmental animal models may also aid in our understanding of the processes leading to the development of schizophrenia during the prodromal phase of the disease and inform the debate on prophylactic treatments aimed at thwarting the progression to psychosis (Powell et al. 2003). Hence, several exciting pharmacological intervention studies have been conducted in which both antipsychotics and antidepressants have shown efficacy as preventive treatments in several neurodevelopmental models (Meyer et al. 2008d; Piontkewitz et al. 2009; Richtand et al. 2006). Specifically, risperidone administered from PND35–56 prevented amphetamine-induced locomotor sensitivity in the nVH lesion model (Richtand et al. 2006). In the PolyI:C model, periadolescent clozapine and fluoxetine (PND35–65) blocked the emergence of schizophrenia-like behavioral profile (PPI, LI, amphetamine-induced hyperlocomotion) in mice (Meyer et al. 2008d), and adolescent (PND34–47) administration of clozapine blocked LI deficits and amphetamine-induced hyperlocomotion in PolyI:C-exposed rats (Piontkewitz et al. 2009). Interestingly, the structural abnormalities observed with prenatal PolyI:C, enlarged lateral ventricles and reduced hippocampal volume as measured by MRI, were also prevented with adolescent clozapine treatment in this same study (Piontkewitz et al. 2009). Thus, neurodevelopmental models may aid in the debate on the efficacy and safety of early preventive treatments during the prodromal phase of illness.

As alluded to in the introductory sections, there are several experimental considerations to take into account when conducting developmental studies such as these, with the two primary concerns being cross-fostering and litter effects. Cross-fostering can be done in one of two ways. One method for determining the effect of the prenatal manipulation on maternal behavior is to give dams of each treatment group litters of both control and prenatally exposed neonates. The other method would be to use control lactating surrogate dams that had not been exposed to either treatment. Indeed, there is evidence in some of the neurodevelopmental models reviewed here that the mother–pup behavioral interaction accounts for some of the observed effects (Meyer et al. 2006c, 2008b). Litter effects can pose statistical problems for the analysis because of the interdependence of individual animals from one litter. Often, the entire litter is used and each animal is treated as an independent observation in the ANOVA. In order to avoid this problem of artificially inflating the sample size, one common practice is to sample only one animal from each litter for the experimental analyses. As Zorilla (Zorrilla 1997) argues, using a representative animal from each litter (or "two-stage sampling") also poses problems for statistical analyses when the within litter variability on a given measure is high. One solution is to average at least two pups per litter, with additional observations not necessarily adding much more power (Zorrilla 1997). Another option is to try to minimize litter effects in the design by using a stud for multiple dams assigned to different treatment groups or using inbred strains (Zorrilla 1997). One easy approach for postnatal manipulations is to randomize treatment conditions within litter. In addition to these experimental design considerations, the effect of litter can be handled somewhat at the analysis step. Covarying for litter in the ANOVA is an option, but it requires at least two littermate observations for each between-litter effect (i.e., two littermates per sex per treatment condition). Perhaps the best way to deal with litter effects is to combine the experimental design suggestions with a within litter statistical analysis in which litter is nested within the treatment condition (reviewed in Zorrilla 1997).

In conclusion, developmental models focused on epidemiological risk factors and neurodevelopmental anomalies have contributed to our understanding of the developing brain, the neuropathology of schizophrenia, and treatment approaches for this debilitating disease. Future studies in this area should continue to examine the 2-hit hypothesis of schizophrenia through the combination of genetic and environmental manipulations and early plus late environmental manipulations. The manipulations used to study secondary or "late hits" could be based on recent findings of putative risk factors that may increase the conversion to psychosis in high-risk individuals (Cannon et al. 2008), many of which occur during adolescence (e.g., social isolation, substance abuse, etc.). Lastly, these models are uniquely suited for epigenetic studies aimed at determining the mechanism by which these manipulations exert long-term effects on brain and behavioral development.

References

Addington J, Addington D, Gasbarre L (1997) Distractibility and symptoms in schizophrenia. J Psychiatry Neurosci 22:180–184

Addington J, Penn D, Woods SW, Addington D, Perkins DO (2008) Social functioning in individuals at clinical high risk for psychosis. Schizophr Res 99:119–124

Aisa B, Tordera R, Lasheras B, Del Rio J, Ramirez MJ (2007) Cognitive impairment associated to HPA axis hyperactivity after maternal separation in rats. Psychoneuroendocrinology 32:256–266

Aisa B, Tordera R, Lasheras B, Del Rio J, Ramirez MJ (2008) Effects of maternal separation on hypothalamic-pituitary-adrenal responses, cognition and vulnerability to stress in adult female rats. Neuroscience 154:1218–1226

Al-Amin HA, Weinberger DR, Lipska BK (2000) Exaggerated MK-801-induced motor hyperactivity in rats with the neonatal lesion of the ventral hippocampus. Behav Pharmacol 11:269–278

Al-Amin HA, Shannon Weickert C, Weinberger DR, Lipska BK (2001) Delayed onset of enhanced MK-801-induced motor hyperactivity after neonatal lesions of the rat ventral hippocampus. Biol Psychiatry 49:528–539

Almeras L, Eyles D, Benech P, Laffite D, Villard C, Patatian A, Boucraut J, Mackay-Sim A, McGrath J, Feron F (2007) Developmental vitamin D deficiency alters brain protein expression in the adult rat: implications for neuropsychiatric disorders. Proteomics 7:769–780

Andersen JD, Pouzet B (2004) Spatial memory deficits induced by perinatal treatment of rats with PCP and reversal effect of D-serine. Neuropsychopharmacology 29:1080–1090

Arnold SE, Hyman BT, Van Hoesen GW, Damasio AR (1991) Some cytoarchitectural abnormalities of the entorhinal cortex in schizophrenia. Arch Gen Psychiatry 48:625–632

Arnold SE, Ruscheinsky DD, Han LY (1997) Further evidence of abnormal cytoarchitecture of the entorhinal cortex in schizophrenia using spatial point pattern analyses. Biol Psychiatry 42:639–647

Ashdown H, Dumont Y, Ng M, Poole S, Boksa P, Luheshi GN (2006) The role of cytokines in mediating effects of prenatal infection on the fetus: implications for schizophrenia. Mol Psychiatry 11:47–55

Asp L, Holtze M, Powell SB, Karlsson H, Erhardt S (2009) Neonatal infection with neurotropic influenza A virus induces the kynurenine pathway in early life and disrupts sensorimotor gating in adult Tap1-/- mice. Int J Neuropsychopharmacol 17:1–11

Ayhan Y, Sawa A, Ross CA, Pletnikov MV (2009) Animal models of gene–environment interactions in schizophrenia. Behav Brain Res 204:274–281

Baharnoori M, Brake WG, Srivastava LK (2009) Prenatal immune challenge induces developmental changes in the morphology of pyramidal neurons of the prefrontal cortex and hippocampus in rats. Schizophr Res 107:99–109

Bakshi VP, Swerdlow NR, Braff DL, Geyer MA (1998) Reversal of isolation rearing-induced deficits in prepulse inhibition by Seroquel and olanzapine. Biol Psychiatry 43:436–445

Barr AM, Powell SB, Markou A, Geyer MA (2006) Iloperidone reduces sensorimotor gating deficits in pharmacological models, but not a developmental model, of disrupted prepulse inhibition in rats. Neuropharmacology 51:457–465

Beasley CL, Zhang ZJ, Patten I, Reynolds GP (2002) Selective deficits in prefrontal cortical GABAergic neurons in schizophrenia defined by the presence of calcium-binding proteins. Biol Psychiatry 52:708–715

Becker RF, Donnell W (1952) Learning behavior in guinea pigs subjected to asphyxia at birth. J Comp Physiol Psychol 45:153–162

Becker A, Grecksch G (2000) Social memory is impaired in neonatally ibotenic acid lesioned rats. Behav Brain Res 109:137–140

Becker A, Grecksch G (2003) Haloperidol and clozapine affect social behaviour in rats postnatally lesioned in the ventral hippocampus. Pharmacol Biochem Behav 76:1–8

Becker A, Grecksch G, Bernstein HG, Hollt V, Bogerts B (1999) Social behaviour in rats lesioned with ibotenic acid in the hippocampus: quantitative and qualitative analysis. Psychopharmacology (Berl) 144:333–338

Becker A, Eyles DW, McGrath JJ, Grecksch G (2005) Transient prenatal vitamin D deficiency is associated with subtle alterations in learning and memory functions in adult rats. Behav Brain Res 161:306–312

Behrens MM, Sejnowski TJ (2009) Does schizophrenia arise from oxidative dysregulation of parvalbumin-interneurons in the developing cortex? Neuropharmacology 57:193–200

Bertolino A, Weinberger DR (1999) Proton magnetic resonance spectroscopy in schizophrenia. Eur J Radiol 30:132–141

Bertrand JB, Langlois JB, Begou M, Volle J, Brun P, d'Amato T, Saoud M, Suaud-Chagny MF (2010) Longitudinal MRI monitoring of brain damage in the neonatal ventral hippocampal lesion rat model of schizophrenia. Hippocampus 20:264–278

Bianchi M, Fone KF, Azmi N, Heidbreder CA, Hagan JJ, Marsden CA (2006) Isolation rearing induces recognition memory deficits accompanied by cytoskeletal alterations in rat hippocampus. Eur J Neurosci 24:2894–2902

Black MD, Selk DE, Hitchcock JM, Wettstein JG, Sorensen SM (1999) On the effect of neonatal nitric oxide synthase inhibition in rats: a potential neurodevelopmental model of schizophrenia. Neuropharmacology 38:1299–1306

Blanc G, Herve D, Simon H, Lisoprawski A, Glowinski J, Tassin JP (1980) Response to stress of mesocortico-frontal dopaminergic neurones in rats after long-term isolation. Nature 284:265–267

Boctor SY, Ferguson SA (2009) Neonatal NMDA receptor antagonist treatments have no effects on prepulse inhibition of postnatal day 25 Sprague-Dawley rats. Neurotoxicology 30:151–154

Boksa P (2004) Animal models of obstetric complications in relation to schizophrenia. Brain Res Brain Res Rev 45:1–17

Boksa P, El-Khodor BF (2003) Birth insult interacts with stress at adulthood to alter dopaminergic function in animal models: possible implications for schizophrenia and other disorders. Neurosci Biobehav Rev 27:91–101

Boksa P, Krishnamurthy A, Brooks W (1995) Effects of a period of asphyxia during birth on spatial learning in the rat. Pediatr Res 37:489–496

Boksa P, Zhang Y, Bestawros A (2002) Dopamine D1 receptor changes due to caesarean section birth: effects of anesthesia, developmental time course, and functional consequences. Exp Neurol 175:388–397

Borrell J, Vela JM, Arevalo-Martin A, Molina-Holgado E, Guaza C (2002) Prenatal immune challenge disrupts sensorimotor gating in adult rats. Implications for the etiopathogenesis of schizophrenia. Neuropsychopharmacology 26:204–215

Bowling SL, Bardo MT (1994) Locomotor and rewarding effects of amphetamine in enriched, social, and isolate reared rats. Pharmacol Biochem Behav 48:459–464

Boyd JH, Pulver AE, Stewart W (1986) Season of birth: schizophrenia and bipolar disorder. Schizophr Bull 12:173–186

Brady AM (2009) Neonatal ventral hippocampal lesions disrupt set-shifting ability in adult rats. Behav Brain Res 205:294–298

Braff DL (2010) Prepulse inhibition (PPI) of the startle reflex: A window on the brain in schizophrenia. In: Swerdlow N (ed) Behavioral Neurobiology of Schizophrenia and Its Treatment. Curr Topics in Behav Neurosci. Springer, Heidelberg

Braff DL, Geyer MA, Swerdlow NR (2001) Human studies of prepulse inhibition of startle: normal subjects, patient groups, and pharmacological studies. Psychopharmacology (Berl) 156:234–258

Brake WG, Boksa P, Gratton A (1997) Effects of perinatal anoxia on the acute locomotor response to repeated amphetamine administration in adult rats. Psychopharmacology (Berl) 133:389–395

Brake WG, Sullivan RM, Gratton A (2000) Perinatal distress leads to lateralized medial prefrontal cortical dopamine hypofunction in adult rats. J Neurosci 20:5538–5543

Broberg BV, Dias R, Glenthoj BY, Olsen CK (2008) Evaluation of a neurodevelopmental model of schizophrenia – early postnatal PCP treatment in attentional set-shifting. Behav Brain Res 190:160–163

Broberg BV, Glenthoj BY, Dias R, Larsen DB, Olsen CK (2009) Reversal of cognitive deficits by an ampakine (CX516) and sertindole in two animal models of schizophrenia – sub-chronic and early postnatal PCP treatment in attentional set-shifting. Psychopharmacology (Berl) 206:631–640

Brown AS, Susser ES (2002) In utero infection and adult schizophrenia. Ment Retard Dev Disabil Res Rev 8:51–57

Brown AS, Susser ES (2008) Prenatal nutritional deficiency and risk of adult schizophrenia. Schizophr Bull 34:1054–1063

Brown AS, Begg MD, Gravenstein S, Schaefer CA, Wyatt RJ, Bresnahan M, Babulas VP, Susser ES (2004a) Serologic evidence of prenatal influenza in the etiology of schizophrenia. Arch Gen Psychiatry 61:774–780

Brown AS, Hooton J, Schaefer CA, Zhang H, Petkova E, Babulas V, Perrin M, Gorman JM, Susser ES (2004b) Elevated maternal interleukin-8 levels and risk of schizophrenia in adult offspring. Am J Psychiatry 161:889–895

Brown AS, Schaefer CA, Quesenberry CP Jr, Liu L, Babulas VP, Susser ES (2005) Maternal exposure to toxoplasmosis and risk of schizophrenia in adult offspring. Am J Psychiatry 162:767–773

Bruins Slot LA, Kleven MS, Newman-Tancredi A (2005) Effects of novel antipsychotics with mixed D(2) antagonist/5-HT(1A) agonist properties on PCP-induced social interaction deficits in the rat. Neuropharmacology 49:996–1006

Buka SL, Tsuang MT, Torrey EF, Klebanoff MA, Wagner RL, Yolken RH (2001) Maternal cytokine levels during pregnancy and adult psychosis. Brain Behav Immun 15:411–420

Burne TH, Becker A, Brown J, Eyles DW, Mackay-Sim A, McGrath JJ (2004a) Transient prenatal vitamin D deficiency is associated with hyperlocomotion in adult rats. Behav Brain Res 154:549–555

Burne TH, Feron F, Brown J, Eyles DW, McGrath JJ, Mackay-Sim A (2004b) Combined prenatal and chronic postnatal vitamin D deficiency in rats impairs prepulse inhibition of acoustic startle. Physiol Behav 81:651–655

Burne TH, O'Loan J, McGrath JJ, Eyles DW (2006) Hyperlocomotion associated with transient prenatal vitamin D deficiency is ameliorated by acute restraint. Behav Brain Res 174:119–124

Cannon M, Jones PB, Murray RM (2002) Obstetric complications and schizophrenia: historical and meta-analytic review. Am J Psychiatry 159:1080–1092

Cannon TD, van Erp TGM, Bearden CE, Loewy R, Thompson P, Toga AW, Huttunnen MO, Keshavan MS, Seidman LJ, Tsuang MT (2003) Early and late neurodevelopmental influences in the prodrome to schizophrenia: contributions of genes, environment, and their interactions. Schizophr Bull 29:653–669

Cannon TD, Hennah W, van Erp TG, Thompson PM, Lonnqvist J, Huttunen M, Gasperoni T, Tuulio-Henriksson A, Pirkola T, Toga AW, Kaprio J, Mazziotta J, Peltonen L (2005) Association of DISC1/TRAX haplotypes with schizophrenia, reduced prefrontal gray matter, and impaired short- and long-term memory. Arch Gen Psychiatry 62:1205–1213

Cannon TD, Cadenhead K, Cornblatt B, Woods SW, Addington J, Walker E, Seidman LJ, Perkins D, Tsuang M, McGlashan T, Heinssen R (2008) Prediction of psychosis in youth at high clinical risk: a multisite longitudinal study in North America. Arch Gen Psychiatry 65:28–37

Chambers RA, Moore J, McEvoy JP, Levin ED (1996) Cognitive effects of neonatal hippocampal lesions in a rat model of schizophrenia. Neuropsychopharmacology 15:587–594

Chudasama Y, Robbins TW (2004) Psychopharmacological approaches to modulating attention in the five-choice serial reaction time task: implications for schizophrenia. Psychopharmacology (Berl) 174:86–98

Cilia J, Reavill C, Hagan JJ, Jones DN (2001) Long-term evaluation of isolation-rearing induced prepulse inhibition deficits in rats. Psychopharmacology (Berl) 156:327–337

Cilia J, Hatcher PD, Reavill C, Jones DN (2005) Long-term evaluation of isolation-rearing induced prepulse inhibition deficits in rats: an update. Psychopharmacology (Berl) 180:57–62

Clancy B, Darlington RB, Finlay BL (2001) Translating developmental time across mammalian species. Neuroscience 105:7–17

Clancy B, Finlay BL, Darlington RB, Anand KJ (2007) Extrapolating brain development from experimental species to humans. Neurotoxicology 28:931–937

Crawley JN (2007) Mouse behavioral assays relevant to the symptoms of autism. Brain Pathol 17:448–459

Cui X, McGrath JJ, Burne TH, Mackay-Sim A, Eyles DW (2007) Maternal vitamin D depletion alters neurogenesis in the developing rat brain. Int J Dev Neurosci 25:227–232

Da Silva NL, Ferreira VM, Carobrez Ade P, Morato GS (1996) Individual housing from rearing modifies the performance of young rats on the elevated plus-maze apparatus. Physiol Behav 60:1391–1396

Dai H, Okuda H, Iwabuchi K, Sakurai E, Chen Z, Kato M, Iinuma K, Yanai K (2004) Social isolation stress significantly enhanced the disruption of prepulse inhibition in mice repeatedly treated with methamphetamine. Ann N Y Acad Sci 1025:257–266

Day-Wilson KM, Jones DNC, Southam E, Cilia J, Totterdell S (2006) Medial prefrontal cortex volume loss in rats with isolation rearing-induced deficits in prepulse inhibition of acoustic startle. Neuroscience 141:1113–1121

Dean K, Bramon E, Murray RM (2003) The causes of schizophrenia: neurodevelopment and other risk factors. J Psychiatr Pract 9:442–454

Del-Bel EA, Joca SR, Padovan CM, Guimaraes FS (2002) Effects of isolation-rearing on serotonin-1A and M1-muscarinic receptor messenger RNA expression in the hipocampal formation of rats. Neurosci Lett 332:123–126

Dieni S, Rees S (2003) Dendritic morphology is altered in hippocampal neurons following prenatal compromise. J Neurobiol 55:41–52

Dieni S, Rees S (2005) BDNF and TrkB protein expression is altered in the fetal hippocampus but not cerebellum after chronic prenatal compromise. Exp Neurol 192:265–273

du Bois TM, Huang XF, Deng C (2008) Perinatal administration of PCP alters adult behaviour in female Sprague-Dawley rats. Behav Brain Res 188:416–419

du Bois TM, Deng C, Han M, Newell KA, Huang XF (2009) Excitatory and inhibitory neurotransmission is chronically altered following perinatal NMDA receptor blockade. Eur Neuropsychopharmacol 19:256–265

Eastvold AD, Heaton RK, Cadenhead KS (2007) Neurocognitive deficits in the (putative) prodrome and first episode of psychosis. Schizophr Res 93:266–277

Eastwood SL, Harrison PJ (1995) Decreased synaptophysin in the medial temporal lobe in schizophrenia demonstrated using immunoautoradiography. Neuroscience 69:339–343

Einon DF, Morgan MJ (1977) A critical period for social isolation in the rat. Dev Psychobiol 10:123–132

El-Khodor BF, Boksa P (1997) Long-term reciprocal changes in dopamine levels in prefrontal cortex versus nucleus accumbens in rats born by Caesarean section compared to vaginal birth. Exp Neurol 145:118–129

El-Khodor BF, Boksa P (1998) Birth insult increases amphetamine-induced behavioral responses in the adult rat. Neuroscience 87:893–904

El-Khodor BF, Boksa P (2000) Transient birth hypoxia increases behavioral responses to repeated stress in the adult rat. Behav Brain Res 107:171–175

Ellenbroek BA, Cools AR (2000) The long-term effects of maternal deprivation depend on the genetic background. Neuropsychopharmacology 23:99–106

Ellenbroek BA, van den Kroonenberg PT, Cools AR (1998) The effects of an early stressful life event on sensorimotor gating in adult rats. Schizophr Res 30:251–260

Elmer GI, Sydnor J, Guard H, Hercher E, Vogel MW (2004) Altered prepulse inhibition in rats treated prenatally with the antimitotic Ara-C: an animal model for sensorimotor gating deficits in schizophrenia. Psychopharmacology 174:177–189

Endo K, Hori T, Abe S, Asada T (2007) Alterations in GABA(A) receptor expression in neonatal ventral hippocampal lesioned rats: comparison of prepubertal and postpubertal periods. Synapse 61:357–366

Engelmann M, Wotjak CT, Landgraf R (1995) Social discrimination procedure: an alternative method to investigate juvenile recognition abilities in rats. Physiol Behav 58:315–321

Eyles D, Brown J, Mackay-Sim A, McGrath J, Feron F (2003) Vitamin D3 and brain development. Neuroscience 118:641–653

Eyles D, Almeras L, Benech P, Patatian A, Mackay-Sim A, McGrath J, Feron F (2007) Developmental vitamin D deficiency alters the expression of genes encoding mitochondrial, cytoskeletal and synaptic proteins in the adult rat brain. J Steroid Biochem Mol Biol 103:538–545

Eyles DW, Feron F, Cui X, Kesby JP, Harms LH, Ko P, McGrath JJ, Burne TH (2009) Developmental vitamin D deficiency causes abnormal brain development. Psychoneuroendocrinology 34(Suppl 1):S247–S257

Falkai P, Schneider-Axmann T, Honer WG (2000) Entorhinal cortex pre-alpha cell clusters in schizophrenia: quantitative evidence of a developmental abnormality. Biol Psychiatry 47:937–943

Fatemi SH, Folsom TD (2009) The neurodevelopmental hypothesis of schizophrenia, revisited. Schizophr Bull 35:528–548

Fatemi SH, Sidwell R, Akhter P, Sedgewick J, Thuras P, Bailey K, Kist D (1998) Human influenza viral infection in utero increases nNOS expression in hippocampi of neonatal mice. Synapse 29:84–88

Fatemi SH, Emamian ES, Kist D, Sidwell RW, Nakajima K, Akhter P, Shier A, Sheikh S, Bailey K (1999) Defective corticogenesis and reduction in Reelin immunoreactivity in cortex and hippocampus of prenatally infected neonatal mice. Mol Psychiatry 4:145–154

Fatemi SH, Earle J, Kanodia R, Kist D, Emamian ES, Patterson PH, Shi L, Sidwell R (2002a) Prenatal viral infection leads to pyramidal cell atrophy and macrocephaly in adulthood: implications for genesis of autism and schizophrenia. Cell Mol Neurobiol 22:25–33

Fatemi SH, Emamian ES, Sidwell RW, Kist DA, Stary JM, Earle JA, Thuras P (2002b) Human influenza viral infection in utero alters glial fibrillary acidic protein immunoreactivity in the developing brains of neonatal mice. Mol Psychiatry 7:633–640

Fatemi SH, Araghi-Niknam M, Laurence JA, Stary JM, Sidwell RW, Lee S (2004) Glial fibrillary acidic protein and glutamic acid decarboxylase 65 and 67 kDa proteins are increased in brains of neonatal BALB/c mice following viral infection in utero. Schizophr Res 69:121–123

Fatemi SH, Pearce DA, Brooks AI, Sidwell RW (2005) Prenatal viral infection in mouse causes differential expression of genes in brains of mouse progeny: a potential animal model for schizophrenia and autism. Synapse 57:91–99

Feldon J, Avnimelech-Gigus N, Weiner I (1990) The effects of pre- and postweaning rearing conditions on latent inhibition and partial reinforcement extinction effect in male rats. Behav Neural Biol 53:189–204

Fendt M, Lex A, Falkai P, Henn FA, Schmitt A (2008) Behavioural alterations in rats following neonatal hypoxia and effects of clozapine: implications for schizophrenia. Pharmacopsychiatry 41:138–145

Ferguson JN, Young LJ, Insel TR (2002) The neuroendocrine basis of social recognition. Front Neuroendocrinol 23:200–224

Feron F, Burne TH, Brown J, Smith E, McGrath JJ, Mackay-Sim A, Eyles DW (2005) Developmental vitamin D3 deficiency alters the adult rat brain. Brain Res Bull 65:141–148

Finamore TL, Port RL (2000) Developmental stress disrupts habituation but spares prepulse inhibition in young rats. Physiol Behav 69:527–530

Finlay BL, Darlington RB (1995) Linked regularities in the development and evolution of mammalian brains. Science 268:1578–1584

Flagstad P, Mork A, Glenthoj BY, van Beek J, Michael-Titus AT, Didriksen M (2004) Disruption of neurogenesis on gestational day 17 in the rat causes behavioral changes relevant to positive and negative schizophrenia symptoms and alters amphetamine-induced dopamine release in nucleus accumbens. Neuropsychopharmacology 29:2052–2064

Flagstad P, Glenthoj BY, Didriksen M (2005) Cognitive deficits caused by late gestational disruption of neurogenesis in rats: a preclinical model of schizophrenia. Neuropsychopharmacology 30:250–260

Flores G, Silva-Gomez AB, Ibanez O, Quirion R, Srivastava LK (2005) Comparative behavioral changes in postpubertal rats after neonatal excitotoxic lesions of the ventral hippocampus and the prefrontal cortex. Synapse 56:147–153

Floresco SB, Zhang Y, Enomoto T (2009) Neural circuits subserving behavioral flexibility and their relevance to schizophrenia. Behav Brain Res 204:396–409

Fone KCF, Porkess MV (2008) Behavioural and neurochemical effects of post-weaning social isolation in rodents – relevance to developmental neuropsychiatric disorders. Neurosci Biobehav Rev 32:1087–1102

Fortier ME, Joober R, Luheshi GN, Boksa P (2004) Maternal exposure to bacterial endotoxin during pregnancy enhances amphetamine-induced locomotion and startle responses in adult rat offspring. J Psychiatr Res 38:335–345

Fortier ME, Luheshi GN, Boksa P (2007) Effects of prenatal infection on prepulse inhibition in the rat depend on the nature of the infectious agent and the stage of pregnancy. Behav Brain Res 181:270–277

Francois J, Ferrandon A, Koning E, Angst MJ, Sandner G, Nehlig A (2009a) Selective reorganization of GABAergic transmission in neonatal ventral hippocampal-lesioned rats. Int J Neuropsychopharmacol 12:1097–1110

Francois J, Koning E, Ferrandon A, Sandner G, Nehlig A (2009b) Metabolic activity in the brain of juvenile and adult rats with a neonatal ventral hippocampal lesion. Hippocampus (Epub ahead of print, July 31)

Frankle WG, Lerma J, Laruelle M (2003) The synaptic hypothesis of schizophrenia. Neuron 39:205–216

Freedman R, Adler LE, Leonard S (1999) Alternative phenotypes for the complex genetics of schizophrenia. Biol Psychiatry 45:551–558

Friston KJ, Frith CD (1995) Schizophrenia: a disconnection syndrome? Clin Neurosci 3:89–97

Geyer MA, Markou A (2002) Animal models of psychiatric disorders. In: Watson S (ed) Psychopharmacology: the fourth generation of progress CD-ROM version 3. Lippincott Williams & Wilkins, Philadelphia

Geyer MA, Moghaddam B (2002) Animal models relevant to schizophrenia disorders. In: Charney D, Coyle J, Davis K, Nemeroff CB (eds) Neuropsychopharmacology: the fifth generation of progress. Lippincott Williams & Wilkins, New York, pp 689–701

Geyer MA, Wilkinson LS, Humby T, Robbins TW (1993) Isolation rearing of rats produces a deficit in prepulse inhibition of acoustic startle similar to that in schizophrenia. Biol Psychiatry 34:361–372

Geyer MA, Krebs-Thomson K, Braff DL, Swerdlow NR (2001) Pharmacological studies of prepulse inhibition models of sensorimotor gating deficits in schizophrenia: a decade in review. Psychopharmacology (Berl) 156:117–154

Grecksch G, Rüthrich H, Höllt V, Becker A (2009) Transient prenatal vitamin D deficiency is associated with changes of synaptic plasticity in the dentate gyrus in adult rats. Psychoneuroendocrinology 34S1:S258–S264

Green MF, Nuechterlein KH, Gold JM, Barch DM, Cohen J, Essock S, Fenton WS, Frese F, Goldberg TE, Heaton RK, Keefe RS, Kern RS, Kraemer H, Stover E, Weinberger DR, Zalcman S, Marder SR (2004) Approaching a consensus cognitive battery for clinical trials in schizophrenia: the NIMH-MATRICS conference to select cognitive domains and test criteria. Biol Psychiatry 56:301–307

Gue M, Bravard A, Meunier J, Veyrier R, Gaillet S, Recasens M, Maurice T (2004) Sex differences in learning deficits induced by prenatal stress in juvenile rats. Behav Brain Res 150:149–157

Hagan JJ, Jones DN (2005) Predicting drug efficacy for cognitive deficits in schizophrenia. Schizophr Bull 31:830–853

Hall FS (1998) Social deprivation of neonatal, adolescent, and adult rats has distinct neurochemical and behavioral consequences. Crit Rev Neurobiol 12:129–162

Hall FS, Humby T, Wilkinson LS, Robbins TW (1997) The effects of isolation-rearing on preference by rats for a novel environment. Physiol Behav 62:299–303

Hall FS, Wilkinson LS, Humby T, Inglis W, Kendall DA, Marsden CA, Robbins TW (1998) Isolation rearing in rats: pre- and postsynaptic changes in striatal dopaminergic systems. Pharmacol Biochem Behav 59:859–872

Harms LR, Eyles DW, McGrath JJ, Mackay-Sim A, Burne TH (2008) Developmental vitamin D deficiency alters adult behaviour in 129/SvJ and C57BL/6J mice. Behav Brain Res 187:343–350

Harrison PJ, Weinberger DR (2005) Schizophrenia genes, gene expression, and neuropathology: on the matter of their convergence. Mol Psychiatry 10:40–68, image 5

Harte MK, Powell SB, Swerdlow NR, Geyer MA, Reynolds GP (2007) Deficits in parvalbumin and calbindin immunoreactive cells in the hippocampus of isolation reared rats. J Neural Transm 114:893–898

Holmes A, le Guisquet AM, Vogel E, Millstein RA, Leman S, Belzung C (2005) Early life genetic, epigenetic and environmental factors shaping emotionality in rodents. Neurosci Biobehav Rev 29:1335–1346

Holtze M, Asp L, Schwieler L, Engberg G, Karlsson H (2008) Induction of the kynurenine pathway by neurotropic influenza A virus infection. J Neurosci Res 86:3674–3683

Huang Z, Liu J, Cheung P-Y, Chen C (2009) Long-term cognitive impairment and myelination deficiency in a rat model of perinatal hypoxic-ischemic brain injury. Brain Res 1301:100–109

Hultman CM, Ohman A, Cnattingius S, Wieselgren IM, Lindstrom LH (1997) Prenatal and neonatal risk factors for schizophrenia. Br J Psychiatry 170:128–133

Hunt PS (2006) Neonatal treatment with a competitive NMDA antagonist results in response-specific disruption of conditioned fear in preweanling rats. Psychopharmacology (Berl) 185:179–187

Huot RL, Plotsky PM, Lenox RH, McNamara RK (2002) Neonatal maternal separation reduces hippocampal mossy fiber density in adult Long Evans rats. Brain Res 950:52–63

Ibi D, Takuma K, Koike H, Mizoguchi H, Tsuritani K, Kuwahara Y, Kamei H, Nagai T, Yoneda Y, Nabeshima T, Yamada K (2008) Social isolation rearing-induced impairment of the hippocampal neurogenesis is associated with deficits in spatial memory and emotion-related behaviors in juvenile mice. J Neurochem 105:921–932

Ibi D, Nagai T, Kitahara Y, Mizoguchi H, Koike H, Shiraki A, Takuma K, Kamei H, Noda Y, Nitta A, Nabeshima T, Yoneda Y, Yamada K (2009) Neonatal polyI:C treatment in mice results in schizophrenia-like behavioral and neurochemical abnormalities in adulthood. Neurosci Res 64:297–305

Ibi D, Nagai T, Koike H, Kitahara Y, Mizoguchi H, Niwa M, Jaaro-Peled H, Nitta A, Yoneda Y, Nabeshima T, Sawa A, Yamada K (2010) Combined effect of neonatal immune activation and mutant DISC1 on phenotypic changes in adulthood. Behav Brain Res 206:32–37

Jakob H, Beckmann H (1986) Prenatal developmental disturbances in the limbic allocortex in schizophrenics. J Neural Transm 65:303–326

Jenkins TA, Harte MK, Stenson G, Reynolds GP (2009) Neonatal lipopolysaccharide induces pathological changes in parvalbumin immunoreactivity in the hippocampus of the rat. Behav Brain Res 205:355–359

Johnstone EC, Ebmeier KP, Miller P, Owens DG, Lawrie SM (2005) Predicting schizophrenia: findings from the Edinburgh High-Risk Study. Br J Psychiatry 186:18–25

Jones GH, Robbins TW, Marsden CA (1989) Isolation-rearing retards the acquisition of schedule-induced polydipsia in rats. Physiol Behav 45:71–77

Jones GH, Marsden CA, Robbins TW (1990) Increased sensitivity to amphetamine and reward-related stimuli following social isolation in rats: possible disruption of dopamine-dependent mechanisms of the nucleus accumbens. Psychopharmacology (Berl) 102:364–372

Jones GH, Hernandez TD, Kendall DA, Marsden CA, Robbins TW (1992) Dopaminergic and serotonergic function following isolation rearing in rats: study of behavioural responses and postmortem and *in vivo* neurochemistry. Pharmacol Biochem Behav 43:17–35

Jones DNC, Gartlon JE, Minassian A, Perry W, Geyer MA (2008) Developing new drugs for schizophrenia: from animals to the clinic. In: McArthur R, Borsini R (eds) Animal and translational models for CNS drug discovery: psychiatric disorders. Elsevier, New York, pp 199–262

Juarez I, De La Cruz F, Zamudio S, Flores G (2005) Cesarean plus anoxia at birth induces hyperresponsiveness to locomotor activity by dopamine D2 agonist. Synapse 58:236–242

Kapoor A, Kostaki A, Janus C, Matthews SG (2009) The effects of prenatal stress on learning in adult offspring is dependent on the timing of the stressor. Behav Brain Res 197:144–149

Kendell RE, Adams W (2002) Exposure to sunlight, vitamin D and schizophrenia. Schizophr Res 54:193–198

Kesby JP, Burne TH, McGrath JJ, Eyles DW (2006) Developmental vitamin D deficiency alters MK 801-induced hyperlocomotion in the adult rat: an animal model of schizophrenia. Biol Psychiatry 60:591–596

Keshavan MS, Hogarty GE (1999) Brain maturational processes and delayed onset in schizophrenia. Dev Psychopathol 11:525–543

Khashan AS, Abel KM, McNamee R, Pedersen MG, Webb RT, Baker PN, Kenny LC, Mortensen PB (2008) Higher risk of offspring schizophrenia following antenatal maternal exposure to severe adverse life events. Arch Gen Psychiatry 65:146–152

Kietzman M, Spring B, Zubin J (1985) Perception, cognition, and information processing. In: Kaplan H, Sadock B (eds) Comprehensive textbook of psychiatry IV. Williams & Williams, Baltimore

Koenig JI (2006) Schizophrenia: a unique translational opportunity in behavioral neuroendocrinology. Horm Behav 50:602–611

Koenig JI, Kirkpatrick B, Lee P (2002) Glucocorticoid hormones and early brain development in schizophrenia. Neuropsychopharmacology 27:309–318

Koenig JI, Elmer GI, Shepard PD, Lee PR, Mayo C, Joy B, Hercher E, Brady DL (2005) Prenatal exposure to a repeated variable stress paradigm elicits behavioral and neuroendocrinological changes in the adult offspring: potential relevance to schizophrenia. Behav Brain Res 156:251–261

Krech D, Rosenzweig MR, Bennett EL (1962) Relations between chemistry and problem-solving among rats raised in enriched and impoverished environments. J Comp Physiol Psychol 55:801–807

Lapiz MD, Mateo Y, Parker T, Marsden C (2000) Effects of noradrenaline depletion in the brain on response on novelty in isolation-reared rats. Psychopharmacology (Berl) 152:312–320

Laruelle M, Abi-Dargham A, van Dyck CH, Gil R, D'Souza CD, Erdos J, McCance E, Rosenblatt W, Fingado C, Zoghbi SS, Baldwin RM, Seibyl JP, Krystal JH, Charney DS, Innis RB (1996) Single photon emission computerized tomography imaging of amphetamine-induced dopamine release in drug-free schizophrenic subjects. Proc Natl Acad Sci USA 93:9235–9240

Le Pen G, Moreau JL (2002) Disruption of prepulse inhibition of startle reflex in a neurodevelopmental model of schizophrenia: reversal by clozapine, olanzapine and risperidone but not by haloperidol. Neuropsychopharmacology 27:1–11

Le Pen G, Grottick AJ, Higgins GA, Martin JR, Jenck F, Moreau JL (2000) Spatial and associative learning deficits induced by neonatal excitotoxic hippocampal damage in rats: further evaluation of an animal model of schizophrenia. Behav Pharmacol 11:257–268

Le Pen G, Kew J, Alberati D, Borroni E, Heitz MP, Moreau JL (2003) Prepulse inhibition deficits of the startle reflex in neonatal ventral hippocampal-lesioned rats: reversal by glycine and a glycine transporter inhibitor. Biol Psychiatry 54:1162–1170

Le Pen G, Gourevitch R, Hazane F, Hoareau C, Jay TM, Krebs MO (2006) Peri-pubertal maturation after developmental disturbance: a model for psychosis onset in the rat. Neuroscience 143:395–405

Lee PR, Brady DL, Shapiro RA, Dorsa DM, Koenig JI (2007) Prenatal stress generates deficits in rat social behavior: reversal by oxytocin. Brain Res 1156:152–167

Lehmann J, Stohr T, Feldon J (2000) Long-term effects of prenatal stress experiences and postnatal maternal separation on emotionality and attentional processes. Behav Brain Res 107:133–144

Lei G, Anastasio NC, Fu Y, Neugebauer V, Johnson KM (2009) Activation of dopamine D1 receptors blocks phencyclidine-induced neurotoxicity by enhancing N-methyl-D-aspartate receptor-mediated synaptic strength. J Neurochem 109:1017–1030

Lemaire V, Koehl M, Le Moal M, Abrous DN (2000) Prenatal stress produces learning deficits associated with an inhibition of neurogenesis in the hippocampus. Proc Natl Acad Sci USA 97:11032–11037

Lena I, Chessel A, Le Pen G, Krebs MO, Garcia R (2007) Alterations in prefrontal glutamatergic and noradrenergic systems following MK-801 administration in rats prenatally exposed to methylazoxymethanol at gestational day 17. Psychopharmacology (Berl) 192:373–383

Lewis DA, Levitt P (2002) Schizophrenia as a disorder of neurodevelopment. Ann Rev Neurosci 25:409–432

Li Q, Cheung C, Wei R, Hui ES, Feldon J, Meyer U, Chung S, Chua SE, Sham PC, Wu EX, McAlonan GM (2009) Prenatal immune challenge is an environmental risk factor for brain and behavior change relevant to schizophrenia: evidence from MRI in a mouse model. PLoS One 4:e6354

Licinio J, Seibyl JP, Altemus M, Charney DS, Krystal JH (1993) Elevated CSF levels of interleukin-2 in neuroleptic-free schizophrenic patients. Am J Psychiatry 150:1408–1410

Lipska BK (2004) Using animal models to test a neurodevelopmental hypothesis of schizophrenia. J Psychiatry Neurosci 29:282–286

Lipska BK, Weinberger DR (1993) Delayed effects of neonatal hippocampal damage on haloperidol-induced catalepsy and apomorphine-induced stereotypic behaviors in the rat. Brain Res Dev Brain Res 75:213–222

Lipska BK, Weinberger DR (1994a) Gonadectomy does not prevent novelty or drug-induced motor hyperresponsiveness in rats with neonatal hippocampal damage. Brain Res Dev Brain Res 78:253–258

Lipska BK, Weinberger DR (1994b) Subchronic treatment with haloperidol and clozapine in rats with neonatal excitotoxic hippocampal damage. Neuropsychopharmacology 10:199–205

Lipska BK, Weinberger DR (2000) To model a psychiatric disorder in animals: schizophrenia as a reality test. Neuropsychopharmacology 23:223–239

Lipska BK, Jaskiw GE, Weinberger DR (1993) Postpubertal emergence of hyperresponsiveness to stress and to amphetamine after neonatal excitotoxic hippocampal damage: a potential animal model of schizophrenia. Neuropsychopharmacology 9:67–75

Lipska BK, Swerdlow NR, Geyer MA, Jaskiw GE, Braff DL, Weinberger DR (1995) Neonatal excitotoxic hippocampal damage in rats causes post-pubertal changes in prepulse inhibition of startle and its disruption by apomorphine. Psychopharmacology (Berl) 122:35–43

Lipska BK, Aultman JM, Verma A, Weinberger DR, Moghaddam B (2002a) Neonatal damage of the ventral hippocampus impairs working memory in the rat. Neuropsychopharmacology 27:47–54

Lipska BK, Halim ND, Segal PN, Weinberger DR (2002b) Effects of reversible inactivation of the neonatal ventral hippocampus on behavior in the adult rat. J Neurosci 22:2835–2842

Lipska BK, Lerman DN, Khaing ZZ, Weickert CS, Weinberger DR (2003) Gene expression in dopamine and GABA systems in an animal model of schizophrenia: effects of antipsychotic drugs. Eur J Neurosci 18:391–402

Lodge DJ, Grace AA (2007) Aberrant hippocampal activity underlies the dopamine dysregulation in an animal model of schizophrenia. J Neurosci 27:11424–11430

Lodge DJ, Grace AA (2009) Gestational methylazoxymethanol acetate administration: a developmental disruption model of schizophrenia. Behav Brain Res 204:306–312

Lodge DJ, Behrens MM, Grace AA (2009) A loss of parvalbumin-containing interneurons is associated with diminished oscillatory activity in an animal model of schizophrenia. J Neurosci 29:2344–2354

Machon RA, Mednick SA, Schulsinger F (1983) The interaction of seasonality, place of birth, genetic risk and subsequent schizophrenia in a high risk sample. Br J Psychiatry 143:383–388

Mallard EC, Rehn A, Rees S, Tolcos M, Copolov D (1999) Ventriculomegaly and reduced hippocampal volume following intrauterine growth-restriction: implications for the aetiology of schizophrenia. Schizophr Res 40:11–21

Mallard C, Loeliger M, Copolov D, Rees S (2000) Reduced number of neurons in the hippocampus and the cerebellum in the postnatal guinea-pig following intrauterine growth-restriction. Neuroscience 100:327–333

Marcelis M, Navarro-Mateu F, Murray R, Selten JP, Van Os J (1998) Urbanization and psychosis: a study of 1942–1978 birth cohorts in The Netherlands. Psychol Med 28:871–879

Marquis JP, Goulet S, Doré FY (2006) Neonatal lesions of the ventral hippocampus in rats lead to prefrontal cognitive deficits at two maturational stages. Neuroscience 140:759–767

Marquis JP, Goulet S, Dore FY (2008) Neonatal ventral hippocampus lesions disrupt extra-dimensional shift and alter dendritic spine density in the medial prefrontal cortex of juvenile rats. Neurobiol Learn Mem 90:339–346

McClellan J, Breiger D, McCurry C, Hlastala SA (2003) Premorbid functioning in early-onset psychotic disorders. J Am Acad Child Adolesc Psychiatry 42:666–672

McGlashan TH, Hoffman RE (2000) Schizophrenia as a disorder of developmentally reduced synaptic connectivity. Arch Gen Psychiatry 57:637–648

McGrath J (1999) Hypothesis: is low prenatal vitamin D a risk-modifying factor for schizophrenia? Schizophr Res 40:173–177

McGrath J, Eyles D, Mowry B, Yolken R, Buka S (2003) Low maternal vitamin D as a risk factor for schizophrenia: a pilot study using banked sera. Schizophr Res 63:73–78

McGrath J, Saari K, Hakko H, Jokelainen J, Jones P, Jarvelin MR, Chant D, Isohanni M (2004) Vitamin D supplementation during the first year of life and risk of schizophrenia: a Finnish birth cohort study. Schizophr Res 67:237–245

McGrath J, Iwazaki T, Eyles D, Burne T, Cui X, Ko P, Matsumoto I (2008) Protein expression in the nucleus accumbens of rats exposed to developmental vitamin D deficiency. PLoS One 3: e2383

McLean S, Grayson B, Harris M, Protheroe C, Bate S, Woolley M, Neill J (2010) Isolation rearing impairs novel object recognition and attentional set-shifting performance in female rats. J Psychopharmacol 24:57–63

Meaney MJ, Mitchell JB, Aitken DH, Bhatnagar S, Bodnoff SR, Iny LJ, Sarrieau A (1991) The effects of neonatal handling on the development of the adrenocortical response to stress: implications for neuropathology and cognitive deficits in later life. Psychoneuroendocrinology 16:85–103

Meaney MJ, Bhatnagar S, Larocque S, McCormick C, Shanks N, Sharma S, Smythe J, Viau V, Plotsky PM (1993) Individual differences in the hypothalamic-pituitary-adrenal stress response and the hypothalamic CRF system. Ann N Y Acad Sci 697:70–85

Mednick SA, Machon RA, Huttunen MO, Bonett D (1988) Adult schizophrenia following prenatal exposure to an influenza epidemic. Arch Gen Psychiatry 45:189–192

Meyer U, Feldon J (2009a) Epidemiology-driven neurodevelopmental animal models of schizophrenia. Prog Neurobiol (Epub ahead of print, Oct 24)

Meyer U, Feldon J (2009b) Neural basis of psychosis-related behaviour in the infection model of schizophrenia. Behav Brain Res 204:322–334

Meyer U, Feldon J, Schedlowski M, Yee BK (2005) Towards an immuno-precipitated neurodevelopmental animal model of schizophrenia. Neurosci Biobehav Rev 29:913–947

Meyer U, Feldon J, Schedlowski M, Yee BK (2006a) Immunological stress at the maternal–foetal interface: a link between neurodevelopment and adult psychopathology. Brain Behav Immun 20:378–388

Meyer U, Nyffeler M, Engler A, Urwyler A, Schedlowski M, Knuesel I, Yee BK, Feldon J (2006b) The time of prenatal immune challenge determines the specificity of inflammation-mediated brain and behavioral pathology. J Neurosci 26:4752–4762

Meyer U, Schwendener S, Feldon J, Yee BK (2006c) Prenatal and postnatal maternal contributions in the infection model of schizophrenia. Exp Brain Res 173:243–257

Meyer U, Murray PJ, Urwyler A, Yee BK, Schedlowski M, Feldon J (2008a) Adult behavioral and pharmacological dysfunctions following disruption of the fetal brain balance between pro-inflammatory and IL-10-mediated anti-inflammatory signaling. Mol Psychiatry 13:208–221

Meyer U, Nyffeler M, Schwendener S, Knuesel I, Yee BK, Feldon J (2008b) Relative prenatal and postnatal maternal contributions to schizophrenia-related neurochemical dysfunction after in utero immune challenge. Neuropsychopharmacology 33:441–456

Meyer U, Nyffeler M, Yee BK, Knuesel I, Feldon J (2008c) Adult brain and behavioral pathological markers of prenatal immune challenge during early/middle and late fetal development in mice. Brain Behav Immun 22:469–486

Meyer U, Spoerri E, Yee BK, Schwarz MJ, Feldon J (2008d) Evaluating early preventive antipsychotic and antidepressant drug treatment in an infection-based neurodevelopmental mouse model of schizophrenia. Schizophr Bull (Epub ahead of print, Oct 8)

Meyer U, Feldon J, Fatemi SH (2009) In-vivo rodent models for the experimental investigation of prenatal immune activation effects in neurodevelopmental brain disorders. Neurosci Biobehav Rev 33:1061–1079

Miller P, Byrne M, Hodges A, Lawrie SM, Owens DG, Johnstone EC (2002) Schizotypal components in people at high risk of developing schizophrenia: early findings from the Edinburgh high-risk study. Br J Psychiatry 180:179–184

Millstein RA, Ralph RJ, Yang RJ, Holmes A (2006) Effects of repeated maternal separation on prepulse inhibition of startle across inbred mouse strains. Genes Brain Behav 5:346–354

Mino Y, Oshima I (2006) Seasonality of birth in patients with schizophrenia in Japan. Psychiatry Clin Neurosci 60:249–252

Mirnics K, Middleton FA, Lewis DA, Levitt P (2001) Analysis of complex brain disorders with gene expression microarrays: schizophrenia as a disease of the synapse. Trends Neurosci 24:479–486

Mittleman BB, Castellanos FX, Jacobsen LK, Rapoport JL, Swedo SE, Shearer GM (1997) Cerebrospinal fluid cytokines in pediatric neuropsychiatric disease. J Immunol 159:2994–2999

Moore H, Jentsch JD, Ghajarnia M, Geyer MA, Grace AA (2006) A neurobehavioral systems analysis of adult rats exposed to methylazoxymethanol acetate on E17: implications for the neuropathology of schizophrenia. Biol Psychiatry 60:253–264

Mouri A, Noda Y, Enomoto T, Nabeshima T (2007) Phencyclidine animal models of schizophrenia: approaches from abnormality of glutamatergic neurotransmission and neurodevelopment. Neurochem Int 51:173–184

Moy SS, Nadler JJ, Perez A, Barbaro RP, Johns JM, Magnuson TR, Piven J, Crawley JN (2004) Sociability and preference for social novelty in five inbred strains: an approach to assess autistic-like behavior in mice. Genes Brain Behav 3:287–302

Murray RM, Lewis SW (1988) Is schizophrenia a neurodevelopmental disorder? Br Med J (Clin Res Ed) 296:263

Nadri C, Belmaker RH, Agam G (2007) Oxygen restriction of neonate rats elevates neuregulin-1alpha isoform levels: possible relationship to schizophrenia. Neurochem Int 51:447–450

Nakatani-Pawlak A, Yamaguchi K, Tatsumi Y, Mizoguchi H, Yoneda Y (2009) Neonatal phencyclidine treatment in mice induces behavioral, histological and neurochemical abnormalities in adulthood. Biol Pharm Bull 32:1576–1583

Nawa H, Takahashi M, Patterson PH (2000) Cytokine and growth factor involvement in schizophrenia – support for the developmental model. Mol Psychiatry 5:594–603

Nestor PG, O'Donnell BF (1998) The mind adrift: attentional dysregulation in schizophrenia. In: Parasuraman R (ed) The attentive brain. MIT Press, Cambridge, MA, pp 527–546

Nolan Y, Vereker E, Lynch AM, Lynch MA (2003) Evidence that lipopolysaccharide-induced cell death is mediated by accumulation of reactive oxygen species and activation of p38 in rat cortex and hippocampus. Exp Neurol 184:794–804

Nudmamud S, Reynolds LM, Reynolds GP (2003) N-acetylaspartate and N-Acetylaspartylglutamate deficits in superior temporal cortex in schizophrenia and bipolar disorder: a postmortem study. Biol Psychiatry 53:1138–1141

Nuechterlein KH, Barch DM, Gold JM, Goldberg TE, Green MF, Heaton RK (2004) Identification of separable cognitive factors in schizophrenia. Schizophr Res 72:29–39

Nyffeler M, Meyer U, Yee BK, Feldon J, Knuesel I (2006) Maternal immune activation during pregnancy increases limbic GABAA receptor immunoreactivity in the adult offspring: implications for schizophrenia. Neuroscience 143:51–62

O'Callaghan E, Sham P, Takei N, Glover G, Murray RM (1991) Schizophrenia after prenatal exposure to 1957 A2 influenza epidemic. Lancet 337:1248–1250

O'Donnell P, Lewis BL, Weinberger DR, Lipska BK (2002) Neonatal hippocampal damage alters electrophysiological properties of prefrontal cortical neurons in adult rats. Cereb Cortex 12:975–982

O'Loan J, Eyles DW, Kesby J, Ko P, McGrath JJ, Burne TH (2007) Vitamin D deficiency during various stages of pregnancy in the rat; its impact on development and behaviour in adult offspring. Psychoneuroendocrinology 32:227–234

Orzack MH, Kornetsky C (1966) Attention dysfunction in chronic schizophrenia. Arch Gen Psychiatry 14:323–326

Owen MJ, Lewis SW (1988) Risk factors in schizophrenia. Br J Psychiatry 153:407

Ozawa K, Hashimoto K, Kishimoto T, Shimizu E, Ishikura H, Iyo M (2006) Immune activation during pregnancy in mice leads to dopaminergic hyperfunction and cognitive impairment in the offspring: a neurodevelopmental animal model of schizophrenia. Biol Psychiatry 59:546–554

Özer S, Ulusahin A, Ulusoy S, Okur H, Coskun T, Tuncall T, Gögüs A, Akarsu AN (2004) Is vitamin D hypothesis for schizophrenia valid? Independent segregation of psychosis in a family with vitamin-D-dependent rickets type IIA. Prog Neuropsychopharmacol Biol Psychiatry 28:255–266

Palmer AA, Printz DJ, Butler PD, Dulawa SC, Printz MP (2004) Prenatal protein deprivation in rats induces changes in prepulse inhibition and NMDA receptor binding. Brain Res 996:193–201

Palmer AA, Brown AS, Keegan D, Siska LD, Susser E, Rotrosen J, Butler PD (2008) Prenatal protein deprivation alters dopamine-mediated behaviors and dopaminergic and glutamatergic receptor binding. Brain Res 1237:62–74

Pascual R, Zamora-Leon SP, Valero-Cabre A (2006) Effects of postweaning social isolation and re-socialization on the expression of vasoactive intestinal peptide (VIP) and dendritic development in the medial prefrontal cortex of the rat. Acta Neurobiol Exp (Wars) 66:7–14

Patterson PH (2009) Immune involvement in schizophrenia and autism: etiology, pathology and animal models. Behav Brain Res 204:313–321

Paulus MP, Bakshi VP, Geyer MA (1998) Isolation rearing affects sequential organization of motor behavior in post-pubertal but not pre-pubertal Lister and Sprague-Dawley rats. Behav Brain Res 94:271–280

Penschuck S, Flagstad P, Didriksen M, Leist M, Michael-Titus AT (2006) Decrease in parvalbumin-expressing neurons in the hippocampus and increased phencyclidine-induced locomotor activity in the rat methylazoxymethanol (MAM) model of schizophrenia. Eur J Neurosci 23:279–284

Pereira LO, Arteni NS, Petersen RC, da Rocha AP, Achaval M, Netto CA (2007) Effects of daily environmental enrichment on memory deficits and brain injury following neonatal hypoxia-ischemia in the rat. Neurobiol Learn Mem 87:101–108

Peters YM, O'Donnell P (2005) Social isolation rearing affects prefrontal cortical response to ventral tegmental area stimulation. Biol Psychiatry 57:1205–1208

Pietropaolo S, Singer P, Feldon J, Yee BK (2008) The postweaning social isolation in C57BL/6 mice: preferential vulnerability in the male sex. Psychopharmacology (Berl) 197:613–628

Piontkewitz Y, Assaf Y, Weiner I (2009) Clozapine administration in adolescence prevents postpubertal emergence of brain structural pathology in an animal model of schizophrenia. Biol Psychiatry 66:1038–1046

Powell SB, Geyer MA (2002) Developmental markers of psychiatric disorders as identified by sensorimotor gating. Neurotox Res 4:489–502

Powell SB, Geyer MA (2007) Overview of animal models of schizophrenia. Curr Protoc Neurosci Chapter 9: Unit 9 24

Powell SB, Risbrough VB, Geyer MA (2003) Potential use of animal models to examine antipsychotic prophylaxis for schizophrenia. Clin Neurosci Res 3:289–296

Powell SB, Zhou X, Geyer MA (2009) Prepulse inhibition and genetic mouse models of schizophrenia. Behav Brain Res 204:282–294

Preece MA, Dalley JW, Theobald DE, Robbins TW, Reynolds GP (2004) Region specific changes in forebrain 5-hydroxytryptamine1A and 5-hydroxytryptamine2A receptors in isolation-reared rats: an *in vitro* autoradiography study. Neuroscience 123:725–732

Pryce CR, Bettschen D, Bahr NI, Feldon J (2001) Comparison of the effects of infant handling, isolation, and nonhandling on acoustic startle, prepulse inhibition, locomotion, and HPA activity in the adult rat. Behav Neurosci 115:71–83

Ranade SC, Rose A, Rao M, Gallego J, Gressens P, Mani S (2008) Different types of nutritional deficiencies affect different domains of spatial memory function checked in a radial arm maze. Neuroscience 152:859–866

Rapoport JL, Addington AM, Frangou S, Psych MRC (2005) The neurodevelopmental model of schizophrenia: update 2005. Mol Psychiatry 10:434–449

Rehn AE, Van Den Buuse M, Copolov D, Briscoe T, Lambert G, Rees S (2004) An animal model of chronic placental insufficiency: relevance to neurodevelopmental disorders including schizophrenia. Neuroscience 129:381–391

Reynolds GP, Beasley CL (2001) GABAergic neuronal subtypes in the human frontal cortex – development and deficits in schizophrenia. J Chem Neuroanat 22:95–100

Reynolds GP, Abdul-Monim Z, Neill JC, Zhang ZJ (2004) Calcium binding protein markers of GABA deficits in schizophrenia – postmortem studies and animal models. Neurotox Res 6:57–61

Rice JE 3rd, Vannucci RC, Brierley JB (1981) The influence of immaturity on hypoxic-ischemic brain damage in the rat. Ann Neurol 9:131–141

Richtand NM, Taylor B, Welge JA, Ahlbrand R, Ostrander MM, Burr J, Hayes S, Coolen LM, Pritchard LM, Logue A, Herman JP, McNamara RK (2006) Risperidone pretreatment prevents elevated locomotor activity following neonatal hippocampal lesions. Neuropsychopharmacology 31:77–89

Robbins TW (2002) The 5-choice serial reaction time task: behavioural pharmacology and functional neurochemistry. Psychopharmacology (Berl) 163:362–380

Romero E, Ali C, Molina-Holgado E, Castellano B, Guaza C, Borrell J (2007) Neurobehavioral and immunological consequences of prenatal immune activation in rats. Influence of antipsychotics. Neuropsychopharmacology 32:1791–1804

Romero E, Guaza C, Castellano B, Borrell J (2008) Ontogeny of sensorimotor gating and immune impairment induced by prenatal immune challenge in rats: implications for the etiopathology of schizophrenia. Mol Psychiatry (Epub ahead of print, Apr 15)

Rothschild DM, O'Grady M, Wecker L (1999) Neonatal cytomegalovirus exposure decreases prepulse inhibition in adult rats: implications for schizophrenia. J Neurosci Res 57:429–434

Rueter LE, Ballard ME, Gallagher KB, Basso AM, Curzon P, Kohlhaas KL (2004) Chronic low dose risperidone and clozapine alleviate positive but not negative symptoms in the rat neonatal ventral hippocampal lesion model of schizophrenia. Psychopharmacology (Berl) 176:312–319

Sahakian BJ, Robbins TW, Morgan MJ, Iversen SD (1975) The effects of psychomotor stimulants on stereotypy and locomotor activity in socially-deprived and control rats. Brain Res 84:195–205

Sahakian BJ, Robbins TW, Iversen SD (1977) The effects of isolation rearing on exploration in the rat. Anim Learn Behav 5:193–198

Sakaue M, Ago Y, Baba A, Matsuda T (2003) The 5-HT1A receptor agonist MKC-242 reverses isolation rearing-induced deficits of prepulse inhibition in mice. Psychopharmacology (Berl) 170:73–79

Sams-Dodd F (1996) Phencyclidine-induced stereotyped behaviour and social isolation in rats: a possible animal model of schizophrenia. Behav Pharmacol 7:3–23

Sams-Dodd F (1998) A test of the predictive validity of animal models of schizophrenia based on phencyclidine and D-amphetamine. Neuropsychopharmacology 18:293–304

Sams-Dodd F, Lipska BK, Weinberger DR (1997) Neonatal lesions of the rat ventral hippocampus result in hyperlocomotion and deficits in social behaviour in adulthood. Psychopharmacology (Berl) 132:303–310

Sandager-Nielsen K, Andersen MB, Sager TN, Werge T, Scheel-Kruger J (2004) Effects of postnatal anoxia on striatal dopamine metabolism and prepulse inhibition in rats. Pharmacol Biochem Behav 77:767–774

Sarter M, Givens B, Bruno JP (2001) The cognitive neuroscience of sustained attention: where top-down meets bottom-up. Brain Res Brain Res Rev 35:146–160

Scaccianoce S, Del Bianco P, Paolone G, Caprioli D, Modafferi AM, Nencini P, Badiani A (2006) Social isolation selectively reduces hippocampal brain-derived neurotrophic factor without altering plasma corticosterone. Behav Brain Res 168:323–325

Schiavone S, Sorce S, Dubois-Dauphin M, Jaquet V, Colaianna M, Zotti M, Cuomo V, Trabace L, Krause KH (2009) Involvement of NOX2 in the development of behavioral and pathologic alterations in isolated rats. Biol Psychiatry 66:384–392

Schrijver NC, Wurbel H (2001) Early social deprivation disrupts attentional, but not affective, shifts in rats. Behav Neurosci 115:437–442

Schrijver NC, Pallier PN, Brown VJ, Wurbel H (2004) Double dissociation of social and environmental stimulation on spatial learning and reversal learning in rats. Behav Brain Res 152:307–314

Schubert MI, Porkess MV, Dashdorj N, Fone KC, Auer DP (2009) Effects of social isolation rearing on the limbic brain: a combined behavioral and magnetic resonance imaging volumetry study in rats. Neuroscience 159:21–30

Schwarz MJ, Muller N, Riedel M, Ackenheil M (2001) The Th2-hypothesis of schizophrenia: a strategy to identify a subgroup of schizophrenia caused by immune mechanisms. Med Hypotheses 56:483–486

Secher T, Berezin V, Bock E, Glenthoj B (2009) Effect of an NCAM mimetic peptide FGL on impairment in spatial learning and memory after neonatal phencyclidine treatment in rats. Behav Brain Res 199:288–297

Segal DS, Geyer MA (1985) Animal models of psychopathology. In: Judd LL, Groves PM (eds) Psychobiological foundations of clinical psychiatry. JB Lippincott Co, Philadelphia, pp 1–14

Segal DS, Geyer MA, Schuckit A (1981) Stimulant-induced psychosis: an evaluation of animal models. In: Youdim MBH, Lovenberg W, Sharman DF, Lagnado JR (eds) Essays in neurochemistry and neuropharmacology. Wiley, New York, pp 95–130

Selten JP, Brown AS, Moons KG, Slaets JP, Susser ES, Kahn RS (1999a) Prenatal exposure to the 1957 influenza pandemic and non-affective psychosis in The Netherlands. Schizophr Res 38:85–91

Selten JP, van der Graaf Y, van Duursen R, Gispen-de Wied CC, Kahn RS (1999b) Psychotic illness after prenatal exposure to the 1953 Dutch flood disaster. Schizophr Res 35:243–245

Shi L, Fatemi SH, Sidwell RW, Patterson PH (2003) Maternal influenza infection causes marked behavioral and pharmacological changes in the offspring. J Neurosci 23:297–302

Silva-Gomez AB, Bermudez M, Quirion R, Srivastava LK, Picazo O, Flores G (2003a) Comparative behavioral changes between male and female postpubertal rats following neonatal excitotoxic lesions of the ventral hippocampus. Brain Res 973:285–292

Silva-Gomez AB, Rojas D, Juarez I, Flores G (2003b) Decreased dendritic spine density on prefrontal cortical and hippocampal pyramidal neurons in postweaning social isolation rats. Brain Res 983:128–136

Sircar R, Rudy JW (1998) Repeated neonatal phencyclidine treatment impairs performance of a spatial task in juvenile rats. Ann N Y Acad Sci 844:303–309

Smith SEP, Li J, Garbett K, Mirnics K, Patterson PH (2007) Maternal immune activation alters fetal brain development through interleukin-6. J Neurosci 27:10695–10702

Son GH, Geum D, Chung S, Kim EJ, Jo JH, Kim CM, Lee KH, Kim H, Choi S, Kim HT, Lee CJ, Kim K (2006) Maternal stress produces learning deficits associated with impairment of NMDA receptor-mediated synaptic plasticity. J Neurosci 26:3309–3318

Son GH, Chung S, Geum D, Kang SS, Choi WS, Kim K, Choi S (2007) Hyperactivity and alteration of the midbrain dopaminergic system in maternally stressed male mice offspring. Biochem Biophys Res Commun 352:823–829

Sorensen HJ, Mortensen EL, Reinisch JM, Mednick SA (2009) Association between prenatal exposure to bacterial infection and risk of schizophrenia. Schizophr Bull 35:631–637

Stefani MR, Moghaddam B (2005) Transient N-methyl-D-aspartate receptor blockade in early development causes lasting cognitive deficits relevant to schizophrenia. Biol Psychiatry 57:433–436

Stevens KE, Johnson RG, Rose GM (1997) Rats reared in social isolation show schizophrenia-like changes in auditory gating. Pharmacol Biochem Behav 58:1031–1036

Susser E, Neugebauer R, Hoek HW, Brown AS, Lin S, Labovitz D, Gorman JM (1996) Schizophrenia after prenatal famine: further evidence. Arch Gen Psychiatry 53:25–31

Swerdlow NR, Geyer MA (1998) Using an animal model of deficient sensorimotor gating to study the pathophysiology and new treatments of schizophrenia. Schizophr Bull 24:285–301

Swerdlow NR, Vaccarino FJ, Amalric M, Koob GF (1986) The neural substrates for the motor-activating properties of psychostimulants: a review of recent findings. Pharmacol Biochem Behav 25:233–248

Swerdlow NR, Braff DL, Taaid N, Geyer MA (1994) Assessing the validity of an animal model of deficient sensorimotor gating in schizophrenic patients. Arch Gen Psychiatry 51:139–154

Swerdlow NR, Light GA, Breier M, Ko D, Mora AB, Shoemaker JM, Saint Marie RL, Neary AC, Geyer MA, Stevens KE, Powell SB (2007) Sensory and sensorimotor gating deficits after neonatal ventral hippocampal lesions in rats. In: 37th Annual meeting of the society for neuroscience, San Diego, CA; Abstract No. 607.9

Swerdlow NR, Weber M, Qu Y, Light GA, Braff DL (2008) Realistic expectations of prepulse inhibition in translational models for schizophrenia research. Psychopharmacology (Berl) 199:331–388

Szuran TF, Pliska V, Pokorny J, Welzl H (2000) Prenatal stress in rats: effects on plasma corticosterone, hippocampal glucocorticoid receptors, and maze performance. Physiol Behav 71:353–362

Takei N, Lewis S, Jones P, Harvey I, Murray RM (1996) Prenatal exposure to influenza and increased cerebrospinal fluid spaces in schizophrenia. Schizophr Bull 22:521–534

Talamini LM, Koch T, Ter Horst GJ, Korf J (1998) Methylazoxymethanol acetate-induced abnormalities in the entorhinal cortex of the rat; parallels with morphological findings in schizophrenia. Brain Res 789:293–306

Talamini LM, Koch T, Luiten PG, Koolhaas JM, Korf J (1999) Interruptions of early cortical development affect limbic association areas and social behaviour in rats; possible relevance for neurodevelopmental disorders. Brain Res 847:105–120

Talamini LM, Ellenbroek B, Koch T, Korf J (2000) Impaired sensory gating and attention in rats with developmental abnormalities of the mesocortex. Implications for schizophrenia. Ann N Y Acad Sci 911:486–494

Thor DH, Holloway WR (1981) Persistence of social investigatory behavior in the male rat: evidence for long-term memory of initial copluatory experience. Anim Learn Behav 9:561–565

Tohmi M, Tsuda N, Watanabe Y, Kakita A, Nawa H (2004) Perinatal inflammatory cytokine challenge results in distinct neurobehavioral alterations in rats: implication in psychiatric disorders of developmental origin. Neurosci Res 50:67–75

Tonkiss J, Galler JR (1990) Prenatal protein malnutrition and working memory performance in adult rats. Behav Brain Res 40:95–107

Tonkiss J, Almeida SS, Galler JR (1998) Prenatally malnourished female but not male rats show increased sensitivity to MK-801 in a differential reinforcement of low rates task. Behav Pharmacol 9:49–60

Torrey EF, Miller J, Rawlings R, Yolken RH (1997) Seasonality of births in schizophrenia and bipolar disorder: a review of the literature. Schizophr Res 28:1–38

Tseng KY, O'Donnell P (2007) Dopamine modulation of prefrontal cortical interneurons changes during adolescence. Cereb Cortex 17:1235–1240

Tseng KY, Lewis BL, Lipska BK, O'Donnell P (2007) Post-pubertal disruption of medial prefrontal cortical dopamine–glutamate interactions in a developmental animal model of schizophrenia. Biol Psychiatry 62:730–738

Tseng KY, Lewis BL, Hashimoto T, Sesack SR, Kloc M, Lewis DA, O'Donnell P (2008) A neonatal ventral hippocampal lesion causes functional deficits in adult prefrontal cortical interneurons. J Neurosci 28:12691–12699

Tseng KY, Chambers RA, Lipska BK (2009) The neonatal ventral hippocampal lesion as a heuristic neurodevelopmental model of schizophrenia. Behav Brain Res 204:295–305

Tsuda N, Tohmi M, Mizuno M, Nawa H (2006) Strain-dependent behavioral alterations induced by peripheral interleukin-1 challenge in neonatal mice. Behav Brain Res 166:19–31

Urakubo A, Jarskog LF, Lieberman JA, Gilmore JH (2001) Prenatal exposure to maternal infection alters cytokine expression in the placenta, amniotic fluid, and fetal brain. Schizophr Res 47:27–36

Vaillancourt C, Boksa P (2000) Birth insult alters dopamine-mediated behavior in a precocial species, the guinea pig. Implications for schizophrenia. Neuropsychopharmacology 23:654–666

van Os J, Selten JP (1998) Prenatal exposure to maternal stress and subsequent schizophrenia. The May 1940 invasion of The Netherlands. Br J Psychiatry 172:324–326

Varty GB, Geyer MA (1998) Effects of isolation rearing on startle reactivity, habituation, and prepulse inhibition in male Lewis, Sprague-Dawley, and Fischer F344 rats. Behav Neurosci 112:1450–1457

Varty GB, Higgins GA (1995) Examination of drug-induced and isolation-induced disruptions of prepulse inhibition as models to screen antipsychotic drugs. Psychopharmacology (Berl) 122:15–26

Varty GB, Marsden CA, Higgins GA (1999) Reduced synaptophysin immunoreactivity in the dentate gyrus of prepulse inhibition-impaired isolation-reared rats. Brain Res 824:197–203

Varty GB, Paulus MP, Braff DL, Geyer MA (2000) Environmental enrichment and isolation rearing in the rat: effects on locomotor behavior and startle response plasticity. Biol Psychiatry 47:864–873

Varty GB, Powell SB, Lehmann-Masten V, Buell MR, Geyer MA (2006) Isolation rearing of mice induces deficits in prepulse inhibition of the startle response. Behav Brain Res 169:162–167

Vohs JL, Chambers RA, Krishnan GP, O'Donnell BF, Hetrick WP, Kaiser ST, Berg S, Morzorati SL (2009) Auditory sensory gating in the neonatal ventral hippocampal lesion model of schizophrenia. Neuropsychobiology 60:12–22

Voikar V, Polus A, Vasar E, Rauvala H (2005) Long-term individual housing in C57BL/6J and DBA/2 mice: assessment of behavioral consequences. Genes Brain Behav 4:240–252

Wakuda T, Matsuzaki H, Suzuki K, Iwata Y, Shinmura C, Suda S, Iwata K, Yamamoto S, Sugihara G, Tsuchiya KJ, Ueki T, Nakamura K, Nakahara D, Takei N, Mori N (2008) Perinatal asphyxia reduces dentate granule cells and exacerbates methamphetamine-induced hyperlocomotion in adulthood. PLoS One 3:e3648

Wang C, McInnis J, West JB, Bao J, Anastasio N, Guidry JA, Ye Y, Salvemini D, Johnson KM (2003) Blockade of phencyclidine-induced cortical apoptosis and deficits in prepulse inhibition by M40403, a superoxide dismutase mimetic. J Pharmacol Exp Ther 304:266–271

Wang CZ, Yang SF, Xia Y, Johnson KM (2008) Postnatal phencyclidine administration selectively reduces adult cortical parvalbumin-containing interneurons. Neuropsychopharmacology 33:2442–2455

Watanabe Y, Hashimoto S, Kakita A, Takahashi H, Ko J, Mizuno M, Someya T, Patterson PH, Nawa H (2004) Neonatal impact of leukemia inhibitory factor on neurobehavioral development in rats. Neurosci Res 48:345–353

Wedzony K, Fijal K, Mackowiak M, Chocyk A, Zajaczkowski W (2008) Impact of postnatal blockade of N-methyl-D-aspartate receptors on rat behavior: a search for a new developmental model of schizophrenia. Neuroscience 153:1370–1379

Weinberger DR (1987) Implications of normal brain development for the pathogenesis of schizophrenia. Arch Gen Psychiatry 44:660–669

Weinberger DR, Berman KF, Suddath R, Torrey EF (1992) Evidence of dysfunction of a prefrontal-limbic network in schizophrenia: a magnetic resonance imaging and regional cerebral blood flow study of discordant monozygotic twins. Am J Psychiatry 149:890–897

Weiner I, Arad M (2009) Using the pharmacology of latent inhibition to model domains of pathology in schizophrenia and their treatment. Behav Brain Res 204:369–386

Weiner I, Schnabel I, Lubow RE, Feldon J (1985) The effects of early handling on latent inhibition in male and female rats. Dev Psychobiol 18:291–297

Weiner I, Lubow RE, Feldon J (1988) Disruption of latent inhibition by acute administration of low doses of amphetamine. Pharmacol Biochem Behav 30:871–878

Weinstock M (2008) The long-term behavioural consequences of prenatal stress. Neurosci Biobehav Rev 32:1073–1086

Weiss IC, Feldon J (2001) Environmental animal models for sensorimotor gating deficiencies in schizophrenia: a review. Psychopharmacology (Berl) 156:305–326

Weiss IC, Domeney AM, Moreau JL, Russig H, Feldon J (2001) Dissociation between the effects of pre-weaning and/or post-weaning social isolation on prepulse inhibition and latent inhibition in adult Sprague-Dawley rats. Behav Brain Res 121:207–218

Weiss IC, Pryce CR, Jongen-Relo AL, Nanz-Bahr NI, Feldon J (2004) Effect of social isolation on stress-related behavioural and neuroendocrine state in the rat. Behav Brain Res 152:279–295

Wiley JL, Buhler KG, Lavecchia KL, Johnson KM (2003) Pharmacological challenge reveals long-term effects of perinatal phencyclidine on delayed spatial alternation in rats. Prog Neuropsychopharmacol Biol Psychiatry 27:867–873

Wilkinson LS, Killcross SS, Humby T, Hall FS, Geyer MA, Robbins TW (1994) Social isolation in the rat produces developmentally specific deficits in prepulse inhibition of the acoustic startle response without disrupting latent inhibition. Neuropsychopharmacology 10:61–72

Winocur G (1991) Functional dissociation of the hippocampus and prefrontal cortex in learning and memory. Psychobiology 19:11–20

Winslow JT, Camacho F (1995) Cholinergic modulation of a decrement in social investigation following repeated contacts between mice. Psychopharmacology (Berl) 121:164–172

Winter C, Djodari-Irani A, Sohr R, Morgenstern R, Feldon J, Juckel G, Meyer U (2009) Prenatal immune activation leads to multiple changes in basal neurotransmitter levels in the adult brain: implications for brain disorders of neurodevelopmental origin such as schizophrenia. Int J Neuropsychopharmacol 12:513–524

Wolff AR, Bilkey DK (2008) Immune activation during mid-gestation disrupts sensorimotor gating in rat offspring. Behav Brain Res 190:156–159

Wong AH, Lipska BK, Likhodi O, Boffa E, Weinberger DR, Kennedy JL, Van Tol HH (2005) Cortical gene expression in the neonatal ventral–hippocampal lesion rat model. Schizophr Res 77:261–270

Wright IK, Upton N, Marsden CA (1991) Resocialisation of isolation-reared rats does not alter their anxiogenic profile on the elevated X-maze model of anxiety. Physiol Behav 50:1129–1132

Wu J, Song TB, Li YJ, He KS, Ge L, Wang LR (2007) Prenatal restraint stress impairs learning and memory and hippocampal PKCbeta1 expression and translocation in offspring rats. Brain Res 1141:205–213

Xu M-Q, Sun W-S, Liu B-X, Feng G-Y, Yu L, Yang L, He G, Sham P, Susser E, St. Clair D, He L (2009) Prenatal malnutrition and adult schizophrenia: further evidence from the 1959–1961 Chinese famine. Schizophr Bull 35:568–576

Yang J, Han H, Cao J, Li L, Xu L (2006) Prenatal stress modifies hippocampal synaptic plasticity and spatial learning in young rat offspring. Hippocampus 16:431–436

Young JW, Powell SB, Risbrough V, Marston HM, Geyer MA (2009) Using the MATRICS to guide development of a preclinical cognitive test battery for research in schizophrenia. Pharmacol Ther 122:150–202

Young JW, Zhou X, Geyer MA (2010) Animal models of schizophrenia. In: Swerdlow N (ed) Behavioral Neurobiology of Schizophrenia and Its Treatment. Curr Topics in Behav Neurosci. Springer, Heidelberg

Zornberg GL, Buka SL, Tsuang MT (2000) The problem of obstetrical complications and schizophrenia. Schizophr Bull 26:249–256

Zorrilla EP (1997) Multiparous species present problems (and possibilities) to developmentalists. Dev Psychobiol 30:141–150

Zuckerman L, Weiner I (2003) Post-pubertal emergence of disrupted latent inhibition following prenatal immune activation. Psychopharmacology (Berl) 169:308–313

Zuckerman L, Weiner I (2005) Maternal immune activation leads to behavioral and pharmacological changes in the adult offspring. J Psychiatr Res 39:311–323

Zuckerman L, Rehavi M, Nachman R, Weiner I (2003) Immune activation during pregnancy in rats leads to a postpubertal emergence of disrupted latent inhibition, dopaminergic hyperfunction, and altered limbic morphology in the offspring: a novel neurodevelopmental model of schizophrenia. Neuropsychopharmacology 28:1778–1789

Part III
Neural substrates of schizophrenia

Prefrontal Cortical Circuits in Schizophrenia

David W. Volk and David A. Lewis

Contents

1 Introduction 486
   1.1 Working Memory Impairments and Dorsolateral
       Prefrontal Cortex Circuitry 486
2 Pathology of DLPFC Circuitry in Schizophrenia 488
   2.1 Abnormalities in Pyramidal Neuron Anatomy and Glutamatergic Signaling 488
   2.2 Abnormalities in GABA Signaling 491
   2.3 Alterations in the Dopamine Neurotransmitter System 492
   2.4 Pathophysiological Consequences of Altered DLPFC Circuitry on Cognitive
       Functioning in Schizophrenia 493
3 Cortical Circuitry Alterations Beyond the DLPFC 494
4 Cannabis Use and Schizophrenia 495
   4.1 Clinical Effects of Cannabis Use in Schizophrenia 495
   4.2 Potential Impact of Cannabis Use of Altered
       Neurotransmitter Systems in Schizophrenia 497
   4.3 Endogenous Cannabinoid System and Schizophrenia 498
5 From Pathology to New Therapeutic Approaches 499
References 500

Abstract Impaired cognitive functioning, including deficits in working memory, is considered to be a core and disabling feature of schizophrenia that is difficult to treat. Deficits in working memory in schizophrenia are attributable, at least in part,

D.W. Volk
Department of Psychiatry, University of Pittsburgh, 3811 O'Hara Street, BST W1653, Pittsburgh, PA 15213, USA
e-mail: volkdw@upmc.edu

D.A. Lewis (✉)
Department of Neuroscience, University of Pittsburgh, 3811 O'Hara Street, BST W1653, Pittsburgh, PA 15213, USA
e-mail: lewisda@upmc.edu

Department of Psychiatry, University of Pittsburgh, 3811 O'Hara Street, BST W1653, Pittsburgh, PA 15213, USA to specific pathological alterations in the neuronal circuitry of the dorsolateral prefrontal cortex that involve, but are not restricted to, disturbances in glutamate, GABA, and dopamine neurotransmission. Cannabis use provides an example of an environmental exposure that may have a deleterious impact on these neurotransmitter systems and thereby contribute to worsening of cognitive functioning in schizophrenia. Increasing knowledge of the nature of the molecular alterations in these cortical circuits may lead to the development of new pathophysiologically informed treatment options for cognitive deficits in schizophrenia.

Keywords Working memory · Pyramidal neurons · GABA · Dopamine · Spine · Thalamus · NMDA receptor · Parvalbumin · GAD67 · Catechol-O-methyltransferase · Gamma-band · Cannabis · CB1 receptor

1 Introduction

A core and disabling feature of schizophrenia involves disturbances in cognitive functioning, including impairments in attention, memory, and executive functions, such as the ability to plan, initiate, and regulate goal-directed behavior (Elvevag and Goldberg 2000). These cognitive deficits are present throughout the entire course of the illness, including the premorbid (Davidson et al. 1999), first-onset (Saykin et al. 1994), and later stages (Breier et al. 1991; Heaton et al. 1994), and consequently do not appear to be attributable to the chronic nature of the illness or to treatment with antipsychotic medications. In addition, cognitive dysfunction in schizophrenia leads to detrimental effects on daily activities and poorer long-term outcomes (Green 1996; Harvey et al. 1998). Furthermore, cognitive deficits are only minimally responsive to available antipsychotic medications (Keefe et al. 2007a, b), indicating a need to develop novel, pathophysiologically informed, treatments for cognitive impairments in schizophrenia (Insel and Scolnick 2006).

1.1 Working Memory Impairments and Dorsolateral Prefrontal Cortex Circuitry

Working memory, typically defined as the ability to transiently maintain and manipulate a limited amount of information in order to guide thought or behavior, has been an area of focus in research on the cognitive impairments in schizophrenia. Working memory involves a series of processes including manipulation of the information being stored, prevention of interference by competing information, and maintenance of goal representations (Barch and Smith 2008). Although individuals with schizophrenia show relatively little impairment on tasks that involve the storage of information in working memory, they consistently show impairments in the manipulation of such information and in the maintenance of goal representations (Barch 2006). These impairments are present in both medicated and unmedicated subjects, occur throughout all stages of the illness, and do not appear to be attributable to other factors such as lack of effort or interest (Tan et al. 2005; Cannon et al. 2005; Barch and Smith 2008).

Working memory is normally associated with activation of dorsolateral prefrontal cortex (DLPFC) circuitry (Fig. 1). DLPFC activation is altered in medication-naïve individuals with schizophrenia but not in subjects with other psychotic disorders

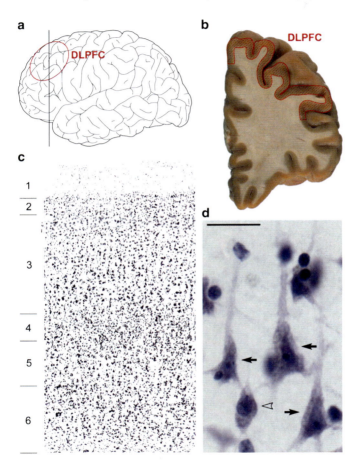

Fig. 1 Anatomy of the prefrontal cortex. (**a**) The relative location of the dorsolateral prefrontal cortex (DLPFC) in the human brain is shown, and the *vertical line* illustrates the approximate location of the coronal slab shown in (**b**). The *highlighted area* of panel (**b**) shows the location of areas 9 and 46 of the DLPFC within the superior and middle frontal gyri. (**c**) A coronal tissue section from the DLPFC stained for Nissl substance and illustrates the presence of six cortical layers. Panel (**d**) from a Nissl-stained tissue section at high power demonstrates the two general classes of neurons in the neocortex: the pyramidal neuron (*solid black arrows*) which is the major excitatory neuron of the cortex, and the GABA interneuron (*open arrow*) which is the major inhibitory neuron of the cortex

(Barch et al. 2001; MacDonald et al. 2003, 2005), suggesting that abnormal DLPFC activation in schizophrenia is not due to the presence of psychosis. Furthermore, varying degrees of altered DLPFC activation are also present in unaffected relatives of individuals with schizophrenia (Barch et al. 2001; MacDonald et al. 2003), which suggests that higher levels of a shared genetic load with an individual with schizophrenia may lead to increasing degrees of abnormalities in the DLPFC, even though the clinical phenotype of schizophrenia may not fully develop. Schizophrenia is not associated with a simple increase or decrease in the degree of DLPFC activation while performing a working memory task; instead, relative to healthy subjects, individuals with schizophrenia exhibit greater DLPFC activation at low working memory loads with intact performance, but reduced activation at high loads with impaired performance (Callicott et al. 2003; Manoach 2003; Tan et al. 2007).

2 Pathology of DLPFC Circuitry in Schizophrenia

Convergent lines of evidence, albeit with varying degrees of replication, have implicated the following components of DLPFC circuitry in the pathology of schizophrenia (1) pyramidal neurons (Figs. 1 and 2), which comprise ~75% of cortical neurons and utilize the excitatory neurotransmitter glutamate; (2) interneurons (Figs. 1 and 2), which comprise ~25% of cortical neurons and utilize the inhibitory neurotransmitter GABA; and (3) projections from dopamine (DA)-containing neurons in the mesencephalon and from the thalamus (Fig. 2).

2.1 Abnormalities in Pyramidal Neuron Anatomy and Glutamatergic Signaling

Pyramidal neurons provide the main source of excitatory signaling in the cerebral cortex, and they provide (and receive) excitatory inputs onto dendritic spines. The total number of DLPFC neurons is not altered in schizophrenia (Thune et al. 2001). However, neuronal density in the DLPFC has been reported to be increased, which appears to reflect a reduction in neuropil (i.e., the number of axon terminals and dendritic spines that occupy the space between neurons; Selemon and Goldman-Rakic 1999). Consistent with this interpretation, shorter dendritic length and the lower density of dendritic spines on pyramidal neurons (Garey et al. 1998; Glantz and Lewis 2000; Black et al. 2004); lower levels of synaptophysin protein, a marker of axon terminals (Perrone-Bizzozero et al. 1996; Glantz and Lewis 1997); and lower levels of gene transcripts that encode proteins present in axon terminals (Mirnics et al. 2000) have been found in the DLPFC of subjects with schizophrenia. Lower spine density is particularly prominent on the basilar dendrites of pyramidal neurons located in deep layer 3 of the DLPFC (Fig. 2; Glantz and Lewis 2000;

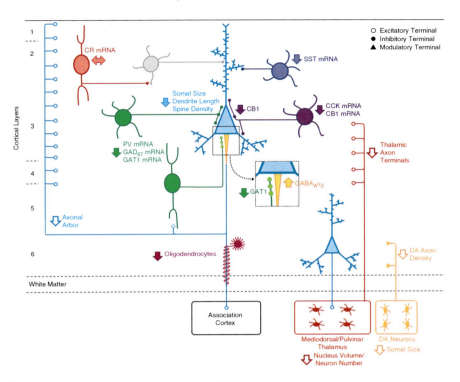

Fig. 2 Summary of putative alterations in DLPFC circuitry in schizophrenia. Pyramidal neurons (*light blue*) in deep layer 3 have smaller somal size, shorter basilar dendrites, lower dendritic spine density, and a reduced axonal arbor in schizophrenia. Altered GABA neurotransmission by parvalbumin (PV)-containing neurons (*green*) is indicated by expression deficits in several gene products as well as by lower GABA membrane transporter (GAT1) protein in the terminals of chandelier neurons and upregulated $GABA_A$ receptor α_2-subunits at their synaptic targets, the axon initial segments of pyramidal neurons (*enlarged square*). Expression of the neuropeptide somatostatin (SST) is decreased in GABA neurons (*dark blue*) that target the distal dendrites of pyramidal neurons. Decreased cholecystokinin (CCK) and cannabinoid receptor 1 (CB1) mRNA levels, and lower CB1 protein in axon terminals, suggest altered regulation of GABA neurotransmission in a subset of basket neurons (*purple*) that target the cell body and proximal dendrites of pyramidal neurons. Gene expression does not seem to be altered in CR-containing GABA neurons (*red*) that primarily target other GABA neurons (*gray*). Putative alterations in thalamic and dopamine (DA) cell bodies and their projections to the DLPFC are also shown. Some studies indicate that the number and/or gene expression in oligodendrocytes is also altered. Not all of the circuitry alterations shown here have been sufficiently replicated or demonstrated to be specific to the disease process of schizophrenia to be considered established "facts"; *solid arrows* indicate abnormalities supported by convergent and/or replicated observations (reprinted from Lewis and Sweet 2009)

Kolluri et al. 2005). Furthermore, the cell size of deep layer 3 pyramidal neurons, which is correlated with the size of a neuron's dendritic tree and axonal arbor (Lewis and Gonzalez-Burgos 2008), is also smaller in subjects with schizophrenia (Fig. 2; Rajkowska et al. 1998; Glantz and Lewis 2000; Pierri et al. 2001). Together, these findings suggest that the number of excitatory inputs to deep layer 3 pyramidal neuron basilar dendrites is reduced in schizophrenia (Figs. 2 and 3).

One source of lower excitatory input to pyramidal neurons in schizophrenia may be a reduced number of thalamic afferents since excitatory projections from the thalamus to the DLPFC synapse primarily on dendritic spines in deep layer 3 and layer 4 (Fig. 2; Erickson and Lewis 2004). Studies of the total number of neurons in the mediodorsal thalamic nucleus, a major source of thalamic projections to the DLPFC, have produced mixed results, with initial studies reporting lower neuron number in schizophrenia but subsequent studies with larger sample sizes failing to detect a difference (reviewed in Dorph-Petersen et al. 2004). However, several postmortem human brain studies have found reduced volume and neuron number in the pulvinar, a thalamic association nucleus that also projects to the DLPFC, in schizophrenia (Byne et al. 2002; Danos et al. 2003; Highley et al. 2003), and these findings are supported by similar observations using MRI in living subjects (Byne et al. 2001; Gilbert et al. 2001; Kemether et al. 2003). Alternatively, the smaller somal volume and lower spine density in deep layer 3 pyramidal neurons could reflect abnormalities intrinsic to this cell class.

The alterations in excitatory inputs to deep layer 3 pyramidal neurons might be developmental, rather than neurodegenerative, in nature given that the densities of excitatory synapses and dendritic spines both normally decline during adolescence in primate DLPFC, with the changes most marked in layer 3 (Bourgeois et al. 1994; Anderson et al. 1995; Huttenlocher and Dabholkar 1997). In humans, this synaptic pruning is thought to underlie the decrease in cortical gray matter thickness that occurs normally during adolescence and to an exaggerated degree in schizophrenia (Rapoport and Gogtay 2008). Current hypotheses hold either that a pre-existing abnormality in glutamate transmission in individuals with schizophrenia is revealed with the pruning of synapses during adolescence, or that the mechanisms of adolescence-related synapse elimination are disturbed which results in excessive synapse pruning and decreased pyramidal neuron spine number (McGlashan and Hoffman 2000).

Excitatory inputs to DLPFC layer 3 pyramidal neurons are mediated by the actions of glutamate on AMPA and NMDA receptors located on dendritic spines, which suggests that glutamatergic signaling is likely abnormal in schizophrenia. Clinical studies have demonstrated that NMDA receptor antagonists (such as phenylcyclidine and/or ketamine) replicate multiple aspects of the clinical syndrome of schizophrenia, including cognitive impairments (Javitt and Zukin 1991; Krystal et al. 1994). Furthermore, NMDA receptor antagonists disrupt working memory in rats (Verma and Moghaddam 1996), and their direct application to the DLPFC impairs working memory performance in monkeys (Dudkin et al. 2001). However, alterations in mRNA or protein levels of NMDA receptor subunits in the DLPFC in postmortem studies of schizophrenia appear to be relatively small and not always replicated (Kristiansen et al. 2007). Interestingly, regulators of NMDA receptor signaling may still be affected in the illness. For example, the modulation of NMDA receptor function by NRG1 signaling pathways has been reported to be reduced in the postmortem DLPFC tissue from subjects with schizophrenia (Hahn et al. 2006).

2.2 Abnormalities in GABA Signaling

Alterations in local inhibitory neurons, or GABA neurons, have also been commonly reported in the DLPFC of subjects with schizophrenia. For example, a reduced mRNA level for GAD_{67}, a principle synthesizing enzyme for GABA, in the DLPFC is perhaps the most widely and consistently replicated observation in postmortem studies of schizophrenia (reviewed in Hashimoto et al. 2008a). At the cellular level, GAD_{67} mRNA expression is not detectable in $\sim$30% of GABA neurons in subjects with schizophrenia, but the remaining GABA neurons exhibit normal levels of GAD_{67} mRNA (Fig. 2; Akbarian et al. 1995; Volk et al. 2000). Similarly, levels of the mRNA for the GABA membrane transporter (GAT1), a protein responsible for reuptake of released GABA into nerve terminals, are also decreased in schizophrenia (Fig. 2; Ohnuma et al. 1999; Hashimoto et al. 2008b), and this decrease is restricted to a minority of GABA neurons (Volk et al. 2001). These abnormalities in GABA neurons appear to be specific to the disease process of schizophrenia because they are not found in subjects with other psychiatric disorders or in monkeys exposed chronically to antipsychotic medications (Volk et al. 2000, 2001; Hashimoto et al. 2003). Taken together, these findings suggest that both the synthesis and reuptake of GABA are lower in a subset of DLPFC neurons in schizophrenia (Fig. 3).

GABA neurons are comprised of diverse subpopulations of neurons with distinct neurochemical, anatomical, physiological, and functional characteristics (Ascoli et al. 2008). The affected neurons in schizophrenia include, but are not restricted to, the $\sim$25% of primate DLPFC GABA neurons that express the calcium-binding protein parvalbumin (PV; Condé et al. 1994), which receive a high number of excitatory inputs from DLPFC layer 3 pyramidal neurons (Melchitzky et al. 2001; Fig. 2). PV neurons include chandelier neurons – whose axon terminals (termed *cartridges*) target the axon initial segments (AIS) of pyramidal neurons – and basket cells, whose axon terminals target the cell bodies of pyramidal neurons. In schizophrenia, the expression of PV mRNA is reduced (Fig. 2; Hashimoto et al. 2003), although the number of PV neurons in the DLPFC appears to be unchanged (Woo et al. 1997; Hashimoto et al. 2003); in addition, approximately half of PV mRNA-containing neurons lack detectable levels of GAD_{67} mRNA (Hashimoto et al. 2003; Fig. 2). In schizophrenia subjects, the chandelier class of PV neurons has lower GAT1 immunoreactivity in their axon cartridges (Fig. 2; Woo et al. 1998); in contrast, immunoreactivity for the $GABA_A$ receptor α_2-subunit, which is present in most $GABA_A$ receptors in the AIS of layer 2–3 pyramidal neurons (Nusser et al. 1996), is markedly increased (Fig. 2; Volk et al. 2002). Several lines of evidence suggest that these pre- (GAT1 and PV) and postsynaptic ($GABA_A$ receptor α_2-subunits) changes may have a compensatory effort for deficient GABA release from chandelier neurons (Lewis et al. 2005; Gonzalez-Burgos and Lewis 2008). For example, lowering PV levels augments GABA release (Vreugdenhil et al. 2003). Thus, the deficit in GAD_{67} mRNA might be a more primary component of the disease process.

Furthermore, recent studies have found lower levels of ankyrin-G at the AIS of DLPFC pyramidal neurons in schizophrenia (Cruz et al. 2009b). Ankyrin-G is an adaptor molecule important for recruiting and stabilizing GABA synapses (Ango et al. 2004) and maintaining other important membrane proteins including voltage-gated sodium channels (Zhou et al. 1998; Susuki and Rasband 2008). Thus, lower levels of ankyrin-G in DLPFC pyramidal neurons may result in improperly developed GABA synapses at the AIS from chandelier cells and, if voltage-gated sodium channels are not sufficiently maintained, a diminished ability to initiate normal action potentials in schizophrenia.

Other populations of DLPFC GABA neurons, such as those that express the neuropeptide somatostatin (Morris et al. 2008) or the neuropeptide cholecystokinin (CCK) and the cannabinoid receptor 1 (CB1R; Eggan et al. 2008), also appear to be disturbed in schizophrenia (Fig. 2). The affected neurons have distinct influences on the function of DLPFC pyramidal neurons. For example, CCK/CB1R- and PV-containing basket cells provide convergent sources of perisomatic inhibition to pyramidal neurons, but play complementary roles in shaping their activity (Freund and Katona 2007; Karson et al. 2009). In contrast, the ~50% of GABA neurons that express the calcium-binding protein calretinin appear to be unaffected in schizophrenia (Hashimoto et al. 2003; Fig. 2). Thus, alterations in certain, critical subpopulations of GABA neurons may contribute to altered regulation of pyramidal neuron activity in schizophrenia (Fig. 3).

2.3 Alterations in the Dopamine Neurotransmitter System

The activity of both pyramidal and GABA neurons in the DLPFC is modulated by inputs from DA-containing neurons located in the ventral mesencephalon. Several lines of evidence point to reduced DA signaling in the DLPFC in schizophrenia (Fig. 2).

First, the DA innervation of the DLPFC appears to be decreased in schizophrenia as indicated by lower levels of markers of DA axons, such as tyrosine hydroxylase (TH; the rate-limiting enzyme in DA synthesis) and the DA transporter (Akil et al. 1999). Although the number of DA neurons is not altered in schizophrenia, some studies suggest that they have smaller somal volumes (Bogerts et al. 1983) and lower levels of TH protein (Perez-Costas et al. 2007). Together, these findings suggest that cortical DA signaling might be diminished in schizophrenia due to a reduced number of axons and/or decreased DA levels per axon.

Second, the availability of extracellular DA in the DLPFC might be reduced in schizophrenia. For example, the DA-degrading enzyme catechol-O-methyltransferase (COMT) is a principal regulator of prefrontal DA levels (Tunbridge et al. 2004b). Although the levels of COMT mRNA and protein do not appear to be altered in schizophrenia (Matsumoto et al. 2003; Tunbridge et al. 2004a), allelic variants (Val158Met) in the *COMT* gene code for enzymes with marked differences in catalytic activity (Chen et al. 2004). For example, the Val-coding allele results in an isoform of COMT protein with higher enzymatic activity. Healthy subjects homozygous for the Val allele have lower levels of cognitive performance, which may be attributable to lower DLPFC DA levels since performance improved with amphetamine-induced increased DA release (Tan et al. 2007). However, studies of the association between *COMT* variants and schizophrenia have been inconclusive (Burmeister et al. 2008).

Third, projections from the cortex are thought to be an essential source of glutamate-mediated excitation to DLPFC-projecting DA neurons (Sesack and Carr 2002; but see Frankle et al. 2006). Thus, reduced excitatory output from DLPFC pyramidal neurons in schizophrenia could lead to persistently decreased activation of DA cells (Fig. 3). Normal working memory performance depends on the optimal level of activation of DA D1 receptors (Goldman-Rakic et al. 2004). Therefore, these deficits in DA signaling in the DLPFC may lead to working memory impairments in schizophrenia.

The combination of decreased DA innervation of the DLPFC, increased DA turnover, and DA cell hypoactivity in the DLPFC in schizophrenia could lead to reduced extracellular DA levels, deficient DA D1 receptor stimulation, and possibly a compensatory upregulation of these receptors. Consistent with this hypothesis, a positron emission tomography study found increased binding of a DA D1 receptor ligand (NNC112) in the DLPFC of drug-free and drug-naïve subjects with schizophrenia (Fig. 2; Abi-Dargham et al. 2002). However, other studies using different ligands have not replicated these results (Okubo et al. 1997; Karlsson et al. 2002) and no changes were found in D1 receptor mRNA levels in postmortem human brain tissue (Meador-Woodruff et al. 1997). Interestingly, preclinical studies indicate that sustained DA depletion differentially affects binding of these DA D1 receptor ligands, and in particular elevates the in vivo binding of NNC112 (Guo et al. 2003). Furthermore, the degree of D1 receptor upregulation in schizophrenia was inversely related to working memory performance (Abi-Dargham et al. 2002), consistent with the idea that D1 upregulation may be a compensatory, but insufficient, response to a DA deficit in the DLPFC.

2.4 Pathophysiological Consequences of Altered DLPFC Circuitry on Cognitive Functioning in Schizophrenia

How could the alterations in DLPFC circuitry summarized in Fig. 2 interact to give rise to the pathophysiology of working memory deficits in schizophrenia (Fig. 3)? One possibility is that hypofunction of NMDA receptors selectively on PV neurons leads to a reduction in GAD_{67} expression and decreased GABA signaling in the DLPFC. The loss of inhibitory regulation of pyramidal neurons may lead to disinhibition of pyramidal neurons and potentially excessive glutamate-mediated activation of non-NMDA receptors that disrupts cortical circuit function (Moghaddam 2004; Lisman et al. 2008). Although aspects of this hypothesis require further explanation, such as the mechanism behind abnormal NMDA receptor function selectively in PV neurons but not in pyramidal neurons (Gonzalez-Burgos and Lewis 2008), it provided the rationale for the development of a novel compound with agonist activity at metabotropic glutamate receptors (mGluR2/3) that reduces glutamate release through a presynaptic mechanism (Moghaddam and Adams 1998). A recent clinical trial demonstrated antipsychotic efficacy of such a compound in schizophrenia, although its effects on cognitive deficits were not reported (Patil et al. 2007).

A second hypothesis posits that an intrinsic deficiency in pyramidal neurons, such as fewer dendritic spines, leads to reduced excitatory output from the DLPFC (Fig. 3). Deficient excitatory output from the DLPFC would produce sustained hypoactivity of, and consequently both morphological and biochemical changes in, cortically projecting DA cells located in the ventral mesencephalon. Impairments in DA cells then lead to reduced DA innervation of the DLPFC and compensatory, but functionally insufficient, upregulation of D1 receptors in pyramidal cells, GABA neurons or both. Because D1 receptor activation increases the activity of PV-positive GABA neurons, reduced D1-mediated signaling might reduce the signaling of these inhibitory neurons in schizophrenia.

A third scenario views the deficit in GAD_{67} mRNA expression as a highly conserved and thus central feature of DLPFC pathology in schizophrenia (Lewis et al. 2005). Because the activity of DLPFC GABA neurons is essential for normal working memory function in monkeys (Rao et al. 2000), reduced GABA signaling from PV-containing neurons to DLPFC pyramidal neurons might contribute to the pathophysiology of working memory impairments. For example, networks of PV-positive GABA neurons appear to be specialized to synchronize the activity of local populations of pyramidal neurons at γ-band frequencies (30–80 Hz; Klausberger et al. 2003; Whittington and Traub 2003; Sohal et al. 2009). Furthermore, γ-band oscillations in the human DLPFC increase in proportion to working memory load (Howard et al. 2003), and the capacity to increase extracellular GABA predicts DLPFC γ-band power during a working memory task in humans (Frankle et al. 2009). In addition, prefrontal γ-band oscillations are reduced during this task in subjects with schizophrenia (Cho et al. 2006). Taken together, these data suggest that a deficit in the synchronization of pyramidal cell firing, resulting from impaired regulation of pyramidal cell networks by PV-positive GABA neurons, could contribute to reduced levels of induced γ-band oscillations, and consequently to impaired working memory in individuals with schizophrenia (Lewis et al. 2005). This hypothesis is supported by recent findings that a novel compound designed to augment GABA neurotransmission selectively at the PV-containing chandelier cell inputs to pyramidal neurons improved both working memory function and prefrontal γ-band oscillations in subjects with schizophrenia (Lewis et al. 2008).

3 Cortical Circuitry Alterations Beyond the DLPFC

Some cortical circuitry alterations in schizophrenia are not restricted to the DLPFC and appear to be relatively widespread across the neocortex. For example, lower dendritic spine density and smaller pyramidal cell somal volumes have been found in other cortical regions such as auditory cortices (Sweet et al. 2003, 2009). Similarly, the same pattern of altered GABA-related gene expression in the DLPFC has also been reported in anterior cingulate, primary visual, and primary motor cortices from subjects with schizophrenia (Hashimoto et al. 2008b). Disturbances in γ-oscillations in schizophrenia have also been observed over a number of cortical regions under different task or stimulus conditions (Spencer et al. 2004; Uhlhaas and Singer 2006; Ferrarelli et al. 2008). Thus, a conserved alteration in GABA neurotransmission across cortical regions could underlie a common abnormality in γ-oscillations that is associated with different clinical features of schizophrenia depending upon the cortical circuits affected.

It is also important to note that alterations in cortical circuitry in schizophrenia are unlikely to be restricted to those discussed earlier. For example, oligodendrocytes, critical mediators of white matter myelination and neuronal development and support, may be dysfunctional and/or reduced in number in schizophrenia (Karoutzou et al. 2008; Fig. 2). These disturbances might contribute to the evidence from functional and structural imaging studies that the connectivity among cortical regions is altered in schizophrenia (Friston and Frith 1995; Fields 2008).

4 Cannabis Use and Schizophrenia

Cannabis use by individuals with schizophrenia (and by adolescents at risk for developing schizophrenia) is becoming recognized as an important public health issue (Moore et al. 2007). Interestingly, cannabis use provides an example of an environmental exposure that may have a deleterious impact on multiple neurotransmitter systems that are already disturbed in schizophrenia and consequently may contribute to worsening of cognitive functioning in the illness. Knowledge of the endogenous cannabinoid (eCB) circuitry that is directly affected by cannabis use may provide additional clues to novel potential pharmacological targets for the disorder.

4.1 Clinical Effects of Cannabis Use in Schizophrenia

Mounting evidence suggests that cannabis use may influence the onset and clinical course of schizophrenia. For example, cannabis use, particularly during adolescence, is associated with an increased risk for developing schizophrenia in a dose-dependent fashion (Zammit et al. 2002; Fergusson et al. 2006; Moore et al. 2007). Furthermore, cannabis use is associated with an earlier onset of psychotic symptoms (Veen et al. 2004; Barnes et al. 2006; Sugranyes et al. 2009), and cannabis use may lead to worse functional outcomes in individuals already afflicted with schizophrenia (Kavanagh et al. 2002; Zammit et al. 2008). Thus, the relatively high prevalence of cannabis use in schizophrenia (Kavanagh et al. 2002) is unlikely

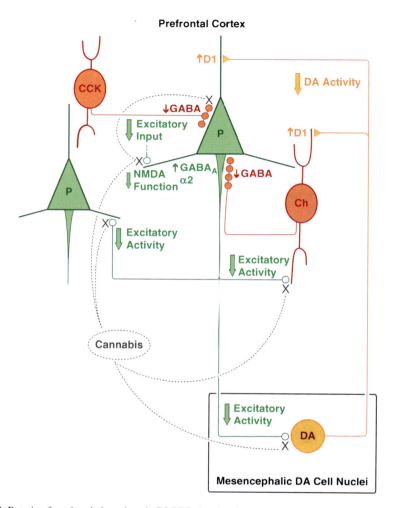

Fig. 3 Putative functional alterations in DLPFC circuitry in schizophrenia and the possible impact of cannabis use. Decreased axonal arbor, somal size, and dendritic spine density (see Fig. 2) of pyramidal neurons (*green*) suggest reduced excitatory input from these neurons onto their primary targets, which include (but are not restricted to) other pyramidal neurons, GABA neurons (*red*), and mesencephalic DA neurons (*orange*). Decreased excitation of midbrain DA neurons may then lead to diminished DA activity in the PFC and a compensatory upregulation of the DA D1 receptor. In addition, chandelier neurons (Ch) and cholecystokinin (CCK)-containing GABA neurons provide deficient inhibitory input to pyramidal neurons due to lower levels of GAD67 (Fig. 2). Since DA normally increases the activity of PV neurons, which include chandelier neurons, a reduction in DA signaling in schizophrenia may further reduce the activity of GABA neurons. Finally, in individuals with schizophrenia, cannabis use, by suppressing neurotransmitter release, may further compound disease-related deficits in GABA, glutamate, and DA neurotransmission and lead to worsening of cognitive deficits in the disorder

to reflect attempts at self-medication and more likely to reflect the presence of shared risk factors (i.e., poverty, less education; Green et al. 2007). Taken together, these data suggest that cannabis abuse may help precipitate schizophrenia in vulnerable individuals and worsen the prognosis in individuals who continue to use cannabis.

Cannabis use results in impairments in multiple cognitive domains relevant to schizophrenia, including working memory (Lichtman et al. 2002; D'Souza et al. 2004). In addition, the cognitive deficits from cannabis use appear to last well beyond the period of acute intoxication and worsen with increasing years of regular use (Solowij et al. 2002), and earlier onset of cannabis use has been linked to greater deficits in cognitive functioning later in life (Ehrenreich et al. 1999; Pope et al. 2003). Cannabis-induced cognitive impairments are also more severe in individuals with schizophrenia compared to control subjects (D'Souza et al. 2005). Thus, DLPFC-related cognitive deficits experienced by individuals with schizophrenia may be particularly susceptible to the deleterious effects of cannabis. Interestingly, the CB1 receptor, which mediates the neural effects of cannabis use, is particularly heavily expressed in certain brain regions in human subjects, including cortical association regions such as the DLPFC (Eggan and Lewis 2007). Thus, cannabis use may result in abnormal activation of the CB1 receptor in an unregulated manner which then disrupts the normal functioning of these brain regions.

4.2 *Potential Impact of Cannabis Use of Altered Neurotransmitter Systems in Schizophrenia*

Exploring the biological basis of the interaction between cannabis use and schizophrenia requires knowledge of the neural circuitry that is directly affected by cannabis: the eCB system. For example, the CB1 receptor is heavily localized to inhibitory axon terminals from CCK-containing neurons in human DLPFC (and several other brain regions; Eggan and Lewis 2007). Repetitive firing of pyramidal neurons, such as during γ-oscillations, results in the synthesis of eCB ligands including 2-arachidonoylglycerol (2-AG) that travel retrogradely to activate CB1 receptors on CCK axon terminals and suppress the release of GABA (Freund et al. 2003). However, cannabis use results in exogenous activation of all CB1 receptors and prolonged suppression of GABA release from all CCK neurons in a manner that lacks spatial and temporal specificity. Thus, cannabis use could substantially disturb synchronized network activity that is normally regulated by CB1 receptor-expressing CCK neurons. These data suggest that one potential link between cannabis use and cognitive impairments in schizophrenia may involve the compounding effects of an environmental insult (cannabis) that further disrupts an already pathologically disturbed GABA system in the PFC (Fig. 3; Volk and Lewis 2005).

Furthermore, recent evidence suggests that eCB signaling can also reduce glutamatergic neurotransmission (Kawamura et al. 2006), although CB1 receptor density appears to be much lower on pyramidal neuron axon terminals than on GABA neuron axon terminals (Kawamura et al. 2006; Katona et al. 2006). Cannabis-induced suppression of glutamatergic activity could further impede intracortical excitatory signaling between pyramidal neurons (Fig. 3), which is already deficient in the disorder (Sect. 2.1). Cannabis-induced suppression of pyramidal neuron activity could then potentially lead to less excitatory drive to GABA neurons, compounding pre-existing problems in GABA signaling in schizophrenia as well (Sect. 2.2).

Endocannabinoid signaling also modulates the activity of DA neurons. For example, under normal conditions, excitatory output from the PFC stimulates midbrain DA neurons (Sect. 2.3), and this excitation also results in the release of 2-AG by the DA neurons (Melis et al. 2004). 2-AG travels retrogradely to the nearby DLPFC axon terminals in the mesencephalon and suppresses glutamate-mediated excitation of DA neurons as a normal feedback regulatory mechanism (Melis et al. 2004). Thus, the use of cannabis may interfere with normal 2-AG-mediated regulation of PFC excitatory inputs to midbrain DA neurons and consequently may excessively suppress an already impaired DLPFC excitation of DA neurons in schizophrenia (Fig. 3; Sect. 2.3).

4.3 Endogenous Cannabinoid System and Schizophrenia

As described earlier, cannabis use in schizophrenia, by exogenously activating CB1 receptor-mediated presynaptic inhibition of neurotransmitter release in an unregulated manner, may aggravate pre-existing deficiencies in glutamatergic, GABAergic, and dopaminergic in schizophrenia, which may in part explain the worsening of cognitive symptoms seen with cannabis use in individuals with the illness (D'Souza et al. 2005). However, additional questions remain.

First, what is the status of eCB signaling in schizophrenia? Lower CB1 receptor mRNA and protein levels have been found in the DLPFC in individuals with schizophrenia regardless of prior cannabis use (Fig. 2), and lower CB1 receptor levels do not appear to be attributable to antipsychotic medications (Eggan et al. 2008). However, do lower CB1 receptor levels reflect an overall deficiency in eCB signaling in the disease, or a downregulation of receptors in response to excessive eCB signaling? Discriminating between these two possibilities requires knowledge of the regulation of the eCB ligands that bind to the CB1 receptor. For example, the eCB 2-AG is synthesized by diacylglycerol lipase (DAGL; Stella et al. 1997) located in pyramidal neurons (Katona et al. 2006; Yoshida et al. 2006) and either activates the CB1 receptor or is degraded by monoglyceride lipase (MGL; Dinh et al. 2002; Gulyas et al. 2004) located in nearby axon terminals (for review, see Placzek et al. 2008). While it is not possible to directly measure 2-AG levels in postmortem human brain tissue (Palkovits et al. 2008), future studies determining the levels of DAGL and/or MGL in the PFC may help elucidate the status of 2-AG signaling in schizophrenia. For example, higher DAGL and/or lower MGL levels would suggest greater 2-AG signaling in schizophrenia, which may further suppress glutamate, GABA, and DA signaling in the disorder, as described earlier.

Second, why does cannabis exposure during adolescence result in an increased risk for developing schizophrenia? During the process of normal adolescent development, the glutamate, GABA, and DA neurotransmitter systems of the DLPFC undergo significant changes (Rosenberg and Lewis 1995; Gonzalez-Burgos et al. 2008; Hashimoto et al. 2009; Cruz et al. 2009a). Perhaps, cannabis use during this sensitive period of cortical development may induce, or worsen pre-existing, alterations in glutamate, GABA, and DA neurotransmitter systems in schizophrenia (Sect. 2). Additional studies, such as long-term administration of cannabis to peripubescent primates, are needed to determine the long-term effects of cannabis exposure on the development of these critical signaling systems in the PFC.

5 From Pathology to New Therapeutic Approaches

The findings reviewed earlier suggest that impaired working memory is attributable, at least in part, to a complex set of alterations in cortical circuitry that involve, but are not restricted to, disturbances in glutamate, GABA, and DA neurotransmission in the DLPFC. The frequency of alterations in these cortical circuits is common enough to be consistently detected in different cohorts of subjects identified by a common set of diagnostic criteria, which suggests that it may be possible to develop novel treatments that are efficacious in many people with schizophrenia, despite the marked heterogeneity in the etiology and clinical phenotype across afflicted individuals. Nonetheless, the extent to which some of these alterations are restricted to only certain types of individuals with schizophrenia remains to be determined. This knowledge is essential for the future advent of personalized approaches to the treatment of schizophrenia. In addition, since the emergence of working memory depends upon more distributed cortical networks, the alterations within local circuits must be considered within the broader organization of the cortex and its connections with subcortical structures.

Any given alteration in DLPFC circuitry could represent a cause (an upstream factor related to the disease pathogenesis), consequence (a deleterious effect of a cause), or compensation (a response to either cause or consequence that helps restore homeostasis; Lewis and Gonzalez-Burgos 2008). Investigating and understanding these distinctions is critical for drug design to determine the appropriate mode of action of the drug. For example, the hypothesis that $GABA_A$ receptors containing α_2-subunits are upregulated in pyramidal neurons due to a deficit in GABA input from chandelier neurons led to the use of a novel, positive allosteric modulator of this receptor subtype that improved both working memory function and prefrontal γ-band oscillations in a small randomized controlled trial of subjects with schizophrenia (Lewis et al. 2008). Similarly, the idea that DA D1 receptors are upregulated to compensate for a deficient DA innervation of the DLPFC has motivated attempts to develop selective approaches for modulating activity at cortical DA D1 receptors (Goldman-Rakic et al. 2004). In this regard, PET-based assessments of the degree of D1 receptor upregulation in individual patients may help in guiding therapy to maximize the likelihood of obtaining optimal levels of D1 receptor stimulation.

Furthermore, recent studies demonstrating the relationship between cannabis use and cognitive impairments in schizophrenia have provided additional clues to novel potential pharmacological targets. For example, since the deleterious effects of cannabis use in schizophrenia may be attributable to further suppression of already impaired glutamate, GABA, and DA signaling in schizophrenia, a novel pharmacological approach might involve an inhibitor of DAGL which would slow 2-AG-mediated suppression of these disturbed neurotransmitter systems. In addition, since the lipophilic 2-AG is produced on-demand and is not stored in synaptic vesicles, DAGL inhibitors would reduce the rate of 2-AG production as it naturally occurs. Importantly, DAGL inhibitors would selectively reduce the synthesis of 2-AG but not other eCB ligands such as anandamide (Stella et al. 1997).

Finally, analyses of pathological circuits might lead in the future to the identification and validation of new types of therapeutic targets beyond the manipulation of neurotransmitter systems. For example, spine-specific kinases whose activity regulates spine size, number, and function might be of potential value as novel targets (Hill et al. 2006; Penzes and Jones 2008). If the adolescence-related pruning of dendritic spines is, as discussed earlier, critical in the emergence of the clinical features of schizophrenia, then such compounds might provide a means for secondary prevention through early intervention in high-risk individuals.

Acknowledgments Cited work conducted by the authors was supported by NIH grants MH043784, MH045156, and MH084053 from the National Institute of Mental Health. The content is solely the responsibility of the authors and does not necessarily represent the official views of the National Institute of Mental Health or the National Institutes of Health. The authors thank Mary Brady for preparation of the figures and Lindsay Karr for preparation of the manuscript.

References

Abi-Dargham A, Mawlawi O, Lombardo I, Gill R, Martinez D, Huang Y, Hwang DR, Keilp J, Kochan L, van Heertum R, Gorman JM, Laruelle M (2002) Prefrontal dopamine D1 receptors and working memory in schizophrenia. J Neurosci 22:3708–3719

Akbarian S, Kim JJ, Potkin SG, Hagman JO, Tafazzoli A, JrWE B, Jones EG (1995) Gene expression for glutamic acid decarboxylase is reduced without loss of neurons in prefrontal cortex of schizophrenics. Arch Gen Psychiatry 52:258–266

Akil M, Pierri JN, Whitehead RE, Edgar CL, Mohila C, Sampson AR, Lewis DA (1999) Lamina-specific alterations in the dopamine innervation of the prefrontal cortex in schizophrenic subjects. Am J Psychiatry 156:1580–1589

Anderson SA, Classey JD, Condé F, Lund JS, Lewis DA (1995) Synchronous development of pyramidal neuron dendritic spines and parvalbumin-immunoreactive chandelier neuron axon terminals in layer III of monkey prefrontal cortex. Neuroscience 67:7–22

Ango F, Di Cristo G, Higashiyama H, Bennett V, Wu P, Huang ZJ (2004) Ankyrin-based subcellular gradient of neurofascin, an immunoglobulin family protein, directs GABAergic innervation at purkinje axon initial segment. Cell 119:257–272

Ascoli GA, Alonso-Nanclares L, Anderson SA, Barrionuevo G, Benavides-Piccione R, Burkhalter A, Buzsaki G, Cauli B, DeFelipe J, Fairen A, Feldmeyer D, Fishell G, Fregnac Y, Freund TF, Gardner D, Gardner EP, Goldberg JH, Helmstaedter M, Hestrin S, Karube F, Kisvarday ZF, Lambolez B, Lewis DA, Marin O, Markram H, Munoz A, Packer A, Petersen CC, Rockland KS, Rossier J, Rudy B, Somogyi P, Staiger JF, Tamas G, Thomson AM, Toledo-Rodriguez M, Wang Y, West DC, Yuste R (2008) Petilla terminology: nomenclature of features of GABAergic interneurons of the cerebral cortex. Nat Rev Neurosci 9:557–568

Barch DM (2006) What can research on schizophrenia tell us about the cognitive neuroscience of working memory? Neuroscience 139:73–84

Barch DM, Smith E (2008) The cognitive neuroscience of working memory: relevance to CNTRICS and schizophrenia. Biol Psychiatry 64:11–17

Barch DM, Carter CS, Braver TS, Sabb FW, MacDonald A, Noll DC, Cohen JD (2001) Selective deficits in prefrontal cortex function in medication-naive patients with schizophrenia. Arch Gen Psychiatry 58:280–288

Barnes TR, Mutsatsa SH, Hutton SB, Watt HC, Joyce EM (2006) Comorbid substance use and age at onset of schizophrenia. Br J Psychiatry 188:237–242

Black JE, Kodish IM, Grossman AW, Klintsova AY, Orlovskaya D, Vostrikov V, Uranova N, Greenough WT (2004) Pathology of layer V pyramidal neurons in the prefrontal cortex of patients with schizophrenia. Am J Psychiatry 161:742–744

Bogerts B, Hantsch J, Herzer M (1983) A morphometric study of the dopamine-containing cell groups in the mesencephalon of normals, Parkinson patients, and schizophrenics. Biol Psychiatry 18:951–969

Bourgeois J-P, Goldman-Rakic PS, Rakic P (1994) Synaptogenesis in the prefrontal cortex of rhesus monkeys. Cereb Cortex 4:78–96

Breier A, Schreiber JL, Dyer J, Pickar D (1991) National Institute of Mental Health longitudinal study of chronic schizophrenia. Arch Gen Psychiatry 48:239–246

Burmeister M, McInnis MG, Zollner S (2008) Psychiatric genetics: progress amid controversy. Nat Rev Genet 9:527–540

Byne W, Buchsbaum MS, Kemether E, Hazlett EA, Shinwari A, Mitropoulou V, Siever LJ (2001) Magnetic resonance imaging of the thalamic mediodorsal nucleus and pulvinar in schizophrenia and schizotypal personality disorder. Arch Gen Psychiatry 58:133–140

Byne W, Buchsbaum MS, Mattiace LA, Hazlett EA, Kemether E, Elhakem SL, Purohit DP, Haroutunian V, Jones L (2002) Postmortem assessment of thalamic nuclear volumes in subjects with schizophrenia. Am J Psychiatry 159:59–65

Callicott JH, Mattay VS, Verchinski BA, Marenco S, Egan MF, Weinberger DR (2003) Complexity of prefrontal cortical dysfunction in schizophrenia: more than up or down. Am J Psychiatry 160:2209–2215

Cannon TD, Glahn DC, Kim J, Van Erp TG, Karlsgodt K, Cohen MS, Nuechterlein KH, Bava S, Shirinyan D (2005) Dorsolateral prefrontal cortex activity during maintenance and manipulation of information in working memory in patients with schizophrenia. Arch Gen Psychiatry 62:1071–1080

Chen J, Lipska BK, Halim N, Ma QD, Matsumoto M, Melhem S, Kolachana BS, Hyde TM, Herman MM, Apud J, Egan MF, Kleinman JE, Weinberger DR (2004) Functional analysis of genetic variation in catechol-O-methyltransferase (COMT): effects on mRNA, protein, and enzyme activity in postmortem human brain. Am J Hum Genet 75:807–821

Cho RY, Konecky RO, Carter CS (2006) Impairments in frontal cortical gamma synchrony and cognitive control in schizophrenia. Proc Natl Acad Sci USA 103:19878–19883

Condé F, Lund JS, Jacobowitz DM, Baimbridge KG, Lewis DA (1994) Local circuit neurons immunoreactive for calretinin, calbindin D-28k, or parvalbumin in monkey prefrontal cortex: distribution and morphology. J Comp Neurol 341:95–116

Cruz DA, Lovallo EM, Stockton S, Rasband M, Lewis DA (2009a) Postnatal development of synaptic structure proteins in pyramidal neuron axon initial segments in monkey prefrontal cortex. J Comp Neurol 514:353–367

Cruz DA, Weaver CL, Lovallo EM, Melchitzky DS, Lewis DA (2009b) Selective alterations in postsynaptic markers of chandelier cell inputs to cortical pyramidal neurons in subjects with schizophrenia. Neuropsychopharmacology 34:2112–2124

Danos P, Baumann B, Krämer A, Bernstein H-G, Stauch R, Krell D, Falkai P, Bogerts B (2003) Volumes of association thalamic nuclei in schizophrenia: a postmortem study. Schizophr Res 60:141–155

Davidson M, Reichenberg A, Rabinowitz J, Weiser M, Kaplan Z, Mark M (1999) Behavioral and intellectual markers for schizophrenia in apparently healthy male adolescents. Am J Psychiatry 156:1328–1335

Dinh TP, Carpenter D, Leslie FM, Freund TF, Katona I, Sensi SL, Kathuria S, Piomelli D (2002) Brain monoglyceride lipase participating in endocannabinoid inactivation. Proc Natl Acad Sci USA 99:10819–10824

Dorph-Petersen K-A, Pierri JN, Sun Z, Sampson AR, Lewis DA (2004) Stereological analysis of the mediodorsal thalamic nucleus in schizophrenia: volume, neuron number, and cell types. J Comp Neurol 472:449–462

D'Souza DC, Perry E, MacDougall L, Ammerman Y, Cooper T, Wu YT, Braley G, Gueorguieva R, Krystal JH (2004) The psychotomimetic effects of intravenous delta-9-tetrahydrocannabinol in healthy individuals: implications for psychosis. Neuropsychopharmacology 29:1558–1572

D'Souza DC, Abi-Saab WM, Madonick S, Forselius-Bielen K, Doersch A, Braley G, Gueorguieva R, Cooper TB, Krystal JH (2005) Delta-9-tetrahydrocannabinol effects in schizophrenia: implications for cognition, psychosis, and addiction. Biol Psychiatry 57:594–608

Dudkin KN, Kruchinin VK, Chueva IV (2001) Neurophysiological correlates of delayed visual differentiation tasks in monkeys: the effects of the site of intracortical blockade of NMDA receptors. Neurosci Behav Physiol 31:207–218

Eggan SM, Lewis DA (2007) Immunocytochemical distribution of the cannabinoid CB1 receptor in the primate neocortex: a regional and laminar analysis. Cereb Cortex 17:175–191

Eggan SM, Hashimoto T, Lewis DA (2008) Reduced cortical cannabinoid 1 receptor messenger RNA and protein expression in schizophrenia. Arch Gen Psychiatry 65:772–784

Ehrenreich H, Rinn T, Kunert HJ, Moeller MR, Poser W, Schilling L, Gigerenzer G, Hoehe MR (1999) Specific attentional dysfunction in adults following early start of cannabis use. Psychopharmacology (Berl) 142:295–301

Elvevag B, Goldberg TE (2000) Cognitive impairment in schizophrenia is the core of the disorder. Crit Rev Neurobiol 14:1–21

Erickson SL, Lewis DA (2004) Cortical connections of the lateral mediodorsal thalamus in cynomolgus monkeys. J Comp Neurol 473:107–127

Fergusson DM, Poulton R, Smith PF, Boden JM (2006) Cannabis and psychosis. BMJ 332:172–175

Ferrarelli F, Massimini M, Peterson MJ, Riedner BA, Lazar M, Murphy MJ, Huber R, Rosanova M, Alexander AL, Kalin N, Tononi G (2008) Reduced evoked gamma oscillations in the frontal cortex in schizophrenia patients: a TMS/EEG study. Am J Psychiatry 165:996–1005

Fields RD (2008) White matter in learning, cognition and psychiatric disorders. Trends Neurosci 31:361–370

Frankle WG, Laruelle M, Haber SN (2006) Prefrontal cortical projections to the midbrain in primates: evidence for a sparse connection. Neuropsychopharmacology 31:1627–1636

Frankle WG, Cho RY, Narendran R, Mason NS, Vora S, Litschge M, Price JC, Lewis DA, Mathis CA (2009) Tiagabine increases [^{11}C]flumazenil binding in cortical brain regions in healthy control subjects. Neuropsychopharmacology 34:624–633

Freund TF, Katona I (2007) Perisomatic inhibition. Neuron 56:33–42
Freund TF, Katona I, Piomelli D (2003) Role of endogenous cannabinoids in synaptic signaling. Physiol Rev 83:1017–1066
Friston KJ, Frith CD (1995) Schizophrenia: a disconnection syndrome? Clin Neurosci 3:89–97
Garey LJ, Ong WY, Patel TS, Kanani M, Davis A, Mortimer AM, Barnes TRE, Hirsch SR (1998) Reduced dendritic spine density on cerebral cortical pyramidal neurons in schizophrenia. J Neurol Neurosurg Psychiatry 65:446–453
Gilbert AR, Rosenberg DR, Harenski K, Spencer S, Sweeney JA, Keshavan MS (2001) Thalamic volumes in patients with first-episode schizophrenia. Am J Psychiatry 158:618–624
Glantz LA, Lewis DA (1997) Reduction of synaptophysin immunoreactivity in the prefrontal cortex of subjects with schizophrenia: regional and diagnostic specificity. Arch Gen Psychiatry 54:943–952
Glantz LA, Lewis DA (2000) Decreased dendritic spine density on prefrontal cortical pyramidal neurons in schizophrenia. Arch Gen Psychiatry 57:65–73
Goldman-Rakic PS, Castner SA, Svensson TH, Siever LJ, Williams GV (2004) Targeting the dopamine D1 receptor in schizophrenia: insights for cognitive dysfunction. Psychopharmacology (Berl) 174:3–16
Gonzalez-Burgos G, Lewis DA (2008) GABA neurons and the mechanisms of network oscillations: implications for understanding cortical dysfunction in schizophrenia. Schizophr Bull 34:944–961
Gonzalez-Burgos G, Kroener S, Zaitsev AV, Povysheva NV, Krimer LS, Barrionuevo G, Lewis DA (2008) Functional maturation of excitatory synapses in layer 3 pyramidal neurons during postnatal development of the primate prefrontal cortex. Cereb Cortex 18:626–637
Green MF (1996) What are the functional consequences of neurocognitive deficits in schizophrenia? Am J Psychiatry 153:321–330
Green AI, Drake RE, Brunette MF, Noordsy DL (2007) Schizophrenia and co-occurring substance use disorder. Am J Psychiatry 164:402–408
Gulyas AI, Cravatt BF, Bracey MH, Dinh TP, Piomelli D, Boscia F, Freund TF (2004) Segregation of two endocannabinoid-hydrolyzing enzymes into pre- and postsynaptic compartments in the rat hippocampus, cerebellum and amygdala. Eur J Neurosci 20:441–458
Guo N, Hwang DR, Lo ES, Huang YY, Laruelle M, Abi-Dargham A (2003) Dopamine depletion and in vivo binding of PET D1 receptor radioligands: implications for imaging studies in schizophrenia. Neuropsychopharmacology 28:1703–1711
Hahn CG, Wang HY, Cho DS, Talbot K, Gur RE, Berrettini WH, Bakshi K, Kamins J, Borgmann-Winter KE, Siegel SJ, Gallop RJ, Arnold SE (2006) Altered neuregulin 1-erbB4 signaling contributes to NMDA receptor hypofunction in schizophrenia. Nat Med 12:734–735
Harvey PD, Howanitz E, Parrella M, White L, Davidson M, Mohs RC, Hoblyn J, Davis KL (1998) Symptoms, cognitive functioning, and adaptive skills in geriatric patients with lifelong schizophrenia: a comparison across treatment sites. Am J Psychiatry 155:1080–1086
Hashimoto T, Volk DW, Eggan SM, Mirnics K, Pierri JN, Sun Z, Sampson AR, Lewis DA (2003) Gene expression deficits in a subclass of GABA neurons in the prefrontal cortex of subjects with schizophrenia. J Neurosci 23:6315–6326
Hashimoto T, Arion D, Unger T, Maldonado-Aviles JG, Morris HM, Volk DW, Mirnics K, Lewis DA (2008a) Alterations in GABA-related transcriptome in the dorsolateral prefrontal cortex of subjects with schizophrenia. Mol Psychiatry 13:147–161
Hashimoto T, Bazmi HH, Mirnics K, Wu Q, Sampson AR, Lewis DA (2008b) Conserved regional patterns of GABA-related transcript expression in the neocortex of subjects with schizophrenia. Am J Psychiatry 165:479–489
Hashimoto T, Nguyen QL, Rotaru D, Keenan T, Arion D, Beneyto M, Gonzalez-Burgos G, Lewis DA (2009) Protracted developmental trajectories of $GABA_A$ receptor $\alpha 1$ and $\alpha 2$ subunit expression in primate prefrontal cortex. Biol Psychiatry 65:1015–1023
Heaton R, Paulsen JS, McAdams LA, Kuck J, Zisook S, Braff D, Harris J, Jeste DV (1994) Neuropsychological deficits in schizophrenics: relationship to age, chronicity, and dementia. Arch Gen Psychiatry 51:469–476

Highley JR, Walker MA, Crow TJ, Esiri MM, Harrison PJ (2003) Low medial and lateral right pulvinar volumes in schizophrenia: a postmortem study. Am J Psychiatry 160:1177–1179

Hill JJ, Hashimoto T, Lewis DA (2006) Molecular mechanisms contributing to dendritic spine alterations in the prefrontal cortex of subjects with schizophrenia. Mol Psychiatry 11:557–566

Howard MW, Rizzuto DS, Caplan JB, Madsen JR, Lisman J, Aschenbrenner-Scheibe R, Schulze-Bonhage A, Kahana MJ (2003) Gamma oscillations correlate with working memory load in humans. Cereb Cortex 13:1369–1374

Huttenlocher PR, Dabholkar AS (1997) Regional differences in synaptogenesis in human cerebral cortex. J Comp Neurol 387:167–178

Insel TR, Scolnick EM (2006) Cure therapeutics and strategic prevention: raising the bar for mental health research. Mol Psychiatry 11:11–17

Javitt DC, Zukin SR (1991) Recent advances in the phencyclidine model of schizophrenia. Am J Psychiatry 148:1301–1308

Karlsson P, Farde L, Halldin C, Sedvall G (2002) PET study of D_1 dopamine receptor binding in neuroleptic-naive patients with schizophrenia. Am J Psychiatry 159:761–767

Karoutzou G, Emrich HM, Dietrich DE (2008) The myelin-pathogenesis puzzle in schizophrenia: a literature review. Mol Psychiatry 13:245–260

Karson MA, Tang AH, Milner TA, Alger BE (2009) Synaptic cross talk between perisomatic-targeting interneuron classes expressing cholecystokinin and parvalbumin in hippocampus. J Neurosci 29:4140–4154

Katona I, Urban GM, Wallace M, Ledent C, Jung KM, Piomelli D, Mackie K, Freund TF (2006) Molecular composition of the endocannabinoid system at glutamatergic synapses. J Neurosci 26:5628–5637

Kavanagh DJ, McGrath J, Saunders JB, Dore G, Clark D (2002) Substance misuse in patients with schizophrenia: epidemiology and management. Drugs 62:743–755

Kawamura Y, Fukaya M, Maejima T, Yoshida T, Miura E, Watanabe M, Ohno-Shosaku T, Kano M (2006) The CB1 cannabinoid receptor is the major cannabinoid receptor at excitatory presynaptic sites in the hippocampus and cerebellum. J Neurosci 26:2991–3001

Keefe RS, Bilder RM, Davis SM, Harvey PD, Palmer BW, Gold JM, Meltzer HY, Green MF, Capuano G, Stroup TS, McEvoy JP, Swartz MS, Rosenheck RA, Perkins DO, Davis CE, Hsiao JK, Lieberman JA (2007a) Neurocognitive effects of antipsychotic medications in patients with chronic schizophrenia in the CATIE trial. Arch Gen Psychiatry 64:633–647

Keefe RS, Sweeney JA, Gu H, Hamer RM, Perkins DO, McEvoy JP, Lieberman JA (2007b) Effects of olanzapine, quetiapine, and risperidone on neurocognitive function in early psychosis: a randomized, double-blind 52-week comparison. Am J Psychiatry 164:1061–1071

Kemether EM, Buchsbaum MS, Byne W, Hazlett EA, Haznedar M, Brickman AM, Platholi J, Bloom R (2003) Magnetic resonance imaging of mediodorsal, pulvinar, and centromedian nuclei of the thalamus in patients with schizophrenia. Arch Gen Psychiatry 60:983–991

Klausberger T, Magill PJ, Marton LF, Roberts JDB, Cobden PM, Buzsaki G, Somogyi P (2003) Brain-state- and cell-type specific firing of hippocampal interneurons in vivo. Nature 421:844–848

Kolluri N, Sun Z, Sampson AR, Lewis DA (2005) Lamina-specific reductions in dendritic spine density in the prefrontal cortex of subjects with schizophrenia. Am J Psychiatry 162:1200–1202

Kristiansen LV, Huerta I, Beneyto M, Meador-Woodruff JH (2007) NMDA receptors and schizophrenia. Curr Opin Pharmacol 7:48–55

Krystal JH, Karper LP, Seibyl JP, Freeman GK, Delaney R, Bremner JD, Heninger GR, Bowers MB Jr, Charney DS (1994) Subanesthetic effects of the noncompetitive NMDA antagonist, ketamine, in humans. Psychotomimetic, perceptual, cognitive, and neuroendocrine responses. Arch Gen Psychiatry 51:199–214

Lewis DA, Gonzalez-Burgos G (2008) Neuroplasticity of neocortical circuits in schizophrenia. Neuropsychopharmacol Rev 33:141–165

Lewis DA, Sweet RA (2009) Schizophrenia from a neural circuitry perspective: advancing toward rational pharmacological therapies. J Clin Invest 119:706–716

Lewis DA, Hashimoto T, Volk DW (2005) Cortical inhibitory neurons and schizophrenia. Nat Rev Neurosci 6:312–324

Lewis DA, Cho RY, Carter CS, Eklund K, Forster S, Kelly MA, Montrose D (2008) Subunit-selective modulation of GABA type A receptor neurotransmission and cognition in schizophrenia. Am J Psychiatry 165:1585–1593

Lichtman AH, Varvel SA, Martin BR (2002) Endocannabinoids in cognition and dependence. Prostaglandins Leukot Essent Fatty Acids 66:269–285

Lisman JE, Coyle JT, Green RW, Javitt DC, Benes FM, Heckers S, Grace AA (2008) Circuit-based framework for understanding neurotransmitter and risk gene interactions in schizophrenia. Trends Neurosci 31:234–242

MacDonald AW III, Pogue-Geile MF, Johnson MK, Carter CS (2003) A specific deficit in context processing in the unaffected siblings of patients with schizophrenia. Arch Gen Psychiatry 60:57–65

MacDonald AW III, Carter CS, Kerns JG, Ursu S, Barch DM, Holmes AJ, Stenger VA, Cohen JD (2005) Specificity of prefrontal dysfunction and context processing deficits to schizophrenia in never-medicated patients with first-episode psychosis. Am J Psychiatry 162:475–484

Manoach DS (2003) Prefrontal cortex dysfunction during working memory performance in schizophrenia: reconciling discrepant findings. Schizophr Res 60:285–298

Matsumoto M, Weickert CS, Beltaifa S, Kolachana B, Chen J, Hyde TM, Herman MM, Weinberger DR, Kleinman JE (2003) Catechol O-methyltransferase (COMT) mRNA expression in the dorsolateral prefrontal cortex of patients with schizophrenia. Neuropsychopharmacology 28:1521–1530

McGlashan TH, Hoffman RE (2000) Schizophrenia as a disorder of developmentally reduced synaptic connectivity. Arch Gen Psychiatry 57:637–648

Meador-Woodruff JH, Haroutunian V, Powchik P, Davidson M, Davis KL, Watson SJ (1997) Dopamine receptor transcript expression in striatum and prefrontal occipital cortex. Arch Gen Psychiatry 54:1089–1095

Melchitzky DS, González-Burgos G, Barrionuevo G, Lewis DA (2001) Synaptic targets of the intrinsic axon collaterals of supragranular pyramidal neurons in monkey prefrontal cortex. J Comp Neurol 430:209–221

Melis M, Perra S, Muntoni AL, Pillolla G, Lutz B, Marsicano G, Di Marzo V, Gessa GL, Pistis M (2004) Prefrontal cortex stimulation induces 2-arachidonoyl-glycerol-mediated suppression of excitation in dopamine neurons. J Neurosci 24:10707–10715

Mirnics K, Middleton FA, Marquez A, Lewis DA, Levitt P (2000) Molecular characterization of schizophrenia viewed by microarray analysis of gene expression in prefrontal cortex. Neuron 28:53–67

Moghaddam B (2004) Targeting metabotropic glutamate receptors for treatment of the cognitive symptoms of schizophrenia. Psychopharmacology (Berl) 174:39–44

Moghaddam B, Adams BW (1998) Reversal of phencyclidine effects by a group II metabotropic glutamate receptor agonist in rats. Science 281:1349–1352

Moore TH, Zammit S, Lingford-Hughes A, Barnes TR, Jones PB, Burke M, Lewis G (2007) Cannabis use and risk of psychotic or affective mental health outcomes: a systematic review. Lancet 370:319–328

Morris HM, Hashimoto T, Lewis DA (2008) Alterations in somatostatin mRNA expression in the dorsolateral prefrontal cortex of subjects with schizophrenia or schizoaffective disorder. Cereb Cortex 18:1575–1587

Nusser Z, Sieghart W, Benke D, Fritschy J-M, Somogyi P (1996) Differential synaptic localization of two major γ-aminobutyric acid type A receptor α subunits on hippocampal pyramidal cells. Proc Natl Acad Sci USA 93:11939–11944

Ohnuma T, Augood SJ, Arai H, McKenna PJ, Emson PC (1999) Measurement of GABAergic parameters in the prefrontal cortex in schizophrenia: focus on GABA content, $GABA_A$

receptor α-1 subunit messenger RNA and human GABA transporter-1 (HGAT-1) messenger RNA expression. Neuroscience 93:441–448

Okubo Y, Suhara T, Suzuki K, Kobayashi K, Inoue O, Terasaki O, Someya Y, Sassa T, Sudo Y, Matsuchima E, Iyo M, Tateno Y, Toru M (1997) Decreased prefrontal dopamine D1 receptors in schizophrenia revealed by PET. Nature 385:634–636

Palkovits M, Harvey-White J, Liu J, Kovacs ZS, Bobest M, Lovas G, Bago AG, Kunos G (2008) Regional distribution and effects of postmortal delay on endocannabinoid content of the human brain. Neuroscience 152:1032–1039

Patil ST, Zhang L, Martenyi F, Lowe SL, Jackson KA, Andreev BV, Avedisova AS, Bardenstein LM, Gurovich IY, Morozova MA, Mosolov SN, Neznanov NG, Reznik AM, Smulevich AB, Tochilov VA, Johnson BG, Monn JA, Schoepp DD (2007) Activation of mGlu2/3 receptors as a new approach to treat schizophrenia: a randomized Phase 2 clinical trial. Nat Med 13:1102–1107

Penzes P, Jones KA (2008) Dendritic spine dynamics – a key role for kalirin-7. Trends Neurosci 31:419–427

Perez-Costas E, Melendez-Ferro M, Gao XM, Conley R, Roberts RC (2007) Decreased expression of tyrosine hydroxylase in the substantia nigra of schizophrenia brains. Schizophr Bull 33:266

Perrone-Bizzozero NI, Sower AC, Bird ED, Benowitz LI, Ivins KJ, Neve RL (1996) Levels of the growth-associated protein GAP-43 are selectively increased in association cortices in schizophrenia. Proc Natl Acad Sci USA 93:14182–14187

Pierri JN, Volk CLE, Auh S, Sampson A, Lewis DA (2001) Decreased somal size of deep layer 3 pyramidal neurons in the prefrontal cortex of subjects with schizophrenia. Arch Gen Psychiatry 58:466–473

Placzek EA, Okamoto Y, Ueda N, Barker EL (2008) Membrane microdomains and metabolic pathways that define anandamide and 2-arachidonyl glycerol biosynthesis and breakdown. Neuropharmacology 55:1095–1104

Pope HG Jr, Gruber AJ, Hudson JI, Cohane G, Huestis MA, Yurgelun-Todd D (2003) Early-onset cannabis use and cognitive deficits: what is the nature of the association? Drug Alcohol Depend 69:303–310

Rajkowska G, Selemon LD, Goldman-Rakic PS (1998) Neuronal and glial somal size in the prefrontal cortex: a postmortem morphometric study of schizophrenia and Huntington disease. Arch Gen Psychiatry 55:215–224

Rao SG, Williams GV, Goldman-Rakic PS (2000) Destruction and creation of spatial tuning by disinhibition: $GABA_A$ blockade of prefrontal cortical neurons engaged by working memory. J Neurosci 20:485–494

Rapoport JL, Gogtay N (2008) Brain neuroplasticity in healthy, hyperactive and psychotic children: insights from neuroimaging. Neuropsychopharmacology 33:181–197

Rosenberg DR, Lewis DA (1995) Postnatal maturation of the dopaminergic innervation of monkey prefrontal and motor cortices: a tyrosine hydroxylase immunohistochemical analysis. J Comp Neurol 358:383–400

Saykin AJ, Shtasel DL, Gur RE, Kester DB, Mozley LH, Stafiniak P, Gur RC (1994) Neuropsychological deficits in neuroleptic naive patients with first-episode schizophrenia. Arch Gen Psychiatry 51:124–131

Selemon LD, Goldman-Rakic PS (1999) The reduced neuropil hypothesis: a circuit based model of schizophrenia. Biol Psychiatry 45:17–25

Sesack SR, Carr DB (2002) Selective prefrontal cortex inputs to dopamine cells: implications for schizophrenia. Physiol Behav 77:513–517

Sohal VS, Zhang F, Yizhar O, Deisseroth K (2009) Parvalbumin neurons and gamma rhythms enhance cortical circuit performance. Nature 459:698–702

Solowij N, Stephens RS, Roffman RA, Babor T, Kadden R, Miller M, Christiansen K, McRee B, Vendetti J (2002) Cognitive functioning of long-term heavy cannabis users seeking treatment. JAMA 287:1123–1131

Spencer KM, Nestor PG, Perlmutter R, Niznikiewicz MA, Klump MC, Frumin M, Shenton ME, McCarley RW (2004) Neural synchrony indexes disordered perception and cognition in schizophrenia. Proc Natl Acad Sci USA 101:17288–17293
Stella N, Schweitzer P, Piomelli D (1997) A second endogenous cannabinoid that modulates long-term potentiation. Nature 388:773–778
Sugranyes G, Flamarique I, Parellada E, Baeza I, Goti J, Fernandez-Egea E, Bernardo M (2009) Cannabis use and age of diagnosis of schizophrenia. Eur Psychiatry 24:282–286
Susuki K, Rasband MN (2008) Spectrin and ankyrin-based cytoskeletons at polarized domains in myelinated axons. Exp Biol Med (Maywood) 233:394–400
Sweet RA, Pierri JN, Auh S, Sampson AR, Lewis DA (2003) Reduced pyramidal cell somal volume in auditory association cortex of subjects with schizophrenia. Neuropsychopharmacology 28:599–609
Sweet RA, Henteleff RA, Zhang W, Sampson AR, Lewis DA (2009) Reduced dendritic spine density in auditory cortex of subjects with schizophrenia. Neuropsychopharmacology 34:374–389
Tan HY, Choo WC, Fones CS, Chee MW (2005) fMRI study of maintenance and manipulation processes within working memory in first-episode schizophrenia. Am J Psychiatry 162:1849–1858
Tan HY, Callicott JH, Weinberger DR (2007) Dysfunctional and compensatory prefrontal cortical systems, genes and the pathogenesis of schizophrenia. Cereb Cortex 17(Suppl 1):i171–i181
Thune JJ, Uylings HBM, Pakkenberg B (2001) No deficit in total number of neurons in the prefrontal cortex in schizophrenics. J Psychiatr Res 35:15–21
Tunbridge E, Burnet PW, Sodhi MS, Harrison PJ (2004a) Catechol-o-methyltransferase (COMT) and proline dehydrogenase (PRODH) mRNAs in the dorsolateral prefrontal cortex in schizophrenia, bipolar disorder, and major depression. Synapse 51:112–118
Tunbridge EM, Bannerman DM, Sharp T, Harrison PJ (2004b) Catechol-o-methyltransferase inhibition improves set-shifting performance and elevates stimulated dopamine release in the rat prefrontal cortex. J Neurosci 24:5331–5335
Uhlhaas PJ, Singer W (2006) Neural synchrony in brain disorders: relevance for cognitive dysfunctions and pathophysiology. Neuron 52:155–168
Veen ND, Selten JP, van der Tweel I, Feller WG, Hoek HW, Kahn RS (2004) Cannabis use and age at onset of schizophrenia. Am J Psychiatry 161:501–506
Verma A, Moghaddam B (1996) NMDA receptor antagonists impair prefrontal cortex function as assessed via spatial delayed alternation performance in rats: modulation by dopamine. J Neurosci 16:373–379
Volk DW, Lewis DA (2005) GABA targets for the treatment of cognitive dysfunction in schizophrenia. Curr Neuropharmacol 3:45–62
Volk DW, Austin MC, Pierri JN, Sampson AR, Lewis DA (2000) Decreased glutamic acid decarboxylase67 messenger RNA expression in a subset of prefrontal cortical gamma-aminobutyric acid neurons in subjects with schizophrenia. Arch Gen Psychiatry 57:237–245
Volk DW, Austin MC, Pierri JN, Sampson AR, Lewis DA (2001) GABA transporter-1 mRNA in the prefrontal cortex in schizophrenia: decreased expression in a subset of neurons. Am J Psychiatry 158:256–265
Volk DW, Pierri JN, Fritschy J-M, Auh S, Sampson AR, Lewis DA (2002) Reciprocal alterations in pre- and postsynaptic inhibitory markers at chandelier cell inputs to pyramidal neurons in schizophrenia. Cereb Cortex 12:1063–1070
Vreugdenhil M, Jefferys JG, Celio MR, Schwaller B (2003) Parvalbumin-deficiency facilitates repetitive IPSCs and gamma oscillations in the hippocampus. J Neurophysiol 89:1414–1422
Whittington MA, Traub RD (2003) Interneuron diversity series: inhibitory interneurons and network oscillations in vitro. Trends Neurosci 26:676–682
Woo T-U, Miller JL, Lewis DA (1997) Schizophrenia and the parvalbumin-containing class of cortical local circuit neurons. Am J Psychiatry 154:1013–1015

Woo T-U, Whitehead RE, Melchitzky DS, Lewis DA (1998) A subclass of prefrontal gamma-aminobutyric acid axon terminals are selectively altered in schizophrenia. Proc Natl Acad Sci USA 95:5341–5346

Yoshida T, Fukaya M, Uchigashima M, Miura E, Kamiya H, Kano M, Watanabe M (2006) Localization of diacylglycerol lipase-alpha around postsynaptic spine suggests close proximity between production site of an endocannabinoid, 2-arachidonoyl-glycerol, and presynaptic cannabinoid CB1 receptor. J Neurosci 26:4740–4751

Zammit S, Allebeck P, Andreasson S, Lundberg I, Lewis G (2002) Self reported cannabis use as a risk factor for schizophrenia in Swedish conscripts of 1969: historical cohort study. BMJ 325:1199–1204

Zammit S, Moore TH, Lingford-Hughes A, Barnes TR, Jones PB, Burke M, Lewis G (2008) Effects of cannabis use on outcomes of psychotic disorders: systematic review. Br J Psychiatry 193:357–363

Zhou DX, Lambert S, Malen PL, Carpenter S, Boland LM, Bennett V (1998) Ankyrin(G) is required for clustering of voltage-gated Na channels at axon initial segments and for normal action potential firing. J Cell Biol 143:1295–1304

Thalamic Pathology in Schizophrenia

Will J. Cronenwett and John Csernansky

Contents

1 Introduction 510
2 The Thalamus Is Uniquely Suited to Modulate Signals Passing to the Cortex 511
3 Syndromes of Thalamic Dysfunction and Their Relevance to Schizophrenia 512
4 Postmortem Evidence: Structural Changes 514
5 Postmortem Evidence: Neurochemical Changes 515
6 Evidence from Neuroimaging 515
    6.1 Lower Thalamic Volume Is Frequently Seen in Schizophrenia: Changes May Be Localized to the MDN, the Anterior Nuclei, and the Pulvinar 515
    6.2 Medication Effects 516
    6.3 Studies at the Onset of Psychosis, and Longitudinal Data 517
    6.4 Imaging Studies in Relatives of Subjects with Schizophrenia 518
    6.5 Other Types of Neuroimaging (fMRI, PET, SPECT, DTI) 518
    6.6 The Limitations of Neuroimaging: Why the Conflicting Results? 521
7 Thalamic Pathology in Schizophrenia: Clinical Correlates 522
8 Dysfunction in Thalamocortical Circuits 523
References 524

Abstract The thalamus plays a critical role in the coordination of information as it passes from region to region within the brain. A disruption of that information flow may give rise to some of the cardinal symptoms of schizophrenia. In support of this hypothesis, schizophrenia-like syndromes emerge when illnesses, such as stroke, selectively damage the thalamus while sparing the rest of the brain.

Evidence from many sources has implicated thalamic dysfunction in schizophrenia. In postmortem studies, several subregions of the thalamus, including the mediodorsal nucleus and the pulvinar, have been shown to have fewer neurons in schizophrenia. Neurochemical disturbances are also seen, with changes in both

W.J. Cronenwett (✉) and J. Csernansky
Psychiatry and Behavioral Sciences, Northwestern University, 446 E. Ontario, Suite 7-200, Chicago, IL 60611, USA
e-mail: w-cronenwett@northwestern.edu

the glutamate and dopamine systems; thalamic glutamate receptor expression is altered in schizophrenia, and dopamine appears to be elevated in thalamic subregions, while evidence exists of an imbalance between dopamine and other neurotransmitters. In vivo studies using magnetic resonance imaging have demonstrated smaller thalamic volumes in schizophrenia, as well as shape deformations suggesting changes in those thalamic regions that are most densely connected to the portions of the brain responsible for executive function and sensory integration. These changes seem to be correlated with clinical symptoms.

The thalamus is a starting point for several parallel, overlapping networks that extend from thalamic nuclei to the cortex. Evidence is emerging that changes in the thalamic nodes of these networks are echoed by changes at other points along the chain; this suggests that schizophrenia might be a disease of disrupted thalamocortical neural networks. This model distributes the pathology throughout the network, but also concentrates attention on the thalamus as a critical structure, especially because of its role in coordinating the flow of information within and between neural networks.

Keywords Thalamus · Thalamorcortical relays · Mediodorsal nucleus · Pulvinar · Dysconnection neural networks · Magnetic resonance imaging

1 Introduction

The thalamus plays a major role in integrating the flow of information to and within the brain. Located roughly in the center of the head, it is the principal component of the diencephalon and consists of a pair of oval-shaped structures between the brainstem and the forebrain. These two symmetrical structures are joined by the massa intermedia in about a third of human beings. Both the right and left thalamus each have two main parts. The larger of these, the dorsal thalamus, is made up of about a dozen individual nuclei which are separated by thin layers of myelinated fibers, and which send excitatory projections to the cerebral cortex. The smaller part, or ventral thalamus, is separated from the nuclei of the dorsal thalamus by the external medullary lamina. It contains the reticular nucleus, which sends GABAergic inhibitory projections to the cells of the dorsal thalamus. The nuclei of the thalamus are richly interconnected with other regions of the brain, which makes the thalamus well suited to its task of processing and propagating signals as they pass into and within the brain. Developmentally, the thalamus also plays a pivotal role in the genesis of the cerebral cortex, with thalamic input being critical for appropriate functional differentiation of the cortex into intercommunicating regions (López-Bendito and Molnár 2003).

Thalamic pathology is clearly associated with schizophrenia. For many decades there has been indirect evidence to support this, in that there are striking similarities between the symptoms of schizophrenia and the syndromes caused by clinical conditions in which the thalamus is exclusively affected. More recently, direct evidence of thalamic involvement in schizophrenia has emerged, from postmortem studies and in vivo neuroimaging. This article will review the role of the thalamus healthy brain functioning, and review the symptoms that arise from thalamic dysfunction in stroke and other illnesses. Next, we will review the accumulated evidence for thalamic changes in schizophrenia. Finally, we will put this into context by discussing the hypothesis that changes seen in the thalamus are but one component of pathology that may in fact be spread widely across thalamocortical neural networks.

2 The Thalamus Is Uniquely Suited to Modulate Signals Passing to the Cortex

All information that ascends to the cortex passes through the thalamus, and all neurons from the periphery synapse there. Thus, the thalamus is appropriately called a major "relay station" of the brain. Its circuitry is intricate, however, and it is capable of a significant amount of dynamic signal processing. Several of the thalamic nuclei, including the lateral geniculate nucleus (LGN) and the ventral portion of the medial geniculate nucleus (MGNv), receive their incoming projections primarily from the peripheral sensory nerves. For example, visual information from the retina reaches the LGN and auditory information reaches the MGNv. Since the neurons that drive these relays have their origins in the periphery, these are said to be first-order relays. However, as many as 95% of the synapses present in first-order relays come not from the peripheral sensory neurons but from interneurons that modulate the firing pattern of the relays (Sherman and Guillery 1998). This heavy overrepresentation of modulating neurons suggests that the thalamus does not relay signals passively to cortex, but instead exerts complex and dynamic control of the information stream. In this way, the thalamus can have an important effect on the conscious perception of stimuli in the environment (Sherman 2004).

One way the thalamus exerts this control is by determining the firing modes of relay cells. The LGN, for example, can fire in both tonic and burst modes (Swadlow and Gusev 2001). In the tonic mode, which is typical of wakeful or attentive states, there is a roughly linear relationship between thalamic input and output. This would appear to maximize the fidelity of continuous transmission of peripheral signals to their cortical destinations. In inattentive states, however, the burst mode predominates. Action potentials arriving at the cortex in bursts are more strongly activating than continuous trains of tonic spikes, rendering such signals more detectable (Swadlow and Gusev 2001). This likely serves to redirect attention towards a novel or changing stimulus. Thus, in addition to its role in the perception of environmental stimuli, the thalamus also helps coordinate the selection and maintenance of attention.

Just as first-order thalamic relays are involved in perception, higher-order relays are involved in the critical appraisal of stimuli, organization and planning of appropriate responses, cognitive flexibility, and inhibition of inappropriate behavior – in other words, the brain executive functions of the brain. These higher-order thalamic relays receive their driver projections from layer 5 of the cortex (Sherman 2005; Sherman and Guillery 1998). The targets of higher-order relays are other cortical regions, implying that these thalamic nuclei facilitate and modulate signals as they pass from one part of cortex to another. In primates, the nuclei involved in cortico-thalamo-cortical communication make up the majority of the thalamus. While the predominant understanding of corticocortical communication had previously been one of hierarchical processing taking place entirely within the cortex (Felleman and Van Essen 1991), experimental evidence of late has demonstrated the involvement of the higher-order relays within the thalamus. This has led investigators to the conclusion that all information destined for a cortical target, even information already contained within the cortex, receives thalamic modulation (Reichova and Sherman 2004). Furthermore, these cortico-thalamo-cortical pathways have a larger diameter and are more heavily myelinated than the corticocortical fibers, allowing for signals to pass more quickly from thalamus to cortex than directly from one region of the cortex to another; thus, the thalamus can "alert" a region of cortex to incoming signals arriving from an interconnected cortical area. The amount of myelination is also variable, such that the time required for the conduction of an impulse remains fairly constant from each thalamic relay to its cortical target regardless of the distance involved. The net conclusion of this is that thalamic signals can arrive at different cortical sites simultaneously (Salami et al. 2003). Finally, the opportunities for thalamocortical signal processing are rich, as the higher-order relays are similar to the first-order relays in that the modulator neurons outnumber the drivers by about ten to one. The picture of the thalamus that emerges is far more than that of a simple relay station. Rather, the thalamus appears well equipped to play an active role in the dynamic processing and coordination of signals both from the periphery and within the cortex affecting sensory perception, executive function, and the active modulation of information flow.

3 Syndromes of Thalamic Dysfunction and Their Relevance to Schizophrenia

One way to illustrate the function of the thalamus is to look at diseases that affect it preferentially. One of the most common of these conditions is cerebrovascular accidents or thalamic strokes, which have been associated with sensorimotor and cognitive deficits since the beginning of the last century. Thalamic infarcts occur when one of the branches of the posterior cerebral artery become occluded. Infarcts of the tuberothalamic artery affect the anterior thalamus and can cause symptoms including executive dysfunction, perseveration, intrusion of unrelated ideas, poverty of speech, and apathy; anterograde amnesia and confabulation are also seen (Carrera and Bogousslavsky 2006; Schmahmann 2003). Strokes in the territory of the paramedian artery frequently affect the mediodorsal nucleus (MDN) of the thalamus and are associated with remarkable behavioral changes that can mimic some symptoms of schizophrenia. These include behavioral disinhibition, apathy, abulia, distractibility, temporal disorientation, impaired autobiographical memory, confabulation, and confusion (Schmahmann 2003). Arousal and attention are often affected in the early phases. These poststroke states can be difficult to distinguish from primary psychiatric disorders. Cyclical, manic-like states have even been reported, in which long periods of apathy and abulia alternate with shorter bursts of racing thoughts, inappropriate joking, and wild confabulations (McGilchrist et al. 1993).

Bilateral paramedian strokes can cause amnesia, severe arousal deficits, akinetic mutism, involuntary movements, and ataxia (Reilly et al. 1992). Strokes in the inferolateral territory are most commonly associated with sensory deficits and the well-documented "thalamic syndrome," in which the patient experiences paroxysmal pain in response to light touch. These patients also have difficulty with learning, naming, and executive function (Annoni et al. 2003). Posterior strokes affecting the pulvinar can lead to sensory loss and gaze disturbance, but in contrast to the syndromes described above, rarely produce behavioral abnormalities (Carrera and Bogousslavsky 2006).

Hallucinatory experiences can be caused by damage to subcortical and brainstem structures. This syndrome, "peduncular hallucinosis," is associated with complex, vivid, naturalistic, multimodal hallucinations, which are often impossible for the patient to distinguish from real experiences. Described only in case reports until recently, this rare condition is thought to arise from the disruption of subcortical control of information flow, leading to severe difficulties integrating sensory information and cognition (Manford and Andermann 1998). Neuropsychological changes such as confusion, disorientation, amnesia, poor insight, and impaired attention can accompany the syndrome. Significantly, this combination of hallucinations and impaired reality monitoring has been seen with lesions confined entirely to the thalamus (Benke 2006).

Finally, the rare prion disease fatal familial insomnia (FFI) rather remarkably demonstrates the role of the thalamus in the maintenance of executive function, attention, and wakefulness. The pathophysiological changes in FFI are limited almost exclusively to the anterior and mediodorsal nuclei of the thalamus; the cause is an inherited mutation in the prion protein gene PRPN on chromosome 20. As the thalamic structures degenerate, the patient becomes unable to sleep. Autonomic hyperactivity is prominent. Behavioral changes include apathy, disinterest, and a marked inability to sustain attention. Intelligence is preserved, but confusion becomes progressively worse as patients lose the ability to arrange memories in proper temporal sequence. Psychiatric symptoms such as depression and hallucinations are common (Gallassi et al. 1996; Nagayama et al. 1996).

These illnesses, in which the damage is limited narrowly to the thalamus, show how thalamic insults can cause cognitive and neuropsychiatric changes resembling some of the symptoms of schizophrenia. In healthy functioning, the thalamus is responsible for coordinating cognitive domains that are known to be impaired in schizophrenia; when the thalamus is damaged, schizophrenia-like syndromes can occur. Therefore, it comes as no surprise that the thalamus has been a target of active schizophrenia research for many years.

4 Postmortem Evidence: Structural Changes

Postmortem investigations of thalamic cytopathology are technically difficult and quite time consuming. For this reason, most studies have looked at specific thalamic nuclei as opposed to the structure as a whole (Clinton and Meador-Woodruff 2004).

Most attention thus far has been paid to the MDN, the anterior nuclei, and the pulvinar. The MDN is the second largest thalamic nucleus behind the pulvinar. Its size makes it somewhat less challenging to study than the smaller thalamic substructures; however, it has also attracted much attention because it is richly connected to areas of the cortex that are involved in integrating multimodal sensory input into a coherent and unified conscious experience (Stein 1998). This is an operation that frequently appears to be defective in schizophrenia. Postmortem descriptions of the MDN are conflicting, but most tend to show smaller neuron numbers without gliosis or any other hallmark signs of neurodegeneration (Heckers 1997). This argues for a process that is not degenerative, but perhaps developmental in origin.

The first high-quality stereotactic cytopathological studies of the thalamus reported lower numbers of both neurons and glial cells in the MDN, independent of medication effects (Pakkenberg 1990, 1992). These studies, and a recent work by Young et al. (2000) found profound reductions in the number of neurons in the MDN when comparing people with schizophrenia to controls. The findings were even more striking because there was virtually no overlap between groups. If consistently replicated, this would represent one of the largest effect sizes in all of schizophrenia research. However, not all investigators have reported results as dramatic as these (Byne et al. 2002; Danos et al. 2005), and one recent study with a large sample size was unable to confirm any differences in number of neurons, MDN volume, or total thalamic volume (Cullen et al. 2003).

Despite inconsistent results, the larger picture suggests that differences in the MDN are in fact present. For example, schizophrenia investigators have found lower neuronal density in the part of the cerebral cortex receiving projections from the MDN. This implies that changes occurring in the thalamus have effects not only within that structure, but in distant interconnected regions as well. In fact, the cortical gray matter of the prefrontal cortex, which receives projections from the MDN, has been shown to have less synaptic connectivity with the thalamus in schizophrenia (Lewis et al. 2001). Selemon et al. (2005) provided experimental support for this process in a study in which they irradiated fetal monkeys at the gestational age corresponding to thalamic neurogenesis; they later noted trend-level reductions in cortical thickness, demonstrating an association between disrupted development of thalamic neurons and the cerebral cortex to which it projects.

Similarly, investigators have repeatedly seen changes in the pulvinar in schizophrenia (reviewed in Byne et al. 2007). The pulvinar shares many cortical targets with the MDN, and in fact its medial parts may best be understood as a continuation of the MDN. The pulvinar has strong connections to the cortical association areas and the limbic system, all of which are affected in schizophrenia (Byne et al. 2002). Danos et al. (2003) found a 22.1% difference in the size of the right medial pulvinar between people with schizophrenia and nonschizophrenic control subjects; in a replication study, Byne et al. (2008a) found 19% difference in volume in the right pulvinar, as well as a 19% difference in neuron number within the medial pulvinar. The anterior nuclei, which also project to the limbic cortex, appear to be similarly reduced in volume and number of neurons (Danos et al. 1998; Young et al. 2000).

5 Postmortem Evidence: Neurochemical Changes

Most neurochemical research on the thalamus has been on the glutaminergic system, for several reasons. Glutamate is the primary neurotransmitter used by the thalamus for sending signals onward to the cortex. NMDA receptor antagonists such as phencyclidine (PCP) and ketamine cause glutamate release in the prefrontal cortex, and have been shown to produce schizophrenia-like syndromes in humans that include hallucinations, delusions, poverty of speech, social withdrawal, disorganized behavior, and cognitive symptoms. These drugs can also cause glutamate-induced cortical neurotoxicity (Javitt 2007).

The thalamus mediates cortical glutaminergic effects. Sharp et al. (2001) demonstrated this in an elegant experiment in which they injected NMDA receptor antagonists directly into the thalamus. While this caused changes in the cortex identical to those seen with systemic injection of the drug, injection into the cortex itself did not. Significantly, the glutaminergic system is known to be dysregulated in schizophrenia, as postmortem work using in-situ hybridization has shown abnormalities in subcortical glutamate receptor expression and in the intracellular signal transduction machinery (Clinton and Meador-Woodruff 2004).

Dopamine has been of great interest since the dawn of antipsychotic psychopharmacology. High performance liquid chromatography has revealed marked elevation of dopamine in the anterior and mediodorsal nuclei of the thalamus, as well as an increase in the ratio of dopamine to norepinephrine. These findings appear to be independent of medication effects (Oke et al. 1988; Oke et al. 1992). Gamma-aminobutyric acid (GABA) is the main inhibitory neurotransmitter in the thalamus and is used in neurons for signal modulation. Abnormalities in the GABA system in schizophrenia have been found in the prefrontal cortex, cingulate, and hippocampus; however, GABAergic abnormalities in the thalamus have not yet been demonstrated (Popken et al. 2002).

6 Evidence from Neuroimaging

6.1 Lower Thalamic Volume Is Frequently Seen in Schizophrenia: Changes May Be Localized to the MDN, the Anterior Nuclei, and the Pulvinar

Many high-quality structural magnetic resonance imaging (MRI) investigations of the thalamus in schizophrenia have been completed. Furthermore, there are several high-quality reviews of the existing literature, including ones by Sim et al. (2006) and Byne et al. (2008b). The most frequent finding to date has been reduced thalamic volume, but this has not been seen in every study. Three-fourths of the studies performed to date report volume loss in the thalamus, with about two-thirds of those findings reaching statistical significance. Four studies have shown reduced thalamic volume in subjects considered by family history to be at high risk for schizophrenia (see below). The magnitude of the differences seen in schizophrenia is highly variable; this is likely explained by the small sample sizes used in relation to the size of the effects being measured. One meta-analysis of absolute and relative (to whole brain) thalamic volume reported combined effect sizes of -0.30 and -0.41, respectively. These are statistically significant but not large (Konick and Friedman 2001).

Intriguingly, some studies that do not report thalamic volume loss nevertheless do describe shape changes in regions of the thalamus that are plausibly affected in schizophrenia. For example, one study failed to detect a change in thalamic volume but did find shape changes in the surface overlying nuclei that project to limbic and association cortices (Hazlett et al. 1999). Another recent study using high-dimensional brain mapping found reduced thalamic volume that vanished when corrected for total cerebral volume, but also found significant and durable shape changes in the anterior and posterior regions. The shape changes they reported were located on the surface of the thalamus overlying the anterior and mediodorsal nuclei (Csernansky et al. 2004). This was partially confirmed in a study of first-episode patients (Coscia et al. 2009), which also found smaller volumes overall, with significant shape changes in the posterior region. This work suggests that differences in thalamic volume may be due to changes within selected subcomponents, instead of being spread evenly throughout the whole structure. Structural MRI investigations of individual thalamic nuclei are problematic because of the small size of the structures and because of the lack of contrast with adjacent tissues. Still, some groups have reported differences in the volumes of the MDN, anterior nuclei, and the pulvinar (Byne et al. 2001; Kemether et al. 2003; see also Sim et al. (2006) for a review); this is in broad agreement with the postmortem findings described above. Changes in these sensory processing and multimodal integration nuclei are plausibly associated with many of the clinical and cognitive deficits of schizophrenia, and this argues for a central role for these parts of the thalamus in the pathophysiology of the disorder.

In sum, the observed structural differences in schizophrenia are small, and not every group of investigators has reported similar findings. Still, the majority of the evidence supports reduced thalamic volume in schizophrenia relative to whole brain size. These volume deficits are likely due at least in part to changes within the MDN, anterior nuclei, and pulvinar.

6.2 Medication Effects

First-generation antipsychotic medications have been associated with thalamic hypertrophy, with a positive correlation between medication dose and increase in size (reported in Gur et al. 1998). This would be consistent with the observation that first-generation antipsychotics are associated with volume recovery in the basal ganglia in schizophrenia (Chakos et al. 1994). In contrast, second-generation medications have been associated with volume deficits (Sullivan et al. 2003), although the Gur study (1998) found that atypical neuroleptics led to volume increase (or volume recovery) in the thalamus. Another small study confirming thalamic volume reduction in schizophrenia also showed volume normalization co-occurring with clinical improvement after 4 weeks of second-generation antipsychotic medications. The subjects in that study who did not respond to medication did not show volume increases (Strungas et al. 2003). Konick and Friedman's large meta-analysis of thalamic size in schizophrenia (2001) did not find medication-related changes, but different types of medications were not analyzed separately.

6.3 Studies at the Onset of Psychosis, and Longitudinal Data

Studies enrolling subjects at the onset of their symptoms are appealing for two reasons. First, these individuals have only a very brief history of treatment (if any at all), and thus in most cases the observed differences are not confounded by medication effects. Additionally, these studies can shed light on the question of whether changes in schizophrenia are progressive. Most first-episode studies, including most childhood and adolescent studies, show reduced total thalamic volume at illness onset without evidence for progressive thalamic volume loss (see Steen et al. (2006) for a review). It is unclear if the lack of observed progression means that changes are stable, or that progression exists but has not been detected yet due to study design, or is undetectable because of small sample size or insufficient technical resolution.

Longitudinal studies of the thalamus in schizophrenia are rare. Two studies with a 2-year follow-up period reported conflicting results: Rapoport et al. (1997) noted progressive decrease in thalamic volume in a population with childhood onset schizophrenia; this change was not present in control subjects. In contrast, James et al. (2004) noted no progression in thalamic volume over time in a group of subjects with adolescent-onset schizophrenia. Volume and shape changes do not appear to correlate with duration of illness (Csernansky et al. 2004), but thalamic volume may be larger in patients whose onset of illness is earlier (Crespo-Facorro et al. 2007). A 5-year follow-up study by van Haren et al. (2008) demonstrated a different trajectory of brain volume changes in subjects with schizophrenia compared to controls, with the velocity of volumetric change in schizophrenia compared to healthy controls being greater in the first decades of illness. It is unclear if these results apply to subcortical gray matter structures as well, but there is at least a suggestion of a more complicated picture than simply either static differences or linear changes over time. Further investigation is needed.

6.4 Imaging Studies in Relatives of Subjects with Schizophrenia

Studies using nonpsychotic relatives of individuals with schizophrenia have the advantage of removing the effects of neuroleptic use as a confounder. Perhaps even more importantly, studies of relatives can shed light on the heritability of the illness. There is strong evidence from twin studies and adoption studies that schizophrenia is a genetic condition (McGuffin et al. 1995). If schizophrenia were similar to other genetically determined conditions, one would expect to see attenuated forms of the pathology in first-degree relatives of affected probands, and indeed this is the case. Many structures affected in schizophrenia are also affected in first-degree relatives (see Boos et al. 2007 for a meta-analysis and review). A study by Staal et al. (1998) showed that the volume of the thalamus was reduced in unaffected first-degree relatives of patients with schizophrenia. Lawrie et al. (2001) replicated these results, and also showed that almost half of the nonschizophrenic relatives who participated in their study had some degree of subthreshold psychotic symptoms. Monozygotic twins with schizophrenia show an interesting trend; twins who are discordant for schizophrenia, as a group, tend to have thalamic volumes that are intermediate between twins who are concordant and matched healthy controls (Ettinger et al. 2007). However, thalamic volume in at-risk individuals has not been shown to predict who will eventually go on to develop schizophrenia (Lawrie et al. 2008).

The volume deficits observed in nonschizophrenic siblings appear to be localized to the anterior and posterior extremes of the thalamus, as has often been seen in their schizophrenic relatives; this adds to the evidence that changes in the thalamus are localized to areas that project to frontotemporal regions of the cortex, and may explain how differences in thalamic structure can be present even when overall volume is unchanged (Csernansky et al. 2004; Harms et al. 2007). Also interestingly, when evaluated using clinical symptom ratings scales and cognitive testing, the nonschizophrenic relatives of affected patients tend to score in an intermediate range between individuals with schizophrenia and healthy control subjects. This is true for positive, negative, and disorganized symptoms, as well as cognitive domains such as episodic and working memory and executive function (Harms et al. 2007). The heritable portion of these changes is unknown, as shared environmental factors may also play a role; however, previous work in twin and adoption studies suggests that the overall contribution of environmental factors to the expression of the disorder is minimal (Kendler et al. 1994). Whether environmental influences can differentially affect various cerebral substructures remains an unanswered question.

6.5 Other Types of Neuroimaging (fMRI, PET, SPECT, DTI)

In the comprehensive review of thalamic neuroimaging by Sim et al. (2006), the authors describe 16 positron emission tomography (PET), single photon emission computed tomography (SPECT), and fMRI studies conducted in a baseline resting state, and a further 19 studies that investigated task-related performance. The task-related studies tend to show decreased thalamic activity, frequently in combination with reduced cortical activation. The changes are present in motor, auditory processing, verbal learning, word fluency, and memory-related tasks. The role of medication is controversial; it has been associated both with increased and decreased thalamic metabolism observed during task performance.

Studies of the resting state reveal a more complicated picture. When looking at thalamic activation at baseline, approximately one-third of studies report no difference between schizophrenia and healthy control subjects, another third report increases in thalamic metabolism, and a final third detect lower metabolism. As is the case with thalamic volume, one explanation for differing findings may be that the functional changes are seen preferentially in certain subcomponents of the thalamus, rather than being spread across the whole structure. One study of never-medicated patients reported significantly lower [^{18}F]-deoxyglucose metabolism in the region of the MDN of the thalamus, with increased metabolism in the pulvinar (Hazlett et al. 2004). Interestingly, they also reported that lower metabolism in the thalamus correlated with poorer verbal memory and increased negative symptoms, while lower overall thalamic metabolism correlated with worsening overall clinical symptoms. fMRI has also shown increased thalamic activation in people at risk for schizophrenia but who have had neither neuroleptic treatment nor any history of a diagnosable mental disorder, but who nonetheless are experiencing some degree of psychotic symptoms (Whyte et al. 2006).

The observed changes elsewhere in the brain that parallel thalamic abnormalities are interesting, and tend to show a distinctive picture. Changes co-occurring with the thalamus are seen, especially during task performance, in the frontal and prefrontal cortices, limbic regions, and cerebellum (Sim et al. 2006). This can be interpreted as further evidence of the dysfunction being distributed along a circuit that includes the thalamus and reciprocally connected regions of cortex.

It has been known for some time that people with schizophrenia, when presented with novel stimuli, display P300 event-related brain potentials that are both delayed and reduced in amplitude (see Turetsky et al. 2007). Until recently, however, the temporal resolution of changes in blood oxygen level dependent (BOLD) signal has not allowed for the measurement of delay in the hemodynamic changes in response to discrete stimuli. This is because the BOLD signal changes within seconds, a time scale that is significantly expanded compared to the neural response, which occurs within milliseconds. Recently, however, using a method described by Henson et al. (2002), one team was able to report delayed hemodynamic response within the thalamus; other affected regions included the basal ganglia and anterior cingulate cortex. There have been no areas in the brain in which the hemodynamic response was seen to occur more quickly in schizophrenia (Ford et al. 2005). This interesting work suggests that neural responses in schizophrenia share both reduced amplitude and increased latency when compared to controls.

The "glutamate hypothesis" of schizophrenia has received much attention in recent years, and evidence implicating glutamate in schizophrenia-related cortical neuroplasticity has begun to appear (see above). It is not possible to measure glutamate concentration or activity in vivo, but magnetic resonance spectroscopy (MRS) is able to demonstrate signal related to glutamine, which is considered a useful marker for glutamate. Thalamic glutamine has been shown on MRS to be elevated in schizophrenia (Théberge et al. 2007) and abnormal in prodromal psychosis (Stone et al. 2009), leading to speculation that glutamate excitotoxicity could be responsible for some of the observed cortical and subcortical volume changes associated with the illness, and perhaps ultimately for the dysfunction of interconnected networks. Supporting a hypothesis that early glutamate toxicity may lead to the appearance of the disorder, glutamine levels have been shown to be elevated at symptom onset and then reduced 30 months later, although not reduced below the levels in healthy matched control subjects. Gray matter volume loss was widespread in that study and was correlated with decreases in thalamic glutamine (Théberge et al. 2007).

Diffusion tensor imaging (DTI) is a relatively new MRI modality that allows for the elaboration of differences within regions of white matter; this is otherwise is difficult because of the uniform appearance of white matter on MRI. Investigators can use DTI to examine fractional anisotropy (FA) of the diffusion of water within a defined volume. This is useful because relatively high FA corresponds to water that is constrained in diffusion by intact cytoarchitecture, while lower FA corresponds to relatively less constrained diffusion of water, as would be seen if axonal continuity were disrupted (Mori et al. 2007). The white matter tracts connecting the MDN of the thalamus to the dorsolateral PFC have been shown to have greater FA in schizophrenia than in healthy controls, arguing for dysfunctional reciprocal connectivity between this major thalamic relay nucleus and its target sites in the cortex (Kito et al. 2009). This would support the "dysconnection hypothesis" of schizophrenia, according to which the pathology is explained by abnormal connectivity that impairs the quality and sequence of communication between brain regions (Friston and Frith 1995).

This type of neuroimaging can also be used to describe the mean diffusivity of water within gray matter structures. The precise interpretation of differences in mean diffusivity are unknown, but one probable explanation is that increased mean diffusivity correlates with increased extracellular space due to altered neuronal architecture, which in turn could imply disrupted or suboptimal neuronal communication (Syková 2004). Mean diffusivity within gray matter was found to be increased in schizophrenia in the mediodorsal and anterior nuclei of the thalamus, as well as in the prefrontal cortex to which those structures are connected, adding to the experimental evidence for thalamic pathology and dysfunctional thalamocortical connectivity (Agarwal et al. 2008; Rose et al. 2006). This may represent in vivo evidence of the postmortem findings of reduced density of neurons or glial cells within the thalamic relay nuclei (see above).

On tasks involving working memory, fMRI has shown differences in activation between relatives and healthy controls in the dorsolateral prefrontal cortex, thalamus, and anterior cingulate gyrus, providing further evidence for thalamocortical dysfunction (Callicott et al. 2003; Thermenos et al. 2004). Similar results appear when brain perfusion maps are created using SPECT, in which alterations in perfusion have been noted to an intermediate degree in relatives of individuals with schizophrenia. The changes also appear to be in the prefrontal and cingulate cortices, as well as in subcortical structures including the thalamus (Blackwood et al. 1999).

6.6 The Limitations of Neuroimaging: Why the Conflicting Results?

Most, but not all, postmortem studies show reduced neuronal and glial density in regions of the thalamus that connect to the limbic and heteromodal association cortices; most, but not all, volumetric studies of the whole thalamus suggest that the size of the thalamus, relative to the whole brain, is reduced in schizophrenia. Many more studies present trend-level data than demonstrate robust statistical significance. One explanation for the discordant results in the structural MRI literature is that the changes might be confined to certain thalamic regions, and these might be differentially affected while the volume of the entire structure remains the same. Several studies have indeed suggested this to be true (see above). Still, it appears even then that the pathological changes are present at a level that is barely detectable given current imaging technology. For studies to be practically feasible, researchers have generally needed to use fairly small samples in proportion to the magnitude of the changes they hope to observe; this leads to highly variable effect sizes across studies, and increases the likelihood that real changes will sometimes remain unobserved (Konick and Friedman 2001). Larger samples would of course be helpful, although this is frequently not possible.

One hopes that technological innovations will increase researchers' ability to discern small differences. But then changing technology may itself bring new problems, as it might not be possible to compare the older results with the newer ones. That said, Wright et al. found in one meta-analysis that slice thickness did not seem to modify effect size, providing reassurance that changes in technology have not much distorted findings thus far (Wright et al. 2000).

There are other limitations that follow from the current resolution of MRI scanning. Even with the smallest voxel size in use today, more than one tissue type can still occur within a given voxel, leading to imprecision when measuring small structures. Human error in the manual delineation of structures is naturally a constant risk, the effects of which are magnified as the differences between health and illness (or the absolute size of structures) get smaller.

In the meta-analysis of brain size in first-episode schizophrenia, Steen et al. (2006) call attention to the multitude of clinical problems that are present in this type of research: patients are typically treated very soon after diagnosis with substantial heterogeneity in medication and dose; also, "first-episode" itself is an imprecise term, since some people may have been symptomatic for a long time before coming to clinical attention. Others may misremember symptoms – or they may simply be unable to reliably self-report when their symptoms began.

This means that terms such as "first-episode" or "never-medicated" correlate to an uncertain degree with duration of illness. Finally, there may be substantial heterogeneity in the illness itself, or indeed there may be different illnesses with different pathophysiological processes that converge on similar clinical syndromes, once again burying the signal in noise.

Given these limitations, it is highly intriguing that the studies of the thalamus in unaffected first-degree relatives of people with schizophrenia tend to be more in agreement with each other than the literature on the thalamus in the patients themselves. This has led some to suggest that the observed differences between studies are perhaps due to a larger-than-anticipated set of confounders that come when studying this clinical population (Wright et al. 2000). Such confounders could include medication history, the unreliability of historical information, recall bias, and the uncertain environmental effects that come from living with such a devastating illness.

7 Thalamic Pathology in Schizophrenia: Clinical Correlates

Clinical correlates of observed thalamic pathology are inconsistent and subject to the same limitations as the structural, functional, and chemical research discussed above. However, other illnesses that affect the thalamus are associated with deficits seen in schizophrenia, such as attention disturbances, memory problems, and executive dysfunction. Therefore, it is plausible that the variations in the thalamus seen in schizophrenia could correlate with some of the clinical and neuropsychological changes seen as well. And indeed, this turns out to be the case.

Lower thalamic volume has been found to correlate with increased thought disorder, bizarre behavior, and hallucinations; correlations to negative symptoms have been mixed (Portas et al. 1998; Preuss et al. 2005). Lower thalamic volume also correlates with reduced performance on the Continuous Performance Test (Salgado-Pineda et al. 2004), and with increased dysfunction in the integration of sensory information (Dazzan et al. 2004).

A recent study of first-episode patients found that volume loss in the pulvinar region correlated with poorer performance in executive functioning and language, but not with the traditional positive and negative symptom domains of schizophrenia (Coscia et al. 2009). This contrasts with early work that found no correlation between clinical symptoms and thalamic volume deficits in first-episode psychosis (Ettinger et al. 2001). Persons at the onset of their illness have been found to have lower thalamic volumes than healthy controls; however, larger thalamic volumes have also been associated with worse cognitive functioning and an increased burden of negative symptoms. One explanation for this is that thalamic enlargement may be present as a compensatory mechanism in the face of functional deficits in interconnected areas (Crespo-Facorro et al. 2007).

fMRI has shown decreased functional activation in memory-related tasks in the MDN of the thalamus, but not the pulvinar (Andrews et al. 2006). Increased glucose metabolism correlates with fewer clinical symptoms measured by the BPRS, while increased metabolism specific to the MD correlated with lower negative symptoms (Hazlett et al. 2004).

8 Dysfunction in Thalamocortical Circuits

Early attempts to characterize the deficits in schizophrenia tried to correlate specific symptoms to specific regions of the brain, and without much success. Thinking has evolved such that the signature pathology of the disorder may be seen as deficiencies not in one region, but in interconnected networks. Schizophrenia, according to this hypothesis, could be a syndrome of neural dysconnection – a disorder associated with disruption in the smooth operation of neural networks, such as are used to integrate sensory information into a unified conscious experience. This leads one to consider distant regions as part of a larger functioning whole; it also focuses attention again on the centrality of the thalamus, which is ideally suited to modulate all information flow from the periphery to the cortex, and from one cortical region to another. Studies using neuronal tracers have described the neuroanatomic basis for these cortico-thalamo-cortical networks (Alexander and Crutcher 1990; Alexander et al. 1986).

If the networks themselves are somehow disrupted, then pathologic changes in the thalamus may be part of a process that is taking place throughout the entire network. Therefore, it is highly interesting that co-occurring changes are in fact seen throughout the reciprocal loop connecting the thalamic relay nuclei with the major cortical multimodal association areas. Andreasen et al. (1994) reported evidence for co-occurring pathology in the thalamus and the white matter tracts leading to the dorsolateral prefrontal cortex; this was replicated later with DTI imaging of the thalamic radiations to the cortex (Kito et al. 2009). Differences in effective connectivity between the thalamus and the dorsolateral prefrontal cortex have been noted during fMRI tasks related to working memory (Schlösser et al. 2003). Differences are also seen in first-episode schizophrenia in resting-state functional connectivity between the dorsolateral prefrontal cortex and thalamus (Zhou et al. 2007). Activation in distributed thalamocortical networks has been seen in actively hallucinating patients (Silbersweig et al. 1995). More recently, studies have attempted to describe the functional integrity of an entire thalamocortical network; for example, Camchong et al. described signal changes in the circuitry linking the thalamus to the prefrontal cortex in delayed response trials (Camchong et al. 2006). Studies continue to find gray matter pathology in distributed nodes in these neuronal circuits (see Ellison-Wright et al. 2008 for a meta-analysis).

In one DTI study, lower FA was seen in the white matter connecting the dorsolateral PFC to the thalamus in both bipolar disorder and schizophrenia, which suggests that the pathology in this circuit may be specific more to psychosis than to schizophrenia (Sussmann et al. 2009).

In any event, such disruptions in thalamocortical circuitry may plausibly produce much of the symptoms of schizophrenia, as the affected cognitive domains would include information processing, prioritizing, and coordination; this has been described as poor mental coordination, or the "cognitive dysmetria" hypothesis of schizophrenia (Andreasen et al. 1998). If this were to be true, then schizophrenia might best be seen as a disease of disrupted thalamocortical neural networks; this model distributes the pathology throughout the network, but also concentrates attention on the thalamus as a common and central node. In any event, the thalamus has attracted the attention of schizophrenia researchers since the advent of modern neuroscience, and it clearly remains an important structure whose entire role has yet to be fully explained.

References

Agarwal N, Rambaldelli G, Perlini C, Dusi N, Kitis O, Bellani M, Cerini R, Isola M, Versace A, Balestrieri M, Gasparini A, Mucelli RP, Tansella M, Brambilla P (2008) Microstructural thalamic changes in schizophrenia: a combined anatomic and diffusion weighted magnetic resonance imaging study. J Psychiatry Neurosci 33:440–448

Alexander GE, Crutcher MD (1990) Functional architecture of basal ganglia circuits: neural substrates of parallel processing. Trends Neurosci 13:266–271

Alexander GE, DeLong MR, Strick PL (1986) Parallel organization of functionally segregated circuits linking basal ganglia and cortex. Annu Rev Neurosci 9:357–381

Andreasen NC, Arndt S, Swayze V, Cizadlo T, Flaum M, O'Leary D, Ehrhardt JC, Yuh WT (1994) Thalamic abnormalities in schizophrenia visualized through magnetic resonance image averaging. Science 266:294–298

Andreasen NC, Paradiso S, O'Leary DS (1998) "Cognitive dysmetria" as an integrative theory of schizophrenia: a dysfunction in cortical–subcortical–cerebellar circuitry? Schizophr Bull 24:203–218

Andrews J, Wang L, Csernansky JG, Gado MH, Barch DM (2006) Abnormalities of thalamic activation and cognition in schizophrenia. Am J Psychiatry 163:463–469

Annoni JM, Khateb A, Gramigna S, Staub F, Carota A, Maeder P, Bogousslavsky J (2003) Chronic cognitive impairment following laterothalamic infarcts: a study of 9 cases. Arch Neurol 60:1439–1443

Benke T (2006) Peduncular hallucinosis: a syndrome of impaired reality monitoring. J Neurol 253:1561–1571

Blackwood DH, Glabus MF, Dunan J, O'Carroll RE, Muir WJ, Ebmeier KP (1999) Altered cerebral perfusion measured by SPECT in relatives of patients with schizophrenia. Correlations with memory and P300. Br J Psychiatry 175:357–366

Boos HB, Aleman A, Cahn W, Pol HH, Kahn RS (2007) Brain volumes in relatives of patients with schizophrenia: a meta-analysis. Arch Gen Psychiatry 64:297–304

Byne W, Buchsbaum MS, Kemether E, Hazlett EA, Shinwari A, Mitropoulou V, Siever LJ (2001) Magnetic resonance imaging of the thalamic mediodorsal nucleus and pulvinar in schizophrenia and schizotypal personality disorder. Arch Gen Psychiatry 58:133–140

Byne W, Buchsbaum MS, Mattiace LA, Hazlett EA, Kemether E, Elhakem SL, Purohit DP, Haroutunian V, Jones L (2002) Postmortem assessment of thalamic nuclear volumes in subjects with schizophrenia. Am J Psychiatry 159:59–65

Byne W, Fernandes J, Haroutunian V, Huacon D, Kidkardnee S, Kim J, Tatusov A, Thakur U, Yiannoulos G (2007) Reduction of right medial pulvinar volume and neuron number in schizophrenia. Schizophr Res 90:71–75

Byne W, Dracheva S, Chin B, Schmeidler JM, Davis KL, Haroutunian V (2008a) Schizophrenia and sex associated differences in the expression of neuronal and oligodendrocyte-specific genes in individual thalamic nuclei. Schizophr Res 98:118–128

Byne W, Hazlett EA, Buchsbaum MS, Kemether E (2008b) The thalamus and schizophrenia: current status of research. Acta Neuropathol 117:347–368

Callicott JH, Egan MF, Mattay VS, Bertolino A, Bone AD, Verchinksi B, Weinberger DR (2003) Abnormal fMRI response of the dorsolateral prefrontal cortex in cognitively intact siblings of patients with schizophrenia. Am J Psychiatry 160:709–719

Camchong J, Dyckman KA, Chapman CE, Yanasak NE, McDowell JE (2006) Basal ganglia-thalamocortical circuitry disruptions in schizophrenia during delayed response tasks. Biol Psychiatry 60:235–241

Carrera E, Bogousslavsky J (2006) The thalamus and behavior: effects of anatomically distinct strokes. Neurology 66:1817–1823

Chakos MH, Lieberman JA, Bilder RM, Borenstein M, Lerner G, Bogerts B, Wu H, Kinon B, Ashtari M (1994) Increase in caudate nuclei volumes of first-episode schizophrenic patients taking antipsychotic drugs. Am J Psychiatry 151:1430–1436

Clinton SM, Meador-Woodruff JH (2004) Thalamic dysfunction in schizophrenia: neurochemical, neuropathological, and in vivo imaging abnormalities. Schizophr Res 69:237–253

Coscia DM, Narr KL, Robinson DG, Hamilton LS, Sevy S, Burdick KE, Gunduz-Bruce H, McCormack J, Bilder RM, Szeszko PR (2009) Volumetric and shape analysis of the thalamus in first-episode schizophrenia. Hum Brain Mapp 30:1236–1245

Crespo-Facorro B, Roiz-Santiáñez R, Pelayo-Terán JM, Rodríguez-Sánchez JM, Pérez-Iglesias R, González-Blanch C, Tordesillas-Gutiérrez D, González-Mandly A, Díez C, Magnotta VA, Andreasen NC, Vázquez-Barquero JL (2007) Reduced thalamic volume in first-episode non-affective psychosis: correlations with clinical variables, symptomatology and cognitive functioning. Neuroimage 35:1613–1623

Csernansky JG, Schindler MK, Splinter NR, Wang L, Gado M, Selemon LD, Rastogi-Cruz D, Posener JA, Thompson PA, Miller MI (2004) Abnormalities of thalamic volume and shape in schizophrenia. Am J Psychiatry 161:896–902

Cullen TJ, Walker MA, Parkinson N, Craven R, Crow TJ, Esiri MM, Harrison PJ (2003) A postmortem study of the mediodorsal nucleus of the thalamus in schizophrenia. Schizophr Res 60:157–166

Danos P, Baumann B, Bernstein HG, Franz M, Stauch R, Northoff G (1998) Schizophrenia and anteroventral thalamic nucleus: selective decrease of parvalbumin-immunoreactive thalamo-cortical projection neurons. Psychiatry Res 82:1–10

Danos P, Baumann B, Krämer A, Bernstein HG, Stauch R, Krell D, Falkai P, Bogerts B (2003) Volumes of association thalamic nuclei in schizophrenia: a postmortem study. Schizophr Res 60:141–155

Danos P, Schmidt A, Baumann B, Bernstein HG, Northoff G, Stauch R, Krell D, Bogerts B (2005) Volume and neuron number of the mediodorsal thalamic nucleus in schizophrenia: a replication study. Psychiatry Res 140:281–289

Dazzan P, Morgan KD, Orr KG, Hutchinson G, Chitnis X, Suckling J, Fearon P, Salvo J, McGuire PK, Mallett RM, Jones PB, Leff J, Murray RM (2004) The structural brain correlates of neurological soft signs in AESOP first-episode psychoses study. Brain 127:143–153

Ellison-Wright I, Glahn DC, Laird AR, Thelen SM, Bullmore E (2008) The anatomy of first-episode and chronic schizophrenia: an anatomical likelihood estimation meta-analysis. Am J Psychiatry 165:1015–1023

Ettinger U, Chitnis XA, Kumari V, Fannon DG, Sumich AL, O'Ceallaigh S, Doku VC, Sharma T (2001) Magnetic resonance imaging of the thalamus in first-episode psychosis. Am J Psychiatry 158:116–118

Ettinger U, Picchioni M, Landau S, Matsumoto K, van Haren NE, Marshall N, Hall MH, Schulze K, Toulopoulou T, Davies N, Ribchester T, McGuire PK, Murray RM (2007) Magnetic resonance imaging of the thalamus and adhesio interthalamica in twins with schizophrenia. Arch Gen Psychiatry 64:401–409

Felleman DJ, Van Essen DC (1991) Distributed hierarchical processing in the primate cerebral cortex. Cereb Cortex 1:1–47

Ford JM, Johnson MB, Whitfield SL, Faustman WO, Mathalon DH (2005) Delayed hemodynamic responses in schizophrenia. Neuroimage 26:922–931

Friston KJ, Frith CD (1995) Schizophrenia: a disconnection syndrome? Clin Neurosci 3:89–97

Gallassi R, Morreale A, Montagna P, Cortelli P, Avoni P, Castellani R, Gambetti P, Lugaresi E (1996) Fatal familial insomnia: behavioral and cognitive features. Neurology 46:935–939

Gur RE, Maany V, Mozley PD, Swanson C, Bilker W, Gur RC (1998) Subcortical MRI volumes in neuroleptic-naive and treated patients with schizophrenia. Am J Psychiatry 155:1711–1717

Harms MP, Wang L, Mamah D, Barch DM, Thompson PA, Csernansky JG (2007) Thalamic shape abnormalities in individuals with schizophrenia and their nonpsychotic siblings. J Neurosci 27:13835–13842

Hazlett EA, Buchsbaum MS, Byne W, Wei TC, Spiegel-Cohen J, Geneve C, Kinderlehrer R, Haznedar MM, Shihabuddin L, Siever LJ (1999) Three-dimensional analysis with MRI and PET of the size, shape, and function of the thalamus in the schizophrenia spectrum. Am J Psychiatry 156:1190–1199

Hazlett EA, Buchsbaum MS, Kemether E, Bloom R, Platholi J, Brickman AM, Shihabuddin L, Tang C, Byne W (2004) Abnormal glucose metabolism in the mediodorsal nucleus of the thalamus in schizophrenia. Am J Psychiatry 161:305–314

Heckers S (1997) Neuropathology of schizophrenia: cortex, thalamus, basal ganglia, and neurotransmitter-specific projection systems. Schizophr Bull 23:403–421

Henson RN, Price CJ, Rugg MD, Turner R, Friston KJ (2002) Detecting latency differences in event-related BOLD responses: application to words versus nonwords and initial versus repeated face presentations. Neuroimage 15:83–97

James AC, James S, Smith DM, Javaloyes A (2004) Cerebellar, prefrontal cortex, and thalamic volumes over two time points in adolescent-onset schizophrenia. Am J Psychiatry 161:1023–1029

Javitt DC (2007) Glutamate and schizophrenia: phencyclidine, N-methyl-D-aspartate receptors, and dopamine–glutamate interactions. Int Rev Neurobiol 78:69–108

Kemether EM, Buchsbaum MS, Byne W, Hazlett EA, Haznedar M, Brickman AM, Platholi J, Bloom R (2003) Magnetic resonance imaging of mediodorsal, pulvinar, and centromedian nuclei of the thalamus in patients with schizophrenia. Arch Gen Psychiatry 60:983–991

Kendler KS, Gruenberg AM, Kinney DK (1994) Independent diagnoses of adoptees and relatives as defined by DSM-III in the provincial and national samples of the Danish Adoption Study of Schizophrenia. Arch Gen Psychiatry 51:456–468

Kito S, Jung J, Kobayashi T, Koga Y (2009) Fiber tracking of white matter integrity connecting the mediodorsal nucleus of the thalamus and the prefrontal cortex in schizophrenia: a diffusion tensor imaging study. Eur Psychiatry 24:269–274

Konick LC, Friedman L (2001) Meta-analysis of thalamic size in schizophrenia. Biol Psychiatry 49:28–38

Lawrie SM, Whalley HC, Abukmeil SS, Kestelman JN, Donnelly L, Miller P, Best JJ, Owens DG, Johnstone EC (2001) Brain structure, genetic liability, and psychotic symptoms in subjects at high risk of developing schizophrenia. Biol Psychiatry 49:811–823

Lawrie SM, McIntosh AM, Hall J, Owens DG, Johnstone EC (2008) Brain structure and function changes during the development of schizophrenia: the evidence from studies of subjects at increased genetic risk. Schizophr Bull 34:330–340

Lewis DA, Cruz DA, Melchitzky DS, Pierri JN (2001) Lamina-specific deficits in parvalbumin-immunoreactive varicosities in the prefrontal cortex of subjects with schizophrenia: evidence for fewer projections from the thalamus. Am J Psychiatry 158:1411–1422

López-Bendito G, Molnár Z (2003) Thalamocortical development: how are we going to get there? Nat Rev Neurosci 4:276–289

Manford M, Andermann F (1998) Complex visual hallucinations. Clinical and neurobiological insights. Brain 121:1819–1840

McGilchrist I, Goldstein LH, Jadresic D, Fenwick P (1993) Thalamo-frontal psychosis. Br J Psychiatry 163:113–115

McGuffin P, Owen MJ, Farmer AE (1995) Genetic basis of schizophrenia. Lancet 346:678–682

Mori T, Ohnishi T, Hashimoto R, Nemoto K, Moriguchi Y, Noguchi H, Nakabayashi T, Hori H, Harada S, Saitoh O, Matsuda H, Kunugi H (2007) Progressive changes of white matter integrity in schizophrenia revealed by diffusion tensor imaging. Psychiatry Res 154:133–145

Nagayama M, Shinohara Y, Furukawa H, Kitamoto T (1996) Fatal familial insomnia with a mutation at codon 178 of the prion protein gene: first report from Japan. Neurology 47:1313–1316

Oke AF, Adams RN, Winblad B, von Knorring L (1988) Elevated dopamine/norepinephrine ratios in thalami of schizophrenic brains. Biol Psychiatry 24:79–82

Oke AF, Putz C, Adams RN, Bird ED (1992) Neuroleptic treatment is an unlikely cause of elevated dopamine in thalamus of schizophrenic subjects. Psychiatry Res 45:203–208

Pakkenberg B (1990) Pronounced reduction of total neuron number in mediodorsal thalamic nucleus and nucleus accumbens in schizophrenics. Arch Gen Psychiatry 47:1023–1028

Pakkenberg B (1992) The volume of the mediodorsal thalamic nucleus in treated and untreated schizophrenics. Schizophr Res 7:95–100

Popken GJ, Leggio MG, Bunney WE, Jones EG (2002) Expression of mRNAs related to the GABAergic and glutamatergic neurotransmitter systems in the human thalamus: normal and schizophrenic. Thalamus Relat Syst 1:349–369

Portas CM, Goldstein JM, Shenton ME, Hokama HH, Wible CG, Fischer I, Kikinis R, Donnino R, Jolesz FA, McCarley RW (1998) Volumetric evaluation of the thalamus in schizophrenic male patients using magnetic resonance imaging. Biol Psychiatry 43:649–659

Preuss UW, Zetzsche T, Jäger M, Groll C, Frodl T, Bottlender R, Leinsinger G, Hegerl U, Hahn K, Möller HJ, Meisenzahl EM (2005) Thalamic volume in first-episode and chronic schizophrenic subjects: a volumetric MRI study. Schizophr Res 73:91–101

Rapoport JL, Giedd J, Kumra S, Jacobsen L, Smith A, Lee P, Nelson J, Hamburger S (1997) Childhood-onset schizophrenia. Progressive ventricular change during adolescence. Arch Gen Psychiatry 54:897–903

Reichova I, Sherman SM (2004) Somatosensory corticothalamic projections: distinguishing drivers from modulators. J Neurophysiol 92:2185–2197

Reilly M, Connolly S, Stack J, Martin EA, Hutchinson M (1992) Bilateral paramedian thalamic infarction: a distinct but poorly recognized stroke syndrome. Q J Med 82:63–70

Rose SE, Chalk JB, Janke AL, Strudwick MW, Windus LC, Hannah DE, McGrath JJ, Pantelis C, Wood SJ, Mowry BJ (2006) Evidence of altered prefrontal-thalamic circuitry in schizophrenia: an optimized diffusion MRI study. Neuroimage 32:16–22

Salami M, Itami C, Tsumoto T, Kimura F (2003) Change of conduction velocity by regional myelination yields constant latency irrespective of distance between thalamus and cortex. Proc Natl Acad Sci USA 100:6174–6179

Salgado-Pineda P, Junqué C, Vendrell P, Baeza I, Bargalló N, Falcón C, Bernardo M (2004) Decreased cerebral activation during CPT performance: structural and functional deficits in schizophrenic patients. Neuroimage 21:840–847

Schlösser R, Gesierich T, Kaufmann B, Vucurevic G, Hunsche S, Gawehn J, Stoeter P (2003) Altered effective connectivity during working memory performance in schizophrenia: a study with fMRI and structural equation modeling. Neuroimage 19:751–763

Schmahmann JD (2003) Vascular syndromes of the thalamus. Stroke 34:2264–2278

Selemon LD, Wang L, Nebel MB, Csernansky JG, Goldman-Rakic PS, Rakic P (2005) Direct and indirect effects of fetal irradiation on cortical gray and white matter volume in the macaque. Biol Psychiatry 57:83–90

Sharp FR, Tomitaka M, Bernaudin M, Tomitaka S (2001) Psychosis: pathological activation of limbic thalamocortical circuits by psychomimetics and schizophrenia? Trends Neurosci 24:330–334

Sherman SM (2004) Interneurons and triadic circuitry of the thalamus. Trends Neurosci 27:670–675

Sherman SM (2005) Thalamic relays and cortical functioning. Prog Brain Res 149:107–126

Sherman SM, Guillery RW (1998) On the actions that one nerve cell can have on another: distinguishing "drivers" from "modulators". Proc Natl Acad Sci USA 95:7121–7126

Silbersweig DA, Stern E, Frith C, Cahill C, Holmes A, Grootoonk S, Seaward J, McKenna P, Chua SE, Schnorr L (1995) A functional neuroanatomy of hallucinations in schizophrenia. Nature 378:176–179

Sim K, Cullen T, Ongur D, Heckers S (2006) Testing models of thalamic dysfunction in schizophrenia using neuroimaging. J Neural Transm 113:907–928

Staal WG, Hulshoff Pol HE, Schnack H, van der Schot AC, Kahn RS (1998) Partial volume decrease of the thalamus in relatives of patients with schizophrenia. Am J Psychiatry 155:1784–1786

Steen RG, Mull C, McClure R, Hamer RM, Lieberman JA (2006) Brain volume in first-episode schizophrenia: systematic review and meta-analysis of magnetic resonance imaging studies. Br J Psychiatry 188:510–518

Stein BE (1998) Neural mechanisms for synthesizing sensory information and producing adaptive behaviors. Exp Brain Res 123:124–135

Stone JM, Day F, Tsagaraki H, Valli I, McLean MA, Lythgoe DJ, O'Gorman RL, Barker G, Mcguire P, OASIS (2009) Glutamate dysfunction in people with prodromal symptoms of psychosis: relationship to gray matter volume. Biol Psychiatry 66:533–539

Strungas S, Christensen JD, Holcomb JM, Garver DL (2003) State-related thalamic changes during antipsychotic treatment in schizophrenia: preliminary observations. Psychiatry Res 124:121–124

Sullivan EV, Rosenbloom MJ, Serventi KL, Deshmukh A, Pfefferbaum A (2003) Effects of alcohol dependence comorbidity and antipsychotic medication on volumes of the thalamus and pons in schizophrenia. Am J Psychiatry 160:1110–1116

Sussmann JE, Lymer GK, McKirdy J, Moorhead TW, Maniega SM, Job D, Hall J, Bastin ME, Johnstone EC, Lawrie SM, McIntosh AM (2009) White matter abnormalities in bipolar disorder and schizophrenia detected using diffusion tensor magnetic resonance imaging. Bipolar Disord 11:11–18

Swadlow HA, Gusev AG (2001) The impact of 'bursting' thalamic impulses at a neocortical synapse. Nat Neurosci 4:402–408

Syková E (2004) Extrasynaptic volume transmission and diffusion parameters of the extracellular space. Neuroscience 129:861–876

Théberge J, Williamson KE, Aoyama N, Drost DJ, Manchanda R, Malla AK, Northcott S, Menon RS, Neufeld RW, Rajakumar N, Pavlosky W, Densmore M, Schaefer B, Williamson PC (2007) Longitudinal grey-matter and glutamatergic losses in first-episode schizophrenia. Br J Psychiatry 191:325–334

Thermenos HW, Seidman LJ, Breiter H, Goldstein JM, Goodman JM, Poldrack R, Faraone SV, Tsuang MT (2004) Functional magnetic resonance imaging during auditory verbal working memory in nonpsychotic relatives of persons with schizophrenia: a pilot study. Biol Psychiatry 55:490–500

Turetsky BI, Calkins ME, Light GA, Olincy A, Radant AD, Swerdlow NR (2007) Neurophysiological endophenotypes of schizophrenia: the viability of selected candidate measures. Schizophr Bull 33:69–94

van Haren NE, Hulshoff Pol HE, Schnack HG, Cahn W, Brans R, Carati I, Rais M, Kahn RS (2008) Progressive brain volume loss in schizophrenia over the course of the illness: evidence of maturational abnormalities in early adulthood. Biol Psychiatry 63:106–113

Whyte MC, Whalley HC, Simonotto E, Flett S, Shillcock R, Marshall I, Goddard NH, Johnstone EC, Lawrie SM (2006) Event-related fMRI of word classification and successful word recognition in subjects at genetically enhanced risk of schizophrenia. Psychol Med 36:1427–1439

Wright IC, Rabe-Hesketh S, Woodruff PW, David AS, Murray RM, Bullmore ET (2000) Meta-analysis of regional brain volumes in schizophrenia. Am J Psychiatry 157:16–25

Young KA, Manaye KF, Liang C, Hicks PB, German DC (2000) Reduced number of mediodorsal and anterior thalamic neurons in schizophrenia. Biol Psychiatry 47:944–953

Zhou Y, Liang M, Jiang T, Tian L, Liu Y, Liu Z, Liu H, Kuang F (2007) Functional dysconnectivity of the dorsolateral prefrontal cortex in first-episode schizophrenia using resting-state fMRI. Neurosci Lett 417:297–302

Hippocampal Pathology in Schizophrenia

Stephan Heckers and Christine Konradi

Contents

1 Introduction 530
2 The Human Hippocampus 530
3 The Hippocampus in Neuropsychiatric Disorders 531
4 Models of Hippocampal Dysfunction in Schizophrenia 532
5 Evidence of Hippocampal Dysfunction in Schizophrenia 534
    5.1 Hippocampal Volume Change in Schizophrenia 534
    5.2 Hippocampal Neurons in Schizophrenia 535
    5.3 Genetic Mechanisms of Hippocampal Pathology in Schizophrenia 538
    5.4 Hippocampal Function and Schizophrenia 540
6 Animal Models 542
7 Critical Review of Findings and Directions for Future Studies 543
References 544

Abstract The hippocampus is abnormal in schizophrenia. Smaller hippocampal volume is the most consistent finding and is present already in the early stages of the illness. The underlying cellular substrate is a subtle, yet functionally significant reduction of hippocampal interneurons. Neuroimaging studies have revealed a pattern of increased hippocampal activity at baseline and decreased recruitment during the performance of memory tasks. Hippocampal lesion models in rodents have replicated some of the pharmacological, anatomical and behavioral phenotype of schizophrenia. Taken together, this pattern of findings points to a disinhibition of

---

S. Heckers (✉)
Department of Psychiatry, Vanderbilt University, 1601 23rd Avenue South, Room 3060, Nashville, TN 37212, USA
e-mail: stephan.heckers@Vanderbilt.Edu

C. Konradi
Department of Pharmacology, Vanderbilt University, 1601 23rd Avenue South, Room 3060, Nashville, TN 37212, USA

hippocampal pyramidal cells and abnormal cortico-hippocampal interactions in schizophrenia.

Keywords Cognitive Deficit · GABA · Glutamate · Hippocampus · Interneurons · Memory · Schizophrenia

1 Introduction

Over the course of the last 20 years, there has been a growing interest in the hippocampus of patients with schizophrenia. The evidence accumulated so far indicates subtle, yet reproducible abnormalities and has given rise to several models of hippocampal dysfunction in schizophrenia. We will provide a brief review of the human hippocampus, including an outline of three neuropsychiatric conditions with known hippocampal pathology. Then, we will review the major models of hippocampal dysfunction in schizophrenia and provide a summary of the main lines of evidence. We will conclude with a critical review of the evidence and a set of recommendations for further studies of the hippocampus in schizophrenia.

2 The Human Hippocampus

The hippocampus is an elongated structure in the ventromedial region of the human temporal lobe. With a volume of 3–5 cm^3 per hemisphere, it occupies less than 1% of the brain. The hippocampus borders the amygdala anteriorly, is covered by the parahippocampal gyrus medially, and extends posteriorly to the pulvinar nucleus of the thalamus. Perpendicular to the anterior–posterior axis, the hippocampus can be divided into five regions: the cornu ammonis (CA) sectors 1–4 and the dentate gyrus (DG) (Fig. 1).

The human hippocampus plays a crucial role in the processing of information (Eichenbaum 2004). Sensory information (e.g., visual, auditory, somatosensory) is relayed from the primary sensory cortices to the multimodal cortices of the prefrontal, parietal, and lateral temporal cortices. These higher-order cortical areas send projections to the entorhinal cortex, a region of the anterior parahippocampal gyrus. The entorhinal cortex sends this highly processed, multimodal sensory information to the hippocampus via two inputs: the direct pathway (to sector CA1) and the indirect pathway (to DG → CA2/3 → CA1) (Witter et al. 2000) (Fig. 1). After processing of the sensory information, the hippocampus sends information back to the entorhinal cortex via the subiculum.

The direct and indirect pathways converge on pyramidal cells in sector CA1 (Fig. 1). The CA1 neurons integrate new sensory information (arriving via the direct pathway) with previously experienced sensory data (retrieved from cortical regions via the entorhinal cortex and the indirect pathway). If these two signals

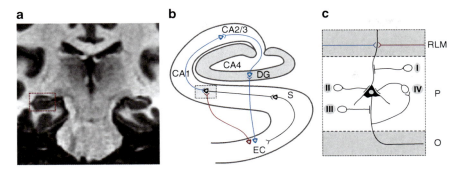

Fig. 1 Anatomy of the hippocampus. (**a**) MRI image of a coronal section through the medial temporal lobe. The *box* highlights the right hippocampus. (**b**) The hippocampus receives input from the entorhinal cortex via the direct (*red*) and indirect (*blue*) pathways. The indirect pathway relays information via the dentate gyrus (DG) and cornu ammonis (CA) sectors 2/3 to sector CA1. The direct pathway projects directly to sector CA1. Neurons in sector CA1 project via the subiculum (S) back to the entorhinal cortex. (**c**) The dendrites of pyramidal neurons in sector CA1 are the target of input from the direct and indirect pathways. Interneurons inhibit pyramidal neurons through axodendritic (I), axosomatic (II), and axoaxonic (III) synapses. Pyramidal neurons dampen their own firing rate through axon collaterals that drive interneurons (IV). The location of the cell bodies and their projections give rise to the three layers of the hippocampus: stratum radiatum/lacunosum/moleculare (RLM), pyramidal cell layer (P), and stratum oriens (O)

coincide, the CA1 neurons send a "match" signal to the cerebral cortex, limbic structures, and the brainstem (Lisman and Grace 2005; Lisman and Otmakhova 2001). The detection of novelty is then followed by the evaluation of valence and salience of new sensory information, through a neural circuit that includes the hippocampus, ventral tegmental area, and nucleus accumbens (Lisman and Grace 2005). The electrophysiological correlates of this information processing are oscillatory rhythms of hippocampal neurons (θ-oscillation at a frequency of 4–8 Hz and γ-oscillation at a frequency of 30–100 Hz) (Tsien 2000). These oscillations are generated by the coordinated firing of excitatory and inhibitory neurons in the hippocampus (Somogyi and Klausberger 2005).

The cellular organization, connections, and electrophysiological properties of the hippocampus as outlined here are crucial for memory function (Eichenbaum 2004) and for the integration of emotion and cognition (Bannerman et al. 2004; McNaughton et al. 2007).

3 The Hippocampus in Neuropsychiatric Disorders

Neuroscientists did not appreciate the crucial role of the hippocampus until the case report of patient H.M., who underwent bilateral medial temporal lobe surgery for intractable seizures in 1953 and subsequently developed amnesia (Corkin 2002;

Scoville and Milner 1957). There is now extensive literature linking three neuropsychiatric conditions, that is, amnesia, dementia, and seizures, to hippocampal pathology.

The most compelling evidence for the importance of the hippocampus comes from case reports of patients with selective hippocampal lesions (Cohen and Eichenbaum 1993). More recently, amnesia patients have provided the foundation for models of hippocampal-specific memory processes. For example, the recollection of entire episodes and the ability to relate individual items to each other has been associated specifically with the hippocampus (Konkel et al. 2008).

While amnesia is typically caused by an acquired lesion of the hippocampus, dementia is associated with a degenerative pathology of the hippocampus. The stages of Alzheimer's disease are defined by the anatomical pattern and severity of neuritic plaque and neurofibrillary tangle depositions and subsequent cell loss in the entorhinal cortex, DG, CA sectors, and subiculum (Braak and Braak 1991; Duyckaerts et al. 2009). Even before autopsy, structural and functional neuroimaging studies can demonstrate smaller hippocampal volume, increased (i.e., inefficient) BOLD signal change, and increased cerebral blood volume (CBV) of the hippocampus in Alzheimer's disease (Bookheimer et al. 2000; Reitz et al. 2009).

Temporal lobe epilepsy is often associated with selective lesions of the hippocampus. A typical scenario is an ischemic lesion of hippocampal neurons, leading to cell loss and a focus of electrical hyperexcitability (Dichter 2009). In addition to acquired seizure syndromes, mouse mutants with selective lesions of hippocampal interneurons display the behavioral, histological, and electrophysiological signs of epilepsy (Cobos et al. 2005).

The evidence for hippocampal pathology in amnesia, dementia, and epilepsy is compelling. In contrast, the emerging literature on hippocampal pathology in schizophrenia is intriguing, but has not provided the basis for an objective test or a neuropathological confirmation of the clinical diagnosis.

4 Models of Hippocampal Dysfunction in Schizophrenia

The rapidly growing evidence for abnormalities of hippocampal structure and function in schizophrenia has given rise to several models of hippocampal dysfunction. In contrast to models of cortical (preliminary prefrontal) dysfunction (Heckers 1997), hippocampal models did not attract serious interest in the schizophrenia research community until the early 1990s (Heckers 2001; Heckers and Konradi 2002). Each hippocampal model emphasizes different aspects of the schizophrenia phenotype (e.g., psychosis vs. memory deficit) and different lines of evidence (e.g., neural circuits vs. neurotransmitters and receptors). For the purpose of this review, we are highlighting five models of hippocampal dysfunction in schizophrenia:

1. Hippocampal pathology drives psychosis. This hypothesis predates any data collected in patients with schizophrenia (Adler and Waldo 1991; Bickford-Wimer et al. 1990; Hemsley 1993; Krieckhaus et al. 1992; Port and Seybold 1995; Roberts 1963; Venables 1992), but has received support from neuroimaging and neural network modeling studies that have linked hippocampal dysfunction to the degree of psychosis in schizophrenia (Friston et al. 1992; Liddle 1992; Siekmeier et al. 2007; Talamini et al. 2005).
2. Hippocampal pathology leads to memory deficits in schizophrenia. While the evidence for memory deficits in schizophrenia is definite (Aleman et al. 1999), it has been surprisingly difficult to find strong clinicopathological correlations of memory deficits and hippocampal abnormalities (Weiss and Heckers 2001). For example, episodic memory deficits in schizophrenia may be due to hippocampal or prefrontal cortex dysfunction and the available literature cannot disambiguate between these two hypotheses (Leavitt and Goldberg 2009; Preston et al. 2005; Reichenberg and Harvey 2007).
3. Disconnection of the hippocampus from multimodal association cortices leads to schizophrenia. This is the anatomical correlate of the psychosis (#1) and memory deficit (#2) models and may serve as the overarching theory of all the hippocampal models of schizophrenia. Since it is unlikely that schizophrenia is a selective lesion of the hippocampus (such as hippocampal amnesia), it is compelling to conceptualize schizophrenia as a perturbation of the reciprocal corticohippocampal pathways (Fletcher 1998). This notion has served as a guide for many neuroimaging studies, which can test the hypothesis of a hippocampal–cortical network dysfunction (Ellison-Wright and Bullmore 2009). Several neural network models have embraced this model as well.
4. Impaired function of glutamatergic (primarily N-methyl-D-aspartate – NMDA) receptors in the hippocampus leads to psychosis. While the glutamate hypothesis of schizophrenia does not make any a priori prediction of localized pathology, the prominent role of hippocampal NMDA receptors in the creation of oscillatory activity and the encoding/retrieval of information has led to several models of hippocampal dysfunction in schizophrenia (Greene 2001; Grunze et al. 1996; Harrison 2004; Harrison et al. 2003; Lisman and Otmakhova 2001; Medoff et al. 2001). These models focus particularly on the integration of the direct and indirect pathways via the NMDA receptor on CA1 pyramidal cells (Greene 2001; Lisman et al. 2008) and the modulation of these neurons by dopamine (Lisman et al. 2008; Lisman and Otmakhova 2001).
5. Insufficient γ-aminobutyric acid (GABA)ergic inhibition of hippocampal neurons leads to schizophrenia (Benes 1999; Benes and Berretta 2001). The abnormal expression of GABAergic genes and proteins and the abnormal activity of the hippocampus observed in studies of glucose metabolism and blood flow have provided compelling evidence for this model (Heckers et al. 1998; Schobel et al. 2009b). Some have proposed that GABAergic dysfunction of the hippocampus is secondary to NMDA receptor hypofunction in schizophrenia (Benes 2009; Lisman et al. 2008; Olney and Farber 1995).

While the models of hippocampal dysfunction in schizophrenia are supported by preliminary evidence, they are still in need of data. This will allow us to decide which model has the greatest explanatory power and how it relates to other models of schizophrenia (e.g., dopamine hypothesis, prefrontal cortex model, thalamic model).

5 Evidence of Hippocampal Dysfunction in Schizophrenia

Several lines of evidence support hippocampal abnormalities in schizophrenia: smaller hippocampal volume, abnormal hippocampal neuron number, abnormal function of genes expressed at high levels in the hippocampus, and abnormal hippocampal activity. While they support some of the models of hippocampal dysfunction in schizophrenia, the studies reviewed below have progressed more or less independently and an integration of their findings is still at an early stage.

5.1 Hippocampal Volume Change in Schizophrenia

The hippocampus is smaller in schizophrenia compared to matched healthy control subjects. This is a very robust finding, supported by many neuroimaging studies and confirmed by several meta-analyses (Honea et al. 2005; Nelson et al. 1998; Steen et al. 2006; Vita et al. 2006; Wright et al. 2000). The effect sizes for smaller hippocampal volume in schizophrenia are 0.8 for all patients and about 0.4 for first-episode patients, putting them at the top of morphometric studies in schizophrenia (Wright et al. 2000). There are, however, several unresolved questions.

The anatomical pattern of hippocampal volume change is not clear. There is some evidence that the volume of the anterior but not the posterior hippocampus is smaller in schizophrenia (Csernansky et al. 1998; Goldman et al. 2007; Rossi et al. 1994; Schobel et al. 2009a; Suddath et al. 1990; Wang et al. 2001), but this anterior/posterior gradient has not been found in all studies (Velakoulis et al. 2001; Weiss et al. 2005). Several high-resolution structural imaging methods have been developed to subdivide the hippocampus into the four CA sectors, which provides another approach to test the hypothesis of regionally selective hippocampal pathology in schizophrenia (Malykhin et al. 2009; Zeineh et al. 1998).

The timing of hippocampal volume change is unknown. Hippocampal volume is already reduced at the time of the first psychotic episode (Steen et al. 2006; Vita et al. 2006). But it is not clear whether hippocampal volume is already reduced in subsyndromal at-risk subjects (Pantelis et al. 2003; Velakoulis et al. 2006; Witthaus et al. 2009).

The quantification of hippocampal volume is not diagnostic for schizophrenia, since it is within the normal range for most patients. Reduced hippocampal volume is also not specific for schizophrenia, since several other psychiatric disorders, such as depression (Campbell and Macqueen 2004; Koolschijn et al. 2009), alcoholism (Geuze et al. 2005), and PTSD (Smith 2005; Woon and Hedges 2008), show a similar pathology. This limits the use of structural imaging as a diagnostic test for schizophrenia, an approach with great promise in the early diagnosis of dementia (Teipel et al. 2008). Complicating the picture even more is the finding that hippocampal volume is also smaller in first-degree relatives of schizophrenia subjects (although not to the same degree as in patients) (Boos et al. 2007; Smith 2005). If this finding holds up in future studies (for concerns, see McDonald et al. 2008), then volume studies need to explore a slope of hippocampal volume from healthy subjects at low genetic risk, through asymptomatic subjects at risk for schizophrenia, to patients with prodromal schizophrenia and ultimately chronic schizophrenia.

5.2 Hippocampal Neurons in Schizophrenia

Each human hippocampus contains approximately 10 million neurons (West and Gundersen 1990). The majority of hippocampal neurons (about 90%) are large, pyramidal-shaped, glutamatergic neurons (principal cells). The remaining 10% of hippocampal neurons are smaller, nonpyramidal, GABAergic neurons (nonprincipal cells) (Freund and Buzsaki 1996; Olbrich and Braak 1985). Despite the relatively small total number of interneurons, they have developed into a highly specialized group of neurons, which differ in their anatomical, biochemical, and electrophysiological properties (Freund and Buzsaki 1996). The two types of hippocampal neurons give rise to an intricate balance of excitation (principal cells) and inhibition (nonprincipal cells). Most hippocampal neurons are located in the pyramidal cell layer (Fig. 1), whereas the two other layers of the hippocampus (i.e., the stratum oriens and the stratum radiatum/lacunosum/moleculare) contain few neurons. Subtle differences in the cellular architecture of the three-layered hippocampus give rise to the four sectors of the cornu ammonis (CA1–4) and the dentate gyrus (Fig. 1).

5.2.1 Hippocampal Neuron Number

The total number of hippocampal neurons in schizophrenia is not reduced to the degree seen in Alzheimer's disease or temporal lobe epilepsy (Falkai and Bogerts 1986; Heckers et al. 1991; Schmitt et al. 2009; Walker et al. 2002). Furthermore, the volume of the pyramidal cell layer and the packing density of cells in the pyramidal cell layer (see Fig. 1) are not decreased in schizophrenia (Heckers et al. 1991; Hurlemann et al. 2005). These studies provide compelling evidence that hippocampal pathology in schizophrenia is distinctly different from that of dementia and epilepsy, both of which are characterized by hippocampal volume reduction due to a significant loss of neurons. It is surprising that the postmortem studies have not corroborated the finding of decreased hippocampal volume in schizophrenia, reported by the large majority of neuroimaging studies (Heckers et al. 1990, 1991; Schmitt et al. 2009; Walker et al. 2002). While postmortem brain volume estimates are prone to substantial bias due to tissue processing, and therefore are of limited value (Braendgaard and Gundersen 1986), we do not have a simple explanation for the discrepancy between the in vivo and ex vivo hippocampal volume estimates in schizophrenia.

5.2.2 Glutamatergic Neurotransmission

Most studies of glutamatergic neurotransmission in schizophrenia have focused on the expression of glutamate receptor complexes, including ionotropic (NMDA, α-amino-3-hydroxy-5-methyl-4-isoxazolepropionic acid – AMPA, and kainate) and G protein-coupled (metabotropic) receptors (Kristiansen et al. 2009; Meador-Woodruff and Healy 2000).

Several studies have reported a decreased expression of the AMPA subunits GluR1 and GluR2 in the hippocampus and the parahippocampal gyrus (Eastwood et al. 1995, 1997; Harrison et al. 1991). In concordance, ligand binding to AMPA receptors was decreased (Kerwin et al. 1990). The kainate receptor subtypes GluR6 and KA2 were also significantly reduced in the hippocampus (Porter et al. 1997). Studies on kainate receptor density, conducted with radiolabeled kainate, demonstrated a decrease in the hippocampus (Deakin et al. 1989; Kerwin et al. 1990).

Initial studies of the NMDA receptor, which focused on the PCP-binding site located inside the ion channel, found no marked changes in the hippocampus in schizophrenia (Kornhuber et al. 1989; Meador-Woodruff and Healy 2000). A study of the NMDA receptor subunits NR1, NR2A, and NR2B found an increase of NR2B mRNA and a decrease of NR1 mRNA in the hippocampus in schizophrenia (Gao et al. 2000), but this has not been replicated in subsequent studies (Kristiansen et al. 2009).

The balance of GABAergic and glutamatergic neurotransmission in the hippocampus in schizophrenia was explored in studies of the expression of the two modulatory proteins: complexin I (presumably reflecting the integrity of GABAergic neurons) and complexin II (presumably reflecting the integrity of glutamatergic neurons). While complexin II expression was found to be more reduced in schizophrenia than complexin I (Eastwood and Harrison 2000; Harrison and Eastwood 1998), this did not correlate with similar changes of the vesicular GABA transporter (vGAT) and vesicular glutamate transporter (vGluT1) (Sawada et al. 2005). However, the complexin II/I ratio correlated inversely with the degree of cognitive impairment antemortem (Sawada et al. 2005), providing intriguing evidence that glutamatergic dysfunction in the hippocampus will lead to cognitive deficits in schizophrenia.

5.2.3 GABAergic Neurons

A growing body of literature deals with abnormalities of GABAergic hippocampal neurons in schizophrenia. Initial studies focused on postsynaptic GABAergic receptors, located on pyramidal and nonpyramidal cells, and revealed a regionally specific upregulation of GABA-A receptor binding in sectors CA2–4, but not CA1 (Benes et al. 1996, 1997). The marked increase of the GABA-A receptor in CA2/3 was found primarily on interneurons, indicating a decreased GABAergic regulation by other interneurons (Benes 1999).

The first evidence for an abnormality of hippocampal interneurons in schizophrenia came from a study of neuron density. Using the shape and staining pattern of pyramidal and nonpyramidal cells as the distinguishing pattern, Benes et al. reported a decrease in the number of nonpyramidal cells, but no changes in the density of pyramidal cells, in schizophrenia and bipolar disorder (Benes et al. 1998). More recent studies have focused on the defining marker of GABAergic neurons, that is, glutamic acid decarboxylase (GAD), the enzyme that converts glutamate to GABA. Two isoforms of *GAD* are known: the gene *GAD1* codes for GAD67 and the gene *GAD2* codes for GAD65. An initial in situ hybridization study of *GAD* mRNA expression in the hippocampus in normal controls, patients with schizophrenia, and patients with bipolar disorder revealed significant decreases of *GAD2* (and to a lesser degree *GAD1*) mRNA expression in bipolar disorder and less significant changes in schizophrenia (Heckers et al. 2002). A subsequent gene expression microarray study confirmed the decreased expression of *GAD1* and *GAD2* in bipolar disorder, but did not find any changes in schizophrenia (Konradi et al. 2004). Finally, a large-scale postmortem study of *GAD1* mRNA expression in 32 patients with schizophrenia and 76 normal control subjects revealed decreased expression in schizophrenia in the dorsolateral prefrontal cortex, but no changes in the hippocampus (Straub et al. 2007). These studies of hippocampal *GAD* mRNA expression in schizophrenia have to be reevaluated in light of a recent study using laser-capture microdissection and microarray profiling, which revealed that changes of hippocampal GAD67 expression in schizophrenia are regionally specific: While the expression was normal in the large sector CA1, it was significantly decreased in sector CA2/3 (Benes et al. 2007).

Additional evidence for selective changes in hippocampal interneurons in schizophrenia comes from the study of calcium-binding proteins, which are differentially expressed in subpopulations of hippocampal interneurons (Freund and Buzsaki 1996; Seress et al. 1993) (Fig. 2). These subpopulations of neurons create a dynamic, spatiotemporal control of hippocampal cell firing, which gives rise to several brain states crucial for normal cognition (Somogyi and Klausberger 2005). An initial study of neuronal density revealed a significantly decreased density of parvalbumin-positive neurons in all hippocampal regions, while the density of calretinin-positive cells was normal (Zhang and Reynolds 2000). The finding of decreased parvalbumin expression has now been corroborated by further studies and provides evidence for a subtype-specific abnormality of interneurons in schizophrenia (Eyles et al. 2002; Torrey et al. 2005).

5.2.4 Other Neurotransmitters

While most studies have explored abnormalities of GABAergic and glutamatergic neurotransmission, additional evidence suggests abnormalities of serotonergic, cholinergic, and dopaminergic neurotransmission in the hippocampus in schizophrenia (Kristiansen et al. 2009). Arguably, the most compelling evidence is the decreased expression of and binding to the α_7 nicotinic receptor and the M1/M4 muscarinic receptors (Kristiansen et al. 2009). In addition, a recent line of evidence has implicated abnormalities of mitochondrial function in both schizophrenia and bipolar disorder. While some have reported abnormal expression of nuclear genes coding for proteins involved in mitochondrial energy metabolism in bipolar disorder but not schizophrenia (Konradi et al. 2004), others have provided evidence for mitochondrial pathology in both disorders (Altar et al. 2005).

Taken together, there is evidence for cellular and molecular abnormalities of the hippocampus in schizophrenia. These changes lead neither to an overall decrease in the number of neurons nor to an overall decrease of either GABAergic or glutamatergic neurotransmission. Rather, hippocampal pathology in schizophrenia seems to be selective for subtypes of neurons and for regions within the CA. Such a pattern of cell- and region-specific pathology could be related to some of the recently identified genetic mechanisms of schizophrenia and could give rise to selective deficits of hippocampal function in patients with schizophrenia.

5.3 Genetic Mechanisms of Hippocampal Pathology in Schizophrenia

The genetic basis of schizophrenia is now firmly established (Owen et al. 2005). However, it is less clear which regions of the brain are affected by changes of DNA sequence or RNA expression (Harrison and Weinberger 2005). Several genes of interest are expressed at high levels in the hippocampus, which makes them sensible targets for the exploration of genetic mechanisms of hippocampal pathology in schizophrenia. Here, we will briefly review the evidence for four genes associated with a risk for schizophrenia, that is, neuregulin-1 (NRG1), disrupted in schizophrenia-1 (DISC1), dystrobrevin-binding protein-1 (DTNBP1), and brain-derived neurotrophic factor (BDNF).

The *NRG1* gene (located on chromosome 8p22) and the gene for one of its receptors, ErbB4 (located on chromosome 2q34), have both been associated with schizophrenia (Harrison and Law 2006; Owen et al. 2005). NRG1 and ErbB4 are expressed in the hippocampus (Law et al. 2004; Mechawar et al. 2007) and regulate GABAergic neurotransmission (Woo et al. 2007). They also affect the function of

α_7 nicotinic receptors (Chang and Fischbach 2006), located on hippocampal interneurons. NRG1 is known to affect long-term potentiation of hippocampal synapses and to modulate dendritic growth and plasticity (Li et al. 2007a). ErbB4 receptors are expressed primarily on hippocampal interneurons and ErbB4 knockout models lead to selective dysfunction of some but not all hippocampal interneurons (Neddens and Buonanno 2009; Vullhorst et al. 2009).

The *DISC1* gene was originally identified in a single pedigree with prominent psychiatric history and has subsequently been associated with several aspects of the schizophrenia phenotype (Roberts 2007). In adult mouse brain, the highest levels of DISC1 mRNA were found in the DG, followed by lower expression in sectors CA1–CA3 (Ma et al. 2002). In a transgenic mouse model, early postnatal induction of mutant C-terminal DISC1 resulted in a cluster of schizophrenia-related phenotypes, including reduced hippocampal dendritic complexity, decreased hippocampal synaptic transmission, and abnormal spatial working memory. This led to the postulation that alterations in DISC1 function during brain development may contribute to the pathogenesis of schizophrenia (Li et al. 2007b). Mice carrying a deletion in the *DISC1* gene that model the schizophrenia-associated translocation showed alterations in the organization of DG neurons, a deficit in short-term plasticity and a selective working memory impairment (Kvajo et al. 2008). Although in schizophrenia the expression of *DISC1* mRNA was not found to be abnormal, the expression of several molecules in the DISC1 pathway was decreased (Lipska et al. 2006). While the exact mechanisms of DISC1 in schizophrenia remain unclear, hippocampal volume and function are under considerable control by the *DISC1* gene (Callicott et al. 2005) and there is tentative evidence that some polymorphisms of the *DISC1* gene contribute to smaller hippocampal volume in schizophrenia (Cannon et al. 2005).

The dystrobrevin-binding protein-1, also known as dysbindin-1, has been associated with schizophrenia in several studies (Harrison and Weinberger 2005). DTNBP1 mRNA is expressed in principal cells of the hippocampus (Talbot et al. 2004). Presynaptic dysbindin-1 expression was reduced in glutamatergic terminals of the hippocampus in schizophrenia. This has been interpreted as contributing to glutamatergic dysfunction in the polysynaptic pathway of the hippocampus and receives support from a mutant mouse model of dysbindin-1 (Talbot 2009).

BDNF plays a major role in brain development and reduced concentrations of the protein have been reported in schizophrenia. In rats, it has been demonstrated that BDNF is vital for hippocampal memory consolidation (Lee et al. 2004). While the association of the *BDNF* gene with schizophrenia is not very strong, several studies have reported that hippocampal volume is larger in individuals with the Val/Val than the Val/Met allele of the most frequently studied single nucleotide, rs6265 (van Haren et al. 2008). Moreover, the Met allele was associated with poorer episodic memory, abnormal hippocampal activation, and lower levels of hippocampal *N*-acetylaspartate in MRI spectroscopy (Egan et al. 2003).

Taken together, it is likely that genetic variations of *NRG1*, *DISC1*, *DTNBP1*, and *BDNF* in schizophrenia affect hippocampal function. This emerging literature on hippocampal effects of schizophrenia risk genes is complemented by the literature on hippocampal pathology in first-degree relatives of patients with schizophrenia. Adult relatives of schizophrenia patients who do not develop schizophrenia (but might show more subtle signs of psychopathology) have smaller hippocampal volumes (Boos et al. 2007; Honea et al. 2005; Seidman et al. 1999, 2002). This indicates that hippocampal pathology in schizophrenia is, at least in part, under the control of genetic factors.

5.4 Hippocampal Function and Schizophrenia

The hippocampus serves a unique role in the encoding and retrieval of memory. It allows the brain to disambiguate relationships between items and to record the sequences of events, making it essential for the creation of relational, episodic, and autobiographical memory (Eichenbaum 2004). Most investigators studying hippocampal function in schizophrenia using positron emission tomography (PET) and functional magnetic resonance imaging (fMRI) have looked at the role of the hippocampus in the encoding and retrieval of memory (Achim and Lepage 2005; Boyer et al. 2007). However, other functions of the hippocampus have also been explored (Bannerman et al. 2004; Bast and Feldon 2003), ranging from sensory gating (Tregellas et al. 2007) to decisional capacity (Eyler et al. 2007).

5.4.1 Hippocampal Activity at Rest in Schizophrenia

Studies of hippocampal activity at rest have found two different patterns in schizophrenia: lower regional cerebral glucose metabolic rates (rCMRglc) (Buchsbaum et al. 1992; Nordahl et al. 1996; Tamminga et al. 1992) and increased regional cerebral blood flow (rCBF) (Friston et al. 1992; Kawasaki et al. 1992, 1996; Lahti et al. 2003; Liddle et al. 1992; Malaspina et al. 2004; Medoff et al. 2001). It is not easy to reconcile these findings. But several studies have demonstrated that increased hippocampal rCBF in schizophrenia is normalized in patients treated with dopamine D2 antagonists (Lahti et al. 2003; Malaspina et al. 2004; Medoff et al. 2001), potentially obscuring abnormal patterns of resting activity. Furthermore, increased left temporal brain metabolism is more prominent in patients with negative symptoms and those with severe delusions and hallucinations (Gur et al. 1995) and resting rCBF values are positively correlated with more severe psychopathology in general (Friston et al. 1992) or with more prominent positive symptoms (delusions and hallucinations) (Liddle et al. 1992). These findings are supported by the few neuroimaging studies that have documented an activation of the hippocampus during auditory hallucinations (Dierks et al. 1999; Silbersweig et al. 1995). Taken together, hippocampal activity has been linked to psychosis, but it remains unclear whether hippocampal activation generates the hallucinatory experience or whether it is involved in the processing, for example, monitoring the source of an auditory representation (Weiss and Heckers 1999).

The studies of hippocampal glucose metabolism and cerebral blood flow in schizophrenia have been complemented by a recent study of CBV (Schobel et al. 2009b). This study of patients with chronic and prodromal schizophrenia revealed increased CBV selectively in hippocampal sector CA1, together with CBV increases in the orbitofrontal cortex and CBV decreases in the dorsolateral prefrontal cortex. The increased CBV in CA1 was interpreted as evidence for a basal hypermetabolic state in the hippocampus in schizophrenia. This was supported by a positive correlation between the degree of CBV and the severity of delusions in the schizophrenia patients.

5.4.2 Hippocampal Activity and Cognitive Function in Schizophrenia

The initial evidence for hippocampal dysfunction during cognitive task performance in schizophrenia came from a PET study of word-stem cued recall (Heckers et al. 1998, 1999) (Fig. 2). While normal subjects activated a right frontal–temporal network to retrieve previously studied words, schizophrenia patients failed to recruit the hippocampus, despite robust and even increased activation of prefrontal regions. Compared to the control group, hippocampal baseline activity was continuously increased in schizophrenia and was not modulated by environmental contingencies. The pattern of increased hippocampal activity at baseline and impaired recruitment during episodic memory retrieval was interpreted as the functional

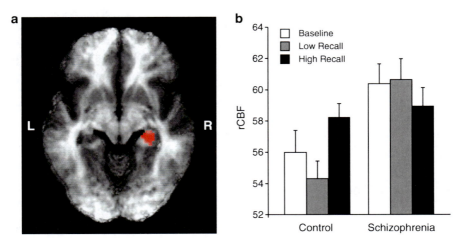

Fig. 2 Abnormal recruitment of the hippocampus during memory retrieval in schizophrenia. (**a**) An axial section through the brain shows the location of abnormal brain activation in the right hippocampus in schizophrenia during a word-stem cued recall experiment (for details of the experimental design, see Heckers et al. 1998). (**b**) Healthy control subjects showed increased regional cerebral blood flow (rCBF) during high accuracy recall when compared with both lexical retrieval at baseline and low accuracy recall. This normal pattern was absent in the schizophrenia group and all three recall conditions were associated with higher rCBF in the hippocampus

correlate of an abnormal corticohippocampal interaction in schizophrenia (in support of the hippocampal model #3 earlier) (Fletcher 1998).

Subsequent studies have extended this initial finding (Hall et al. 2009; Jessen et al. 2003; Leavitt and Goldberg 2009; Ongur et al. 2006; Ragland et al. 2001; Weiss et al. 2003, 2004). First, patients with schizophrenia relied less on the recruitment of the hippocampus and showed more widespread activation of the prefrontal cortex during the retrieval of previously learned information (Weiss et al. 2003). Second, the ability to classify new items as previously not experienced was impaired in schizophrenia (i.e., a higher false alarm rate to new items) and was associated with decreased activation and smaller volume of the hippocampus (Weiss et al. 2004). Third, hippocampal recruitment in schizophrenia was impaired during a relational, but not during a nonrelational memory task (Ongur et al. 2006). In addition to these findings of impaired hippocampal activation during memory retrieval in chronic patients with schizophrenia, a study of first-episode psychosis patients has revealed a selective deficit to engage hippocampal-dependent relational binding, resulting in poorer subsequent recognition performance (Achim et al. 2007).

Abnormal activation of the hippocampus in schizophrenia is not limited to memory function. For example, patients with schizophrenia demonstrate significantly greater activation of the hippocampus while passively viewing facial expressions (Holt et al. 2006). Furthermore, healthy subjects demonstrate significant habituation of hippocampal activity to the repeated presentation of fearful faces, whereas patients with schizophrenia do not demonstrate such habituation (Holt et al. 2005).

In summary, functional neuroimaging studies have reported increased blood flow in the hippocampus in schizophrenia, which is associated with higher levels of psychopathology and psychosis (i.e., delusions and hallucinations). The evidence of abnormal hippocampal activity is particularly strong for the domain of memory, with several studies revealing specific abnormalities of hippocampal recruitment during the performance of memory tasks. Future studies need to explore whether such abnormalities of hippocampal function, demonstrated so far in small samples of subjects, can explain the memory deficits in most patients with schizophrenia and whether they can explain the social dysfunction resulting from memory deficits in schizophrenia (Eyler et al. 2007; Green 1996).

6 Animal Models

Several animal models of hippocampal pathology in schizophrenia have supported the evidence from postmortem and in vivo studies in patients with schizophrenia (Feldon and Weiner 2009; Sawa 2009).

A substantial number of experiments in rodents have shown that a neonatal ventral hippocampal lesion induces several of the pharmacological and behavioral features of schizophrenia (Lipska 2004; Tseng et al. 2009). While the lesion does not provide a good model for the subtle abnormalities of volume and cell number reported for the hippocampus in schizophrenia, it does provide evidence that an early developmental lesion of the hippocampus perturbs the normal regulation of dopamine release and the proper function of the cerebral cortex.

The methylazoxymethanol (MAM)-G17 model employs the administration of a mitotoxin, MAM, on gestational day 17 to pregnant rats, to induce a developmental disruption of the hippocampus (Lodge and Grace 2009). The MAM model shows a loss of parvalbumin-containing interneurons (especially in the ventral hippocampus), leading to diminished oscillatory activity (Lodge et al. 2009).

The infusion of picrotoxin, a noncompetitive antagonist of the GABA-A receptor, into the basolateral complex of the amygdala is a model of perturbed amygdala–hippocampal interaction. It leads to an increased flow of excitatory activity into stratum oriens of hippocampal sectors CA2 and CA3, resulting in a selective reduction of GABAergic interneurons containing parvalbumin, calbindin, and calretinin (Berretta et al. 2009).

The blockade of NMDA receptors in the hippocampus (especially in sector CA1) leads to a decreased activity of parvalbumin-positive interneurons, which in turn leads to a disinhibition of hippocampal pyramidal cells (Behrens et al. 2007; Bickel and Javitt 2009; Greene 2001; Kinney et al. 2006; Lisman et al. 2008). In addition, an NR1 knockdown mouse model shows a marked deficit in the phase coupling between θ- and γ-oscillations, indicating abnormal integration of hippocampal–cortical interactions (Ramsey 2009).

Taken together, these rodent models of environmental or genetic lesions of the hippocampus replicate some of the core findings in patients with schizophrenia, that is, loss of parvalbumin-containing interneurons, increased neural activity in the hippocampus, and memory deficits. These models may serve as a crucial bridge between the studies in humans and the study of basic hippocampal mechanisms of schizophrenia, potentially leading to the development of better pharmacological treatments of schizophrenia.

7 Critical Review of Findings and Directions for Future Studies

The literature on hippocampal pathology in schizophrenia is rapidly growing. The main body of literature, that is, studies of hippocampal structure and function in patients with schizophrenia, is supported by postmortem and animal studies of cellular and molecular mechanisms. However, the significance of hippocampal pathology in schizophrenia is still unknown. Here, we will highlight three important aspects that need to be addressed in future studies: regional specificity, timing, and mechanisms of hippocampal pathology.

The anatomical pattern of hippocampal pathology in schizophrenia needs to be studied more thoroughly. Some investigators have proposed regionally specific pathology in sector CA1, others in sectors CA2 and CA3. The two sectors contribute uniquely to different stages of memory formation (e.g., pattern separation, pattern completion, novelty detection) (Cutsuridis et al. 2010; Lisman and Grace 2005; Neves et al. 2008). In addition, some investigators have reported abnormalities in the anterior but not posterior hippocampus in schizophrenia. The anterior hippocampus is more closely connected with limbic structures and medial prefrontal cortical areas, whereas the posterior hippocampus has prominent reciprocal connections to the dorsolateral prefrontal cortex (Barbas and Blatt 1995; Goldman-Rakic et al. 1984; Lepage et al. 1998; Strange and Dolan 2006; Strange et al. 1999). Regionally specific pathology of the hippocampus should, therefore, predict distinct patterns of hippocampal dysfunction in schizophrenia.

The developmental profile of hippocampal pathology in schizophrenia is unknown. While neuroimaging studies have clearly shown that hippocampal volume changes are present at the time of the first psychotic episode, it is not clear whether hippocampal structure is already abnormal during early stages of development or during the asymptomatic at-risk state (Lawrie 2007; Tebartz van Elst et al. 2007). The issue of vulnerability for smaller hippocampal volume and its progression throughout the illness are an important issue for further study, since the finding of smaller hippocampal volume in first-degree relatives of schizophrenia probands indicates a genetic liability. Similarly, the role of hippocampal dysfunction in the emergence of clinical features, such as psychosis and cognitive deficits, needs to be clarified. Studies of the timing of hippocampal pathology in schizophrenia should include studies of genetic and environmental factors that can affect hippocampal development and postmaturational integrity (Arango et al. 2001; Phillips et al. 2006).

The mechanisms of hippocampal pathology in schizophrenia remain unclear. Several models have been proposed, highlighting abnormalities of interneurons, NMDA receptors, and corticohippocampal connections. To advance our understanding of the hippocampus in schizophrenia, we need to move to strong hypothesis testing, resulting in the exclusion of some of the current hypotheses. This needs to include a test of the specificity of hippocampal pathology. For example, if NMDA receptor hypofunction and a loss of parvalbumin-containing interneurons are at the core of hippocampal pathology in schizophrenia, how is it different from the pathology in the cerebral cortex, where similar changes have been described (Lewis and Hashimoto 2007)?

Ultimately, we want to understand how subtle perturbations in a small, yet crucial region of the medial temporal lobe contribute to schizophrenia. This may include diagnostic tests to predict occurrence (Davatzikos et al. 2005), targets for the development of new drugs (Dhikav and Anand 2007; Newton and Duman 2007), and, eventually, strategies to prevent the development of schizophrenia.

References

Achim AM, Lepage M (2005) Episodic memory-related activation in schizophrenia: meta-analysis. Br J Psychiatry 187:500–509

Achim AM, Bertrand MC, Sutton H, Montoya A, Czechowska Y, Malla AK, Joober R, Pruessner JC, Lepage M (2007) Selective abnormal modulation of hippocampal activity during memory formation in first-episode psychosis. Arch Gen Psychiatry 64:999–1014

Adler LE, Waldo MC (1991) Counterpoint: a sensory gating–hippocampal model of schizophrenia. Schizophr Bull 17:19–24

Aleman A, Hijman R, de Haan EHF, Kahn RS (1999) Memory impairment in schizophrenia: a meta-analysis. Am J Psychiatry 156:1358–1366

Altar CA, Jurata LW, Charles V, Lemire A, Liu P, Bukhman Y, Young TA, Bullard J, Yokoe H, Webster MJ, Knable MB, Brockman JA (2005) Deficient hippocampal neuron expression of proteasome, ubiquitin, and mitochondrial genes in multiple schizophrenia cohorts. Biol Psychiatry 58:85–96

Arango C, Kirkpatrick B, Koenig J (2001) At issue: stress, hippocampal neuronal turnover, and neuropsychiatric disorders. Schizophr Bull 27:477–480

Bannerman DM, Rawlins JN, McHugh SB, Deacon RM, Yee BK, Bast T, Zhang WN, Pothuizen HH, Feldon J (2004) Regional dissociations within the hippocampus – memory and anxiety. Neurosci Biobehav Rev 28:273–283

Barbas H, Blatt GJ (1995) Topographically specific hippocampal projections target functionally distinct prefrontal areas in the rhesus monkey. Hippocampus 5:511–533

Bast T, Feldon J (2003) Hippocampal modulation of sensorimotor processes. Prog Neurobiol 70:319–345

Behrens MM, Ali SS, Dao DN, Lucero J, Shekhtman G, Quick KL, Dugan LL (2007) Ketamine-induced loss of phenotype of fast-spiking interneurons is mediated by NADPH-oxidase. Science 318:1645–1647

Benes FM (1999) Evidence for altered trisynaptic circuitry in schizophrenic hippocampus. Biol Psychiatry 46:589–599

Benes FM (2009) Neural circuitry models of schizophrenia: is it dopamine, GABA, glutamate, or something else? Biol Psychiatry 65:1003–1005

Benes FM, Berretta S (2001) GABAergic interneurons: implications for understanding schizophrenia and bipolar disorder. Neuropsychopharmacology 25:1–27

Benes FM, Khan Y, Vincent SL, Wickramasinghe R (1996) Differences in the subregional and cellular distribution of GABAA receptor binding in the hippocampal formation of schizophrenic brain. Synapse 22:338–349

Benes FM, Wickramasinghe R, Vincent SL, Khan Y, Todtenkopf M (1997) Uncoupling of GABA(A) and benzodiazepine receptor binding activity in the hippocampal formation of schizophrenic brain. Brain Res 755:121–129

Benes FM, Kwok EW, Vincent SL, Todtenkopf MS (1998) A reduction of nonpyramidal cells in sector CA2 of schizophrenics and manic depressives. Biol Psychiatry 44:88–97

Benes FM, Lim B, Matzilevich D, Walsh JP, Subburaju S, Minns M (2007) Regulation of the GABA cell phenotype in hippocampus of schizophrenics and bipolars. Proc Natl Acad Sci USA 104:10164–10169

Berretta S, Gisabella B, Benes FM (2009) A rodent model of schizophrenia derived from postmortem studies. Behav Brain Res 204:363–368

Bickel S, Javitt DC (2009) Neurophysiological and neurochemical animal models of schizophrenia: focus on glutamate. Behav Brain Res 204:352–362

Bickford-Wimer PC, Nagamoto H, Johnson R, Adler LE, Egan M, Rose GM, Freedman R (1990) Auditory sensory gating in hippocampal neurons: a model system in the rat. Biol Psychiatry 27:183–192

Bookheimer SY, Strojwas MH, Cohen MS, Saunders AM, Pericak-Vance MA, Mazziotta JC, Small GW (2000) Patterns of brain activation in people at risk for Alzheimer's disease. N Engl J Med 343:450–456

Boos HB, Aleman A, Cahn W, Pol HH, Kahn RS (2007) Brain volumes in relatives of patients with schizophrenia: a meta-analysis. Arch Gen Psychiatry 64:297–304

Boyer P, Phillips JL, Rousseau FL, Ilivitsky S (2007) Hippocampal abnormalities and memory deficits: new evidence of a strong pathophysiological link in schizophrenia. Brain Res Rev 54:92–112

Braak H, Braak E (1991) Neuropathological stageing of Alzheimer-related changes. Acta Neuropathol 82:239–259

Braendgaard H, Gundersen HJG (1986) The impact of recent stereological advances on quantitative studies of the nervous system. J Neurosci Methods 18:39–78

Buchsbaum MS, Haier RJ, Potkin SG, Nuechterlein K, Bracha HS, Katz M, Lohr J, Wu J, Lottenberg S, Jerabek PA, Trenary M, Tafalla R, Reynolds C, Bunney WE (1992) Frontostriatal disorder of cerebral metabolism in never-medicated schizophrenics. Arch Gen Psychiatry 49:935–942

Callicott JH, Straub RE, Pezawas L, Egan MF, Mattay VS, Hariri AR, Verchinski BA, Meyer-Lindenberg A, Balkissoon R, Kolachana B, Goldberg TE, Weinberger DR (2005) Variation in DISC1 affects hippocampal structure and function and increases risk for schizophrenia. Proc Natl Acad Sci USA 102:8627–8632

Campbell S, Macqueen G (2004) The role of the hippocampus in the pathophysiology of major depression. J Psychiatry Neurosci 29:417–426

Cannon TD, Hennah W, van Erp TG, Thompson PM, Lonnqvist J, Huttunen M, Gasperoni T, Tuulio-Henriksson A, Pirkola T, Toga AW, Kaprio J, Mazziotta J, Peltonen L (2005) Association of DISC1/TRAX haplotypes with schizophrenia, reduced prefrontal gray matter, and impaired short- and long-term memory. Arch Gen Psychiatry 62:1205–1213

Chang Q, Fischbach GD (2006) An acute effect of neuregulin 1 beta to suppress alpha 7-containing nicotinic acetylcholine receptors in hippocampal interneurons. J Neurosci 26:11295–11303

Cobos I, Calcagnotto ME, Vilaythong AJ, Thwin MT, Noebels JL, Baraban SC, Rubenstein JL (2005) Mice lacking Dlx1 show subtype-specific loss of interneurons, reduced inhibition and epilepsy. Nat Neurosci 8:1059–1068

Cohen NJ, Eichenbaum H (1993) Memory, amnesia and the hippocampal system. MIT, Cambridge, MA

Corkin S (2002) What's new with the amnesic patient H.M.? Nat Rev Neurosci 3:153–160

Csernansky JG, Joshi S, Wang L, Haller JW, Gado M, Miller JP, Grenander U, Miller MI (1998) Hippocampal morphometry in schizophrenia by high dimensional brain mapping. Proc Natl Acad Sci USA 95:11406–11411

Cutsuridis V, Cobb S, Graham BP (2010) Encoding and retrieval in a model of the hippocampal CA1 microcircuit. Hippocampus 20(3):423–446

Davatzikos C, Shen D, Gur RC, Wu X, Liu D, Fan Y, Hughett P, Turetsky BI, Gur RE (2005) Whole-brain morphometric study of schizophrenia revealing a spatially complex set of focal abnormalities. Arch Gen Psychiatry 62:1218–1227

Deakin JFW, Slater P, Simpson MDC, Gilchrist AC, Skan WJ, Royston MC, Reynolds GP, Cross AJ (1989) Frontal cortical and left temporal glutamatergic dysfunction in schizophrenia. J Neurochem 52:1781–1786

Dhikav V, Anand KS (2007) Is hippocampal atrophy a future drug target? Med Hypotheses 68:1300–1306

Dichter MA (2009) Emerging concepts in the pathogenesis of epilepsy and epileptogenesis. Arch Neurol 66:443–447

Dierks T, Linden DE, Jandl M, Formisano E, Goebel R, Lanfermann H, Singer W (1999) Activation of Heschl's gyrus during auditory hallucinations. Neuron 22:615–621

Duyckaerts C, Delatour B, Potier MC (2009) Classification and basic pathology of Alzheimer disease. Acta Neuropathol 118:5–36

Eastwood SL, Harrison PJ (2000) Hippocampal synaptic pathology in schizophrenia, bipolar disorder and major depression: a study of complexin mRNAs. Mol Psychiatry 5:425–432

Eastwood SL, McDonald B, Burnet PWJ, Beckwith JP, Kerwin RW, Harrison PJ (1995) Decreased expression of mRNAs encoding non-NMDA glutamate receptors GluR1 and GluR2 in medial temporal lobe neurons in schizophrenia. Mol Brain Res 29:211–223

Eastwood SL, Burnet PW, Harrison PJ (1997) GluR2 glutamate receptor subunit flip and flop isoforms are decreased in the hippocampal formation in schizophrenia: a reverse transcriptase–polymerase chain reaction (RT–PCR) study. Brain Res Mol Brain Res 44:92–98

Egan MF, Kojima M, Callicott JH, Goldberg TE, Kolachana BS, Bertolino A, Zaitsev E, Gold B, Goldman D, Dean M, Lu B, Weinberger DR (2003) The BDNF val66met polymorphism affects activity-dependent secretion of BDNF and human memory and hippocampal function. Cell 112:257–269

Eichenbaum H (2004) Hippocampus: cognitive processes and neural representations that underlie declarative memory. Neuron 44:109–120

Ellison-Wright I, Bullmore E (2009) Meta-analysis of diffusion tensor imaging studies in schizophrenia. Schizophr Res 108:3–10

Eyler LT, Olsen RK, Nayak GV, Mirzakhanian H, Brown GG, Jeste DV (2007) Brain response correlates of decisional capacity in schizophrenia: a preliminary FMRI study. J Neuropsychiatry Clin Neurosci 19:137–144

Eyles DW, McGrath JJ, Reynolds GP (2002) Neuronal calcium-binding proteins and schizophrenia. Schizophr Res 57:27–34

Falkai P, Bogerts B (1986) Cell loss in the hippocampus of schizophrenics. Eur Arch Psychiatry Neurol Sci 236:154–161

Feldon J, Weiner I (2009) Editorial: special issue on modeling schizophrenia. Behav Brain Res 204:255–257

Fletcher P (1998) The missing link: a failure of fronto-hippocampal integration in schizophrenia. Nat Neurosci 1:266–267

Freund TF, Buzsaki G (1996) Interneurons of the hippocampus. Hippocampus 6:347–470

Friston KJ, Liddle PF, Frith CD, Hirsch SR, Frackowiak RS (1992) The left medial temporal region and schizophrenia. A PET study. Brain 115:367–382

Gao X-M, Sakai K, Roberts RC, Conley RR, Dean B, Tamminga CA (2000) Ionotropic glutamate receptors and expression of N-methyl-D-aspartate receptor subunits in subregions of human hippocampus: effects of schizophrenia. Am J Psychiatry 157:1141–1149

Geuze E, Vermetten E, Bremner JD (2005) MR-based in vivo hippocampal volumetrics: 2. Findings in neuropsychiatric disorders. Mol Psychiatry 10:160–184

Goldman MB, Torres IJ, Keedy S, Marlow-O'Connor M, Beenken B, Pilla R (2007) Reduced anterior hippocampal formation volume in hyponatremic schizophrenic patients. Hippocampus 17:554–562

Goldman-Rakic PS, Selemon LD, Schwartz ML (1984) Dual pathways connecting the dorsolateral prefrontal cortex with the hippocampal formation and parahippocampal cortex in the rhesus cortex. Neuroscience 12:719–743

Green M (1996) What are the functional consequences of neurocognitive deficits in schizophrenia? Am J Psychiatry 153:321–330

Greene R (2001) Circuit analysis of NMDAR hypofunction in the hippocampus, in vitro, and psychosis of schizophrenia. Hippocampus 11:569–577

Grunze HC, Rainnie DG, Hasselmo ME, Barkai E, Hearn EF, McCarley RW, Greene RW (1996) NMDA-dependent modulation of CA1 local circuit inhibition. J Neurosci 16:2034–2043

Gur RE, Mozley PD, Resnick SM, Mozley LH, Shtasel DL, Gallacher F, Arnold SE, Karp JS, Alavi A, Reivich M, Gur RC (1995) Resting cerebral glucose metabolism in first-episode and previously treated patients with schizophrenia relates to clinical features. Arch Gen Psychiatry 52:657–667

Hall J, Whalley HC, Marwick K, McKirdy J, Sussmann J, Romaniuk L, Johnstone EC, Wan HI, McIntosh AM, Lawrie SM (2009) Hippocampal function in schizophrenia and bipolar disorder. Psychol Med [Epub ahead of print, Sept 7]

Harrison PJ (2004) The hippocampus in schizophrenia: a review of the neuropathological evidence and its pathophysiological implications. Psychopharmacology (Berl) 174:151–162

Harrison PJ, Eastwood SL (1998) Preferential involvement of excitatory neurons in medial temporal lobe in schizophrenia. Lancet 352:1669–1673

Harrison PJ, Law AJ (2006) Neuregulin 1 and schizophrenia: genetics, gene expression, and neurobiology. Biol Psychiatry 60:132–140

Harrison PJ, Weinberger DR (2005) Schizophrenia genes, gene expression, and neuropathology: on the matter of their convergence. Mol Psychiatry 10:40–68

Harrison PJ, McLaughlin D, Kerwin RW (1991) Decreased hippocampal expression of a glutamate receptor gene in schizophrenia. Lancet 337:450–452

Harrison PJ, Law AJ, Eastwood SL (2003) Glutamate receptors and transporters in the hippocampus in schizophrenia. Ann NY Acad Sci 1003:94–101

Heckers S (1997) Neuropathology of schizophrenia: cortex, thalamus, basal ganglia, and neurotransmitter-specific projection systems. Schizophr Bull 23:403–421

Heckers S (2001) Neuroimaging studies of the hippocampus in schizophrenia. Hippocampus 11:520–528

Heckers S, Konradi C (2002) Hippocampal neurons in schizophrenia. J Neural Transm 109:891–905

Heckers S, Heinsen H, Heinsen YC, Beckmann H (1990) Limbic structures and lateral ventricle in schizophrenia. A quantitative postmortem study. Arch Gen Psychiatry 47:1016–1022

Heckers S, Heinsen H, Geiger B, Beckmann H (1991) Hippocampal neuron number in schizophrenia. A stereological study. Arch Gen Psychiatry 48:1002–1008

Heckers S, Rauch SL, Goff D, Savage CR, Schacter DL, Fischman AJ, Alpert NM (1998) Impaired recruitment of the hippocampus during conscious recollection in schizophrenia. Nat Neurosci 1:318–323

Heckers S, Goff D, Schacter DL, Savage CR, Fischman AJ, Alpert NM, Rauch SL (1999) Functional imaging of memory retrieval in deficit vs nondeficit schizophrenia. Arch Gen Psychiatry 56:1117–1123

Heckers S, Stone D, Walsh J, Shick J, Koul P, Benes FM (2002) Differential hippocampal expression of glutamic acid decarboxylase 65 and 67 messenger RNA in bipolar disorder and schizophrenia. Arch Gen Psychiatry 59:521–529

Hemsley DR (1993) A simple (or simplistic?) cognitive model for schizophrenia. Behav Res Ther 31:633–45

Holt DJ, Weiss AP, Rauch SL, Wright CI, Zalesak M, Goff DC, Ditman T, Welsh RC, Heckers S (2005) Sustained activation of the hippocampus in response to fearful faces in schizophrenia. Biol Psychiatry 57:1011–1019

Holt DJ, Kunkel L, Weiss AP, Goff DC, Wright CI, Shin LM, Rauch SL, Hootnick J, Heckers S (2006) Increased medial temporal lobe activation during the passive viewing of emotional and neutral facial expressions in schizophrenia. Schizophr Res 82:153–162

Honea R, Crow TJ, Passingham D, Mackay CE (2005) Regional deficits in brain volume in schizophrenia: a meta-analysis of voxel-based morphometry studies. Am J Psychiatry 162:2233–2245

Hurlemann R, Tepest R, Maier W, Falkai P, Vogeley K (2005) Intact hippocampal gray matter in schizophrenia as revealed by automatized image analysis postmortem. Anat Embryol (Berl) 210:513–517

Jessen F, Scheef L, Germeshausen L, Tawo Y, Kockler M, Kuhn KU, Maier W, Schild HH, Heun R (2003) Reduced hippocampal activation during encoding and recognition of words in schizophrenia patients. Am J Psychiatry 160:1305–1312

Kawasaki Y, Suzuki M, Maeda Y, Urata K, Yamaguchi N, Matsuda H, Hisada K, Suzuki M, Takashima T (1992) Regional cerebral blood flow in patients with schizophrenia. A preliminary report. Eur Arch Psychiatry Clin Neurosci 241:195–200

Kawasaki Y, Maeda Y, Sakai N, Higashima M, Yamaguchi N, Koshino Y, Hisada K, Suzuki M, Matsuda H (1996) Regional cerebral blood flow in patients with schizophrenia: relevance to symptom structures. Psychiatry Res 67:49–58

Kerwin R, Patel S, Meldrum B (1990) Quantitative autoradiographic analysis of glutamate binding sites in the hippocampal formation in normal and schizophrenic brain post mortem. Neuroscience 39:25–32

Kinney JW, Davis CN, Tabarean I, Conti B, Bartfai T, Behrens MM (2006) A specific role for NR2A-containing NMDA receptors in the maintenance of parvalbumin and GAD67 immunoreactivity in cultured interneurons. J Neurosci 26:1604–1615

Konkel A, Warren DE, Duff MC, Tranel DN, Cohen NJ (2008) Hippocampal amnesia impairs all manner of relational memory. Front Hum Neurosci 2:15

Konradi C, Eaton M, MacDonald ML, Walsh J, Benes FM, Heckers S (2004) Molecular evidence for mitochondrial dysfunction in bipolar disorder. Arch Gen Psychiatry 61:300–308

Koolschijn PC, van Haren NE, Lensvelt-Mulders GJ, Hulshoff Pol HE, Kahn RS (2009) Brain volume abnormalities in major depressive disorder: a meta-analysis of magnetic resonance imaging studies. Hum Brain Mapp 30:3719–3735

Kornhuber J, Mack-Burkhardt F, Riederer P, Hebenstreit GF, Reynolds GP, Andrews HB, Beckmann H (1989) [3H]MK-801 binding sites in postmortem brain regions of schizophrenic patients. J Neural Transm Gen Sect 77:231–236

Krieckhaus EE, Donahoe JW, Morgan MA (1992) Paranoid schizophrenia may be caused by dopamine hyperactivity of CA1 hippocampus. Biol Psychiatry 31:560–570

Kristiansen LV, Cowell RM, Biscaia M, McCullumsmith RE, Meador-Woodruff JH (2009) Alterations of neurotransmitter receptors in schizophrenia: evidence from postmortem studies. In: Javitt DC, Kantrowitz J (eds) Handbook of neurochemistry and molecular neurobiology: schizophrenia, 3rd edn. Springer, New York, NY, pp 443–492

Kvajo M, McKellar H, Arguello PA, Drew LJ, Moore H, MacDermott AB, Karayiorgou M, Gogos JA (2008) A mutation in mouse Disc1 that models a schizophrenia risk allele leads to specific alterations in neuronal architecture and cognition. Proc Natl Acad Sci USA 105:7076–7081

Lahti AC, Holcomb HH, Weiler MA, Medoff DR, Tamminga CA (2003) Functional effects of antipsychotic drugs: comparing clozapine with haloperidol. Biol Psychiatry 53:601–608

Law AJ, Shannon Weickert C, Hyde TM, Kleinman JE, Harrison PJ (2004) Neuregulin-1 (NRG-1) mRNA and protein in the adult human brain. Neuroscience 127:125–136

Lawrie S (2007) Distinguishing vulnerability, prediction, and progression in the preschizophrenic brain. Arch Gen Psychiatry 64:250–251, author reply 252–253

Leavitt VM, Goldberg TE (2009) Episodic memory in schizophrenia. Neuropsychol Rev 19:312–323

Lee JL, Everitt BJ, Thomas KL (2004) Independent cellular processes for hippocampal memory consolidation and reconsolidation. Science 304:839–843

Lepage M, Habib R, Tulving E (1998) Hippocampal PET activations of memory encoding and retrieval: the HIPER model. Hippocampus 8:313–322

Lewis DA, Hashimoto T (2007) Deciphering the disease process of schizophrenia: the contribution of cortical GABA neurons. Int Rev Neurobiol 78:109–131

Li B, Woo RS, Mei L, Malinow R (2007a) The neuregulin-1 receptor erbB4 controls glutamatergic synapse maturation and plasticity. Neuron 54:583–597

Li W, Zhou Y, Jentsch JD, Brown RA, Tian X, Ehninger D, Hennah W, Peltonen L, Lonnqvist J, Huttunen MO, Kaprio J, Trachtenberg JT, Silva AJ, Cannon TD (2007b) Specific developmental disruption of disrupted-in-schizophrenia-1 function results in schizophrenia-related phenotypes in mice. Proc Natl Acad Sci USA 104:18280–18285

Liddle PF (1992) Syndromes of schizophrenia on factor analysis. Br J Psychiatry 161:860–861

Liddle PF, Friston KJ, Frith CD, Jones T, Hirsch SR, Frackowiak RSJ (1992) Patterns of cerebral blood flow in schizophrenia. Br J Psychiatry 160:179–186

Lipska BK (2004) Using animal models to test a neurodevelopmental hypothesis of schizophrenia. J Psychiatry Neurosci 29:282–286

Lipska BK, Peters T, Hyde TM, Halim N, Horowitz C, Mitkus S, Weickert CS, Matsumoto M, Sawa A, Straub RE, Vakkalanka R, Herman MM, Weinberger DR, Kleinman JE (2006) Expression of DISC1 binding partners is reduced in schizophrenia and associated with DISC1 SNPs. Hum Mol Genet 15:1245–1258

Lisman JE, Grace AA (2005) The hippocampal–VTA loop: controlling the entry of information into long-term memory. Neuron 46:703–713

Lisman JE, Otmakhova NA (2001) Storage, recall, and novelty detection of sequences by the hippocampus: elaborating on the SOCRATIC model to account for normal and aberrant effects of dopamine. Hippocampus 11:551–568

Lisman JE, Coyle JT, Green RW, Javitt DC, Benes FM, Heckers S, Grace AA (2008) Circuit-based framework for understanding neurotransmitter and risk gene interactions in schizophrenia. Trends Neurosci 31:234–242

Lodge DJ, Grace AA (2009) Gestational methylazoxymethanol acetate administration: a developmental disruption model of schizophrenia. Behav Brain Res 204:306–312

Lodge DJ, Behrens MM, Grace AA (2009) A loss of parvalbumin-containing interneurons is associated with diminished oscillatory activity in an animal model of schizophrenia. J Neurosci 29:2344–2354

Ma L, Liu Y, Ky B, Shughrue PJ, Austin CP, Morris JA (2002) Cloning and characterization of Disc1, the mouse ortholog of DISC1 (disrupted-in-schizophrenia 1). Genomics 80:662–672

Malaspina D, Harkavy-Friedman J, Corcoran C, Mujica-Parodi L, Printz D, Gorman JM, Van Heertum R (2004) Resting neural activity distinguishes subgroups of schizophrenia patients. Biol Psychiatry 56:931–937

Malykhin NV, Lebel RM, Coupland NJ, Wilman AH, Carter R (2009) In vivo quantification of hippocampal subfields using 4.7 T fast spin echo imaging. NeuroImage 49:1224–1230

McDonald C, Dineen B, Hallahan B (2008) Meta-analysis of brain volumes in unaffected first-degree relatives of patients with schizophrenia overemphasizes hippocampal deficits. Arch Gen Psychiatry 65:603–604, author reply 604–605

McNaughton N, Kocsis B, Hajos M (2007) Elicited hippocampal theta rhythm: a screen for anxiolytic and procognitive drugs through changes in hippocampal function? Behav Pharmacol 18:329–346

Meador-Woodruff JH, Healy DJ (2000) Glutamate receptor expression in schizophrenic brain. Brain Res Brain Res Rev 31:288–294

Mechawar N, Lacoste B, Yu WF, Srivastava LK, Quirion R (2007) Developmental profile of neuregulin receptor ErbB4 in postnatal rat cerebral cortex and hippocampus. Neuroscience 148:126–139

Medoff DR, Holcomb HH, Lahti AC, Tamminga CA (2001) Probing the human hippocampus using rCBF: contrasts in schizophrenia. Hippocampus 11:543–550

Neddens J, Buonanno A (2009) Selective populations of hippocampal interneurons express ErbB4 and their number and distribution is altered in ErbB4 knockout mice. Hippocampus [Epub ahead of print, Aug 4]

Nelson MD, Saykin AJ, Flashman LA, Riordan HJ (1998) Hippocampal volume reduction in schizophrenia as assessed by magnetic resonance imaging: a meta-analytic study. Arch Gen Psychiatry 55:433–440

Neves G, Cooke SF, Bliss TV (2008) Synaptic plasticity, memory and the hippocampus: a neural network approach to causality. Nat Rev Neurosci 9:65–75

Newton SS, Duman RS (2007) Neurogenic actions of atypical antipsychotic drugs and therapeutic implications. CNS Drugs 21:715–725

Nordahl TE, Kusubov N, Carter C, Salamat S, Cummings AM, O'Shora-Celaya L, Eberling J, Robertson L, Huesman RH, Jagust W, Budinger TF (1996) Temporal lobe metabolic differences in medication-free outpatients with schizophrenia via the PET-600. Neuropsychopharmacology 15:541–554

Olbrich HG, Braak H (1985) Ratio of pyramidal cells versus non-pyramidal cells in sector CA1 of the human Ammon's horn. Anat Embryol 173:105–110

Olney JW, Farber NB (1995) Glutamate receptor dysfunction and schizophrenia. Arch Gen Psychiatry 52:998–1007

Ongur D, Cullen TJ, Wolf DH, Rohan M, Barreira P, Zalesak M, Heckers S (2006) The neural basis of relational memory deficits in schizophrenia. Arch Gen Psychiatry 63:1268–1277

Owen MJ, Craddock N, O'Donovan MC (2005) Schizophrenia: genes at last? Trends Genet 9:518–525

Pantelis C, Velakoulis D, McGorry PD, Wood SJ, Suckling J, Phillips LJ, Yung AR, Bullmore ET, Brewer W, Soulsby B, Desmond P, McGuire PK (2003) Neuroanatomical abnormalities before and after onset of psychosis: a cross-sectional and longitudinal MRI comparison. Lancet 361:281–288

Phillips LJ, McGorry PD, Garner B, Thompson KN, Pantelis C, Wood SJ, Berger G (2006) Stress, the hippocampus and the hypothalamic–pituitary–adrenal axis: implications for the development of psychotic disorders. Aust NZ J Psychiatry 40:725–741

Port RL, Seybold KS (1995) Hippocampal synaptic plasticity as a biological substrate underlying episodic psychosis. Biol Psychiatry 37:318–324

Porter RH, Eastwood SL, Harrison PJ (1997) Distribution of kainate receptor subunit mRNAs in human hippocampus, neocortex and cerebellum, and bilateral reduction of hippocampal GluR6 and KA2 transcripts in schizophrenia. Brain Res 751:217–231

Preston AR, Shohamy D, Tamminga CA, Wagner AD (2005) Hippocampal function, declarative memory, and schizophrenia: anatomic and functional neuroimaging considerations. Curr Neurol Neurosci Rep 5:249–256

Ragland JD, Gur RC, Raz J, Schroeder L, Kohler CG, Smith RJ, Alavi A, Gur RE (2001) Effect of schizophrenia on frontotemporal activity during word encoding and recognition: a PET cerebral blood flow study. Am J Psychiatry 158:1114–1125

Ramsey AJ (2009) NR1 knockdown mice as a representative model of the glutamate hypothesis of schizophrenia. Prog Brain Res 179:51–58

Reichenberg A, Harvey PD (2007) Neuropsychological impairments in schizophrenia: integration of performance-based and brain imaging findings. Psychol Bull 133:833–858

Reitz C, Brickman AM, Brown TR, Manly J, DeCarli C, Small SA, Mayeux R (2009) Linking hippocampal structure and function to memory performance in an aging population. Arch Neurol 66:1385–1392

Roberts DR (1963) Schizophrenia and the brain. J Neuropsychiatry 5:71–79

Roberts RC (2007) Schizophrenia in translation: disrupted in schizophrenia (DISC1): integrating clinical and basic findings. Schizophr Bull 33:11–15

Rossi A, Stratta P, Mancini F, Gallucci M, Mattei P, Core L, Di Michele V, Casacchia M (1994) Magnetic resonance imaging findings of amygdala–anterior hippocampus shrinkage in male patients with schizophrenia. Psychiatry Res 52:43–53

Sawa A (2009) Genetic models of schizophrenia. Prog Brain Res 179:3–6

Sawada K, Barr AM, Nakamura M, Arima K, Young CE, Dwork AJ, Falkai P, Phillips AG, Honer WG (2005) Hippocampal complexin proteins and cognitive dysfunction in schizophrenia. Arch Gen Psychiatry 62:263–272

Schmitt A, Steyskal C, Bernstein HG, Schneider-Axmann T, Parlapani E, Schaeffer EL, Gattaz WF, Bogerts B, Schmitz C, Falkai P (2009) Stereologic investigation of the posterior part of the hippocampus in schizophrenia. Acta Neuropathol 117:395–407

Schobel SA, Kelly MA, Corcoran CM, Van Heertum K, Seckinger R, Goetz R, Harkavy-Friedman J, Malaspina D (2009a) Anterior hippocampal and orbitofrontal cortical structural brain abnormalities in association with cognitive deficits in schizophrenia. Schizophr Res 114:110–118

Schobel SA, Lewandowski NM, Corcoran CM, Moore H, Brown T, Malaspina D, Small SA (2009b) Differential targeting of the CA1 subfield of the hippocampal formation by schizophrenia and related psychotic disorders. Arch Gen Psychiatry 66:938–946

Scoville WB, Milner B (1957) Loss of recent memory after bilateral hippocampal lesions. J Neurol Neurosurg Psychiatry 20:11–21

Seidman LJ, Faraone SV, Goldstein JM, Goodman JM, Kremen WS, Toomey R, Tourville J, Kennedy D, Makris N, Caviness VS, Tsuang MT (1999) Thalamic and amygdala–hippocampal volume reductions in first-degree relatives of patients with schizophrenia: an MRI-based morphometric analysis. Biol Psychiatry 46:941–954

Seidman LJ, Faraone SV, Goldstein JM, Kremen WS, Horton NJ, Makris N, Toomey R, Kennedy D, Caviness VS, Tsuang MT (2002) Left hippocampal volume as a vulnerability indicator for schizophrenia. Arch Gen Psychiatry 59:839–849

Seress L, Gulyas AI, Ferrer I, Tunon T, Soriano E, Freund TF (1993) Distribution, morphological features, and synaptic connections of parvalbumin- and calbindin D28k-immunoreactive neurons in the human hippocampal formation. J Comp Neurol 337:208–230

Siekmeier PJ, Hasselmo ME, Howard MW, Coyle J (2007) Modeling of context-dependent retrieval in hippocampal region CA1: implications for cognitive function in schizophrenia. Schizophr Res 89:177–190

Silbersweig DA, Stern E, Frith C, Cahill C, Holmes A, Grootoonk S, Seaward J, McKenna P, Chua SE, Schnorr L, Jones T, Frackowiak RSJ (1995) A functional neuroanatomy of hallucinations in schizophrenia. Nature 378:176–179

Smith ME (2005) Bilateral hippocampal volume reduction in adults with post-traumatic stress disorder: a meta-analysis of structural MRI studies. Hippocampus 15:798–807

Somogyi P, Klausberger T (2005) Defined types of cortical interneurone structure space and spike timing in the hippocampus. J Physiol 562:9–26

Steen RG, Mull C, McClure R, Hamer RM, Lieberman JA (2006) Brain volume in first-episode schizophrenia: systematic review and meta-analysis of magnetic resonance imaging studies. Br J Psychiatry 188:510–518

Strange BA, Dolan RJ (2006) Anterior medial temporal lobe in human cognition: memory for fear and the unexpected. Cogn Neuropsychiatry 11:198–218

Strange BA, Fletcher PC, Henson RN, Friston KJ, Dolan RJ (1999) Segregating the functions of human hippocampus. Proc Natl Acad Sci USA 96:4034–4039

Straub RE, Lipska BK, Egan MF, Goldberg TE, Callicott JH, Mayhew MB, Vakkalanka RK, Kolachana BS, Kleinman JE, Weinberger DR (2007) Allelic variation in GAD1 (GAD(67)) is associated with schizophrenia and influences cortical function and gene expression. Mol Psychiatry 12:854–869

Suddath RL, Christison GW, Torrey EF, Casanova MF, Weinberger DR (1990) Anatomical abnormalities in the brains of monozygotic twins discordant for schizophrenia. N Engl J Med 322:789–794

Talamini LM, Meeter M, Elvevag B, Murre JM, Goldberg TE (2005) Reduced parahippocampal connectivity produces schizophrenia-like memory deficits in simulated neural circuits with reduced parahippocampal connectivity. Arch Gen Psychiatry 62:485–493

Talbot K (2009) The sandy (sdy) mouse: a dysbindin-1 mutant relevant to schizophrenia research. Prog Brain Res 170:87–94

Talbot K, Eidem WL, Tinsley CL, Benson MA, Thompson EW, Smith RJ, Hahn CG, Siegel SJ, Trojanowski JQ, Gur RE, Blake DJ, Arnold SE (2004) Dysbindin-1 is reduced in intrinsic, glutamatergic terminals of the hippocampal formation in schizophrenia. J Clin Invest 113:1353–1363

Tamminga CA, Thaker GK, Buchanan R, Kirkpatrick B, Alphs LD, Chase TN, Carpenter WT (1992) Limbic system abnormalities identified in schizophrenia using positron emission tomography with fluorodeoxyglucose and neocortical alterations with deficit syndrome. Arch Gen Psychiatry 49:522–530

Tebartz van Elst L, Ebert D, Hesslinger B (2007) Amygdala volume status might reflect dominant mode of emotional information processing. Arch Gen Psychiatry 64:251–252, author reply 252–253

Teipel SJ, Meindl T, Grinberg L, Heinsen H, Hampel H (2008) Novel MRI techniques in the assessment of dementia. Eur J Nucl Med Mol Imaging 35(Suppl 1):S58–S69

Torrey EF, Barci BM, Webster MJ, Bartko JJ, Meador-Woodruff JH, Knable MB (2005) Neurochemical markers for schizophrenia, bipolar disorder, and major depression in postmortem brains. Biol Psychiatry 57:252–260

Tregellas JR, Davalos DB, Rojas DC, Waldo MC, Gibson L, Wylie K, Du YP, Freedman R (2007) Increased hemodynamic response in the hippocampus, thalamus and prefrontal cortex during abnormal sensory gating in schizophrenia. Schizophr Res 92:262–272

Tseng KY, Chambers RA, Lipska BK (2009) The neonatal ventral hippocampal lesion as a heuristic neurodevelopmental model of schizophrenia. Behav Brain Res 204:295–305

Tsien JZ (2000) Linking Hebb's coincidence-detection to memory formation. Curr Opin Neurobiol 10:266–273

van Haren NE, Bakker SC, Kahn RS (2008) Genes and structural brain imaging in schizophrenia. Curr Opin Psychiatry 21:161–167

Velakoulis D, Stuart GW, Wood SJ, Smith DJ, Brewer WJ, Desmond P, Singh B, Copolov D, Pantelis C (2001) Selective bilateral hippocampal volume loss in chronic schizophrenia. Biol Psychiatry 50:531–539

Velakoulis D, Wood SJ, Wong MT, McGorry PD, Yung A, Phillips L, Smith D, Brewer W, Proffitt T, Desmond P, Pantelis C (2006) Hippocampal and amygdala volumes according to psychosis stage and diagnosis: a magnetic resonance imaging study of chronic schizophrenia, first-episode psychosis, and ultra-high-risk individuals. Arch Gen Psychiatry 63:139–149

Venables PH (1992) Hippocampal function and schizophrenia. Experimental psychological evidence. Ann NY Acad Sci 658:111–127

Vita A, De Peri L, Silenzi C, Dieci M (2006) Brain morphology in first-episode schizophrenia: a meta-analysis of quantitative magnetic resonance imaging studies. Schizophr Res 82:75–88

Vullhorst D, Neddens J, Karavanova I, Tricoire L, Petralia RS, McBain CJ, Buonanno A (2009) Selective expression of ErbB4 in interneurons, but not pyramidal cells, of the rodent hippocampus. J Neurosci 29:12255–12264

Walker MA, Highley JR, Esiri MM, McDonald B, Roberts HC, Evans SP, Crow TJ (2002) Estimated neuronal populations and volumes of the hippocampus and its subfields in schizophrenia. Am J Psychiatry 159:821–828

Wang L, Joshi SC, Miller MI, Csernansky JG (2001) Statistical analysis of hippocampal asymmetry in schizophrenia. NeuroImage 14:531–545

Weiss AP, Heckers S (1999) Neuroimaging of hallucinations: a review of the literature. Psychiatry Res: Neuroimaging 92:61–74

Weiss AP, Heckers S (2001) Neuroimaging of declarative memory in schizophrenia. Scand J Psychol 42:239–250

Weiss AP, Schacter DL, Goff DC, Rauch SL, Alpert NM, Fischman AJ, Heckers S (2003) Impaired hippocampal recruitment during normal modulation of memory performance in schizophrenia. Biol Psychiatry 53:48–55

Weiss AP, Zalesak M, DeWitt I, Goff D, Kunkel L, Heckers S (2004) Impaired hippocampal function during the detection of novel words in schizophrenia. Biol Psychiatry 55:668–675

Weiss AP, DeWitt I, Goff D, Ditman T, Heckers S (2005) Anterior and posterior hippocampal volumes in schizophrenia. Schizophr Res 73:103–112

West MJ, Gundersen HJ (1990) Unbiased stereological estimation of the number of neurons in the human hippocampus. J Comp Neurol 296:1–22

Witter MP, Wouterlood FG, Naber PA, Van Haeften T (2000) Anatomical organization of the parahippocampal–hippocampal network. In: Scharfman HE, Witter MP, Schwarcz R (eds) The parahippocampal region: implications for neurological and psychiatric diseases. Annals of the New York Academy of Sciences, New York, NY, pp 1–24

Witthaus H, Kaufmann C, Bohner G, Ozgurdal S, Gudlowski Y, Gallinat J, Ruhrmann S, Brune M, Heinz A, Klingebiel R, Juckel G (2009) Gray matter abnormalities in subjects at ultra-high risk for schizophrenia and first-episode schizophrenic patients compared to healthy controls. Psychiatry Res 173:163–169

Woo RS, Li XM, Tao Y, Carpenter-Hyland E, Huang YZ, Weber J, Neiswender H, Dong XP, Wu J, Gassmann M, Lai C, Xiong WC, Gao TM, Mei L (2007) Neuregulin-1 enhances depolarization-induced GABA release. Neuron 54:599–610

Woon FL, Hedges DW (2008) Hippocampal and amygdala volumes in children and adults with childhood maltreatment-related posttraumatic stress disorder: a meta-analysis. Hippocampus 18:729–736

Wright IC, Rabe-Hesketh S, Woodruff PW, David AS, Murray RM, Bullmore ET (2000) Meta-analysis of regional brain volumes in schizophrenia. Am J Psychiatry 157:16–25

Zeineh MM, Engel SA, Bookheimer SY (1998) Segmentation and unfolding of the human hippocampus and parahippocampal gyrus with projection of functional MRI. NeuroImage 7:S693

Zhang ZJ, Reynolds GP (2000) A selective deficit in the relative density of parvalbumin-immunoreactive neurons in the hippocampus in schizophrenia. Schizophr Res 49:65

Integrative Circuit Models and Their Implications for the Pathophysiologies and Treatments of the Schizophrenias

Neal R. Swerdlow

Contents

1 Introduction 556
2 Distributed Neural Dysfunction: The "Hole" Thing Is Wrong 558
3 Now That We Know This, What Do We Ask? 562
   3.1 Primary Versus Secondary? 562
   3.2 Clinical Correlates? 563
   3.3 Different Etiologies? 563
   3.4 Risk Markers? 564
   3.5 Which Target? 566
4 Where Does This Lead Us? 568
   4.1 The Fourth Option 569
   4.2 Old News, New Urgency 571
5 Conclusion 573
References 574

Abstract A preponderance of evidence indicates that the heterogeneous group of schizophrenias is accompanied by disturbances in neural elements distributed throughout multiple levels of interconnected cortico-striato-pallido-thalamic circuitry. These disturbances include a substantial loss of, or failure to develop, both cells and/or appropriate cellular connections in regions that include at least portions of the hippocampus, parahippocampal gyrus, entorhinal cortex, amygdala, prefrontal and anterior cingulate cortex, superior and transverse temporal gyri, and mediodorsal, anterior, and pulvinar nuclei of the thalamus; they appear to reflect failures of early brain maturation, that become codified into dysfunctional circuit

N.R. Swerdlow
School of Medicine, University of California, San Diego, 9500 Gilman Dr., La Jolla, CA 92093-0804, USA
e-mail: nswerdlow@ucsd.edu

properties, that in the opinion of this author cannot be "undone" or even predictably remediated in any physiological manner by existing pharmacotherapies. These circuit disturbances are variable across individuals with schizophrenia, perhaps reflecting the interaction of multiple different risk genes and multiple different epigenetic events. Evidence for these complex circuit disturbances has significant implications for many areas of schizophrenia research, and for future efforts toward developing more effective therapeutic approaches for this group of disorders. The conclusion of this chapter is that such future efforts should focus on further developing and refining medications that target nodal or convergent circuit points within the limbic–motor interface, with the goal of constraining the scope and severity of psychotic exacerbations, to be used in concert with systematic rehabilitative psychotherapies designed to engage healthy neural systems to compensate for and replace dysfunctional higher circuit elements. This strategy should be applied in both preventative and treatment settings, and disseminated for community delivery via an evidence-based manualized format. In contrast to alternative treatment strategies that range from complex polypharmacy to gene therapies to psychosurgical interventions, the use of combined medication plus targeted cognitive and behavioral psychotherapy has both common sense and time-tested documented efficacy with numerous other neuropsychiatric disorders.

Keywords Hippocampus · Neurorehabilitation · Prefrontal cortex · Schizophrenia · Striatum · Thalamus

1 Introduction

The chapters in this text are testimony to the fact that our field has made major gains toward understanding the biology of the group of schizophrenias. In the elegance of this research and the biological models it produces, we do not lose sight of the tragedy of schizophrenia. While we do not yet know its causes, nor is there a clear consensus on the best ways to treat schizophrenia, the wish to reduce the suffering created by this disorder is a unifying force behind the painstaking and often lifelong research undertaken by each of the authors of this text, and by so many others in our field. The symptoms of this "complex phenotype" are both captivating and devastating. As clinicians, we know that the first presentation of schizophrenia – often in a previously healthy, vivacious teen – foretells decades of suffering and significant impairment that may span virtually all areas of life function. Family members of schizophrenia patients tell us, "it's like we lost him, and he never came back."

The disturbances of thought, affect, and behavior in the schizophrenias are so severe, and their life consequences so profound, that it has been the understandable hope of scientists and clinicians that a single cause and pathophysiology would link most forms of this illness, and that one class of intervention could reverse or prevent it. In fact, the severity of the typical "first break," its accompanying step down in global function, and subsequent sustained functional deficits is one "phenotype" that separates the schizophrenias from all but a small handful of other mental disorders. This distinctive pattern of onset and course gave rise to hope that, despite subtle differences in presentation, the schizophrenias all reflected one common event or biological process.

But as the elegant summaries in this text detail, decades of careful investigation have made it clear that no single "hole" or brain lesion accounts for the symptoms of this disorder, nor does one gene code for all of its aberrant neural substrates. The heterogeneity of clinical symptoms of the schizophrenias reflects abnormal activity in multiple, distributed, interacting brain circuits, with a differing involvement of these circuits across individuals. As we explicate the detailed cell–cell interactions both within and across components of these circuits, we must consider the likelihood that no medications will be able to restore to full working order these intricate synaptic interconnections. And, even with the growing list of *receptor* targets detailed by Kim and Stahl (2010), the *clinical* targets of schizophrenia polypharmacy will likely remain symptom control, fewer and shorter psychotic exacerbations, and optimistically, modest gains in cognitive, social, and vocational function. Indeed, in most individuals, achievement of these goals via pharmacological interventions will, for the foreseeable future, be viewed as a major success.

We are only now beginning to understand the structural and functional properties of limbic cortical and subcortical circuits that are conveyed through programmed cell migration, pre- and postnatal synaptic reorganization, and apoptosis across *normal* development (cf. Tau and Peterson 2010; Bouwmeester et al. 2002). This intricate order is turned to chaos by the pathological processes underlying schizophrenia and the resultant or compensatory changes stimulated and then hard-wired by tightly choreographed, interdependent developmental events. The problem is not like diabetes, where a single hormone can replace one lost. It is not like hypertension, where therapeutics are geared less toward the myriad causes and more toward a small number of final common pathways. It is certainly not like Parkinson's disease (PD), where impaired motor initiation reflects the loss of predominantly one neuron type, whose role is to supply a single chemical to cells in a way that can be somewhat mimicked by orally administering the precursor for that chemical; this is possible in PD *because the organization of the postsynaptic circuitry has developed normally, and for a substantial portion of adulthood has retained the detailed interconnections intended in the "normal" design*. In schizophrenia, the root cause appears to be a developmental interruption and tangling of neural connections that are orders of magnitude too complex to restore or replace, and which in their complexity regulate not a motor function but rather the psychological identity of the individual (cf. Nelson et al. 2009; Eack et al. 2008; Levesque et al. 2003; Ochsner et al. 2002). Subtle differences in a pancreas or substantia nigra might contribute to modest individual differences in the presentation and course of diabetes or PD; by contrast, the interindividual differences in limbic cortical connectivity imparted first by multiple different and potentially interacting "risk" genes, and next by differently timed insults across development are orders of magnitude greater, and from this follows even greater variability in the schizophrenia phenotype.

Let us pick one component of the complex circuitry implicated in schizophrenia: the prefrontal cortex (PFC). This is not a relatively homogeneous group of cells, like a normally developed pancreas or even substantia nigra, but rather a hub for extremely complex circuit interactions, within which pathology triggers pre- and postsynaptic compensatory changes among many functionally distinct subregions and cell types, and convergent influences of GABA, glutamate, dopamine, and a long list of peptides and other neuromodulators, all within adjacent lamina. Calculate the possible permutations of synaptic interactions in the simplest cartoon schematic, the number of different risk genes and epigenetic events, and multiply by several orders of magnitude, and you begin to see the level of chaos into which we as clinicians introduce medications. Whether it will ever be feasible to develop medications that selectively target individual elements of this complex circuitry in a manner that predictably improves cognition is an empirical question being tested at this time. However, in this author's opinion, *it is neither scientifically rational nor clinically helpful to rely on the fact that pharmacology will in the foreseeable future be able to reach backwards two decades through a variable web of absent and misguided neural connections, replace missing and improper ones with healthy ones, and thereby disentangle schizophrenia from the self.* It will be argued here that prefrontal and limbic corticothalamic pathology in schizophrenia is too widely distributed, complex and variable to be predictably engaged with medications, and that our field should therefore consider alternative future strategies for understanding and treating the schizophrenias.

2 Distributed Neural Dysfunction: The "Hole" Thing Is Wrong

Much as the concept of wave–particle duality did for quantum mechanics, the dual concepts of localization versus distribution of central nervous function have had both divisive and unifying effects on our field. There are a handful of brain disorders in which pathology can be localized primarily to one cell type or region. For the formative years of our field, these disorders – particularly PD and Huntington's disease (HD) – served as models for how many brain disorders should be conceptualized and studied. Over time, simple models of "localized" pathology for PD and HD revealed themselves to be inadequate, and neurologists like Penny and Young (1983) and Mink (2006) advanced models of distributed interacting cortico-striato-pallido-thalamic (CSPT) circuits. Despite this, models for neuropsychiatric disorders like schizophrenia continued to focus primarily on one or two brain regions – for example, the prefrontal or mesial temporal cortex – taking the perspective of the "hole" and not the "whole." Alheid and Heimer (1988), Stevens (1973), Nauta (1982), and others recognized that the "anatomy" of schizophrenia must involve interactions among cortical and subcortical circuits at the "limbic–motor interface," rather than a single localized "lesion," and this concept eventually made its way into the psychiatric literature (e.g., Swerdlow and Koob 1987), finding traction in models of several other disorders, including obsessive–compulsive disorder (OCD; Modell et al. 1989). The specific limitations of localizationist approaches to schizophrenia were the focus of my 1991 essay (Swerdlow 1991), titled, in part, "The 'hole' thing is wrong."

To some degree, the inherent tension between localized and distributed models of pathology of schizophrenia is a product of the ways in which we study the disorder. If we focus with increasing power on specific localized substrates in schizophrenia – for example, in Lewis and colleagues' elegant postmortem studies of GABA signaling in dorsolateral PFC chandelier neurons (Lewis et al. 2004a, b, 2005; Akil et al. 1999) – it becomes increasingly difficult to incorporate models of more distributed circuit dysfunction. However, as their chapter in this text describes, (Volk and Lewis 2010) even at the "localized" level, there are complex circuit interactions, involving multiple neural elements, each with different spatial, temporal, chemical, and electrophysiological properties. Other levels of analysis, such as the neurocognitive, functional neuroimaging and electrophysiological approaches described in chapters by Kalkstein et al. (2010), Brown and Thompson (2010), Braff (2010), Rissling and Light (2010), and others in this text may be more able to identify disturbances in coordinated neural activity across distributed brain systems, but lack the level of neural resolution needed to detect the organizational disturbances at the localized, microcircuit level. So, one can conceptualize schizophrenia as a disorder of either localized or system-level pathology, depending on how you study it.

But regardless of how one conceptualizes it, the evidence for distributed neural dysfunction in schizophrenia is compelling. Let us first look only at areas where *structural* abnormalities are reported (and not, e.g., areas activated abnormally during hallucinations, as described by Silbersweig et al. (1995) and updated in the chapter by Brown and Thompson (2010)), or under stress or cognitive demands, e.g., Fig. 2 by Heckers and Konradi (2010). A preponderance of findings in different schizophrenia cohorts support significant volumetric and/or morphometric abnormalities in brain regions that include at least portions of the hippocampus, parahippocampal gyrus, entorhinal cortex, amygdala, prefrontal, anterior cingulate and inferior parietal cortex, superior and transverse temporal gyri, mediodorsal, anterior and pulvinar nuclei of the thalamus, and portions of the cerebellum (see citations in chapters by Levitt et al. (2010): particularly the outstanding summary tables and Fig. 1; Volk and Lewis (2010); Cronenwett and Csernansky (2010); Heckers and Konradi (2010); and in several other recent reviews: cf. Pakkenberg et al. 2009; Ellison-Wright et al. 2008; Eisenberg et al. 2010; Benes 2010).

But it is not simply that regions are the wrong size or shape. Rather, volumetric and morphometric abnormalities must reflect the more complex biological basis for perturbations in the number, size or shape of cells, fibers, or extraparenchymal elements of all of these brain regions in schizophrenia. Indeed, a long list of papers report laminar- and subregion-specific reductions and other abnormalities in the number of neurons, the length of their dendrites, the density of their dendritic spines, varicosities and levels of corresponding cellular proteins and mRNA in prefrontal, mesial temporal and auditory cortex, striatum, and thalamus, and even the cerebellum and midbrain dopamine (DA) nuclei, among other regions (cf. Akil et al. 1999; Davidsson et al. 1999; Katsetos et al. 1997; Aparacio-Legarza et al. 1997; Beasley and Reynolds 1997; Benes 1999; Benes et al. 1991; Conrad et al. 1991; Glantz and Lewis 2000; Harrison and Eastwood 1998; Jakob and Beckmann 1986; Kolluri et al. 2005; Pakkenberg 1990; Paz et al. 2006; Pierri et al. 1999; Sweet et al. 2009; Vawter et al. 1999).

Studies also document abnormalities in the number or distribution of neurotransmitter receptors in these and other brain regions, which may reflect a primary loss of cells that support these receptors, a secondary response to abnormalities of the fibers that innervate these receptors or the chemicals they deliver, or combinations thereof. Many of these findings are summarized by Urban and Abi-Dargham (2010), Volk and Lewis (2010), and in several recent scholarly reviews (cf. Cruz et al. 2009; Dean et al. 2009; Howes et al. 2009; Roberts et al. 2009; Lewis et al. 2008; Gur et al. 2007). For example, the numbers of prefrontal cortical cells supporting receptors for cholecystokinin, somatostatin, and endocannabinoids are diminished in schizophrenia (cf. Lewis et al. 2008). Meta-analyses also document increased DA D2 receptor density in the striatum (Kestler et al. 2001; Weinberger 2001; Laruelle 1998; Zakzanis and Hansen 1998), accompanied paradoxically by increased DA release (Abi-Dhargam et al. 1998). Others report abnormalities in serotonin (5HT) receptors in PFC and GABA-A receptors in hippocampus (cf. Urban and Abi-Dhargam 2010; Heckers and Konradi 2010).

The long list of distributed neural disturbances that can be documented in studies of large samples of schizophrenia patients is really the "tip of the iceberg." For several reasons, we can be confident that the "real" list of disturbances and combinations thereof is likely to be much more extensive. These reasons are outlined below.

First, studies of neural circuit abnormalities in schizophrenia have been relatively circumscribed in their targets. To some degree, we have been like the man looking for his lost keys under the lamp post. But emerging evidence shows cortical abnormalities to exist well beyond the prefrontal and mesial temporal regions, suggesting more generalized neurodevelopmental disturbances that might impact brain regions in schizophrenia patients that have not yet been carefully examined. In one such example, Sweet et al. (2007, 2009) have presented compelling evidence for GABAergic interneuron disturbances in auditory cortex, similar to those previously described in PFC.

Second, disturbances in neuronal number, size, shape, and connectivity give rise to perturbations in neurochemical transmission, cellular metabolism, signal transduction molecules, gene expression, and many other levels of the machinery required for normal neural function. Some of these "cascading" disturbances are known, as reported by Kvajo et al. (2010) and Benes (2010). These many neuronal changes demonstrated under "static" (e.g., postmortem) and low-load (e.g., "resting state") conditions can, under "high-load" conditions of stress or cognitive challenge, give rise to additional functional deficits in patterns of transmitter release or neuronal activation (Dolan et al. 1995; Heckers et al. 1998; Kumari et al. 2003; Volz et al. 1999) some of these issues are discussed by Brown and Thompson (2010) and others in this text.

Third, identifiable disturbances in one neuronal element can translate into widely distributed dysfunction within "intact" or otherwise normal brain circuits efferent from, or projecting to, the "damaged" element Bertolino et al. (1997, 2000). For example, pathology that impairs the normal "γ-band" synchronization of discharges from large populations of normal cortical neurons is thought to impair the efficiency of information processing among those "normal" cells and the circuits that they form (cf. Uhlhaas and Singer 2006). Thus, the loss or disturbance of a single cell type can have multiplier effects on downstream circuit dysfunction, even among circuits that – in postmortem analyses or resting state imaging measures – *might have normal structural and morphological properties.*

Fourth, the *variance* across and within studies for each of these reported disturbances is substantial, suggesting that the specific patterns of neural deficits in schizophrenia may be as heterogeneous as the patterns of clinical expression. For as many disturbances as are seen in these lists of "group mean" differences, there are many more combinations and permutations expressed in any patient. In two individuals with schizophrenia, the same brain region may be relatively normal in one and grossly abnormal in another. Furthermore, among the list of regions known to be statistically different in cohorts of patients versus comparison subjects, any given patient might exhibit some but not all of these regional abnormalities. And with any given CSPT locus, reduced volumetric measures in two different schizophrenia patients might reflect the loss or absence of two very different cell populations, resulting in two very different patterns of abnormal efferent projections and innervation. We do not know which of these many different abnormalities are interrelated versus independent because (as discussed earlier) most studies focus on one or a small number of measures or neural elements. In one example, for all of our hypotheses and empirical basic studies related to hippocampal–frontal interactions, we do not know what percentage of schizophrenia patients with reduced mesial temporal volume also have reduced prefrontal volume, nor how these two (of the long list of possible) deficits are related in causality, let alone the clinical implications thereof; studies of these issues are in process both with hippocampal–frontal (cf. Gur et al. 2007) and with thalamo-frontal circuitry that are yielding highly informative findings (cf. Cronenwett and Csernansky 2010, Sect. 7). There are several examples of studies that have correlated metabolic activity across different levels of CSPT circuitry in normals and patients to identify aberrant patterns of functional connections or interdependence in schizophrenia and other disorders (e.g., Wu et al. 1990; Schwartz et al. 1996; Calhoun et al. 2009). Still, analyses of the interdependence or independence of abnormalities in size, shape, metabolism, receptor measures, etc., in multiple different structures within CSPT or related circuitry are rare and should yield an understanding of the illness not afforded by analyses of individual measures at a single locus.

3 Now That We Know This, What Do We Ask?

For scientists studying the pathogenesis and treatments of schizophrenia, the long list of distributed neural deficits in this disorder raises many questions, of which only a few will be mentioned herein. Are some of these deficits "primary" versus "secondary," that is, a direct part of the schizophrenia pathology versus a downstream consequence of changes elsewhere in the brain? Are particular neural disturbances associated with particular clinical manifestations of schizophrenia, or predictive of specific therapeutic interventions? Are the various neural disturbances differentially associated with particular etiologies, for example, some with epigenetic/environmental influences versus specific disorder genes (e.g. Stefanis et al. 1999; Wood et al. 2005)? Are some of these neural disturbances "vulnerability" markers, rather than causes of the symptoms per se? Which of these distributed disturbances should we prioritize to study as medication targets, and why?

3.1 Primary Versus Secondary?

It is almost certain that most patients with schizophrenia have multiple affected loci within limbic CSPT circuitry, and further, that these multiple disturbances do not arise as independent events. It thus seems reasonable to ask whether one of these disturbances is most often primary, that is, a direct result of the root cause of schizophrenia, versus secondary, that is, a consequence of aberrant neural function elsewhere in the brain. Certainly, there is no reason to believe that the symptoms of schizophrenia (which are almost certainly "downstream" events) result from a disturbance that is primary rather than secondary. Nonetheless, if we can identify the primary disturbance, and study biological processes closer to the genesis of the illness, we will better be able to narrow the list of etiologies, deduce ways to detect individuals at greatest risk for developing the illness, and even perhaps design interventions that blunt or prevent the progression of disturbances to secondary and tertiary loci.

While this issue may be relevant to studies of the etiology of schizophrenia, the distinction between "primary" and "secondary" brain disturbances is perhaps less relevant to the clinician treating schizophrenia, and certainly to the patient and family suffering from this disorder. The treatment at age 20 will not be different, whether the symptom-causing neural disturbance is "primary" (e.g., the loss of neuron "A" very early in development as the result of an immune response to viral exposure) versus "secondary" (e.g., the loss or improper migration of neuron "B" somewhat later in development resulting from a loss of trophic factors normally supplied or stimulated by neuron "A"). This argument changes somewhat when considering earlier preventative interventions – for example, in a 10-year old with identified biomarkers that predict an increased risk for later developing schizophrenia – but I suggest below that the conclusion does not. Certainly, prenatal interventions – analogous to the use of folic acid to dramatically reduce the incidence of neural tube defects (cf. Wolff et al. 2009) – might conceivably be applied as preventative measure should a "primary" disturbance be identified in schizophrenia, but it is hard to imagine that such preventative measures will be based on information emerging from our current approaches to neural circuit analyses in this disorder.

3.2 Clinical Correlates?

If different circuit elements are impacted to differing degrees within individual schizophrenia patients (a conjecture supported by the variability in regional disturbances both within and across studies), and if function can be "localized" to some degree within this circuitry, it is reasonable to ask whether disturbances at particular circuit nodes are associated with specific profiles of dysfunction in schizophrenia. If this is the case, one might argue that we could use clinical/neurocognitive profiles, in combination with neuroimaging or electrophysiological data, to predict patterns of circuit disturbances in any given patient, and thereby use treatments to target a more specific biology (a circuit element) rather than a less specific symptom profile. For example, if we can determine that a specific neurocognitive pattern is associated with a relative loss of PFC receptor "X," and we can identify a molecular target that is specifically associated with PFC receptor "X" signaling, we would have a basis for rational drug design to treat this neurocognitive deficit and its resulting functional impairment. This is the general rationale behind ongoing clinical trials of several novel and potentially "procognitive" agents in schizophrenia, which might be "personalized" through the use of neurocognitive performance or imaging data to predict drug-sensitive individuals. A less ambitious approach is suggested by evidence that antipsychotic responsivity is associated with increased number and density of striatal dopamine receptors, particularly at axodendritic dopamine synapses (Roberts et al. 2009). While we cannot reasonably expect in the foreseeable future the widespread use of receptor imaging to select individualized medications, it is possible that more easily accessible markers for elevated striatal dopaminergic receptors could be identified and used as surrogates for receptor studies. Of course, attempts to use symptom profiles and biomarkers to predict treatment sensitivity are not new to psychiatry, and as yet remains a strategy of unproven value.

3.3 Different Etiologies?

Given the heterogeneity of circuit disturbances in the schizophrenias, it is reasonable to consider whether different patterns of disturbances reflect distinct etiologies; that is, whether specific circuit disturbances could be used as "endophenotypes" for identifying vulnerability genes (an approach similar to that taken by Meyer-Lindenberg 2009), or as grouping factors in characterizing epigenetic antecedents to schizophrenia (e.g., maternal smoking, viral exposure, malnutrition, low birth weight, low Apgar scores, etc.). Analyses could take advantage of existing large databases of normative regional volumetric and morphometric information (e.g., Mazziotta et al. 2001), to generate profiles of disturbances at multiple nodes within CSPT circuits for individual patients, and use this information to form meaningful subgroups to test hypotheses. For example, we might determine that the presence of particular hippocampal and PFC disturbances (e.g., >1 SD below the mean volume in both hippocampal subregions and PFC) are associated with specific vulnerability genes, or even with specific epigenetic events; such a possibility is suggested by preclinical findings, in which prenatal viral exposure (Borrell et al. 2002) or postnatal hippocampal and PFC NMDA receptor knockdown (Belforte et al. 2010) in preclinical models results in both hippocampal and PFC dysfunction and schizophrenia-related phenotypes of reduced prepulse inhibition (PPI; see Braff 2010, this text).

3.4 Risk Markers?

An expanding literature is now documenting that hippocampus, amygdala, and other structures, including anterior cingulate cortex, are reduced in volume and perhaps functionally impaired in asymptomatic first-degree relatives of schizophrenia probands, and in "ultra-high-risk" individuals (e.g., asymptomatic individuals from families with multiple affected probands; cf. Pantelis et al. 2009; Ho and Magnotta 2010; Boos et al. 2007; Cannon et al. 2003). In some cases, volumetric and functional disturbances in unaffected individuals are associated with specific genetic polymorphisms that also appear to convey a risk for schizophrenia (Kempf et al. 2008; Esslinger et al. 2009; Hall et al. 2008; Gruber et al. 2008; Callicott et al. 2005; Honea et al. 2009). One implication of these findings is that while these circuit disturbances are associated with an increased vulnerability for schizophrenia, that is familial and likely genetic, they are not sufficient to produce the disorder. This could reflect a need for multiple "hits" (e.g., the inherited neural dysfunction plus epigenetic events) or alternatively, it could reflect the resilience conveyed by "protective" factors in unaffected relatives. Even if these phenotypes (and potential endophenotypes) are not sufficient to produce the illness, they are nonetheless important biological markers of increased risk that can inform us about both etiologies and potential preventative interventions.

The oft-cited example of colorectal cancer, for which genes were identified through the risk marker of familial adenomatous polyposis (Groden et al. 1991), provides a model for the successful use of endophenotypes as biomarkers to understand the genetics of complex disorders. The vulnerability to colon cancer is conveyed predominantly by a single phenotype – adenomatous colon polyps. The biology of colon cancer therefore reflects a common foundation ("polyp biology"), acted upon by different "second hits" (diet, smoking, other risk genes, etc.) to generate the disorder (cancer). While it is too early to reach firm conclusions, initial findings suggest that – like the distributed CSPT "lesions" in schizophrenia probands – the neural circuit "vulnerability markers" in *unaffected relatives* of schizophrenia probands are also at multiple different loci, which may be associated with several different genetic polymorphisms and perhaps rare gene variants. There is no a priori reason to assume that two different individuals carrying different predispositions to schizophrenia based on reduced hippocampal versus anterior cingulate volumes, respectively, would be vulnerable to the same epigenetic events. Thus, unlike the biology of colon cancer, and the genetic models that it suggests, the heterogeneity in schizophrenia may *not* reflect the impact of "second hits" on a common foundation of a single "vulnerable" phenotype. Rather, this heterogeneity may approximate the mathematical product of the different possible circuit vulnerability phenotypes and the different epigenetic modifying events. This, of course, suggests a more complex model, both in terms of its application to schizophrenia genetics and in terms of the possible different underlying neurobiological substrates of the disorder.

But for the sake of argument, let us suppose that a list of neural circuit phenotypes, epigenetic events, and genetic markers could be identified in clinically normal 10-year-old children, which conveyed a fractionally increased risk for the later development of schizophrenia. Surely, this list would include many "false positives," that is, items that are common in a population maturing to normal adulthood. What would we, as clinicians, do with this information? Would we widely administer prophylactic drugs (were they to ever become available) to asymptomatic school children, to prevent or forestall the development of schizophrenia in a small percentage of these children? It is hard enough to get parents to vaccinate children for measles (Omer et al. 2009; Yarwood et al. 2005) do we think that there would be widespread acceptance of preventative drugs for schizophrenia, or that such drugs – as neuroactive agents – would be innocuous in a pediatric population? Would we stratify children (or fetuses) based on genetic testing? Not likely: genetic markers could at best suggest a statistical increase in the risk for developing schizophrenia. Dozens of genes have been identified via a variety of genetic and statistical approaches that appear to be associated with an increased risk for schizophrenia, and/or specific neurocognitive deficits in this disorder (cf. SchizophreniaGene: http://www.schizophreniaforum.org/res/sczgene/default.asp; Eisenberg et al. 2010); indeed, emerging findings suggest that this long list grossly underestimates the number of rare highly penetrant gene mutations that might contribute to different forms of this illness (Walsh et al. 2008). Even the most highly predictive genetic testing for brain disorders – independent of any effective preventative intervention – is controversial, and may not necessarily be widely utilized in an asymptomatic population, based on experience with testing for frontotemporal dementia (Riedijk et al. 2009). Are we – as some suggest (Lesch 1999) – anticipating gene therapies, in a polygenic disorder where individual genes account for only a small fraction of the variance in the diagnosis? We cannot know what the future holds, but I do not see this as a realistic option. Perhaps "protective" drug interventions would be reserved for high-risk individuals or those exhibiting prodromal symptoms. But evidence indicates that substantial neural and neurocognitive dysfunction is already present by the time of the schizophrenia prodrome, suggesting that interventions at this point might not be fully "preventative." The use of antipsychotics and antidepressants in prodromal populations are currently being studied (Kobayashi et al. 2009; Walker et al. 2009), and the strategy of limiting the potentially damaging effects of untreated psychosis on the adolescent or young adult brain may be very important (Barnes et al. 2008; Farooq et al. 2009); this approach makes clinical sense and is being applied independent of any other biological markers.

3.5 Which Target?

Perhaps, the most basic question raised by the "target-rich" environment of distributed neural disturbances in schizophrenia is which ones do we study as potential targets for medications? One might argue that despite evidence for distributed neural deficits in the schizophrenias, some aspects of "normal brain function" can still be localized; hence, it is reasonable to select specific brain regions to study, based on the convergence of known functional localization and known functional deficits in schizophrenia patients. Thus, to understand and potentially remediate specific neurocognitive deficits in schizophrenia, perhaps we should study regions to which specific neurocognitive functions can be localized, for example, executive functions within the PFC. But the PFC is innervated by both the mediodorsal thalamus (MD) and hippocampal complex; both of these PFC inputs (and many others) are significantly disturbed in schizophrenia, and the ability of the PFC to normally regulate executive function depends on the normal microcircuit patterns of MD and hippocampal efferents onto specific laminar and sublaminar PFC targets. We cannot reasonably expect medications to replace in any physiological manner (e.g., synchronized with moment-to-moment load demands) the specificity and complexity of this convergent PFC input, and thereby restore normal neurocognitive function. And, because some of the MD and hippocampal cells that normally form these PFC afferents appear to be missing or dysfunctional both prior to and after the onset of the illness (cf. Early et al. 1987; Ellison-Wright et al. 2008; Wright et al. 2000; Wang et al. 2008; Velakoulis et al. 1999; Van Erp et al. 2004; Weiss et al. 2005; Pantelis et al. 2009; Ho and Magnotta 2010; Harms et al. 2007; Qiu et al. 2009; Sismanlar et al. 2010; Witthaus et al. 2009) does it make sense to study or intervene within either of these regions, as a means to normalize PFC activity and thus executive function? And what of the inputs to the MD and hippocampus, arising from other regions thought to be disturbed in schizophrenia (Menon et al. 2001; Wang et al. 2008; Ellison-Wright et al. 2008)? In deference to Kim and Stahl (2010), picking a point to study and intervene, faced with evidence of distributed cellular deficits within a reverberating circuit, brings to mind a different elephant parable: "it's elephants all the way down."[1]

In her landmark paper, "An anatomy of schizophrenia," Janice Stevens (1973) suggested that a logical point of intervention within reverberating limbic CSPT circuitry is at a nodal point created by the funnel-neck compression of limbic and frontal corticostriatal projections descending through the ventral striatum. Information with cortical-level complexity descends onto striatal targets, is distributed among functional subcircuits for specialized processing, and then re-expanded in striatal projections across pallidal targets (Flaherty and Graybiel 1994). At this point of striatal convergence, ascending dopaminergic projections terminate on the necks of dendritic spines distributed over Spiny I GABAergic striatal projection neurons, and are thereby positioned to "gate" information passed from descending limbic cortical and PFC pyramidal cells via projections terminating upon the heads of these same dendritic spines (cf. Sesack and Grace 2010). This critical positioning of DAergic inputs and receptors, combined with the subcortical hyperdopaminergic state in schizophrenia created by both excessive D2 receptors and presynaptic DA release, is likely responsible for some of the antipsychotic properties of functional DA antagonists, and excessive amounts of these axodendritic dopamine synapses appear to predict antipsychotic sensitivity (Roberts et al. 2009).

Clinicians have long known – and the CATIE study validated (Lieberman et al. 2005) – that current DA antagonists are not "antischizophrenia" drugs, nor do they have more than modest therapeutic impact in "real-life" clinical settings. The use of antipsychotic drugs approximates the "final common pathway" model used in the treatment of hypertension, and there is no way that drugs acting at this final convergence point will restore normal organization to thought structure defined by complex interconnections at the level of the PFC, anterior cingulate, amygdala, or hippocampus. It nonetheless remains possible that the right combination of pharmacological effects targeting a "nodal" point within an aberrant reverberating circuit might offer "feed-forward" therapeutic benefits – with each pass through the CSPT loop – sensoring or blunting the intrusion of aberrant cortical activity into consciousness; this could have the effect of diminishing the severity and/or frequency of psychotic thought content. The goal of this treatment is essentially "spin control": constraining the chaos created by reverberating misinformation. The impact of such therapy is quantitative more than qualitative: that is, it primarily reduces the "volume" (amount and intensity) of disruptive and disorganized cortical information, more so than directly impacting cognition (the latter being a function of intrinsic cortical circuits). This is precisely what patients tell us ("the voices get quieter and easier to ignore"), what we as clinicians experience (delusions become less complex and systematic, often leaving behind

---

[1]Child: "What holds up the earth?"; Wise man: "The earth sits atop an elephant." Child: "What holds up the elephant?"; Wise man: "It's elephants all the way down."

a "core" belief, suspicion, or idiosyncratic logic), and what laboratory tests confirm ("gating" measures improve more so than neurocognitive functions; Swerdlow et al. 2006; Keefe et al. 2007). Still, it is reasonable to suggest (and empirical data support) that the right drugs targeting the right circuit nodes will help limit psychotic exacerbations; this is the rationale for efforts to design drugs that are free from the many real-life adverse effects of current antipsychotic medications (e.g., see Meyer 2010), and are deliverable (and tolerated) in a manner that is practical and reliable (e.g., see Rabin and Siegel 2010). In the view of this author, the development of efficacious and tolerable drugs that act at the "nodal point" of the limbic–motor interface to reduce the severity and frequency of psychotic exacerbations remains an achievable goal, but because these drugs do not correct higher cortical disturbances in neural function, they will not likely be sufficient to generate sustainable clinical gains in the absence of a second level of intervention in many schizophrenia patients.

4 Where Does This Lead Us?

Our field has fully embraced the imperative to understand the biology of the schizophrenias, in hopes that this knowledge will be applied via existing or evolving interventional models – pharmacologic, immunologic, surgical, or genetic – toward the prevention or treatment of this group of disorders. What we have learned about the neural circuit and genetic complexities of this heterogeneous condition, in my opinion, now raises the conundrum that these interventional models may not be appropriate for schizophrenia. Indeed, we are coming to grips with the fact that this circuitry, its development, and regulation by genes and molecules are so complex that we still do not have optimal models for studying it (cf. Benes 2010), let alone for treating it. This is not a failure of translational neuroscience: *while we have not "solved" schizophrenia and its treatment, it is very important that we have learned enough to make key refinements to the expectations of our interventional models.* From a neural circuit perspective, this chapter has superficially addressed interventions at different nodes within a dysfunctional CSPT circuit (1) at the "highest" cortical levels, where (I suggest) drugs offer little hope of predictably recreating in any physiological manner the complexity of healthy synaptic dynamics; (2) elsewhere within higher corticothalamic or corticocortical connections, that also appears to be intrinsically perturbed beyond reach of even the "smartest" drugs; or (3) at striatal or striatopallidal targets "downstream" from the aberrant cortical activity, blunting the impact of disorganized cortical information but not restoring order to cognitive function. A *fourth option* is to intervene within healthy cortical circuitry, using the intact complexity of intrinsic healthy circuits to compensate for, and potentially restore or subsume the function of damaged circuit elements, or even protect this circuitry from future damage. Indeed, this fourth option is already being applied successfully via many different forms of cognitive remediation or rehabilitation for schizophrenia.

4.1 The Fourth Option

Among the most important findings in the evolution of modern psychiatry is that psychotherapy (and particularly cognitive and behavioral therapy) changes the brain (Baxter et al. 1992; Schwartz et al. 1996; Saxena et al. 2009; Mundo 2006; Paquette et al. 2003; Schwartz 1998). Justice cannot be done here to the wealth of physiological, psychological, and philosophical issues associated with this set of observations, and for these, I must refer readers elsewhere (e.g., Schwartz and Begley 2002; Schwartz et al. 2005). Perhaps what is so intriguing about these findings is that something so obvious and predictable would generate such controversy; much of our field so stridently championed the biological basis of mental illness that there has been a tendency to view with skepticism the notion that psychological interventions were anything other than labor-intensive ways to increase medication compliance. But the evidence is clear that psychotherapy changes thoughts and behavior, and the organ responsible for thoughts and behavior is the brain. *How* psychotherapy changes the brain, and the extent to which these changes reflect processes from gene expression up to the organization of circuits and systems, are questions of ongoing scientific investigation (de Lange et al. 2008; Fox 2009; Keller and Just 2009; Korosi and Baram 2009; Porto et al. 2009; Saxena et al. 2009). As it relates to the topic of this text, perhaps the most relevant question is: to what extent can we expect cognitive and behavioral measures to prevent or compensate for neural dysfunction in schizophrenia, where the structure of cognition (a primary tool in this intervention) is fundamentally impaired?

There are many examples of effective neurorehabilitative strategies in severe brain disorders, which have provided important models for the use of these strategies in schizophrenia. In stroke and brain trauma rehabilitation, behavioral and cognitive tools engage parallel circuits to compensate for dysfunction, enabling the restoration of motor and speech abilities, among others (cf. Taub et al. 2002; Kelley and Borazanci 2009; Laatsch et al. 2007; Musso et al. 1999). Compensatory plasticity is perhaps greatest early in development (cf. Kirton et al. 2007; Johnston et al. 2009), suggesting that cognitive "compensation" or rehabilitation may be most effective in premorbid children or adolescents at risk for developing schizophrenia. In Habit Reversal Therapy for tics in Tourette syndrome (TS), patients willfully suppress aberrant motor activity that arises within dysfunctional neural circuits, and substitute or engage less disruptive behaviors (Wilhelm et al. 2003; Deckersbach et al. 2006; Himle et al. 2006). TS is an inherited disorder of childhood onset, suggesting that genetically programmed pathology impacting early forebrain connectivity is amenable to psychotherapeutic interventions in adulthood. The example of OCD is particularly important, both because the age of onset is not so dissimilar from that of schizophrenia and because cognition in OCD – like schizophrenia – is severely impacted by intrusive and disturbing information. Cognitive and behavioral therapies (CBT) are highly effective in OCD, and successful CBT is associated with normalized metabolism in frontal, striatal, and thalamic regions, and with normalized patterns of correlated metabolism across multiple CSPT regions (Schwartz et al. 1996). Thus, there is ample clinical evidence from neuropsychiatric disorders to lead us to predict that CBT should have rehabilitative effects in schizophrenia.

Indeed, despite the difficulties in studying and utilizing cognitive therapy in schizophrenia (e.g., Braff 1992), the evidence that a variety of different forms of cognitive, behavioral, and social rehabilitation can reduce symptoms and improve function in schizophrenia patients is now substantial. While effect sizes vary across different forms of treatment, treatment cohorts, and outcome measures, a number of meta-analyses document clear safety, feasibility, acceptance, and efficacy of a number of different remediation/rehabilitation therapies in schizophrenia (Klingberg et al. 2009; Medalia and Choi 2009; McGurk et al. 2007), with sustained benefits in many cases lasting years (e.g., Granholm et al. 2007; McGurk et al. 2009; Sellwood et al. 2007; Eack et al. 2009). The details of these interventions, and the studies that have assessed them, can be found in several recent comprehensive reviews (e.g., Medalia and Choi 2009; Tai and Turkington 2009; Demily and Franck 2008; Wykes et al. 2008). Response predictors, including demographic, neurocognitive, and structural and functional imaging measures (Granholm et al. 2008; Kurtz et al. 2009; Brabban et al. 2009; Kumari et al. 2009; Premkumar et al. 2009) are being identified; treatments that target specific functional outcome measures (e.g., vocation: McGurk et al. 2009) and specific symptoms (e.g., hallucinations: Penn et al. 2009 or negative symptoms: Klingberg et al. 2009) are also being developed. Rehabilitative treatments under study are being delivered by nurses (Malik et al. 2009), families (Sellwood et al. 2007), individual and group therapists (Saksa et al. 2009; Lecardeur et al. 2009), and computers (Cavallaro et al. 2009), in inpatient (Lindenmayer et al. 2008), outpatient (Eack et al. 2009), and residential/home settings (Mizuno et al. 2005; Sellwood et al. 2007; Velligan et al. 2009) several attempts have been made to manualize these therapies (e.g., Davis et al. 2005; Klingberg et al. 2009; Roberts and Penn 2009). It will be important to identify and extract the features among these many different therapies that are responsible for their beneficial effects, so that a smaller number of standardized and maximally effective treatments can emerge for detailed study and general clinical application. Another challenge will be to disseminate and apply these therapies as part of standard clinical practice: data suggest that while cognitive and behavioral therapy for schizophrenia is endorsed by the National Health Service and widely and successfully used in the United Kingdom, its acceptance is much lower among practitioners in the United States (Kuller et al. 2010).

Can cognitive, behavioral, and psychosocial interventions be used in preventative approaches to schizophrenia (Morrison et al. 2004)? Compared to preventative medications or vaccines for schizophrenia, we might expect greater public acceptance for preventative measures that involve cognitive and behavioral paradigms or measures to promote trust and healthy socialization. Such paradigms might be identified that activate components of developing CSPT circuitry in a manner that offers supportive or neuroprotective effects for these developing circuits. Presumably, activity within developing synaptic arrays – *stimulated or sustained endogenously by self-initiated and contextually appropriate effects of cognitive, behavioral, or social interventions* – would better approximate normal physiological properties, compared to those stimulated by the relatively passive, exogenous administration of medications. Tasks with high prefrontal demands, for example, might be part of a special educational regimen for high-risk children who have a particular neurocognitive or genetic profile. Therapies that engender appropriate trust in human interactions, and nurture areas of identified cognitive or emotional strengths (and their underlying limbic and frontal neural substrates), might be protective against pathological processes that would otherwise later lead to paranoia and social isolation. The development and optimal implementation of such tasks would not be a trivial undertaking, but could draw substantially from paradigms successfully developed for applications ranging from early childhood development programs (Anderson et al. 2003) to pediatric brain injury (Ylvisaker et al. 2005). It is not hard to imagine that a society would balk at the widespread use of drugs or vaccines in school children to prevent a small percentage from developing schizophrenia, but would embrace the notion of providing evidence-based educational and social enrichment to children with special needs. And, while the use of cognitive, behavioral, and social regimens as "neuroprotective" interventions to prevent or mitigate the development of brain disorders has not been systematically tested, there are substantial preclinical and clinical data supporting the benefits of environmental enrichment on neural development and cognitive function (Hack et al. 1995; Diamond et al. 1976; Briones et al. 2009; Sweatt 2009) and potentially in the creation of a protective "cognitive reserve" (Mandolesi et al. 2008; Dhanushkodi and Shetty 2008).

4.2 Old News, New Urgency

As medical students many decades ago, my classmates and I were taught that *mental illness is better treated with medications and psychotherapy delivered together, than with either treatment delivered alone*. In fact, it is increasingly difficult to demonstrate greater-than-placebo clinical benefits of either antipsychotics or antidepressants in the absence of concomitant psychotherapy (Gelenberg et al. 2008; Leucht et al. 2009). But lacking clear evidence-based psychotherapeutic interventions for the schizophrenias (and in part in response to the past use of inappropriate psychotherapies for this disorder; Lehman and Steinwachs 1998a, b), the standard treatment of schizophrenia in the United States in the last decades of the twentieth century devolved into monthly brief medication "checks." This model was fueled by complex social, economic, political, and psychodynamic (e.g., countertransference) reasons, including the potent influence of the pharmaceutical and health insurance industries on treatment models in psychiatry. Despite growing evidence for the inadequacy of this "med-check" model, cognitive rehabilitation strategies remain relatively underused in the United States (Kuller et al. 2010). Perhaps, most alarming in this regard is the re-emerging interest in psychosurgical interventions for severe mental illness (cf. Kuhn et al. 2009), including schizophrenia (Mikell et al. 2009);

this movement has developed in tandem with the increasing and relatively unregulated financial incentives to "thought leaders" in our field by medical device manufacturers. Lest we repeat the tragic failures of our psychosurgical past, we must recall a second lesson from our early medical training: *Primum non nocere.*

It is important to underscore that this "fourth option" – using the normal physiological and anatomical properties of healthy brain circuits to restore or subsume the function of damaged circuit elements – does *not* diminish the need for effective antipsychotic pharmacotherapy. In fact, one could argue that – faced with the potential for real functional gains via cognitive, behavioral, and social interventions in schizophrenia – there is even a *greater* urgency to control psychosis via optimized medication treatment. The cognitive, behavioral, and social *demands* of neurorehabilitative therapies are not compatible with severe psychotic states, and hence severe, uncontrolled psychosis will almost certainly impede these therapies. The clear lasting negative functional impact of uncontrolled psychosis is mentioned earlier in this chapter, and suggests that preventative cognitive, behavioral, and social interventions might also be best implemented in concert with prophylactic medications in high-risk or prodromal populations. Furthermore, the need for effective antipsychotic treatment also underscores the imperative to identify positive predictors/biomarkers of medication response (discussed earlier) and to minimize the adverse consequences of current antipsychotic options that contribute to negative health sequelae and poor medication adherence (see Meyer 2010; Rabin and Siegel 2010). Presumably, the positive impact of effective antipsychotic medications on neurorehabilitation will be met in kind by the positive effects of functional, vocational, and social stability on psychotic exacerbations, medication adherence, and dose requirements. Medications that enhance specific cognitive functions (e.g. working memory) might reasonably be expected to amplify the clinical benefits of cognitive and rehabilitative therapies, even if they appear to be clinically ineffective when administered without the demands of cognitive interventions. This true synergy of medication and nonmedication therapies might potentially be enhanced by fully integrating medication use into these psychotherapies, not just as "treatment-as-usual" add-ons, but in a manner that further engages our patients' neural substrates for willful intent (Beauregard et al. 2001). While this strategy has been applied to nonpsychotic disorder such as OCD (e.g., Schwartz et al. 1996), the focus in schizophrenia has instead emphasized the advantages of dependable, long-acting but passively administered medications. Optimistically, the integration of successful neurorehabilitation and pharmacotherapy can restore our patients' capacity for informed agency, and empower them to take more active roles in their treatment.

There are, however, broader implications to the awareness that the schizophrenias reflect widely distributed and complex neural disturbances that are present early in life and reflect multiple and variable neurodevelopmental and genetic events. While we can never predict how future findings might cause a "paradigm shift" in the treatment of brain disorders, in my opinion *we can neither afford nor justify a single-minded pursuit of the genetic and molecular bases of a disorder for which such information will not (in the foreseeable future) lead to practical clinical interventions*. Resources for studying any brain disorder are finite and nonrenewable, both in terms of research funding and its impact on the critical mass of focused intellectual energy required to solve extremely complicated problems. In my opinion, we will best serve our patients and their families by using the substantial knowledge gained about schizophrenia as an impetus to refocus our field away from interventional models that no longer make sense, and toward the use of evidence-based medication and psychotherapies in ways that are both biologically informed and clinically rational.

5 Conclusion

Neural disturbances in schizophrenia are widely distributed and variable across limbic CSPT circuitry, including but not limited to the prefrontal and anterior cingulate cortex, hippocampus, amygdala, at least two thalamic nuclei, and the striatum. Dysfunction that might be less readily detected by existing imaging and postmortem tissue approaches likely extends well beyond these structures, and dysregulation of synchronized cortical activity may have widespread deleterious effects on the efficiency of information processing. The complex intrinsic circuitry of many CSPT loci in schizophrenia, and the unpredictable and variable rearrangement of this circuitry by multiple different forms of early developmental pathology and subsequent compensatory responses, forces us to be highly skeptical of the notion that even the "smartest" medications will be able to significantly restore this circuitry to proper functioning order. This awareness leads us to a number of research strategies for better understanding the neurobiology and genetics of this disorder, some of which are addressed in this chapter. From a clinical standpoint, I argue that the most parsimonious approach to such distributed circuit and microcircuit disturbances is to intervene within healthy cortical circuitry, using the normal anatomical and physiological organization of intrinsic healthy circuits to compensate for, and potentially restore or subsume the function of damaged circuit elements. Toward this end, I suggest that our focus for future treatments of schizophrenia should be on the continued development of standardized, evidence-based neurorehabilitative psychotherapies, delivered together with medications that enhance cognitive functions utilized in these psychotherapies, and with better-tolerated antipsychotic medications targeting specific circuit nodes, which are capable of limiting the number and duration of psychotic exacerbations. Research should be directed toward identifying and manualizing the most effective forms of rehabilitative psychotherapies, to facilitate their delivery in community settings. This treatment model may involve ongoing and perhaps lifelong psychological and medication management in schizophrenia, but gains in clinical stability will reduce the frequency, intensity, and cost of more extreme clinical interventions. The goals for "next-generation" drugs will be focused substantially, and this should lead to a reduction in pharmaceutical costs, accompanied by clear economic and broader societal benefits of improved treatments for schizophrenia.

Acknowledgments While the opinions expressed here are attributed to the author, they were formed and shaped through the process of discussions with, and mentorship from, many individuals, among whom are Drs. David Braff, Jeffrey Schwartz, and Nancy Downs. The author also acknowledges the outstanding administrative assistance by Ms. Maria Bongiovanni in the preparation of this chapter, and the support of his family as he used more than a few hours of "family time" to develop his ideas and compose this chapter.

References

Abi-Dargham A, Gil R, Krystal J, Baldwin RM, Seibyl JP, Bowers M, van Dyck CH, Charney DS, Innis RB, Laruelle M (1998) Increased striatal dopamine transmission in schizophrenia: confirmation in a second cohort. Am J Psychiatry 155:761–767

Akil M, Pierri JN, Whitehead RE, Edger CL, Mohila C, Sampson AR, Lewis DA (1999) Lamina-specific alterations in the dopamine innervation of the prefrontal cortex in schizophrenic subjects. Am J Psychiatry 156:1580–1589

Alheid GF, Heimer L (1988) New perspectives in basal forebrain organization of special relevance for neuropsychiatric disorders: the striatopallidal, amygdaloid, and corticopetal components of substantia innominata. Neuroscience 27:1–39

Anderson LM, Shinn C, Fullilove MT, Scrimshaw SC, Fielding JE, Normand J, Carande-Kulis VG, Task Force on Community Preventive Services (2003) The effectiveness of early childhood development programs. A systematic review. Am J Prev Med 24:32–46

Aparacio-Legarza MI, Cutts AJ, Davis B, Reynolds GP (1997) Deficits of [3H]D-aspartate binding to glutamate uptake sites in striatal and accumbens tissue in patients with schizophrenia. Neurosci Lett 232:13–16

Barnes TR, Leeson VC, Mutsatsa SH, Watt HC, Hutton SB, Joyce EM (2008) Duration of untreated psychosis and social function: 1-year follow-up study of first-episode schizophrenia. Br J Psychiatry 193:203–209

Baxter LR Jr, Schwartz JM, Bergman KS, Szuba MP, Guze BH, Mazziotta JC, Alazraki A, Selin CE, Ferng HK, Munford P, Phelps ME (1992) Caudate glucose metabolic rate changes with both drug and behavior therapy for obsessive–compulsive disorder. Arch Gen Psychiatry 49:681–689

Beasley CL, Reynolds GP (1997) Parvalbumin-immunoreactive neurons are reduced in the prefrontal cortex of schizophrenics. Schizophr Res 24:349–355

Beauregard M, Lévesque J, Bourgouin P (2001) Neural correlates of the conscious self-regulation of emotion. J Neurosci 21:1–6

Belforte JE, Zsiros V, Sklar ER, Jiang Z, Yu G, Li Y, Quinlan EM, Nakazawa K (2010) Postnatal NMDA receptor ablation in corticolimbic interneurons confers schizophrenia-like phenotypes. Nat Neurosci 13:76–83

Benes FM (1999) Evidence for altered trisynaptic circuitry in schizophrenic hippocampus. Biol Psychiatry 46:589–599

Benes FM (2010) Amygdalocortical circuitry in schizophrenia: from circuits to molecules. Neuropsychopharmacology 35:239–257

Benes FM, McSparren J, Bird ED, SanGiovanni JP, Vincent SL (1991) Deficits in small interneurons in prefrontal and cingulate cortices of schizophrenic and schizoaffective patients. Arch Gen Psychiatry 48:996–1001

Bertolino A, Saunders RC, Mattay VS, Bachevalier J, Frank JA, Weinberger DR (1997) Altered development of prefrontal neurons in rhesus monkeys with neonatal mesial temporo-limbic lesions: a proton magnetic resonance spectroscopic imaging study. Cereb Cortex 7:740–748

Bertolino A, Breier A, Callicott JH, Adler C, Mattay VS, Shapiro M, Frank JA, Pickar D, Weinberger DR (2000) The relationship between dorsolateral prefrontal neuronal N-acetylaspartate and evoked release of striatal dopamine in schizophrenia. Neuropsychopharmacology 22:125–132

Bogerts B, Ashtari M, Degreef G, Alvir JM, Bilder RM, Lieberman JA (1990) Reduced temporal limbic structure volumes on magnetic resonance images in first episode schizophrenia. Psychiatry Res Neuroimaging 35:1–13

Boos HB, Aleman A, Cahn W, Hulshoff Pol H, Kahn RS (2007) Brain volumes in relatives of patients with schizophrenia: a meta-analysis. Arch Gen Psychiatry 64:297–304

Borrell J, Vela JM, Arevalo-Martin A, Molina-Holgado E, Guaza C (2002) Prenatal immune challenge disrupts sensorimotor gating in adult rats. Implications for the etiopathogenesis of schizophrenia. Neuropsychopharmacology 26:204–215

Bouwmeester H, Wolterink G, van Ree JM (2002) Neonatal development of projections from the basolateral amygdala to prefrontal, striatal, and thalamic structures in the rat. J Comp Neurol 442:239–249

Braff DL (1992) Reply to cognitive therapy and schizophrenia. Schizophr Bull 18:37–38

Braff DL (2010) Prepulse inhibition (PPI) of the startle reflex: A window on the brain in schizophrenia. In: Swerdlow N (ed) Behavioral Neurobiology of Schizophrenia and Its Treatment. Curr Top Behav Neurosci

Brabban A, Tai S, Turkington D (2009) Predictors of outcome in brief cognitive behavior therapy for schizophrenia. Schizophr Bull 35:859–864

Briones TL, Rogozinska M, Woods J (2009) Environmental experience modulates ischemia-induced amyloidogenesis and enhances functional recovery. J Neurotrauma 26:613–625

Brown GG, Thompson WK (2010) Functional brain imaging in schizophrenia: Selected results and methods. In: Swerdlow N (ed) Behavioral Neurobiology of Schizophrenia and Its Treatment. Curr Top Behav Neurosci

Calhoun VD, Eichele T, Pearlson G (2009) Functional brain networks in schizophrenia: a review. Front Hum Neurosci 3:17

Callicott JH, Straub RE, Pezawas L, Egan MF, Mattay VS, Hariri AR, Verchinski BA, Meyer-Lindenberg A, Balkissoon R, Kolachana B, Goldberg TE, Weinberger DR (2005) Variation in DISC1 affects hippocampal structure and function and increases risk for schizophrenia. Proc Natl Acad Sci USA 102:8627–8632

Cannon TD, van Erp TG, Bearden CE, Loewy R, Thompson P, Toga AW, Huttunen MO, Keshavan MS, Seidman LJ, Tsuang MT (2003) Early and late neurodevelopmental influences in the prodrome to schizophrenia: contributions of genes, environment, and their interactions. Schizophr Bull 29:653–669

Cavallaro R, Anselmetti S, Poletti S, Bechi M, Ermoli E, Cocchi F, Stratta P, Vita A, Rossi A, Smeraldi E (2009) Computer-aided neurocognitive remediation as an enhancing strategy for schizophrenia rehabilitation. Psychiatry Res 169:191–196

Conrad AJ, Abebe T, Austin R, Forsythe S, Scheibel AB (1991) Hippocampal pyramidal cell disarray in schizophrenia as a bilateral phenomenon. Arch Gen Psychiatry 48:413–417

Cronenwett WJ, Csernansky J (2010) Thalamic pathology in schizophrenia. In: Swerdlow N (ed) Behavioral Neurobiology of Schizophrenia and Its Treatment. Current Topics in Behavioral Neuroscience. Springer, Heidelberg

Cruz DA, Weaver CL, Lovallo EM, Melchitzky DS, Lewis DA (2009) Selective alterations in postsynaptic markers of chandelier cell inputs to cortical pyramidal neurons in subjects with schizophrenia. Neuropsychopharmacology 34:2112–2124

Davidsson P, Gottfries J, Bogdanovic N, Ekman R, Karlsson I, Gottfries CG, Blennow K (1999) The synaptic-vesicle-specific proteins rab3a and synaptophysin are reduced in thalamus and related cortical brain regions in schizophrenic brains. Schizophr Res 40:23–29

Davis LW, Lysaker PH, Lancaster RS, Bryson GJ, Bell MD (2005) The Indianapolis Vocational Intervention Program: a cognitive behavioral approach to addressing rehabilitation issues in schizophrenia. J Rehabil Res Dev 42:35–45

De Lange FP, Koers A, Kalkman JS, Bleijenberg G, Hagoort P, van der Meer JW, Toni I (2008) Increase in prefrontal cortical volume following cognitive behavioural therapy in patients with chronic fatigue syndrome. Brain 131:2172–2180

Dean B, Boer S, Gibbons A, Money T, Scarr E (2009) Recent advances in postmortem pathology and neurochemistry in schizophrenia. Curr Opin Psychiatry 22:154–160

Deckersbach T, Rauch S, Buhlmann U, Wilhelm S (2006) Habit reversal versus supportive psychotherapy in Tourette's disorder: a randomized controlled trial and predictors of treatment response. Behav Res Ther 44:1079–1090

Demily C, Franck N (2008) Cognitive remediation: a promising tool for the treatment of schizophrenia. Expert Rev Neurother 8:1029–1036

Dhanushkodi A, Shetty AK (2008) Is exposure to enriched environment beneficial for functional post-lesional recovery in temporal lobe epilepsy? Neurosci Biobehav Rev 32:657–674

Diamond MC, Ingham CA, Johnson RE, Bennett EL, Rosenzweig MR (1976) Effects of environment on morphology of rat cerebral cortex and hippocampus. J Neurobiol 7:75–85

Dolan RJ, Fletcher P, Frith CD, Friston KJ, Frackowiak RS, Grasby PM (1995) Dopaminergic modulation of impaired cognitive activation in the anterior cingulate cortex in schizophrenia. Nature 378:180–182

Eack SM, George MM, Prasad KM, Keshavan MS (2008) Neuroanatomical substrates of foresight in schizophrenia. Schizophr Res 103:62–70

Eack SM, Greenwald DP, Hogarty SS, Cooley SJ, DiBarry AL, Montrose DM, Keshavan MS (2009) Cognitive enhancement therapy for early-course schizophrenia: effects of a two-year randomized controlled trial. Psychiatr Serv 60:1468–1476

Early TS, Relman EM, Raichle ME, Spitznagel EL (1987) Left globus pallidus abnormality in newly medicated patients with schizophrenia. Proc Natl Acad Sci USA 84:561–563

Eisenberg DP, Sarpal D, Kohn PD, Meyer-Lindenberg A, Wint D, Kolachana B, Apud J, Weinberger DR, Berman KF (2010) Catechol-O-Methyltransferase Valine(158)Methionine genotype and resting regional cerebral blood flow in medication-free patients with schizophrenia. Biol Psychiatry 67:287–290

Ellison-Wright I, Glahn DC, Laird AR, Thelen SM, Bullmore E (2008) The anatomy of first-episode and chronic schizophrenia: an anatomical likelihood estimation meta-analysis. Am J Psychiatry 165:1015–1023

Esslinger C, Walter H, Kirsch P, Erk S, Schnell K, Arnold C, Haddad L, Mier D, Opitz von Boberfeld C, Raab K, Witt SH, Rietschel M, Cichon S, Meyer-Lindenberg A (2009) Neural mechanisms of a genome-wide supported psychosis variant. Science 324:605

Farooq S, Large M, Nielssen O, Waheed W (2009) The relationship between the duration of untreated psychosis and outcome in low-and-middle income countries: a systematic review and meta analysis. Schizophr Res 109:15–23

Flaherty AW, Graybiel AM (1994) Input–output organization of the sensorimotor striatum in the squirrel monkey. J Neurosci 74:599–610

Floresco SB, Grace AA (2003) Gating of hippocampal-evoked activity in prefrontal cortical neurons by inputs from the mediodorsal thalamus and ventral tegmental area. J Neurosci 23:3930–3943

Fox K (2009) Experience-dependent plasticity mechanisms for neural rehabilitation in somatosensory cortex. Philos Trans R Soc Lond B Biol Sci 364:369–381

Gelenberg AJ, Thase ME, Meyer RE, Goodwin FK, Katz MM, Kraemer HC, Potter WZ, Shelton RC, Fava M, Khan A, Trivedi MH, Ninan PT, Mann JJ, Bergeson S, Endicott J, Kocsis JH, Leon AC, Manji HK, Rosenbaum JF (2008) The history and current state of antidepressant clinical trial design: a call to action for proof-of-concept studies. J Clin Psychiatry 69:1513–1528

Glantz LA, Lewis DA (2000) Decreased dendritic spine density on prefrontal cortical pyramidal neurons in schizophrenia. Arch Gen Psychiatry 57:65–73

Granholm E, McQuaid JR, McClure FS, Link PC, Perivoliotis D, Gottlieb JD, Patterson TL, Jeste DV (2007) Randomized controlled trial of cognitive behavioral social skills training for older people with schizophrenia: 12-month follow-up. J Clin Psychiatry 68:730–737

Granholm E, McQuaid JR, Link PC, Fish S, Patterson T, Jeste DV (2008) Neuropsychological predictors of functional outcome in cognitive behavioral social skills training for older people with schizophrenia. Schizophr Res 100:133–143

Groden J, Thliveris A, Samowitz W, Carlson M, Gelbert L, Albertsen H, Joslyn G, Stevens J, Spirio L, Robertson M, Sargeant L, Krapcho K, Wolff E, Randall Burt R, Hughes JP, Warrington J, McPherson J, Wasmuth J, Le Paslier D, Abderrahim H, Cohen D, Leppert M, White R (1991) Identification and characterization of the familial adenomatous polyposis coli gene. Cell 66:589–600

Gruber O, Falkai P, Schneider-Axmann T, Schwab SG, Wagner M, Maier W (2008) Neuregulin-1 haplotype HAP(ICE) is associated with lower hippocampal volumes in schizophrenic patients and in non-affected family members. J Psychiatr Res 43:1–6

Gur RE, Keshavan MS, Lawrie SM (2007) Deconstructing psychosis with human brain imaging. Schizophr Bull 33:921–931

Hack M, Klein NK, Taylor HG (1995) Long-term developmental outcomes of low birth weight infants. Future Child 5:176–196

Hall J, Whalley HC, Moorhead TW, Baig BJ, McIntosh AM, Job DE, Owens DG, Lawrie SM, Johnstone EC (2008) Genetic variation in the DAOA (G72) gene modulates hippocampal function in subjects at high risk of schizophrenia. Biol Psychiatry 64:428–433

Harms MP, Wang L, Mamah D, Barch DM, Thompson PA, Csernansky JG (2007) Thalamic shape abnormalities in individuals with schizophrenia and their nonpsychotic siblings. J Neurosci 27:13835–13842

Harrison PJ, Eastwood SL (1998) Preferential involvement of excitatory neurons in medial temporal lobe in schizophrenia. Lancet 352:1669–1673

Harte MK, Powell SB, Reynolds LM, Swerdlow NR, Geyer MA, Reynolds GP (2004) Reduced N-acetylaspartate in the temporal cortex of rats reared in isolation. Biol Psychiatry 56:296–299

Heckers S, Konradi C (2010) Hippocampal pathology in schizophrenia. Curr Top Behav Neurosci. doi: 10.1007/7854_2010_43

Heckers S, Rauch SL, Goff D, Savage CR, Schacter DL, Fischman AJ, Alpert NM (1998) Impaired recruitment of the hippocampus during conscious recollection in schizophrenia. Nat Neurosci 1:318–323

Himle MB, Woods DW, Piacentini JC, Walkup JT (2006) Brief review of habit reversal training for Tourette syndrome. J Child Neurol 21:719–725

Ho BC, Magnotta V (2010) Hippocampal volume deficits and shape deformities in young biological relatives of schizophrenia probands. NeuroImage 49:3385–3393

Honea R, Verchinski BA, Pezawas L, Kolachana BS, Callicott JH, Mattay VS, Weinberger DR, Meyer-Lindenberg A (2009) Impact of interacting functional variants in COMT on regional gray matter volume in human brain. NeuroImage 45:44–51

Howes OD, Egerton A, Allan V, McGuire P, Stokes P, Kapur S (2009) Mechanisms underlying psychosis and antipsychotic treatment response in schizophrenia: insights from PET and SPECT imaging. Curr Pharm Des 15:2550–2559

Jakob H, Beckmann H (1986) Prenatal developmental disturbances in the limbic allocortex in schizophrenics. J Neural Transm 65:303–326

Johnston MV, Ishida A, Ishida WN, Matsushita HB, Nishimura A, Tsuji M (2009) Plasticity and injury in the developing brain. Brain Dev 31:1–10

Kalkstein S, Hurford I, Gur RC (2010) Neurocognition in schizophrenia. Curr Top Behav Neurosci. doi: 10.1007/7854_2010_42

Katsetos CD, Hyde TM, Herman MM (1997) Neuropathology of the cerebellum in schizophrenia – an update: 1996 and future directions. Biol Psychiatry 42:213–224

Kaur T, Cadenhead KS (2010) Treatment implications of the schizophrenia prodrome. In: Swerdlow N (ed) Behavioral Neurobiology of Schizophrenia and Its Treatment. Curr Top Behav Neurosci

Keefe RS, Bilder RM, Davis SM, Harvey PD, Palmer BW, Gold JM, Meltzer HY, Green MF, Capuano G, Stroup TS, McEvoy JP, Swartz MS, Rosenheck RA, Perkins DO, Davis CE, Hsiao JK, Lieberman JA, CATIE Investigators, Neurocognitive Working Group (2007)

Neurocognitive effects of antipsychotic medications in patients with chronic schizophrenia in the CATIE Trial. Arch Gen Psychiatry 64:633–647

Kestler LP, Walker E, Vega EM (2001) Dopamine receptors in the brains of schizophrenia patients: a meta-analysis of the findings. Behav Pharmacol 12:355–371

Keller TA, Just MA (2009) Altering cortical connectivity: remediation-induced changes in the white matter of poor readers. Neuron 64:624–631

Kelley RE, Borazanci AP (2009) Stroke rehabilitation. Neurol Res 31:832–840

Kempf L, Nicodemus KK, Kolachana B, Vakkalanka R, Verchinski BA, Egan MF, Straub RE, Mattay VA, Callicott JH, Weinberger DR, Meyer-Lindenberg A (2008) Functional polymorphisms in PRODH are associated with risk and protection for schizophrenia and fronto-striatal structure and function. PLoS Genet 4:e1000252

Kim DH, Stahl SM (2010) Antipsychotic drug development. Curr Top Behav Neurosci. doi: 10.1007/7854_2010_47

Kirton A, Westmacott R, deVeber G (2007) Pediatric stroke: rehabilitation of focal injury in the developing brain. NeuroRehabilitation 22:371–382

Klingberg S, Wittorf A, Herrlich J, Wiedemann G, Meisner C, Buchkremer G, Frommann N, Wölwer W (2009) Cognitive behavioural treatment of negative symptoms in schizophrenia patients: study design of the TONES study, feasibility and safety of treatment. Eur Arch Psychiatry Clin Neurosci 259:S149–S154

Kobayashi H, Morita K, Takeshi K, Koshikawa H, Yamazawa R, Kashima H, Mizuno M (2009) Effects of aripiprazole on insight and subjective experience in individuals with an at-risk mental state. J Clin Psychopharmacol 29:421–425

Kolluri N, Sun Z, Sampson AR, Lewis DA (2005) Lamina-specific reductions in dendritic spine density in the prefrontal cortex of subjects with schizophrenia. Am J Psychiatry 162:1200–1202

Korosi A, Baram TZ (2009) The pathways from mother's love to baby's future. Front Behav Neurosci 3:27

Kuhn J, Gaebel W, Klosterkoetter J, Woopen C (2009) Deep brain stimulation as a new therapeutic approach in therapy-resistant mental disorders: ethical aspects of investigational treatment. Eur Arch Psychiatry Clin Neurosci 259(Suppl 2):S135–S141

Kuller AM, Ott BD, Goisman RM, Wainwright LD, Rabin RJ (2010) Cognitive behavioral therapy and schizophrenia: a survey of clinical practices and views on efficacy in the United States and United Kingdom. Community Ment Health J 46(1):2–9

Kumari V, Gray JA, Geyer MA, Ffytche D, Soni W, Mitterschiffthaler MT, Vythelingum GN, Simmons A, Williams SC, Sharma T (2003) Neural correlates of tactile prepulse inhibition: a functional MRI study in normal and schizophrenic subjects. Psychiatry Res 122:99–113

Kumari V, Peters ER, Fannon D, Antonova E, Premkumar P, Anilkumar AP, Williams SC, Kuipers E (2009) Dorsolateral prefrontal cortex activity predicts responsiveness to cognitive–behavioral therapy in schizophrenia. Biol Psychiatry 66:594–602

Kurtz MM, Seltzer JC, Fujimoto M, Shagan DS, Wexler BE (2009) Predictors of change in life skills in schizophrenia after cognitive remediation. Schizophr Res 107:267–274

Kvajo M, McKellar H, Joseph A, Gogos JA (2010) Molecules, signaling and schizophrenia. Curr Top Behav Neurosci. doi: 10.1007/7854_2010_41

Laatsch L, Harrington D, Hotz G, Marcantuono J, Mozzoni MP, Walsh V, Hersey KP (2007) An evidence-based review of cognitive and behavioral rehabilitation treatment studies in children with acquired brain injury. J Head Trauma Rehabil 22:248–256

Laruelle M (1998) Imaging dopamine transmission in schizophrenia. A review and meta-analysis. Q J Nucl Med 42:211–221

Lecardeur L, Stip E, Giguere M, Blouin G, Rodriguez JP, Champagne-Lavau M (2009) Effects of cognitive remediation therapies on psychotic symptoms and cognitive complaints in patients with schizophrenia and related disorders: a randomized study. Schizophr Res 111:153–158

Lehman AF, Steinwachs DM (1998a) Translating research into practice: the schizophrenia patient outcomes research team (PORT) treatment recommendations. Schizophr Bull 24:1–10

Lehman AF, Steinwachs DM (1998b) Patterns of usual care for schizophrenia: initial results from the schizophrenia patient outcomes research team (PORT) client survey. Schizophr Bull 24:11–20
Lesch KP (1999) Gene transfer to the brain: emerging therapeutic strategy in psychiatry? Biol Psychiatry 45:247–253
Leucht S, Arbter D, Engel RR, Kissling W, Davis JM (2009) How effective are second-generation antipsychotic drugs? A meta-analysis of placebo-controlled trials. Mol Psychiatry 14:429–447
Lévesque J, Joanette Y, Paquette V, Mensour B, Beaudoin G, Leroux J-M, Bourgouin P, Beauregard M (2003) Neural circuitry underlying voluntary self-regulation of sadness. Biol Psychiatry 53:502–510
Levitt JJ, Bobrow L, Lucia D, Srinivasan P (2010) A selective review of volumetric and morphometric imaging in schizophrenia. In: Swerdlow N (ed) Behavioral Neurobiology of Schizophrenia and Its Treatment. Curr Top Behav Neurosci
Lewis DA, Cruz D, Eggan S, Erickson S (2004a) Postnatal development of prefrontal inhibitory circuits and the pathophysiology of cognitive dysfunction in schizophrenia. Ann NY Acad Sci 1021:64–76
Lewis DA, Volk DW, Hashimoto T (2004b) Selective alterations in prefrontal cortical GABA neurotransmission in schizophrenia: a novel target for the treatment of working memory dysfunction. Psychopharmacology 174:143–150
Lewis DA, Hashimoto T, Volk DW (2005) Cortical inhibitory neurons and schizophrenia. Nat Rev Neurosci 6:312–324
Lewis DA, Hashimoto T, Morris HM (2008) Cell and receptor type-specific alterations in markers of GABA neurotransmission in the prefrontal cortex of subjects with schizophrenia. Neurotox Res 14:237–248
Lieberman JA, Stroup TS, McEvoy JP, Swartz MS, Rosenheck RA, Perkins DO, Keefe RS, Davis SM, Davis CE, Lebowitz BD, Severe J, Hsiao JK, Clinical Antipsychotic Trials of Intervention Effectiveness (CATIE) Investigators (2005) Effectiveness of antipsychotic drugs in patients with chronic schizophrenia. N Engl J Med 353:1209–1223
Lindenmayer JP, McGurk SR, Mueser KT, Khan A, Wance D, Hoffman L, Wolfe R, Xie H (2008) A randomized controlled trial of cognitive remediation among inpatients with persistent mental illness. Psychiatr Serv 59:241–247
Malik N, Kingdon D, Pelton J, Mehta R, Turkington D (2009) Effectiveness of brief cognitive–behavioral therapy for schizophrenia delivered by mental health nurses: relapse and recovery at 24 months. J Clin Psychiatry 70:201–207
Mandolesi L, De Bartolo P, Foti F, Gelfo F, Federico F, Leggio MG, Petrosini L (2008) Environmental enrichment provides a cognitive reserve to be spent in the case of brain lesion. J Alzheimers Dis 15:11–28
Mazziotta J, Toga A, Evans A, Fox P, Lancaster J, Zilles K, Woods R, Paus T, Simpson G, Pike B, Holmes C, Collins L, Thompson P, MacDonald D, Iacoboni M, Schormann T, Amunts K, Palomero-Gallagher N, Geyer S, Parsons L, Narr K, Kabani N, Le Goualher G, Boomsma D, Cannon T, Kawashima R, Mazoyer B (2001) A probabilistic atlas and reference system for the human brain: international consortium for brain mapping (ICBM). Philos Trans R Soc Lond B Biol Sci 356:1293–1322
McGurk SR, Twamley EW, Sitzer DI, McHugo GJ, Mueser KT (2007) A meta-analysis of cognitive remediation in schizophrenia. Am J Psychiatry 164:1791–1802
McGurk SR, Mueser KT, DeRosa TJ, Wolfe R (2009) Work, recovery, and comorbidity in schizophrenia: a randomized controlled trial of cognitive remediation. Schizophr Bull 35:319–335
Medalia A, Choi J (2009) Cognitive remediation in schizophrenia. Neuropsychol Rev 19:353–364
Menon V, Anagnoson RT, Glover GH, Pfefferbaum A (2001) Functional magnetic resonance imaging evidence for disrupted basal ganglia function in schizophrenia. Am J Psychiatry 158:646–649
Meyer JM (2010) Antipsychotics and metabolics in the post-CATIE era. Curr Top Behav Neurosci. doi: 10.1007/7854_2010_45

Meyer-Lindenberg A (2009) Neural connectivity as an intermediate phenotype: brain networks under genetic control. Hum Brain Mapp 30:1938–1946

Mikell CB, McKhann GM, Segal S, McGovern RA, Wallenstein MB, Moore H (2009) The hippocampus and nucleus accumbens as potential therapeutic targets for neurosurgical intervention in schizophrenia. Stereotact Funct Neurosurg 87:256–265

Mink JW (2006) Neurobiology of basal ganglia and Tourette syndrome: basal ganglia circuits and thalamocortical outputs. Adv Neurol 99:89–98

Mizuno M, Sakuma K, Ryu Y, Munakata S, Takebayashi T, Murakami M, Falloon IR, Kashima H (2005) The Sasagawa project: a model for deinstitutionalisation in Japan. Keio J Med 54:95–101

Modell JG, Mountz JM, Curtis GC, Greden JF (1989) Neurophysiologic dysfunction in basal ganglia/limbic striatal and thalamocortical circuits as a pathogenetic mechanism of obsessive–compulsive disorder. J Neuropsychiatry Clin Neurosci 1:27–36

Morrison AP, French P, Malford L, Lewis SW, Kilcommons A, Green J, Parker S, Bentall RP (2004) Cognitive therapy for the prevention of psychosis in people at ultra-high risk. Br J Psychiatry 185:291–297

Mundo E (2006) Neurobiology of dynamic psychotherapy: an integration possible? J Am Acad Psychoanal Dyn Psychiatry 34:679–691

Musso M, Weiller C, Kiebel S, Muller SP, Bulau P, Rijntjes M (1999) Training-induced brain plasticity in aphasia. Brain 122:1781–1790

Nauta WJ (1982) Limbic innervation of the striatum. Adv Neurol 35:41–47

Nelson B, Fornito A, Harrison BJ, Yücel M, Sass LA, Yung AR, Thompson A, Wood SJ, Pantelis C, McGorry PD (2009) A disturbed sense of self in the psychosis prodrome: linking phenomenology and neurobiology. Neurosci Biobehav Rev 33:807–817

O'Driscoll GA, Florencio PS, Gagnon D, Wolff AV, Benkelfat C, Mikula L, Lal S, Evans AC (2001) Amygdala–hippocampal volume and verbal memory in first-degree relatives of schizophrenic patients. Psychiatry Res 107:75–85

Ochsner KN, Bunge SA, Gross JJ, Gabrieli JDE (2002) Rethinking feelings: an fMRI study of the cognitive regulation of emotion. J Cogn Neurosci 14:1215–1229

Omer SB, Salmon DA, Orenstein WA, deHart MP, Halsey N (2009) Vaccine refusal, mandatory immunization, and the risks of vaccine-preventable diseases. N Engl J Med 360:1981–1988

Pakkenberg B (1990) Pronounced reduction of total neuron number in mediodorsal thalamic nucleus and nucleus accumbens in schizophrenics. Arch Gen Psychiatry 47:1023–1028

Pakkenberg B, Scheel-Kruger J, Kristiansen LV (2009) Schizophrenia; from structure to function with special focus on the mediodorsal thalamic prefrontal loop. Acta Psychiatr Scand 120:345–354

Pantelis C, Yucel M, Bora E, Fornito A, Testa R, Brewer WJ, Velakoulis D, Wood SJ (2009) Neurobiological markers of illness onset in psychosis and schizophrenia: the search for a moving target. Neuropsychol Rev 19:385–398

Paquette V, Lévesque J, Mensour B, Leroux J-M, Beaudoin G, Bourgouin P, Beauregard M (2003) "Change the mind and you change the brain": effects of cognitive–behavioral therapy on the neural correlates of spider phobia. NeuroImage 18:401–409

Paz RD, Andreasen NC, Daoud SZ, Conley R, Roberts R, Bustillo J, Perrone-Bizzozero NI (2006) Increased expression of activity-dependent genes in cerebellar glutamatergic neurons of patients with schizophrenia. Am J Psychiatry 163:1829–1831

Penn DL, Meyer PS, Evans E, Wirth RJ, Cai K, Burchinal M (2009) A randomized controlled trial of group cognitive–behavioral therapy vs. enhanced supportive therapy for auditory hallucinations. Schizophr Res 109:52–59

Penny JB, Young AB (1983) Speculations on the functional anatomy of basal ganglia disorders. Annu Rev Neurosci 6:73–94

Pierri JN, Chaudry AS, Woo TU, Lewis DA (1999) Alterations in chandelier neuron axon terminals in the prefrontal cortex of schizophrenic subjects. Am J Psychiatry 156:1709–1719

Porto PR, Oliveira L, Mari J, Volchan E, Figueira I, Ventura P (2009) Does cognitive behavioral therapy change the brain? A systematic review of neuroimaging in anxiety disorders. J Neuropsychiatry Clin Neurosci 21:114–125

Premkumar P, Fannon D, Kuipers E, Peters ER, Anilkumar AP, Simmons A, Kumari V (2009) Structural magnetic resonance imaging predictors of responsiveness to cognitive behaviour therapy in psychosis. Schizophr Res 115:146–155

Qiu A, Wang L, Younes L, Harms MP, Ratnanather JT, Miller MI, Csernansky JG (2009) Neuroanatomical asymmetry patterns in individuals with schizophrenia and their non-psychotic siblings. NeuroImage 47:1221–1229

Rabin CR, Siegel SJ (2010) Antipsychotic dosing and drug delivery. Curr Top Behav Neurosci. doi: 10.1007/7854_2010_46

Riedijk SR, Niermeijer MFN, Dooijes D, Tibben A (2009) A decade of genetic counseling in frontotemporal dementia affected families: few counseling requests and much familial opposition to testing. J Genet Counsel 18:350–356

Rissling AJ, Light GA (2010) Neurophysiological measures of sensory registration, stimulus discrimination and selection in schizophrenia patients. In: Swerdlow N (ed) Behavioral Neurobiology of Schizophrenia and Its Treatment. Curr Top Behav Neurosci

Roberts DL, Penn DL (2009) Social cognition and interaction training (SCIT) for outpatients with schizophrenia: a preliminary study. Psychiatry Res 166:141–147

Roberts RC, Roche JK, Conley RR, Lahti AC (2009) Dopaminergic synapses in the caudate of subjects with schizophrenia: relationship to treatment response. Synapse 63:520–530

Saksa JR, Cohen SJ, Srihari VH, Woods SW (2009) Cognitive behavior therapy for early psychosis: a comprehensive review of individual vs. group treatment studies. Int J Group Psychother 59:357–383

Saxena S, Gorbis E, O'Neill J, Baker SK, Mandelkern MA, Maidment KM, Chang S, Salamon N, Brody AL, Schwartz JM, London ED (2009) Rapid effects of brief intensive cognitive–behavioral therapy on brain glucose metabolism in obsessive–compulsive disorder. Mol Psychiatry 14:197–205

Schwartz JM (1998) Neuroanatomical aspects of cognitive–behavioural therapy response in obsessive–compulsive disorder: an evolving perspective on brain and behavior. Br J Psychiatry 173(Suppl 35):39–45

Schwartz JM, Begley S (2002) The mind and the brain: neuroplasticity and the power of mental force. Harper Collins, New York, NY

Schwartz JM, Stoessel PW, Baxter LR Jr, Martin KM, Phelps ME (1996) Systematic changes in cerebral glucose metabolic rate after successful behavior modification treatment of obsessive–compulsive disorder. Arch Gen Psychiatry 53:109–113

Schwartz JM, Stapp HP, Beauregard M (2005) Quantum physics in neuroscience and psychology: a neurophysical model of mind–brain interaction. Philos Trans R Soc Lond B Biol Sci 360:1309–1327

Sellwood W, Wittkowski A, Tarrier N, Barrowclough C (2007) Needs-based cognitive–behavioural family intervention for patients suffering from schizophrenia: 5-year follow-up of a randomized controlled effectiveness trial. Acta Psychiatr Scand 116:447–452

Sesack SR, Grace AA (2010) Cortico-basal ganglia reward network: microcircuitry. Neuropsychopharmacology 35:27–47

Shenton ME, Kirkinis R, Jolesz FA, Pollak SD, LeMay M, Wible CG, Hokama H, Martin J, Metcalf D, Coleman M et al (1992) Abnormalities of the left temporal lobe and thought disorder in schizophrenia. N Engl J Med 327:604–612

Silbersweig DA, Stern E, Frith C, Cahill C, Holmes A, Grootoonk S, Seaward J, McKenna P, Chua SE, Schnorr L, Jones T, Frackowiak RSJ (1995) A functional neuroanatomy of hallucinations in schizophrenia. Nature 378:176–179

Sişmanlar SG, Anik Y, Coşkun A, Ağaoğlu B, Karakaya I, Yavuz CI (2010) The volumetric differences of the fronto-temporal region in young offspring of schizophrenic patients. Eur Child Adolesc Psychiatry 19:151–157

Stefanis N, Frangou S, Yakeley J, Sharma T, O'Connell P, Morgan K, Sigmudsson T, Taylor M, Murray R (1999) Hippocampal volume reduction in schizophrenia: effects of genetic risk and pregnancy and birth complications. Biol Psychiatry 46:697–702

Stevens JR (1973) An anatomy of schizophrenia? Arch Gen Psychiatry 29:177–189

Sweatt JD (2009) Experience-dependent epigenetic modifications in the central nervous system. Biol Psychiatry 65:191–197

Sweet RA, Bergen SE, Sun Z, Marcsisin MJ, Sampson AR, Lewis DA (2007) Anatomical evidence of impaired feedforward auditory processing in schizophrenia. Biol Psychiatry 61:854–864

Sweet RA, Henteleff RA, Zhang W, Sampson AR, Lewis DA (2009) Reduced dendritic spine density in auditory cortex of subjects with schizophrenia. Neuropsychopharmacology 34:374–389

Swerdlow NR (1991) Neuropsychology of schizophrenia: the "hole" thing is wrong. Behav Brain Sci 14:51–53

Swerdlow NR, Koob GF (1987) Dopamine, schizophrenia, mania and depression: toward a unified hypothesis of cortico-striato-pallido-thalamic function. Behav Brain Sci 10:197–245

Swerdlow NR, Light GA, Cadenhead KS, Sprock J, Hsieh MH, Braff DL (2006) Startle gating deficits in a large cohort of patients with schizophrenia: relationship to medications, symptoms, neurocognition, and level of function. Arch Gen Psychiatry 63:1325–1335

Tai S, Turkington D (2009) The evolution of cognitive behavior therapy for schizophrenia: current practice and recent developments. Schizophr Bull 35:865–873

Tau GZ, Peterson BS (2010) Normal development of brain circuits. Neuropsychopharmacology 35:147–168

Taub E, Uswatte G, Elbert T (2002) New treatments in neurorehabilitation founded on basic research. Nat Rev Neurosci 3:228–236

Uhlhaas PJ, Singer W (2006) Neural synchrony in brain disorders: relevance for cognitive dysfunctions and pathophysiology. Neuron 52:155–168

Urban N, Abi-Dargham A (2010) Neurochemical imaging in schizophrenia. Curr Top Behav Neurosci. doi: 10.1007/7854_2010_37

van Erp TG, Saleh PA, Huttunen M, Lonnqvist J, Kaprio J, Salonen O, Valanne L, Poutanen VP, Standertskjold-Nordenstam CG, Cannon TD (2004) Hippocampal volumes in schizophrenic twins. Arch Gen Psychiatry 61:346–353

Vawter MP, Howard AL, Hyde TM, Kleinman JE, Freed WJ (1999) Alterations of hippocampal secreted N-AM in bipolar disorder and synaptophysin in schizophrenia. Mol Psychiatry 4:467–475

Velakoulis D, Pantelis C, McGorry PD, Dudgeon P, Brewer W, Cook M, Desmond P, Bridle N, Tierney P, Murrie V, Singh B, Copolov D (1999) Hippocampal volume in first-episode psychoses and chronic schizophrenia: a high-resolution magnetic resonance imaging study. Arch Gen Psychiatry 56:133–141

Velligan DI, Draper M, Stutes D, Maples N, Mintz J, Tai S, Turkington D (2009) Multimodal cognitive therapy: combining treatments that bypass cognitive deficits and deal with reasoning and appraisal biases. Schizophr Bull 35:884–893

Volk DW, Lewis DA (2010) Prefrontal cortical circuits in schizophrenia. Curr Top Behav Neurosci. doi: 10.1007/7854_2010_44

Volz H, Gaser C, Hager F, Rzanny R, Ponisch J, Mentzel H, Kaiser WA, Sauer H (1999) Decreased frontal activation in schizophrenics during stimulation with the Continuous Performance Test – a functional magnetic resonance imaging study. Eur Psychiatry 14:17–24

Walker EF, Cornblatt BA, Addington J, Cadenhead KS, Cannon TD, McGlashan TH, Perkins DO, Seidman LJ, Tsuang MT, Woods SW, Heinssen R (2009) The relation of antipsychotic and antidepressant medication with baseline symptoms and symptom progression: a naturalistic study of the North American Prodrome Longitudinal Sample. Schizophr Res 115:50–57

Walsh T, McClellan JM, McCarthy SE, Addington AM, Pierce SB, Cooper GM, Nord AS, Kusenda M, Malhotra D, Bhandari A, Stray SM, Rippey CF, Roccanova P, Makarov V,

Lakshmi B, Findling RL, Sikich L, Stromberg T, Merriman B, Gogtay N, Butler P, Eckstrand K, Noory L, Gochman P, Long R, Chen Z, Davis S, Baker C, Eichler EE, Meltzer PS, Nelson SF, Singleton AB, Lee MK, Rapoport JL, King MC, Sebat J (2008) Rare structural variants disrupt multiple genes in neurodevelopmental pathways in schizophrenia. Science 320:539–543

Wang L, Mamah D, Harms MP, Karnik M, Price JL, Gado MH, Thompson PA, Barch DM, Miller MI, Csernansky JG (2008) Progressive deformation of deep brain nuclei and hippocampal–amygdala formation in schizophrenia. Biol Psychiatry 64:1060–1068

Weiss AP, Dewitt I, Goff D, Ditman T, Heckers S (2005) Anterior and posterior hippocampal volumes in schizophrenia. Schizophr Res 73:103–112

Wilhelm S, Deckersbach T, Coffey BJ, Bohne A, Peterson AL, Baer L (2003) Habit reversal versus supportive psychotherapy for Tourette's disorder: a randomized controlled trial. Am J Psychiatry 160:1175–1177

Witthaus H, Kaufmann C, Bohner G, Ozgürdal S, Gudlowski Y, Gallinat J, Ruhrmann S, Brüne M, Heinz A, Klingebiel R, Juckel G (2009) Gray matter abnormalities in subjects at ultra-high risk for schizophrenia and first-episode schizophrenic patients compared to healthy controls. Psychiatry Res 173:163–169

Wolff T, Witkop CT, Miller T, Syed SB, U.S. Preventive Services Task Force (2009) Folic acid supplementation for the prevention of neural tube defects: an update of the evidence for the U.S. Preventive Services Task Force. Ann Intern Med 150:632–639

Wood SJ, Yucel M, Velakoulis D, Phillips LJ, Yung AR, Brewer W, McGorry PD, Pantelis C (2005) Hippocampal and anterior cingulate morphology in subjects at ultra-high risk for psychosis: the role of family history of psychotic illness. Schizophr Res 75:295–301

Wright IC, Rabe-Hesketh S, Woodruff PW, David AS, Murray RM, Bullmore ET (2000) Meta-analysis of regional brain volumes in schizophrenia. Am J Psychiatry 157:16–25

Wu JC, Segal BV Jr, Haier RJ, Buchsbaum MS (1990) Testing the Swerdlow/Koob model of schizophrenia pathophysiology using positron emission tomography. Behav Brain Sci 13:169–170

Wykes T, Steel C, Everitt B, Tarrier N (2008) Cognitive behavior therapy for schizophrenia: effect sizes, clinical models, and methodological rigor. Schizophr Bull 34:523–537

Yarwood J, Noakes K, Kennedy D, Campbell H, Salisbury D (2005) Tracking mothers attitudes to childhood immunisation 1991–2001. Vaccine 23:5670–5687

Ylvisaker M, Adelson PD, Braga LW, Burnett SM, Glang A, Feeney T, Moore W, Rumney P, Todis B (2005) Rehabilitation and ongoing support after pediatric TBI: twenty years of progress. J Head Trauma Rehabil 20:95–109

Zakzanis KK, Hansen KT (1998) Dopamine D2 densities and the schizophrenic brain. Schizophr Res 32:201–206

Part IV
Genetic and molecular substrates of schizophrenia

Experimental Approaches for Identifying Schizophrenia Risk Genes

Kiran K. Mantripragada, Liam S. Carroll, and Nigel M. Williams

Contents

1 Introduction 588
2 Linkage Mapping 588
3 Association Mapping 591
4 Identifying Genes Through Structural Chromosomal Variations 597
5 Re-Sequencing 602
References 603

Abstract Schizophrenia is a severe, debilitating and common psychiatric disorder, which directly affects ~1% of the population worldwide. Although previous studies have unequivocally shown that schizophrenia has a strong genetic component, our understanding of its pathophysiology remains limited. The precise genetic architecture of schizophrenia remains elusive and is likely to be complex. It is believed that multiple genetic variants, with each contributing a modest effect on disease risk, interact with environmental factors resulting in the phenotype. In this chapter, we summarise the main molecular genetic approaches that have been utilised in identifying susceptibility genes for schizophrenia and discuss the advantages and disadvantages of each approach. First, we detail the findings of linkage mapping in pedigrees (affected families), which analyse the co-segregation of polymorphic genetic markers with disease phenotype. Second, the contribution of targeted and genome-wide association studies, which compare differential allelic frequencies in schizophrenia cases and matched controls, is presented. Third, we discuss about the identification of susceptibility genes through analysis of chromosomal structural variation (gains and losses of genetic material). Lastly, we introduce the concept of re-sequencing, where the entire genome/exome is sequenced both in affected and

K.K. Mantripragada (✉), L.S. Carroll, and N.M. Williams (✉)
Department of Psychological Medicine and Neurology, MRC Centre in Neuropsychiatric Genetics and Genomics, Cardiff University School of Medicine, Cardiff, UK
e-mails: MantripragadaKK@cardiff.ac.uk; WilliamsNM@cardiff.ac.uk

unaffected individuals. This approach has the potential to provide a clarified picture of the majority of the genetic variation underlying disease pathogenesis.

Keywords Association analysis · Copy number variation · Linkage · Resequencing · Schizophrenia · Structural variation

1 Introduction

Schizophrenia is a severe, debilitating and common psychiatric disorder, which directly affects ∼1% of the population worldwide (Owen et al. 2002). Despite years of search for biological markers, our understanding of the pathophysiology of schizophrenia remains limited, and therefore the diagnosis of schizophrenia is still solely based on the assessment of clinical criteria (Owen et al. 2002). Evidence from family, twin and adoption studies shows unequivocally that schizophrenia has a strong genetic component with estimates of the heritability, defined as the proportion of variance of the phenotype that can be explained by genetic variation, as high as ∼80% (Owen et al. 2002). Therefore, although it is clear that unidentified environmental factors also influence risk to develop schizophrenia, identifying the genetic risk factors remains a key challenge in understanding the pathogenesis of schizophrenia.

The precise genetic architecture of schizophrenia is largely unknown and is likely to be complex, with the phenotype being a product of and/or interactions between multiple genetic and environmental factors. Moreover, we expect each variant to have a modest effect on disease risk (Risch 1990b). In addition, despite the use of highly reliable diagnostic rating scales, the general lack of biological validity means we will inevitably be faced with a level of aetiological heterogeneity among the diagnostic groups. These are the main difficulties that have to be overcome by molecular genetic mapping studies, which aim to identify each susceptibility gene from approximately 30,000 others in the human genome. Here, we will detail the main approaches that are used in these studies (Table 1).

2 Linkage Mapping

Linkage analysis is a well established front-line approach to map genes (Botstein and Risch 2003). Essentially polymorphic genetic markers are typed in pedigrees and analysed for co-segregation with the disease phenotype. Ideally, this takes the form of a genome-wide analysis using several hundred markers evenly distributed throughout the genome. Regions containing markers that co-segregate with disease more often than expected by chance are said to be linked, and if the linkage is real, it will contain a susceptibility gene for the disorder under investigation.

Table 1 Gene mapping strategies

Strategy	Genome wide	Positional/functional
Linkage	+ Relatively inexpensive (compared with association analysis) + No assumptions of gene function or risk variant + Ability to detect rare alleles (<1%) + Multiple tools for analysis available − Large pedigrees segregating disease are rare − Allele sharing methods require very large samples − Many failures to replicate regions of proposed linkage − Common designs with marker intervals of 5–10 cM will prove insufficiently informative	+ Relatively inexpensive + Ability to detect rare alleles (<1%) + Multiple tools for analysis available − Prior knowledge of gene position or function required − Denser mapping may be limited by low resolution of genetic maps. Linkage analysis will be severely inhibited by incorrect marker order − Many failures to replicate regions of proposed linkage
Association	+ No assumptions of gene function or risk variant + Moderately sized samples will have greater power to detect common alleles (>5%) of modest effect + Do not require large multiply affected families + DNA pooling strategies can be applied − Expensive − High density marker maps required − Not suited to detect rare risk variants (<1%) − Multiple testing issues	+ Inexpensive + Moderately sized samples will have greater power to detect common alleles (>5%) of modest effect + Do not require large multiply affected families + DNA pooling strategies can be applied + Multiple tools for analysis available − Prior knowledge of gene position or function required − May be susceptible to population stratification − Not suited to detect rare risk variants (<1%)
Resequencing	+ Sufficient power to detect rare risk variants (<1%) + Do not require large multiply affected families − Prohibitively expensive and laborious − Prior hypothesis required based on variant function − Not yet attempted	+ Inexpensive + Sufficient power to detect rare risk variants (<1%) + Do not require large multiply affected families − Prior hypothesis required based on variant function − Prior knowledge of gene position or function required

Linkage analysis was applied with considerable success in studies of Mendelian disorders, where there is by nature a single major disease locus with rare highly penetrant alleles (Badano and Katsanis 2002). The archetypal form of linkage analysis requires the specification of a number of genetic parameters (parametric linkage analysis) for the disease being studied, such as a precise genetic model detailing the mode of inheritance, gene frequencies, penetrance rate (the frequency at which the disease phenotype manifests itself in the presence of the analysed gene) and phenocopy rate (the frequency at which the disease phenotype occurs in the absence of the analysed gene) (Lander and Schork 1994).

Some of the most successful linkage studies in psychiatry utilised this approach and have effectively reduced the complexity of phenotypic heterogeneity by focusing on pedigrees segregating rare Mendelian sub-forms of the phenotype. They hypothesised that such pedigrees will segregate genes of major effect, which linkage analysis will have high power to detect. Most notably successes include studies of familial Alzheimer's disease (FAD, an unusually early onset of Alzheimer's transmitted in an autosomal dominant fashion), which led to the identification of the susceptibility genes encoding amyloid precursor protein (APP), presenilin 1 and presenilin 2 (PSEN1, PSEN2) (Campion et al. 1995). Although this form of analysis is potentially powerful, multiply affected pedigrees segregating schizophrenia in a classical Mendelian manner have proved to be vanishingly rare. Schizophrenia does not have the aforementioned characteristics of Mendelian disorders, and consequently it is much more difficult to accurately define a genetic model (Risch 1990a). To overcome this, studies of schizophrenia have typically performed parametric linkage analysis using large numbers of multiply affected pedigrees under several estimated genetic models. While this remains a powerful gene mapping approach, it creates the potential of false positive results because of multiple testing and false negative results if the data are analysed under the wrong genetic parameters.

Given the inherent difficulty of parametric linkage analysis, researchers have attempted to map genes using allele sharing methods of linkage analysis (Williams et al. 2003), the premise of which is to identify marker alleles that are shared by affected individuals within pedigrees more often than is expected by chance. Allele sharing approaches do not require the prior specification of a genetic model (non-parametric) and, on this basis, are typically considered more robust than parametric linkage methods for complex disorders (Risch 1990b). Although methods exist to perform allele sharing linkage analysis in large pedigrees, it is their suitability for analysing small nuclear pedigrees that some researchers have argued makes them particularly attractive to map schizophrenia genes. In this scenario, it is possible to collect and limit analysis to large numbers of narrowly defined small nuclear pedigrees (Risch 1990b). In comparison to the parametric linkage studies with the correctly specified model, allele sharing methods require much larger sample sizes to achieve the equivalent power. For a common disease such as schizophrenia, it is possible to collect such a large sample.

Whole genome linkage studies have implicated a large number of putative schizophrenia susceptibility loci; however, four regions have received both genome-wide evidence for linkage as well as supportive evidence from other studies; (6p24-22 (Straub et al. 1995), 1q21-22 (Brzustowicz et al. 2000) and 13q32-34 (Blouin et al. 1998) and 6q (Lindholm et al. 2001; Lerer et al. 2003). However, despite these observations the results of linkage findings in schizophrenia usually fall short of genome-wide significance and failures to replicate are common. Failure to replicate true linkages to loci of moderate effect are to be expected, given that the typical samples of between 20 and 100 families are well short of what is required for adequate power (Suarez et al. 1994; Göring et al. 2001). It is, therefore, difficult to separate the true positive finding from the (presumed) false positives. One solution, albeit imperfect (Levinson et al. 2003) is meta-analysis. In a meta-analysis of 20 schizophrenia genome scans (Lewis et al. 2003), the number of loci meeting aggregate criteria for significance was much greater than the number of loci expected by chance ($p < 0.001$), revealing an encouraging level of consistency. The authors concluded that a number of schizophrenia loci are highly likely to be present in some, perhaps even all, of those detected at chromosome 2p-q, 5q, 3p, 11q, 6p, 1p-q, 22q11-12, 8p 20p, and 14pter-q13. Other regions located on chromosome 16p-q, 18q, 10p, 15q, 6q and 17q were also implicated; however, the evidence was somewhat weaker.

Overall, schizophrenia linkage studies have proved to be limited; studies providing compelling evidence for linkage are rare and failures to replicate proposed linkages are common. This is typical for disorders likely to involve the combined action of several genes of modest effect, and though this can be addressed by studying much larger samples, the sample sizes required to achieve sufficient power to identify genes of the effect size expected in schizophrenia are prohibitive (Risch and Merikangas 1996).

3 Association Mapping

In its simplest form, an association study is a test of differences in allele frequency between a set of cases and controls matched for age, sex and ethnicity. Association studies have a number of practical advantages over linkage studies: they do not require multiply affected families; the analysis requires no prior definition of the disease model and importantly they can have sufficient power to detect common genetic variants that contribute only modestly to disease using much smaller sample sizes than those required by linkage analysis (Risch and Merikangas 1996). For example, a sample of 700 well-matched cases and controls should have sufficient power to detect an allele contributing to 10% of the disease in a population, compared with over 5,000 affected-sibling-pairs that would be required to detect a significant linkage.

Largely because of historical, technical and economical limitations, the majority of association studies performed to date have typically been focussed on candidate genes for schizophrenia (O'Donovan et al. 2003). A locus is typically defined as a candidate for association analysis either because of its physical location within a region previously identified by linkage analysis (positional candidate) or because the function of its protein product fits into a proposed model explaining the physiology of the phenotype (hypothesis-driven candidate). Although such studies have yielded some positive evidence for association to schizophrenia (Chumakov et al. 2002; Shifman et al. 2002; Chowdari et al. 2002; Straub et al. 2002), because of their typically modest sample sizes, they have on the whole been limited in power, and this is reflected in the inherent lack of replication between independent studies. In addition, our general ignorance of the biological underpinnings of schizophrenia mean that there are thousands of potential candidate genes for schizophrenia, and consequently the prior odds of one being a true susceptibility locus is low.

Until recently, because of the requirement of prohibitively large number of genotypes, the concept of genome-wide association studies (GWAS) has not been feasible. However, over recent years, we have witnessed a series of significant technological advances, which have made GWAS an exciting powerful alternative to whole genome linkage analysis (Kruglyak 2008; Hirschhorn and Daly 2005). First, a series of studies identified and mapped large numbers of single nucleotide polymorphisms (SNPs) present in the human genome (Miller et al. 2005; Thorisson and Stein 2003). This collection currently exceeds 16 million SNPs, the details of which are publicly accessible at http://www.ncbi.nlm.nih.gov/projects/SNP/. Using current genotyping technologies, it is not, however, economically viable to analyse every one of these polymorphisms; therefore, this high density SNP map has become the focus of studies that have aimed to identify a highly informative set of SNPs spanning the human genome (Gabriel et al. 2002; Hinds et al. 2005). It has long been recognised that alleles at a genetic locus do not occur independently but often show some degree of correlation; i.e. they occur on the same haplotype (Gabriel et al. 2002). This is the phenomenon of linkage disequilibrium (LD).

The International HapMap Project (International HapMap Consortium 2005; Frazer et al. 2007) has assayed over four million SNPs spanning the human genome in a series of samples that are representative of more than four distinct populations. By assessing the patterns of interSNP LD in these samples, it has been possible to establish that the genotypes of many SNPs are indeed correlated and that consequently a large percentage of the common haplotypic diversity present in a given population can be captured by a relatively small number of highly informative 'tag' SNPs spanning the genome (Gabriel et al. 2002). This observation makes it possible to perform association analysis on most of the common genetic variation present in a given population without directly genotyping all the polymorphisms, and this has empowered GWAS.

SNP genotyping arrays have been developed to enable the simultaneous and efficient analysis of large numbers of SNPs. At the time of writing, typical arrays carry 500,000 to 2 million SNPs and capture over 80% of the common variation (SNPs with a minor allele frequency >0.1) present in the Caucasian population (Li et al. 2008). The analysis of so many SNPs has, even more so than with linkage studies, incurred a statistical price to distinguish between the true and false positives. Typically, the corresponding p-value to 0.05 for a single marker test is $<10^{-7}$ in a GWAS (Ioannidis et al. 2009). The size of the association sample is directly related to the power of the study to significantly detect an association signal (Fig. 1) and moreover, our experience from linkage and candidate gene association studies informs us that we can expect that most schizophrenia disease risk alleles are likely to impart a very small risk to carriers (Lewis et al. 2003). It is, therefore, clear that in order for a GWAS to robustly detect risk alleles of such small effect and meet the very low level of genome wide significance, we require sample sizes in the tens of thousands (Fig. 1) (Risch and Merikangas 1996; Ioannidis et al. 2009).

It is possible to ascertain cases that are enriched for specific disease risk alleles by focussing on the collection of those with a known family history or even those with an extreme phenotype (e.g. an early age of onset). As these approaches can

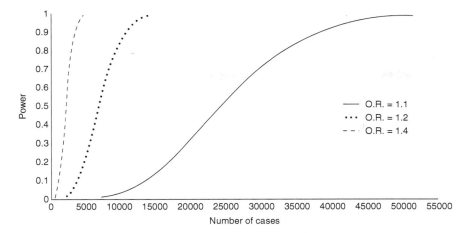

Fig. 1 The relationship between the size of sample and genetic effect on power: The relationship between sample size and the power to significantly detect an allelic association ($p = 0.0001$) at a risk allele with minor allele frequency 0.2. Power is calculated assuming a threefold excess of controls to cases and at a range of odd ratio (O.R.) to demonstrate the increase in sample size required to detect genes of small effect

potentially result in a smaller number of cases being genotyped, they can be financially attractive. However, largely because our knowledge of the genetic architecture of schizophrenia remains limited, the increased value of performing GWAS on samples such as these remains unknown. Instead, to date most GWAS have analysed large, well diagnosed sets of cases.

Studies performing case–control association analysis have historically been hampered by concerns that the presence of any latent population stratification can potentially inflate the type I error rate, and this remains a concern in focussed candidate gene association studies (Lander and Schork 1994; Voight and Pritchard 2005). While it remains essential to collect controls from the same broad ethnic background as the cases, for GWAS it is now possible to identify and exclude individuals (both cases and controls) who have a substantially different genetic background and also correct the association results for any residual substructure present in the sample (Zheng et al. 2006; Price et al. 2006). Although these procedures correct for the variation in the average genome-wide measures of ethnic admixture in the sample, it is possible that any specific markers that strongly reflect population differences may remain in the analysis. Nevertheless, when these statistical tools are appropriately applied to a GWAS, the overall inflation of type I error is typically modest (Zheng et al. 2006; Price et al. 2006). As a result of this, though parent-proband trio designs have been advocated to eliminate the risk of population stratification, it is now generally accepted to be more efficient in terms of both time and economy to perform GWAS using a case–control design (McCarthy et al. 2008). Furthermore, a number of studies have demonstrated that for common diseases such as schizophrenia, there is no benefit in performing

GWAS using a 'hyper-normal' control group (The Wellcome Trust Case Control Consortium 2007; Moskvina et al. 2005). Researchers instead compare the cases to a large population-based control sample on the premise that any loss of power due to misclassification bias (a small number of control individuals meeting the criteria for being a case) can be overcome by increasing the sample size (The Wellcome Trust Case Control Consortium 2007).

The accuracy of GWAS is highly dependent on robust genotype data and in order to obtain this, each stage of the molecular and statistical analysis has to be performed with meticulous attention to detail. First, it is essential that the DNA is of high quality, since even subtle differences in the DNA extraction procedures potentially produce highly significant differences in the association signal between the cases and controls (Clayton et al. 2005). Given the vast number of genotypes generated in GWAS, it is clearly not practical to manually identify each SNP and hence genotypes are ascertained via automated computer algorithms. Although these generate high quality data, it is, nonetheless, essential that they are scrutinised by stringent quality control procedures to obtain a dataset ready for association analysis. Therefore, each study has to find a compromise between setting a high enough QC threshold so that the data have a sufficiently low type I error rate while avoiding over-stringent QC, which would introduce an unacceptable number of spurious association signals. Typically QC procedures will assess the integrity of each sample by testing the overall SNP call rate, molecularly confirming the gender, identifying duplicate, swapped, contaminated and even related samples (McCarthy et al. 2008). Additionally, SNPs will be discarded if they fail to genotype in too many samples or have an extreme measure of Hardy Weinberg equilibrium in the controls (McCarthy et al. 2008). It is, therefore, expected that following standardised QC procedures, a subset of samples and SNPs will be excluded to generate a high quality 'clean' dataset on which the final tests of association will be performed.

Several GWAS publications now exist for schizophrenia, each following a slightly different design and therefore each being differently powered to detect alleles for schizophrenia. The first notable GWAS was reported by Sullivan and colleagues (Sullivan et al. 2008), who analysed a total of 650,000 SNPs in a sample of 738 cases, 733 controls. This study was underpowered to detect risk alleles of small effect; therefore, its failure to identify any markers that met the criteria for genome wide significance confirmed that schizophrenia disease risk alleles are likely to impart a very small risk to carriers.

This limitation was addressed by O'Donovan and colleagues (O'Donovan et al. 2008) who employed a 'two-stage design' where, following GWAS on a limited size 'discovery sample', they increased their power by genotyping the subset of SNPs showing strongest evidence for association in a much larger set of samples. In this case, the initial analysis of over 360,000 SNPs in the initial sample of 479 cases and 2,937 controls from the UK yielded no SNPs that met the criteria for genome wide significance. However, after testing the most promising SNPs for association in a further 16,726 subjects, O'Donovan and colleagues did obtain strong evidence for association ($p = 1.61 \times 10^{-7}$) with an intronic SNP at the gene *ZNF804A* (O'Donovan et al. 2008).

Very few independent researchers have been able to collect the size of samples that are required to perform a well powered GWAS. Consequently, researchers have combined their resources and formed international consortia to perform GWAS. In several recently published studies (Stefansson et al. 2009; Shi et al. 2009; Purcell et al. 2009), these well powered samples have started to yield some intriguing results. In the smallest of these collaborative studies, the Molecular Genetics of Schizophrenia Consortium (MGS) analysed a discovery sample of 2,681 cases and 2,653 controls of European descent and also 1,286 cases and 973 controls of African-American origin (Shi et al. 2009). These were analysed both in isolation and in combination. However, no SNPs exceeded their threshold for genome-wide significance (Shi et al. 2009). Using a slightly larger sample of 3,322 cases and 3,587 controls from Europe, the International Schizophrenia Consortium (ISC) were able to identify evidence for genome-wide significant association at two independent loci: a SNP within the first intron of the myosin XVIIIB (*MYO18B*) gene ($p = 3.4 \times 10^{-7}$) on chromosome 22; and a SNP (rs3130297, $p = 4.79 \times 10^{-8}$) located at the *MHC* (major histocompatibility complex) locus on chromosome 6 (Purcell et al. 2009). Despite a large discovery sample (2,663 cases and 13,498 controls from Europe), the SGENE-plus consortium initially failed to identify any makers that exceeded their threshold of genome wide significance (Stefansson et al. 2009). Following the analysis of the top 1,500 SNPs in an additional 4,999 cases and 15,555 controls, this study was able to identify genome-wide significant evidence for association at a single SNP at the neurogranin (*NRGN*) locus on chromosome 11 (rs12807809, $p = 5.0 \times 10^{-7}$) and also with multiple SNPs again at the *MHC* locus. Finally, the most striking evidence was obtained with a combined analysis of all 12,945 European cases and 34,591 European controls from all three studies, which was able to demonstrate highly significant evidence for association with schizophrenia with SNPs at the *MHC* locus (strongest evidence at rs6932590, $p = 1.4 \times 10^{-12}$), *NRGN* (rs12807809, $p = 2.4 \times 10^{-9}$) and additionally at the transcription factor 4 (*TCF4*) locus on chromosome 18 (rs9960767, $p = 4.1 \times 10^{-9}$) (Stefansson et al. 2009). These findings provide us with an unprecedented insight into the molecular pathogenesis of schizophrenia, in particular the *MHC* findings are consistent with there being an immune component to schizophrenia risk, while the associations at *NRGN* and *TCF4* indicate the importance of molecular pathways involved in brain development, memory and cognition.

The schizophrenia GWAS performed to date have taught us lessons that we should bear in mind when interpreting these data. Possibly because of variation in the genetic effect across two different datasets, we should not be surprised when the findings of two apparently equally powered studies are not entirely concordant. This could also be explained by the 'winners curse' effect (Zollner and Pritchard 2007) whereby the inflated effect size of the initial observation of genome wide significant evidence for association is not seen in subsequent studies. Replication studies, therefore, have to be sufficiently powered to account for this effect. GWAS findings are therefore typically followed up not by replication but by sequentially combining additional independent datasets (Zeggini et al. 2007). Although this is not entirely straightforward, it can be achieved through appropriately combining p values or effect sizes according to standard meta-analysis methods (Ioannidis et al. 2009). It can therefore be expected that the results of schizophrenia GWAS will be continually updated by the cumulative meta-analysis of additional datasets once they become available (Zeggini et al. 2007).

When considering the experimental approaches used to discover the causative variants, one should take into account the expected pattern of risk alleles in an individual or a population. There are several possible allelic architectures of schizophrenia and given that the highly informative sets of SNPs present on arrays are powered to detect association with common alleles, it is important to appreciate that the basic premise underlying GWAS is the common disease-common variant (CDCV) hypothesis (Pritchard 2001; Reich and Lander 2001). This assumes that schizophrenia risk variants are ancient, have undergone little negative evolutionary selection and therefore occur at high frequency (>0.01) in patient cohorts. Arguments supporting this notion include the rapid expansion of the human population (Tishkoff and Verrelli 2003) and the similar risk of developing schizophrenia between populations (Shih et al. 2004).

In contrast, the common disease-rare variant (CDRV) model (Pritchard 2001) proposes that common diseases reflect the action of multiple rare variants (each having a minor allele frequency less than 1%). Proponents of this model utilise evidence that the majority of schizophrenia patients have no close affected relatives, paternal age is associated with risk of schizophrenia (paternal age is associated with an increased mutation rate) and decreased fertility of patients is consistent with many recently occurring mutations causing increased susceptibility (McClellan et al. 2007). There are now examples from complex diseases that show a wide scale of allelic diversity from weak acting common alleles (O'Donovan et al. 2008; Purcell et al. 2009; Stefansson et al. 2009; Shi et al. 2009) to rare variants that confer a greater risk (Lusis and Pajukanta 2008; Kathiresan et al. 2009). Most now accept that the current wave of schizophrenia GWAS are underpowered to rigorously detect common alleles with a very small effect size that we expect are increasing risk to schizophrenia (Terwilliger and Hiekkalinna 2006; Manolio et al. 2009). Nevertheless, there is a growing opinion that the explanation for the relatively few positive signals identified by GWAS is that a large number of very rare, possibly larger effect, risk alleles confer risk to schizophrenia (Manolio et al. 2009). An objective view is that both common and rare variants are likely to explain most instances of schizophrenia. However, given that even GWAS using large samples are underpowered to detect rare risk alleles, it is now widely accepted that we need to adopt alternative approaches to detect low frequency risk variants (Manolio et al. 2009; Kruglyak 2008). In light of this, a number of recent studies have sought and identified evidence that rare risk variants are associated with schizophrenia. Interestingly in these situations, the source has not been SNPs but rather structural chromosomal variants.

4 Identifying Genes Through Structural Chromosomal Variations

It has long been recognised that physically, the human genome is a dynamic entity and that structural chromosomal variants can locally disrupt our standard view that the 'normal' genome contains two copies of all autosomal genes and one copy of gonosomal (X and Y) genes. It is well known that individuals carrying a small interstitial deletion of approximately 1.5–3 Mb at chromosome 22q11 have approximately 20- to 30-fold increased risk of psychosis, especially schizophrenia (Murphy et al. 1999; Papolos et al. 1996; Pulver et al. 1994; Shprintzen et al. 1992). Although no single gene has been unambiguously implicated in the increased risk to develop these psychiatric manifestations, a few candidate genes such as *COMT* and *PRODH* have been considered important (Shifman et al. 2002; Liu et al. 2002). Similarly, a (1;11) (q42;q14.3) balanced reciprocal translocation was found to co-segregate with schizophrenia (LOD 3.6) and other psychiatric disorders (LOD 7.1) in a single Scottish family (Blackwood et al. 2001). The translocation site on chromosome 1 disrupts a gene which, because of its unknown function, was named Disrupted in schizophrenia 1 (DISC1) (Millar et al. 2000). Although the pathogenic mechanisms in this family remain unknown, subsequent molecular studies have implicated the *DISC1* protein as well as its binding partners, such as *PDE4B* protein and microtubule-associated dynein motor complex, in mechanisms important in brain development (Fatemi et al. 2008; Pickard et al. 2007).

Over the past decade, the application of high resolution technologies has revealed that our genomes carry a much larger number of smaller copy number variants (CNVs) (Redon et al. 2006). These variants are defined as DNA fragments for which varying copies have been observed between two or more genomes within the same species. Typically CNVs manifest in the form of deletions, copy gains (duplications) or complex rearrangements, ranging in size from 1 kilobase (kb) to several megabases (Mb), and can encompass none, one or even a large number of contiguous genes. Given that the rate of locus-specific new nucleotide mutation in CNVs is three to four times higher than for SNPs (van Ommen 2005), they are clearly an important source of genetic variation. To date, ~29,000 CNVs have been catalogued in the Database of Genomic Variants (http://projects.tcag.ca/variation/) affecting a substantial number of annotated genes.

The efficiency in detection of CNVs is directly proportional to the resolution at which the genome can be investigated. For this purpose, identification of copy number alterations using comparative genomic hybridisation (CGH) and SNP genotyping platforms has proven to be instrumental. In CGH, differentially labelled 'test' and 'reference' DNA are competitively hybridised to selected probes on a glass slide. The ratio of the fluorescence intensities detected is indicative of the relative DNA copy number in test vs. reference DNA (Fig. 2a). Initially, metaphase chromosomes were utilised as probes on a glass slide, which limited the resolution of analysis to few megabases at the best. However, the advent of array-CGH, which allowed small genomic fragments to be examined, has resulted in identification of

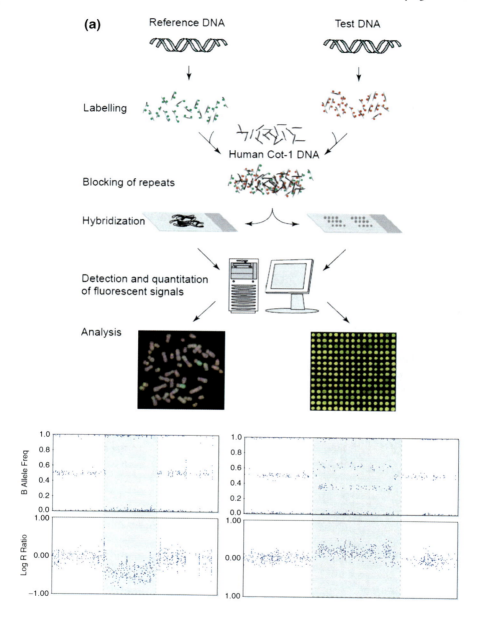

Fig. 2 (**a**). Genome wide analysis of DNA copy number using comparative genomic hybridization (CGH) to metaphase spreads and microarrays. Reproduced with permission from Trends in Genetics Volume 20, Issue 2, February 2004, Pages 87–94. (**b**). Analysis of DNA copy number using Illumina 660W SNP genotyping platform. As shown in the graph, a shift in the log R ratios (*top*) and the B-allele frequency (*bottom*) for a set of contiguous probes (*highlighted by broken rectangles*) would indicate the DNA dosage variation at the locus. The *left panel* displays a deletion and the *right panel* represents a gain in copy number, using Illumina's genome viewer software

submicroscopic copy number changes (up to 1 kb) with high throughput and at high resolution. A wide variety of probes have been used in array-CGH including genomic clones such as bacterial artificial clones (BACs), cDNA clones, PCR products and oligonucleotides.

In addition to CGH, SNP genotyping platforms have also been used in CNV detection. In the last decade, there has been considerable development in technologies for performing SNP genotyping across entire genomes. The signal intensities from each probe could be used for inference of genotypes and to calculate the relative copy number of maternal and paternal alleles in the genome (Fig. 2b). Therefore, using these GWAS platforms in congruence with sophisticated bioinformatic algorithms, a large amount of CNV data is now available through the SNP studies.

There are a number of ways that CNVs can have a deleterious effect on the function of a dosage sensitive gene, including loss of function, unmasking recessive functional alleles, producing novel gene fusion products or even by perturbing a functional regulatory element by means of a position effect (Fig. 3). Genes encompassed by CNVs are therefore good candidates for research in disease susceptibility and have been shown to increase risk to a wide range of phenotypes; *CCL3L1* (17q11.2) dosage in HIV susceptibility (Gonzalez et al. 2005) and copy gain of *CYP2D6* (22q13.1) in drug metabolism (Johansson et al. 1993). In addition, CNVs have also been documented in neurodevelopmental disorders such as autism, mental retardation and seizures (Sebat et al. 2007; Abrahams and Geschwind 2008; Cook and Scherer 2008). However, despite their obvious functional relevance, the identification of CNVs in unaffected controls indicates that many can have apparently inconsequential phenotypic effects (Feuk et al. 2006; McCarroll et al. 2005; Conrad et al. 2005, 2009). Nevertheless, it has also been shown that CNVs with no apparent functional affect in a carrier can cause genomic instability in future generations (Feuk et al. 2006). Consequently, interpretation of CNV screens can be potentially challenging. Encouragingly, recent screens of schizophrenia cohorts have yielded strong evidence that some CNVs can increase risk to develop schizophrenia.

The large majority of studies to date have performed microarray analysis to identify large, rare CNVs that are associated with schizophrenia. Two such studies screened a series of cases and controls using high-resolution array-CGH (Walsh et al. 2008; Kirov et al. 2008) and were able to identify deletions intersecting the neurexin 1 gene (*NRXN1*) located at chromosome 2p16.3 in schizophrenia cases. These findings were particularly intriguing as the *NRXN1* protein plays a vital role in GABAergic and glutamatergic synaptic differentiation (Craig and Kang 2007), making it an excellent functional candidate for schizophrenia. Moreover, analogous mutations at the *NRXN1* locus as well as in other functionally related genes such as contactin-associated protein-like 2 precursor (*CNTNAP2*) and neuroligins (Lawson-Yuen et al. 2008; Alarcón et al. 2008; Friedman et al. 2008; Cook and Scherer 2008) have been identified by others in autism patients. These studies therefore advocate the possible role of neurexins, their related genes and biological pathways in schizophrenia pathogenesis.

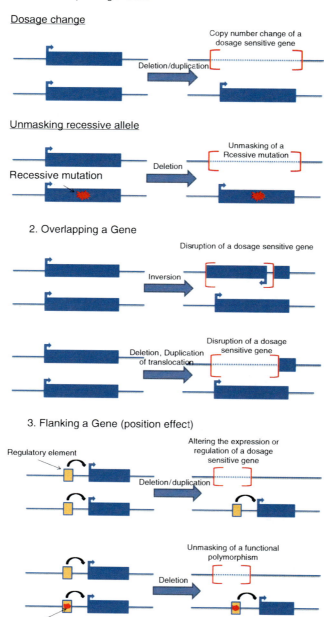

Fig. 3 Deleterious effect of CNVs on dosage sensitive genes

The development of analytical techniques that enable SNP arrays to accurately detect large CNVs has resulted in a number of key studies whereby large schizophrenia GWAS have been interrogated to identify specific loci that are enriched for large, rare CNVs in schizophrenia. The simultaneous publication of two such studies demonstrated the power of this approach, with both studies identifying evidence for significant association for CNVs at 22q11.21, 1q21.1 and 15q13.3 (Stefansson et al. 2008; The International Schizophrenia Consortium 2008). In addition, significant enrichment of CNVs at 15q11.2 was identified by just one study (Stefansson et al. 2008). All of these regions harbour a large number of contiguous genes; however, though not reported as part of the main findings, both of these studies also identified a predominance of rare deletions at the *NRXN1* locus in schizophrenia patients. More recently, by carefully chelating several large datasets, two studies have now demonstrated that schizophrenia patients carry a highly significant excess of CNVs at both chromosome 16p13 ($p = 0.00010$) and also at 16p11 ($p = 4.8 \times 10^{-7}$) (Ingason et al. 2009; McCarthy et al. 2009). Although the confidence intervals are large, the odds ratio for carrying a CNV at each of these regions are potentially exciting (McCarthy et al. 2009; Ingason et al. 2009). Moreover, the presence of these CNVs in unaffected family members suggests that they are not highly penetrant Mendelian mutations. Finally, it is particularly intriguing that the CNVs showing association with schizophrenia do not conform to classic nosological disease boundaries, each also conferring risk to other psychiatric diseases (Cook and Scherer 2008).

Despite showing early signs of success, studies analysing CNVs in schizophrenia have been limited to analysing rare CNVs (typically less than 1% population frequency) largely because of limitations in the resolution of many of the arrays used. The analysis of more common, polymorphic CNVs will require them to be screened in very large cohorts and called with the accuracy expected of GWAS. This will require arrays with increased density and higher accuracy than can be currently achieved as well as the development of new analysis algorithms (Wang et al. 2007; McCarroll 2008; McCarroll et al. 2008). However, a recent study has demonstrated that as well as capturing the association signal of proxy SNPs, an informative set of tag SNPs can also capture the signals of polymorphic common CNVs (Conrad et al. 2009). It is therefore possible that the current wave of GWAS have already indirectly tested the majority of common CNVs for association with schizophrenia.

The analysis of CNVs in schizophrenia is still in its early stages; however, important candidate genes and associated pathways have already emerged as strong candidates in schizophrenia pathogenesis. As well as the potential of testing common polymorphic CNVs for association with schizophrenia, it is essential that large schizophrenia cohorts are screened using higher density arrays and ultimately by next generation sequencing (see further) to establish the effect of smaller, rare CNVs have on schizophrenia pathogenesis.

5 Re-Sequencing

Schizophrenia is the result of a complex interplay between both genetic and environment factors. Linkage and association studies have identified a catalogue of genes/loci that are likely to contribute to schizophrenia pathogenesis. As discussed, these approaches are particularly powered for identifying genes with susceptibility alleles that are common (minor allele $>1\%$). It is reasonable to assume that schizophrenia will be contributed by rare as well as common alleles. The genetic variants identified so far should therefore, at best, be considered as supportive in the pathogenesis of schizophrenia, and alternative approaches will be required to map the remaining susceptibility loci, which are likely to have low frequency risk alleles (Hirschhorn and Altshuler 2002; Manolio et al. 2009).

Ideally, we would obtain a clarified picture of underlying genetic contributions to schizophrenia by screening the entire genomic sequence of all disease-carrying individuals and unaffected controls. However, although the reductions in cost and improvements in reliability will eventually make this possible via whole genome DNA sequencing, it is currently impractical. In the more immediate future, it will be possible to sequence the entire exonic content of a large number of cases and controls and proof of principle studies have recently emerged (Ng et al. 2009).

As with other common genetic disorders, it is possible that there may be additional rare sequence variants present in the genes implicated by GWAS (Kathiresan et al. 2009). Consequently, a viable strategy that is approachable with available resources is the re-sequencing of a large set of appropriate candidate genes that have been implicated by GWAS in a sample of both patients and controls or in individuals at the extremes of a distribution (e.g. re-sequencing 120 cases and 120 controls would have $>90\%$ power to detect risk variants of frequency $<1\%$). This should include their coding regions, $5'$ and $3'$ UTRs, splice sites, promoters and conserved intronic regions, but a major challenge will be distinguishing functional from benign variants and appropriately assessing their relevance by association analysis (Li and Leal 2008). All significant and marginally significant findings from initial studies could be followed up in larger case–control samples, which would allow identification of risk and protective genotypes for schizophrenia aetiology. To date, when this approach has been applied, it has been generally limited to screening the coding sequence of only a small number of candidate genes; however, over the coming years we will witness an increase in the number of studies applying this approach.

Finally, the advent of next generation sequencing technologies is transforming our approach to understanding complex genomes (Ng et al. 2008, 2009; Levy et al. 2007). These technologies provide an immense potential in comprehensively investigating a range of cellular processes at single base resolution and depth and provide a holistic view of genomic and transcriptional regulation as well as variation. With the future application of these massively parallel sequencing technologies, there seems little doubt that more rare variants (SNPs and CNVs), which might present appreciable effect size to schizophrenia pathogenesis, will be uncovered.

References

Abrahams BS, Geschwind DH (2008) Advances in autism genetics: on the threshold of a new neurobiology. Nat Rev Genet 9:341–355

Alarcón M, Abrahams BS, Stone JL, Duvall JA, Perederiy JV, Bomar JM, Sebat J, Wigler M, Martin CL, Ledbetter DH, Nelson SF, Cantor RM, Geschwind DH (2008) Linkage, association, and gene-expression analyses identify CNTNAP2 as an autism-susceptibility gene. Am J Hum Genet 82:150–159

Badano JL, Katsanis N (2002) Beyond Mendel: an evolving view of human genetic disease transmission. Nat Rev Genet 3:779–789

Blackwood DHR, Fordyce A, Walker MT, St Clair DM, Porteous DJ, Muir WJ (2001) Schizophrenia and affective disorders – cosegregation with a translocation at chromosome 1q42 that directly disrupts brain-expressed genes: clinical and P300 findings in a family. Am J Hum Genet 69:428–433

Blouin JL, Dombroski BA, Nath SK, Lasseter VK, Wolyniec PS, Nestadt G, Thornquist M, Ullrich G, McGrath J, Kasch L, Lamacz M, Thomas MG, Gehrig C, Radhakrishna U, Snyder SE, Balk KG, Neufeld K, Swartz KL, DeMarchi N, Papadimitriou GN, Dikeos DG, Stefanis CN, Chakravarti A, Childs B, Housman DE, Kazazian HH, Antonarakis S, Pulver AE (1998) Schizophrenia susceptibility loci on chromosomes 13q32 and 8p21. Nat Genet 20:70–73

Botstein D, Risch N (2003) Discovering genotypes underlying human phenotypes: past successes for Mendelian disease, future approaches for complex disease. Nat Genet 33 Suppl:228–237

Brzustowicz LM, Hodgkinson KA, Chow EW, Honer WG, Bassett AS (2000) Location of a major susceptibility locus for familial schizophrenia on chromosome 1q21–q22. Science 288:678–682

Campion D, Flaman JM, Brice A, Hannequin D, Dubois B, Martin C, Moreau V, Charbonnier F, Didierjean O, Tardieu S, Penet C, Puel M, Pasquier F, Le Doze F, Bellis G, Calenda A, Heilig R, Martinez M, Mallet J, Bellis M, Clerget-Darpoux F, Agid Y, Frebourg T (1995) Mutations of the presenilin I gene in families with early-onset Alzheimer's disease. Hum Mol Genet 4:2373–2377

Chowdari KV, Mirnics K, Semwal P, Wood J, Lawrence E, Bhatia T, Deshpande SN, Thelma BK, Ferrell RE, Middleton FA, Devlin B, Levitt P, Lewis DA, Nimgaonkar VL (2002) Association and linkage analyses of RGS4 polymorphisms in schizophrenia. Hum Mol Genet 11:1373–1380

Chumakov I, Blumenfeld M, Guerassimenko O, Cavarec L, Palicio M, Abderrahim H, Bougueleret L, Barry C, Tanaka H, La Rosa P, Puech A, Tahri N, Cohen-Akenine A, Delabrosse S, Lissarrague S, Picard FP, Maurice K, Essioux L, Millasseau P, Grel P, Debailleul V, Simon AM, Caterina D, Dufaure I, Malekzadeh K, Belova M, Luan JJ, Bouillot M, Sambucy JL, Primas G, Saumier M, Boubkiri N, Martin-Saumier S, Nasroune M, Peixoto H, Delaye A, Pinchot V, Bastucci M, Guillou S, Chevillon M, Sainz-Fuertes R, Meguenni S, Aurich-Costa J, Cherif D, Gimalac A, Van Duijn C, Gauvreau D, Ouellette G, Fortier I, Raelson J, Sherbatich T, Riazanskaia N, Rogaev E, Raeymaekers P, Aerssens J, Konings F, Luyten W, Macciardi F, Sham PC, Straub RE, Weinberger DR, Cohen N, Cohen D (2002) Genetic and physiological data implicating the new human gene G72 and the gene for D-amino acid oxidase in schizophrenia. Proc Natl Acad Sci USA 99:13675–13680

Clayton DG, Walker NM, Smyth DJ, Pask R, Cooper JD, Maier LM, Smink LJ, Lam AC, Ovington NR, Stevens HE, Nutland S, Howson JM, Faham M, Moorhead M, Jones HB, Falkowski M, Hardenbol P, Willis TD, Todd JA (2005) Population structure, differential bias and genomic control in a large-scale, case-control association study. Nat Genet 37:1243–1246

Conrad DF, Andrews TD, Carter NP, Hurles ME, Pritchard JK (2005) A high-resolution survey of deletion polymorphism in the human genome. Nat Genet 38:75–81

Conrad DF, Pinto D, Redon R, Feuk L, Gokcumen O, Zhang Y, Aerts J, Andrews TD, Barnes C, Campbell P, Fitzgerald T, Hu M, Ihm CH, Kristiansson K, Macarthur DG, Macdonald JR, Onyiah I, Pang AW, Robson S, Stirrups K, Valsesia A, Walter K, Wei J, The Wellcome Trust

Case Control Consortium, Tyler-Smith C, Carter NP, Lee C, Scherer SW, Hurles ME (2009) Origins and functional impact of copy number variation in the human genome. Nature (Epub ahead of print, Oct 7) Available at: http://www.ncbi.nlm.nih.gov/pubmed/19812545 (Accessed October 15, 2009)

Cook EH, Scherer SW (2008) Copy-number variations associated with neuropsychiatric conditions. Nature 455:919–923

Craig AM, Kang Y (2007) Neurexin–neuroligin signaling in synapse development. Curr Opin Neurobiol 17:43–52

Fatemi SH, King DP, Reutiman TJ, Folsom TD, Laurence JA, Lee S, Fan YT, Paciga SA, Conti M, Menniti FS (2008) PDE4B polymorphisms and decreased PDE4B expression are associated with schizophrenia. Schizophr Res 101:36–49

Feuk L, Carson AR, Scherer SW (2006) Structural variation in the human genome. Nat Rev Genet 7:85–97

Frazer KA, Ballinger DG, Cox DR, Hinds DA, Stuve LL, Gibbs RA, Belmont JW, Boudreau A, Hardenbol P, Leal SM, Pasternak S, Wheeler DA, Willis TD, Yu F, Yang H, Zeng C, Gao Y, Hu H, Hu W, Li C, Lin W, Liu S, Pan H, Tang X, Wang J, Wang W, Yu J, Zhang B, Zhang Q, Zhao H, Zhao H, Zhou J, Gabriel SB, Barry R. Blumenstiel B, Camargo A, Defelice M, Faggart M, Goyette M, Gupta S, Moore J, Nguyen H, Onofrio RC, Parkin M, Roy J, Stahl E, Winchester E, Ziaugra L, Altshuler D, Shen Y, Yao Z, Huang W, Chu X, He Y, Jin L, Liu Y, Shen Y, Sun W, Wang H, Wang Y, Wang Y, Xiong X, Xu L, Waye MM, Tsui SK, Xue H, Wong JT, Galver LM, Fan JB, Gunderson K, Murray SS, Oliphant AR, Chee MS, Montpetit A, Chagnon F, Ferretti V, Leboeuf M, Olivier JF, Phillips MS, Roumy S, Sallée C, Verner A, Hudson TJ, Kwok PY, Cai D, Koboldt DC, Miller RD, Pawlikowska L, Taillon-Miller P, Xiao M, Tsui LC, Mak W, Song YQ, Tam PK, Nakamura Y, Kawaguchi T, Kitamoto T, Morizono T, Nagashima A, Ohnishi Y, Sekine A, Tanaka T, Tsunoda T, Deloukas P, Bird CP, Delgado M, Dermitzakis ET, Gwilliam R, Hunt S, Morrison J, Powell D, Stranger BE, Whittaker P, Bentley DR, Daly MJ, de Bakker PI, Barrett J, Chretien YR, Maller J, McCarroll S, Patterson N, Pe'er I, Price A, Purcell S, Richter DJ, Sabeti P, Saxena R, Schaffner SF, Sham PC, Varilly P, Altshuler D, Stein LD, Krishnan L, Smith AV, Tello-Ruiz MK, Thorisson GA, Chakravarti A, Chen PE, Cutler DJ, Kashuk CS, Lin S, Abecasis GR, Guan W, Li Y, Munro HM, Qin ZS, Thomas DJ, McVean G, Auton A, Bottolo L, Cardin N, Eyheramendy S, Freeman C, Marchini J, Myers S, Spencer C, Stephens M, Donnelly P, Cardon LR, Clarke G, Evans DM, Morris AP, Weir BS, Tsunoda T, Mullikin JC, Sherry ST, Feolo M, Skol A, Zhang H, Zeng C, Zhao H, Matsuda I, Fukushima Y, Macer DR, Suda E, Rotimi CN, Adebamowo CA, Ajayi I, Aniagwu T, Marshall PA, Nkwodimmah C, Royal CD, Leppert MF, Dixon M, Peiffer A, Qiu R, Kent A, Kato K, Niikawa N, Adewole IF, Knoppers BM, Foster MW, Clayton EW, Watkin J, Gibbs RA, Belmont JW, Muzny D, Nazareth L, Sodergren E, Weinstock GM, Wheeler DA, Yakub I, Gabriel SB, Onofrio RC, Richter DJ, Ziaugra L, Birren BW, Daly MJ, Altshuler D, Wilson RK, Fulton LL, Rogers J, Burton J, Carter NP, Clee CM, Griffiths M, Jones MC, McLay K, Plumb RW, Ross MT, Sims SK, Willey DL, Chen Z, Han H, Kang L, Godbout M, Wallenburg JC, L'Archevêque P, Bellemare G, Saeki K, Wang H, An D, Fu H, Li Q, Wang Z, Wang R, Holden AL, Brooks LD, McEwen JE, Guyer MS, Wang VO, Peterson JL, Shi M, Spiegel J, Sung LM, Zacharia LF, Collins FS, Kennedy K, Jamieson R, Stewart J (2007) A second generation human haplotype map of over 3.1 million SNPs. Nature 449:851–861

Friedman JI, Vrijenhoek T, Markx S, Janssen IM, van der Vliet WA, Faas BH, Knoers NV, Cahn W, Kahn RS, Edelmann L, Davis KL, Silverman JM, Brunner HG, van Kessel AG, Wijmenga C, Ophoff RA, Veltman JA (2008) CNTNAP2 gene dosage variation is associated with schizophrenia and epilepsy. Mol Psychiatry 13:261–266

Gabriel SB, Schaffner SF, Nguyen H, Moore JM, Roy J, Blumenstiel B, Higgins J, DeFelice M, Lochner A, Faggart M, Liu-Cordero SN, Rotimi C, Adeyemo A, Cooper R, Ward R, Lander ES, Daly MJ, Altshuler D (2002) The structure of haplotype blocks in the human genome. Science 296:2225–2229

Gonzalez E, Kulkarni H, Bolivar H, Mangano A, Sanchez R, Catano G, Nibbs RJ, Freedman BI, Quinones MP, Bamshad MJ, Murthy KK, Rovin BH, Bradley W, Clark RA, Anderson SA, O'connell RJ, Agan BK, Ahuja SS, Bologna R, Sen L, Dolan MJ, Ahuja SK (2005) The influence of CCL3L1 gene-containing segmental duplications on HIV-1/AIDS susceptibility. Science 307:1434–1440

Göring HH, Terwilliger JD, Blangero J (2001) Large upward bias in estimation of locus-specific effects from genomewide scans. Am J Hum Genet 69:1357–1369

Hinds DA, Stuve LL, Nilsen GB, Halperin E, Eskin E, Ballinger DG, Frazer KA, Cox DR (2005) Whole-genome patterns of common DNA variation in three human populations. Science 307:1072–1079

Hirschhorn JN, Altshuler D (2002) Once and again-issues surrounding replication in genetic association studies. J Clin Endocrinol Metab 87(10):4438–41

Hirschhorn JN, Daly MJ (2005) Genome-wide association studies for common diseases and complex traits. Nat Rev Genet 6:95–108

Ingason A, Rujescu D, Cichon S, Sigurdsson E, Sigmundsson T, Pietiläinen OP, Buizer-Voskamp JE, Strengman E, Francks C, Muglia P, Gylfason A, Gustafsson O, Olason PI, Steinberg S, Hansen T, Jakobsen KD, Rasmussen HB, Giegling I, Möller HJ, Hartmann A, Crombie C, Fraser G, Walker N, Lonnqvist J, Suvisaari J, Tuulio-Henriksson A, Bramon E, Kiemeney LA, Franke B, Murray R, Vassos E, Toulopoulou T, Mühleisen TW, Tosato S, Ruggeri M, Djurovic S, Andreassen OA, Zhang Z, Werge T, Ophoff RA; GROUP Investigators, Rietschel M, Nöthen MM, Petursson H, Stefansson H, Peltonen L, Collier D, Stefansson K, Clair DM (2009) Copy number variations of chromosome 16p13.1 region associated with schizophrenia. Mol Psychiatry (Epub ahead of print, Sept 29) Available at: http://www.ncbi.nlm.nih.gov/pubmed/19786961 (Accessed September 30, 2009)

International HapMap Consortium (2005) A haplotype map of the human genome. Nature 437: 1299–1320

Ioannidis JP, Thomas G, Daly MJ (2009) Validating, augmenting and refining genome-wide association signals. Nat Rev Genet 10:318–329

Johansson I, Lundqvist E, Bertilsson L, Dahl ML, Sjöqvist F, Ingelman-Sundberg M (1993) Inherited amplification of an active gene in the cytochrome P450 CYP2D locus as a cause of ultrarapid metabolism of debrisoquine. Proc Natl Acad Sci USA 90:11825–11829

Kathiresan S, Willer CJ, Peloso GM, Demissie S, Musunuru K, Schadt EE, Kaplan L, Bennett D, Li Y, Tanaka T, Voight BF, Bonnycastle LL, Jackson AU, Crawford G, Surti A, Guiducci C, Burtt NP, Parish S, Clarke R, Zelenika D, Kubalanza KA, Morken MA, Scott LJ, Stringham HM, Galan P, Swift AJ, Kuusisto J, Bergman RN, Sundvall J, Laakso M, Ferrucci L, Scheet P, Sanna S, Uda M, Yang Q, Lunetta KL, Dupuis J, de Bakker PI, O'Donnell CJ, Chambers JC, Kooner JS, Hercberg S, Meneton P, Lakatta EG, Scuteri A, Schlessinger D, Tuomilehto J, Collins FS, Groop L, Altshuler D, Collins R, Lathrop GM, Melander O, Salomaa V, Peltonen L, Orho-Melander M, Ordovas JM, Boehnke M, Abecasis GR, Mohlke KL, Cupples LA (2009) Common variants at 30 loci contribute to polygenic dyslipidemia. Nat Genet 41:56–65

Kirov G, Gumus D, Chen W, Norton N, Georgieva L, Sari M, O'Donovan MC, Erdogan F, Owen MJ, Ropers HH, Ullmann R (2008) Comparative genome hybridization suggests a role for NRXN1 and APBA2 in schizophrenia. Hum Mol Genet 17:458–465

Kruglyak L (2008) The road to genome-wide association studies. Nat Rev Genet 9:314–318

Lander ES, Schork NJ (1994) Genetic dissection of complex traits. Science 265:2037–2048

Lawson-Yuen A, Saldivar JS, Sommer S, Picker J (2008) Familial deletion within NLGN4 associated with autism and Tourette syndrome. Eur J Hum Genet 16:614–618

Lerer B, Segman RH, Hamdan A, Kanyas K, Karni O, Kohn Y, Korner M, Lanktree M, Kaadan M, Turetsky N, Yakir A, Kerem B, Macciardi F (2003) Genome scan of Arab Israeli families maps a schizophrenia susceptibility gene to chromosome 6q23 and supports a locus at chromosome 10q24. Mol Psychiatry 8:488–498

Levinson DF, Levinson MD, Segurado R, Lewis CM (2003) Genome scan meta-analysis of schizophrenia and bipolar disorder, part I: Methods and power analysis. Am J Hum Genet 73:17–33

Levy S, Sutton G, Ng PC, Feuk L, Halpern AL, Walenz BP, Axelrod N, Huang J, Kirkness EF, Denisov G, Lin Y, MacDonald JR, Pang AW, Shago M, Stockwell TB, Tsiamouri A, Bafna V, Bansal V, Kravitz SA, Busam DA, Beeson KY, McIntosh TC, Remington KA, Abril JF, Gill J, Borman J, Rogers YH, Frazier ME, Scherer SW, Strausberg RL, Venter JC (2007) The diploid genome sequence of an individual human. PLoS Biol 5:e254

Lewis CM, Levinson DF, Wise LH, DeLisi LE, Straub RE, Hovatta I, Williams NM, Schwab SG, Pulver AE, Faraone SV, Brzustowicz LM, Kaufmann CA, Garver DL, Gurling HM, Lindholm E, Coon H, Moises HW, Byerley W, Shaw SH, Mesen A, Sherrington R, O'Neill FA, Walsh D, Kendler KS, Ekelund J, Paunio T, Lönnqvist J, Peltonen L, O'Donovan MC, Owen MJ, Wildenauer DB, Maier W, Nestadt G, Blouin JL, Antonarakis SE, Mowry BJ, Silverman JM, Crowe RR, Cloninger CR, Tsuang MT, Malaspina D, Harkavy-Friedman JM, Svrakic DM, Bassett AS, Holcomb J, Kalsi G, McQuillin A, Brynjolfson J, Sigmundsson T, Petursson H, Jazin E, Zoëga T, Helgason T (2003) Genome scan meta-analysis of schizophrenia and bipolar disorder, part II: Schizophrenia. Am J Hum Genet 73:34–48

Li B, Leal SM (2008) Methods for detecting associations with rare variants for common diseases: application to analysis of sequence data. Am J Hum Genet 83:311–321

Li M, Li C, Guan W (2008) Evaluation of coverage variation of SNP chips for genome-wide association studies. Eur J Hum Genet 16:635–643

Lindholm E, Ekholm B, Shaw S, Jalonen P, Johansson G, Pettersson U, Sherrington R, Adolfsson R, Jazin E (2001) A schizophrenia-susceptibility locus at 6q25, in one of the world's largest reported pedigrees. Am J Hum Genet 69:96–105

Liu H, Heath SC, Sobin C, Roos JL, Galke BL, Blundell ML, Lenane M, Robertson B, Wijsman EM, Rapoport JL, Gogos JA, Karayiorgou M (2002) Genetic variation at the 22q11 PRODH2/DGCR6 locus presents an unusual pattern and increases susceptibility to schizophrenia. Proc Natl Acad Sci USA 99:3717–3722

Lusis AJ, Pajukanta P (2008) A treasure trove for lipoprotein biology. Nat Genet 40:129–130

Manolio TA, Collins FS, Cox NJ, Goldstein DB, Hindorff LA, Hunter DJ, McCarthy MI, Ramos EM, Cardon LR, Chakravarti A, Cho JH, Guttmacher AE, Kong A, Kruglyak L, Mardis E, Rotimi CN, Slatkin M, Valle D, Whittemore AS, Boehnke M, Clark AG, Eichler EE, Gibson G, Haines JL, Mackay TF, McCarroll SA, Visscher PM (2009) Finding the missing heritability of complex diseases. Nature 461:747–753

McCarroll SA (2008) Extending genome-wide association studies to copy-number variation. Hum Mol Genet 17:R135

McCarroll SA, Hadnott TN, Perry GH, Sabeti PC, Zody MC, Barrett JC, Dallaire S, Gabriel SB, Lee C, Daly MJ, Altshuler DM, International HapMap Consortium (2005) Common deletion polymorphisms in the human genome. Nat Genet 38:86–92

McCarroll SA, Kuruvilla FG, Korn JM, Cawley S, Nemesh J, Wysoker A, Shapero MH, de Bakker PI, Maller JB, Kirby A, Elliott AL, Parkin M, Hubbell E, Webster T, Mei R, Veitch J, Collins PJ, Handsaker R, Lincoln S, Nizzari M, Blume J, Jones KW, Rava R, Daly MJ, Gabriel SB, Altshuler D (2008) Integrated detection and population-genetic analysis of SNPs and copy number variation. Nat Genet 40:1166–1174

McCarthy MI, Abecasis GR, Cardon LR, Goldstein DB, Little J, Ioannidis JP, Hirschhorn JN (2008) Genome-wide association studies for complex traits: consensus, uncertainty and challenges. Nat Rev Genet 9:356–369

McCarthy SE, Makarov V, Kirov G, Addington AM, McClellan J, Yoon S, Perkins DO, Dickel DE, Kusenda M, Krastoshevsky O, Krause V, Kumar RA, Grozeva D, Malhotra D, Walsh T, Zackai EH, Kaplan P, Ganesh J, Krantz ID, Spinner NB, Roccanova P, Bhandari A, Pavon K, Lakshmi B, Leotta A, Kendall J, Lee YH, Vacic V, Gary S, Iakoucheva LM, Crow TJ, Christian SL, Lieberman JA, Stroup TS, Lehtimäki T, Puura K, Haldeman-Englert C, Pearl J, Goodell M, Willour VL, Derosse P, Steele J, Kassem L, Wolff J, Chitkara N, McMahon FJ,

Malhotra AK, Potash JB, Schulze TG, Nöthen MM, Cichon S, Rietschel M, Leibenluft E, Kustanovich V, Lajonchere CM, Sutcliffe JS, Skuse D, Gill M, Gallagher L, Mendell NR, Wellcome Trust Case Control Consortium, Craddock N, Owen MJ, O'Donovan MC, Shaikh TH, Susser E, Delisi LE, Sullivan PF, Deutsch CK, Rapoport J, Levy DL, King MC, Sebat J (2009) Microduplications of 16p11.2 are associated with schizophrenia. Nat Genet 41:1223–1227

McClellan JM, Susser E, King M (2007) Schizophrenia: a common disease caused by multiple rare alleles. Br J Psychiatry 190:194–199

Millar JK, Wilson-Annan JC, Anderson S, Christie S, Taylor MS, Semple CA, Devon RS, St Clair DM, Muir WJ, Blackwood DH, Porteous DJ (2000) Disruption of two novel genes by a translocation co-segregating with schizophrenia. Hum Mol Genet 9:1415–1423

Miller RD, Phillips MS, Jo I, Donaldson MA, Studebaker JF, Addleman N, Alfisi SV, Ankener WM, Bhatti HA, Callahan CE, Carey BJ, Conley CL, Cyr JM, Derohannessian V, Donaldson RA, Elosua C, Ford SE, Forman AM, Gelfand CA, Grecco NM, Gutendorf SM, Hock CR, Hozza MJ, Hur S, In SM, Jackson DL, Jo SA, Jung SC, Kim S, Kimm K, Kloss EF, Koboldt DC, Kuebler JM, Kuo FS, Lathrop JA, Lee JK, Leis KL, Livingston SA, Lovins EG, Lundy ML, Maggan S, Minton M, Mockler MA, Morris DW, Nachtman EP, Oh B, Park C, Park CW, Pavelka N, Perkins AB, Restine SL, Sachidanandam R, Reinhart AJ, Scott KE, Shah GJ, Tate JM, Varde SA, Walters A, White JR, Yoo YK, Lee JE, Boyce-Jacino MT, Kwok PY, SNP Consortium Allele Frequency Project (2005) High-density single-nucleotide polymorphism maps of the human genome. Genomics 86:117–126

Moskvina V, Holmans P, Schmidt KM, Craddock N (2005) Design of case-controls studies with unscreened controls. Ann Hum Genet 69:566–576

Murphy KC, Jones LA, Owen MJ (1999) High rates of schizophrenia in adults with velo-cardio-facial syndrome. Arch Gen Psychiatry 56:940–945

Ng PC, Levy S, Huang J, Stockwell TB, Walenz BP, Li K, Axelrod N, Busam DA, Strausberg RL, Venter JC (2008) Genetic variation in an individual human exome. PLoS Genet 4:e1000160

Ng SB, Turner EH, Robertson PD, Flygare SD, Bigham AW, Lee C, Shaffer T, Wong M, Bhattacharjee A, Eichler EE, Bamshad M, Nickerson DA, Shendure J (2009) Targeted capture and massively parallel sequencing of 12 human exomes. Nature 461:272–276

O'Donovan MC, Williams NM, Owen MJ (2003) Recent advances in the genetics of schizophrenia. Hum Mol Genet 12 (Spec No 2):R125–R133

O'Donovan MC, Craddock N, Norton N, Williams H, Peirce T, Moskvina V, Nikolov I, Hamshere M, Carroll L, Georgieva L, Dwyer S, Holmans P, Marchini JL, Spencer CC, Howie B, Leung HT, Hartmann AM, Möller HJ, Morris DW, Shi Y, Feng G, Hoffmann P, Propping P, Vasilescu C, Maier W, Rietschel M, Zammit S, Schumacher J, Quinn EM, Schulze TG, Williams NM, Giegling I, Iwata N, Ikeda M, Darvasi A, Shifman S, He L, Duan J, Sanders AR, Levinson DF, Gejman PV, Cichon S, Nöthen MM, Gill M, Corvin A, Rujescu D, Kirov G, Owen MJ, Buccola NG, Mowry BJ, Freedman R, Amin F, Black DW, Silverman JM, Byerley WF, Cloninger CR, Molecular Genetics of Schizophrenia Collaboration (2008) Identification of loci associated with schizophrenia by genome-wide association and follow-up. Nat Genet 40:1053–1055

Owen MJ, O'Donovan MC, Gottesman II (2002) Schizophrenia. In: McGuffin P, Owen MJ, Gotesman II (eds) Psychiatric genetics and genomics. Medical Publications, Oxford, pp 247–266

Papolos DF, Faedda GL, Veit S, Goldberg R, Morrow B, Kucherlapati R, Shprintzen RJ (1996) Bipolar spectrum disorders in patients diagnosed with velo-cardio-facial syndrome: does a hemizygous deletion of chromosome 22q11 result in bipolar affective disorder? Am J Psychiatry 153:1541–1547

Pickard BS, Thomson PA, Christoforou A, Evans KL, Morris SW, Porteous DJ, Blackwood DH, Muir WJ (2007) The PDE4B gene confers sex-specific protection against schizophrenia. Psychiatr Genet 17:129–133

Price AL, Patterson NJ, Plenge RM, Weinblatt ME, Shadick NA, Reich D (2006) Principal components analysis corrects for stratification in genome-wide association studies. Nat Genet 38:904–909

Pritchard JK (2001) Are rare variants responsible for susceptibility to complex diseases? Am J Hum Genet 69:124–137

Pulver AE, Nestadt G, Goldberg R, Shprintzen RJ, Lamacz M, Wolyniec PS, Morrow B, Karayiorgou M, Antonarakis SE, Housman D, Kucherlapati R (1994) Psychotic illness in patients diagnosed with velo-cardio-facial syndrome and their relatives. J Nerv Ment Dis 182: 476–478

Purcell SM, Wray NR, Stone JL, Visscher PM, O'Donovan MC, Sullivan PF, Sklar P (2009) Common polygenic variation contributes to risk of schizophrenia and bipolar disorder. Nature 460:748–752

Redon R, Ishikawa S, Fitch KR, Feuk L, Perry GH, Andrews TD, Fiegler H, Shapero MH, Carson AR, Chen W, Cho EK, Dallaire S, Freeman JL, González JR, Gratacòs M, Huang J, Kalaitzopoulos D, Komura D, MacDonald JR, Marshall CR, Mei R, Montgomery L, Nishimura K, Okamura K, Shen F, Somerville MJ, Tchinda J, Valsesia A, Woodwark C, Yang F, Zhang J, Zerjal T, Zhang J, Armengol L, Conrad DF, Estivill X, Tyler-Smith C, Carter NP, Aburatani H, Lee C, Jones KW, Scherer SW, Hurles ME (2006) Global variation in copy number in the human genome. Nature 444:444–454

Reich DE, Lander ES (2001) On the allelic spectrum of human disease. Trends Genet 17:502–510

Risch N (1990a) Linkage strategies for genetically complex traits. I. Multilocus models. Am J Hum Genet 46:222–228

Risch N (1990b) Linkage strategies for genetically complex traits. II. The power of affected relative pairs. Am J Hum Genet 46:229–241

Risch N, Merikangas K (1996) The future of genetic studies of complex human diseases. Science 273:1516–1517

Sebat J, Lakshmi B, Malhotra D, Troge J, Lese-Martin C, Walsh T, Yamrom B, Yoon S, Krasnitz A, Kendall J, Leotta A, Pai D, Zhang R, Lee YH, Hicks J, Spence SJ, Lee AT, Puura K, Lehtimäki T, Ledbetter D, Gregersen PK, Bregman J, Sutcliffe JS, Jobanputra V, Chung W, Warburton D, King MC, Skuse D, Geschwind DH, Gilliam TC, Ye K, Wigler M (2007) Strong association of de novo copy number mutations with autism. Science 316:445–449

Shi J, Levinson DF, Duan J, Sanders AR, Zheng Y, Pe'er I, Dudbridge F, Holmans PA, Whittemore AS, Mowry BJ, Olincy A, Amin F, Cloninger CR, Silverman JM, Buccola NG, Byerley WF, Black DW, Crowe RR, Oksenberg JR, Mirel DB, Kendler KS, Freedman R, Gejman PV (2009) Common variants on chromosome 6p22.1 are associated with schizophrenia. Nature 460:753–757

Shifman S, Bronstein M, Sternfeld M, Pisanté-Shalom A, Lev-Lehman E, Weizman A, Reznik I, Spivak B, Grisaru N, Karp L, Schiffer R, Kotler M, Strous RD, Swartz-Vanetik M, Knobler HY, Shinar E, Beckmann JS, Yakir B, Risch N, Zak NB, Darvasi A (2002) A highly significant association between a COMT haplotype and schizophrenia. Am J Hum Genet 71: 1296–1302

Shih RA, Belmonte PL, Zandi PP (2004) A review of the evidence from family, twin and adoption studies for a genetic contribution to adult psychiatric disorders. Int Rev Psychiatry 16:260–283

Shprintzen RJ, Goldberg R, Golding-Kushner KJ, Marion RW (1992) Late-onset psychosis in the velo-cardio-facial syndrome. Am J Med Genet 42:141–142

Stefansson H, Rujescu D, Cichon S, Pietiläinen OP, Ingason A, Steinberg S, Fossdal R, Sigurdsson E, Sigmundsson T, Buizer-Voskamp JE, Hansen T, Jakobsen KD, Muglia P, Francks C, Matthews PM, Gylfason A, Halldorsson BV, Gudbjartsson D, Thorgeirsson TE, Sigurdsson A, Jonasdottir A, Jonasdottir A, Bjornsson A, Mattiasdottir S, Blondal T, Haraldsson M, Magnusdottir BB, Giegling I, Möller HJ, Hartmann A, Shianna KV, Ge D, Need AC, Crombie C, Fraser G, Walker N, Lonnqvist J, Suvisaari J, Tuulio-Henriksson A, Paunio T, Toulopoulou T, Bramon E, Di Forti M, Murray R, Ruggeri M, Vassos E, Tosato S, Walshe M, Li T, Vasilescu C, Mühleisen TW, Wang AG, Ullum H, Djurovic S, Melle I, Olesen J, Kiemeney LA, Franke B, GROUP, Sabatti C, Freimer NB, Gulcher JR, Thorsteinsdottir U,

Kong A, Andreassen OA, Ophoff RA, Georgi A, Rietschel M, Werge T, Petursson H, Goldstein DB, Nöthen MM, Peltonen L, Collier DA, St Clair D, Stefansson K (2008) Large recurrent microdeletions associated with schizophrenia. Nature 455:232–236

Stefansson H, Ophoff RA, Steinberg S, Andreassen OA, Cichon S, Rujescu D, Werge T, Pietiläinen OP, Mors O, Mortensen PB, Sigurdsson E, Gustafsson O, Nyegaard M, Tuulio-Henriksson A, Ingason A, Hansen T, Suvisaari J, Lonnqvist J, Paunio T, Børglum AD, Hartmann A, Fink-Jensen A, Nordentoft M, Hougaard D, Norgaard-Pedersen B, Böttcher Y, Olesen J, Breuer R, Möller HJ, Giegling I, Rasmussen HB, Timm S, Mattheisen M, Bitter I, Réthelyi JM, Magnusdottir BB, Sigmundsson T, Olason P, Masson G, Gulcher JR, Haraldsson M, Fossdal R, Thorgeirsson TE, Thorsteinsdottir U, Ruggeri M, Tosato S, Franke B, Strengman E, Kiemeney LA, Genetic Risk and Outcome in Psychosis (GROUP), Melle I, Djurovic S, Abramova L, Kaleda V, Sanjuan J, de Frutos R, Bramon E, Vassos E, Fraser G, Ettinger U, Picchioni M, Walker N, Toulopoulou T, Need AC, Ge D, Yoon JL, Shianna KV, Freimer NB, Cantor RM, Murray R, Kong A, Golimbet V, Carracedo A, Arango C, Costas J, Jönsson EG, Terenius L, Agartz I, Petursson H, Nöthen MM, Rietschel M, Matthews PM, Muglia P, Peltonen L, St Clair D, Goldstein DB, Stefansson K, Collier DA (2009) Common variants conferring risk of schizophrenia. Nature 460:744–747

Straub RE, MacLean CJ, O'Neill FA, Burke J, Murphy B, Duke F, Shinkwin R, Webb BT, Zhang J, Walsh D, Kendler KS (1995) A potential vulnerability locus for schizophrenia on chromosome 6p24-22: evidence for genetic heterogeneity. Nat Genet 11:287–293

Straub RE, Jiang Y, MacLean CJ, Ma Y, Webb BT, Myakishev MV, Harris-Kerr C, Wormley B, Sadek H, Kadambi B, Cesare AJ, Gibberman A, Wang X, O'Neill FA, Walsh D, Kendler KS (2002) Genetic variation in the 6p22.3 gene DTNBP1, the human ortholog of the mouse dysbindin gene, is associated with schizophrenia. Am J Hum Genet 71:337–348

Suarez B, Hampe C, VanEerdewegh P (1994) Problems of replicating linkage claims in psychiatry. In: Genetic approaches to mental disorders. American Psychiatric Press, Washington DC, pp 23–46

Sullivan PF, Lin D, Tzeng JY, van den Oord E, Perkins D, Stroup TS, Wagner M, Lee S, Wright FA, Zou F, Liu W, Downing AM, Lieberman J, Close SL (2008) Genomewide association for schizophrenia in the CATIE study: results of stage 1. Mol Psychiatry 13:570–584

Terwilliger JD, Hiekkalinna T (2006) An utter refutation of the "fundamental theorem of the HapMap". Eur J Hum Genet 14:426–437

The International Schizophrenia Consortium (2008) Rare chromosomal deletions and duplications increase risk of schizophrenia. Nature 455:237–241

The Wellcome Trust Case Control Consortium (2007) Genome-wide association study of 14,000 cases of seven common diseases and 3,000 shared controls. Nature 447:661–678

Thorisson GA, Stein LD (2003) The SNP Consortium website: past, present and future. Nucleic Acids Res 31:124–127

Tishkoff SA, Verrelli BC (2003) Patterns of human genetic diversity: implications for human evolutionary history and disease. Annu Rev Genomics Hum Genet 4:293–340

van Ommen GB (2005) Frequency of new copy number variation in humans. Nat Genet 37:333–334

Voight BF, Pritchard JK (2005) Confounding from cryptic relatedness in case-control association studies. PLoS Genet 1:e32

Walsh T, McClellan JM, McCarthy SE, Addington AM, Pierce SB, Cooper GM, Nord AS, Kusenda M, Malhotra D, Bhandari A, Stray SM, Rippey CF, Roccanova P, Makarov V, Lakshmi B, Findling RL, Sikich L, Stromberg T, Merriman B, Gogtay N, Butler P, Eckstrand K, Noory L, Gochman P, Long R, Chen Z, Davis S, Baker C, Eichler EE, Meltzer PS, Nelson SF, Singleton AB, Lee MK, Rapoport JL, King MC, Sebat J (2008) Rare structural variants disrupt multiple genes in neurodevelopmental pathways in schizophrenia. Science 320:539–543

Wang K, Li M, Hadley D, Liu R, Glessner J, Grant SF, Hakonarson H, Bucan M (2007) PennCNV: an integrated hidden Markov model designed for high-resolution copy number variation detection in whole-genome SNP genotyping data. Genome Res 17:1665–1674

Williams NM, Norton N, Williams H, Ekholm B, Hamshere ML, Lindblom Y, Chowdari KV, Cardno AG, Zammit S, Jones LA, Murphy KC, Sanders RD, McCarthy G, Gray MY, Jones G, Holmans P, Nimgaonkar V, Adolfson R, Osby U, Terenius L, Sedvall G, O'Donovan MC, Owen MJ (2003) A systematic genomewide linkage study in 353 sib pairs with schizophrenia. Am J Hum Genet 73:1355–1367

Zeggini E, Weedon MN, Lindgren CM, Frayling TM, Elliott KS, Lango H, Timpson NJ, Perry JR, Rayner NW, Freathy RM, Barrett JC, Shields B, Morris AP, Ellard S, Groves CJ, Harries LW, Marchini JL, Owen KR, Knight B, Cardon LR, Walker M, Hitman GA, Morris AD, Doney AS, Wellcome Trust Case Control Consortium (WTCCC), McCarthy MI, Hattersley AT (2007) Replication of genome-wide association signals in UK samples reveals risk loci for type 2 diabetes. Science 316:1336–1341

Zheng G, Freidlin B, Gastwirth JL (2006) Robust genomic control for association studies. Am J Hum Genet 78:350–356

Zollner S, Pritchard JK (2007) Overcoming the winner's curse: estimating penetrance parameters from case-control data. Am J Hum Genet 80:605–615

Epigenetics of Schizophrenia

Schahram Akbarian

Contents

1 Introduction 612
2 Histone Modifications and DNA Methylation 613
3 Findings in Schizophrenia Postmortem Brain 615
4 Reproducibility of Epigenetic Alterations in Schizophrenia Postmortem Brain 617
5 Cellular Specificity of Epigenetic Markings 618
6 Epigenetic Markings in Brain: How Stable? 619
7 Implications for the Neurobiology of Schizophrenia 620
8 Chromatin Remodeling and Antipsychotic Medication 621
9 Synopsis and Outlook 622
References 622

Abstract Epigenetic regulators of gene expression including DNA cytosine methylation and posttranslational histone modifications could play a role for some of the molecular alterations associated with schizophrenia. For example, in prefrontal cortex of subjects with schizophrenia, abnormal DNA or histone methylation at sites of specific genes and promoters is associated with changes in RNA expression. These findings are of interest from a neurodevelopmental perspective because there is increasing evidence that epigenetic markings for a substantial portion of genes and loci are highly regulated during the first years of life. Furthermore, there is circumstantial evidence that a subset of antipsychotic drugs, including the atypical, Clozapine, interfere with chromatin remodeling mechanisms. Challenges for the field include (1) no clear consensus yet regarding disease-associated changes, (2) the lack of cell-specific chromatin assays which makes it difficult to ascribe epigenetic alterations to specific cell populations, and (3) lack of knowledge about the stability or turnover of epigenetic markings at specific loci in (brain) chromatin.

S. Akbarian
Department of Psychiatry, Brudnick Neuropsychiatric Research Institute, University of Massachusetts Medical School, Worcester, MA 01604, USA
e-mail: schahram.akbarian@umassmed.edu

Despite these shortcomings, the study of DNA and histone modifications in chromatin extracted from diseased and control brain tissue is likely to provide valuable insight into the genomic risk architecture of schizophrenia, particularly in the large majority of cases for which a straightforward genetic cause still remains elusive.

Keywords Epigenome · Gene expression · Nucleosome · Postmortem brain · Psychosis

1 Introduction

Schizophrenia and related disorders such as bipolar illness and autism are considered complex and heterogeneous in their etiologies, as reflected by concordance rates of less than 70% in monozygotic twins and non-Mendelian inheritance patterns (Kaminsky et al. 2006). Important advancements toward a better understanding of the underlying genetics have been achieved recently, and the list of copy number variations, microdeletions, and polymorphisms associated with genetic risk for schizophrenia is steadily increasing. Still, straightforward genetic causes are lacking for a very large majority of affected individuals (The International Schizophrenia Consortium 2008; Lupski 2008; Bertolino and Blasi 2009; Delisi 2009; Smith 2009).

Given that current genetic approaches do not sufficiently resolve the mechanisms of disease, it had been hypothesized that "epigenetic" mechanisms, perhaps in interaction with genetic factors, could play a pivotal role in the pathophysiology of schizophrenia (Abdolmaleky et al. 2005b; Kaminsky et al. 2006). Traditionally, the term "epigenetic" is applied in the context of phenotypes and mechanisms thought to reflect (or result from) heritable changes in gene expression and function without any alterations of the DNA sequence (Jablonka and Lamb 2002; Berger et al. 2009). A well-known example of this scenario is the inactivation pattern of the second X chromosome in female somatic cells which is faithfully passed on from mother to daughter cells after the earliest stages of intrauterine development (Payer and Lee 2008). In addition, the term "epigenetic" is in the fields of neuroscience and translational medicine applied to a wide range of molecular mechanism focused on chromatin structure and function both in dividing and in postmitotic cells (Tsankova et al. 2007; Jiang et al. 2008a; Akbarian and Huang 2009). In this context, as outlined in more detail below, the study of epigenetic markings offers an attractive perspective for postmortem brain research, in which chromatin assays typically are combined with quantification of the RNA and proteins in order to obtain a measure for gene expression activity.

Schizophrenia and several other of the major psychiatric diseases including bipolar illness and some cases in the autism spectrum typically lack a defining cellular pathology but often are associated with alterations in levels of RNAs encoding a wide range of transcripts involved in inhibitory or excitatory neurotransmission, myelination, and metabolism, among others (Hof et al. 2002; Davis and Haroutunian 2003; Fatemi 2005; Kristiansen et al. 2007; Connor and Akbarian 2008; Akbarian and Huang 2009). Both DNA methylation at sites of CpG dinucleotides, and some of the covalent histone modifications at gene promoters (discussed in detail further below) are involved in the regulation of transcription. Therefore, these "epigenetic" markings are often studied in schizophrenia postmortem brain in conjunction with quantification of the corresponding RNA molecules (Abdolmaleky et al. 2005a; Grayson et al. 2005; Huang et al. 2007). The typical working hypothesis implies that a change in promoter DNA methylation or histone modification is a potential indicator for altered gene expression activity at that locus, which in turn leads to an alteration in the level of the corresponding RNA.

Several recent excellent reviews summarize our state of knowledge on epigenetic regulation in psychiatric disorders, with the majority of studies published to date primarily focused on preclinical models (Levenson and Sweatt 2005; Tsankova et al. 2007; Abel and Zukin 2008; Monteggia and Kavalali 2009; Sweatt 2009). The purpose of this review will be to provide an overview of the chromatin markings that to date were studied in brains of subjects with schizophrenia, including the potential implications for the neurobiology of the disease, and to discuss current limitations and challenges the field needs to overcome. Because practically all schizophrenia studies explored either DNA or histone methylation markings, the present review is mostly focused on these types of modifications. It must be noted, however, that the "language" of chromatin is far more complex, and involves a complex set of histone modifications including lysine acetylation, methylation, ubiquitinylation, and SUMOylation, as well as arginine methylation, serine phosphorylation, and proline isomerization (Berger 2007). Many of these modifications remain to be explored in animal models and human postmortem studies.

2 Histone Modifications and DNA Methylation

Histones are, together with DNA that wraps around them, the fundamental structural unit of eukaryote chromatin and thus regulate gene expression, DNA repair, and chromosome segregation, among others. The backbone of chromatin is provided by the core histones H2A, H2B, H3, and H4 that together form the nucleosome, a drum-like structure with 147 bp of genomic DNA wrapped around it (Wolffe 1992; Hayes and Hansen 2001). Dynamic changes in chromatin structure and accessibility of transcription factors are mediated by two molecular mechanisms: chromatin remodeling and histone tail modifications (Felsenfeld and Groudine 2003). Chromatin remodeling is defined by the mobilization of genomic DNA wrapped around the nucleosome core particle (Peterson 2002). Covalent modifications of specific arginine, lysine, and serine residues at the histone N-terminal tails on the nucleosome surface define a "histone code" that is differentially regulated in chromatin at sites of active gene expression, in comparison to inactive and silenced chromatin (Jenuwein and Allis 2001; Berger 2002; Iizuka and Smith 2003; Peterson and Laniel 2004; Wang et al. 2004). The highly basic histone tails are predicted to be less structured than the histone-fold regions and are believed to interact with the negatively charged DNA backbone or with other chromatin-associated proteins. In particular, the acetylation of specific lysine residues in the N-terminal tails of H3 and H4 is associated with active gene expression, while deacetylation of these residues results in transcriptional silencing, chromatin condensation, and heterochromatin formation (Jenuwein and Allis 2001; Berger 2002; Iizuka and Smith 2003; Peterson and Laniel 2004; Wang et al. 2004). In addition, differential methylation and phosphorylation of specific serine, lysine, and arginine residues at the N-terminal end of H3 and H4 is associated with transcriptional regulation in eukaryotes, including humans (Jenuwein and Allis 2001; Berger 2002; Fischle et al. 2003; Iizuka and Smith 2003; Peterson and Laniel 2004; Wang et al. 2004). Studies conducted on postmortem brain tissue provided evidence that this "histone code," which differentiates chromatin at sites with active gene expression from silenced chromatin (Jenuwein and Allis 2001; Berger 2002; Iizuka and Smith 2003; Peterson and Laniel 2004; Wang et al. 2004), contributes to transcriptional regulation in the human brain (Stadler et al. 2005; Huang et al. 2006, 2007). In addition, recent animal studies demonstrated that transcriptional regulation in a diverse set of learning- or drug-induced plasticity models is accompanied by dynamic histone modification changes (Swank and Sweatt 2001; Guan et al. 2002; Alarcon et al. 2004; Korzus et al. 2004; Li et al. 2004; Tsankova et al. 2004; Kumar et al. 2005; Levine et al. 2005; Tsankova et al. 2006). Therefore, there is little doubt that the principles of chromatin remodeling and histone N-terminal tail modifications, which are traditionally explored in dividing cells (Jenuwein and Allis 2001; Berger 2002; Iizuka and Smith 2003; Peterson and Laniel 2004; Wang et al. 2004), also apply to the central nervous system and its postmitotic constituents, including neurons (Martin and Sun 2004; Colvis et al. 2005; Jiang et al. 2008a).

In vertebrates, the methylation of C5 carbon of cytosine residues at CpG dinucleotides positioned within transcribed sequences is loosely associated with gene expression activity but when found within regulatory sequences surrounding transcription start sites and proximal promoters, DNA methylation is often (but not always) thought to be involved in transcriptional repression (Klose and Bird 2006; Suzuki and Bird 2008). DNA methylation plays an important role for genome stability, X-chromosome inactivation, and the regulation of imprinted gene loci that show parent-of-origin-specific gene expression (Hajkova et al. 2002; Morgan et al. 2005; Waggoner 2007). There is, not surprisingly, ample evidence that DNA methylation and covalent histone modifications often are coregulated and functionally interconnected. To mention just one example, it has been suggested that histone H3-lysine 9 (H3K9) methylation, a repressive chromatin mark, is a recruitment signal for DNA methyltransferase enzymes, which catalyze methylation of CpG dinucleotides which in turn are recognized by methyl CpG-binding proteins associated with repressive chromatin remodeling and histone deacetylase (HDAC) activity (Fuks 2005).

3 Findings in Schizophrenia Postmortem Brain

To date, most studies exploring DNA methylation changes in schizophrenia utilized postmortem tissue from cerebral cortex. A number of pioneering studies, focusing on less than a handful of candidate genes often implied in cortical dysfunction of schizophrenia, reported various degrees of aberrant CpG hyper- or hypomethylation in regulatory gene sequences and promoters. These include the glycoprotein *REELIN* (Abdolmaleky et al. 2005a; Grayson et al. 2005), catechyl-*O*-methyltransferase (*COMT*) (Abdolmaleky et al. 2006), and *SOX10*, encoding a transcription factor important for myelination and oligodendrocyte function (Iwamoto et al. 2005). It is encouraging that these disease-related hyper(hypo)methylation changes were associated with decreased (increased) levels of the corresponding RNA, consistent with the repressive function of CpG methylation around transcription start sites. On the other hand, the disease-related *REELIN* and *COMT* DNA methylation changes could not be replicated in independent studies (Mill et al. 2008; Tochigi et al. 2008). Numerous factors could have contributed to these disparate findings. For example, DNA CpG methylation in a variety of tissues, including CNS, could be sensitive to the social environment (Rampon et al. 2000; Weaver et al. 2004), ischemia (Endres et al. 2000), environmental toxins (Desaulniers et al. 2005; Bollati et al. 2007; Salnikow and Zhitkovich 2008), nicotine (Satta et al. 2007, 2008), exposure to alcohol (Marutha Ravindran and Ticku 2004), psychostimulants (Numachi et al. 2004, 2007), and antipsychotic drugs (Shimabukuro et al. 2006; Cheng et al. 2008; Mill et al. 2008). It is likely that these factors also play a role in shaping DNA methylation patterns in the human brain, thereby contributing to sample variability in postmortem studies.

Recently, a more comprehensive DNA methylation screen for approximately 8,000 genes was conducted on the frontal cortex of schizophrenia and bipolar subjects; altogether approximately 100 loci were identified that showed, in comparison to controls, significant methylation changes in the schizophrenia and bipolar cohorts. These loci, taken together, comprised a highly heterogeneous group of a diverse set of genes, which were thought to regulate glutamatergic and GABAergic neurotransmission and neurodevelopment (Mill et al. 2008). That schizophrenia and bipolar disorder showed methylation changes of similar magnitude and direction for a number of loci resonates with some of the clinical, genetic, and neurochemical features common to both disorders (Craddock and Owen 2007). Strikingly, the list of genes showing altered methylation in Mill et al. includes at least one locus, dysbindin, which confers genetic risk for psychosis (Straub et al. 2002; Raybould et al. 2005; Pae et al. 2007). This is interesting because evidence is emerging that for a number of risk loci, epigenetic modifications in brain chromatin are influenced by local sequence polymorphisms and other genetic variations (Polesskaya et al. 2006; Huang and Akbarian 2007).

These studies on the frontal lobe of schizophrenia patients indicate that the number of loci showing, in comparison to controls, DNA hyper- versus hypomethylation, was roughly equal, which makes it very unlikely that schizophrenia is associated with a generalized drift toward increased (or decreased) DNA methylation in brain. A similar conclusion was reached by an earlier study on the temporal cortex of schizophrenia, reporting no significant DNA methylation changes for the entire set of 50 genes examined (Siegmund et al. 2007).

Notably, for each of the DNA methylation studies reporting positive findings, the overall magnitude of the disease-related changes, in comparison to controls, were surprisingly subtle. For example, in the Mill et al. study, disease-related DNA methylation levels even for one of the most significantly changed genes (*WDR18*) were reported as 17 versus 25% (Mill et al. 2008). While it is difficult, even for single copy genes, to extrapolate the percentages of methylated DNA extracted from a tissue homogenate to the number of cell nuclei affected, the evidence so far would support the view that many of the DNA methylation changes associated with schizophrenia are likely to involve only a small subset of cells residing in the cerebral cortex.

To date, few studies have explored histone modification changes at sites of specific genes in schizophrenia, using either nucleated peripheral blood cells (Gavin et al. 2009) or postmortem brain tissue (Huang and Akbarian 2007; Huang et al. 2007; Mellios et al. 2008) as input material. For the latter, preference was given to the study of lysine methylation which apparently is less affected by tissue autolysis and other postmortem confounds (Akbarian et al. 2005; Huang et al. 2006). Like the methylation of CpG dinucleotides in genomic DNA, the methylation of lysine and arginine residues of the histone proteins plays an important role in the complex regulation of chromatin structure and function, whereby the specific position of the modified residue within the histone protein, and even the numbers of (methyl) groups added to the residue, reflect distinct chromatin states differentiating transcriptionally active ("open") and silenced or repressed loci (Kouzarides 2002; Berger 2007; Akbarian and Huang 2009). The side chain of lysine residues can carry up to three methyl groups, while arginines could be mono- or dimethylated. In particular, coding and noncoding sequences transcribed by the RNA polymerase II complex are typically defined by sharp peaks of trimethylated histone H3-lysine 4 (H3K4) in nucleosomes positioned around transcription start sites and a more broad distribution of trimethylated H3K36, or dimethylated H3K79, along the transcribed region (Klose et al. 2007; Seila et al. 2008; Guttman et al. 2009). Furthermore, trimethylated H3K9, H3K27, and H4K20 are associated with repressive chromatin, in striking contrast to monomethylated H3K9 and H4K20 which in human cells are associated with transcription (Barski et al. 2007).

Focusing on the distribution of trimethyl-H3K4 and -H3K27, two chromatin marks differentiating between open and repressive chromatin, respectively, we recently showed that some subjects with schizophrenia are affected by a shift from the H3K4me3 to the H3K27me3 mark in prefrontal chromatin surrounding the transcription start site of *GAD1*, encoding 67 kDa glutamic acid decarboxylase GABA synthesis enzyme (Huang et al. 2007). Of note, downregulation of GAD67 RNA affects widespread portions of the cerebral cortex, the hippocampus, and other areas of forebrain, and even the cerebellar cortex (Guidotti et al. 2000; Dracheva et al. 2004; Fatemi et al. 2005; Benes et al. 2008; Bullock et al. 2008; Hashimoto et al. 2008). Indeed, the shift from H3K4 to H3K27 methylation at the *GAD1* gene in the schizophrenia cases was accompanied by a deficit in GAD1(GAD67) RNA, in comparison to the matched control (Huang et al. 2007). This study thus illustrates again the present approach by the field, which is to study epigenetic markings of a gene of interest for which a change in RNA level had been described. Comparison of these two molecular phenotypes then potentially informs about a change in gene expression activity in the diseased tissue.

4 Reproducibility of Epigenetic Alterations in Schizophrenia Postmortem Brain

Given that, to date, only few studies explored chromatin modifications in schizophrenia postmortem brain, it is difficult to predict whether or not the "epigenetic approach" will contribute to a critical advancement of our knowledge about the mechanisms of disease. Obviously, independent replication or confirmation of reported alterations in diseased tissue is a prerequisite. However, as one might expect from an epigenetic marking, recent studies suggest that DNA methylation patterns show, on a genome-wide scale, considerable heterogeneity between individuals, with larger differences between dizygotic as compared to monozygotic twins (Kaminsky et al. 2009). Furthermore, the similarities in DNA methylation and histone acetylation patterns among monozygotic twin pairs tend to diminish with age (Fraga et al. 2005). Given this apparent variability in chromatin markings between individuals, which only increases during the course of aging, it comes as no surprise that to date, the field still awaits independent replication for the reported alterations in schizophrenia postmortem studies. For example, DNA methylation analysis for the *REELIN* promoter (encoding a glycoprotein essential for orderly brain development and neuronal connectivity) either reported no change (Mill et al. 2008; Tochigi et al. 2008) in the cerebral cortex of schizophrenics or identified aberrant hypermethylation at different portions of the gene (Abdolmaleky et al. 2005a; Grayson et al. 2005). Likewise, DNA hypomethylation and increased gene expression for *COMT* (encoding a key enzyme for catecholamine metabolism) was reported in one study (Abdolmaleky et al. 2006) while another group reported no methylation changes for the same gene (Mill et al. 2008). Besides of interindividual differences, numerous factors could have contributed to these disparate findings. Importantly, the activity of DNA methyltransferases (the enzymes which methylate DNA) and DNA CpG methylation in a variety of tissues, including CNS, are sensitive to the social environment (Rampon et al. 2000; Weaver et al. 2004), ischemia (Endres et al. 2000), environmental toxins (Desaulniers et al. 2005; Bollati et al. 2007; Salnikow and Zhitkovich 2008), nicotine (Satta et al. 2007, 2008), alcohol (Marutha Ravindran and Ticku 2004), psychostimulants (Numachi et al. 2004, 2007), and antipsychotic drugs (Shimabukuro et al. 2006; Cheng et al. 2008; Mill et al. 2008). It is likely that these factors also play a role in shaping DNA methylation patterns in the human brain, thus further contributing to sample variability in postmortem studies. Given that, as mentioned earlier, the differences between schizophrenia and control subjects for most of the epigenetic markings studied so far are comparatively subtle, some caution in the interpretation of the postmortem data is required. From a skeptics' viewpoint, it is very well possible that many of disease-associated changes published so far will not be consistently replicated in subsequent postmortem studies, and additional work will be required to obtain a clearer picture on the role of epigenetic mechanisms in schizophrenia.

5 Cellular Specificity of Epigenetic Markings

To date, most assays designed to detect and quantify DNA methylation and histone modifications require an input material between 10^3 and 10^8 nuclei, and this lack of cellular resolution poses a challenge because brain tissue is comprised of an extremely heterogeneous mixture of different cell types. To date, many studies exploring epigenetic dysregulation of gene expression in schizophrenia utilized tissue homogenates for their DNA and histone modification assays, while the gene (s) of interest often are expressed only in a select subpopulation of neurons or other cells (Abdolmaleky et al. 2005a, 2006; Grayson et al. 2005; Iwamoto et al. 2005; Huang et al. 2007; Siegmund et al. 2007; Tamura et al. 2007; Mill et al. 2008; Tochigi et al. 2008). Some of these changes, such as the hypermethylation of the *REELIN* promoter DNA in cortical layer I of subjects with schizophrenia (Veldic et al. 2004, 2005; Grayson et al. 2005), or the shift from open chromatin-associated (H3-trimethyl-K4) to repressive histone methylation (H3-trimethyl-K27) at the *GAD1* gene promoter could indicate an epigenetic defect in the population of inhibitory interneurons (Huang et al. 2007). This is a very attractive working hypothesis that could potentially explain the observed deficits in the corresponding REELIN and GAD1(GAD67) RNAs (Guidotti et al. 2009). A more conclusive test of this hypothesis will await the development of technologies that allow the efficient sorting of GABAergic neuron chromatin directly from postmortem tissue. To this end, some progress has been made. For example, it is possible to purify, immunotag (with anti-NeuN antibody), and sort efficiently 10^7–10^8 neuronal and nonneuronal nuclei from less than 1 g of postmortem cerebral cortex in a single day, thereby enabling separate processing of neuronal and nonneuronal chromatin (Jiang et al. 2008b; Matevossian and Akbarian 2008). These approaches should be, in principle, also applicable to selected subpopulations of neurons and other cell types. As an alternative, it could be feasible to sort chromatin from tissue homogenates with a set of anti-methyl-histone antibodies that differentiate between open chromatin at sites of actual or potential gene expression, and repressive chromatin, and then assay DNA methylation and other types of modifications separately for these different fractions. This approach was used to show that DNA methylation alterations at the *GAD1* promoter in schizophrenia selectively affect the fraction of repressive chromatin (Huang and Akbarian 2007). However, until the cellular resolution of chromatin immunoprecipitation, DNA methylation, and other "epigenetic" assays becomes critically refined, the study of epigenetic markings in GABAergic and other genes in human (or, for that matter, animal) brain tissue homogenates will face the challenge to relate the readout from these experiments back to the subpopulation of cells for which these changes would be most relevant.

6 Epigenetic Markings in Brain: How Stable?

The rationale to explore certain types of epigenetic modifications in postmortem brain of subjects with schizophrenia is, as mentioned earlier, primarily driven by hypotheses that try to associate a change in RNA levels to alterations in DNA or histone modifications at the corresponding gene (promoter). Often it is implied, explicitly or implicitly, that any changes in RNA levels observed in postmortem tissue reflect changes in steady-state RNA levels, and that any changes in DNA or histone modifications reflect a stable and long-lasting epigenetic alteration in the brain of subjects with schizophrenia. Little is known, however, about the stability, or dynamic turnover, of epigenetic markings in human brain. Cell culture studies suggest that DNA methylation undergoes bidirectional changes – on the scale of minutes to hours – in response to DNA methyltransferase inhibitors (Levenson et al. 2006; Kundakovic et al. 2007) and depolarization (Martinowich et al. 2003; Nelson et al. 2008). In addition, molecular machineries that mediate demethylation of CpG dinucleotides or of histone lysine residues are now well characterized, notwithstanding the fact that DNA and histone methylation were not long ago considered to be potentially irreversible (Klose and Zhang 2007; Ooi and Bestor 2008). Indeed, studies in rodent hippocampus suggest that neuronal activity drives the expression of Gadd45b and other molecules regulating DNA methylation (Ma et al. 2009). Furthermore, learning and memory is associated with highly dynamic DNA methylation changes at *PP1*, *REELIN*, and gene promoters regulating synaptic plasticity (Ma et al. 2009). Therefore, one could assume that at least some of the epigenetic alterations reported in schizophrenia postmortem brain are not necessarily stable imprints that exist for very long periods of time, and instead, could potentially reflect a pathophysiological mechanism that operates on a much shorter time scale, from hours to minutes. On the other hand, a recent study monitoring DNA methylation levels at 50 gene loci across the lifespan of the human cerebral cortex reported for a subset of genes highly dynamic changes, even exponential increases during late pre- or early postnatal development, but then levels remained nearly unchanged during the course of subsequent maturation and aging (Siegmund et al. 2007). Another set of genes showed a steady, almost linear increase in DNA methylation levels with increasing age (Siegmund et al. 2007). Taken all these findings together, it is likely that the precise dynamics of DNA methylation changes during normal development and aging are variable, depending on the specific locus and other factors including cell type.

7 Implications for the Neurobiology of Schizophrenia

As discussed earlier, the study of DNA and histone modifications at defined promoter sequences in schizophrenia postmortem brain is attractive because this approach could potentially uncover epigenetic mechanisms of dysregulated gene expression in that disorder. However, presently it remains an open question whether some of the aforementioned disease-related promoter DNA and histone methylation changes reflect indeed long-lasting and sustained defects in the regulation of gene expression.

Notably, the reported histone methylation changes in prefrontal GAD1 chromatin of schizophrenia subjects (Huang et al. 2007) are also of interest from the viewpoint of neurodevelopment. There is evidence that H3K4 methylation is progressively upregulated in human prefrontal cortex at a select set of GABAergic gene promoters (GAD1/GAD2/NPY/SST) during the transitions from the fetal period to childhood and then to adulthood (Huang et al. 2007). This extended course of chromatin regulation during the first two decades of life was paralleled by similar increases on the RNA level (Huang et al. 2007). Taken together, these findings imply that chromatin remodeling and transcriptional mechanisms involved in the prefrontal dysfunction of schizophrenia also play a role during normal development. Whether the GAD1 chromatin alterations in schizophrenia indeed reflect disordered neurodevelopment, or a type of pathophysiology occurring later in life, remains to be determined in future studies.

From a broader perspective, evidence is emerging that epigenetic markings are highly regulated particularly during late prenatal and early postnatal human brain development. For example, when DNA methylation changes were assayed at 50 gene promoters across the lifespan in human temporal cortex, more than one half of this select set of genes showed age-related methylation changes, which were particularly pronounced during the transition from the fetal period and early childhood years to later stages of development (Siegmund et al. 2007). Likewise, maturation of cerebellar cortex was associated with histone methylation changes in chromatin surrounding glutamate receptor gene chromatin (Stadler et al. 2005). Future studies that map epigenetic markings on a comprehensive, genome-wide scale during normal human brain development will be necessary to find out whether or not these developmental changes are representative for a significant portion of the genome and whether or not schizophrenia-related risk loci and genes show age-related changes. Given the emerging link between childhood adversity, epigenetic dysregulation of gene expression, and adult psychopathology (McGowan et al. 2008, 2009), it is not unlikely that a neurodevelopmental approach will also advance our current understanding of chromatin-associated alterations in brains of subjects with schizophrenia.

8 Chromatin Remodeling and Antipsychotic Medication

As mentioned earlier, DNA methylation changes at selected gene loci have been linked to a wide range of drugs including nicotine (Satta et al. 2007, 2008), alcohol (Marutha Ravindran and Ticku 2004), and psychostimulants (Numachi et al. 2004, 2007). Interestingly, studies dating back in the 1960s and 1970s observed that subjects with schizophrenia treated with the amino acid methionine – which functions as a methyl donor in a wide range of biochemical pathways – frequently exhibited a worsening of their symptoms (Grayson et al. 2009). Based on preclinical studies on GABAergic gene promoters, it is possible that excessive intake of methionine may have led to aberrant DNA (and perhaps also histone) methylation events at selected loci in the CNS of the study participants, leading to transcriptional dysregulation and ultimately further worsening of psychosis (Grayson et al. 2009).

If the fine-tuning of DNA and histone methylation activities indeed is important from the viewpoint of the neurobiology of psychosis, then one could expect that these mechanisms are affected by antipsychotic drugs. Interestingly, one study reported that global DNA brain methylation levels showed a slight decrease in animals exposed to haloperidol (Shimabukuro et al. 2006), which is an interesting observation given that the amino acid and methyl donor, methionine, potentially worsens psychosis (Grayson et al. 2009). Furthermore, there is increasing evidence that changes the dopamine D_2-system and other receptors targeted by antipsychotics are coupled by various signaling pathways to chromatin remodeling, including DNA methylation and histone modification activities in the nucleus (Li et al. 2004; Dong et al. 2008; Costa et al. 2009). Furthermore, in the comprehensive study of Mill et al. (2008), DNA methylation levels in one out of approximately 8,000 promoters, mitogen-activated protein kinase I (*MEK1*), showed a significant (inverse) association with lifetime exposure to antipsychotic medication. Interestingly, phosphorylation of the MEK1 protein itself is also changed in animal brain exposed to the atypical antipsychotic Clozapine (Browning et al. 2005), which would indicate multiple layers by which certain antipsychotic drugs affect MEK1 expression activity. How MEK1 could be involved in the neurobiology of psychosis remains unclear, albeit the MEK1 and other components of the mitogen-activated protein kinase (MAPK) signaling regulate a plethora of mechanisms in the developing and mature nervous system (Rosen et al. 1994; Sweatt 2001; Bozon et al. 2003). Furthermore, there is evidence from animal models and even postmortem tissue that expression and activity of the histone H3K4-specific methyltransferase, mixed-lineage leukemia 1 (MLL1), is in human prefrontal cortex upregulated after exposure to Clozapine (Huang and Akbarian 2007; Huang et al. 2007). This effect does not appear to be the result of a direct interaction between Clozapine and histone methyltransferase, because the drug's effect requires intact brain circuitry and is not observed in the cell culture preparation (Huang and Akbarian 2007; Huang et al. 2007).

9 Synopsis and Outlook

The study of epigenetic regulators of gene expression offers an attractive perspective to schizophrenia research, given that genetic factors remain difficult to identify in many affected cases. Postmortem brain studies, conducted on schizophrenia and matched control cohorts, explored primarily DNA and histone methylation changes, but so far no consensus has emerged regarding which, if any, genomic sequence(s) are frequently (let alone consistently) changed in brains of subjects with schizophrenia. The evidence so far indicates that epigenetic alterations observed in the brains of schizophrenia patients, when compared to controls, are typically subtle in nature. However, further optimization and development of chromatin and methylation assays will be necessary to reach a more definitive conclusion, because to date practically all schizophrenia-related studies were limited to tissue homogenates mostly from the frontal cortex. Many of the studies reporting positive findings show that DNA and histone methylation changes often are associated with altered expression of the corresponding gene transcripts. Therefore, these epigenetic regulators could contribute to dysregulated gene expression in schizophrenia. Following on the heels of these pioneering studies, future work, using so-called "next-generation" (or "massively parallel" or "deep") sequencing approaches, which process hundreds of thousands or millions of DNA templates in parallel (Kharchenko et al. 2008), is likely to provide a wealth of novel information about chromatin regulation. By applying these assays to defined cell populations extracted from human brain (Jiang et al. 2008b), we can expect to significantly advance our knowledge of the epigenetic landscape in schizophrenia on a genome-wide scale. Furthermore, it is likely that the field will go beyond the study of DNA and histone modifications, and explore the role of noncoding RNA and other molecules that emerge as key regulators of chromatin function and epigenetic memory (Mattick and Makunin 2006; Sessa et al. 2007). Therefore, it could be expected that the study of epigenetic mechanisms will move the frontiers in schizophrenia research, like it already has done in many other areas of medicine (Feinberg 2007; Szyf 2009).

Acknowledgments Research conducted by the author is supported by grants from the National Institutes of Health and the Staglin Family Music Festival Schizophrenia Research Award/NARSAD. The author reports no conflicts of interest.

References

Abdolmaleky HM, Cheng KH, Russo A, Smith CL, Faraone SV, Wilcox M, Shafa R, Glatt SJ, Nguyen G, Ponte JF, Thiagalingam S, Tsuang MT (2005a) Hypermethylation of the reelin (RELN) promoter in the brain of schizophrenic patients: a preliminary report. Am J Med Genet B Neuropsychiatr Genet 134:60–66

Abdolmaleky HM, Thiagalingam S, Wilcox M (2005b) Genetics and epigenetics in major psychiatric disorders: dilemmas, achievements, applications, and future scope. Am J Pharmacogenomics 5:149–160

Abdolmaleky HM, Cheng KH, Faraone SV, Wilcox M, Glatt SJ, Gao F, Smith CL, Shafa R, Aeali B, Carnevale J, Pan H, Papageorgis P, Ponte JF, Sivaraman V, Tsuang MT, Thiagalingam S (2006) Hypomethylation of MB-COMT promoter is a major risk factor for schizophrenia and bipolar disorder. Hum Mol Genet 15:3132–3145

Abel T, Zukin RS (2008) Epigenetic targets of HDAC inhibition in neurodegenerative and psychiatric disorders. Curr Opin Pharmacol 8:57–64

Akbarian S, Huang HS (2009) Epigenetic regulation in human brain-focus on histone lysine methylation. Biol Psychiatry 65:198–203

Akbarian S, Ruehl MG, Bliven E, Luiz LA, Peranelli AC, Baker SP, Roberts RC, Bunney WE Jr, Conley RC, Jones EG, Tamminga CA, Guo Y (2005) Chromatin alterations associated with down-regulated metabolic gene expression in the prefrontal cortex of subjects with schizophrenia. Arch Gen Psychiatry 62:829–840

Alarcon JM, Malleret G, Touzani K, Vronskaya S, Ishii S, Kandel ER, Barco A (2004) Chromatin acetylation, memory, and LTP are impaired in CBP+/− mice: a model for the cognitive deficit in Rubinstein–Taybi syndrome and its amelioration. Neuron 42:947–959

Barski A, Cuddapah S, Cui K, Roh TY, Schones DE, Wang Z, Wei G, Chepelev I, Zhao K (2007) High-resolution profiling of histone methylations in the human genome. Cell 129:823–837

Benes FM, Lim B, Matzilevich D, Subburaju S, Walsh JP (2008) Circuitry-based gene expression profiles in GABA cells of the trisynaptic pathway in schizophrenics versus bipolars. Proc Natl Acad Sci USA 105:20935–20940

Berger SL (2002) Histone modifications in transcriptional regulation. Curr Opin Genet Dev 12:142–148

Berger SL (2007) The complex language of chromatin regulation during transcription. Nature 447:407–412

Berger SL, Kouzarides T, Shiekhattar R, Shilatifard A (2009) An operational definition of epigenetics. Genes Dev 23:781–783

Bertolino A, Blasi G (2009) The genetics of Schizophrenia. Neuroscience 164:288–299

Bollati V, Baccarelli A, Hou L, Bonzini M, Fustinoni S, Cavallo D, Byun HM, Jiang J, Marinelli B, Pesatori AC, Bertazzi PA, Yang AS (2007) Changes in DNA methylation patterns in subjects exposed to low-dose benzene. Cancer Res 67:876–880

Bozon B, Kelly A, Josselyn SA, Silva AJ, Davis S, Laroche S (2003) MAPK, CREB and zif268 are all required for the consolidation of recognition memory. Philos Trans R Soc Lond B Biol Sci 358:805–814

Browning JL, Patel T, Brandt PC, Young KA, Holcomb LA, Hicks PB (2005) Clozapine and the mitogen-activated protein kinase signal transduction pathway: implications for antipsychotic actions. Biol Psychiatry 57:617–623

Bullock WM, Cardon K, Bustillo J, Roberts RC, Perrone-Bizzozero NI (2008) Altered expression of genes involved in GABAergic transmission and neuromodulation of granule cell activity in the cerebellum of schizophrenia patients. Am J Psychiatry 165:1594–1603

Cheng MC, Liao DL, Hsiung CA, Chen CY, Liao YC, Chen CH (2008) Chronic treatment with aripiprazole induces differential gene expression in the rat frontal cortex. Int J Neuropsychopharmacol 11:207–216

Colvis CM, Pollock JD, Goodman RH, Impey S, Dunn J, Mandel G, Champagne FA, Mayford M, Korzus E, Kumar A, Renthal W, Theobald DE, Nestler EJ (2005) Epigenetic mechanisms and gene networks in the nervous system. J Neurosci 25:10379–10389

Connor CM, Akbarian S (2008) DNA methylation changes in schizophrenia and bipolar disorder. Epigenetics 3:55–58

Costa E, Chen Y, Dong E, Grayson DR, Kundakovic M, Maloku E, Ruzicka W, Satta R, Veldic M, Zhubi A, Guidotti A (2009) GABAergic promoter hypermethylation as a model to study the neurochemistry of schizophrenia vulnerability. Expert Rev Neurother 9:87–98

Craddock N, Owen MJ (2007) Rethinking psychosis: the disadvantages of a dichotomous classification now outweigh the advantages. World Psychiatry 6:84–91

Davis KL, Haroutunian V (2003) Global expression-profiling studies and oligodendrocyte dysfunction in schizophrenia and bipolar disorder. Lancet 362:758

Delisi LE (2009) Searching for the true genetic vulnerability for schizophrenia. Genome Med 1:14

Desaulniers D, Xiao GH, Leingartner K, Chu I, Musicki B, Tsang BK (2005) Comparisons of brain, uterus, and liver mRNA expression for cytochrome p450s, DNA methyltransferase-1, and catechol-*o*-methyltransferase in prepubertal female Sprague–Dawley rats exposed to a mixture of aryl hydrocarbon receptor agonists. Toxicol Sci 86:175–184

Dong E, Nelson M, Grayson DR, Costa E, Guidotti A (2008) Clozapine and sulpiride but not haloperidol or olanzapine activate brain DNA demethylation. Proc Natl Acad Sci USA 105:13614–13619

Dracheva S, Elhakem SL, McGurk SR, Davis KL, Haroutunian V (2004) GAD67 and GAD65 mRNA and protein expression in cerebrocortical regions of elderly patients with schizophrenia. J Neurosci Res 76:581–592

Endres M, Meisel A, Biniszkiewicz D, Namura S, Prass K, Ruscher K, Lipski A, Jaenisch R, Moskowitz MA, Dirnagl U (2000) DNA methyltransferase contributes to delayed ischemic brain injury. J Neurosci 20:3175–3181

Fatemi SH (2005) Reelin glycoprotein in autism and schizophrenia. Int Rev Neurobiol 71:179–187

Fatemi SH, Stary JM, Earle JA, Araghi-Niknam M, Eagan E (2005) GABAergic dysfunction in schizophrenia and mood disorders as reflected by decreased levels of glutamic acid decarboxylase 65 and 67 kDa and Reelin proteins in cerebellum. Schizophr Res 72:109–122

Feinberg AP (2007) Phenotypic plasticity and the epigenetics of human disease. Nature 447:433–440

Felsenfeld G, Groudine M (2003) Controlling the double helix. Nature 421:448–453

Fischle W, Wang Y, Allis CD (2003) Histone and chromatin cross-talk. Curr Opin Cell Biol 15:172–183

Fraga MF, Ballestar E, Paz MF, Ropero S, Setien F, Ballestar ML, Heine-Suner D, Cigudosa JC, Urioste M, Benitez J, Boix-Chornet M, Sanchez-Aguilera A, Ling C, Carlsson E, Poulsen P, Vaag A, Stephan Z, Spector TD, Wu YZ, Plass C, Esteller M (2005) Epigenetic differences arise during the lifetime of monozygotic twins. Proc Natl Acad Sci USA 102:10604–10609

Fuks F (2005) DNA methylation and histone modifications: teaming up to silence genes. Curr Opin Genet Dev 15:490–495

Gavin DP, Kartan S, Chase K, Jayaraman S, Sharma RP (2009) Histone deacetylase inhibitors and candidate gene expression: an *in vivo* and *in vitro* approach to studying chromatin remodeling in a clinical population. J Psychiatr Res 43:870–876

Grayson DR, Jia X, Chen Y, Sharma RP, Mitchell CP, Guidotti A, Costa E (2005) Reelin promoter hypermethylation in schizophrenia. Proc Natl Acad Sci USA 102:9341–9346

Grayson DR, Chen Y, Dong E, Kundakovic M, Guidotti A (2009) From trans-methylation to cytosine methylation: evolution of the methylation hypothesis of schizophrenia. Epigenetics 4:144–149

Guan Z, Giustetto M, Lomvardas S, Kim JH, Miniaci MC, Schwartz JH, Thanos D, Kandel ER (2002) Integration of long-term-memory-related synaptic plasticity involves bidirectional regulation of gene expression and chromatin structure. Cell 111:483–493

Guidotti A, Auta J, Davis JM, Di-Giorgi-Gerevini V, Dwivedi Y, Grayson DR, Impagnatiello F, Pandey G, Pesold C, Sharma R, Uzunov D, Costa E (2000) Decrease in reelin and glutamic acid decarboxylase67 (GAD67) expression in schizophrenia and bipolar disorder: a postmortem brain study. Arch Gen Psychiatry 57:1061–1069

Guidotti A, Dong E, Kundakovic M, Satta R, Grayson DR, Costa E (2009) Characterization of the action of antipsychotic subtypes on valproate-induced chromatin remodeling. Trends Pharmacol Sci 30:55–60

Guttman M, Amit I, Garber M, French C, Lin MF, Feldser D, Huarte M, Zuk O, Carey BW, Cassady JP, Cabili MN, Jaenisch R, Mikkelsen TS, Jacks T, Hacohen N, Bernstein BE, Kellis M, Regev A, Rinn JL, Lander ES (2009) Chromatin signature reveals over a thousand highly conserved large non-coding RNAs in mammals. Nature 458:223–227

Hajkova P, Erhardt S, Lane N, Haaf T, El-Maarri O, Reik W, Walter J, Surani MA (2002) Epigenetic reprogramming in mouse primordial germ cells. Mech Dev 117:15–23

Hashimoto T, Bazmi HH, Mirnics K, Wu Q, Sampson AR, Lewis DA (2008) Conserved regional patterns of GABA-related transcript expression in the neocortex of subjects with schizophrenia. Am J Psychiatry 165:479–489
Hayes JJ, Hansen JC (2001) Nucleosomes and the chromatin fiber. Curr Opin Genet Dev 11:124–129
Hof PR, Haroutunian V, Copland C, Davis KL, Buxbaum JD (2002) Molecular and cellular evidence for an oligodendrocyte abnormality in schizophrenia. Neurochem Res 27:1193–1200
Huang HS, Akbarian S (2007) GAD1 mRNA expression and DNA methylation in prefrontal cortex of subjects with schizophrenia. PLoS ONE 2:e809
Huang HS, Matevossian A, Jiang Y, Akbarian S (2006) Chromatin immunoprecipitation in postmortem brain. J Neurosci Methods 156:284–292
Huang HS, Matevossian A, Whittle C, Kim SY, Schumacher A, Baker SP, Akbarian S (2007) Prefrontal dysfunction in schizophrenia involves mixed-lineage leukemia 1-regulated histone methylation at GABAergic gene promoters. J Neurosci 27:11254–11262
Iizuka M, Smith MM (2003) Functional consequences of histone modifications. Curr Opin Genet Dev 13:154–160
Iwamoto K, Bundo M, Yamada K, Takao H, Iwayama-Shigeno Y, Yoshikawa T, Kato T (2005) DNA methylation status of SOX10 correlates with its downregulation and oligodendrocyte dysfunction in schizophrenia. J Neurosci 25:5376–5381
Jablonka E, Lamb MJ (2002) The changing concept of epigenetics. Ann NY Acad Sci 981:82–96
Jenuwein T, Allis CD (2001) Translating the histone code. Science 293:1074–1080
Jiang Y, Langley B, Lubin FD, Renthal W, Wood MA, Yasui DH, Kumar A, Nestler EJ, Akbarian S, Beckel-Mitchener AC (2008a) Epigenetics in the nervous system. J Neurosci 28:11753–11759
Jiang Y, Matevossian A, Huang HS, Straubhaar J, Akbarian S (2008b) Isolation of neuronal chromatin from brain tissue. BMC Neurosci 9:42
Kaminsky Z, Wang SC, Petronis A (2006) Complex disease, gender and epigenetics. Ann Med 38:530–544
Kaminsky ZA, Tang T, Wang SC, Ptak C, Oh GH, Wong AH, Feldcamp LA, Virtanen C, Halfvarson J, Tysk C, McRae AF, Visscher PM, Montgomery GW, Gottesman II, Martin NG, Petronis A (2009) DNA methylation profiles in monozygotic and dizygotic twins. Nat Genet 41:240–245
Kharchenko PV, Tolstorukov MY, Park PJ (2008) Design and analysis of ChIP-seq experiments for DNA-binding proteins. Nat Biotechnol 26:1351–1359
Klose RJ, Bird AP (2006) Genomic DNA methylation: the mark and its mediators. Trends Biochem Sci 31:89–97
Klose RJ, Zhang Y (2007) Regulation of histone methylation by demethylimination and demethylation. Nat Rev Mol Cell Biol 8:307–318
Klose RJ, Gardner KE, Liang G, Erdjument-Bromage H, Tempst P, Zhang Y (2007) Demethylation of histone H3K36 and H3K9 by Rph1: a vestige of an H3K9 methylation system in *Saccharomyces cerevisiae*? Mol Cell Biol 27:3951–3961
Korzus E, Rosenfeld MG, Mayford M (2004) CBP histone acetyltransferase activity is a critical component of memory consolidation. Neuron 42:961–972
Kouzarides T (2002) Histone methylation in transcriptional control. Curr Opin Genet Dev 12:198–209
Kristiansen LV, Huerta I, Beneyto M, Meador-Woodruff JH (2007) NMDA receptors and schizophrenia. Curr Opin Pharmacol 7:48–55
Kumar A, Choi KH, Renthal W, Tsankova NM, Theobald DE, Truong HT, Russo SJ, Laplant Q, Sasaki TS, Whistler KN, Neve RL, Self DW, Nestler EJ (2005) Chromatin remodeling is a key mechanism underlying cocaine-induced plasticity in striatum. Neuron 48:303–314
Kundakovic M, Chen Y, Costa E, Grayson DR (2007) DNA methyltransferase inhibitors coordinately induce expression of the human reelin and glutamic acid decarboxylase 67 genes. Mol Pharmacol 71:644–653

Levenson JM, Sweatt JD (2005) Epigenetic mechanisms in memory formation. Nat Rev Neurosci 6:108–118

Levenson JM, Roth TL, Lubin FD, Miller CA, Huang IC, Desai P, Malone LM, Sweatt JD (2006) Evidence that DNA (cytosine-5) methyltransferase regulates synaptic plasticity in the hippocampus. J Biol Chem 281:15763–15773

Levine AA, Guan Z, Barco A, Xu S, Kandel ER, Schwartz JH (2005) CREB-binding protein controls response to cocaine by acetylating histones at the fosB promoter in the mouse striatum. Proc Natl Acad Sci USA 102:19186–19191

Li J, Guo Y, Schroeder FA, Youngs RM, Schmidt TW, Ferris C, Konradi C, Akbarian S (2004) Dopamine D2-like antagonists induce chromatin remodeling in striatal neurons through cyclic AMP-protein kinase A and NMDA receptor signaling. J Neurochem 90:1117–1131

Lupski JR (2008) Schizophrenia: incriminating genomic evidence. Nature 455:178–179

Ma DK, Jang MH, Guo JU, Kitabatake Y, Chang ML, Pow-Anpongkul N, Flavell RA, Lu B, Ming GL, Song H (2009) Neuronal activity-induced Gadd45b promotes epigenetic DNA demethylation and adult neurogenesis. Science 323:1074–1077

Martin KC, Sun YE (2004) To learn better, keep the HAT on. Neuron 42:879–881

Martinowich K, Hattori D, Wu H, Fouse S, He F, Hu Y, Fan G, Sun YE (2003) DNA methylation-related chromatin remodeling in activity-dependent BDNF gene regulation. Science 302:890–893

Marutha Ravindran CR, Ticku MK (2004) Changes in methylation pattern of NMDA receptor NR2B gene in cortical neurons after chronic ethanol treatment in mice. Brain Res Mol Brain Res 121:19–27

Matevossian A, Akbarian S (2008) Neuronal nuclei isolation from human postmortem brain tissue. J Vis Exp 20:pii: 914. doi: 10.3791/914

Mattick JS, Makunin IV (2006) Non-coding RNA. Hum Mol Genet 15(Spec No. 1):R17–R29

McGowan PO, Sasaki A, Huang TC, Unterberger A, Suderman M, Ernst C, Meaney MJ, Turecki G, Szyf M (2008) Promoter-wide hypermethylation of the ribosomal RNA gene promoter in the suicide brain. PLoS ONE 3:e2085

McGowan PO, Sasaki A, D'Alessio AC, Dymov S, Labonte B, Szyf M, Turecki G, Meaney MJ (2009) Epigenetic regulation of the glucocorticoid receptor in human brain associates with childhood abuse. Nat Neurosci 12:342–348

Mellios N, Huang HS, Baker SP, Galdzicka M, Ginns E, Akbarian S (2008) Molecular determinants of dysregulated GABAergic gene expression in the prefrontal cortex of subjects with schizophrenia. Biol Psychiatry 65:1006–1014

Mill J, Tang T, Kaminsky Z, Khare T, Yazdanpanah S, Bouchard L, Jia P, Assadzadeh A, Flanagan J, Schumacher A, Wang SC, Petronis A (2008) Epigenomic profiling reveals DNA-methylation changes associated with major psychosis. Am J Hum Genet 82:696–711

Monteggia LM, Kavalali ET (2009) Rett syndrome and the impact of MeCP2 associated transcriptional mechanisms on neurotransmission. Biol Psychiatry 65:204–210

Morgan HD, Santos F, Green K, Dean W, Reik W (2005) Epigenetic reprogramming in mammals. Hum Mol Genet 14(Spec No 1):R47–R58

Nelson ED, Kavalali ET, Monteggia LM (2008) Activity-dependent suppression of miniature neurotransmission through the regulation of DNA methylation. J Neurosci 28:395–406

Numachi Y, Yoshida S, Yamashita M, Fujiyama K, Naka M, Matsuoka H, Sato M, Sora I (2004) Psychostimulant alters expression of DNA methyltransferase mRNA in the rat brain. Ann NY Acad Sci 1025:102–109

Numachi Y, Shen H, Yoshida S, Fujiyama K, Toda S, Matsuoka H, Sora I, Sato M (2007) Methamphetamine alters expression of DNA methyltransferase 1 mRNA in rat brain. Neurosci Lett 414:213–217

Ooi SK, Bestor TH (2008) The colorful history of active DNA demethylation. Cell 133:1145–1148

Pae CU, Serretti A, Mandelli L, Yu HS, Patkar AA, Lee CU, Lee SJ, Jun TY, Lee C, Paik IH, Kim JJ (2007) Effect of 5-haplotype of dysbindin gene (DTNBP1) polymorphisms for the susceptibility to bipolar I disorder. Am J Med Genet B Neuropsychiatr Genet 144:701–703

Payer B, Lee JT (2008) X chromosome dosage compensation: how mammals keep the balance. Annu Rev Genet 42:733–732
Peterson CL (2002) Chromatin remodeling: nucleosomes bulging at the seams. Curr Biol 12: R245–R247
Peterson CL, Laniel MA (2004) Histones and histone modifications. Curr Biol 14:R546–R551
Polesskaya OO, Aston C, Sokolov BP (2006) Allele C-specific methylation of the 5-HT2A receptor gene: evidence for correlation with its expression and expression of DNA methylase DNMT1. J Neurosci Res 83:362–373
Rampon C, Jiang CH, Dong H, Tang YP, Lockhart DJ, Schultz PG, Tsien JZ, Hu Y (2000) Effects of environmental enrichment on gene expression in the brain. Proc Natl Acad Sci USA 97:12880–12884
Raybould R, Green EK, MacGregor S, Gordon-Smith K, Heron J, Hyde S, Caesar S, Nikolov I, Williams N, Jones L, O'Donovan MC, Owen MJ, Jones I, Kirov G, Craddock N (2005) Bipolar disorder and polymorphisms in the dysbindin gene (DTNBP1). Biol Psychiatry 57:696–701
Rosen LB, Ginty DD, Weber MJ, Greenberg ME (1994) Membrane depolarization and calcium influx stimulate MEK and MAP kinase via activation of Ras. Neuron 12:1207–1221
Salnikow K, Zhitkovich A (2008) Genetic and epigenetic mechanisms in metal carcinogenesis and cocarcinogenesis: nickel, arsenic, and chromium. Chem Res Toxicol 21:28–44
Satta R, Maloku E, Costa E, Guidotti A (2007) Stimulation of brain nicotinic acetylcholine receptors (nAChRs) decreases DNA methyltransferase 1 (DNMT1) expression in cortical and hippocampal GABAergic neurons of Swiss albino mice. Soc Neurosci Abstr no. 60.7
Satta R, Maloku E, Zhubi A, Pibiri F, Hajos M, Costa E, Guidotti A (2008) Nicotine decreases DNA methyltransferase 1 expression and glutamic acid decarboxylase 67 promoter methylation in GABAergic interneurons. Proc Natl Acad Sci USA 105:16356–16361
Seila AC, Calabrese JM, Levine SS, Yeo GW, Rahl PB, Flynn RA, Young RA, Sharp PA (2008) Divergent transcription from active promoters. Science 322:1849–1851
Sessa L, Breiling A, Lavorgna G, Silvestri L, Casari G, Orlando V (2007) Noncoding RNA synthesis and loss of polycomb group repression accompanies the colinear activation of the human HOXA cluster. RNA 13:223–239
Shimabukuro M, Jinno Y, Fuke C, Okazaki Y (2006) Haloperidol treatment induces tissue- and sex-specific changes in DNA methylation: a control study using rats. Behav Brain Funct 2:37
Siegmund KD, Connor CM, Campan M, Long TI, Weisenberger DJ, Biniszkiewicz D, Jaenisch R, Laird PW, Akbarian S (2007) DNA methylation in the human cerebral cortex is dynamically regulated throughout the life span and involves differentiated neurons. PLoS ONE 2:e895
Smith M (2009) The year in human and medical genetics. Highlights of 2007–2008. Ann NY Acad Sci 1151:1–21
Stadler F, Kolb G, Rubusch L, Baker SP, Jones EG, Akbarian S (2005) Histone methylation at gene promoters is associated with developmental regulation and region-specific expression of ionotropic and metabotropic glutamate receptors in human brain. J Neurochem 94:324–336
Straub RE, Jiang Y, MacLean CJ, Ma Y, Webb BT, Myakishev MV, Harris-Kerr C, Wormley B, Sadek H, Kadambi B, Cesare AJ, Gibberman A, Wang X, O'Neill FA, Walsh D, Kendler KS (2002) Genetic variation in the 6p22.3 gene DTNBP1, the human ortholog of the mouse dysbindin gene, is associated with schizophrenia. Am J Hum Genet 71:337–348
Suzuki MM, Bird A (2008) DNA methylation landscapes: provocative insights from epigenomics. Nat Rev Genet 9:465–476
Swank MW, Sweatt JD (2001) Increased histone acetyltransferase and lysine acetyltransferase activity and biphasic activation of the ERK/RSK cascade in insular cortex during novel taste learning. J Neurosci 21:3383–3391
Sweatt JD (2001) The neuronal MAP kinase cascade: a biochemical signal integration system subserving synaptic plasticity and memory. J Neurochem 76:1–10
Sweatt JD (2009) Experience-dependent epigenetic modifications in the central nervous system. Biol Psychiatry 65:191–197

Szyf M (2009) Epigenetics, DNA methylation, and chromatin modifying drugs. Annu Rev Pharmacol Toxicol 49:243–263

Tamura Y, Kunugi H, Ohashi J, Hohjoh H (2007) Epigenetic aberration of the human REELIN gene in psychiatric disorders. Mol Psychiatry 12:593–600

The International Schizophrenia Consortium (2008) Rare chromosomal deletions and duplications increase risk of schizophrenia. Nature 455:237–241

Tochigi M, Iwamoto K, Bundo M, Komori A, Sasaki T, Kato N, Kato T (2008) Methylation status of the reelin promoter region in the brain of schizophrenic patients. Biol Psychiatry 63:530–533

Tsankova NM, Kumar A, Nestler EJ (2004) Histone modifications at gene promoter regions in rat hippocampus after acute and chronic electroconvulsive seizures. J Neurosci 24:5603–5610

Tsankova NM, Berton O, Renthal W, Kumar A, Neve RL, Nestler EJ (2006) Sustained hippocampal chromatin regulation in a mouse model of depression and antidepressant action. Nat Neurosci 9:519–525

Tsankova N, Renthal W, Kumar A, Nestler EJ (2007) Epigenetic regulation in psychiatric disorders. Nat Rev Neurosci 8:355–367

Veldic M, Caruncho HJ, Liu WS, Davis J, Satta R, Grayson DR, Guidotti A, Costa E (2004) DNA-methyltransferase 1 mRNA is selectively overexpressed in telencephalic GABAergic interneurons of schizophrenia brains. Proc Natl Acad Sci USA 101:348–353

Veldic M, Guidotti A, Maloku E, Davis JM, Costa E (2005) In psychosis, cortical interneurons overexpress DNA-methyltransferase 1. Proc Natl Acad Sci USA 102:2152–2157

Waggoner D (2007) Mechanisms of disease: epigenesis. Semin Pediatr Neurol 14:7–14

Wang Y, Fischle W, Cheung W, Jacobs S, Khorasanizadeh S, Allis CD (2004) Beyond the double helix: writing and reading the histone code. Novartis Found Symp 259:3–17, discussion 17–21, 163–169

Weaver IC, Cervoni N, Champagne FA, D'Alessio AC, Sharma S, Seckl JR, Dymov S, Szyf M, Meaney MJ (2004) Epigenetic programming by maternal behavior. Nat Neurosci 7:847–854

Wolffe AP (1992) New insights into chromatin function in transcriptional control. FASEB J 6:3354–3361

Molecules, Signaling, and Schizophrenia

Mirna Kvajo, Heather McKellar, and Joseph A. Gogos

Contents

1	Introduction	630
2	*AKT1* Gene and the AKT/GSK3β Signaling Pathway	631
	2.1 Neuronal Plasticity	632
	2.2 Dopamine Signaling	633
	2.3 GSK3β Signaling	634
3	*PPP3CC* Gene and the Calcineurin Signaling Pathway	635
	3.1 Dopamine Signaling	636
	3.2 Transcriptional Regulation	637
	3.3 Synaptic Plasticity and Neurotransmission	637
	3.4 Nitric Oxide Signaling	638
	3.5 Vesicle Trafficking	638
	3.6 Cytoskeletal Phosphorylation	639
4	*DISC1* Gene and the cAMP and GSK3/Wnt Signaling Pathways	639
	4.1 Disc1 and cAMP Signaling	641
	4.2 Disc1 and GSK3β Signaling	646
5	General Summary	646
References		648

J.A. Gogos (✉)
Department of Physiology and Cellular Biophysics, College of Physicians and Surgeons, Columbia University, New York, NY 10032, USA
e-mail: jag90@columbia.edu
Department of Neuroscience, Columbia University, New York, NY 10032, USA

M. Kvajo
Department of Physiology and Cellular Biophysics, College of Physicians and Surgeons, Columbia University, New York, NY 10032, USA
Department of Psychiatry, College of Physicians and Surgeons, Columbia University, New York, NY 10032, USA

H. McKellar
Integrated Program in Cellular, Molecular, and Biophysical Studies, Columbia University Medical Center, New York, NY 10032, USA

Abstract Schizophrenia is one of the most common psychiatric disorders, but despite some progress in identifying the genetic factors implicated in its development, the molecular mechanisms underlying its etiology and pathogenesis remain poorly understood. However, accumulating evidence suggests that regardless of the underlying genetic complexity, the mechanisms of the disease may impact a small number of common signaling pathways. In this review, we discuss the evidence for a role of schizophrenia susceptibility genes in intracellular signaling cascades by focusing on three prominent candidate genes: *AKT*, *PPP3CC* (calcineurin), and *DISC1*. We describe the regulation of a number of signaling cascades by AKT and calcineurin through protein phosphorylation and dephosphorylation, and the recently uncovered functions of DISC1 in cAMP and GSK3β signaling. In addition, we present independent evidence for the involvement of their downstream signaling pathways in schizophrenia. Finally, we discuss evidence supporting an impact of these susceptibility genes on common intracellular signaling pathways and the convergence of their effects on neuronal processes implicated in schizophrenia.

Keywords AKT · Calcineurin · Candidate genes · DISC1 · Schizophrenia · Signaling pathways

1 Introduction

Schizophrenia represents one of the most common and devastating psychiatric disorders but the molecular mechanisms underlying its etiology and pathogenesis remain to be fully elucidated. The complexity of schizophrenia is such that no consistent physiological correlates of the disease symptoms have yet been identified, and the mechanisms behind the therapeutic effects of most psychotropic drugs are still shrouded in mystery. Despite the large number of candidate genes that have been so far implicated in schizophrenia, our understanding of their specific contributions to the underlying molecular and cellular abnormalities is still in its infancy, even for the few most robust of these genetic findings. However, regardless of this genetic complexity, accumulating evidence suggests that the underlying molecular mechanisms may converge into relatively few distinct intracellular signaling pathways. Indeed, uncovering the contributions of common molecular targets known to regulate multiple signaling cascades is an exceptionally attractive possibility and particularly relevant for drug development efforts.

Evidence for a role of aberrant intracellular signaling in the etiology and pathogenesis of schizophrenia comes from multiple lines of research. As discussed in detail later, changes in levels and activity of key components of signaling cascades have been noted early on in patients with schizophrenia. Furthermore, pharmacological studies showed that psychotropic medications directly target key components of signaling cascades and have immediate and long-term effects on these pathways. Even more importantly, the very nature of intracellular signaling provides a mechanism by which even minor variations in signaling can have large and encompassing effects on a wide range of complex functions. Indeed, our current understanding suggests that different aspects of pathogenesis may be caused by simultaneous dysregulation of multiple pathways and processes, including neuronal development and adult brain plasticity. This concept was validated by recent studies utilizing a combination of biochemical and behavioral approaches in mouse models, showing that mutations in signaling components can, through multiple mechanisms, exert major effects on complex neuronal functions as well as behavior.

In this review we discuss the roles of intracellular signaling pathways in the pathophysiology of psychiatric illnesses by showcasing the contributions of three prominent schizophrenia candidate genes: *AKT1*, *PPP3CC*, and *DISC1*. AKT and calcineurin modulate signaling pathways through their well-established roles in phosphorylation and dephosphorylation, while DISC1 has been recently implicated in the control of cAMP-dependent signaling and GSK3β activation. As described in the pertinent sections, despite their distinct molecular functions the downstream effects of these genes converge to common signaling pathways that regulate essential cellular processes. Thus, they not only affect common processes such as neuronal development and survival, synaptic plasticity, and memory and learning, but also often do so by contributing to the same signaling cascades.

Overall, the dynamic regulation of complex, interacting signaling pathways may be an important underlying factor contributing to the development of psychiatric disorders, and genes which are uniquely positioned to control such signaling may represent promising targets for drug development strategies.

2 *AKT1* Gene and the AKT/GSK3β Signaling Pathway

AKT, also known as protein kinase B (PKB), is a key regulator of signal transduction processes mediated by protein phosphorylation and a central molecule in cell signaling downstream of growth factors, cytokines, and other external stimuli (Brazil et al. 2004). AKT is a serine/threonine kinase activated by 3′-phospholipids, the products of phosphoinositide 3-kinase (PI3K) activity. Several extracellular and intracellular factors that act through receptor tyrosine kinases or G protein-coupled receptors lead to enhanced PI3K activity and production of the membrane phospholipids which, in turn, recruit AKT from the cytosol to the cell membrane. This transition causes its activation through phosphorylation at multiple sites, most prominently at Thr308 and Ser473, allowing substrate binding and elevated rate of catalysis. Upon activation, AKT phosphorylates a variety of substrate proteins (Brazil et al. 2004), thereby controlling diverse biological responses such as programmed cell death, cell proliferation, migration, and metabolic processes.

Since the initial report in 2004 (Emamian et al. 2004), accumulating evidence suggests that impaired AKT signaling plays a role in the pathogenesis of schizophrenia. This study and several subsequent ones uncovered association between *AKT1* genetic variants and schizophrenia in case/control samples and family cohorts (an updated compilation of such studies is now available; Allen et al. 2008). In addition, a number of studies provided convergent evidence of a decrease in *AKT1* mRNA, protein, and activity levels in the prefrontal cortex and hippocampus, as well as in peripheral blood of some individuals with schizophrenia (Emamian et al. 2004; Thiselton et al. 2008; Zhao et al. 2006). The activity of major AKT1 targets such as GSK3β was also found to be altered in individuals with schizophrenia (Emamian et al. 2004). Negative findings have also been reported, but overall, an increasingly strong link between a dysregulation of AKT signaling and schizophrenia is emerging. Pharmacological evidence further supports a role for *AKT* in psychiatric disorders. For instance, the common antipsychotic haloperidol and several other drugs used in the management of psychosis were found to affect AKT phosphorylation (Beaulieu et al. 2007; Emamian et al. 2004), suggesting that this may be one mechanism of action of these commonly used medications.

The *AKT* gene family includes three members (*AKT1*, *AKT2*, and *AKT3*) which, although encoded by distinct genes, have a high structural homology (Brazil et al. 2004). While the three isoforms are expected to have partially overlapping functions, they also have independent physiological roles as demonstrated by the differential phenotypes of Akt-deficient mice (Yang et al. 2004). Mice lacking Akt1 have increased perinatal mortality and reduced body weight, while Akt2-deficient mice primarily exhibit a diabetes-like syndrome. Akt3-deficient mice have decreased brain size but maintain normal body weights and glucose homeostasis. Although genetic evidence at present points primarily at *AKT1* variants in the pathogenesis of schizophrenia, all three isoforms may mediate disease-relevant pathways.

Due to its critical roles in apoptosis and cell survival, the functions of Akt in the nervous system have been initially mainly studied in the context of neurodegenerative disorders, malignancies, and injury. However, it is evident that at least some of the diverse roles attributed to Akt may also contribute to the fundamental functions of the nervous system. Its high expression in the neurogenic zones of the developing brain is indicative of a role in proliferation and migration during brain morphogenesis, while its localization in the mature brain where it is restricted to select cellular populations such as the cerebellar Purkinje cells and the hippocampal pyramidal neurons suggests a role in neuron-specific physiological processes (Owada et al. 1997; Znamensky et al. 2003). Indeed, as described in the following sections, Akt has been firmly established as a key player in a range of neuronal functions.

2.1 Neuronal Plasticity

Studies in *Drosophila* and a mouse model of Akt deficiency revealed its critical role in long-term plasticity and establishment of dendritic architecture (Guo and Zhong 2006; Lai et al. 2006). Multiple upstream and downstream Akt effectors have been implicated in these processes. Indeed, several extracellular activators of Akt including BDNF, PDGF, bFGF, EGF, NGF, and NMDA have critical roles in neuronal plasticity and memory formation (Brazil et al. 2004). In fact, BDNF-dependent phosphorylation of Akt1 in the hippocampus is correlated with spatial reference and working memory formation (Mizuno et al. 2003), and in the amygdala increased phosphorylation of Akt was observed after long-term potentiation (LTP) and fear conditioning (Lin et al. 2001).

Recent studies uncovered the diverse Akt substrates that mediate these processes. Akt signaling downstream of Ras controls the insertion of the GluR1 subunits of AMPA receptors, thereby regulating AMPA receptor-mediated synaptic responses (Qin et al. 2005), while its direct phosphorylation of the β_2 subunit of $GABA_A$ receptors regulates their insertion into the plasma membrane, modulating the strength of inhibitory synapses (Wang et al. 2003b). Akt has also been associated with estrogen-dependent local protein synthesis and synaptogenesis in hippocampal spines (Akama and McEwen 2003), and has a role in the establishment of dendritic arborization and axonal polarity through at least two independent downstream effectors: mTOR (Jaworski et al. 2005) and TIAM (Tolias et al. 2005). Akt signaling also controls the activation of the transcription factor CREB (cAMP response element-binding protein) (Du and Montminy 1998). CREB regulates production of targets as diverse as growth factors, transcriptional factors, and neurotransmitter receptors. Through activation of an array of proteins, CREB can affect mood and addiction, growth factor-dependent cell survival, and learning and memory (Carlezon et al. 2005). Indeed, Akt1-deficient mice display a prominent alteration in gene expression, with enrichment in genes controlling synaptic function, neuronal development, myelination, and the actin cytoskeleton (Lai et al. 2006).

2.2 *Dopamine Signaling*

A recently described role of Akt in dopaminergic transmission is particularly relevant for its involvement in psychiatric disorders. Dopamine has critical roles in the modulation of glutamate- and GABA-mediated neurotransmission and is involved in brain functions such as movement, emotion, reward, and affect (Gainetdinov and Caron 2003). Dopaminergic dysfunction has been extensively linked to schizophrenia through genetic, pharmacological, and biochemical evidence (Gainetdinov and Caron 2003; Lang et al. 2007). As more extensively discussed later in this review, signaling downstream of D_1 and D_2 dopamine receptors has been classically associated with the "canonical" cAMP-dependent pathway. However, in a series of elegant experiments, Caron and colleagues recently established Akt as a novel and key signaling intermediate downstream of D_2 dopamine receptors (Beaulieu et al. 2005, 2007). Indeed, Akt activation by D_2 receptors appears to be promoted by mechanisms that were previously implicated in the termination of canonical dopamine transmission. Following stimulation, dopamine receptors are rapidly inactivated through recruitment of scaffolding proteins called arrestins, which leads to termination of signaling and induces the assembly of an internalization complex. Strikingly, activation of D_2 receptors results also in the formation of a complex that along with arrestins contains protein phosphatase 2A (PP2A) and Akt, causing deactivation of Akt by PP2A-mediated dephosphorylation and the subsequent stimulation of downstream GSK3-mediated signaling.

Behavioral analysis of Akt1 mutant mice provided further evidence for a role of Akt1 in dopaminergic signaling. These mice have a normal basic behavioral profile without widespread deficits; however, upon neurochemical challenge of the dopaminergic system, they display abnormal working memory retention. Strikingly, pharmacological activation of dopamine D_2 but not D_1 receptors results in worse working memory performance (Lai et al. 2006), underscoring the findings of Caron et al. implicating D_2 receptors in Akt signaling. These results were expanded by a recent report linking AKT-dependent dopaminergic transmission with working memory in human subjects (Tan et al. 2008). The authors found that several polymorphisms in the *AKT* sequence which have been previously associated with schizophrenia also contribute to cognitive processes in healthy subjects. Moreover, one polymorphism which was also correlated with reduced AKT expression was also associated with inefficient execution of a task which critically depends on proper dopamine levels in the PFC. Subjects with this variant that also carried polymorphisms in the *COMT* gene, which is responsible for breaking down dopamine, had disproportionally inefficient PFC processing, suggesting a genetic interaction between these two genes and an important role for the dopamine–AKT signaling pathway in cognitive processes.

2.3 GSK3β Signaling

GSK3, a serine/threonine protein kinase, is a canonical Akt target that has been implicated in fundamental cellular processes such as cell proliferation, metabolism, transcriptional control, and determination of cell fate through phosphorylation of a wide array of target proteins (Frame and Cohen 2001). The two isoforms of GSK3, α and β, are encoded by distinct genes and, similarly to Akt, have separate, but overlapping roles. Akt and other kinases regulate the activity of GSK3 isoforms through phosphorylation of a serine residue within the N-terminal domain (Ser21 in GSK3α and Ser9 in GSK3β), which leads to its inhibition and the activation of its many substrates. GSK3β has been first placed in the spotlight with the discovery that drugs effective in mood disorders, in particular lithium, affect its phosphorylation and activity (Jope and Roh 2006). Extensive genetic and biochemical studies indeed correlated GSK3β with pathologies such as depression and bipolar disorder (Jope and Roh 2006). Despite the absence of genetic association with schizophrenia, its role in dopaminergic transmission suggests that it may be also implicated in this disorder. Indeed, studies in human tissue revealed changes in GSK3β levels and activity in patients (Emamian et al. 2004; Jope and Roh 2006). Results from these assays are somewhat contradictory, with some studies showing a decrease in levels and activity, and others suggesting an increase in function. Degradation of postmortem tissue may account for these discrepancies, especially as independent evidence showing increased D_2 dopamine receptors in patients with schizophrenia (Abi-Dargham et al. 2000), and pharmacological data indicating that typical antipsychotics act on D_2 receptors and inhibit GSK3 activity (Beaulieu et al. 2007; Emamian et al. 2004), support the notion of increased GSK3β signaling in schizophrenia.

A different perspective on the role of GSK3β dysregulation in psychiatric disorders came from its association with the Wnt signaling pathway. Wnt is a well-known regulator of brain development (Malaterre et al. 2007), and GSK3β plays a central role in its signaling as an essential kinase which regulates the stability of a downstream effector, β-catenin, and the transcription of a multitude of genes (Frame and Cohen 2001). In this signaling cascade, GSK3β activity is regulated by association with other pathway members rather then by the canonical phosphorylation at Ser9. Thus, changes in GSK3β levels and its interactions with other proteins, rather than its phosphorylation status, can affect its downstream signaling. Association of several members of this pathway with schizophrenia has been reported (for review see Lovestone et al. 2007), making it a particularly attractive candidate to account for the neurodevelopmental changes that have been linked with the disorder. This is further emphasized by recent findings associating GSK3β to Disc1 function (Mao et al. 2009), which is discussed in the later sections of this review.

3 *PPP3CC* Gene and the Calcineurin Signaling Pathway

Calcineurin, formerly known as Protein Phosphatase 2B, is a calcium/calmodulin-dependent Ser/Thr phosphatase whose abundance in the brain, regulatory properties, and large number of targets give it a unique position in the regulation of neuronal processes. Calcineurin is a heterodimer composed of a catalytic (CaN A) and a regulatory (CaN B) subunit. The activity of the catalytic domain is suppressed at basal conditions, but upon calcium and calmodulin binding to the regulatory subunit there is a dramatic increase in its activity. Calcineurin dependence upon calcium/calmodulin thus provides a useful link between depolarization-induced Ca^{2+} influx and phosphorylation state. With a wide range of substrates, calcineurin plays a critical role in maintaining the balance of phosphorylation and thus activity for a diverse set of proteins which affect a wide range of processes such as endocytosis, cytoskeletal stability, synaptic plasticity, and neurotransmitter receptor regulation (Shibasaki et al. 2002).

Calcineurin is highly enriched in the hippocampus, the striatum, and the substantia nigra, and is found in the cortex, the cerebellum, and the amygdala to a lesser degree (Eastwood et al. 2005; Goto et al. 1986). Three isoforms of CaN A, with distinct expression patterns, have been identified (Eastwood et al. 2005). CaN Aα and CaN Aβ are strongly expressed from late embryonic stages until adulthood, whereas CaN Aγ expression is higher during embryonic development and decreases in adulthood. Although the distinct functions of these isoforms are not well understood, these expression patterns suggest differential roles during development and in the adult brain. Subcellular localization studies show that in the adult brain, calcineurin localizes to the neuronal soma, spines, postsynaptic density, and axon terminals, suggesting a role in neuronal transmission (Goto et al. 1986).

Calcineurin was first linked to schizophrenia when Gerber et al. (2003) found an association between the gene encoding for the CaN Aγ isoform (*PPP3CC*) and schizophrenia in two independent populations. Further genetic studies have supported *PPP3CC* as a risk gene for schizophrenia (Goto et al. 1986; Horiuchi et al. 2007; Liu et al. 2007; Yamada et al. 2007) and bipolar disease (Mathieu et al. 2008). Several studies failed to show an association between common *PPP3CC* variants and schizophrenia, possibly due to the power of the studies and the level of risk associated with altered calcineurin signaling (Kinoshita et al. 2005; Sanders et al. 2008; Xi et al. 2007).

Animal models provide strong evidence that altered calcineurin signaling may underlie many schizophrenia-associated endophenotypes. Mice with disrupted calcineurin activity display behavioral deficits reminiscent of the pathological alterations in schizophrenia. They also show a decrease in long-term depression (LTD) and the modification threshold for LTD/LTP in the hippocampus (Miyakawa et al. 2003; Zeng et al. 2001). Since LTP and LTD are critical for learning and memory, it follows that deficits in these processes could contribute to the cognitive deficits observed in schizophrenia. The link to schizophrenia is further strengthened because chronic administration of antipsychotic drugs to rats decreased calcineurin protein and mRNA levels in the striatum and prefrontal cortex, regions implicated in schizophrenia. Interestingly, calcineurin activity was actually increased, suggesting that the decrease in expression may reflect a compensatory response (Rushlow et al. 2005), and further emphasizing the importance of balanced calcineurin signaling in neuronal processes.

3.1 Dopamine Signaling

Calcineurin regulates a diverse range of processes, many of which have been suggested to be altered in schizophrenia. As previously discussed, dopaminergic signaling plays an important role in the pathology of psychiatric disease and calcineurin has been implicated in its control on multiple levels. Calcineurin-mediated dephosphorylation regulates the activity of DARPP32 (Dopamine- and cAMP-regulated phosphoprotein of 32 kDa), also called Protein Phosphatase 1 regulatory (inhibitor) subunit 1B (PP1R1B), one of the central regulators of dopamine signaling. DARPP32 plays a critical role in the integration of signaling pathways through selective phosphorylation and dephosphorylation by several kinases and phosphatases (Svenningsson et al. 2004). DARPP32 is phosphorylated and activated by protein kinase A (PKA) via a cAMP-mediated pathway that is activated through dopamine D_1 receptors, adenosine A_{2A} receptors, and vasoactive intestinal protein (VIP) receptors. In its phosphorylated state, DARPP32 is a potent inhibitor of Protein Phosphatase 1 (PP1), a serine/threonine protein phosphatase with wide-ranging targets, from voltage-gated channels to neurotransmitter receptors, which is critical for the integration of signaling of various neurotransmitters (Svenningsson et al. 2004). Calcineurin-induced dephosphorylation of DARPP32 following Ca^{2+} influx through a number of receptors including NMDA, AMPA, and dopamine D_2 receptors inhibits this pathway, leading to an increase in PP1 activity (Nishi et al. 2002). Interestingly, DARPP32 has also been implicated in psychiatric disorders, although the genetic and biochemical evidence is at present controversial (Feldcamp et al. 2008; Lang et al. 2007; Yoshimi et al. 2008). Calcineurin can also directly modulate dopamine signaling through D_1 and D_2 receptors. Activation of D_2 receptors leads to an increase in calcineurin activity and a dephosphorylation of D_1 receptors with an associated decrease in D_1 activity (Adlersberg et al. 2004).

3.2 Transcriptional Regulation

Through inhibition and activation of transcription factors such as CREB and NFATc (nuclear factor of activated T cells), calcineurin has the ability to regulate a large number of pathways. CREB activity is attenuated after dephosphorylation by PP1, which is normally inhibited by DARPP32 or inhibitor-1; however, these are both inactivated by calcineurin, leading to an increase in PP1 activity and a decrease in transcription of the numerous CREB targets (Hagiwara et al. 1992). While there is extensive evidence that CREB may be involved in depression (Carlezon et al. 2005), one study also found variants in the promoter region of CREB in patients with schizophrenia (Kawanishi et al. 1999). Additionally, calcineurin activates the transcription factor NFAT following netrin- or neurotrophin-induced increases in calcium (Nguyen and Di 2008). This pathway has been implicated in neuronal growth and axonal guidance during development, as shown by severe neurodevelopmental defects of NFATc2/c3/c4 triple mutants. Interestingly, inhibition of calcineurin during embryonic development leads to similar defects (Graef et al. 2001). Activation of NFATc4 following increased calcium flux through L-type Ca^{2+} channels or neuronal depolarization induces inositol 1,4,5-trisphosphate receptor (IP_3R) expression in a calcineurin-dependent manner. This may be important for synaptic plasticity and memory because calcium release from IP_3-sensitive intracellular stores is critical for the induction of LTD (Graef et al. 1999).

3.3 Synaptic Plasticity and Neurotransmission

Many of the targets of calcineurin are responsible for aspects of synaptic plasticity, particularly induction of LTD. Calcineurin plays a role in limiting LTP, which could be important for maintaining a balance when processing new memories. Reversible expression of a calcineurin inhibitor actually facilitates LTP *in vitro* and *in vivo* and enhances learning and strengthens both long- and short-term memory (Malleret et al. 2001). Importantly, overexpression of a constitutively active truncated form of calcineurin inhibits the transition of short- to long-term memory (Mansuy et al. 1998). These effects may be mediated by several neurotransmitter systems. Small influxes of calcium through NMDA receptors induce an increase in PP1 activity through a calcineurin-dependent pathway and this leads to an induction of LTD by modulating neurotransmitter receptors (Snyder et al. 2003). Calcium influx through NMDA receptors also leads to a dephosphorylation of the GluR1 subunit of the AMPA receptor and an inhibition of its activity (Snyder et al. 2003) as well as endocytosis of the AMPA-R itself (Beattie et al. 2000). Calcineurin also affects inhibitory GABA signaling during the induction of LTD through the dephosphorylation of $GABA_A$ receptors (Wang et al. 2003a), and independent genetic and biochemical evidence also suggest that GABA signaling may be affected in schizophrenia (Allen et al. 2008).

3.4 Nitric Oxide Signaling

Nitric oxide synthase (NOS), the enzyme which catalyzes the production of nitric oxide (NO), is a substrate of calcineurin. NO signaling through its receptor soluble guanylate cyclase and its second messenger cGMP affects numerous neuronal processes including synaptic plasticity, mRNA stability, and transcription factor activity (Calabrese et al. 2007). Dephosphorylation of NOS by calcineurin leads to its activation and increases in NO levels. NOS is directly associated with NMDA receptors and upon glutamate binding to the receptor, NOS catalyzes the production of NO (Bernstein et al. 2005). This can lead to LTP; however with high levels of activation, the generation of nitric oxide free radicals can be neurotoxic (Calabrese et al. 2007). Mice lacking nNOS have cognitive impairments which suggest that disruption of the signaling pathway in schizophrenia could contribute to the learning and memory deficits seen in the disease (Kirchner et al. 2004). Both increases and decreases in protein, mRNA, and activity levels of NOS have been seen in patients with schizophrenia. Genetic association in several populations has also been found, although some of these findings were not replicated (for review, see Bernstein et al. 2005).

3.5 Vesicle Trafficking

Dynamin 1 and the other dephosphins (amphiphysin 1 and 2, synaptojanin, AP180, epsin, Eps15) are calcineurin substrates which are activated upon Ca^{2+} influx and induce clathrin-mediated synaptic vesicle endocytosis (Cousin and Robinson 2001). Endocytosis is a critical process that functions to maintain plasma membrane area and to traffic receptors from the synapse to control reactivity. The bidirectional regulation of receptor trafficking is thought to be critical for synaptic plasticity, and deficits in this process could be involved in abnormal functioning in neuropsychiatric disorders (Lau and Zukin 2007). Calcineurin and the dephosphins have also been implicated in the control of bulk endocytosis which is critical for recovering large volumes of synaptic membrane following periods of strong stimulation (Clayton et al. 2007). The function of synapsin 1, another component of vesicle-trafficking machinery, is also regulated by calcineurin. Calcineurin-dependent dephosphorylation of synapsin is necessary for maintaining the reserve pool of neurotransmitter vesicles at the synapse. Inhibition of calcineurin increases glutamate release because phosphorylated synapsin leads to trafficking of the vesicles for neurotransmitter release (Chi et al. 2003). Synapsin 1 mRNA and protein levels have also been shown to be altered in brains of some patients with schizophrenia, suggesting that this protein may be important in the disease (Browning et al. 1993; Tcherepanov and Sokolov 1997). Thus, calcineurin affects a number of proteins that are necessary for a neuron to continually respond to stimuli and prepare for future episodes of activity.

3.6 Cytoskeletal Phosphorylation

Cytoskeletal proteins tubulin, tau, MAP2, and F-actin are dephosphorylated by calcineurin (Goto et al. 1985; Halpain et al. 1998) and this plays a role in axonal outgrowth (Ferreira et al. 1993), synaptic development and remodeling (Quinlan and Halpain 1996), and spine stability (Halpain et al. 1998). The role of calcineurin in the regulation of cytoskeletal proteins has been investigated mainly in the context of neurodegenerative disorders such as Alzheimer's disease; however, there has been accumulating evidence that changes in the cytoskeleton may play a role in the pathogenesis of schizophrenia as well (Benitez-King et al. 2004). Patients with schizophrenia show decreases in spine density (Glantz and Lewis 2001; Hill et al. 2006; Kolluri et al. 2005) and neuronal complexity (Broadbelt et al. 2002; Kalus et al. 2002). Moreover, not only abnormal expression of cytoskeletal proteins MAP2 and MAP5 (Arnold et al. 1991), but also altered phosphorylation of MAP2 (Cotter et al. 1997, 2000) has been found in brains from patients. Finally, morphological changes in neurons have also been observed in several mouse models of schizophrenia (Kvajo et al. 2008; Mukai et al. 2008; Stark et al. 2008), suggesting that the stability of these structures may be important for the disease pathogenesis.

4 *DISC1* Gene and the cAMP and GSK3/Wnt Signaling Pathways

Disrupted In Schizophrenia-1 (*DISC1*) is a prominent candidate gene for schizophrenia and bipolar disorder. In 1990, St. Clair et al. (1990) reported a pedigree of a large Scottish family with many individuals displaying a variety of psychiatric diagnoses including schizophrenia, bipolar disorder, recurrent depression, and conduct disorder. Genomic analysis revealed a presence of a $t(1;11)(q42;q14)$ balanced chromosomal translocation which cosegregated with mental illness in affected individuals, and lead to a disruption of two genes, Disrupted In Schizophrenia-1 (*DISC1*) and Disrupted In Schizophrenia-2 (*DISC2*) (Millar et al. 2000). Only *DISC1* was found to have an open reading frame, while the *DISC2* transcripts lack any protein-coding potential and their functions have not been experimentally addressed.

Since its identification as a genetic risk factor for major mental illness, the *DISC1* locus has been also implicated (albeit in a controversial manner) in psychiatric disorders in karyotypically normal patient populations. Linkage and association studies linked *DISC1* variants to schizophrenia, schizoaffective disorder, and major depression (Allen et al. 2008). Several studies also showed that healthy subjects and patients with schizophrenia carrying *DISC1* risk alleles display changes in gray matter volume in the hippocampus, prefrontal cortex, and other brain regions (Callicott et al. 2005; Cannon et al. 2005; Di Giorgio et al. 2008; Szeszko et al. 2008), as well as deficits in working, short-term, and long-term memory (Burdick et al. 2005; Cannon et al. 2005; Hennah et al. 2005), attention (Hennah et al. 2005; Liu et al. 2006), and hippocampal and prefrontal cortical activation (Callicott et al. 2005; Prata et al. 2008). It should be noted that like other associations studies, these studies of *DISC1* have not produced any statistically unequivocal results, suggesting that the well-defined mutation in the Scottish family is currently the only disease-associated allele within the *DISC1* gene.

The *DISC1* gene gives rise to at least four well-defined splice variants: L (Long; 90 kDa), Lv (Long Variant; 95 kDa), S (Short; 70 kDa), and Es (Extremely Short; 41 kDa) (Taylor et al. 2003), but their individual functions and contribution to the disease pathogenesis are not well understood. The Scottish translocation occurs between exons 8 and 9 of the gene, truncating the C-terminal region of the major (L, Lv, and S) isoforms (Millar et al. 2000). Notably, in mice carrying a truncating lesion in the endogenous *Disc1* orthologue that models the *DISC1* risk allele the mutation leads to a loss of the major splice variants suggesting that the truncated isoforms are unstable and become rapidly degraded (Koike et al. 2006; Kvajo et al. 2008). The Es Disc1 isoforms are present in the Scottish translocation carriers, indicating that the phenotypic consequences of *DISC1* disruption are dictated by its effects on select splice variants.

Studies of the effects of the mutation on Disc1 function are hindered by the fact that its protein sequence has no significant homology to other proteins. Functional analysis indicates the presence of an N-terminal domain with no clear structure, and a long helical C-terminal domain containing multiple regions with high coiled-coil forming potential (Millar et al. 2000). This region has a capacity for binding multiple proteins, and *in vitro* assays revealed that Disc1 associates with a number of proteins involved in different cellular processes. Indeed, Disc1 has been found localized to several cellular compartments including the centrosome (Morris et al. 2003), mitochondria (James et al. 2004), spines and dendrites (Bradshaw et al. 2008; Millar et al. 2005), growth cones and axon terminals in neurons and neuronal-like cells (Taya et al. 2007), as well as stress granules (Ogawa et al. 2005) and the nucleus (Sawamura et al. 2005). While the majority of Disc1 binding partners have still not been confirmed *in vivo*, the multiplicity of its binding and localizations lead to a suggestion that Disc1 may act as a molecular scaffold and serve as an assembly point for large multiprotein complexes.

In the brain Disc1 expression shows a strict spatial and temporal regulation with highest levels during development when Disc1 mRNA can be detected in all hippocampal subfields, the cortex, and the thalamus. In the adult brain, Disc1 displays a more restricted localization with a high expression retained only in the dentate gyrus formation of the hippocampus, and, to a lesser degree in the CA1 region (Austin et al. 2004). Studies in Disc1-deficient mice revealed region-specific phenotypes, including altered organization and complexity of newly born and mature neurons of the dentate gyrus, as well as changes in short-term plasticity in CA3/CA1 synapses (Kvajo et al. 2008). These mice also displayed a highly selective impairment in working memory (Koike et al. 2006; Kvajo et al. 2008), suggesting that Disc1 deficiency affects neural circuits within the hippocampus as well as the prefrontal cortex. Mouse models overexpressing truncated versions of the human DISC1 protein under various exogenous promoters also provided evidence for anatomical and behavioral abnormalities (Hikida et al. 2007; Pletnikov et al. 2007, 2008), whereas those carrying point mutations in the *Disc1* gene demonstrated that allelic heterogeneity at the *Disc1* locus can lead to distinct behavioral phenotypes (Clapcote et al. 2007).

4.1 Disc1 and cAMP Signaling

In recent years, the role of Disc1 in the control of cAMP signaling has received particular attention. Millar et al. showed that Disc1 binds phosphodiesterase 4b (PDE4b), an enzyme which inactivates the key intracellular second messenger adenosine 3′,5′-monophosphate (cAMP), and is believed to be the sole means of cAMP degradation in the brain (Millar et al. 2005). While this study fell short of showing an effect of Disc1 on PDE4b activity, it demonstrated that their interaction appeared to be dynamically regulated by intracellular cAMP levels. An elevation in cAMP led to a dissociation of the complex in a PKA activity-dependent manner. Since previous studies showed that a raise in cAMP levels induces a PKA-dependent phosphorylation of PDE4b and an increase in its catalytic activity (MacKenzie et al. 2002), it was hypothesized that Disc1 sequesters PDE4b in its low-activity, unphosphorylated form until it is required to switch off cAMP signaling, whereupon PKA-phosphorylated, high-activity PDE4b is released from Disc1.

PDE4b binds to Disc1 through a highly conserved UCR2 domain which is present in all genes of the PDE4 family of phosphodiesterases (Millar et al. 2005), and Disc1 was able to form complexes with representative isoforms from all four (A, B, C, and D) *PDE4* gene families. The N-terminal domain of Disc1 is responsible for PDE4 binding, suggesting that all known Disc1 splice variants can participate in this complex. Indeed, a biochemical analysis revealed the complexity of this interaction, showing that Disc1 contains several PDE4 binding sites some of which were specific for PDE4b, while others could bind additional PDE4 isoforms (Murdoch et al. 2007). Complexes between various PDE4 isoforms and Disc1 splice variants showed differential sensitivity to cAMP, possibly owing to the differences in PDE4 binding sites. However, the cAMP-mediated disruption of PDE4–Disc1 complexes could not be attributed to PKA phosphorylation of either PDE4 or Disc1. Therefore, the exact mechanism of the feedback loop controlling this association is yet unknown, and may be based on the association with other PDE4 binding factors.

While this interaction has been well documented in vitro, little is known about the cellular roles of Disc1 in PDE4 regulation. One clue comes from the specific spatial and temporal regulation of local cAMP levels by PDEs within cell compartments. PDE4 isoforms are discreetly regulated in response to cAMP levels and targeted to different cellular compartments through inclusion of regulatory regions and unique N-terminal sequences (Lynch et al. 2006). Disc1 may act as a scaffold for PDE4, and the differential interaction between isoforms may convey an additional layer of specificity in regulating cAMP levels. Indeed, Disc1 and PDE4 isoforms colocalize at a wide array of cellular locations including the centrosome and mitochondria in cell lines, and the dendrites and dendritic spines in primary hippocampal neurons (Bradshaw et al. 2008; Millar et al. 2005). At the centrosome, PDE4b appears to be interacting with Disc1, NDE1, NDEL1, LIS1, and dynein (Bradshaw et al. 2008), suggesting that Disc1 assembles a large protein complex whose interactions and functions may be modulated by PDE4b-dependent control of cAMP signaling.

First indications of the physiological effects of Disc1 on PDE4 function came from a recent analysis of two mouse lines with point mutations in the *Disc1* gene introduced by *N*-nitroso-*N*-ethylurea (ENU) mutagenesis (Clapcote et al. 2007). These missense mutations, changing a glutamine to leucine at position 31 (Q31L) and a leucine to proline at position 100 (L100P), were located in exon 2, which is present in all known splice isoforms and encodes most of the Disc1 N-terminal domain. The mutations did not affect the production or stability of major Disc1 isoforms, but the mice carrying them displayed an array of behavioral deficits that were suggestive of schizophrenia in L100P mice, and depression in Q31L mice. *In vitro* analysis showed that both mutations decreased the binding of PDE4b to Disc1, with the L100P mutation showing a stronger effect than the Q31L. Based on the *in vitro* findings, the decreased binding in mutant mice should result in increased PDE4 activity. Instead, a 50% decrease in activity was observed in the Q31L mice, with no change in L100P, implying that *in vivo* a reduced Disc1–PDE4b association does necessarily result in increased PDE4b activation. These enzymatic findings were backed up by pharmacology because treatment with rolipram, an inhibitor of PDE4b which also acts as an antidepressant, was not able to reverse the behavioral deficits of Q31L mice. While these mouse models showed an *in vivo* interplay between PDE4 and Disc1, difficulties in predicting the effects of these mutations on Disc1 structure and function will require additional effort to dissect out the mechanisms behind their interaction.

4.1.1 PDE4 and Psychiatric Disorders

Although their interaction with DISC1 highlighted their potential contribution to the genetic basis of schizophrenia and related psychiatric disorders, phosphodiesterases have been in the spotlight as effectors of neuronal signaling cascades for several decades, since a mutation in the *Drosophila melanogaster* orthologue of the *PDE4* gene was correlated with impaired memory and learning (Davis 1996). This mutation disrupted PDE4 function, leading to increased cAMP levels and abnormalities in ion channel function and nerve excitability, synaptic transmission, plasticity, axonal growth cone motility, and neuronal arborization. The evidence for a role of phosphodiesterases in psychiatric disorders first came from pharmacological studies showing that PDE4 inhibitors have an antidepressant (Houslay and Adams 2003), memory-enhancing and antipsychotic-like effect (Kanes et al. 2007). Subsequent studies showed that PDE4 activity underlies the mechanisms of other antidepressant drugs, further strengthening the role of cAMP signaling in psychiatric disorders (Renau 2004).

Genetic evidence for *PDE4* in psychiatric disease is at present promising but preliminary, as it is based on a limited number of studies. Analysis of a family with a $t(1;16)$ chromosomal translocation and a history of psychosis in two family members carrying this translocation found a disruption of the *PDE4B* gene on chromosome 1 and the gene encoding cadherin 8 (*CDH8*) on chromosome 16 (Millar et al. 2005). While PDE4B expression was reduced in a lymphoblastoid cell line from these patients, this translocation does not provide an unequivocal evidence for a role of the *PDE4B* gene in psychosis because cadherins have been also linked with psychiatric and neurodevelopmental disorders (Wang et al. 2009). A handful of studies in karyotypically normal populations associated polymorphisms in *PDE4B* and *PDE4D* with increased susceptibility to schizophrenia (Fatemi et al. 2008a; Numata et al. 2008; Tomppo et al. 2009), with other variants conferring a protective effect (Pickard et al. 2007; Tomppo et al. 2009). Changes in PDE4 expression were also found in patients with schizophrenia (Fatemi et al. 2008a), bipolar disorder (Fatemi et al. 2008b), depression (Numata et al. 2009), and autism (Braun et al. 2007). Due to the paucity of these findings, further studies are needed to establish their relevance and substantiate the involvement of *PDE4B* in the etiology of psychiatric disorders.

PDE4 isoforms show a complex expression pattern in the adult brain, further emphasizing the intricacy of cAMP regulation. Analysis of PDE4 protein and mRNA revealed that PDE4a, PDE4b, and PDE4d are widely expressed in the adult brain, while PDE4c is mainly expressed in peripheral tissues (Cherry and Davis 1999; Perez-Torres et al. 2000). PDE4 isoforms exhibit isoform-specific localization patterns suggesting that, analogous to cellular compartmentalization, each isoform may have a specific role within particular brain structures. PDE4a displayed the most restricted expression pattern, being abundant in limbic and olfactory regions as well as in the cortical layers associated with motor control. PDE4b was found to be highly expressed in the nucleus accumbens, the hypothalamus, ventral striatum, and locus coeruleus, regions that have been implicated in anxiety-related behaviors and addiction. Its levels in the hippocampus were not as high, indicating that its role in hippocampal-dependent learning may be limited (Cherry and Davis 1999; Perez-Torres et al. 2000). Indeed, a behavioral analysis of a PDE4b mouse mutant supported a role in the regulation of anxiety and striatal function, but not memory and learning (Siuciak et al. 2008; Zhang et al. 2008). PDE4d is the isoform most prominently expressed in the hippocampus, where it was found in pyramidal cells of the CA2/CA3 region, and its transcript was detected in the dentate gyrus. Behavioral analysis of PDE4d-deficient mice revealed a divergent role from PDE4b, as these mice displayed a robust antidepressant-like effect, suggesting that PDE4d may be the major isoform mediating the antidepressant effects of rolipram (Zhang et al. 2002). Interestingly, all three PDE4 isoforms appear to be particularly prominent in axonal projection fibers and in synaptic compartments (Cherry and Davis 1999), indicating that PDE4 activity in many regions may be presynaptic.

The expression of PDE4 isoforms in the hippocampus is intriguing, particularly considering their putative roles in adult neurogenesis of the dentate gyrus. PDE4b-deficient mice have increased adult neurogenesis (Zhang et al. 2008) showing, with other studies (Duman et al. 2001; Nakagawa et al. 2002), the importance of cAMP signaling for this process. Deficits in adult neurogenesis have been linked to psychiatric disorders such as depression and schizophrenia, and it is well documented that long-term administration of rolipram and other antidepressants increases adult neurogenesis, and that this may mediate their therapeutic effects (Duman et al. 2001). Disc1 has been implicated in the control of adult neurogenesis, affecting the production, migration, and morphological development of adult born neurons (Duan et al. 2007; Kvajo et al. 2008; Mao et al. 2009). While its effects on neurogenesis have been recently attributed to its interaction with GSK3β (Mao et al. 2009), it is tempting to speculate that its effects may be at least in part mediated through PDE4 regulation.

4.1.2 Adenylyl Cyclase and Psychiatric Disorders

Adenylyl cyclase (AC) is an ATP-pyrophosphate lyase that synthesizes cAMP and thus has a critical role in the control of cAMP-dependent signaling (Sadana and Dessauer 2009). Several mammalian ACs have been described with distinct expression patterns and regulatory properties. While at present there is no genetic evidence linking ACs with psychiatric disorders, studies of mouse knockout models revealed their importance for pertinent neurobiological processes including long-term memory, response to stress, and sensitivity to drugs of abuse (Sadana and Dessauer 2009). In a study of affective behaviors, a double knockout for AC1 and AC8 displayed complex behavioral changes suggestive of a depressive-like phenotype. On the other hand, a deficiency in AC5, which has been previously implicated in opiate dependency and pain responses, induced strong antidepressant and anxiolytic phenotypes, underlining the multifaceted effects of cAMP regulation (Krishnan et al. 2008).

Additional evidence for the involvement of cAMP signaling in psychiatric disorders came from studies of adenylyl cyclase-regulating molecules. AC activity is primarily controlled through binding to subunits of G protein receptors, which couple several neurotransmitter systems to cAMP production and downstream signaling (Sadana and Dessauer 2009). Altered levels and activity of G protein subunits were reported in patients with schizophrenia and bipolar disorder (Nishino et al. 1993; Okada et al. 1994; Schreiber and Avissar 2000; Yang et al. 1998), and a genetic polymorphism potentially resulting in increased mRNA expression of Gαs, a subunit which activates adenylyl cyclases, has been associated with psychosis (Minoretti et al. 2006). Mice expressing a constitutively active isoform of Gαs exhibit sensorimotor, cognitive, biochemical, and neuroanatomical deficits similar to those in schizophrenia (Kelly et al. 2007, 2009). Intriguingly, another AC-regulating protein, PACAP (pituitary adenylate cyclase-activating polypeptide), which activates adenylyl cyclase through binding to G protein-coupled receptors, has been implicated in schizophrenia (Hashimoto et al. 2007; Ishiguro et al. 2001). PACAP-deficient mice also showed several schizophrenia-related behaviors including hyperactivity, increased novelty-seeking behavior, cognitive deficits, and disrupted sensorimotor gating (Hashimoto et al. 2001).

4.1.3 cAMP-Dependent Downstream Signaling and Schizophrenia

cAMP-dependent downstream signaling has been associated with a variety of cellular processes that have been implicated in the pathology of schizophrenia. Its role in PKA activation is of particular interest, as it affects multiple processes through phosphorylation of key targets such as the transcription factor CREB and components of the dopamine signaling pathway (Nguyen and Woo 2003). While we already touched on several of these processes earlier in the review, in this section we will focus on the remaining cAMP-dependent signaling targets that are most relevant for the pathology of psychiatric disorders.

Dopamine Signaling

The critical role of cAMP signaling downstream of dopaminergic transmission firmly established its importance in the pathogenesis of psychiatric disorders. As described earlier in the review, canonical dopamine signaling is tightly coupled to cAMP-mediated processes (Gainetdinov and Caron 2003). The two major classes of dopamine receptors, D_1 and the D_2, are prototypical G protein-coupled receptors, and their stimulation results in the regulation of adenylyl cyclase activity through G protein subunits such as Gαs, leading to changes in cAMP levels. As described earlier, PKA activation by cAMP leads to a phosphorylation of a number of substrates including DARPP32, a key regulator of dopamine transmission (Svenningsson et al. 2004).

EPAC-Mediated Signaling

EPAC (exchange factor directly activated by cAMP), also called RAPGEF3/4 (Rap guanine nucleotide exchange factor (GEF) 3/4), a guanine exchange factor regulated by cAMP, recently emerged as a candidate target for cAMP dysregulation in schizophrenia. In mice overexpressing Gαs, the G protein subunit that stimulates adenylyl cyclase activity, compensatory changes lead to decreased cAMP levels which are correlated with schizophrenia-like phenotypes (Kelly et al. 2007, 2009). Pharmacological stimulation of EPAC exhibited antipsychotic-like properties, suggesting that the detrimental effects of decreased cAMP may be at least partially mediated through EPAC. EPAC is on its own right a relevant candidate gene. It has been implicated in developmentally relevant processes such as cell proliferation, survival, and differentiation, as well as plasticity and long-term memory formation (Gelinas et al. 2008; Ouyang et al. 2008). Furthermore, a recent genetic study found an association of variants in its sequence with anxiety and depression (Middeldorp et al. 2009). Thus, EPAC may represent a prototype for target genes that mediate cAMP-dependent deficits in schizophrenia.

4.2 *Disc1 and GSK3β Signaling*

A recent study implicated Disc1 in the control of GSK3β signaling (Mao et al. 2009), suggesting that this interaction plays a role in neurogenesis during brain development and adulthood. As previously described, GSK3β is a key player downstream of both the Akt and Wnt signaling pathways (Frame and Cohen 2001). Disc1 directly binds to and inhibits GSK3β activity, affecting its ability to phosphorylate β-catenin, as well as its autophosphorylation at Y216, a noncanonical phosphorylation site which has been implicated in GSK3β activation by upstream kinases such as Akt. Suppression of Disc1 levels in individual neurons resulted in decreased neurogenesis which was mediated by β-catenin signaling (Mao et al. 2009). Surprisingly, Disc1-mediated inhibition modified only selective GSK3β substrates with several of the well-characterized targets apparently not being affected. While this report offers tantalizing indications of the convergence of signaling pathways in schizophrenia, the complex outcomes of Disc1 binding on GSK3β signaling as well as absence of some critical controls in this study pose a challenge to understanding the place of this interaction within the framework of GSK3β signaling.

5 General Summary

In this review, we discuss the contributions of schizophrenia susceptibility genes *AKT*, *PPP3CC*, and *DISC1* to intracellular signaling pathways and neuronal processes underlying the etiology and pathogenesis of schizophrenia. The growing body of evidence behind these genes stems from a wide range of genetic, biochemical, and animal studies which delineate their involvement in cellular and behavioral processes ultimately related to well-established schizophrenia endophenotypes.

The three schizophrenia susceptibility genes described here have distinct cellular functions. AKT is a wide-range kinase which modulates intracellular signaling cascades activated by growth factors, cytokines, and other external stimuli, and is thus critically implicated in the regulation of a diverse set of processes, from cell survival to neuronal plasticity. Through its phosphatase activity, calcineurin also controls the phosphorylation state of numerous proteins in signaling pathways, and due to its specific regulatory mechanisms, it is considered a major player in the regulation of neuronal-specific processes. Although the functions of *DISC1* are the least understood, this candidate gene has also been implicated in diverse processes

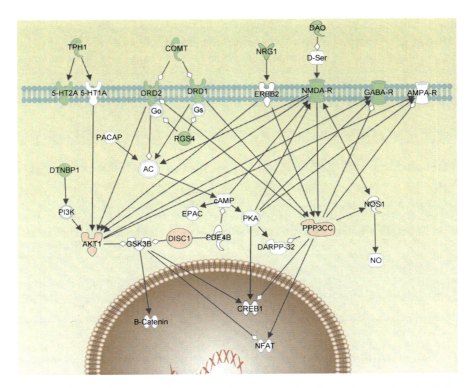

Fig. 1 Schizophrenia susceptibility genes and intracellular signaling pathways. Depicted is the interaction of intracellular signaling pathways involving AKT, calcineurin (PPP3CC), and DISC1. The scheme also shows their intersection with other candidate susceptibility genes (*green*) included in the top 30 candidates genes from the schizophrenia gene database meta-analyses (http://www.szgene.com; Allen et al. 2008). *Lines with a solid arrowhead* indicate activation, whereas *lines with an open diamond* indicate inhibition or deactivation. The schematic was prepared using the Ingenuity Systems software

such as cellular proliferation, neurogenesis, neurite outgrowth, and migration. In this review we touch on its regulation of two distinct cellular pathways with critical roles in neuronal function; the effect on cAMP signaling through the interaction with PDE4B, and its recently described regulation of GSK3β.

Although AKT, calcineurin, and DISC1 have different targets, they ultimately converge on a number of intracellular signaling cascades (Fig. 1) and affect several of the same neuronal processes. For example, a large body of evidence implicates both AKT and calcineurin in direct and indirect control of dopaminergic signaling, transcriptional control, as well as synaptic plasticity. DISC1 is placed at the intersection of these signaling cascades through its regulation of PDE4 and GSK3β activity. Our review of the existing literature focused primarily on three genes and the associated signaling pathways. However, as depicted in Fig. 1, when we consider other candidate susceptibility genes extracted in an unbiased way from the top 30 candidate genes from the schizophrenia gene database meta-analyses (http://www.szgene.com; Allen et al. 2008), several of them appear to be key players in the same pathways, underscoring the importance of these intracellular signaling in understanding the etiology and pathogenesis of schizophrenia. As more is understood about the mechanisms of actions of these and other schizophrenia susceptibility genes, common intracellular signaling pathways are likely to emerge affording not only a greater understanding of the etiology and pathogenesis of the disease, but also promising drug targets for treatment of its symptoms.

References

Abi-Dargham A, Rodenhiser J, Printz D, Zea-Ponce Y, Gil R, Kegeles LS, Weiss R, Cooper TB, Mann JJ, Van Heertum RL, Gorman JM, Laruelle M (2000) Increased baseline occupancy of D2 receptors by dopamine in schizophrenia. Proc Natl Acad Sci USA 97:8104–8109

Adlersberg M, Hsiung SC, Glickstein SB, Liu KP, Tamir H, Schmauss C (2004) Regulation of dopamine D-receptor activation*in vivo*by protein phosphatase 2B (calcineurin). J Neurochem 90:865–873

Akama KT, McEwen BS (2003) Estrogen stimulates postsynaptic density-95 rapid protein synthesis via the Akt/protein kinase B pathway. J Neurosci 23:2333–2339

Allen NC, Bagade S, McQueen MB, Ioannidis JP, Kavvoura FK, Khoury MJ, Tanzi RE, Bertram L (2008) Systematic meta-analyses and field synopsis of genetic association studies in schizophrenia: the SzGene database. Nat Genet 40:827–834

Arnold SE, Lee VM, Gur RE, Trojanowski JQ (1991) Abnormal expression of two microtubule-associated proteins (MAP2 and MAP5) in specific subfields of the hippocampal formation in schizophrenia. Proc Natl Acad Sci USA 88:10850–10854

Austin CP, Ky B, Ma L, Morris JA, Shughrue PJ (2004) Expression of Disrupted-In-Schizophrenia-1, a schizophrenia-associated gene, is prominent in the mouse hippocampus throughout brain development. Neuroscience 124:3–10

Beattie EC, Carroll RC, Yu X, Morishita W, Yasuda H, ZM von, Malenka RC (2000) Regulation of AMPA receptor endocytosis by a signaling mechanism shared with LTD. Nat Neurosci 3:1291–1300

Beaulieu JM, Sotnikova TD, Marion S, Lefkowitz RJ, Gainetdinov RR, Caron MG (2005) An Akt/ beta-arrestin 2/PP2A signaling complex mediates dopaminergic neurotransmission and behavior. Cell 122:261–273

Beaulieu JM, Tirotta E, Sotnikova TD, Masri B, Salahpour A, Gainetdinov RR, Borrelli E, Caron MG (2007) Regulation of Akt signaling by D2 and D3 dopamine receptors in vivo. J Neurosci 27:881–885

Benitez-King G, Ramirez-Rodriguez G, Ortiz L, Meza I (2004) The neuronal cytoskeleton as a potential therapeutical target in neurodegenerative diseases and schizophrenia. Curr Drug Targets CNS Neurol Disord 3:515–533

Bernstein HG, Bogerts B, Keilhoff G (2005) The many faces of nitric oxide in schizophrenia. A review. Schizophr Res 78:69–86

Bradshaw NJ, Ogawa F, ntolin-Fontes B, Chubb JE, Carlyle BC, Christie S, Claessens A, Porteous DJ, Millar JK (2008) DISC1, PDE4B, and NDE1 at the centrosome and synapse. Biochem Biophys Res Commun 377:1091–1096

Braun NN, Reutiman TJ, Lee S, Folsom TD, Fatemi SH (2007) Expression of phosphodiesterase 4 is altered in the brains of subjects with autism. NeuroReport 18:1841–1844

Brazil DP, Yang ZZ, Hemmings BA (2004) Advances in protein kinase B signalling: AKTion on multiple fronts. Trends Biochem Sci 29:233–242

Broadbelt K, Byne W, Jones LB (2002) Evidence for a decrease in basilar dendrites of pyramidal cells in schizophrenic medial prefrontal cortex. Schizophr Res 58:75–81

Browning MD, Dudek EM, Rapier JL, Leonard S, Freedman R (1993) Significant reductions in synapsin but not synaptophysin specific activity in the brains of some schizophrenics. Biol Psychiatry 34:529–535

Burdick KE, Hodgkinson CA, Szeszko PR, Lencz T, Ekholm JM, Kane JM, Goldman D, Malhotra AK (2005) DISC1 and neurocognitive function in schizophrenia. NeuroReport 16:1399–1402

Calabrese V, Mancuso C, Calvani M, Rizzarelli E, Butterfield DA, Stella AM (2007) Nitric oxide in the central nervous system: neuroprotection versus neurotoxicity. Nat Rev Neurosci 8:766–775

Callicott JH, Straub RE, Pezawas L, Egan MF, Mattay VS, Hariri AR, Verchinski BA, Meyer-Lindenberg A, Balkissoon R, Kolachana B, Goldberg TE, Weinberger DR (2005) Variation in DISC1 affects hippocampal structure and function and increases risk for schizophrenia. Proc Natl Acad Sci USA 102:8627–8632

Cannon TD, Hennah W, van Erp TG, Thompson PM, Lonnqvist J, Huttunen M, Gasperoni T, Tuulio-Henriksson A, Pirkola T, Toga AW, Kaprio J, Mazziotta J, Peltonen L (2005) Association of DISC1/TRAX haplotypes with schizophrenia, reduced prefrontal gray matter, and impaired short- and long-term memory. Arch Gen Psychiatry 62:1205–1213

Carlezon WA Jr, Duman RS, Nestler EJ (2005) The many faces of CREB. Trends Neurosci 28:436–445

Cherry JA, Davis RL (1999) Cyclic AMP phosphodiesterases are localized in regions of the mouse brain associated with reinforcement, movement, and affect. J Comp Neurol 407:287–301

Chi P, Greengard P, Ryan TA (2003) Synaptic vesicle mobilization is regulated by distinct synapsin I phosphorylation pathways at different frequencies. Neuron 38:69–78

Clapcote SJ, Lipina TV, Millar JK, Mackie S, Christie S, Ogawa F, Lerch JP, Trimble K, Uchiyama M, Sakuraba Y, Kaneda H, Shiroishi T, Houslay MD, Henkelman RM, Sled JG, Gondo Y, Porteous DJ, Roder JC (2007) Behavioral phenotypes of Disc1 missense mutations in mice. Neuron 54:387–402

Clayton EL, Evans GJ, Cousin MA (2007) Activity-dependent control of bulk endocytosis by protein dephosphorylation in central nerve terminals. J Physiol 585:687–691

Cotter D, Kerwin R, Doshi B, Martin CS, Everall IP (1997) Alterations in hippocampal non-phosphorylated MAP2 protein expression in schizophrenia. Brain Res 765:238–246

Cotter D, Wilson S, Roberts E, Kerwin R, Everall IP (2000) Increased dendritic MAP2 expression in the hippocampus in schizophrenia. Schizophr Res 41:313–323

Cousin MA, Robinson PJ (2001) The dephosphins: dephosphorylation by calcineurin triggers synaptic vesicle endocytosis. Trends Neurosci 24:659–665

Davis RL (1996) Physiology and biochemistry of *Drosophila* learning mutants. Physiol Rev 76:299–2317

Di Giorgio A, Blasi G, Sambataro F, Rampino A, Papazacharias A, Gambi F, Romano R, Caforio G, Rizzo M, Latorre V, Popolizio T, Kolachana B, Callicott JH, Nardini M, Weinberger DR, Bertolino A (2008) Association of the SerCys DISC1 polymorphism with human hippocampal formation gray matter and function during memory encoding. Eur J Neurosci 28:2129–2136

Du K, Montminy M (1998) CREB is a regulatory target for the protein kinase Akt/PKB. J Biol Chem 273:32377–32379

Duan X, Chang JH, Ge S, Faulkner RL, Kim JY, Kitabatake Y, Liu XB, Yang CH, Jordan JD, Ma DK, Liu CY, Ganesan S, Cheng HJ, Ming GL, Lu B, Song H (2007) Disrupted-In-Schizophrenia 1 regulates integration of newly generated neurons in the adult brain. Cell 130:1146–1158

Duman RS, Nakagawa S, Malberg J (2001) Regulation of adult neurogenesis by antidepressant treatment. Neuropsychopharmacology 25:836–844

Eastwood SL, Salih T, Harrison PJ (2005) Differential expression of calcineurin A subunit mRNA isoforms during rat hippocampal and cerebellar development. Eur J Neurosci 22:3017–3024

Emamian ES, Hall D, Birnbaum MJ, Karayiorgou M, Gogos JA (2004) Convergent evidence for impaired AKT1-GSK3beta signaling in schizophrenia. Nat Genet 36:131–137

Fatemi SH, King DP, Reutiman TJ, Folsom TD, Laurence JA, Lee S, Fan YT, Paciga SA, Conti M, Menniti FS (2008a) PDE4B polymorphisms and decreased PDE4B expression are associated with schizophrenia. Schizophr Res 101:36–49

Fatemi SH, Reutiman TJ, Folsom TD, Lee S (2008b) Phosphodiesterase-4A expression is reduced in cerebella of patients with bipolar disorder. Psychiatr Genet 18:282–288

Feldcamp LA, Souza RP, Romano-Silva M, Kennedy JL, Wong AH (2008) Reduced prefrontal cortex DARPP-32 mRNA in completed suicide victims with schizophrenia. Schizophr Res 103:192–200

Ferreira A, Kincaid R, Kosik KS (1993) Calcineurin is associated with the cytoskeleton of cultured neurons and has a role in the acquisition of polarity. Mol Biol Cell 4:1225–1238

Frame S, Cohen P (2001) GSK3 takes centre stage more than 20 years after its discovery. Biochem J 359:1–16

Gainetdinov RR, Caron MG (2003) Monoamine transporters: from genes to behavior. Annu Rev Pharmacol Toxicol 43:261–284

Gelinas JN, Banko JL, Peters MM, Klann E, Weeber EJ, Nguyen PV (2008) Activation of exchange protein activated by cyclic-AMP enhances long-lasting synaptic potentiation in the hippocampus. Learn Mem 15:403–411

Gerber DJ, Hall D, Miyakawa T, Demars S, Gogos JA, Karayiorgou M, Tonegawa S (2003) Evidence for association of schizophrenia with genetic variation in the 8p21.3 gene, PPP3CC, encoding the calcineurin gamma subunit. Proc Natl Acad Sci USA 100:8993–8998

Glantz LA, Lewis DA (2001) Dendritic spine density in schizophrenia and depression. Arch Gen Psychiatry 58:203

Goto S, Yamamoto H, Fukunaga K, Iwasa T, Matsukado Y, Miyamoto E (1985) Dephosphorylation of microtubule-associated protein 2, tau factor, and tubulin by calcineurin. J Neurochem 45:276–283

Goto S, Matsukado Y, Mihara Y, Inoue N, Miyamoto E (1986) The distribution of calcineurin in rat brain by light and electron microscopic immunohistochemistry and enzyme-immunoassay. Brain Res 397:161–172

Graef IA, Mermelstein PG, Stankunas K, Neilson JR, Deisseroth K, Tsien RW, Crabtree GR (1999) L-type calcium channels and GSK-3 regulate the activity of NF-ATc4 in hippocampal neurons. Nature 401(703):70–78

Graef IA, Chen F, Chen L, Kuo A, Crabtree GR (2001) Signals transduced by Ca(2+)/calcineurin and NFATc3/c4 pattern the developing vasculature. Cell 105:863–875

Guo HF, Zhong Y (2006) Requirement of Akt to mediate long-term synaptic depression in *Drosophila*. J Neurosci 26:4004–4014

Hagiwara M, Alberts A, Brindle P, Meinkoth J, Feramisco J, Deng T, Karin M, Shenolikar S, Montminy M (1992) Transcriptional attenuation following cAMP induction requires PP-1-mediated dephosphorylation of CREB. Cell 70:105–113

Halpain S, Hipolito A, Saffer L (1998) Regulation of F-actin stability in dendritic spines by glutamate receptors and calcineurin. J Neurosci 18:9835–9844

Hashimoto H, Shintani N, Tanaka K, Mori W, Hirose M, Matsuda T, Sakaue M, Miyazaki J, Niwa H, Tashiro F, Yamamoto K, Koga K, Tomimoto S, Kunugi A, Suetake S, Baba A (2001) Altered psychomotor behaviors in mice lacking pituitary adenylate cyclase-activating polypeptide (PACAP). Proc Natl Acad Sci USA 98:13355–13360

Hashimoto R, Hashimoto H, Shintani N, Chiba S, Hattori S, Okada T, Nakajima M, Tanaka K, Kawagishi N, Nemoto K, Mori T, Ohnishi T, Noguchi H, Hori H, Suzuki T, Iwata N, Ozaki N, Nakabayashi T, Saitoh O, Kosuga A, Tatsumi M, Kamijima K, Weinberger DR, Kunugi H, Baba A (2007) Pituitary adenylate cyclase-activating polypeptide is associated with schizophrenia. Mol Psychiatry 12:1026–1032

Hennah W, Tuulio-Henriksson A, Paunio T, Ekelund J, Varilo T, Partonen T, Cannon TD, Lonnqvist J, Peltonen L (2005) A haplotype within the DISC1 gene is associated with visual memory functions in families with a high density of schizophrenia. Mol Psychiatry 10:1097–1103

Hikida T, Jaaro-Peled H, Seshadri S, Oishi K, Hookway C, Kong S, Wu D, Xue R, Andrade M, Tankou S, Mori S, Gallagher M, Ishizuka K, Pletnikov M, Kida S, Sawa A (2007) From the Cover: dominant-negative DISC1 transgenic mice display schizophrenia-associated phenotypes detected by measures translatable to humans. Proc Natl Acad Sci USA 104:14501–14506

Hill JJ, Hashimoto T, Lewis DA (2006) Molecular mechanisms contributing to dendritic spine alterations in the prefrontal cortex of subjects with schizophrenia. Mol Psychiatry 11:557–566

Horiuchi Y, Ishiguro H, Koga M, Inada T, Iwata N, Ozaki N, Ujike H, Muratake T, Someya T, Arinami T (2007) Support for association of the PPP3CC gene with schizophrenia. Mol Psychiatry 12:891–893

Houslay MD, Adams DR (2003) PDE4 cAMP phosphodiesterases: modular enzymes that orchestrate signalling cross-talk, desensitization and compartmentalization. Biochem J 370:1–18

Ishiguro H, Ohtsuki T, Okubo Y, Kurumaji A, Arinami T (2001) Association analysis of the pituitary adenyl cyclase activating peptide gene (PACAP) on chromosome 18p11 with schizophrenia and bipolar disorders. J Neural Transm 108:849–854

James R, Adams RR, Christie S, Buchanan SR, Porteous DJ, Millar JK (2004) Disrupted in Schizophrenia 1 (DISC1) is a multicompartmentalized protein that predominantly localizes to mitochondria. Mol Cell Neurosci 26:112–122

Jaworski J, Spangler S, Seeburg DP, Hoogenraad CC, Sheng M (2005) Control of dendritic arborization by the phosphoinositide-3′-kinase-Akt-mammalian target of rapamycin pathway. J Neurosci 25:11300–11312

Jope RS, Roh MS (2006) Glycogen synthase kinase-3 (GSK3) in psychiatric diseases and therapeutic interventions. Curr Drug Targets 7:1421–1434

Kalus P, Bondzio J, Federspiel A, Muller TJ, Zuschratter W (2002) Cell-type specific alterations of cortical interneurons in schizophrenic patients. NeuroReport 13:713–717

Kanes SJ, Tokarczyk J, Siegel SJ, Bilker W, Abel T, Kelly MP (2007) Rolipram: a specific phosphodiesterase 4 inhibitor with potential antipsychotic activity. Neuroscience 144:239–246

Kawanishi Y, Harada S, Tachikawa H, Okubo T, Shiraishi H (1999) Novel variants in the promoter region of the CREB gene in schizophrenic patients. J Hum Genet 44:428–430

Kelly MP, Isiegas C, Cheung YF, Tokarczyk J, Yang X, Esposito MF, Rapoport DA, Fabian SA, Siegel SJ, Wand G, Houslay MD, Kanes SJ, Abel T (2007) Constitutive activation of Galphas within forebrain neurons causes deficits in sensorimotor gating because of PKA-dependent decreases in cAMP. Neuropsychopharmacology 32:577–588

Kelly MP, Stein JM, Vecsey CG, Favilla C, Yang X, Bizily SF, Esposito MF, Wand G, Kanes SJ, Abel T (2009) Developmental etiology for neuroanatomical and cognitive deficits in mice overexpressing Galphas, a G-protein subunit genetically linked to schizophrenia. Mol Psychiatry 14:398–415

Kinoshita Y, Suzuki T, Ikeda M, Kitajima T, Yamanouchi Y, Inada T, Yoneda H, Iwata N, Ozaki N (2005) No association with the calcineurin A gamma subunit gene (PPP3CC) haplotype to Japanese schizophrenia. J Neural Transm 112:1255–1262

Kirchner L, Weitzdoerfer R, Hoeger H, Url A, Schmidt P, Engelmann M, Villar SR, Fountoulakis M, Lubec G, Lubec B (2004) Impaired cognitive performance in neuronal nitric oxide synthase knockout mice is associated with hippocampal protein derangements. Nitric Oxide 11:316–330

Koike H, Arguello PA, Kvajo M, Karayiorgou M, Gogos JA (2006) Disc1 is mutated in the 129S6/SvEv strain and modulates working memory in mice. Proc Natl Acad Sci USA 103:3693–3697

Kolluri N, Sun Z, Sampson AR, Lewis DA (2005) Lamina-specific reductions in dendritic spine density in the prefrontal cortex of subjects with schizophrenia. Am J Psychiatry 162:1200–1202

Krishnan V, Graham A, Mazei-Robison MS, Lagace DC, Kim KS, Birnbaum S, Eisch AJ, Han PL, Storm DR, Zachariou V, Nestler EJ (2008) Calcium-sensitive adenylyl cyclases in depression and anxiety: behavioral and biochemical consequences of isoform targeting. Biol Psychiatry 64:336–343

Kvajo M, McKellar H, Arguello PA, Drew LJ, Moore H, Macdermott AB, Karayiorgou M, Gogos JA (2008) A mutation in mouse Disc1 that models a schizophrenia risk allele leads to specific alterations in neuronal architecture and cognition. Proc Natl Acad Sci USA 105:7076–7081

Lai WS, Xu B, Westphal KG, Paterlini M, Olivier B, Pavlidis P, Karayiorgou M, Gogos JA (2006) Akt1 deficiency affects neuronal morphology and predisposes to abnormalities in prefrontal cortex functioning. Proc Natl Acad Sci USA 103:16906–16911

Lang UE, Puls I, Muller DJ, Strutz-Seebohm N, Gallinat J (2007) Molecular mechanisms of schizophrenia. Cell Physiol Biochem 20:687–702

Lau CG, Zukin RS (2007) NMDA receptor trafficking in synaptic plasticity and neuropsychiatric disorders. Nat Rev Neurosci 8:413–426

Lin CH, Yeh SH, Lin CH, Lu KT, Leu TH, Chang WC, Gean PW (2001) A role for the PI-3 kinase signaling pathway in fear conditioning and synaptic plasticity in the amygdala. Neuron 31:841–851

Liu YL, Fann CS, Liu CM, Chen WJ, Wu JY, Hung SI, Chen CH, Jou YS, Liu SK, Hwang TJ, Hsieh MH, Ouyang WC, Chan HY, Chen JJ, Yang WC, Lin CY, Lee SF, Hwu HG (2006) A single nucleotide polymorphism fine mapping study of chromosome 1q42.1 reveals the vulnerability genes for schizophrenia, GNPAT and DISC1: association with impairment of sustained attention. Biol Psychiatry 60:554–562

Liu YL, Fann CS, Liu CM, Chang CC, Yang WC, Hung SI, Yu SL, Hwang TJ, Hsieh MH, Liu CC, Tsuang MM, Wu JY, Jou YS, Faraone SV, Tsuang MT, Chen WJ, Hwu HG (2007) More evidence supports the association of PPP3CC with schizophrenia. Mol Psychiatry 12:966–974

Lovestone S, Killick R, Di FM, Murray R (2007) Schizophrenia as a GSK-3 dysregulation disorder. Trends Neurosci 30:142–149

Lynch MJ, Hill EV, Houslay MD (2006) Intracellular targeting of phosphodiesterase-4 underpins compartmentalized cAMP signaling. Curr Top Dev Biol 75:225–259

MacKenzie SJ, Baillie GS, McPhee I, MacKenzie C, Seamons R, McSorley T, Millen J, Beard MB, HG H, van MD (2002) Long PDE4 cAMP specific phosphodiesterases are activated by protein kinase A-mediated phosphorylation of a single serine residue in Upstream Conserved Region 1 (UCR1). Br J Pharmacol 136:421–433

Malaterre J, Ramsay RG, Mantamadiotis T (2007) Wnt-Frizzled signalling and the many paths to neural development and adult brain homeostasis. Front Biosci 12:492–506

Malleret G, Haditsch U, Genoux D, Jones MW, Bliss TV, Vanhoose AM, Weitlauf C, Kandel ER, Winder DG, Mansuy IM (2001) Inducible and reversible enhancement of learning, memory, and long-term potentiation by genetic inhibition of calcineurin. Cell 104:675–686

Mansuy IM, Mayford M, Jacob B, Kandel ER, Bach ME (1998) Restricted and regulated overexpression reveals calcineurin as a key component in the transition from short-term to long-term memory. Cell 92:39–49

Mao Y, Ge X, Frank CL, Madison JM, Koehler AN, Doud MK, Tassa C, Berry EM, Soda T, Singh KK, Biechele T, Petryshen TL, Moon RT, Haggarty SJ, Tsai LH (2009) Disrupted in schizophrenia 1 regulates neuronal progenitor proliferation via modulation of GSK3beta/beta-catenin signaling. Cell 136:1017–1031

Mathieu F, Miot S, Etain B, El Khoury MA, Chevalier F, Bellivier F, Leboyer M, Giros B, Tzavara ET (2008) Association between the PPP3CC gene, coding for the calcineurin gamma catalytic subunit, and bipolar disorder. Behav Brain Funct 4:2

Middeldorp CM, Vink JM, Hettema JM, de Geus EJ, Kendler KS, Willemsen G, Neale MC, Boomsma DI, Chen X (2009) An association between Epac-1 gene variants and anxiety and depression in two independent samples. Am J Med Genet B Neuropsychiatr Genet 153:214–219

Millar JK, Wilson-Annan JC, Anderson S, Christie S, Taylor MS, Semple CA, Devon RS, Clair DM, Muir WJ, Blackwood DH, Porteous DJ (2000) Disruption of two novel genes by a translocation co-segregating with schizophrenia. Hum Mol Genet 9:1415–1423

Millar JK, Pickard BS, Mackie S, James R, Christie S, Buchanan SR, Malloy MP, Chubb JE, Huston E, Baillie GS, Thomson PA, Hill EV, Brandon NJ, Rain JC, Camargo LM, Whiting PJ, Houslay MD, Blackwood DH, Muir WJ, Porteous DJ (2005) DISC1 and PDE4B are interacting genetic factors in schizophrenia that regulate cAMP signaling. Science 310:1187–1191

Minoretti P, Politi P, Coen E, Di VC, Bertona M, Bianchi M, Emanuele E (2006) The T393C polymorphism of the GNAS1 gene is associated with deficit schizophrenia in an Italian population sample. Neurosci Lett 397:159–163

Miyakawa T, Leiter LM, Gerber DJ, Gainetdinov RR, Sotnikova TD, Zeng H, Caron MG, Tonegawa S (2003) Conditional calcineurin knockout mice exhibit multiple abnormal behaviors related to schizophrenia. Proc Natl Acad Sci USA 100:8987–8992

Mizuno M, Yamada K, Takei N, Tran MH, He J, Nakajima A, Nawa H, Nabeshima T (2003) Phosphatidylinositol 3-kinase: a molecule mediating BDNF-dependent spatial memory formation. Mol Psychiatry 8:217–224

Morris JA, Kandpal G, Ma L, Austin CP (2003) DISC1 (Disrupted-In-Schizophrenia 1) is a centrosome-associated protein that interacts with MAP1A, MIPT3, ATF4/5 and NUDEL: regulation and loss of interaction with mutation. Hum Mol Genet 12:1591–1608

Mukai J, Dhilla A, Drew LJ, Stark KL, Cao L, Macdermott AB, Karayiorgou M, Gogos JA (2008) Palmitoylation-dependent neurodevelopmental deficits in a mouse model of 22q11 microdeletion. Nat Neurosci 11:1302–1310

Murdoch H, Mackie S, Collins DM, Hill EV, Bolger GB, Klussmann E, Porteous DJ, Millar JK, Houslay MD (2007) Isoform-selective susceptibility of DISC1/phosphodiesterase-4 complexes to dissociation by elevated intracellular cAMP levels. J Neurosci 27:9513–9524

Nakagawa S, Kim JE, Lee R, Malberg JE, Chen J, Steffen C, Zhang YJ, Nestler EJ, Duman RS (2002) Regulation of neurogenesis in adult mouse hippocampus by cAMP and the cAMP response element-binding protein. J Neurosci 22:3673–3682

Nguyen T, Di GS (2008) NFAT signaling in neural development and axon growth. Int J Dev Neurosci 26:141–145

Nguyen PV, Woo NH (2003) Regulation of hippocampal synaptic plasticity by cyclic AMP-dependent protein kinases. Prog Neurobiol 71:401–437

Nishi A, Bibb JA, Matsuyama S, Hamada M, Higashi H, Nairn AC, Greengard P (2002) Regulation of DARPP-32 dephosphorylation at PKA- and Cdk5-sites by NMDA and AMPA receptors: distinct roles of calcineurin and protein phosphatase-2A. J Neurochem 81:832–841

Nishino N, Kitamura N, Hashimoto T, Kajimoto Y, Shirai Y, Murakami N, Nakai T, Komure O, Shirakawa O, Mita T (1993) Increase in [3H]cAMP binding sites and decrease in Gi alpha and Go alpha immunoreactivities in left temporal cortices from patients with schizophrenia. Brain Res 615:41–49

Numata S, Ueno S, Iga J, Song H, Nakataki M, Tayoshi S, Sumitani S, Tomotake M, Itakura M, Sano A, Ohmori T (2008) Positive association of the PDE4B (phosphodiesterase 4B) gene with schizophrenia in the Japanese population. J Psychiatr Res 43:7–12

Numata S, Iga J, Nakataki M, Tayoshi S, Taniguchi K, Sumitani S, Tomotake M, Tanahashi T, Itakura M, Kamegaya Y, Tatsumi M, Sano A, Asada T, Kunugi H, Ueno S, Ohmori T (2009) Gene expression and association analyses of the phosphodiesterase 4B (PDE4B) gene in major depressive disorder in the Japanese population. Am J Med Genet B Neuropsychiatr Genet 150B:527–534

Ogawa F, Kasai M, Akiyama T (2005) A functional link between Disrupted-In-Schizophrenia 1 and the eukaryotic translation initiation factor 3. Biochem Biophys Res Commun 338: 771–776

Okada F, Tokumitsu Y, Takahashi N, Crow TJ, Roberts GW (1994) Reduced concentrations of the alpha-subunit of GTP-binding protein Go in schizophrenic brain. J Neural Transm Gen Sect 95:95–104

Ouyang M, Zhang L, Zhu JJ, Schwede F, Thomas SA (2008) Epac signaling is required for hippocampus-dependent memory retrieval. Proc Natl Acad Sci USA 105:11993–11997

Owada Y, Utsunomiya A, Yoshimoto T, Kondo H (1997) Expression of mRNA for Akt, serine–threonine protein kinase, in the brain during development and its transient enhancement following axotomy of hypoglossal nerve. J Mol Neurosci 9:27–33

Perez-Torres S, Miro X, Palacios JM, Cortes R, Puigdomenech P, Mengod G (2000) Phosphodiesterase type 4 isozymes expression in human brain examined by in situ hybridization histochemistry and [3H]rolipram binding autoradiography. Comparison with monkey and rat brain. J Chem Neuroanat 20:349–374

Pickard BS, Thomson PA, Christoforou A, Evans KL, Morris SW, Porteous DJ, Blackwood DH, Muir WJ (2007) The PDE4B gene confers sex-specific protection against schizophrenia. Psychiatr Genet 17:129–133

Pletnikov MV, Ayhan Y, Nikolskaia O, Xu Y, Ovanesov MV, Huang H, Mori S, Moran TH, Ross CA (2007) Inducible expression of mutant human DISC1 in mice is associated with brain and behavioral abnormalities reminiscent of schizophrenia. Mol Psychiatry 13(2):173–186

Pletnikov MV, Ayhan Y, Xu Y, Nikolskaia O, Ovanesov M, Huang H, Mori S, Moran TH, Ross CA (2008) Enlargement of the lateral ventricles in mutant DISC1 transgenic mice. Mol Psychiatry 13:115

Prata DP, Mechelli A, Fu CH, Picchioni M, Kane F, Kalidindi S, McDonald C, Kravariti E, Toulopoulou T, Miorelli A, Murray R, Collier DA, McGuire PK (2008) Effect of disrupted-in-schizophrenia-1 on pre-frontal cortical function. Mol Psychiatry 13:915–917

Qin Y, Zhu Y, Baumgart JP, Stornetta RL, Seidenman K, Mack V, van AL, Zhu JJ (2005) State-dependent Ras signaling and AMPA receptor trafficking. Genes Dev 19:2000–2015

Quinlan EM, Halpain S (1996) Emergence of activity-dependent, bidirectional control of microtubule-associated protein MAP2 phosphorylation during postnatal development. J Neurosci 16:7627–7637

Renau TE (2004) The potential of phosphodiesterase 4 inhibitors for the treatment of depression: opportunities and challenges. Curr Opin Investig Drugs 5:34–39

Rushlow WJ, Seah YH, Belliveau DJ, Rajakumar N (2005) Changes in calcineurin expression induced in the rat brain by the administration of antipsychotics. J Neurochem 94:587–596

Sadana R, Dessauer CW (2009) Physiological roles for G protein-regulated adenylyl cyclase isoforms: insights from knockout and overexpression studies. Neurosignals 17:5–22

Sanders AR, Duan J, Levinson DF, Shi J, He D, Hou C, Burrell GJ, Rice JP, Nertney DA, Olincy A, Rozic P, Vinogradov S, Buccola NG, Mowry BJ, Freedman R, Amin F, Black DW, Silverman JM, Byerley WF, Crowe RR, Cloninger CR, Martinez M, Gejman PV (2008) No significant association of 14 candidate genes with schizophrenia in a large European ancestry sample: implications for psychiatric genetics. Am J Psychiatry 165:497–506

Sawamura N, Sawamura-Yamamoto T, Ozeki Y, Ross CA, Sawa A (2005) A form of DISC1 enriched in nucleus: altered subcellular distribution in orbitofrontal cortex in psychosis and substance/alcohol abuse. Proc Natl Acad Sci USA 102:118711–118792

Schreiber G, Avissar S (2000) G proteins as a biochemical tool for diagnosis and monitoring treatments of mental disorders. Isr Med Assoc J 2(Suppl):86–91

Shibasaki F, Hallin U, Uchino H (2002) Calcineurin as a multifunctional regulator. J Biochem 131:1–15

Siuciak JA, McCarthy SA, Chapin DS, Martin AN (2008) Behavioral and neurochemical characterization of mice deficient in the phosphodiesterase-4B (PDE4B) enzyme. Psychopharmacology (Berl) 197:115–126

Snyder GL, Galdi S, Fienberg AA, Allen P, Nairn AC, Greengard P (2003) Regulation of AMPA receptor dephosphorylation by glutamate receptor agonists. Neuropharmacology 45:703–713

St Clair CD, Blackwood D, Muir W, Carothers A, Walker M, Spowart G, Gosden C, Evans HJ (1990) Association within a family of a balanced autosomal translocation with major mental illness. Lancet 336:13–16

Stark KL, Xu B, Bagchi A, Lai WS, Liu H, Hsu R, Wan X, Pavlidis P, Mills AA, Karayiorgou M, Gogos JA (2008) Altered brain microRNA biogenesis contributes to phenotypic deficits in a 22q11-deletion mouse model. Nat Genet 40:751–760

Svenningsson P, Nishi A, Fisone G, Girault JA, Nairn AC, Greengard P (2004) DARPP-32: an integrator of neurotransmission. Annu Rev Pharmacol Toxicol 44:269–296

Szeszko PR, Hodgkinson CA, Robinson DG, DeRosse P, Bilder RM, Lencz T, Burdick KE, Napolitano B, Betensky JD, Kane JM, Goldman D, Malhotra AK (2008) DISC1 is associated with prefrontal cortical gray matter and positive symptoms in schizophrenia. Biol Psychol 79:103–110

Tan HY, Nicodemus KK, Chen Q, Li Z, Brooke JK, Honea R, Kolachana BS, Straub RE, Meyer-Lindenberg A, Sei Y, Mattay VS, Callicott JH, Weinberger DR (2008) Genetic variation in AKT1 is linked to dopamine-associated prefrontal cortical structure and function in humans. J Clin Invest 118:2200–2208

Taya S, Shinoda T, Tsuboi D, Asaki J, Nagai K, Hikita T, Kuroda S, Kuroda K, Shimizu M, Hirotsune S, Iwamatsu A, Kaibuchi K (2007) DISC1 regulates the transport of the NUDEL/LIS1/14-3-3epsilon complex through kinesin-1. J Neurosci 27:15–26

Taylor MS, Devon RS, Millar JK, Porteous DJ (2003) Evolutionary constraints on the disrupted in schizophrenia locus. Genomics 81:67–77

Tcherepanov AA, Sokolov BP (1997) Age-related abnormalities in expression of mRNAs encoding synapsin 1A, synapsin 1B, and synaptophysin in the temporal cortex of schizophrenics. J Neurosci Res 49:639–644

Thiselton DL, Vladimirov VI, Kuo PH, McClay J, Wormley B, Fanous A, O'Neill FA, Walsh D, Van den Oord EJ, Kendler KS, Riley BP (2008) AKT1 is associated with schizophrenia across multiple symptom dimensions in the Irish study of high density schizophrenia families. Biol Psychiatry 63:449–457

Tolias KF, Bikoff JB, Burette A, Paradis S, Harrar D, Tavazoie S, Weinberg RJ, Greenberg ME (2005) The Rac1-GEF Tiam1 couples the NMDA receptor to the activity-dependent development of dendritic arbors and spines. Neuron 45:525–538

Tomppo L, Hennah W, Lahermo P, Loukola A, Tuulio-Henriksson A, Suvisaari J, Partonen T, Ekelund J, Lonnqvist J, Peltonen L (2009) Association between genes of disrupted in schizophrenia 1 (DISC1) interactors and schizophrenia supports the role of the DISC1 pathway in the etiology of major mental illnesses. Biol Psychiatry 65:1055–1062

Wang J, Liu S, Haditsch U, Tu W, Cochrane K, Ahmadian G, Tran L, Paw J, Wang Y, Mansuy I, Salter MM, Lu YM (2003a) Interaction of calcineurin and type-A GABA receptor gamma 2 subunits produces long-term depression at CA1 inhibitory synapses. J Neurosci 23:826–836

Wang Q, Liu L, Pei L, Ju W, Ahmadian G, Lu J, Wang Y, Liu F, Wang YT (2003b) Control of synaptic strength, a novel function of Akt. Neuron 38:915–928

Wang K, Zhang H, Ma D, Bucan M, Glessner JT, Abrahams BS, Salyakina D, Imielinski M, Bradfield JP, Sleiman PM, Kim CE, Hou C, Frackelton E, Chiavacci R, Takahashi N, Sakurai T, Rappaport E, Lajonchere CM, Munson J, Estes A, Korvatska O, Piven J, Sonnenblick LI, Varez Retuerto AI, Herman EI, Dong H, Hutman T, Sigman M, Ozonoff S, Klin A, Owley T, Sweeney JA, Brune CW, Cantor RM, Bernier R, Gilbert JR, Cuccaro ML, McMahon WM, Miller J, State MW, Wassink TH, Coon H, Levy SE, Schultz RT, Nurnberger JI, Haines JL, Sutcliffe JS, Cook EH, Minshew NJ, Buxbaum JD, Dawson G, Grant SF, Geschwind DH, Pericak-Vance MA, Schellenberg GD, Hakonarson H (2009) Common genetic variants on 5p14.1 associate with autism spectrum disorders. Nature 459:528–533

Xi Z, Yu L, Shi Y, Zhang J, Zheng Y, He G, He L, Wei Q, Yao W, Zhang K, Gu N, Feng G, Zhu S (2007) No association between PPP3CC and schizophrenia in the Chinese population. Schizophr Res 90:357–359

Yamada K, Gerber DJ, Iwayama Y, Ohnishi T, Ohba H, Toyota T, Aruga J, Minabe Y, Tonegawa S, Yoshikawa T (2007) Genetic analysis of the calcineurin pathway identifies members of the EGR gene family, specifically EGR3, as potential susceptibility candidates in schizophrenia. Proc Natl Acad Sci USA 104:2815–2820

Yang CQ, Kitamura N, Nishino N, Shirakawa O, Nakai H (1998) Isotype-specific G protein abnormalities in the left superior temporal cortex and limbic structures of patients with chronic schizophrenia. Biol Psychiatry 43:12–19

Yang ZZ, Tschopp O, Baudry A, Dummler B, Hynx D, Hemmings BA (2004) Physiological functions of protein kinase B/Akt. Biochem Soc Trans 32:350–354

Yoshimi A, Takahashi N, Saito S, Ito Y, Aleksic B, Usui H, Kawamura Y, Waki Y, Yoshikawa T, Kato T, Iwata N, Inada T, Noda Y, Ozaki N (2008) Genetic analysis of the gene coding for DARPP-32 (PPP1R1B) in Japanese patients with schizophrenia or bipolar disorder. Schizophr Res 100:334–341

Zeng H, Chattarji S, Barbarosie M, Rondi-Reig L, Philpot BD, Miyakawa T, Bear MF, Tonegawa S (2001) Forebrain-specific calcineurin knockout selectively impairs bidirectional synaptic plasticity and working/episodic-like memory. Cell 107:617–629

Zhang HT, Huang Y, Jin SL, Frith SA, Suvarna N, Conti M, O'donnell JM (2002) Antidepressant-like profile and reduced sensitivity to rolipram in mice deficient in the PDE4D phosphodiesterase enzyme. Neuropsychopharmacology 27:587–595

Zhang HT, Huang Y, Masood A, Stolinski LR, Li Y, Zhang L, Dlaboga D, Jin SL, Conti M, O'donnell JM (2008) Anxiogenic-like behavioral phenotype of mice deficient in phosphodiesterase 4B (PDE4B). Neuropsychopharmacology 33:1611–1623

Zhao Z, Ksiezak-Reding H, Riggio S, Haroutunian V, Pasinetti GM (2006) Insulin receptor deficits in schizophrenia and in cellular and animal models of insulin receptor dysfunction. Schizophr Res 84:1–14

Znamensky V, Akama KT, McEwen BS, Milner TA (2003) Estrogen levels regulate the subcellular distribution of phosphorylated Akt in hippocampal CA1 dendrites. J Neurosci 23:2340–2347

Index

A
Abstract reasoning, 378
ACG. *See* Anterior cingulate gyrus
Adenylyl cyclase, 145, 151, 152, 644–646
Adult neurogenesis, 644
Affective processing, 382
Agency, 572
Akt
   animal models, 636
   and dopaminergic signaling, 634, 648
   expression, 632–634
   gene family, 632
   GSK3β signaling, 631–635
   and neuronal plasticity, 632–633, 647
   in schizophrenia, 632–635
   in transcriptional regulation, 637
$\alpha 7$, 416, 419
α-adrenergic receptors, 149, 150
α-amino-3-hydroxy-5-methyl-4-isoxazolepropionic acid (AMPA), 536
α_7 nicotinic receptor, 538, 539
Amantadine, 71, 76
AMG. *See* Amygdala
Amino acid methionine, 621
AMPA. *See* α-amino-3-hydroxy-5-methyl-4-isoxazolepropionic acid
AMPA receptor modulators, 61, 66
Amygdala (AMG), 245, 246, 250–254, 271, 275
Analysis of covariance, 198
Animal models, 391–420, 542–543
Anterior cingulate gyrus (ACG), 245, 246, 254, 256, 260, 263, 271
Antidepressants, 107, 112, 114
Antipsychotic medications, in PPI, 354, 358–360
Antipsychotics, 459, 517, 563, 566–568, 571–573
Apoptosis, 108
Ara-C. *See* Cytosine arabinocide
Association analysis, 591–593
Atomoxetine, 71, 76
Attention, 376–378, 384, 441
Attentional set-shifting task, 401–402
Atypical antipsychotic, 143, 145
Autonomy, 146

B
Bacteria, 446
Behavioral tasks
   auditory oddball, 201, 202, 204
   delayed response, 194, 195
   N-back, 194–196, 198, 202
   sentence completion, 201
   Sternberg's item recognition task, 196, 203
   verbal fluency (word production), 200, 201, 205
β-catenin, 635, 646
Biodegradable, 155–159, 163
Bonn Scale for the Assessment of Basic Symptoms (BSABS), 100
Brain-derived neurotrophic factor (BDNF), 538, 539
Brain networks
   small world networks, 204
   task negative networks, 203
   task positive networks, 203
Brain regions
   amygdala, 185
   anterior cingulate, 185, 189
   caudate nucleus, 185, 196
   frontal lobe
      dorsolateral prefrontal cortex (DLPFC), 195, 196, 200–203, 205

inferior frontal gyrus, 185
medial frontal, 201
middle frontal gyrus, 196, 201
orbitofrontal, 189
prefrontal, 189, 198, 205
occipital-parietal regions, 191
occipital-temporal region, 191
temporal lobe
  entorhinal cortex, 196
  hippocampus, 202
  insula, 185
  middle temporal gyrus, 185, 201
  parahippocampal gyrus, 202
  perirhinal cortex, 196
  superior temporal gyrus, 184, 190
BSABS. *See* Bonn Scale for the Assessment of Basic Symptoms

C

CAARMS. *See* Comprehensive Assessment of At-Risk Mental State
Calcineurin
  animal models, 636
  in cytoskeletal phosphorylation, 639
  in dopaminergic signaling, 648
  expression, 635–637
  in nitric oxide signaling, 638
  in schizophrenia, 636–639
  in synaptic plasticity, 637–638
  in transcriptional regulation, 637
  in vesicle trafficking, 638–639
cAMP. *See* Cyclic adenosine monophosphate
Cannabis, 495–500
Cardiovascular risk, 26–29, 31–33
Catechol-O-methyl-transferase (COMT), 414–416, 418–419, 492, 493
CATIE. *See* Clinical Antipsychotic Trials of Intervention Effectiveness
Caudate nucleus 32, 244, 264–268
CB1 receptor, 497, 498
CBT. *See* Cognitive and behavioral therapy
CBV. *See* Cerebral blood volume
Cerebral blood flow, 183, 192
Cerebral blood volume (CBV), 532, 541
Cesarean section (C-section), 450, 451
Chlorpromazine equivalents, 146–149
5-Choice serial reaction-time task, 401
Cholinergic agents, cognition
  cholinesterase inhibitors, 47, 51
  muscarinic receptors, 55–60
  nicotine and nicotinic receptor agonists, 51, 54–59

Cholinergic hypothesis, 405
Cholinergic receptors, 149
Cholinesterase inhibitors
  donepezil studies in, 47–49
  galantamine studies in, 47, 51–53
  rivastigmine studies in, 47, 50
Chromatin, 612–622
Clinical Antipsychotic Trials of Intervention Effectiveness (CATIE) Schizophrenia Trial
  baseline cardiometabolic profile of subjects, 30
  cardiometabolic outcomes, 25
  design, 25
  study, 567
Clinical staging criteria, 111–112, 114
CNTRICS. *See* Cognitive Neuroscience Treatment Research to Improve Cognition in Schizophrenia
Cognition, 283
Cognition, pharmacological strategies
  cholinergic agents, 46–60
  control tasks, 68
  dopaminergic agents, 70–77
  executive control, 45, 47, 54
  gamma-aminobutyric acid modulating agents, 67–69
  glutamatergic agents, 60–67
  improvement in, 71, 76, 83
  mifepristone, 80
  minocycline, 80
  mirtazapine, 80
  mixed/unique agent studies, 81–82
  modafinil, 77–79
  pregnenolone, 80
  research on, 44–45
  working memory, 68, 70, 71
Cognitive and behavioral therapy (CBT), 569, 570
Cognitive behavioral therapy in prodrome, 105
Cognitive deficits, 441–442, 453, 458
Cognitive Neuroscience Treatment Research to Improve Cognition in Schizophrenia (CNTRICS), 400–402
Cognitive symptoms
  attention, 400, 401
  reasoning and problem solving, 400
  social cognition, 400
  speed of processing, 400
  verbal learning and memory, 400
  visual learning and memory, 400, 403
  working memory, 402–403
Competence distinction, 5

Index

Comprehensive Assessment of At-Risk Mental State (CAARMS), 99
Computerized neuropsychological testing, 384
COMT. *See* Catechol-O-methyl-transferase (COMT)
Confrontational naming, 380
Constructional skills, 381
Construct validity, 7, 438
Conversion rate, 101, 102, 112, 114
Copy number variations (CNVs), 363, 597
Cornu ammonis (CA), 530, 531, 535
Cortico-striato-pallido-thalamic (CSPT), 558, 561, 562, 564, 565, 567, 568, 570, 573
Countertransference, 571
CpG dinucleotides, 613, 616, 619
C-reactive protein (CRP), 26
CREB, 633, 637, 645
Critical brain regions, 183, 191
Cross-fostering, 463
CRP. *See* C-reactive protein
C-section. *See* Cesarean section
CSPT. *See* Cortico-striato-pallido-thalamic
Cyclic adenosine monophosphate (cAMP), signaling and schizophrenia, 645–646
Cytoarchitectural abnormalities, 437
Cytokines, 447
Cytosine, 614
Cytosine arabinocide (Ara-C), 461
Cytoskeleton, 633, 639

D

DARPP32, 636, 637, 645
Decanoate, 155, 161, 164
Default mode network (DMN), 201, 203, 204
Dementia, 532, 535
Dementia praecox, 182
Demethylation, 619
Dentate gyrus (DG), 530, 531, 535
Depots, 154–155, 161, 162, 164, 169, 170
Depression, 535
DG. *See* Dentate gyrus
Diffusion tensor imaging (DTI), 518–521, 523
3-[(2,4-Dimethoxy)benzylidene]anabaseine (DMXB-A), 59
Disrupted-in-schizophrenia-1 (Disc1), 414, 417–418, 448, 538, 539
 binding to PDE4, 641, 642
 and cAMP signaling, 641–646
 cellular localization, 642
 expression, 641
 and GSK3β signaling, 646
 mouse models, 641, 642

 in schizophrenia, 639, 640, 642
 splice variants, 640–642
DMN. *See* Default mode network
DNA methylation, 613–615, 618–621
Donepezil, 47–49
Dopamine (DA), 488, 489, 510, 515
 extrastriatal, 220, 224–225, 230, 231
 prepulse inhibition (PPI), startle reflex, 356–357
 release, 221, 228
 striatal, 220, 222–223
 synthesis, 223–224
 transporters, 222
Dopamine D_2 partial agonism, 127
Dopamine D_2 receptors, 125–129
Dopamine D_3 receptors, 129
Dopamine hyper-reactivity, 455
Dopamine hypothesis, 143–144, 405
Dopamine receptor type 2 (D_2) antagonists, 143
Dopaminergic agents
 atomoxetine and amantadine, 71, 76
 dihydrexidine (DAR–0100), 76–77
 indirect agonists, 71
 modulation studies in, 72–75
Dopaminergic receptors, 150
Dopaminergic signaling
 and AKT, 634, 648
 and calcineurin, 648
 and cAMP, 645
Duration of untreated psychosis, 99–101
Dysconnection hypothesis of scizophrenia, 520
Dysconnectivity hypothesis, 200
Dystrobrevin-binding protein-1 (DTNBP1), 538, 539

E

Early identification center, 100–102, 112
Ecological validity, 7, 11
EEG. *See* Electroencephalography
Effective connectivity analysis, signaling and schizophrenia, 645–646
Effort, 14
Electroencephalography (EEG), 286–288
Encoding and retrieval of memory, 540
Endophenotypes, 111, 336, 374, 383–384, 412, 415
Entorhinal cortex (EC), 245, 246, 250, 271
Environmental factors, 7
EPAC, 646
Epidemiological studies, 437

ErbB4, 538, 539
Estrogen, 108
Event-related potentials, 284, 286–287
Executive functioning, 375–379, 384
Experience sampling methods, 14
Experimental design, 464
Expressive language, 380
External validity, 13
Extraretinal processes, 331–333
Eye tracking dysfunction, 311–337

F

Face validity, 438
Fatal familial insomnia (FFI), 513
Fine motor skills, 381
First-episode studies, 517
First psychotic episode, 534, 544
fMRI. *See* Functional magnetic resonance imaging
Frontal lobe deficit hypothesis, 191
Functional brain image methods
  blood oxygen level dependent (BOLD), 197
  electroencephalography (EEG), 199, 201, 205
  functional magnetic resonance imaging (fMRI), 183, 197, 199, 201–204
  magnetoencephalography (MEG), 199
  positron emission tomography (PET), 183, 191, 193, 199–202
  single photon emission computed tomography (SPECT), 183, 193
Functional capacity, 5
Functional connectivity analysis
  coherence, 199, 201, 205
  eigenimage analysis, 202
  independent components analysis (ICA), 199, 201, 203, 204
  zero-order correlation, 199, 200, 202
Functional decline, 99, 110, 114
Functional heterogeneity, 8
Functional magnetic resonance imaging (fMRI), 518–523, 540
Functional outcomes, 5
Functional performance, 5
Functioning, 295, 300, 301
Fusiform gyrus, 268, 270, 271

G

(GABA)ergic, 533, 535–538, 543
GABA. *See* Gamma-amino butyric acid
GABA interneurons, 456

GAD. *See* Glutamic acid decarboxylase
GAD67. *See* Glutamic acid decarboxylase 67
Galantamine, 47, 51–53
(Gamma) "γ"-band synchronization, 561
Gamma-amino butyric acid (GABA), 227–228, 460, 462, 487–489, 491–500, 515
Gamma-aminobutyric acid modulating agents
  lorazepam and flumazenil, 68
  MK-0777, 68
  modulation studies in, 69
Gamma-band, 494, 499
Gating, 440
Gene–environment interactions, 448
Genetic model, 398, 412–416
Genetics, 313, 314, 336–337
Genome wide association studies (GWAS), 361
Genome wide linkage and association studies, 413
Glutamate, 228, 510, 515, 519, 520
Glutamate hypothesis of schizophrenia, 519
Glutamatergic, 150
Glutamatergic agents
  AMPA receptor modulators, 61, 66
  glycine allosteric modulators, 60–61
  modulation studies in, 62–65
  NMDA receptor antagonists, 66–67
  phosphodiesterase 5 inhibitors, 66
Glutamatergic hypothesis, 392
Glutamatergic receptors, 132–133
Glutamic acid decarboxylase 67 (GAD67), 491, 493, 494, 496
Glutamic acid decarboxylase (GAD), 537
Glycine, 150
Glycine agonists, 134
Glycine allosteric modulators, 60–61
Glycogen synthase kinase-3β (GSK3β), 631, 632, 634–635, 639–646
G-protein coupled receptor, 151, 152
G-protein receptor subunits, 645
GSK-3β. *See* Glycogen synthase kinase-3β

H

Hallucinations
  auditory/verbal, 184, 185, 189
  visual, 185
Heschl's gyrus, 268, 271, 273, 274
Hippocampus, 245–247, 250–254, 271, 277, 458, 459, 530–544
Histone modifications, 613–614, 616, 618–622

Index

2-Hit, 463, 464
Hope, 9
HPA axis, 453, 454
Hyperfrontality, 195
Hyperlocomotion, 463
Hypofrontality, 191, 193, 195

I
Immune, 443
Implantable, 161–165, 168, 170
Increased sensitivity to amphetamine, 462
Infections, 443
Inferior frontal gyrus, 257, 260, 261, 271
Inferior temporal gyrus, 270, 271
Influenza, 443, 445
Information processing, 284, 286, 287, 301
Injectable, 155, 163, 165
Intellectual functioning, 375–376
Intelligence, 375, 382
Interneurons, 511, 531, 532, 535, 537, 538, 543, 544
Intra-cranial self-stimulation, 399, 400
Intrauterine hypoxia, 451
Ionotropic N-methyl-D-aspartate (NMDA) glutamate receptors, 132
Iontophoretic, 167–168
IQ, 375, 378, 385
Isolation, 454
Isolation rearing, 455–457

K
Kainite, 536
Kraepelin, E., 182, 190, 200
KYNA. *See* Kynurenic acid
Kynurenic acid (KYNA), 447

L
Lamotrigine, 66–67
Language, 378, 380–381
Latent inhibition (LI), 440
Lateral geniculate nucleus (LGN), 511
LI. *See* Latent inhibition
Linkage, 588–592, 602
Lipopolysaccharide (LPS), 445, 446
Litter, 464
Litter effects, 438, 439, 463
Localized and distributed models, 559
Locomotion, 440
Long-term depression, 636
Long-term potentiation, 633

LPS. *See* Lipopolysaccharide
Lysine methylation, 616

M
Magnetic resonance spectroscopy (MRS), 520
Malnutrition, 448
MAM-G17 model. *See* Methylazoxymethanol-G17 model
Maternal behavior, 439
Maternal malnutrition, 449
Maternal separation, 454
MATRICS. *See* Measurement and Treatment Research to Improve Cognition in Schizophrenia
MATRICS battery, 375, 386
MATRICS-CT, 6
Measurement and Treatment Research to Improve Cognition in Schizophrenia (MATRICS), 394, 400, 401
Medial geniculate nucleus, 511
Medication-enhanced cognitive therapy, 572, 573
Mediodorsal nucleus, of thalamus, 512
Memantine, 67
Memory, 375, 379–380, 384
Memory scanning, 186
Mental flexibility, 378, 379, 384
Meta-analysis, 320, 321
Meta-analytic studies, 193, 195
Metabolic syndrome
  criteria, 28, 29
  dyslipidemia, 27, 30
Metabotropic glutamate, 150, 170
Metabotropic glutamate receptors (mGluRs), 133
Methylazoxymethanol acetate (MAM), 461
Methylazoxymethanol-G17 model, 543
mGluR2/3, 150
mGluRs. *See* Metabotropic glutamate receptors
Microparticles, 155, 157, 162
Middle frontal gyrus, 257, 260, 261, 271
Middle temporal gyrus, 269, 271
Mifepristone, 80
Minocycline, 80
Mirtazapine, 80
Mismatch negativity (MMN), 284, 295
M1/M4, 538
MMN. *See* Mismatch negativity
Modafinil, 77–79
Model of schizophrenia, 375, 385
Morris water maze, 403

Motion processing, 322, 325, 326, 328, 330, 331, 333, 335–337
Muscarinic receptors, 55–60, 538

N
N100, 291, 292
NAPLS. *See* North American Prodromal Longitudinal Studies
Naturalistic observation, 15
Negative symptoms, 190–199, 203, 204, 393, 399–400, 407–409
Neonatal hypoxia, 451
Neonatal infections, 447
Neonatal ventral hippocampal lesion (nVH), 458–460, 542
Neuregulin 1 gene (NRG1), 362, 413, 414, 416, 417, 538, 539
Neuroanatomical abnormalities, 437
Neurochemical imaging techniques
positron emission tomography (PET), 216–220, 222–229, 231, 233
single photon emission tomography (SPECT), 216, 217, 219, 220, 222, 224, 227–230
Neurocognition in schizophrenia, 373–386
Neurodegenerative, 385
Neurodevelopmental hypothesis, 437
Neurodevelopmental models, 462
Neurodevelopmental models of schizophrenia, 385
Neuroimaging, 334–337
Neuropathology, 183, 194
Neuroprotection, 107–109
Neuropsychological deficits, 374, 375, 377, 384–385
Neurorehabilitation, 572
Nicotine
and nicotinic receptor agonists, 51, 54–59
prepulse inhibition (PPI), startle reflex, 357–358
Nitric oxide signaling, 638
NMDA. *See* N-methyl-D-aspartate
NMDA-R antagonist, 461
NMDA receptor antagonists, 515
lamotrigine, 66–67
memantine, 67
NMDA receptors, 462, 490, 493
N-methyl-D-aspartate (NMDA), 533, 536, 543, 544
Nonbiodegradable, 156–157
North American Prodromal Longitudinal Studies (NAPLS), 102
Novel object recognition, 403
Novelty, 531, 543
NRG1. *See* Neuregulin 1 gene

O
Obstetric complications, 450
Occupancy, 228–232
Odor span task (OST), 402–403, 419
Omega 3 fatty acid, 108, 112, 114
Optimized antagonist, 129
Orbitofrontal cortex, 245, 257, 258
OST. *See* Odor span task

P
P300, 284, 287, 288, 293, 297–300
PACAP. *See* Pituitary adenyl cyclase activating peptide gene
Parahippocampal cortex, 246
Parenteral, 163, 164
Parvalbumin, 489, 491, 537, 543, 544
PCL. *See* Polycaprolactone
PDE4. *See* Phosphodiesterase 4
Peduncular hallucinosis, 513
Performance distinction, 5
Perinatal NMDA receptor antagonist, 462
Peripheral blood cells, 616
PET. *See* Positron emission tomography
PFC. *See* Prefrontal cortex
Pharmacological isomorphism, 393, 396, 397
Pharmacological models, 443
acute, 406
repeated/escalating, 406–407
subchronic, 405
Pharmacological studies, of PPI
dopamine, 356–357
nicotine, 357–358
Pharmacologic treatment in prodrome, 98, 100, 104
Phase I, 168, 169
Phase III, 169
Phosphodiesterase 4 (PDE4)
and adult neurogenesis, 644
expression, 643, 644
mouse models, 642
in psychiatric disorders, 643–644
Phosphodiesterase 5 (PDE5) inhibitors, 66
Picrotoxin, 543
Pituitary adenyl cyclase activating peptide gene (PACAP), 645
Placental insufficiency, 452
Planning, 378, 384

Planum temporale (PT), 268, 271, 273, 275, 276
Plasma half-life, 151
PLGA. *See* Poly(lactic-co-glycolic acid)
Poly(lactic-co-glycolic acid) (PLGA), 156–159, 163, 165
Polycaprolactone (PCL), 157–159
PolyI:C, 445
Polypharmacy, 125
Positive symptoms, 143–145, 170, 184–190, 202, 393, 398–400
Positron emission tomography (PET), 518–521, 540, 541
Postmortem brain, 612–620, 622
Post-traumatic stress disorder (PTSD), 535
Predictive validity, 438
Prefrontal cortex (PFC), 244, 252, 256–264, 271, 459, 460, 462, 558
Pregnenolone, 80
Premorbid period, 98
Prenatal, 446
Prenatal stress, 452, 453
Prenatal toxin, 461
Prepulse inhibition (PPI), 397, 403–404, 440, 454, 456, 462, 463
   animal models, 353, 355, 359, 361
   antipsychotic medications, 354, 358–360, 364
   dopamine (DA), 356–357
   endophenotype, 354, 359, 361–363
   environmental vectors, 352
   gating disorders, 353
   gating model, 353–354
   gating processes, 352
   genomic influences, 359, 361–363
   information processing, 350, 353
   nicotine, 357–358
   paradigm, 351
   plasticity, 350
   PubMed citations, 351
   variables and, relationship, 355
President's New Freedom Commission on Mental Health, 9
Preventative intervention, 562, 564, 565
Preventive treatments, 463
Primary *versus* secondary neural dysfunction, 562–563
Primum non nocere, 572
Prion disease, 513
Processing speed, 377, 383, 384
Prodromal period, 98, 109
Protein deficiency, 449
Provigil. *See* Modafinil

Pruning, 108
Psychological stress, 452
Psychosis, 532, 533, 540, 542, 544
Psychosurgery, 571, 572
Psychotherapy, 556, 569, 571
Psychotomimetics, 398, 407, 409
PT. *See* Planum temporale
PTSD. *See* Post-traumatic stress disorder
Pulvinar, 513–516, 519, 522
Pursuit initiation, 315, 316, 322, 330–331, 333
PV interneurons, 457
Pyramidal neurons, 487–494, 496, 498, 499

R

Radial arm maze (RAM), 402, 409, 411, 412, 419
rCBF. *See* Regional cerebral blood flow
rCMRglc. *See* Regional cerebral glucose metabolic rates
Recall, 379, 380
Receptive language, 380
Receptor tautology, 406
Recognition, 379, 380, 382, 384
Recovery movement, 7, 9
Regional cerebral blood flow (rCBF), 540, 541
Regional cerebral glucose metabolic rates (rCMRglc), 540
Relapse, 153, 154, 161, 162, 170
Resequencing, 602
Resting CBF or metabolism
   absolute values, 192
   resting values, 192
Reverse translational, 398
Risk-benefit ratio, 114
Risk marker, 564–566
Rivastigmine, 47, 50

S

Saccades, 312, 314, 316–319, 321, 323, 330, 331, 333
Schizophrenia, 283–301, 587–602
   metabolic dysfunction, 24, 35
   mortality, 24
Self-report, 5
Sensory gating, 441
Serotonergic receptors, 149
Serotonin
   receptors, 225, 227, 231–232
   transporters, 225
Serotonin 5-HT_{1A} receptors, 130–131
Serotonin 5-HT_{2A} receptors, 130

Serotonin 5-HT$_{2c}$ receptors, 132
Serotonin 5-HT$_7$ receptors, 132
Side effects, 134–135
Single photon emission computer tomography (SPECT), 518–521
SIPS. *See* Structured Interview of Prodromal Symptoms
Smooth pursuit eye movements, 312, 313, 315, 319
Social cognition, 382–383, 386
Social interaction, 442
Social isolation, 455, 457
Social withdrawal, 442, 454
Sp^4, 415
Speech
  emotional prosody, 185
  frontotemporal speech circuits, 185, 200
  internal, 184, 185, 189, 190
  self-monitoring of, 189, 190
Spine, 488–490, 494, 496, 500
Startle, 440
Startle reflex. *See* Prepulse inhibition (PPI)
Step-ramp paradigms, 316, 325
Stereotypy, 440
Stigma, 9
Straight gyrus, 257, 258, 260, 262, 263, 271
Structural variation, 587
Structured Interview of Prodromal Symptoms (SIPS), 99, 112, 114
Superior frontal gyrus, 257, 260, 271
Superior temporal gyrus, 271, 272
Susceptibility genes, 412–417
Sustained attention task, 401
Synaptic plasticity, 631, 635, 637–638, 648

T
Temporal lobe epilepsy, 532, 535
Temporal pole, 268, 269, 271
Test of grocery shopping skills (TOGSS), 13
Thalamic syndrome, 513
Thalamus, 488, 490
  infarcts, 512
  magnetic resonance imaging of, 515
  modulates attention, 523
  as relay station, 511
  role in integrating information, 510
  role in signal processing, 511
  volume loss in schizophrenia, 516
Theory of mind, 382, 383
TOGSS. *See* Test of grocery shopping skills
Transdermal, 165–168, 170

Transgenics, 392, 416, 418, 419
Treatment Units for Research On Neurocognition and Schizophrenia (TURNS), 400
Triglycerides, nonfasting, 29, 31, 34
Tropisetron, 59
TURNS. *See* Treatment Units for Research On Neurocognition and Schizophrenia

U
University of California, San Diego Performance-Based Skills Assessment (UPSA), 14

V
VALERO. *See* Validation of Everyday Real-World Outcomes
Validation
  construct validity
    convergent, 397
  etiological validity, 397
  face validity, 396
  predictive validity, 396
Validation of Everyday Real-World Outcomes (VALERO), 6
Validity
  construct, 194, 197
  face, 195, 197
Verbal memory, 379, 384
Veridicality, 12
Verisimilitude, 12
Vesicle trafficking, 638–639
Vigilance, 376, 383
Visual memory, 379, 380
Visuospatial processing, 381
Vitamin D, 449, 450
Vocabulary, 380

W
Wernicke, C., 184, 199
Willful intent, 572
Wnt, 635, 639–646
Working memory (WM), 68, 70, 71, 193–198, 202, 205–207, 376, 384, 486–488, 490, 493, 494, 497, 499

X
Xanomeline, 60